AF765175

CARBOHYDRATES IN DRUG DISCOVERY AND DEVELOPMENT

CARBOHYDRATES IN DRUG DISCOVERY AND DEVELOPMENT

SYNTHESIS AND APPLICATION

Edited by

VINOD KUMAR TIWARI

Associate Professor, Banaras Hindu University, Varanasi, India

ELSEVIER

Elsevier
Radarweg 29, PO Box 211, 1000 AE Amsterdam, Netherlands
The Boulevard, Langford Lane, Kidlington, Oxford OX5 1GB, United Kingdom
50 Hampshire Street, 5th Floor, Cambridge, MA 02139, United States

Library of Congress Cataloging-in-Publication Data

A catalog record for this book is available from the Library of Congress

British Library Cataloguing-in-Publication Data

A catalogue record for this book is available from the British Library

ISBN: 978-0-12-816675-8

Publisher: Susan Dennis
Acquisitions Editor: Emily M. McCloskey
Editorial Project Manager: Mona Zahir
Production Project Manager: Bharatwaj Varatharajan
Designer: Miles Hitchen

Typeset by Thomson Digital

Contents

Contributors

Anand K. Agrahari
Department of Chemistry, Institute of Science, Banaras Hindu University, Varanasi, India

Indrapal Singh Aidhen
Department of Chemistry, Indian Institute of Technology Madras, Chennai, India

Priyanka Bose
Department of Chemistry, Institute of Science, Banaras Hindu University, Varanasi, India

Maude Cloutier
Centre Armand-Frappier Santé Biotechnologie, Institut national de la recherche scientifique (INRS), boul. des Prairies, Laval, QC, Canada

Rituparna Das
Sweet Lab, Department of Chemical Sciences, Indian Institute of Science Education and Research (IISER) Kolkata, India

Ashutosh K. Dash
School of Pharmaceutical Sciences, Department of Pharmaceutical Chemistry, Shoolini University, Solan, Himachal Pradesh, India

Charles Gauthier
Centre Armand-Frappier Santé Biotechnologie, Institut national de la recherche scientifique (INRS), boul. des Prairies, Laval, QC, Canada

Xuefei Huang
Department of Chemistry, Michigan State University, East Lansing, MI; Department of Biomedical Engineering, Michigan State University, East Lansing, MI; Institute for Quantitative Health Science and Engineering, Michigan State University, East Lansing, MI, United States

Nazar Hussain
Academy of Scientific and Innovative Research (AcSIR), CSIR-Indian Institute of Integrative Medicine, Jammu, Jammu and Kashmir; Natural Product Chemistry Division, CSIR-Indian Institute of Integrative Medicine, Jammu, Jammu and Kashmir, India

Manoj K. Jaiswal
Department of Chemistry, Institute of Science, Banaras Hindu University, Varanasi, India

Vineet Kumar
Chemistry and Bioprospecting Division, Forest Research Institute, Dehradun, India

Priti Kumari
Department of Chemistry, Institute of Science, Banaras Hindu University, Varanasi, India

Divya Kushwaha
Department of Chemistry (MMV); Institute of Science, Banaras Hindu University, Varanasi, India

Rajeswara Reddy Mannem
Department of Chemistry, Indian Institute of Technology Madras, Chennai, India

Nidhi Mishra
Department of Chemistry, Institute of Science, Banaras Hindu University, Varanasi, India

Debaraj Mukherjee
Academy of Scientific and Innovative Research (AcSIR), CSIR-Indian Institute of Integrative Medicine, Jammu, Jammu and Kashmir; Natural Product Chemistry Division, CSIR-Indian Institute of Integrative Medicine, Jammu, Jammu and Kashmir, India

Balaram Mukhopadhyay
Sweet Lab, Department of Chemical Sciences, Indian Institute of Science Education and Research (IISER) Kolkata, India

Saddam M. Muthana
Department of Chemistry, College of Sciences & General Studies, Alfaisal University, Riyadh, Kingdom of Saudi Arabia

Shipra Nagar
Chemistry and Bioprospecting Division, Forest Research Institute, Dehradun; National Centre for Polar and Ocean Research, Ministry of Earth Sciences, Goa, India

Chintam Narayana
Department of Chemistry, Institute of Science, Banaras Hindu University, Varanasi, India.

Balaji Olety
Department of Molecular Cell and Cancer Biology, University of Massachusetts Medical School, Worcester, MA, United States

Nivedita T. Patil
Division of Organic Chemistry, National Chemical Laboratory, Pune, India

Roland J. Pieters
Department of Chemical Biology & Drug Discovery, Utrecht Institute for Pharmaceutical Sciences, Utrecht University, Utrecht, The Netherlands

Ashok K. Prasad
Bioorganic Laboratory, Department of Chemistry, University of Delhi, Delhi, India

Namakkal G. Ramesh
Department of Chemistry, Indian Institute of Technology Delhi, New Delhi, India

Ram Sagar
Department of Chemistry, Institute of Science, Banaras Hindu University, Varanasi, India

Richard R. Schmidt
Department of Chemistry, Universitat Konstanz, Germany

Pradeep Sharma
Chemistry and Bioprospecting Division, Forest Research Institute, Dehradun, Uttarakhand, India

Vivek K. Sharma
RNA Therapeutics Institute, University of Massachusetts Medical School, Worcester, MA, United States

Mysore S. Shashidhar
Division of Organic Chemistry, National Chemical Laboratory, Pune; The Academy of Scientific and Innovative Research, Ghaziabad, India

Anoop S. Singh
Department of Chemistry, Institute of Science, Banaras Hindu University, Varanasi, India

Girija S. Singh
Chemistry Department, University of Botswana, Gaborone, Botswana

Sumit K. Singh
Department of Chemistry, Institute of Science, Banaras Hindu University, Varanasi, India

Naveenkumar Thoti
Department of Chemistry, Indian Institute of Technology Madras, Chennai, India

Ghanshyam Tiwari
Department of Chemistry, Institute of Science, Banaras Hindu University, Varanasi, India

Vinod K. Tiwari
Department of Chemistry, Institute of Science, Banaras Hindu University, Varanasi, India

Jordi van Heteren
Department of Chemical Biology & Drug Discovery, Utrecht Institute for Pharmaceutical Sciences, Utrecht University, Utrecht, The Netherlands

Jicheng Zhang
Department of Chemistry, Michigan State University, East Lansing, MI, United States

Foreword

Carbohydrates are the most abundant group of natural products. They can be found almost everywhere on Earth and they are present in all living organisms. For more than a century carbohydrates are already known as structural components of biological systems and to be essential for the cellular metabolism. The more recent finding that carbohydrates mediate various physiological processes was highly inspiring for glycoscience in general, this way stimulating chemical, biological, medicinal, and pharmaceutical studies in this field. The most prominent characteristics of carbohydrates are the unparalleled structural diversity, that is, due to presence of several hydroxy groups and different anomeric linkages with varying stereo-orientation. Thus, gigantic possibilities for structural, as well as for linkage generation between sugar residues are available. These possibilities have been used by nature for diverse biological demands. For instance, such structural variations permit corresponding variations of the overall topology, of the hydrophilicity and hydrophobicity, and of the presentation of functional groups. Understanding of such phenomena is a firm basis for the rationalization of scientific results and particularly for rational drug design.

The emerging understanding of the important structural and functional roles of glycosides, oligosaccharides, and the various types of glycoconjugates (glycolipids, glycopeptides and - proteins, peptidoglycans, glycophospholipids, glycolipidated glycoproteins, etc.) and in derived medicinal studies created an urgent need for pure samples of such materials. Hence, efficient methodologies for the synthesis of such compounds have been meanwhile developed. Yet, improvements of the methodologies, particularly by introducing catalytic synthetic procedures, is still in progress.

Interactions between molecules are decisive for cellular recognition and communication events in living systems. Thus, transient dynamic cellular interactions are available that provide all requirements for molecular recognition and, this way, for information exchange, adaptation to the environment, and for self-regulation. Evolution of this process is accomplished in nature by generating (1) strong individual interactions between receptor and ligand, or (2) by variation of the number of specific interactions of the same type, thus leading to what is termed "multivalency." Therapeutically important inhibitors of glycosidases or glycosyltransferases are generally of the first type, whereas molecular recognition via carbohydrate-mediated association or self-association of proteins into large assemblies follows the second type. Hence, multivalency is decisive for tissue-differentiation, cell-cell recognition and communication, host interactions with pathogens (for instance, in bacterial and viral infections and with secreted toxins), cancer metastasis, inflammation, and protein targeting and clearing. Hence, multivalency plays also an important role for the structure design of effectors in medicinal studies in order to mediate pathological ligand-receptor interactions. Therefore, understanding the underlying binding effects should propel significant progress in the development of therapeutic agents.

As the well-being of any living system depends on a carefully controlled metabolism, on transport, recognition and signal transduction events, as well as on faithful DNA-replication, RNA-transscription, and derived protein synthesis, the life-maintaining processes were regarded to depend essentially on proteins. Not surprisingly, medicinal chemistry has traditionally targeted proteins by employing (small) organic molecules as enzyme inhibitors, receptor or channel blockers, or interferers in the protein synthesis machinery. This view is supported by the new drugs approved in the last decades. However, as the knowledge of the complexity of intra- and intercellular events has dramatically increased, other molecular species, and particularly carbohydrates and carbohydrate-conjugates, emerge as therapeutic agents as well.

This book on "Carbohydrates in Drug Discovery and Development" appears at a very fortuitous time, as the field is in a period of immense progress and the promises for future progression are spectacular. This view is underscored by the many newly founded companies devoted to this field. The authors of this book are experts in their field and they cover important subjects and aspects of the structure and function of the "glycome," a term that comprises all carbohydrates and carbohydrate-containing structures in an organism and the available knowledge of their function.

Richard R. Schmidt
University of Konstanz, Germany
February 2020

Preface

The chemistry and biology of carbohydrates are considered to be a highly promising area of interest since long back. Awesomely, the application of carbohydrates in different disciplines of science is growing exponentially. Their role in several important biological processes, notably energy storage, transport, modulation of protein function, intercellular adhesion, malignant transformation, signal transduction, viral and bacterial cell surface recognition formulate this moiety to be an exceedingly considerable scaffold for the development of new chemical entities of pharmacological importance. Moreover the readily availability, high functionality and chiral-pool character are the additional few fascinating structural features of carbohydrates, which may further enhance the utilities. In this respect, we herein present a judicious endeavor to communicate the impact of carbohydrates in drug discovery and development to the readers.

The main objective of this book is to introduce a concept of multidisciplinary opportunity with a critical discussion about diverse emerging relevance of carbohydrates in biology and medicine. This book covers sixteen different chapters that would greatly illustrate the scope of carbohydrates in drug discovery research. Several leading experts who are very active in this field have contributed to this Elsevier venture and I am grateful to all of them for their excellent contribution.

In the introductory Chapter 1, Schmidt and my group describe the general consideration about recent trends, opportunity, and challenges on carbohydrates-based molecular scaffolding. Xuefei Huang presented a detailed discussion about the heparin and their derivatives as molecular scaffold in drug discovery along with the significance of heparin mimetics including oligosaccharides, noncarbohydrate-based small molecules, and multivalent glycopolymers and dendrimers as the possible tools for modulation of biology and therapy in Chapter 2. Indarpal Singh presented the recent developments in synthesis and bioactivity of C-glycosylated compounds with future perspectives in Chapter 3. In Chapter 4, Charles and coworker presented the chemistry and antibacterial activity of 3-deoxy-D-manno-oct-2ulosonic acid and their derivatives. Chapter 5 covers the impact of sialic acids with their chemical diversity, a common strategy for sialic acid modification, their impact in chemical biology in addition to neurobiology by our group. Opportunity and challenges of glycan microarrays for the study of carbohydrate-protein interaction useful in drug discovery program is presented by Saddam Muthana in Chapter 6.

Some interesting common synthetic routes to biologically active inositol and their derivatives is well presented by Shashidhar and coworkers in Chapter 7. Impact of the diverse range of imino sugars in drug discovery program is nicely presented by N. G. Ramesh in Chapter 8. Roland J. Peters presents an interesting aspect of carbohydrate-protein interaction and their opportunity for the designed mimics in Chapter 9. Kushwaha and Tiwari described the scope of Cu-catalyzed Click chemistry (CuAAC) with their scope in drug discovery development in Chapter 10. Ramsagar and coworker briefly presented the impact of glycohybrid molecules in the pharmaceutical industry in Chapter 11. Mukherjee and coworkers describe the scope and challenges of biologically relevant carbohydrate-containing macrocycles in Chapter 12. A brief discussion about carbohydrate-based antibiotics is described by Girija Singh in Chapter 13. Mukhopadhayay and Roy presented about carbohydrate-based antibacterial and anticancer vaccines in Chapter 14. In Chapter 15, a general consideration about a few selected plant oligosaccharides and polysaccharides with their opportunity in drug development is described by Kumar and Nagar. And at the last, Sharma and Prasad presented about *N*-acetylgalactosamine–conjugates and their potential application in delivering oligonucleotide drugs to the liver.

Thus, the topics included the basic and advanced aspect of carbohydrates starting from a brief introduction, the biological behavior of carbohydrates in living cells and further their biomedical applications. I sincerely thank all the contributors who are eminent scientists actively working on various aspect of carbohydrate chemistry. Further, words are insufficient to convey my special thanks to all the learned reviewers who provide constructive comments and suggestion to make our book even more interesting to readers and useful for researchers actively working on medicinal chemistry/carbohydrate chemistry/microarray/chemical and enzymatic synthesis/ and drug development.

I would like to express my heartfelt thanks and deep sense of appreciation to em. Prof. Richard R. Schmidt, a renowned scientist, and Professor at Universitat Konstanz, Germany for his consistent support, suggestions and keen interest in the chapters and moreover for writing the venerated foreword to the Elsevier book. I personally thank Dr. Rama P. Tripathi, Prof. Srinivasan Chandrasekaran, and Prof. Xi Chen for their useful suggestion, discussion, and help and also my PhD students, in particular Mr. Anand

K. Agrahari and also Dr. Sanchayita Rajkhowa, both for proofread during reviewing and typesetting stage of this book.

Last but not least, I would like to express my deep sense of appreciations to Dr. Joseph Hayton, publishing director, and the entire editorial team associated with Elsevier Inc. in particular, Ms. Carr Hilary, Emily McCloskey, Hess Anneka, Bharatwaj Varatharajan, and most notable Mona Zahir for their keen interest and throughout timely support in publishing this theme with the high standard of publication maintained in bringing out this volume on "*Carbohydrates in Drug Discovery and Development*" with Elsevier Inc.

Vinod Kumar Tiwari
Associate Professor, Organic Chemistry
Department of Chemistry, Institute of Science
Banaras Hindu University, Varanasi-221005, INDIA
Vinod.Tiwari@bhu.ac.in
3 February, 2020

Abbreviations

AAP	American Academy of Pediatrics
Ac	Acetyl
ACGs	Apigenin *C*-glycosides
ADME	Absorption distribution metabolism and excretion
AFM	Atomic force microscopy
AFP-L3	Anti α-fetoprotein fraction L3
AG	Arabinogalactan
AIDS	Acquired immunodeficiency syndrome
AIV	Avian influenza virus
ALL	Acute lymphoblastic leukemia
AML	Acute myeloid leukemia
AMP-DNJ	*N*-[5-(adamantan-1-yl-methoxy)-pentyl]-1-deoxynojirimycin
APC	Antigen presenting cells
ASI	α-syn aggregation inhibitor
ATIII	Antithrombin III
AZA	Acetazolamide
AZT	3′-Azido-3′-deoxythymidine
Bn	Benzyl
Bz	Benzoyl
BLI	Biolayer interferometry
BBB	Blood-brain barrier
NBS	*N*-bromosuccinimide
9-BBN	9-Borabicyclo[3.3.1]nonane
BLG	Bovine liver galactosidase
NB-DNJ	*N*-butyl-1-deoxynojirimycin
BLI	Biolayer interferometry
Boc	*tert*-butyloxycarbonyl
BSA	Bovine serum albumin
BVDV	Bovine viral diarrhea virus
CA	Carbonic anhydrases
CD	Crohn's disease
CDI	*Clostridium defficile* infection
CHB	Chronic hepatitis B
CML	Cronic myelogenous leukemia
CMP-Neu5NAc	Cytidine 5′-monophospho-*N*-acetylneuraminic acid
CMT	Chaperone-mediated therapy
CPE	Cytopathic effect
CPI-GC	Ceramide phosphoinositol glycan core
CPS	Capsular polysaccharide
CRD	Carbohydrate recognizing domains
CSA	Camphorsulfonic acid
CT	Cholera toxin
CTP	Cytidine triphosphate

CuAAC	Cu(I)-catalyzed azide-alkyne cycloaddition
CuANCR	Copper catalyzed alkyne-nitrone cycloaddition
CUTI	Complicated urinary tract infections
CVDs	Cardiovascular diseases
DAB-1	1,4-Dideoxy-1,4-imino-D-arabinitol
DANA	2,3-Dehydro-*N*-acetylneuraminic acid
3DG	3-deoxyglucosone
2-DOS	2-deoxystreptamine
DAST	Diethylaminosulfur trifluoride
DBU	1,8-Diazabicyclo[5.4.0]undec-7-ene
DCC	*N,N′*-Dicychlohexylcarbodiimide
DEAD	Diethyl azodicarboxylate
DENV	Dengue virus
DGDP	2,5-Dideoxy-2,5-imino-D-glucitol
DGJ	1-Deoxygalactonojrimycin
DIBAL-H	Diisobutylaluminium hydride
DIX	1,5-Dideoxy-1,5-iminoxylitol
DLS	Dynamic light scattering
DMAP	*N,N*-Dimethylaminopyridine
DMDP	2,5-Dihydroxymethyl-3,4-dihydroxypyrrolidine
DMNB	Dimethoxy nitrobenzyl
DNJ	1-Deoxynojirimycin
DP	Degrees of polymerization
DPA	Decaprenol phosphoarabinose
DPB	Diffuse pan bronchiolitis
DTBMP	2,6-Di-*tert*-butyl-4-methylpyridine
ECG	*Escherichia coli* galactosidase
ECM	Extracellular matrix
EGC	Endoglycoceramidase (II)
ELISA	Enzyme-linked immunosorbent assay
ELLA	Enzyme-linked lectin assay
EM	Electron microscopy
ENDOR	Electron-nuclear double resonance
EPS	Exopolysaccharide
ER	Endoplasmic reticulum
ESI-MS	Electrospray ionization mass spectrometry
FAK	Focal adhesion kinase
FDA	Food and Drug Administration
FGFs	Fibroblast growth factors
Fuc	Fucose
α-1,3-FucT	α-1,3-Fucosyltransferase
α-1,6-FucT	α-1,6-Fucosyltransferase
GA	18*β*-Glycerrhetinic acid
GAGs	Glycosaminoglycans
Gal	Galactose
GalNAc	*N*-acetylgalactosamine
GBP	Glycan-binding proteins

GCase	Lysosomal glucocerebrosidase
GCS	Glucosylceramide synthase
GDP	Fucose-guanosine diphosphate *β*-L-fucose
GEMA	Glucosyloxyethyl methacrylate
GH	Glycoside hydrolase
Glc	Glucose
GlcA	Glucuronic acid
GlcN	Glucosamine
β-GCase	β-Glucocerebrosidase
GlcNAc	*N*-acetylglucosamine
GLUTs	Glucose transporters
GNB	Gram-negative bacteria
GP	Glycogen phosphorylase
GPIs	Glycosylphosphatidylinositols
GR	Glutathione reductase
GSL	Glycosphingolipids
GST	Glutathione-S-transferase
GTs	Glycosyl transferases
Gul	Gulose
GXM	Glucuronoxylomnnan
β-GCase	β-Glucocerebrosidase
HA	Hemagglutinin
HAART	Highly active anti-retroviral therapy
hCA	Humancarbonic anhydrase
HCC	Hepatocelluar carcinoma
HCV	Hepatitis C virus
HeLa	Human cervical cell line
HGJ	Homogalactonojirimycin
HIA	Hemagglutinin inhibition assay
HiB	Haemophilus influenzae type B
HIT	Heparin-induced thrombocytopenia
HMPA	Hexamethylphosphoramide
HIV	Human immunodeficiency virus
HMMPO	2-(Hydroxymethyl)-2-methyl-3,4-dihydro-2*H*-pyrrole 1-oxide
CYB-3	2-Hydroxymethyl-3-hydroxypyrrolidine
hNeu2	Human neuraminidase 2
hPIV-3HN	Human parainfluenza viruses type 3 hemagglutinin–neuraminidase
HPLC	High-performance liquid chromatography
HRP	Horseradish peroxidase
HS	Heparan sulfate
HSA	Human serum albumin
HSPGs	Heparan sulfate proteoglycans
HSV	Herpes simplex viruses
HTS	High-throughput screening
HWE	Horner-Wadsworth-Emmons reaction
2-OH-ABK	2-Hydroxyarbekacin
$(DHQ)_2$-AQN	Hydroquinine anthraquinone-1,4-diyl diether

IA	Invasive aspergillosis
IAV	Influenza A virus
IBD	Inflammatory bowel disease
IgG	Immunoglobulin G
IgM	Immunoglobulin M
IP_3	D-myo-inositol 1,4,5-trisphosphate
ITC	Isothermal titration calorimetry
IV	Intravenous
JbαMan	Jack bean α-mannosidase
KDH	3-Deoxy-α-D-*lyxo*-hept-2-ulosonic acid
Kdo	3-Deoxy-D-*manno*-2-octulosonic acid
$KdoN_3$	8-Azido-8-deoxy-Kdo
$KdoNH_2$	8-Amino-8-deoxy-Kdo
KdsA	3-Deoxy-α-D-*manno*-2-octulosonate-8-phosphate
KHMDS	Potassium bis(trimethylsilyl)amine
KLH	Keyhole limpet hemocyanin
LAM	Lipoarabinomannan
LDA	Lithium diisopropylamide
LEAPT	Lectin directed enzyme-activated prodrug therapy
LMWH	Low-molecular weight heparin
LPH	Lactase-phlorizin hydrolase
LPS	Lipopolysaccharide
LRP	Luciferase reporter phage
LSD	Lysosomal storage disorders
MAC	*Mycobacterium avium* complex
mAGP	Mycolyl-arabinogalactan peptidoglycan
MALDI-TOF	Matrix-assisted laser desorption/ionization-time of flight
Man	Mannose
ManA	Mannuronic acid
ManN	Mannosamine
ManNAc	*N*-Acetylmannosamine
m-CPBA	*meta*-chloroperoxybenzoic acid
MCs	Macrocyclic carbohydrates
MDR	Multi-drug-resistant
MeOTf	Trifluoromethane sulfonate
MGL	Metabolic glycan labeling
MHC	Minimal hemolytic concentration
MIC	Minimal inhibitory concentration
MJ	Mannojirimycin
MOM	Methoxymethyl
mRNA	Messenger ribonucleic acid
MRSA	Methicillin- resistant *Staphylococcus aureus*
Ms	Mesyl
MS	Multiple sclerosis
Nap	2-Naphthylmethyl
*N*AP-DNJ	*N*-(6′-[4″-azido-2″-nitrophenylamino]hexyl)-1-deoxynojirimycin
NAs	Neuraminidases

NDGA	Nordihydroguaiaretic acid
neoPGs	Neoproteoglycans
Neu5Ac	*N*-acetylneuraminic acid
Neu5Ac2en	*N*-Acetyl-2,3-dehydro-2-deoxyneuraminic acid
Neu5Ac9Lt	9-*O*-Lactoyl-*N*-acetylneuraminic acid
NHS	*N*-Hydroxysuccinimide
NHTP	*N*-Hydroxy-2-thiopyridone
NIS	*N*-Iodosuccinimide
NJ	Nojirimycin
NMO	*N*-Methylmorpholine *N*-oxide
NPs	Natural products
OA	Oleanolic acid
OGlcNAc	*O*-linked β-*N*-acetylglucosamine
OM	Outer membrane
OTf	Trifluoromethanesulfonate
OVA	Oval albumin
PAGE	Polyacrylamide gel electrophoresis
PCC	Pyridinium chloroformate
PCs	Pharmacological chaperones
3-PG	3-Phosphoglyceric acid
PDT	Photodynamic therapy
PGTs	Peptidoglycan glycosyl tranferases
PIMs	Phosphatidylinositol mannosides
PI-PLC	Phosphatidyl inositol-specific phospholipase C
Piv	Pivalate
PM	Propionyloxymethyl
PNA	Peanut agglutinin
PPTS	Pyridinium *p*-toluenesulfonate
PTC	Papillary thyroid cancer
PTPs	Protein tyrosine phosphatase
PTSA	*para*-toluenesulfonic acid
RCM	Ring closing metathesis
Rha	Rhamnose
RMGP	Rabbit muscle glycogen phosphorylase
RRMS	Relapsing-remitting multiple sclerosis
rRNA	Ribosomal ribonucleic acid
SAC	Saccharin
SAE	Sharpless asymmetric epoxidation
SAG	Sweet almond glucosidase
SARS-CoV	SARS-coronavirus
SC	Subcutaneous
SGLT2	Sodium glucose co-transporter 2
sLe^a	Sialyl Lewis a
sLe^x	Sialyl Lewis x
α-2,3ST	α-2,3-(*N*)-Sialyltransferase
α-2,6ST	α-2,6-(*N*)-Sialyltransferase
SPAAC	Strain-promoted alkyne-azide cycloaddition

SPR	Surface plasmon resonance
SQAP	Sulfoquinovosylacylpropanediol
SQMG	Sulfoquinovosyl-monoacylglycerol
SRB	Sulforhodamine B
SRT	Substrate reduction therapy
ST	Sialyltransferase
Sulfo-SIAB	Sulfosuccinimidyl (4-iodoacetylamino) benzoate
SV-AUC	Sedimentation velocity analytical ultracentrifugation
TACAs	Tumor-associated carbohydrate antigens
TAMRA	Tetramethylrhodamine
TBAF	Tetra-*N*-butylammonium fluoride
TBAHS	Tetrabutylammonium hydrogen sulfate
TBAI	Tetrabutylammonium iodide
TBS	*tert*-butyldimethylsilyl
TBSCl	*tert*-butyldimethylsilyl chloride
TCA	Trichloroacetimidate
TCPTP	T-cell protein tyrosine phosphatase
TcTS	*Trypanosoma cruzi trans-sialidase*
TEAC	Trolox equivalent antioxidant capacity
TFA	Trifluoroacetic acid
thio-dGTP	6-Thioguanosine 5′-triphosphate
TIPS	triisopropylsilyl group
TMS	Trimethylsilyl group
TMSCl	Trimethylsilyl chloride
TMSCN	Trimethylsilyl cyanide
TMSOTf	Trimethylsilyl triflate
TPHB	Triphenylphosphorane hydrobromide
Troc	2,2,2-Trichloroethoxycarbonyl
Ts	Tosyl
TT	Tetanus toxoid
Tyc	Tyrocidine
UC	Ulcerative colitis
UFH	Unfractionated heparin
ULMWH	Ultra-low molecular weight heparin
VCNA	*Vibrio cholerae* neuraminidase
VEGF	Vascular endothelial growth factor
VL	Visceral Leishmaniasis
VRE	Vancomycine-resistant enterococci
VRSA	Vancomycin-resistant *Staphylococcus aureus*
VTE	Venous thromboembolism
VZV	Varicellazoster virus
WA	Weinreb-amide
WHO	World Health Organization
ZBG	Zinc binding group
α-*C*-Man-Trp	α-*C*-Mannosyltryptophan
β4GalT	β-1,4-galactosyltransferase I
γ-Gcase	γ-glutamyl cysteine synthetase

CHAPTER ONE

Recent trends and challenges on carbohydrate-based molecular scaffolding: general consideration toward impact of carbohydrates in drug discovery and development

Nidhi Mishra[a], Vinod K. Tiwari[a], Richard R. Schmidt[b]
[a]Department of Chemistry, Institute of Science, Banaras Hindu University, Varanasi, India
[b]Department of Chemistry, Universitat Konstanz, Germany

1 Introduction

Design and synthesis of novel compounds having desired pharmacological traits is the main target of drug discovery. In this aspect, molecular scaffolding has been emerged as the most efficient way for designing of potential drug candidates. Molecular scaffold, a distinguished and one of the most imperative notions in medicinal chemistry, is defined as the core structure of a molecule with preferable bioactive properties.[1] Carbohydrates, being a biologically important class of molecules, provide a suitable choice for drug discovery. The exceptional advantages associated with this class of molecules make them appropriate for molecular scaffolding. Carbohydrates are the most abundant and easily accessible natural products. Although, their extraordinary chemical significance which lies in their unique structural traits, such as, presence of multiple hydroxy groups, simple yet firm structures, chirality and easily modifiable architecture, have always been an interesting subject for chemistry community,[2] but, carbohydrates are incomparable when it comes to their importance in biological world. Sugars are great source of energy and serve as energy storage molecules. Their importance in survival of living world makes them unavoidable component of our food. Besides this, carbohydrates are essential part of many biological systems and processes. For example, ribose sugar is an

Carbohydrates in Drug Discovery and Development. http://dx.doi.org/10.1016/B978-0-12-816675-8.00001-4

essential part of the genetic materials (nucleic acid), carbohydrates serve as antigens and play a major role in deciding the blood group, heparin serves as natural anticoagulant in blood, carbohydrates are present in plasma membrane in form of glycolipids, constitute the structural framework of cell in plants (cellulose), fungi (chitin), bacteria (peptidoglycan) and exoskeleton of arthropods (chitin). Carbohydrate-protein interaction is essential for many fundamental cell functions, such as, signal transduction, cell adhesion, inflammation, stabilization of protein structure and host-pathogen recognition.[3]

Although the significance of carbohydrates in cellular physiology has been acknowledged by biochemists for more than two centuries but, their biological importance was under estimated as compared to nucleic acids and proteins until their crucial role in intercellular and cell-pathogen interactions was accounted. Since then glycobiology has emerged as one of the most significant field of biochemistry with new discoveries explaining role of carbohydrate moieties in the transmission of biological information.[4] Some of the carbohydrate-mediated important biological processes play crucial role in various disease functions and hence are medicinally important too.[5] As a result, the study of the functions of carbohydrates, specially oligosaccharides and galectins, in biological systems is an essential tool for development of new therapeutics for many human diseases.[6] These studies have assisted the development of promising drugs, vaccines and novel drug delivery systems based on carbohydrate molecules to cure conditions related to diabetes, arthritis, thrombosis, viral and bacterial infections, Alzheimer's disease, tuberculosis, cancer, etc.[7–12]

Carbohydrate-based drug discovery has gained much attention during last few decades. Many naturally occurring carbohydrates as well their synthetic modifications have shown their potency as drug leads. Various marketed drugs and vaccines are based on carbohydrates and several are under clinical trials. Also, many recent researches have presented new carbohydrate-based molecules as potential leads towards drug discovery. Researchers have also proved carbohydrates to be promising scaffolds to develop new effective drug delivery systems. Moreover, the binding property of carbohydrate moiety with metal ions has promoted the discovery and development of carbohydrate-based metallodrugs. This chapter focuses on the importance of carbohydrates scaffolds in drug discovery, development of carbohydrate derivatives as drug systems and vaccines, new leads in carbohydrate-drug discovery and the future perspectives of carbohydrate-based therapeutics.

2 Biological significance of carbohydrates and their impact in drug discovery

2.1 Carbohydrates on cell surface

The significant roles played by carbohydrates in biological processes make them an interesting scaffold for drug discovery to start with. The presence of carbohydrates on cell surface in form of oligosaccharides, glycoproteins and glycolipids on cell surfaces makes them a key contributor in regulation of cell-cell, cell-effectors and cell-extra cellular matrix interactions which expresses their major role in cellular recognition and signal transduction (Fig. 1.1). The participation of carbohydrates in these two important cellular processes exhibits their significance in diagnosis and therapy of diseases related to immunity, neurodegenerative disease, oncology, infection, and inflammation including neurodegeneration (multiple sclerosis, Alzheimer's), vascular diseases (diabetes, thrombosis, atherosclerosis), orthopedic ailments (arthritis), and cancer.

The exemplary role of carbohydrates in cell-cell recognition can be observed in determination of ABO blood group. The ABO blood types are determined on the basis of antigens found attached with lipids and proteins on the surface of erythrocytes. These antigens are polysaccharides in which arrangement of sugar units describes the type of antigen (A, B or O). For instance, the O antigen has a saccharide chain following —Lipid—Glucose—Galactose—*N*-acetylglucosamine—Galactose—Fucose sequence (Fig. 1.2). This sequence and hence the O antigen is the most

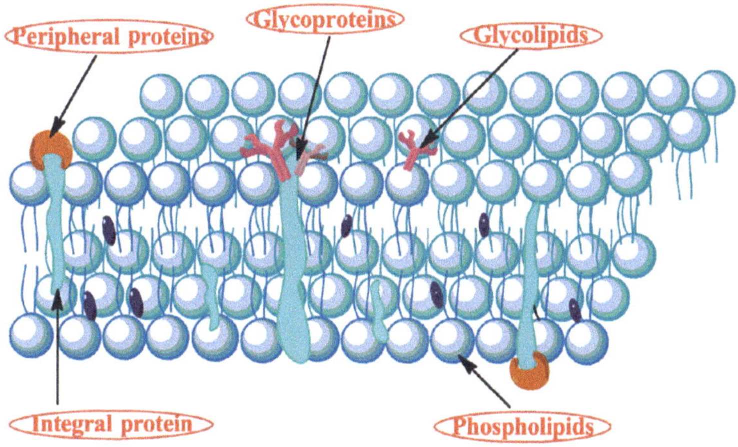

Figure 1.1 *Carbohydrates in plasma membrane.*

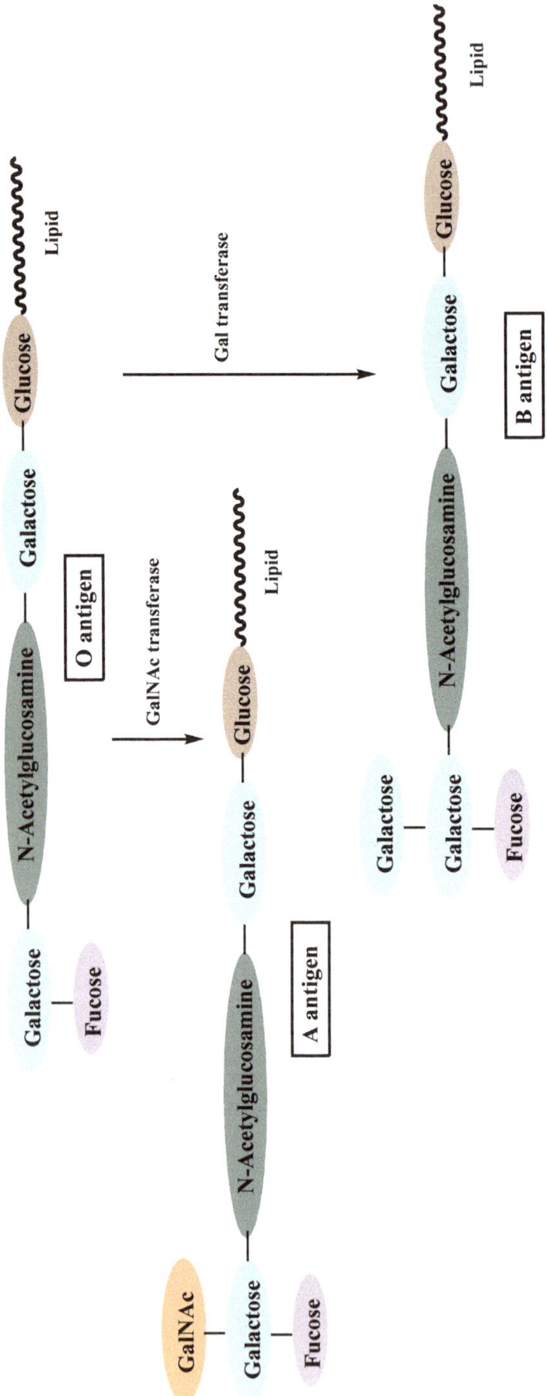

Figure 1.2 *Role of carbohydrates in determination of ABO blood type.*

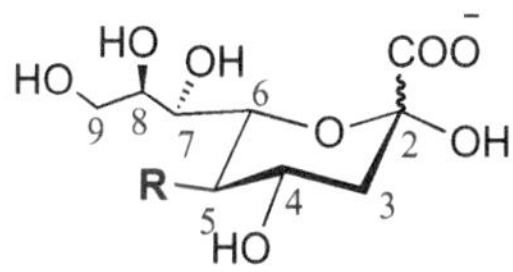

R = NHAc, Neu5Ac (**1**)
R = NHGc, Neu5Gc (**2**)
R = OH, KDN (**3**)

Figure 1.3 ***Structure of sialic acids (N-acetylneuraminic acid (Neu5Ac), N-glycolylneuraminic acid (Neu5Gc), and deaminoneuraminic acid (KDN).***

basic oligosaccharide attachment which is found in all the three types of blood groups. A antigen has additional *N*-acetylgalactosamine (GalNAc), whereas, the B antigen has an added galactose unit. Hence, the O blood type contains only O antigen, the type A has an additional GalNAc attached to O antigen with a glycosidic bond and type B has an additional galactose bonded through glycosidic bond to the O antigen.[13]

Sialic acids (**1-3**), which are *O*- or *N*-substituted derivatives of neuraminic acids (Fig. 1.3) are found in plasma membrane in form of glycolipids and glycoproteins. These acidic sugars, consisting of a nine-carbon framework, decorate the outermost end of glycan chain on cell surface.[14] They play crucial role in many physiological functions and pathological processes. The hydrophilicity and negative charge of sialic acids make them apposite for many modulatory and structural roles. They also provide binding sites to pathogens and toxins.[15]

Maximum concentration of sialic acids is found in human brain where these carbohydrates assist neural transmission.[16] Sialic acids act as receptor in microbial binding of host cell to viruses, such as, influenza viruses (*Orthomyxoviridae*), rotaviruses (*Reoviridae*) and adenoviruses (*Adenoviridae*). Sialic acids being the key components of selectin ligands, significantly assist the regulation of immune response. A high density of sialoglycoproteins is expressed by metastatic cancer cells. A correlation of regulation of ST6GAL1(ST6 β-galactoside α-2,6-sialyltransferase 1) gene with sialic acids has been analyzed in various carcinomas which clearly indicates the role of sialic acids in They play a significant role in advancement of human malignancies. The pathological functions of sialic acids have instigated the development of several drugs by modification in sialic acid structures.[17–19] The role of sialic acid in neurobiology and its importance as a scaffold in drug development has been elaborately discussed in **Tiwari Chapter** of this book.

Carbohydrates or glycans and their conjugates are involved in wide range of biological processes and play an important role in various diseases, however investigation of their biological significances is really challenging.

Although, the emerging of glycan microarrays as high-throughput technology for knowledge about carbohydrate interactions have overcome some of issues, and have greatly contributed to our understanding of the biological roles of carbohydrates. Interestingly, glycan microarrays offer new applications in biomedical research. The notable developments and essential role of glycan microarrays in drug discovery and development has been described in **Prof. Saddam Muthana Chapter** of this book.

Inflammatory response and immune regulation are mainly governed by carbohydrate-lectin interactions in which lectins are responsible for recognition of specific carbohydrates.[20,21] Hence, these carbohydrate-protein interactions attract medical science for drug research related to inflammation and immune-related ailments. Understanding of glycans, which are recognition targets of carbohydrate-binding proteins,[22] as well as, the study of carbohydrate-protein interactions has emerged as aid to develop glycomimetics- an advancement towards carbohydrate-based drugs.[23] A change in the structure of glycan is considered to be a characteristic of advancement of a disease related to inflammation.[24]

Abnormal glycosylation apprises about diseased state and is considered as a prognostic and diagnostic biomarker for tumors, autoimmunity, congenital disorders and cancers.[25] Many neuronal processes, such as, neurite proturbence, are governed by glycosylation which modulates protein function resulting into modulation in cell signalling which further governs memory consolidation. Researches reveal that developmental defects and change in cognitive behaviour can be caused due to genetically removal of enzyme responsible for glycosylation.[26]

In addition to their remarkable biological significance of sialic acids, a negatively charged nine-carbon sugars found in the terminal end of cell surface, 3-deoxy-D-*manno*-oct-2-ulosonic acid (Kdo) is also very important sugar having eight carbon known for their great pharmacological activities. As a highly conserved and essential element of (LPS), 3-deoxy-D-*manno*-oct-2-ulosonic acid (Kdo) are considered as a significant element of Gram-negative bacteria lipopolysaccharides and hence, the importance of Kdo and their derivatives as antibacterial agents confirms their worth in antibiotic drug development. **Prof. Charles Gauthier Chapter** briefly describes the importance of Kdo and total synthesis of Kdo and its derivatives.

Furthermore, inositols and their derivatives has been considered as a very interesting class of molecules occur in plants and animals, where phosphorylated inositol derivatives are known to involved in various cellular processes in eukaryotic cells. The D-myo-inositol 1,4,5-trisphosphate

(IP_3) is released from the phosphatidyl inositol 4,5-bisphosphate through enzymatic hydrolysis mediated by phospholipase C is known to mobilize Ca^{2+} from the intracellular stores of mostmammalian cells. The detail information about recent developments in this area particularly synthetic methods for an easy access of enantiomeric inositol and their derivatives and improved the pace of description of the biological roles played by phosphoinositols hass been discussed in **Prof. M S Shashidhar Chapter** of this book.

2.2 Carbohydrate for genetic information

Nucleic acids are the most important biomolecules for living world in which all the genetic information are encoded. Polynucleotide strands contained in nucleic acids are composed of monomer nucleotides consisting of three components: nitrogenous bases, a phosphate group and a furanose sugar (Ribose in RNA and Deoxyribose in DNA). Furanose sugar units build up a backbone for nucleic acids by formation of covalent bonds with phosphates of other units. This special arrangement bestows a firm architecture by resisting cleavage.

Oligonucleotide-based therapeutics can target any candidates in a specific manner which was once considered 'undruggable' and thus they could be a highly promising drug development platform. A simple covalent conjugation of a carbohydrate moiety, for example *N*-acetylgalactosamine (Gal*N*Ac), to the oligonucleotides facilitates receptor-mediated uptake of oligonucleotides specifically to hepatocytes in the liver and also found to be many folds more potent than its unconjugated counterpart. This approach particularly conjugates targeting other cell and tissue types towards oligonucleotide delivery for liver diseases has been well described in **Drs. Sharma, Olety, and Prasad chapter** of this book.

2.3 Importance of furanose structure

Furanose form of sugar offers apt conformation which in turn helps the desired biological response. For instance, when Gly-Gly portion in Leu-enkaphalin (**4**) (an endogenous opioid peptide neurotransmitter) is replaced by a furanose sugar counterpart, it induces notable analgesic activity by providing suitable conformation consistent with its binding domain. Hence, carbopeptide with sugar substitution (**5**) shows analgesic activity while corresponding carbopeptide without any furanose sugar part does not exhibit any such biological activity (Fig. 1.4).[27]

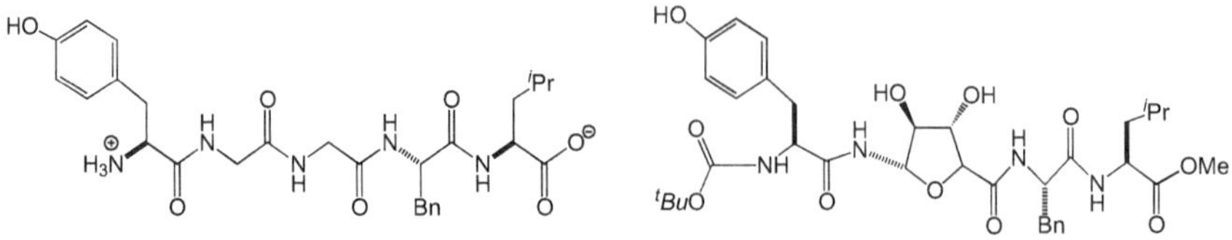

Leu- enkephalin 4
(No Analgesic activity)

5
Shows Analgesic activity

Figure 1.4 ***Effect on biological activity by replacing aminoacid in Leu-encephalin by a furanose sugar.***

2.4 Heparin: a natural anticoagulant

Heparin (**6**), found in animal tissues in form of heparin sulphate (**7**) (Fig. 1.5), is a naturally occurring polysaccharide which contains a highly sulfated glycosaminoglycan framework. It possesses highest negative charge density as compared to other biomolecules which makes Heparin suitable for cell encapsulation. The presence of heparin sulfate on cell surface or extra cellular protein matrix assists various biological functions, such as, cell surface reception, differentiation, migration, proliferation, cancer metathesis.

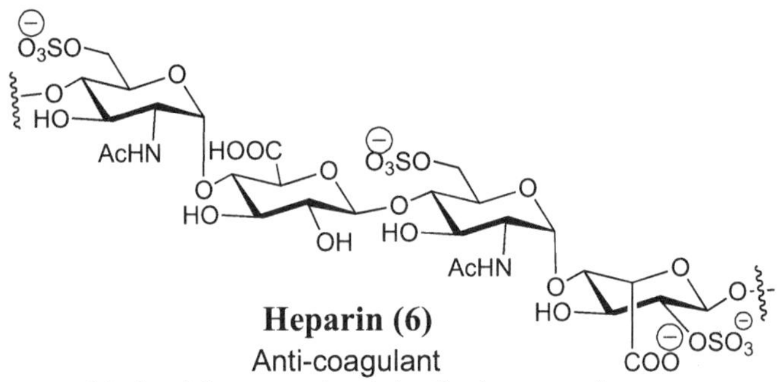

Heparin (6)
Anti-coagulant
(derived from porcine intestinal mucosa)

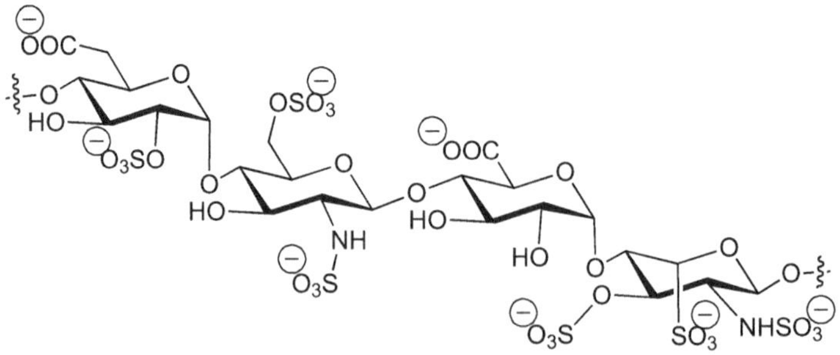

Heparin sulfate (7)

Figure 1.5 ***Structure of heparin and heparin sulfate: natural anticoagulant.***

The major and best known function of heparin is anticoagulation of blood. Heparin binds toserine protease inhibitor antithrombin III (AT) which results in activation of AT. The activated AT blocks thrombin and inactivates factors Xa and IIa which are responsible for blood coagulation.[28]

The impact of heparin and their derivatives in drug discovery and development has been elaborately discussed in **Prof. Xuefei Huang Chapter** of this book.

3 Carbohydrate-based drugs

The biological significance of carbohydrates in various pharmacological processes has compelled the drug discovery community to think over the development of carbohydrate-based therapeutic agents. Also the structural traits and functionality of carbohydrates assist in attaining desired structural modifications required for inducing biological activity in novel drug leads, as well as, enhancement of activity of existing drugs. Moreover, the stereochemistry of carbohydrates and their existence in nature in non-racemic enantiomerically pure forms encourages the carbohydrate-based drug designs with specific stereo requirements for a better binding with target proteins of specific orientations.

3.1 Naturally occurring carbohydrates and their derivatives as drug candidates

Most of the carbohydrate-derived pharmaceuticals which are well established marketed drugs are naturally occurring saccharides originating from plant extracts, animal tissues or bacteria.[29] Heparin, which has been discussed previously in this chapter, is globally sourced from porcine intestinal mucosa in form of its sodium salt. Heparin is used as a medication mainly as an anticoagulant, in treatment of unstable angina and cardiac attacks.[30] Another carbohydrate-based drug, Acarbose (**8**) is derived from cultures of *Actinoplanes utahensis*. This drug is an α-glucosidase inhibitor used in the treatment of type 2 diabetes mellitus. The mode of action of acarbose follows the reversible binding to α-glucosidase in the brush border of intestinal mucosa and α-amylase in pancreas. Inhibition of these enzymes curbs the digestion of complex carbohydrates resulting into decrease in glucose level in blood.[31]

Ancer 20 (**9**, Fig. 1.6) another naturally occurring carbohydrate derivative is derived from *tubercle bacillum* and used for the treatment of leucopenia.[32] Sodium hyaluronate (**10**), a glycosaminoglycan, is derived from animal

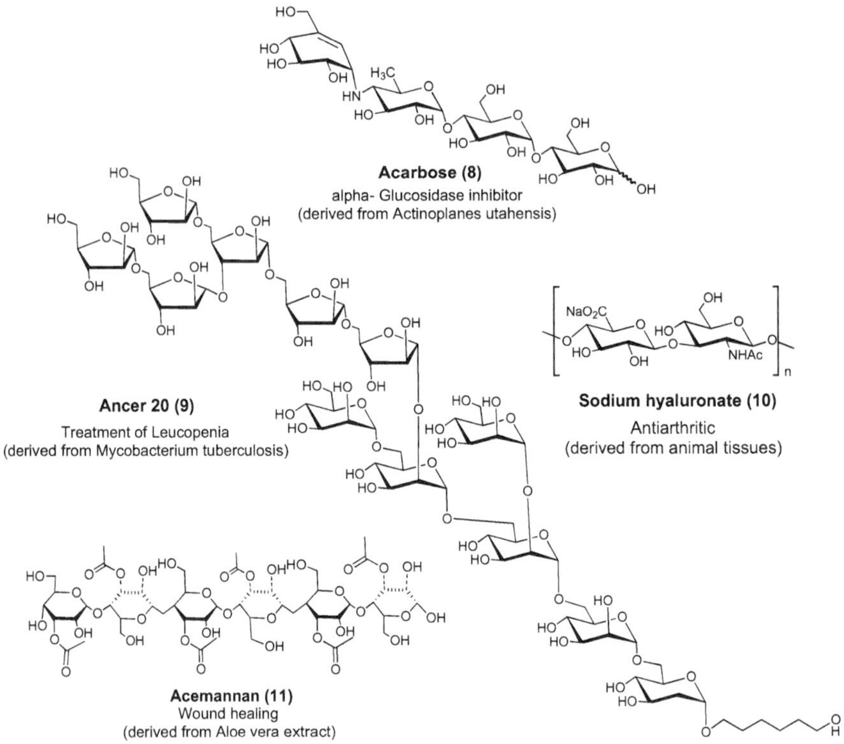

Figure 1.6 ***Structure of some carbohydrate-based marketed drugs*** **(8-11)** ***derived from natural origins.***

tissues. It acts as a tissue lubricant and assists the modulation of tissue interactions. The viscoelastic solution formed by sodium hyaluronate with water provides mechanical protection and protective buffer for tissues. The drug is used to cure arthritis.[33]

Acemannan (**11**), derived from aloe-vera extract is a polymanno-galacto acetate withβ-1,4-glycosidic bond linkage. This hydrophilic polymer is known to induce macrophages to secrete cytokines, such as, interferon, interleukins and tumor necrosis factor-α which have been identified to cause inflammation. Presently this medication is being used for the clinical management and cure of fibrosarcoma in dogs and cats.[34]

Carbohydrates pose as a promising candidate for vaccine development, which is considered a highly priority area. The principle behind the development glycoconjugate anti-bacterial, anti-tumor and anti- cancer vaccines along with their synthetic steps and commercialization has been described in great detail in **Prof. Balaram Mukhopadhyay** Chapter of this book.

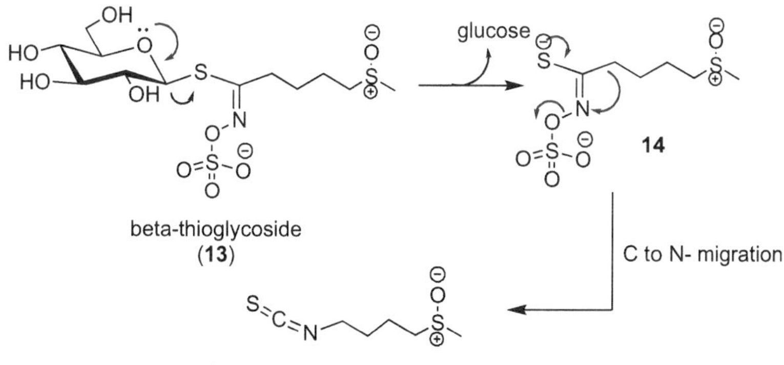

Figure 1.7 ***Anti-cancer drug sulforaphane* 12 *resulted from thio-glycoside found in green vegetables, like Broccoli.***

Sulforaphane (**13**), a potential anti-cancer molecule, is derived from respective thio-glycoside found in green vegetables, such as, broccoli and cabbage.[35] β-Thioglycoside **14** that undergoes Hoffman type C-N migration followed by subsequent removal of sulfate as the leaving group during the course of reaction and finally leads to the formation of sulforaphane (**12**) (Fig. 1.7). Green vegetables contain several thioglycosides[36] and therefore can be beneficial to prevent cancer if adequately used in daily diet.

Carbohydrate-based flavones, such as, Baicalin (**15**) and Astilbin (**16**) are known to have various biological activities which provide medicinal value to their source plants. Baicalin is extracted from *Scutellaria baicalensis* which is a well known Chinese herbal medicine to cure dysentery, diarrhea, hemorrhaging, hypertension, respiratory infections, insomnia and inflammation.[37] Recent reports support the exceptional pharmacological activities of Baicalin including antiviral activity against H1N1 influenza virus,[38] anti-inflammatory activity,[39] protection against UVC-induced cytotoxicity,[40] anti-apoptotic activity against oxidative neuronal damage,[41] antioxidant property,[42] and antidiabeticactivity[43] (in combination with metmorfin) (Fig. 1.8). On the other

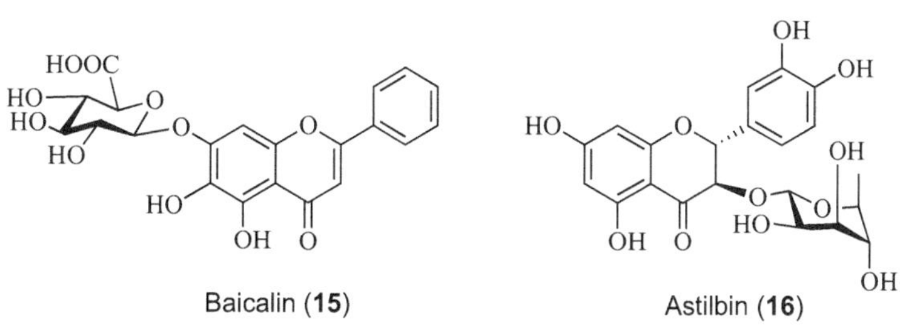

Figure 1.8 ***Biologically active flavonoids Baicalin* (15) *and Astilbin* (16).**

hand, astilbin (**16**) is a flavonol extracted from *Smilax glabra, Smilax china L., Hypericum perforatum Engelhardtia chrysolepis* and some other plants. Biological evaluation of astilbin advocates its insecticidal activity against *Spodoptera frugiperda* and *Anticarsia gemmatalis,*[44] immunoregulatory and anti-inflammatory activity,[45] and antibacterial activity.[46]

A number of recent researches have presented the promising anti-tumor and anti-cancer properties of some naturally occurring glycosides isolated from various plant extracts. For example, some of the new cytotoxic glycosides (aquaterins) **17–23** (Fig. 1.9) found in water spinach (*Ipomoea aquatic*) showed cytotoxicity against human breast carcinoma cell lines MCF-7 and MCF-7/ADR, human hepatoma celllines SMMC-7721and HepG2, and human osteosarcoma cell lines U2-OS and MG-63.[47] Another plant from

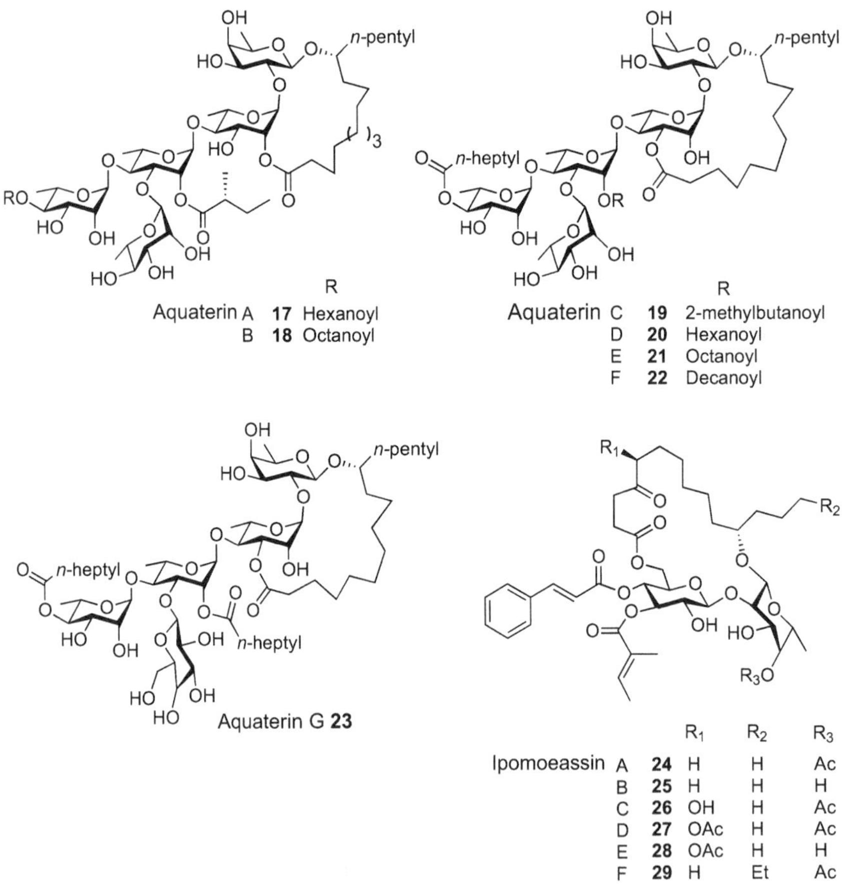

Figure 1.9 *Cytoxic glycosides extracted from* **Ipomoea aquatic** *and* **Ipomoea squamosal.**

the same genus, *Ipomoea squamosa* was found to have cytotoxic properties against ovarian cancer. Glycosides identified in the leaf extract of *Ipomoea squamosa* (ipomoeassins) **24–29** exhibited cytotoxicity in the nanomolar range against ovarian cancer cell lines.[48] Fascinated by the exceptional anticancer activity of ipomoeassins, their total synthesis was achieved and reported by different research groups during years 2007–15.[49] The structure-activity relationship (SAR) of Ipomoeassin F **29** has been recently reported by Zong et al. which shows the importance of acylation pattern for bioactivity advocating a possible interaction of α,β-unsaturated esters in the receptor. Also, it was found that while the high cytotoxic activities of Ipomoeassins is favourable for further explore them for anticancer therapy, but, their poor selectivity between cancer cells and normal cells curb the further possibilities.[50]

Recently, the isolation of six different di-acetylated lactonic sophorolipids from the culture of *Starmerella bombicola* CGMCC 1576 has been reported by Li et al.The cytotoxic evaluation of these sulphorolipid against cervical carcinoma cell lines CaSki and HeLa revealed that among all the isolated sulphorolipids, compound **30** (Fig. 1.10) showed highest cytotoxicity with IC_{50} value of 25.35 mg/mL and 12.23 mg/mL respectively.[51] This sulphorolipid also exhibited activity against breast cancer cell line MDA-MB-231 with IC_{50} value of 21.8 mM.[52]

A detailed study of carbohydrate-containing macrocyclic compounds with their biological activities, common synthetic routes and their impact in drug discovery and development has been greatly described in **Prof. D Mukherjee** Chapter.

Glucose transporter GLUT1 and fructose transporter GLUT5 are expressed in normal cell metabolism as well as diseased state such as cancer

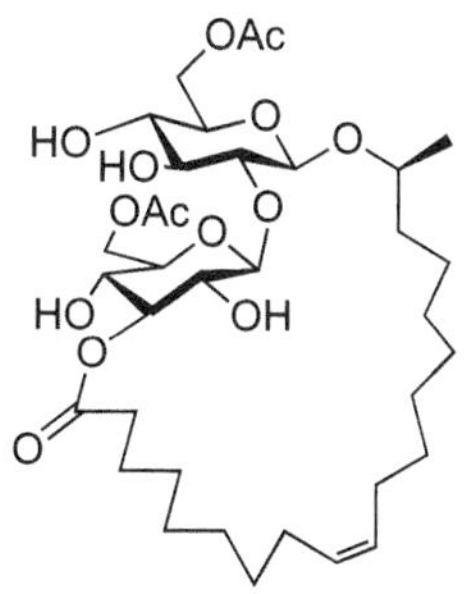

30

Figure 1.10 ***Cytotoxic sulphorolipid 30 extracted from Starmerella bombicola CGMCC 1576.***

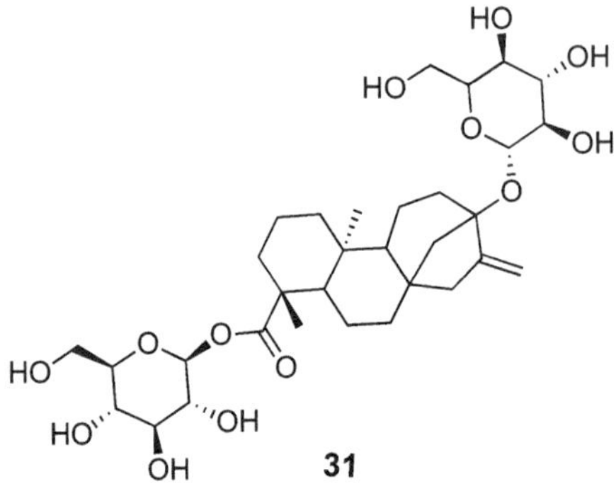

31

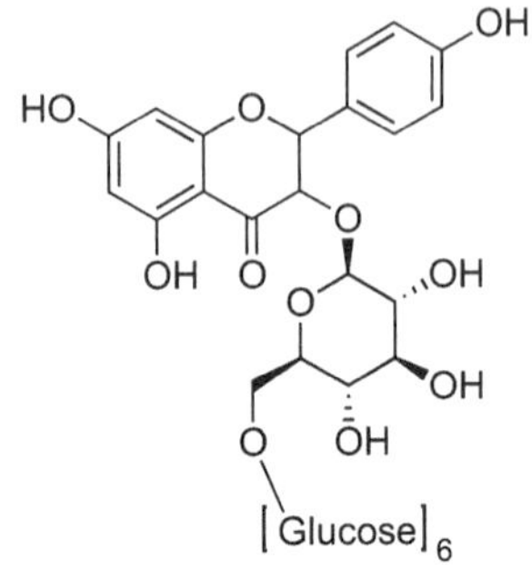

32

Figure 1.11 ***GLUT inhibitors extracted from plants.***

and diabetes. Thompson et al. reported two natural products extracted from plants as GLUT inhibitors. Rubusoside (**31**), extracted from *Rubus suavissimus,* was found to have inhibitory activity against human GLUT1 and GLUT5, whereas, astragalin-6-glucoside (**32**, Fig. 1.11), a glycosylated derivative of astragalin extracted from *Phytolacca Americana*, showed inhibitory activity against human GLUT5. Traditionally *Rubus suavissimus* is used as a traditional medicine for weight loss, while *Phytolacca Americana* is a traditional medicine for cancer therapy.[53]

Oligosaccharides and polysaccharides extracted from various plant sources have been used as traditional medicines for centuries. New researches also advocate the medicinal properties of such phytochemicals on the basis of their excellent biological activity assays. **Dr. Vineet Kumar Chapter** of this book has elaborately discussed the plant derived oligosaccharides and polysaccharide, their biological activities and their role in development of drug leads.

3.2 Carbohydrate scaffolds in nature and their impact in drug development

Design and synthesis of carbohydrate-based drugs has been inspired from nature itself. Modification in carbohydrate structures in form of iminosugars, sugar amino acids and polycyclic derivatives to be used as scaffolds to develop bioactive compounds for drug development was enthused by natural carbohydrate scaffolds, such as, peptidoglycan in which a link between a sugar unit and a peptide chain is established by *N*-acetyl-muramic acid, sialyl Lewis X, in which galactose acts as linker to attach two sugar units, and lipopolysaccharide (LPS) in which a glucosamine scaffold works as a linker to link complex saccharide chain to other glucosamine unit, as well as, a phosphate unit to lipidic chains (Fig. 1.12).[54]

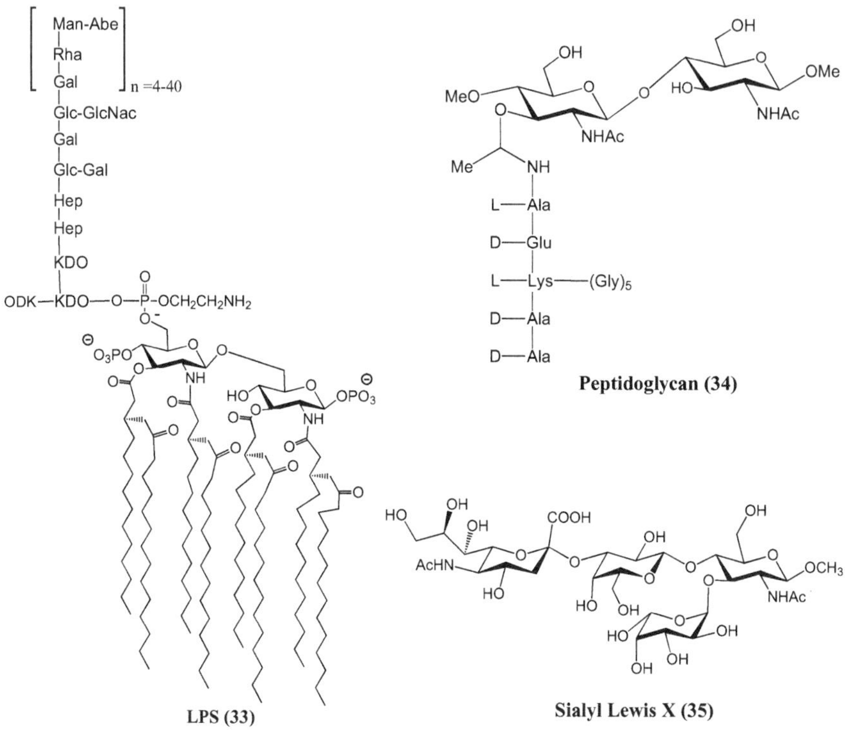

Figure 1.12 ***Carbohydrate scaffolds in nature.***

Due to natural abundance of monosaccharides, they provide a cheap and promising precursor for synthesis of molecules showing biological activity. Mimics of biomolecules are used to enhance stability and biological activity and facilitate understanding of physiological functions and their mechanism which in turn assists drug discovery.[55] Glucose-based somatotropin release inhibitory factor mimic (SRIF-mimic) **36** and **37** were reported by Hirschmann et al. which bind to somatostatin receptors on AtT-20 cells. The work was extended by Papageorgiou and co-workers by introducing a tetrasubstituted xylofuranose **38** as a non-peptide mimic of somatostatin.[56] Another non-peptide SRIF-mimic **39** was developed by Hirschmann and co-workers using *D*-glucose backbone for Cyclo(Arg-Gly-Asp-D-Phe-Val) (cRGDFV), an antagonist vitronectin (Fig. 1.13).[57]

The biological significance of glycosides in nature appeals the researchers for their study and development of their modified mimics with desirable activities for specific targets. This further enhances their importance in the field of drug discovery. The importance of naturally occurring glycosides

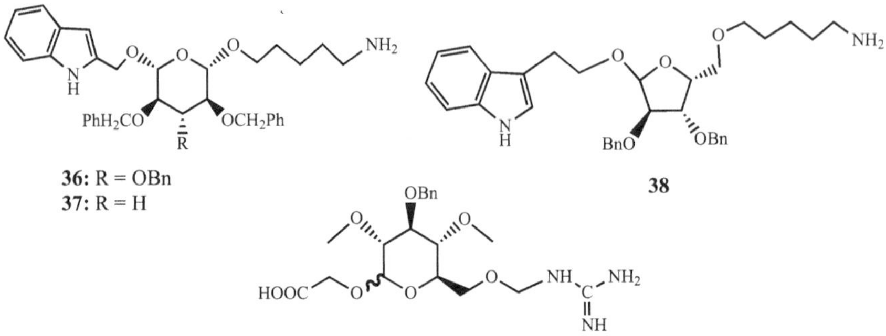

36: R = OBn
37: R = H
38

Carbohydrated based mimic of cRGDfV (39)

Figure 1.13 ***Structure of few carbohydrate-based mimics.***

(especially C-glycosides) and their impact on drug discovery has been discussed in **Prof. IPS Aidhen Chapter** in this book.

3.3 Carbohydrate-based antibiotics

Antibioticis defined as an antibacterial substance which is used to check the growth of bacteria. Antibiotics are mainly known as substances produced by microorganisms, such as, bacteria, lichens, fungi, moulds, algae, and some other plants. In past few years, semi synthetic and synthetic antibiotics have shown their potency in form of a substitute to the antibiotics obtained from natural sources.[58] Their success has encouraged the designing and development of new molecules with antibiotic activity which can overcome the drug resistance developed by microorganisms against existing antibiotics. Majority of the carbohydrate-based antibiotics are extracted from microorganisms, such as, bacteria and fungi and are known to affect the protein synthesis (translation). These antibiotics can be divided into three major categories. First category contains cyclitol antibiotics in which carbohydrate unit links to cyclitols or aminocyclitols through a glycosidic linkage. These antibiotics are called aminogylocoside antibiotics. Examples of such antibiotics are Streptomycin, Amikacin and Kanamycin. The second category of carbohydrate-based antibiotics contains a carbohydrate-nucleotide linkage. Examples of this category of antibiotics are Mureidomycins, Tunicamycin and Liposidomycin. The third group of carbohydrate-based antibiotics, known as macrolide antibiotics, includes carbohydrate linkage to a non-carbohydrate unit (for example, macrocyclic lactone) *via* a glycosidic bond. Examples of some carbohydrate-based macrolide antibiotics are Azithromycin and Erythromycin A.[59]

Aminoglycosides, which contain aminosugar substructures, are known as a class of antibiotics which deals with mainly the treatment of aerobic gram-negative bacilli infections. These molecules bind to 30S ribosomal proteins in an irreversible manner, whereas, the peptide elongation is blocked by microlides by binding to 50S ribosomal unit reversibly. Some of the oldest examples of such antibiotics are Streptomycin (**40**), Neomycin (**41**) and Gentamicin (**42**) (Fig. 1.14). Streptomycin (**40**), which comes from antinobacterium *Streptomyces grieseus*, is a known medication for management of tuberculosis. The mode of action of Streptomycin consists of binding to small 16S rRNA of bacterial ribosome to inhibit protein

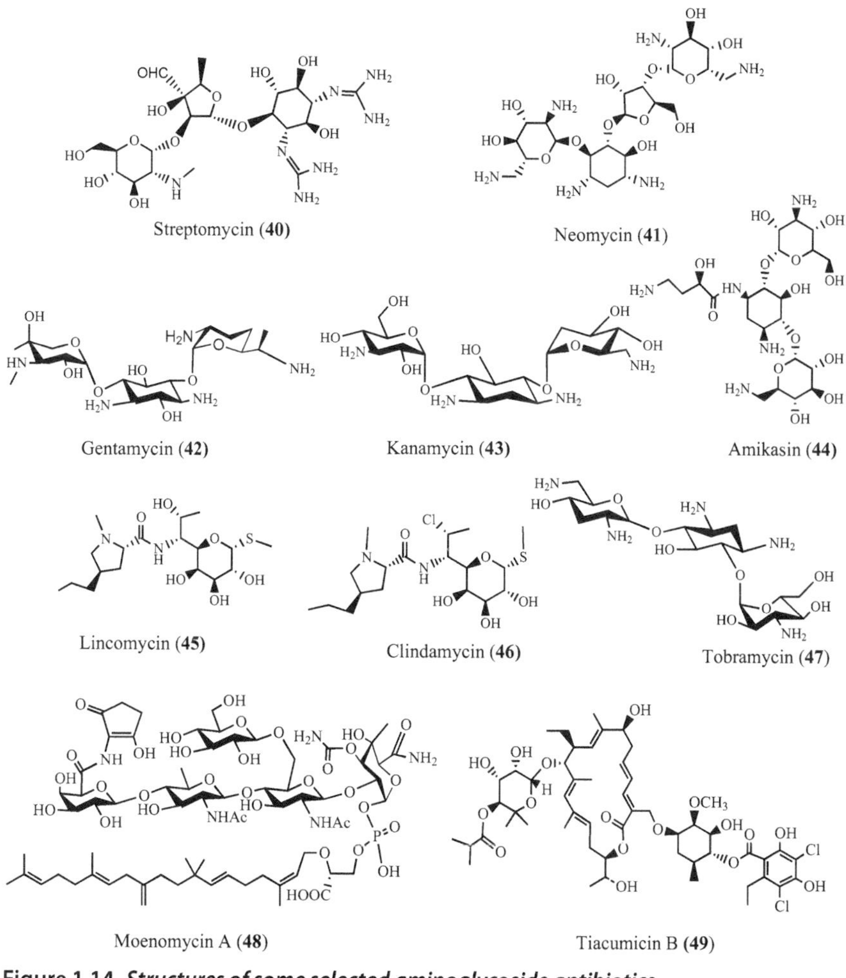

Figure 1.14 ***Structures of some selected aminoglycoside antibiotics.***

synthesis which results into interference with formyl-methionyl-tRNA binding to 30s subunit.[60] Neomycin (**41**) is generally used as a topical medicine to inhibit bacterial growth. The mode of action of Neomycin is similar to that of Streptomycin.[61] Gentamicin (**42**) is used for treatment of various types of bacterial infections including urinary tract infection, bone infection, pelvic inflammatory disease, pneumonia, meningitis, endocarditis and sepsis.[62] Kanamycin (**43**) is another aminoglycoside, which is derived from the bacterium *Streptomyces kanamyceticus* effective against *Proteus* infections; derived from bacterium *Streptomyces kanamyceticus.* Kanamycin works against *Proteus* infections by inhibiting the activity of a range of gram-positive and gram-negative bacteria. Moreover, common side effects related to aminoglycoside medication, such as, nephrotoxicity and ototoxicity are seldom observed with Kanamycin usage which makes this drug a successful antibiotic for various bacterial infections. Umezawa and co-workers analyzed the occurrence of O-phosphorylation in Kanamycin resistant organisms which is mainly responsible for the inactivation of Kanamycin. This drawback related to Kanamycin was removed by modifying the structure of Kanamycin in form of a new semi synthetic antibiotic Amikasin (**44**). Amikasin was developed by acylation of amino group attached to third carbon of 2-deoxystreptamine in Kanamycin by L-4-amino-2-hydroxybutyric acid. Amikasin was found to show enhanced activity against *Pseudomonas aeruginosa*, *Staphylococcus aureus* and *Enterobacter* species isolates. Lincomycin (**45**), a lincosamide antibiotic, is derived from actinomycete *Streptomyces lincolnensis*, whereas its semi synthetic derivative Clindamycin (**46**) is synthesized by replacement of 7-hydroxy group of Lincomycin **45** with a chlorine atom with inversion in stereochemistry. Lincomycin and Clindamycin both are used for oral administration of microbial infections. Tobramycin (**47**), another carbohydrate-based antibiotic, comes from *Streptomyces tenebrarius*. This antibiotic is used as medication against infections caused by *pseudomonas* species. The mechanism of action of Tobramycin follows the certain portions of mRNA to prevent formation of 70S complex which further inhibits the synthesis of protein. This drug is inefficient to pass gastrointestinal track, so the only mode of administration is intramuscular or intravenous.[63]

Moenomycin A (**48**) is a naturally occurring drug of Moenomycin family of phosphoglycolipid antibiotics which are metabolites of the bacterial genus *Streptomyces.* Moenomycinswork by direct inhibition of peptidoglycan glycosyl tranferases (PGTs) in the last step of synthesis of bacterial cell wall.[64] Presence of 3-Phosphoglyceric acid (3-PG) with atypical isoprenoid chain at the second hydroxyl group and the reducing end of polysaccharide linked with a phosphate group are some unique structural features associated with the Moenomycin family.[65] A 18-membered macrolactone Tiacumicin B (**49**) is used as a medication for the treatment of *Clostridium defficile* infection (CDI). It acts as a RNA synthesis inhibitor which was approved in 2011by FDA.[66]

Nucleoside antibiotics is a large group of natural products having different antibiotic activities. In general, nucleoside antibiotic combat the bacterial growth by inhibiting the biosynthesis of peptidoglycan layer which are essential part of bacterial cell wall. Nucleoside antibiotics inhibit phospho-Mur*N*Ac-pentapeptide translocase that further restricts biosynthesis of peptidoglycan layer.[67] The only known clinically used nucleoside antibiotic is Ramoplanin (**50**, Fig. 1.15) which works by inhibition of *O*-linked *N*-acetyleglucosamine (GlcNAc) transferase gene.[68]

Ramoplanin (**50**) is a glycolipodepsipeptide antibiotic which is extracted from the strain ATCC 33076 of *Actinoplanes.* It is being administered under phase-2 trials for cure of multiple antibiotic-resistant *Clostridium difficile* infection of the gastrointestinal tract,The mode of action of this drug is follows formation of a U-shaped structure that inhibits of the bacterial cell wall synthesis which in turn captures and binds the specific intermediate Lipid II (C35-MurNAc-peptide-GlcNAc) during formation of membrane.[69,70]

Macrolide is another class of antibiotics which are derived from natural sources, mainlyfungal metabolites (Fig. 1.16). Macrolides are consisting of a

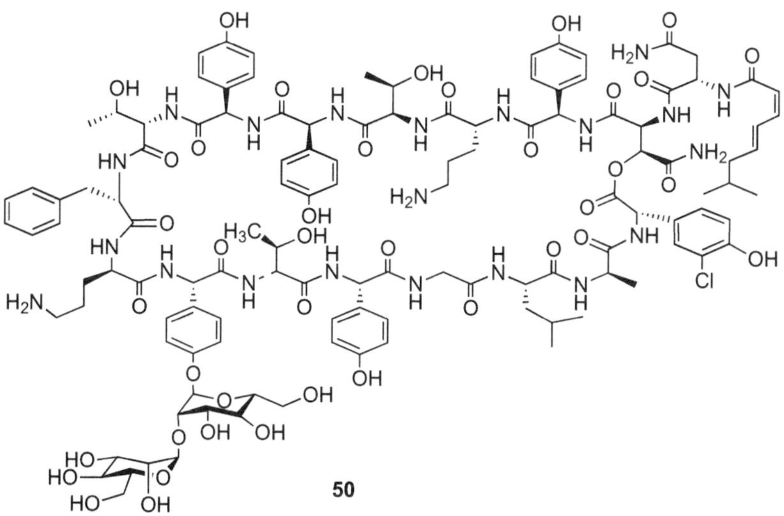

50

Figure 1.15 ***Structure of Ramoplanin* (50)*, a carbohydrate-based nucleoside antibiotic.***

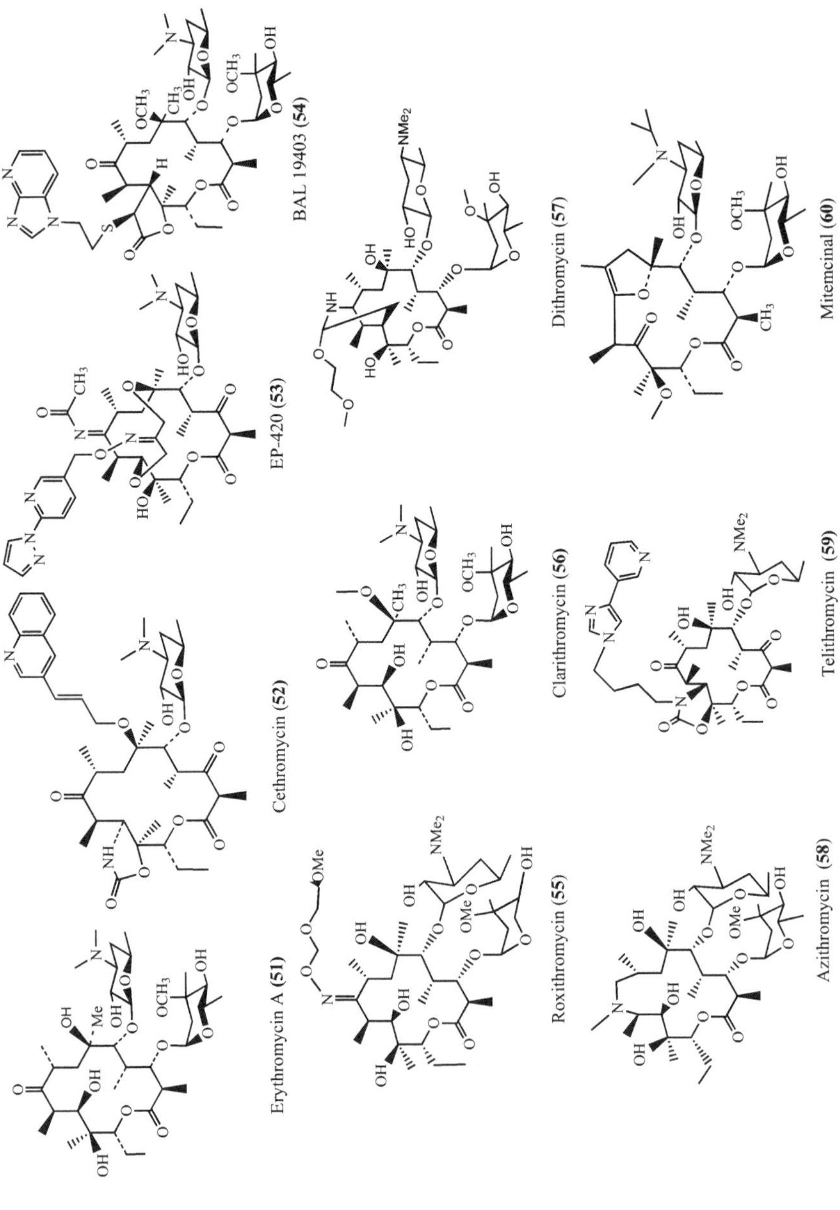

Figure 1.16 ***Structure of few selected macrolide glycosides antibiotics.***

macrocyclic lactone unit attached to one or two sugar units or amino sugars.[71] In general, macrolides are clinically used against bacterial diseases caused by gram positive bacteria. Example of a well known macrolide is Erythromycin A (**51**), which is derived from *Streptomyces erythreus.* Eryhthromycin A is very effective against main respiratory pathogens, such as, *Mycoplasma* and *Legionellia.* The mechanism of this antibiotic follows binding to 50S ribosomal subunit through peptidyltransferase site and inhibition of the protein synthesis. Other derivatives of this antibiotic, such as, Cerythromycin (**52**), EP-420 (**53**) and BAL-19403 (**54**), are under clinical trials. Cerythromycin **(52)**, is a ketolide antibiotic, considered as clinically effective against plague, tularaemia, and anthrax. This drug has been labeled as an 'orphan drug' for cure of tularaemia and anthrax. EP-420 **(53)** is a bridged bicyclolide which is under phase II trials as medication against acquired pneumonia (CAP). Another macrolide BAL-19403 **(54)** has shown its potency in clinical trials against acne. BAL-19403 is considered to facilitate the mutations in *Propionibacterium acnes* in the 2057 to 2059 region of 23S rRNA.[72–74]

A second generation of macrolides, such as, Roxithromycin **(55)**, Clarithromycin **(56)**, Dithromycin **(57)**, Azithromycin **(58)**, and Telithromycin **(59)** have been introduced as anti-infective agents with a range of activities after finding limitations of Erythromycin regarding solubility in acidic medium and limited spectrum of activity. With enhanced physicochemical profiles and improved pharmacokinetics, these macrolides show reduced side effects as compared to Erythromycin A.[75] Telithromycin **(59),** a semi synthetic derivative of Eryhthromycin A, is the first FDA approved antibiotic to be clinically deployed for cure of respiratory tract infections mainly acquired pneumonia (CAP). Mitemcinal **(60)** is a non-peptide motilin receptor agonist with add on properties like acid-resistance and orally active. This drug has shown clinical potency in Phase II trials for improvement of gastric emptying by promoting antroduodenal motility which further commences gastric contractions in diabetic and idiopathic gastroparesis patients.

Liposidomycin (**61**) is clinically found to potentially inhibit the biosynthesis of bacterial peptidoglycan.[76] Mureidomycin A (**62**) is a peptidylnucleoside which shows antipseudomonal activity. This antibiotic inhibits the synthesis of peptidoglycan and formation of lipid-intermediate from UDP-N-acetylglucosamine and UDP-N-acetylmuramyl (MurNAc)-peptide.[77] Tunicamycin (**53**) (Fig. 1.17) is extracted from *Streptomyces lysosuperificus* and is known to inhibit *N*-linked glycosylation. This drug inhibits protein folding by restricting the addition of oligosaccharide to nacent polypeptides.[78]

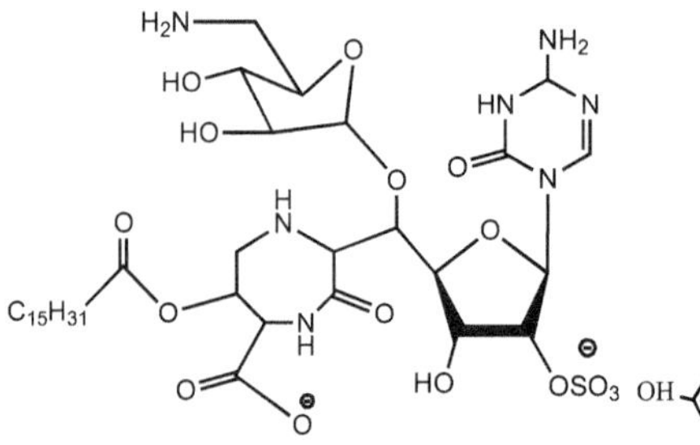

Liposidomycin (**61**)

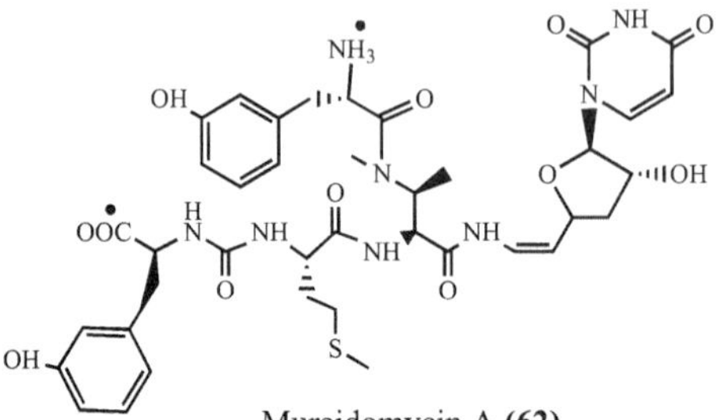

Mureidomycin A (**62**)

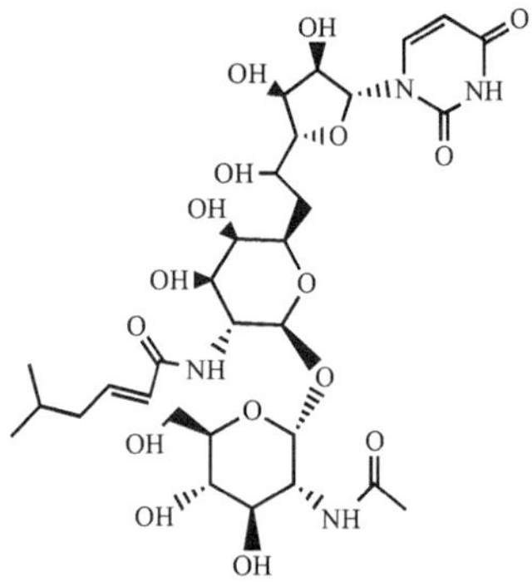

Tunicamycin (**63**)

Figure 1.17 ***Structure of peptidoglycan inhibitors Liposidomycin and Mureidomycin A, and protein folding inhibitor Tunicamycin.***

Vancomycin (**64**) is used intravenously as an antibiotic for treatment of a various bacterial infections, such as, blood stream infections, skin infections, bone and joint infections, endocarditis, and meningitis. Vancomycin and related antibiotics show specific activity against Gram-positive bacteria and show inefficacy in passing through the outer membrane of Gram-negative bacteria making them inactive against Gram-negative bacteria. Telavancin (**65**), a derivative of Vancomycin, is used for the treatment of treatment of infections related to skin. This drug exhibits dual mode of action against Gram-positive bacteria by restricting transglycosylation and further transpeptidation. With regards to inhibition of peptidoglycan synthesis in intact Methicillin- resistant *Staphylococcus aureus* (MRSA) cells, Telavancin shows more efficacy than Vancomycin.[79,80] Another semisynthetic derivative of Vancomycin, Tiecoplanin A40926 (**66**) was introduced to clinical trials in 1959 and found to have anti-Neisseria activity. Dalbavancin (**67**) is a synthetic derivative of Teicoplanin A40926 is another lipoglycopeptide antibiotic for cure of bacterial skin infections which has been recently approved by FDA in 2014.[81,82] (Fig. 1.18)

Oritavancin (**68**), was developed by a Eli Lilly in 2001 as a semi-synthetic glycopeptides derivative of Chloroerythromycin (**69,** Fig. 1.18).

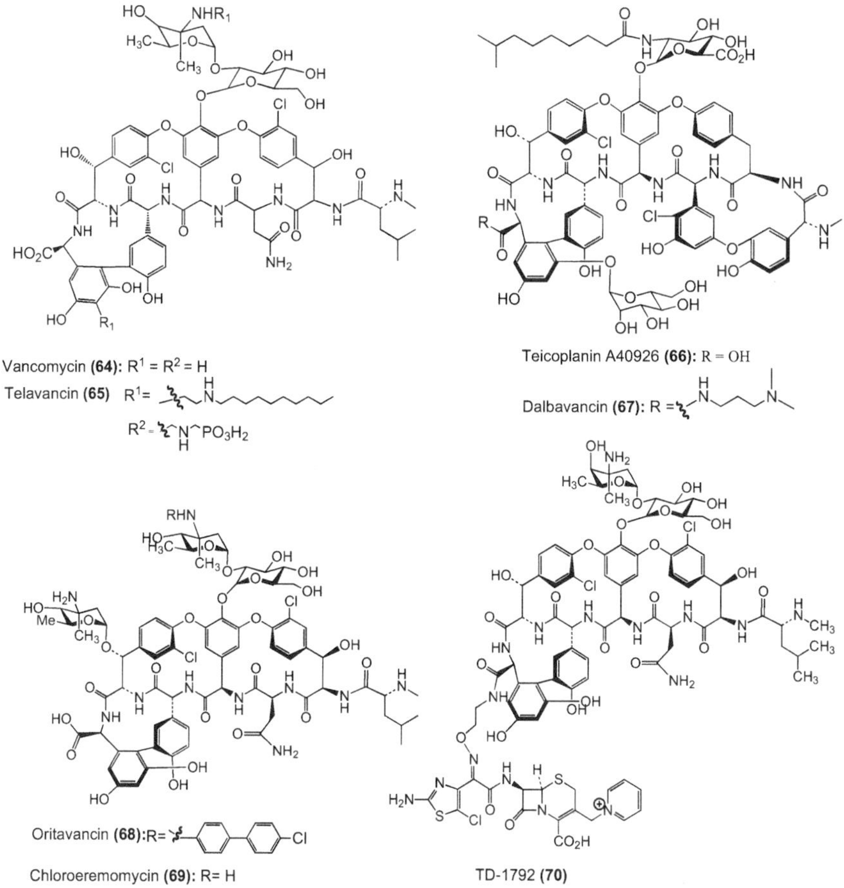

Figure 1.18 ***Clinically used carbohydrate-based macrolide antibiotics.***

Oritavancin follows a mechanism of action similar to that of Telavancin to combat the Gram-positive bacteria. Non-availability of required data related to Oritavacin caused declination of its approval by FDA. TD-1792 (**70**), an improved glycopeptides-cephalosporin heterodimer derivative of Vancomycin is found to be effective against complicated skin and skin structure infections including Methicillin-resistant Staphylococcus aureusin Phase-II trials and proved better efficacy as compared to Vancomycin (**64**).[83]

Nicalaou and coworkers opted to modify the carbohydrate unit in Vancomycin for better results in terms of enhanced activity and reduced side effects. Applying this approach, they developed a library of Vancomycin derivatives by treatment of vancomycin analogs (**71**) (Fig. 1.19) with a variety of 3,4-disubstituted benzaldehydes. The outcome of this strategy was fruitful with numerous synthesized derivatives showing more potency against vancomycin resistant strains.[84]

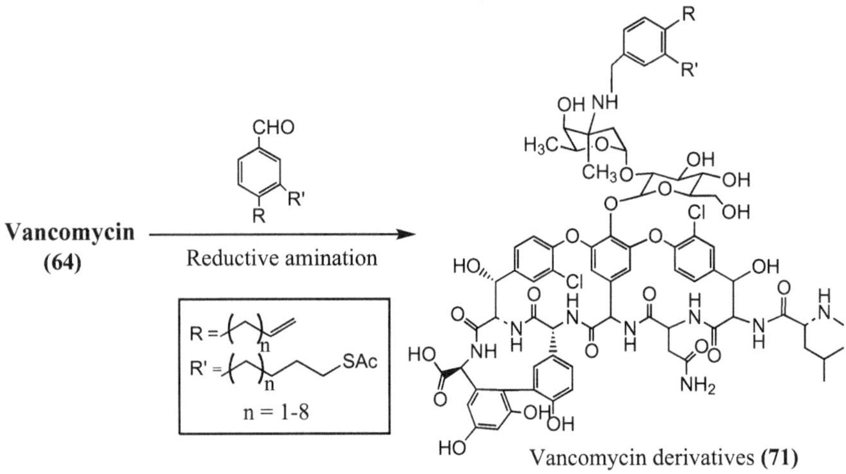

Figure 1.19 ***Synthesis of vancomycin derivatives* 71.**

Due to significant outcomes of carbohydrate derivatives as antibiotics, drug discovery and pharma research in the field of antibiotics has welcomed carbohydrate moiety with open hands. In last few years many carbohydrate-based molecules have reported to have excellent antibiotic properties and many of them are under clinical trials. A detailed account of these development and advancements in carbohydrate-based antibiotic drug discovery has been given in **Prof. Girija S Singh Chapter**.

Cu(I)-catalyzed Click chemistry (CuAAC) to generate 1-4-disubstituted triazoles with a great ease is receiving great success in past few years, particularly in carbohydrate chemistry.[85] In the field of drug discovery and development, this protocol in combination with combinatorial technique, is considered to be an unparalleled approach for the target based structural design and development of new leads. This combination has led facile development various novel therapeutic agents and pharmaco-mimics.[86] Carbo Click offers an elaborated description of significance of carbo-click reaction in drug discovery and this section is greatly discussed in Kushwaha and Tiwari Chapter. This useful click strategy in combination with thioesterase mediated macrocyclization was applied by Walsh and Lin to develop a library of 247 glycopeptides with a motive to attain a reduced toxicity as compared to the existing antibiotics. Biological assessment of the developed compounds exhibited good results with two compounds Tyc4PG-14 **(72)** and Tyc4PG-15 **(73)** showing anti-bacterial properties equivalent to native Tyrocidine, and having six folds better therapeutic index (Fig.1.20).[87]

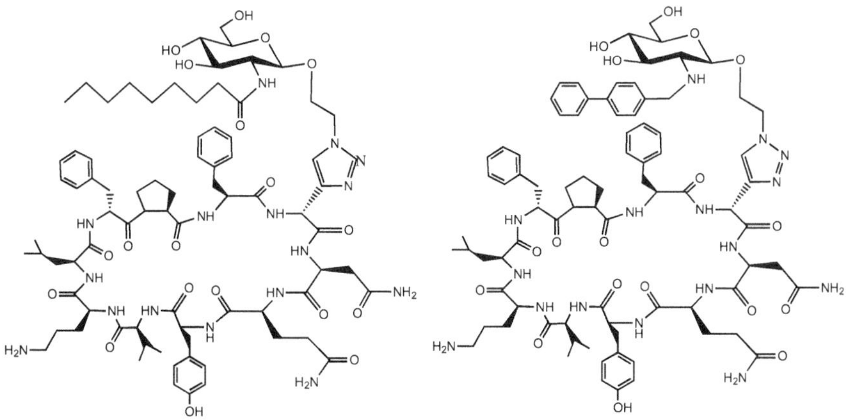

Tyc4PG-14 **(72)** Tyc4PG-15 **(73)**

Figure 1.20 ***Synthetic glycopeptides* 72, 73 *with anti-bacterial activity.***

Hybrid molecules combine two distinct biologically active molecules that act at different targets, into one new molecule/chemical entity to combine the effects of each molecule. Impact of glycohybrids molecules in drug discovery and development has been discussed in **Prof. Ram Sagar Chapter** of this book.

Despite of extensive development in this field, carbohydrate-containing molecules are still not truly well represented, mainly due their poor absorption, *in vivo* stability, overall polar character, and also their poor affinity. In recent years, Click inspired multivalency principle has been used to enhance the affinities. Role of statistical rebinding through multivalency effects in order to enhance our understanding towards Carbohydrate -protein Interactions has been nicely presented in **Prof. Roland J. Pieters Chapter.**

3.4 Carbohydrate-based anti-cancer agents

Tumor-specific carbohydrate antigens (TACAs) are over expressed on the surface of cancer cells as a consequence of aberrant glycosylation. These carbohydrate antigens (TACAs) are conventional biomarkers for cancer detection and are concerned when it comes to development of anti-cancer therapies. An increased number of lectins and glucose transporters (GLUTs) are found on the membrane of cancer cells which helps in binding and transportation of carbohydrate moieties. GLUTs help in fulfilment of the increased demand of glucose for energy in cancer cells proliferation. Thus, cancer cells show a higher uptake of glucose than normal cells.[88] Thus a sugar-based targeted drug delivery is considered a fruitful approach for development of

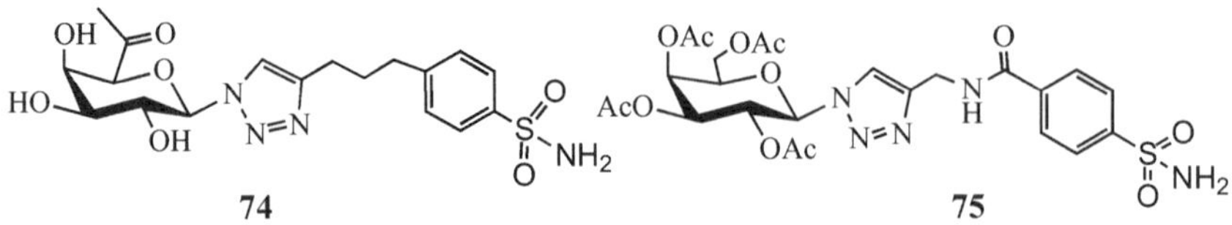

74 75

Figure 1.21 ***Carbohydrate-based benzene sulphonamides as carbonic anhydrase inhibitors.***

anti-cancer therapies. Carbonic anhydrases (CAs) are enzymes which catalyze the reversible hydration of carbon dioxide which results in production of a proton and bicarbonate. Continuous division of cancer cells creates hypoxic conditions which results into an acidic microenvironment. CAs help in maintaining the physiological intracellular pH for cancer cells and make an acidic extracellular pH which helps the cancer cell survival, invasion and metastatis. Hence, carbonic anhydrase inhibitors are supposed to provide a therapy lead against cancer cells.[89] A conjugated support, such as, polymers, nanoparticles, liposomes and antibodies, enhance the activity of anti-cancer agents by providing a suitable drug-delivery. This strategy has been explored by pharmacists to develop new therapeutic agents.

A carbohydrate appendage offers a useful mean in development of CA-Isozyme selective moieties. A series of benzene sulphonamide glycoconjugates was developed by Wilkinson et al. employing Click chemistry. The developed compounds were tested for the anti-cancer activity by Pastoreková et al. who reported compounds **74** and **75** (Fig. 1.21) to be active against the CA-IX.[90,91]

Recently, Mishra et al. have developed a series of noscapine glycoconjugates from noscapine (**76**) by utilizing Click chemistry (Fig. 1.22),[92] where

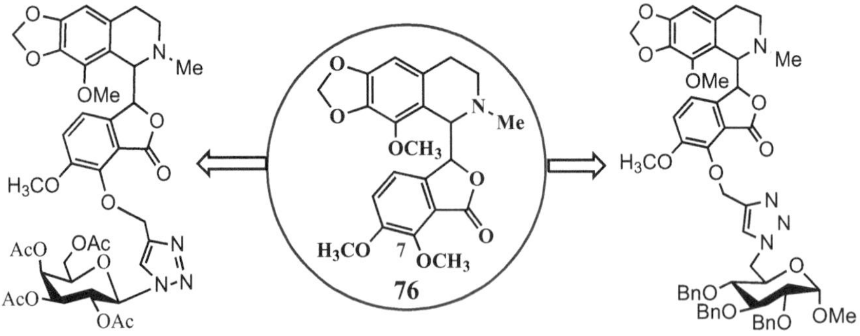

77: Anti-cancer activity, (IC_{50} = 29.2μM)

78: Anti-leishmanial, (IC_{50} = 8.36 μM)

Figure 1.22 ***Click inspired biologically relevant noscapine glycoconjugates.***

few compounds exhibited significant anti-cancer activity (compound **77** displayed IC_{50} = 29.2 μM, where 5-flourouracil was used as standard having IC_{50} = 49.2 μM). In addition, these molecules were evaluated for the anti-leishmanial activity, and few of them showed significant bioactivity. For example, compound **78** displayed IC_{50} = 8.36 μM and Miltefosine was used as standard reference drug for both anti-leishmanial and cell toxicity evaluations.

Neocarzinostatin (**79,** Fig. 1.23) is an antitumor agent which is derived from *Streptomyces macromomyceticus.* It contains a macrocyclic chromoprotein enediyne structure which is consisted of two components, first a 113 amino acid protein, and second a labile chromophore non-covalantly bound with the protein part with high affinity. Two molecules of polystyrene-co-maleic acid and one molecule of neocarzinostatin chromoprotein conjugate to form Zinostatin stimalamer which is a lipophilic intra-arterial therapeutic agent, and is used to treat hepatocellular carcinoma (HCC).[90] Gemtuzumab ozogamicin (**80**) is a conjugate of calicheamicin and a humanized anti-CD33 monoclonal antibody and is used for treatment of acute mylogenous leukemia.[91]

Inotuzumab ozogamicin (CMC-544) (**81,** Fig. 1.23) is a drug recently approved by European Commission and FDA in 2017 for the treatment of relapsed acute lymphoblastic leukemia (ALL). This drug consists humanized IgG4 anti-CD22 acid sensitive 4(4-acetylphenoxy) butanoic acid attached to an *N*-acetyl gamma calicheamicin dimethyl hydrazide by covalent bonds and marketed by Pfizer/Wyeth under the trade name Besponsa.[93–96]

Anthracyclines are anti tumor antibiotics which are used in chemotherapy of cancers. The action of these drugs involves the interference of the enzyme sengaged in DNA replication. Most of the members of this class of anticancer drugs contain a deoxy amino sugar unit in their structure. For example, Doxorubicin (**82**) is an anthracycline anti-tumor drug which is used for carcinomas of breast, ovary, liver, lung, and thyroid, lymphomas and lukemias under the trade nameAdriamycin.[97] The action of Doxorubicin follows the stabilisation of the topoisomerase II complex which is responsible for the relaxation of DNA supercoils during transcription. Hence the stabilization of topoisomerase II enzyme ceases the DNA replication.[98,99] Crystallographic studies of doxorubicin reveals that aromatic chromophoric moiety of doxorubicin inserts between the base pairs of the DNA, whereas, daunosamin sugar present on minor grooves of DNA interact with the flanking base pairs.[100] Another prototypical drug of this class is Daunorubicin (**83**) which is used to treat chronic myelogenous leukemia (CML), acute

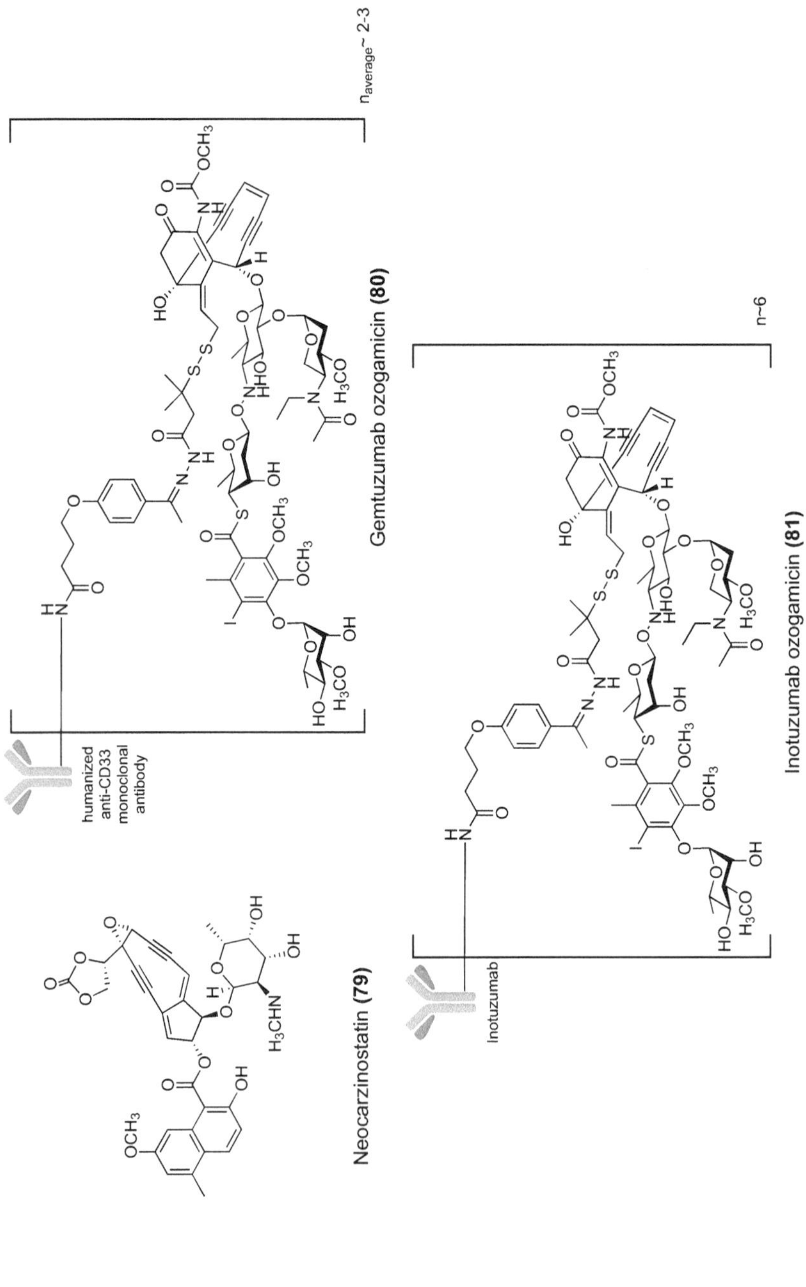

Figure 1.23 ***NP-Antibody glycoconjugates as anti-cancer drugs.***

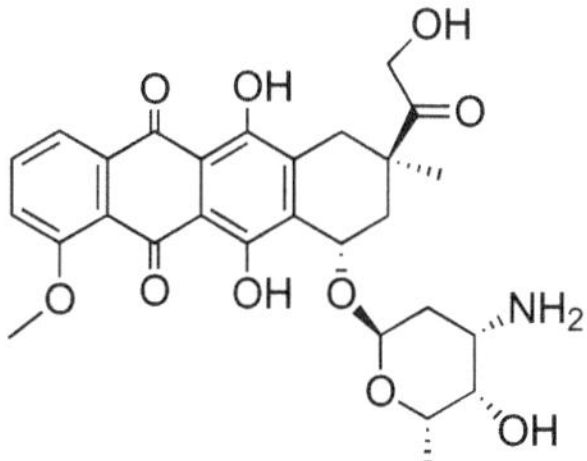

Doxorubicin (**82**)

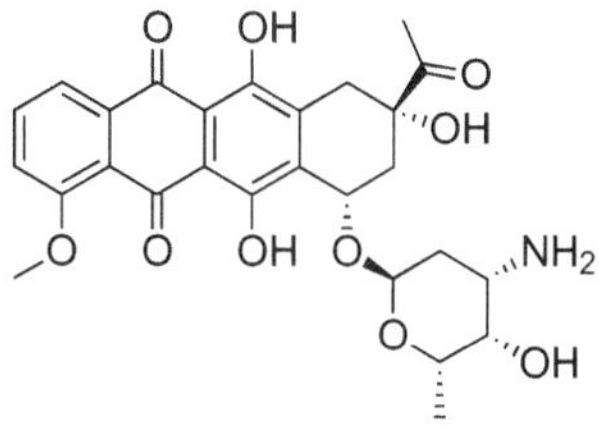

Daunorubicin (**83**)

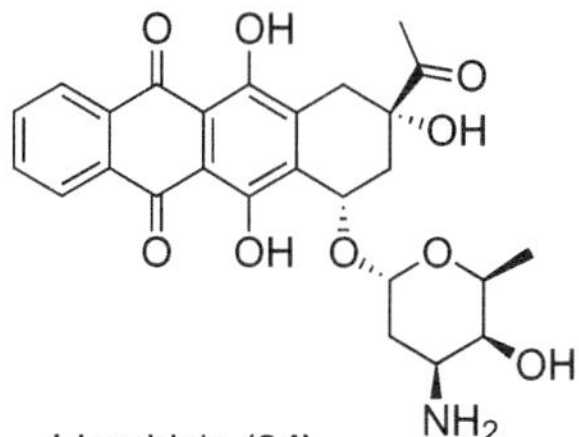

Idarubicin (**84**)

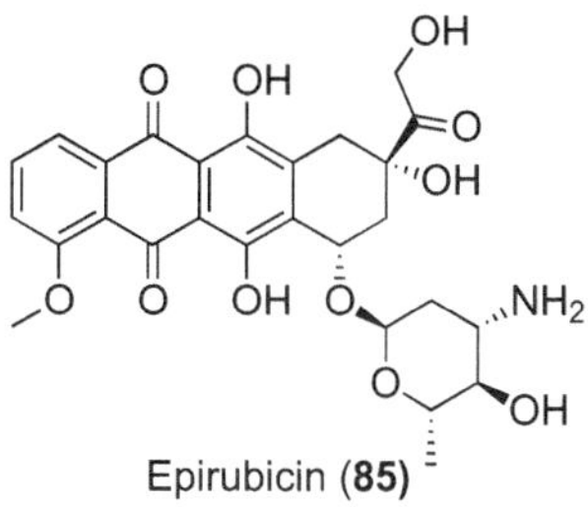

Epirubicin (**85**)

Figure 1.24 ***Carbohydrate-based anthracyclines for chemotherapy.***

lumphocytic leukemia (ALL), acute myeloid leukemia (AML), and Kaposi's sarcoma.[101] Idarubicin (**84**) and Epirubicin (**85**) (Fig. 1.24) are two anthracyclines which have got acceptance for worldwide chemotherapeutic use. Idarubicin is a fat soluble variation of Daunorubicin which is orally bioavailable therapy for treatment of acute myeloid leukemia (AML).[102] Epirubicin shows mode of action and activity similar to Doxorubicin though has lesser cardiotoxic side effects.[103]

Latest reports have revealed the importance of glycolipids as pharmacophore scaffolds for development of compounds with pronounced pharmaceutical use in cure of ailments related to aberrant regulation of protein synthesis and cell proliferation such as cancer. In a recent patent by Bottley et al. compound **86** has been reported to show 90% inhibition against rapidly progressive breast carcinoma cell lines MDA-MB-231 and MCF-7 at a concentration of 6 μg/mL for 96 hours treatment. On the other hand, for slow growing lung cancer cell line A549 which is cisplatin resistant, a 1 μM combination of cisplatin with compound **C** was found able to sensitize the carcinoma cells with complete inhibition at 8 μg/mL concentration. In a similar way, compounds **87**, **88** and **89** (Fig. 1.25) having a *D*-glucopyranosyl or *D*-mannopyranosyl residue in place of *D*-galactopyranose moiety and 2-hydroxymethyl-1,3-propanediol group instead of glycerol residue of **86**,

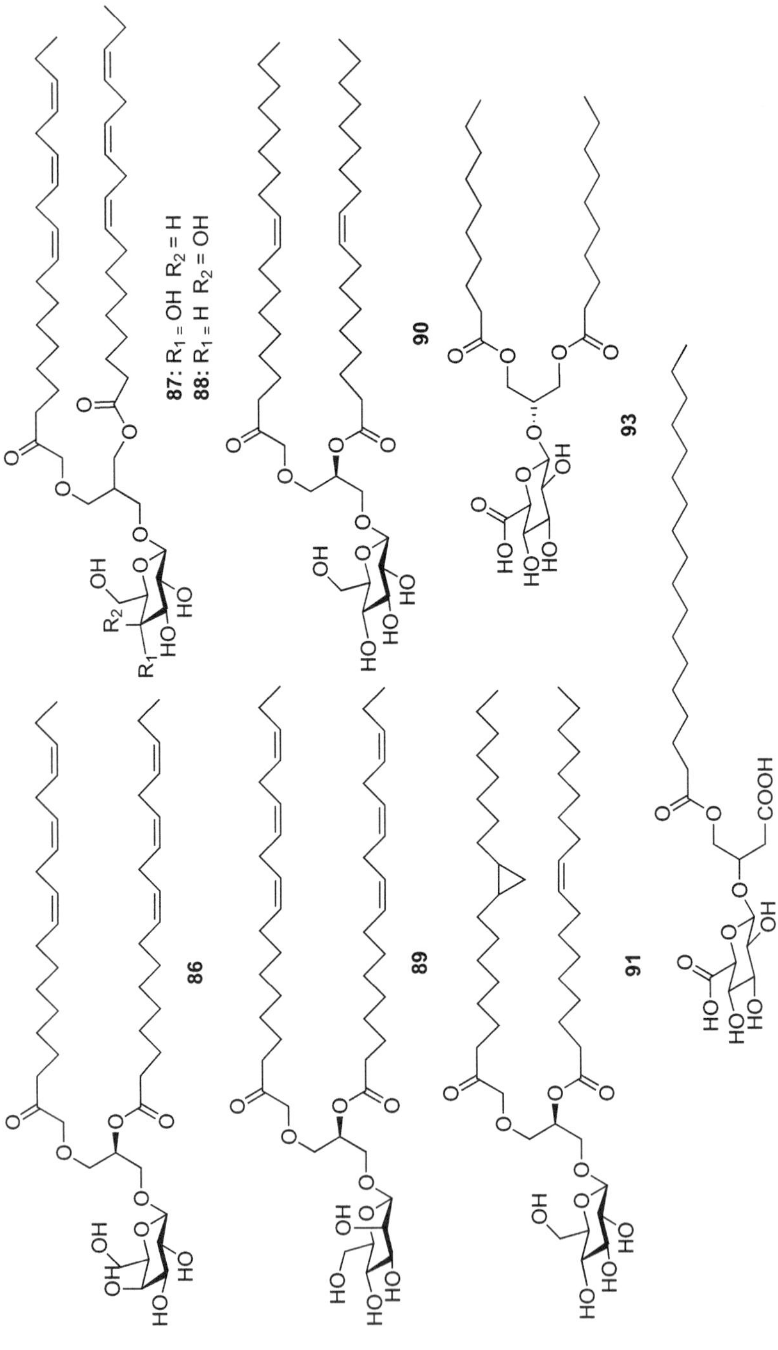

Figure 1.25 ***Glycolipid with significant cytotoxic properties.***

also showed good results for inhibition of lung carcinoma cell line A549 with complete inhibition at 20–40 μM concentration. These compounds with a combination of very low dose of cisplatin also sensitized A549 cisplatin resistant lung cancer cell line.[104] Glycolipid analogs **90** and **91** developed by Srikanth et al. exhibited cytotoxic properties against human cancer cell lines A549, MCF-7, DU145, and SKOV3,[105] while **92** and **93** were reported by Vetro et al. to have promising cytotoxic activity against IGROV-1 ovarian carcinoma cell line.[106]

Compounds which comprise sulfur containing functionalities are acknowledged to exhibit excellent anticancer activities, whether they are naturally occurring or synthetically developed.[107] Kazunari and co-workers have recently developed an analog of sulfoquinovosylmonoacylglycerol(SQMG) (18:0) in which glycerol moiety was replaced by propanediol moiety. The developed compound sulfoquinovosylacylpropanediol (SQAP) **94** was found a potent agent for radiation therapy of prostate cancer DU145 cells in mice.[108] SQAP assisted radiotherapy exhibited promising results for human urothelial carcinoma cell lines UM-UC-3 and 5637.[109] Fumio and co-workers tried to apprehend the mechanism of action of SQAP. In their study focal adhesion kinase (FAK), which plays an important role in angiogenesis and cell turnover, was found to be one of the five target proteins identified.[110] Another natural sulfonated glycolipid sulfoquinovosyl diacylglycerols (SQDG) inspired Costa et al. for the development of synthetic sulfoglycolipid **95** as potential serine/threonine protein kinase (AKT) inhibitor for anti-cancer therapy. Compound **95** (Fig. 1.26) showed

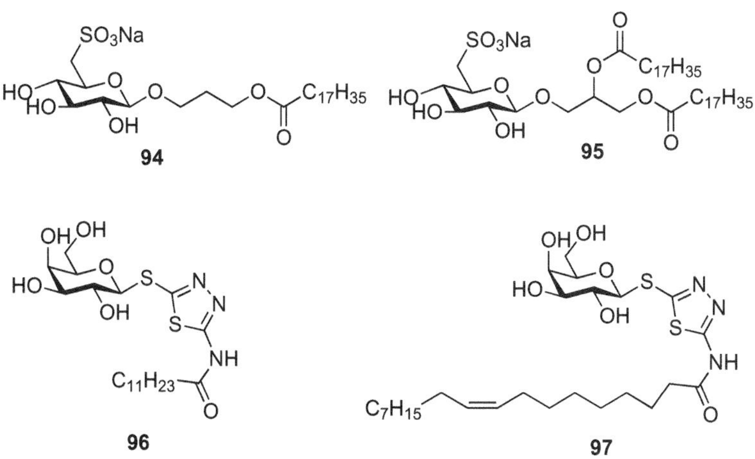

Figure 1.26 ***Carbohydrate-thio derivatives with anticancer activity.***

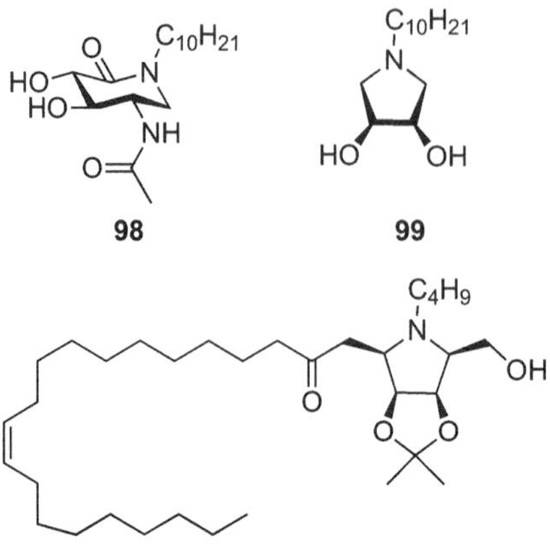

Figure 1.27 ***Azasugar derivatives with promising cytotoxic properties.***

concentration dependent inhibition of human papillary thyroid cancer (PTC) cell line NIM-1 with IC_{50} 3 ± 1.6 μM and dose dependent inhibition of ovarian cancer cell line IGROV-1 with IC_{50} value 2 ± 1.4 μM.[111] Very recently,Vudhgiri and co-workers have developed a series of cytotoxic thioglycosides of 5-acylamido-1,3,4-thiadiazole-2-thiol. The most notable cytotoxicity was exhibited by compounds **96** and **97** against cervical carcinoma cell line HeLa with IC50 value of 8.7 and 8.8 mM, respectively.[112]

Certain derivatives of azasugars have been found to possess exceptional cytotoxicity against various carcinoma cell lines. For example, compound **98** and **99** (Fig. 1.27) have shown promising activity against K562 cell line of chronic myeloid leukemia[113], whereas, compound **100** exhibited cytotoxicity against pancreatic carcinoma lines PK9 and MiaPaCa2.[114]

Photodynamic therapy (PDT) is a treatment for cancers which uses a drug (a photosensitizer or photosensitizing agent) which when exposed to a particular light wavelength, produces a form of oxygen which in turn kills cancer cells. Since, porphyrin core of the dendrimer acts as a photosensitizer and produces lethal singlet oxygen to kill cancer cells while the carbohydrate part of the dendrimer provides increased tumor cell specificity as well the water solubility, thus glycodendrimers with porphyrin core may found remarkable application in PDT. In this context, we developed porphyrin-cored glycodendrimers which were studied for their absorption-emission behavior,[115] however their use in PDT is still under investigation.

A number of triazolyl glycoconjugates with promising anticancer activities have been developed during last eighteen years by employing the facile synthetic approach of click-chemistry. This section has been well discussed in **Kushwaha and Tiwari Chapter** of this book.

3.5 Carbohydrate-based anti-diabetic agents

Diabetes mellitus has a globally spread health issue which has emerged as a great cause of concern because of related mortality rate. Type II diabetes mellitus which is the most common type of diabetes is described by a condition in which the course of processing of blood sugar (glucose) by the body is affected. Type II diabetes mellitus causes declines in insulin secretion, in addition to insulin resistance in peripheral tissues which in turn causes hyperglycemia, an increase in blood glucose level. Therapy often involves the identification of specific targets that involve and affect glucose metabolism and their modulation.

3.5.1 Glycosidase inhibitors with anti-diabetic effects

We recently presented the scope of carbohydrate containing molecules in the management ofType II diabetes.[116] A number of sugar-based glycosidase inhibitors and sodium glucose co-transporter 2 (SGLT2) inhibitors were considered as the major active classes of molecules for the management of diabetes. Many glycosidase enzymes assist the catalytic hydrolysis of glycosidic bonds. Therefore glycosidase inhibitors are helpful in controlling the blood sugar level.

The α-glucosidase inhibitors with anti-diabetic activity are projected with a number of common structural aspects with a family of new chemical entities that perform as mimic of sugar skeletons. In general they bear ring structure with amine functionality and also few hydroxyl groups set in three dimensional manners analogous to basic carbohydrate framework. These structural similarities are often used as a tool to modify functional aspects of mimicking entities, where the concept is incredibly envisaged in case of carbasugars (**101**) having carbocyclic framework and also iminosugars (**102**) (Fig. 1.28) having *N*- as heteroatom in ring with various related functionalities. On account of their promising bioactivity and notable enzymatic actions, they can generate a family of glucosidase inhibitors aiming to target diabetes.[117\]

Acarbose (**8**), a pseudotetrasaccharide extracted from culture broths of *Actinoplanes* strain SE 50, is the first approved anti-diabetic drug. The drug acts by reversible binding to oligosaccharide binding site of α-glucosidase which in turn inhibits the hydrolysis of oligosaccharides and decreases the level of glucose in blood.[117] Another α- glucosidase inhibitor Miglitol (**104**, Fig. 1.29),

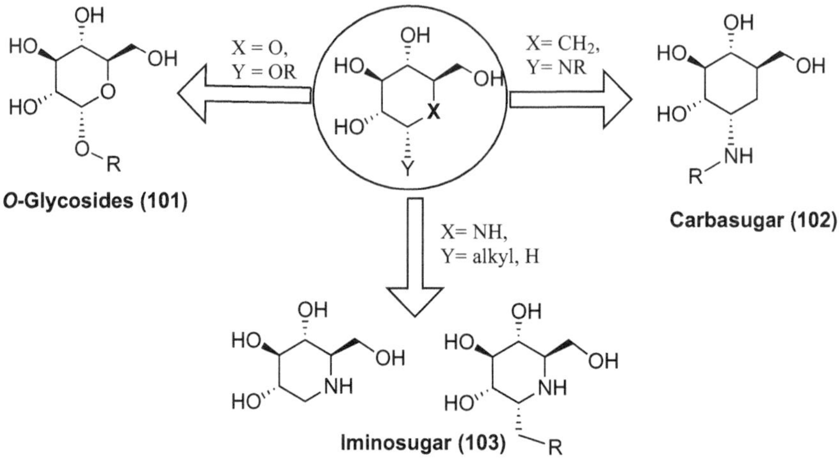

Figure 1.28 ***Strategic modification in carbohydrate framework to develop α-glucosidase inhibitors.***

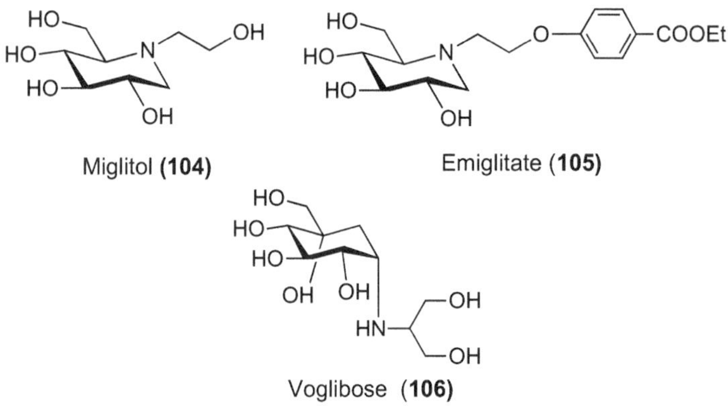

Figure 1.29 ***Some carbohydrate mimetics as approved glycosidase inhibitors with anti-diabetic activity.***

which is an approved anti-diabetic agent and prevents the hydrolysis of carbohydrates into simpler forms.[118] Emiglitate (**105**) was developed as an analog of Miglitol having more lipophilic nature but lower inhibitory profile.[119] Voglibose (**106**) is another α-glucosidase inhibitor used for patients of diabetes mellitus to lower the level of blood glucose in postprandial condition.[120]

3.5.2 Sodium dependent glucose co-transporter inhibitors

Sodium dependent glucose co-transporter (SGLT2) has become known to be a novel and extensive target for therapy of Type II diabetes mellitus. The

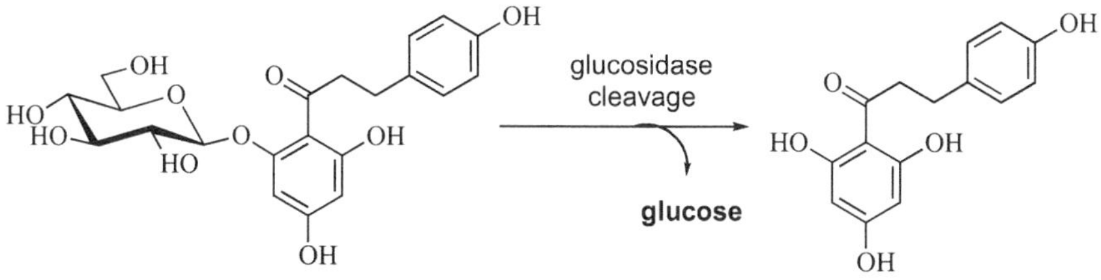

Phlorizin **(107)** Phloretin **(108)**

Scheme 1.1 ***General scheme for glycosidic cleavage of Phlorizin producing Phloretin.***

first SGLT inhibitor was Phlorizin **(107)**, a naturally occurring *O*-aryl glycoside. It was found a potential drug for treatment of type-2 diabetes but its metabolic instability compelled to develop more promising and more selective analogs.[121]

Phlorizin (**107**) is known to be the first *O*-aryl glycoside SGLT2 inhibitor, however, because of its unstable metabolism and vulnerability to β-glycosidic cleavage, this *O*-aryl glycoside couldn't be approved as a safe anti-diabetic drug. A general scheme for glycosidic cleavage of phlorizin (**107**) leading to formation of phloretin (**108)** is depicted in Scheme 1.1. Further developments in this direction offered a number of related *O*-glycosides with sufficient metabolic stability and also improved efficacy than phlorizin.[122,123]

One of the synthetic analog of Phlorizin (107) is Empagliflozin (**109**) which is used to lower blood sugar level of type-2 diabetes patients.[124] Canagliflozin **(110)** is also an analog of Phlorizin (**107**) which was developed as SGLT2 inhibitor for treatment of type-2 diabetes mellitus.This drug got approval from FDA in 2013as an orally administered and safe drug.This drug showed efficacy in treatment of obesity too.[125] Another example is Dapagliflozin (**111**)**,** a C-aryl glycoside which has recently got approval by FDA in 2014 as a SGLT2 inhibitior.The ameliorating effect of Dapagliflozin **111** on both postprandial and hypoglycemic state made it superior and beneficial to other medications used for glycemic control.[126–129]

To develop novel potent SGLT2 inhibitors in form of enhanced selectivity and activity, many derivatives of Dapagliflozin have been developed. For example, in order to get better binding affinity with SGLT2 by decreasing the negative charge density of oxygen in the ring, a series of fluorodapagliflozins (**112–115**) with high electron withdrawing difluoro subsituent were synthesized. Better SGLT2 inhibitory activities were shown by fluorodapagliflozins (**112–115**) (Fig. 1.30) as compared to Dapagliflozin

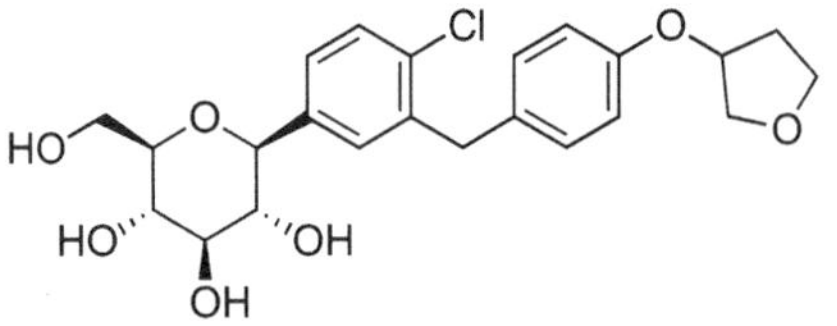

Empagliflozin (**109**)

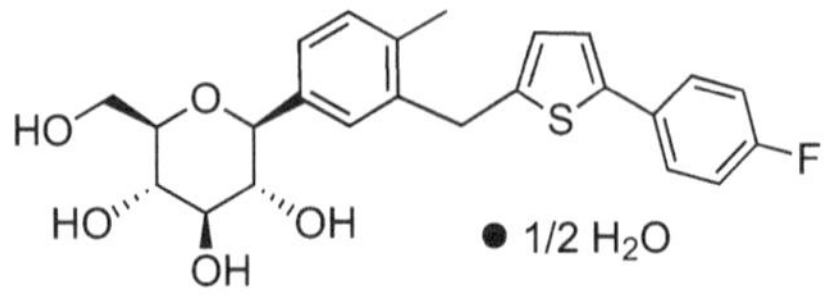

Canagliflozin hydrate (**110**)

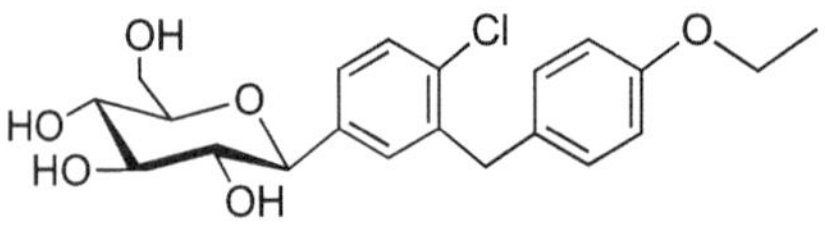

Dapagliflozin (**111**)

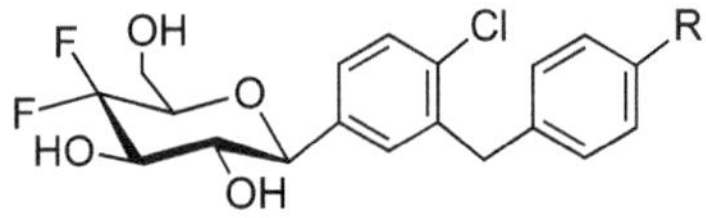

Fluorodapagliflozin

112: R = Et; **113**: R = Me

114: R = O^{i}-Pr; **115**: R = OMe

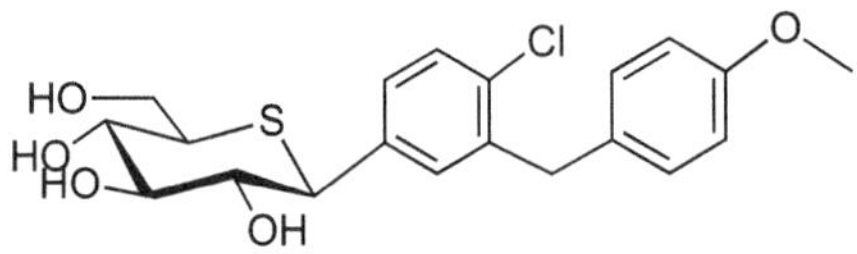

Thiodapagliflozin (**116**)

Figure 1.30 ***Recently developedcarbohydrate-based SGLT2 inhibitors.***

(**111**) with a 3.6- to 5.6- folds increase in potency and IC_{50} values less than 0.60 nm.[130] Thiodapagliflozin (**116**) is a potent SGLT2 inhibitor designed by replacing sugar moiety of Dapagliflozin with a thio-sugar analog. Promising SGLT2 inhibitory activity was shown by Thiodapagliflozin with IC_{50} value 1.68 nm.[130]

Likewise, a novel class of macrocyclic *C*-aryl glycoside was developed in order to get a promising SGLT2 inhibitor simply with an aid of macrocyclization to join carbohydrate moiety with proximal cyclic structure. Macrocycle **117**and **118** (Fig. 1.31) displayed significant *in vitro* activity

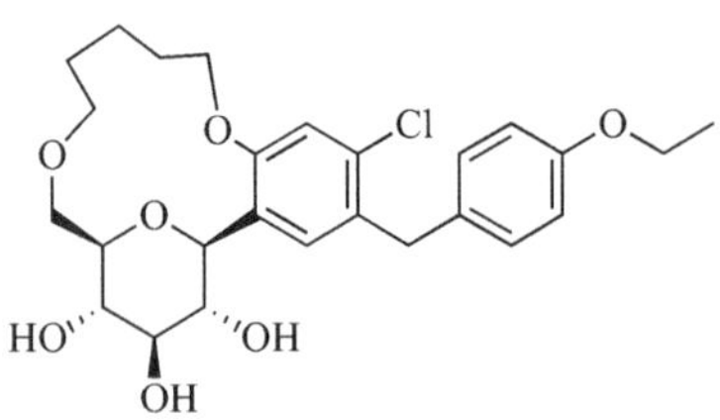

117

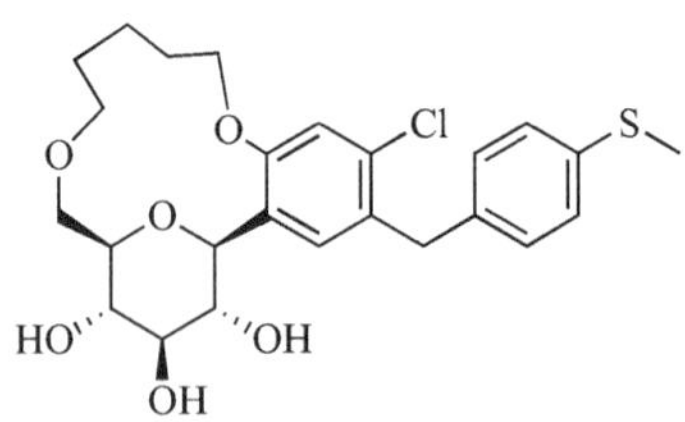

118

Figure 1.31 ***Structure of macrocyclic C-aryl glycosides as potential SGLT2 inhibitors.***

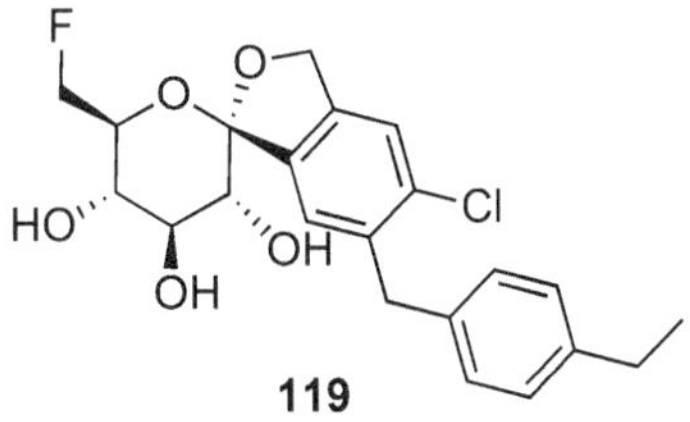

119

Figure 1.32 ***Novel O-spiroketal C-arylglucoside as potential SGLT2 inhibitor.***

withIC_{50} value 0.778 nM and 0.899 nM against hSGLT2, respectively.[131] *O*-spiroketal C-arylglucoside (**119**) (Fig. 1.32) has also been reported to show remarkable SGLT2 inhibitor activity.[132]

Ertugliflozin is an oral SGLT2 inhibitor discovered by Pfizer and Merck. This drug has been approved recently by USFDA in 2017 and then by the European Medicines Agency in 2018 for the treatment of type II diabetes mellitus.[133] The process-scale synthetic approach to the metabolic drug Ertugliflozin-L-pyroglutamic acid **(121)** has been depicted in Scheme 1.2.[95]

The scale-up synthesis of Ertugliflozin begins from benzylated glucose derivative (**122**), which was first subjected to oxidation using DMSO/Ac_2O and gluconolactone thus obtained was further reacted with *N*-methylpiperazine to give respective amide **123**. Oxidation of the 5-hydroxyl group of compound **123** using SO_3-pyridine in presence of DIPEA in DMSO gave respective ketone **124** which as such on reaction with organomagnesium reagent, derived from iodomethyl pivalate as hydroxymethyl anion equivalent, gave compound **125** with key C1 quaternary stereo-center as mixture of diastereomer (95:5 ratio). Compound **125** on pivalate removal using NaOMe in toluene afforded compound **126** in same diasteromeric ratio, that is, with 90% *de*. Diol **126** first underwent isopropylidene protection using 2,2-dimethoxypropane catalyzed by methanesulfonic acid followed by treatment with oxalic acid/MTBE afforded oxalate salt **127** as solid. Compound **127** was reacted with aqueous $NaHCO_3$ in toluene to get the free base **128** isolated from the organic layer after evaporated azeotropically. This compound is required in completely dry condition for the addition of the aryl anion via metalation using *n*-hexyl lithium at -15°C to afford respective ketone **128**. This compound on acetonide removal using TFA followed by reductive hydrogenolysis of the benzyl ether gave compound **129** as diastereomeric mixture. Finally, debenzylation was done under hydrogenation condition (H_2/Pd-C in acidic

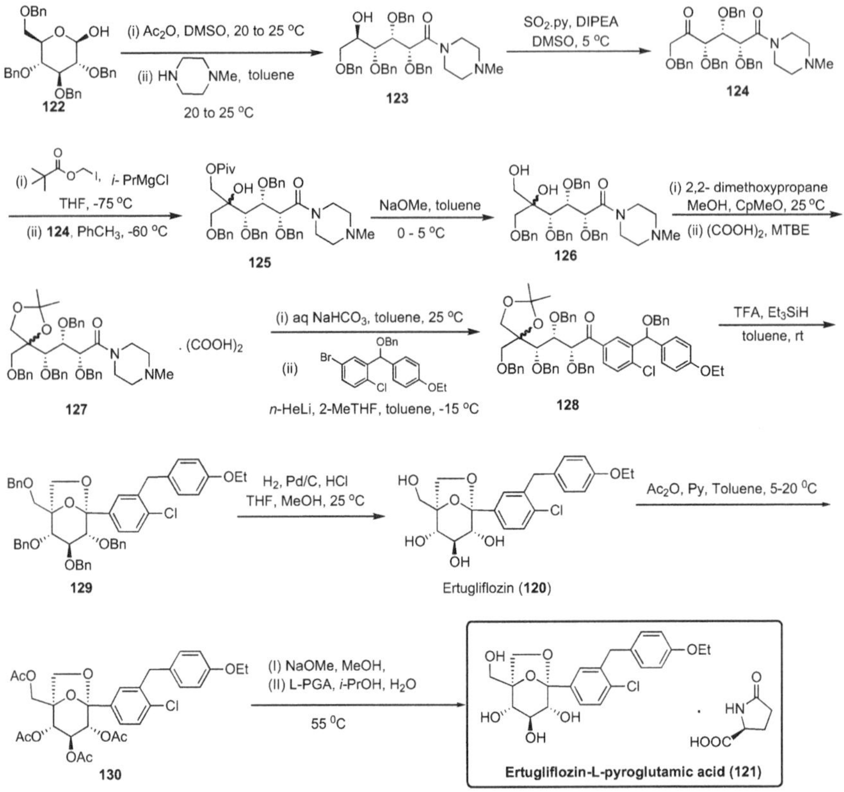

Scheme 1.2 ***Process-scale-up synthesis of metabolic drug Ertugliflozin-L- pyroglutamic acid*** **(121).**

condition) to afford the desired ertugliflozin **120**. For the final co-crystal formation of ertugliflozin, this drug was subjected to peracetylation to have a highly crystalline form, that is, compound **130**, and then acetate removal under standard reaction condition to provide an ertugliflozin solution with purity. This was finally concentrated, taken up in isopropanol, and heated to 60°C then treated with H_2O and at the end treated with of *L*-pyroglutamic solution in water to afford high yield of ertugliflozin-L- pyroglutamic acid (**121**) as crystalline compound (m.p. = 142.5°C).[95]

3.6 Carbohydrate-based anti-tubercular agents

A number of novel specific target-based anti-tubercular drugs have been developed so far which target biosynthesis of nucleic acids or proteins, DNA topoisomerases, enzymes and genes.[136] Enzymes and genes, which play crucial role in biosynthesis of saccharides on cell wall of TB bacteria, are the most interesting targets for anti-tubercular drug development and instigate to develop novel drugs to fight TB.[137] Interestingly, the presence of cell wall in bacterial cell and the absence of cell wall in the host cell make biosynthesis of mycobacteial cell wall as an important molecular target for designing and development of anti-mycobacterial agents. Polysaccharides present in the plasma membrane of the cell wall *M. tuberculosis* are largely in their furanose form. Mycolyl-arabinogalactan peptidoglycan (mAGP) complex, which assists the synthesis of lipoarabinomannan (LAM) and arabinogalactan (AG) with furanose form of arabinose and galactose residue, respectively, is vital for synthesis of mycobacterial cell wall. Restriction of the biosynthesis of LAM and AG is the target factor that can help in designing and development of novel anti-tuberculosis drugs.[136–138] A number of novel specific target-based anti-tubercular drugs have been developed so far, for example, Rifampicin, Pyrazinamide, Isoniazid, Ethambutol, Ofloxacin, etc., which target biosynthesis of nucleic acids or proteins, DNA topoisomerases, enzymes and genes,[136] where few simple carbohydrate derivatives have also shown notable activity against tuberculosis. A β-glycosyl amino acid have exhibited notable *in vitro* anti-tubercular activitywith MIC value 3.12 μg/mL, envisaged to imitate the enzymes *D-alanine racemase and D-alanine synthetase* known to play key role in biosynthesis of peptidoglycan.[139] Ethambutol (EMB, **131**), a synthetic β-aminoalcohol showed activity against *M.* strains with bactericidal mechanism involving the inhibition of biosynthesis of arabinan in both AG and LAM.[140] Thus, considering the anti-tubercular activity of EMB, Tripathi Group at CSIR-Central Drug Research Institute have developed a series of β-glycosyl amino alcohols and diglycosylated diaminoalcohols (n = 3, 7, 10, 12; **132-135**), where these simple molecules displayed anti-tubercular activity against *M. tuberculosis* $H_{37}Ra$ and $H_{37}Rv$ strains with MIC 6.25–3.12 μg/mL in both virulent and avirulent strains.[141] Likewise, from a library of galactopyranosyl amino alcohols, N^1,N^n-digalactopyranosylated amino alcohol (**136,** Fig. 1.33) showed potent *in vitro* activity against *M. tuberculosis* $H_{37}Rv$ in addition to remarkable activity against MDR TB.[142]

A cofactor, mainly NAD^+ and ATP, is required to assist DNA ligases (LigA) in catalyzing the phosphodiester bond formation. Thus NAD^+-dependent DNA ligases appeared to be a prominent target for drug development against bacterial infections. In this regard, glycosyl amino alcohol **132** was identified to inhibit NAD^+ binding site of the *M. tuberculosis* enzyme (*Mtu*LigA).[143]

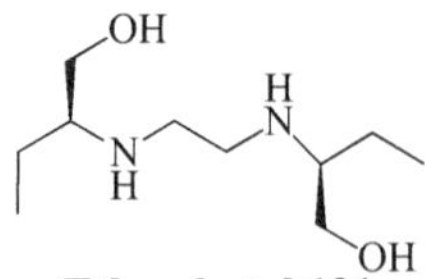

Ethambutol 131
(anti-tubercular drug)

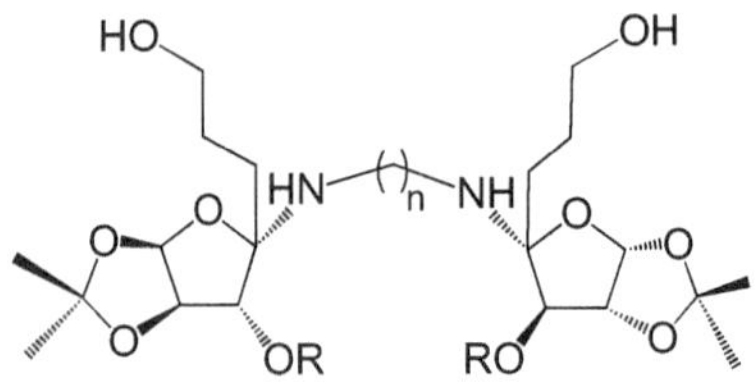

Glycosyl amino alcohols (132-135)
n = 3, R = Bn as MtuLigA inhibitor
(IC50 (μM) = 46.2 ± 1)

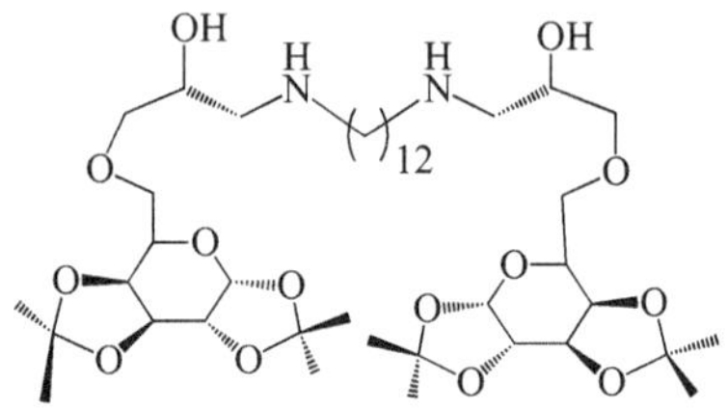

136

Figure 1.33 ***Carbohydrate-based diamino alcohols* (132-136) *as anti-tubercular agent.***

3.7 Carbohydrate-based anti-parasitic agents

Antiparasitics is a category of drugs which are used for the cure against parasitic diseases, like those caused by protozoa, ectoparasites, amoeba, helminths, and parasitic fungi. Many carbohydrate-derivatives have been found active against diseases caused by parasites, such as, leishmaniasis, malaria, and filariasis.

Leishmaniasis is a disease caused by Liesmania parasites of more than 20 species which are spread by the bite of phlebotomine sand flies.[144] Available anti-leishmanial therapies include a combination of paromomycin, pentavalent miltefosine, antimonialsand liposomal Amphotericin B (**137**) (Fig. 1.34). Amphotericin B is a carbohydrate-linked macrocyclic polyene

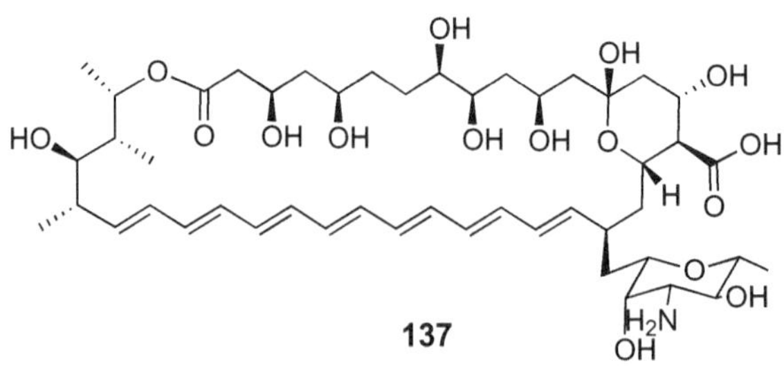

137

Figure 1.34 ***Structure of Amphotericin B, a drug for visceral leishmaniasis.***

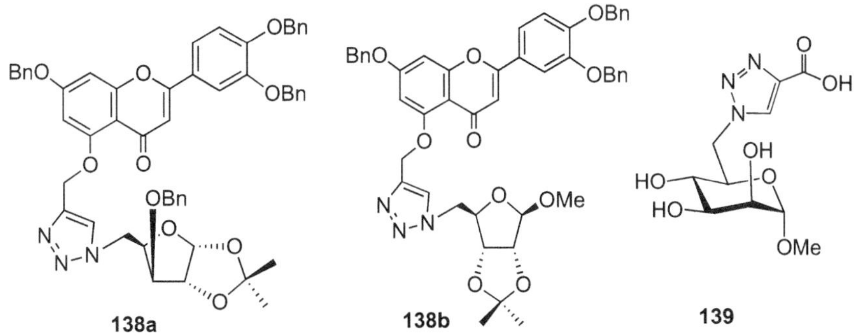

138a 138b 139

Figure 1.35 ***Structure of click inspired triazolyl glycoconjugates with anti-leishmanial activity.***

which is isolated from *Streptomyces nodosus.*[145] However, new drug for the effective treatment of leishmaniasisis urgently required with enhanced activity than existing drugs without their related limitations such as, side effects, non-oral admintration, etc.

Click chemistry as the most successful protocol in glyco-chemistry has been widely used in development of novel glycoconjugates of pharmacological significance. A series of triazolyl *O*-benzylquercetin glycoconjugates were developed *via* click coupling of *O*-benzylquercetin and deoxy-*azido* sugars. Results of the evaluation of anti-leishmanial activity of these compounds revealed that compounds **138a** and **138b** (Fig. 1.35) exhibited remarkable *in vitro* activity against intramacrophase amastigote and promastigotes forms of *Leishmania donovani.*[146] Likewise, Click inspired glycoconjugate **139** have demonstrated their bioactivity as inhibitor of *Leishmania.*[147] Various other triazolyl glycoconjugates with significant biological activities have been listed in Carbo-Click chapter in this book.

A number of carbohydrate-containing molecules displayed promising anti-malarial activities. Malaria is caused by *Plasmodiumfalciparum,* a protozoan parasite and causes thousands of deaths every year in tropical and subtropical regions of the world. In spite of a number of existing antimalarial drugs, such as, pyrimethanamine, chloroquine, sulfonamides etc. the developed drug resistance against these drugs compels to discover novel anti-malarial drugs. As erythrocytic phase of the malaria parasite involves Fe^{+3} ions, thus, iron chelators like hydroxamates can be a potential annihilator of malarial parasites. Among many attempts hydroxamic acids into picture. Anti-malarial activity of glycosyl hydroxamates **(140, 141)** revealed that the amine functionality is essential for biactivity and compound having glucofuranose moiety is even more active than galactopyranose ring.[148]

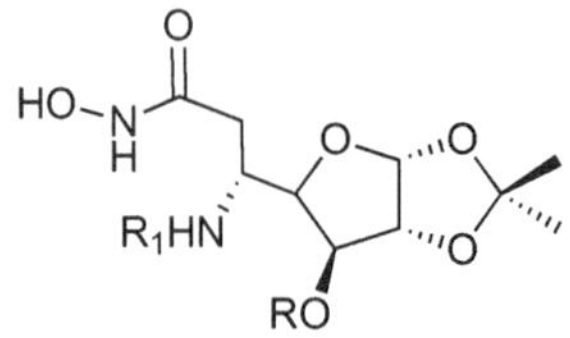

140

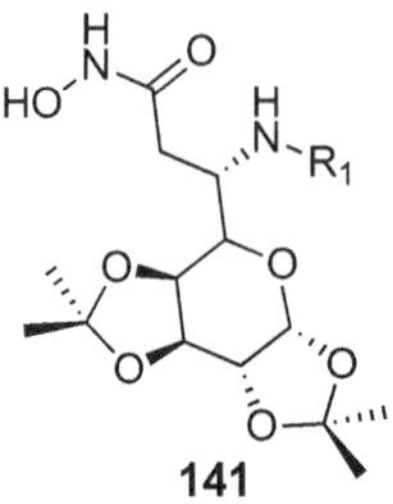

141

Glycosyl hydroxamates as potential anti-malarial agents:

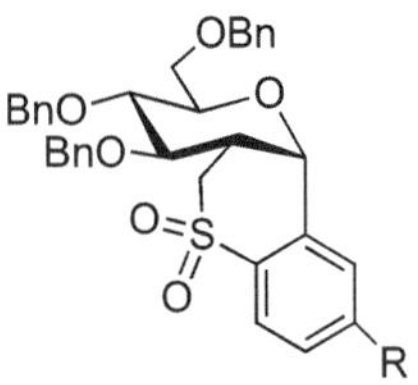

Sulfoxide derivatives
142a: R= Me
142b: R= OMe

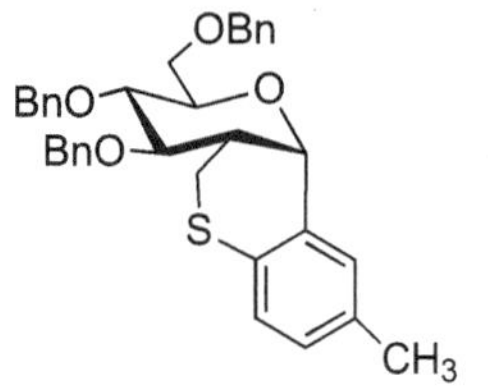

ThiochromanSulphide derivative
143a: R= Me
143b: R= OMe

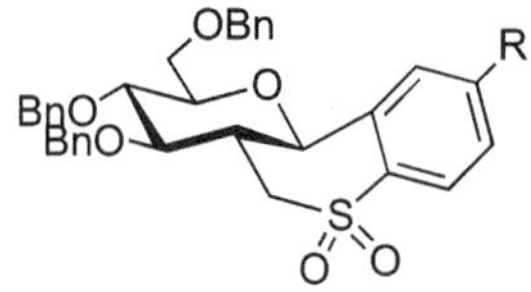

Epimerized sulfoxides
144a: R= Me
144b: R= OMe

Glycosyl thiocromans with anti-malarial activities:

Figure 1.36 ***Structure of some simple carbohydrate derivatives as potential anti-malarial agents.***

Sulfoxide and sulfone derivatives of carbohydrate-fused thiochromans were developed as potential anti-malarial agents (Fig. 1.36). Results of the biological assessment of developed compounds for antimalarial activity against chloroquine resistant FCR3 and chloroquine sensitive 3D7 strains revealed that sulfones (**142a,b**) are more active against malaria parasite as compared to corresponding sulfides (**143a,b**). Also, epimerized sulfones (**144a,b**) showed higher anti-malarial activity than the corresponding thiochroman sulfide derivatives.[149]

A number of carbohydrate-containing molecules displayed promising anti-filarial activities. Filariasis is caused by Filarioide a roundworm which is spread by mosquitoes and black flies. Various anti-filarial therapies have been developed so far targeting different target sites of parasite, such as, nicotinic acetylcholine receptors, GABA receptor channel, β-tubulins, Filarial DNA and DNA Topoisomerases etc. Antioxidant defence mechanism, involving Glutathione (a tripeptide) enzyme, is the main target for anti-filarial drugs. Diamines and polyamines are considered as reliable architectures for development of anti-parasitic agents. In this context, a series of N^1,N^n-glycosylated diaminoalkanes (**145, 146**) (Fig. 1.37) showed

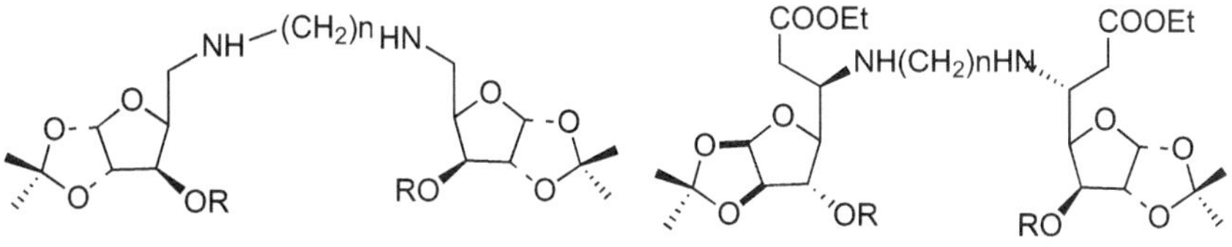

145 R = Me, Bn; n = 3, 7, 10, 12 **146**

Figure 1.37 ***Diaminoalkane glycoconjugates with anti-filarial activity.***

remarkable anti-filarial activity. The *in vitro* assessment against Glutathione metabolism were revealed by evaluation of their effect on intracellular GSH segment of bovine filarial worms along with effect upon glutathione-S-transferase (GST), glutathione reductase (GR) and γ-glutamyl cysteine synthetase (γ-Gcase). Compound **146** was found active enough to affect MTT and motility of filarial worms *B. malayi* with a 50 mg/kg oral administration and a few of them displayed mild macrofilaricidal, microfilaricidal, and sternlization effect.[150]

3.8 Carbohydrate-containing molecules as anti-HIV agents

Acquired immunodeficiency syndrome (AIDS) is one of the most fatal diseases with highest mortality rate. The cause of AIDS is human immunodeficiency virus (HIV) by which helper T-cells are affected which in turnbreaks down the immune system and the carrier becomes vulnerable to a wide array of infections.[151] Even after substantial studies of various immunologic, virologic, and genetic aspects of HIV, its treatment is still a challenge before pharma community.[152,153] The main target of potential anti-HIV agents is reverse transcriptase (HIV-RT) enzyme which is involved in viral replication. A library of benzyl 1,2,3- triazole derivatives was developed and assessed biologically for activity against HIV-RT. Several developed compounds exhibited notable anti-HIV-RT activity among which compounds (**147-149,** Fig. 1.38) highest activity with 60% to 65% inhibition at 50 μM.[154]

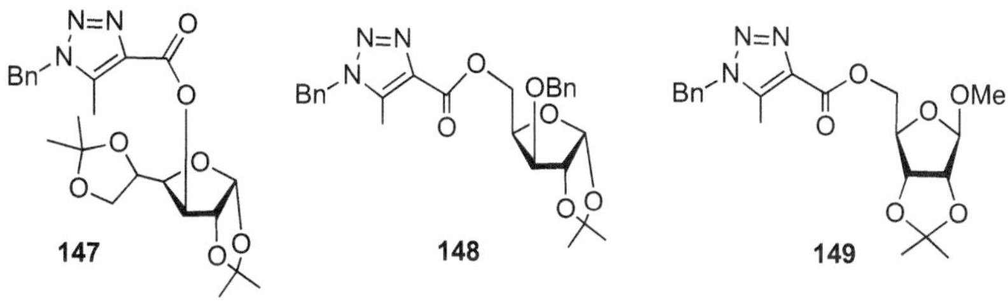

147 **148** **149**

Figure 1.38 ***Anti HIV-RT glycosyl triazoles.***

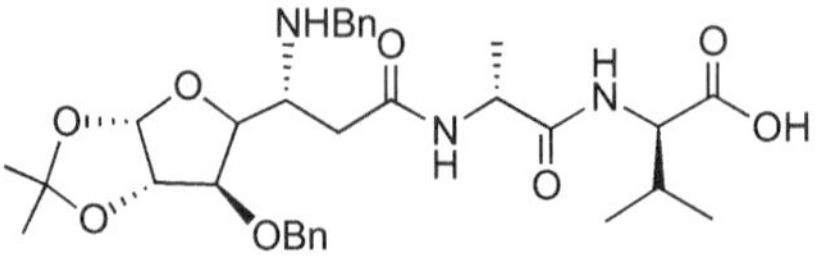

150

IC_{50} (nM): 760 - 0.02

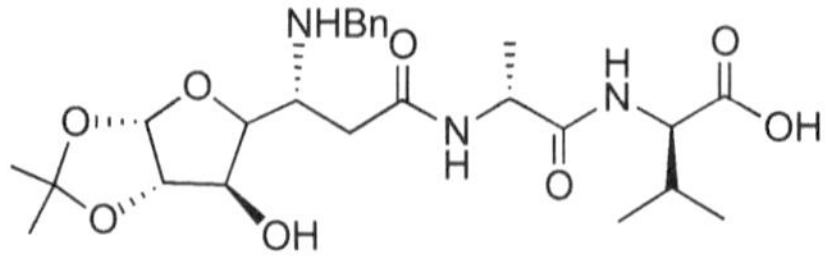

151

IC_{50} (nM): 710 - 0.04

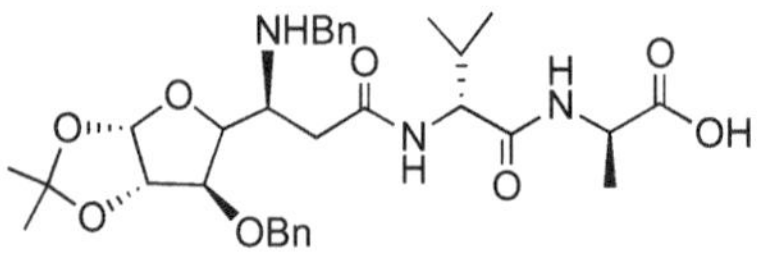

152

IC_{50} (nM): 1090 - 0.02

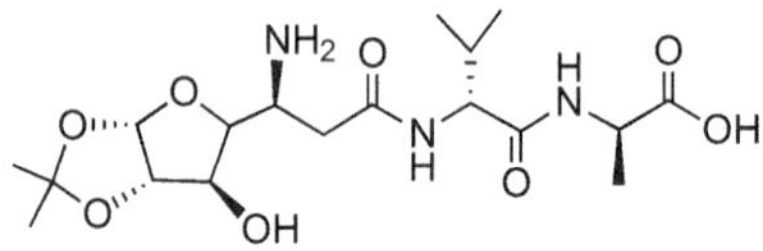

153

IC_{50} (nM): 950 - 0.07

Figure 1.39 ***Carbohydrate peptidomimetics as HIV-1 protease inhibitor.***

Integrases and proteases are other crucial enzymes which play important role in viral replication and hence are major target of anti-HIV therapies. The inhibition of Protease enzyme is the main approach in HIV treatment named as highly active anti-retroviral therapy (HAART). Carbohydrate-based carbapeptides were developed as potential Protease inhibitors and assessed *in vitro* for inhibitory activity against HIV- 1 protease. Among various developed peptidomimetics **150–153** (Fig. 1.39) displayed notable inhibitory activity against HIV-1 protease.[155]

3.9 Carbohydrate-based anti-coagulants

We have previously discussed the significance of heparin and heparin sulfate as anti-coagulant of blood. An architecture containing uronic acid linked to repeated disaccharide glucosamine units is found in both heparin and heparin sulfate, however, more *L*-iduronic acid units are found in heparin along with maximum *N*-sulfated glucosaminyl fragments.[156]

Heparin is administered as intravenous infusion or subcutaneous injection to prevent blood coagulation in some medical conditions. Side effects related to heparin, like thrombocytopenia, are due to its ability to activate blood proteins including platelet factor 4. This leads to a need of modification in heparin structure to attain a desired anti-coagulant activity without related side effects. One of these modifications was achieved in form of depolymerization of heparin forming low molecular weight heparin (LMWH). New strategies with advanced techniques have managed the generation of ultra-low molecular weight heparin (ULMWH) AVE5026 which is under clinical trials of Phase III for thromboembolism related to

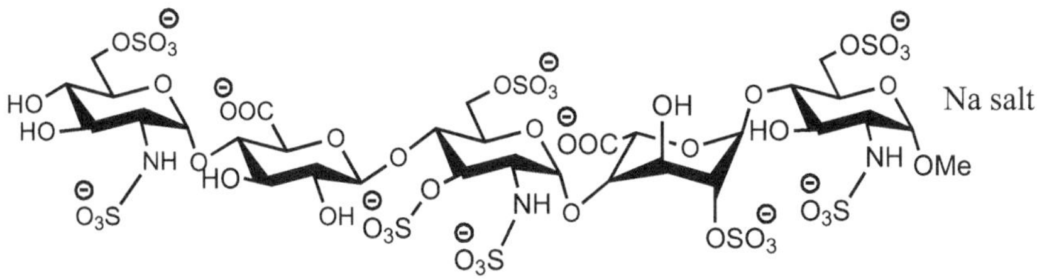

Fondaparinux (**154**)

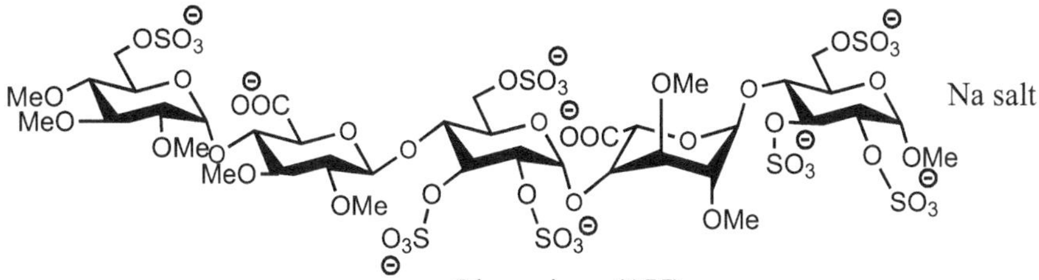

Idraparinux (**155**)

Figure 1.40 ***Synthetic analogs of antithrobin binding sugar.***

veins. LMWH and ULMWH exhibited improved tolerance in living systems, as well as, better pharmacological activities.[157]

New heparin mimetics have emerged as the potential therapeutic aids for coagulation related conditions, as well as, for development of anti-inflammatory, anti-HIV and anti-cancer agents. Hadri et al. reported a synthetic analog of natural antithrombin binding sugar (Fig. 1.40), Fondaparinux (**154**), which is chemically related to LMWH. In 2001, FDA approved this drug under the trade name Arixtra for the prevention of venous thromoembolism. A synthetic analog of Fondraparinux, Idraparinux (**155**) is more effective drug for prevention, as well as, treatment of venous thromboembolism.[158]

A detailed discussion about heparin as molecular scaffold in drug discovery along with the significance of heparin mimetics including oligosaccharides, non-carbohydrate based small molecules, and multivalent glycopolymers and dendrimers as tools for modulation of biology and therapy has been discussed in **Prof. Xuefei Huang** Chapter.

3.10 Carbohydrate-mimetics as potential sialyltransferase inhibitors

Many biological processes, such as, cellular recognition, tumor metastases, and immune response are adhered to sialylation of glycoconjugates at their non-reducing end. This process is governed by sialyltransferase (ST) enzymes, thus inhibition of these enzymes has significant biochemical and physiological applications. ST inhibition has emerged as a potential strategy for cure of ailments like various carcinomas, viral infection, inflammation, immune response, and neurological disorders.[159]

Since sialyltransferases assist the transfer of sialyl residue from a specific sugar nucleotide donor to a glycoonjugate acceptor with specific terminal structure of the sugar residue. Hence, sugar-acceptor analogs can be strategically developed as potential sialyltransferase inhibitors. Kajihara et al. developed *N*-acetyl-*β*-lactosamide derivatives (**156-159**) and reported them to exhibit modest inhibition against α-(2-6)-sialyltransferase from rat liver with *K*i range 0.76–4.14 mM.[160]

Derivatives of methyl 5a'-carba-*β*-lactoside (**160-162**), developed by Okazaki and co-workers were found to show inhibitory activity against rST3Gal I and ST6Gal I in micromolar range.[161] (Fig. 1.41)

Designing of donor-based sialyltransferase inhibitors is rationally done on the basis of structural specifications of cytidine monophosphate *N*-acetylneuraminic acid (**163**) (CMP-Neu5Ac). CMP-Neu5Ac is consists of nucleotide moiety CMP and carbohydrate residue Neu5Ac and is the common donor sugar nucleotide of sialyltransferases. Some examples of donor analogs of sialyltransferase inhibitors have been shown in Fig. 1.42. CMP-quinic acid derivative **164** shows inhibitory activity against α(2-6)-sialyltransferase from rat liver with *K*i value of 20 μM.[162] Another CMP-Neu5Ac analog which showed strong inhibition is **165** with *K*i value of 44 μM, equivalent to that of CMP.[163]

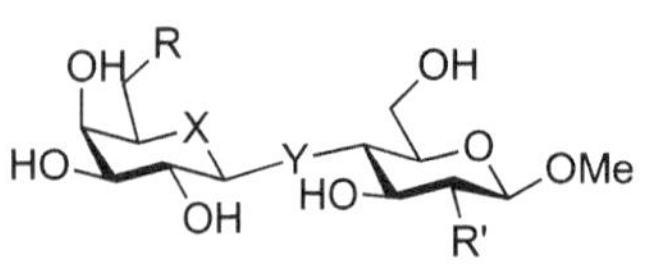

	R	R'	X	Y	*Ki (μM)*
156	H	NHAc	O	O	760 (ST6Gal I)
157	SH	NHAc	O	O	3780 (ST6Gal I)
158	OTHP	NHAc	O	O	4140 (ST6Gal I)
159	H	OH	C	O	419 (rST3Gal I) 903 (rST6Gal I)
160	H	OH	C	NH	185 (rST3Gal I) 533 (rST6Gal I)
161	H	NHAc	C	NH	245 (rST3Gal I) 651 (rST6Gal I)

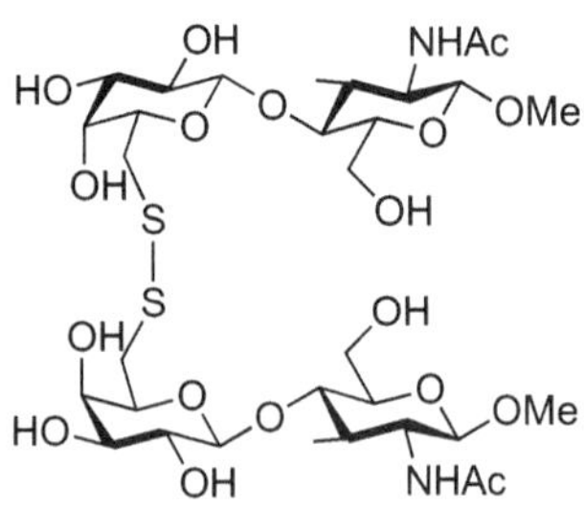

162, *Ki*= 2000 *μM* (ST6Gal I)

Figure 1.41 ***Carbohydrate-based acceptor analogs of ST-inhibitor.***

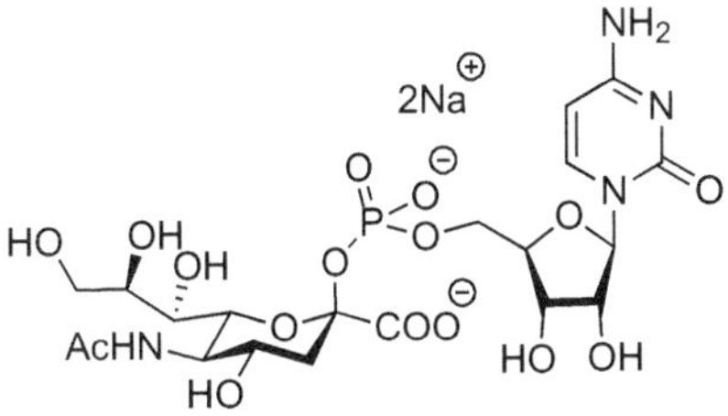

CMP-Neu5Ac
163, *Ki*= 46 μM

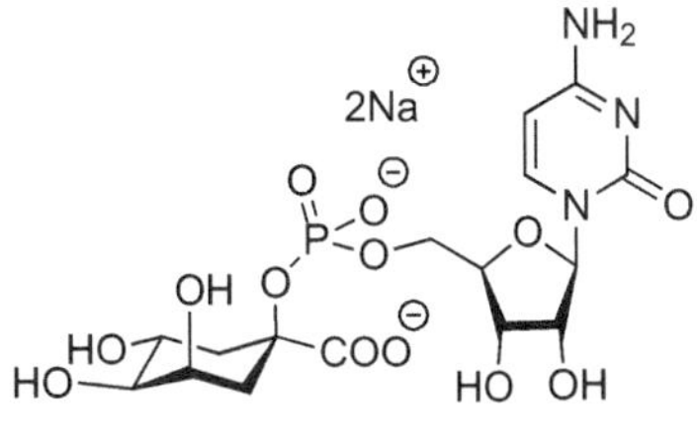

CMP-quinic acid
164, *Ki*= 44 μM

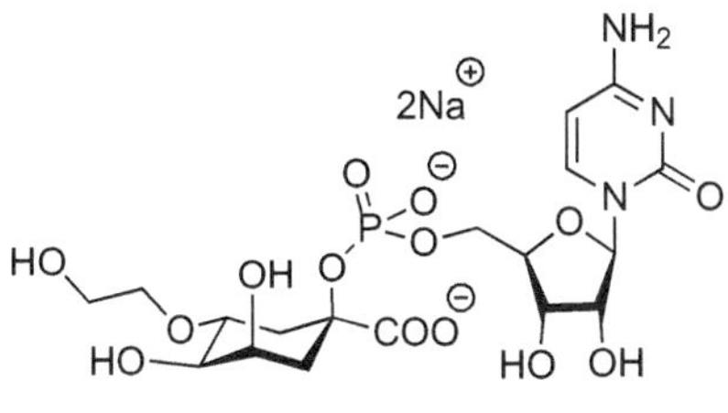

165, *Ki*= 20 μM

Figure 1.42 ***Examples of carbohydrate-based donor analogs of ST-inhibitor.***

Amann et al. developed three transition state analogs of CMP-Neu5Ac out of which, analog **166** which consists the linkage of CMP residue to C-1 of 2,3-dehydro-N-acetylneuramin-1-yl phosphonate, showed excellent inhibitory activity (*Ki* = 350 nm) against α(2-6)-sialyltransferase which is 130-fold stronger than that of CMPNeu5Ac.[164] Muller *et al.* developed a series of potential sialyltransferases based on transition state analogs among which **167** and **168** (Fig. 1.43) showed better inhibitory activity than that of **166** while **169** exhibited strongest inhibition with Ki value of 40 nM.[165]

2-Benzoyloxyethyl *N*-acetylglucosamine β-glycoside derivatives **170–173** (Fig. 1.44) exhibited excellent inhibition of α(2-6)-sialyltransferase with *Ki* value ranging from 29 nM-69 nM.[166]

3.11 Carbohydrate-based potential glycosidase inhibitors

Glycoside hydrolase, also known as glycosidase, are a class of enzymes which are crucially involved in the catalytic hydrolysis of the glycosidic bond and play vital role in a number of important biological functions like lysosomal catabolism of glycoconjugates, intestinal digestion and posttranslational modification of glycoproteins. Inhibition of glycosidase enzymes affects the secretion, maturation and transport of glycoproteins which in turn promotes the use of glycosidase inhibitors for treatment of viral infections,

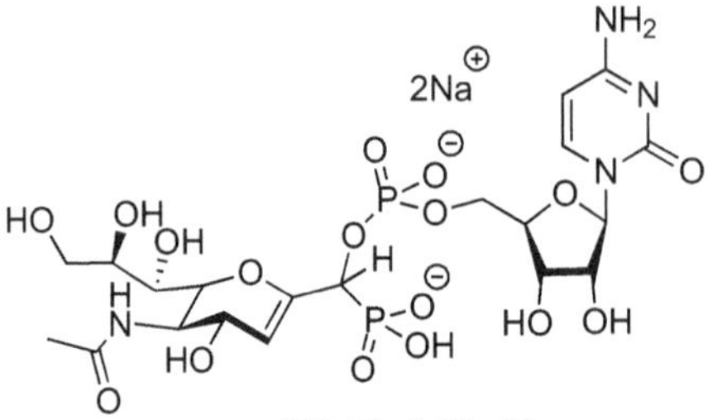

166, *Ki*= 0.35 μM

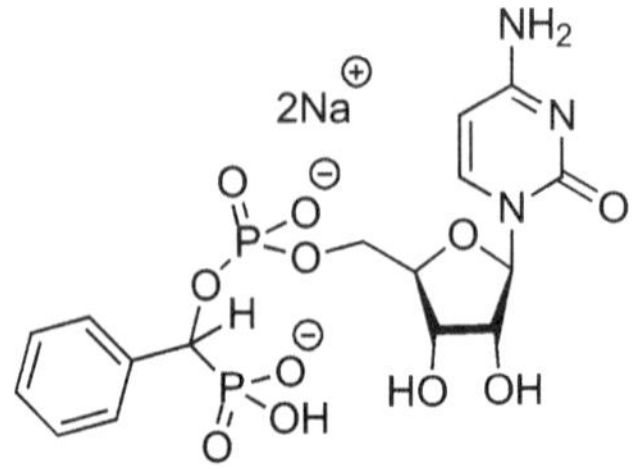

167, *Ki*= 0.20 μM

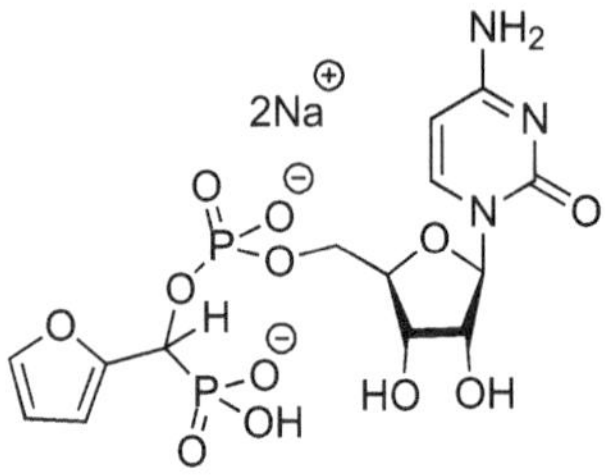

168, *Ki*= 0.28 μM

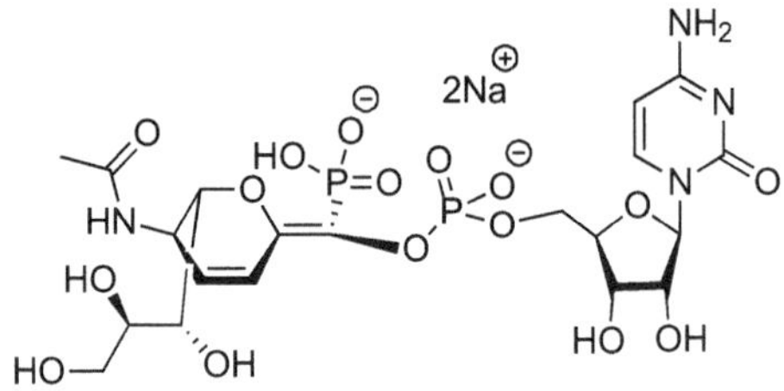

169, *Ki*= 0.04 μM

Figure 1.43 ***Carbohydrate-based transition state analogs of CMP-Neu5Ac with ST-inhibitor activity.***

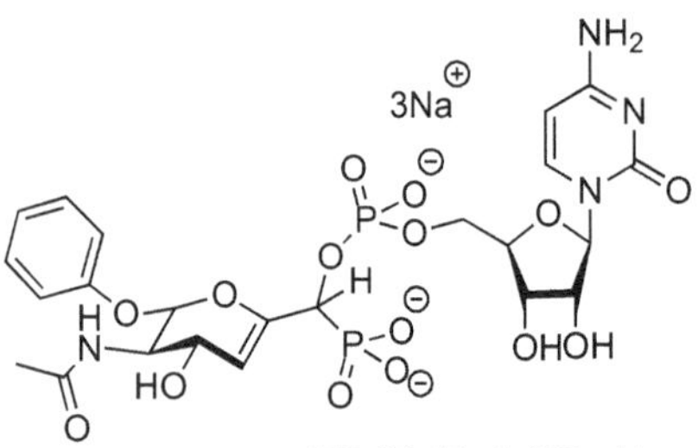

170 (h), *Ki*= 0.029 μM
171 (l), *Ki*= 0.69 μM

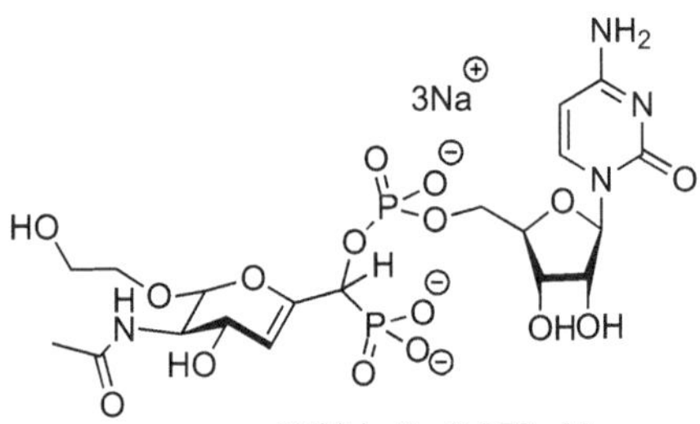

172(h), *Ki*= 0.059 μM
173(l), *Ki*= 0.038 μM

Figure 1.44 ***N-Acetylglucosamine derivatives with excellent ST-inhibitor activity.***

cancer, diabetes and genetic disorders.[167] Iminosugars have revealed their potential as therapeutic agents and they have also emerged as a promising moiety to act as glycosidase inhibitors. D'Alonzo et al. studied the biochemistry of *L*-iminosugar and found that it shows interactions and inhibitory action against specific enzymes.[168] The first iminosugar glycosidase inhibitor was Miglustat (*N*-butyl-deoxynojirimycin, **174**) which was extracted from microorganisms and plants. Miglustat shows inhibitory activity against glucosylceramide synthase enzyme and in turn helps in

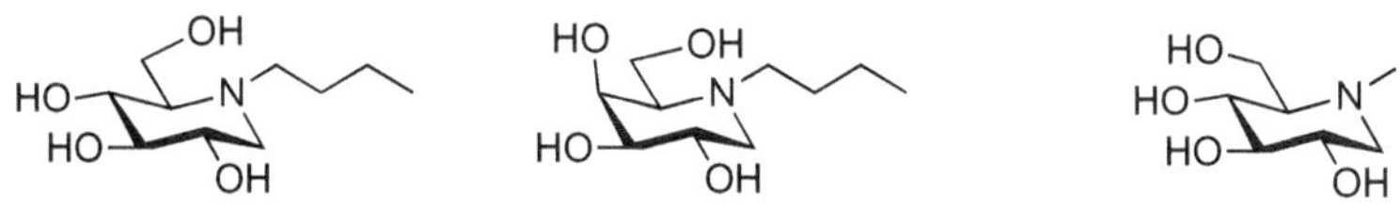

Miglustat **174** *N*-Butyl-1-deoxy-galactonojirimycin **175** NMDNJ **176**

Figure 1.45 ***Carbohydrate mimetics for treatment of lysosomal storage diseases.***

reduction of biosynthesis of glucosylceramide from ceramide which results into reduction in glycosphingolipids (GSL) synthesis. Miglustat has been also found an effective inhibitor of α-glucosidase I and II, sucrase, maltase and glucocerebrosidases (lysosomal and non- lysosomal). Miglustat (**174**) is used for treatment of cases of mild to moderate type I Gaucher disease in which enzyme replacement therapy cannot be done due to medical risks. FDA approved this drug in 2003 for treatment of Type 1 Gaucher disease. Miglustat is also approved for treatment of neurological problems linked with Niemann- Pick type C disease (Lysosomal storage disease) in Canada, Japanand Europe.[169] Other sugar mimics which were found efficient glycosidase inhibitors for treatment of Gaucher disease and Fabry disease are *N*-butyl-1-deoxy-galactonojirimycin (**175**) and *N*-methyl-1-deoxy-nojirimycin (**176**) (Fig. 1.45).[170–174]

A detailed study of imino sugar-based drugs and significance of imino sugar scaffolds in drug discovery, its advantages and prospects has been described in **Prof. NG Ramesh Chapter**.

Neuraminidases, a class of glycosidase enzymes catalyzes the hydrolysis of glycosidic bond of neuraminic acid and viral neuraminidases are best known as target for prevention of influenza viral infection. Neu5Ac2en (**177**), a neuraminic acid derivative, showed inhibitory activity against sialidases (neuraminidase which hydrolyze sialic acids). Two more potential neuraminidase inhibitor Oseltamivir (Tamiflu, **178**) and Zanamivir (Relenza, **179)** (Fig. 1.46) were developed as drugs for treatment of infection caused by influenza A and Influenza B virus.[175] The mechanism of action of these drugs involves binding to viral glycoproteins through sialic acid binding site to check the binding of influenza virus to host cell which in turn prevents the infection spread.[176]A triazolyl derivative of Zanamivir with a triazole ring at C-4 position (**180**) exhibited notable *in vitro*glycosidase inhibitor activity against Avian influenza virus with IC_{50} value 6.4 μM.[177]

Details of scalable synthesis of neuraminidase inhibitors Zanamivir and Oseltamivir and the pharmaceutical significance of other sialic acid derivatives have been elaborately discussed in **Tiwari Chapter** of this book.

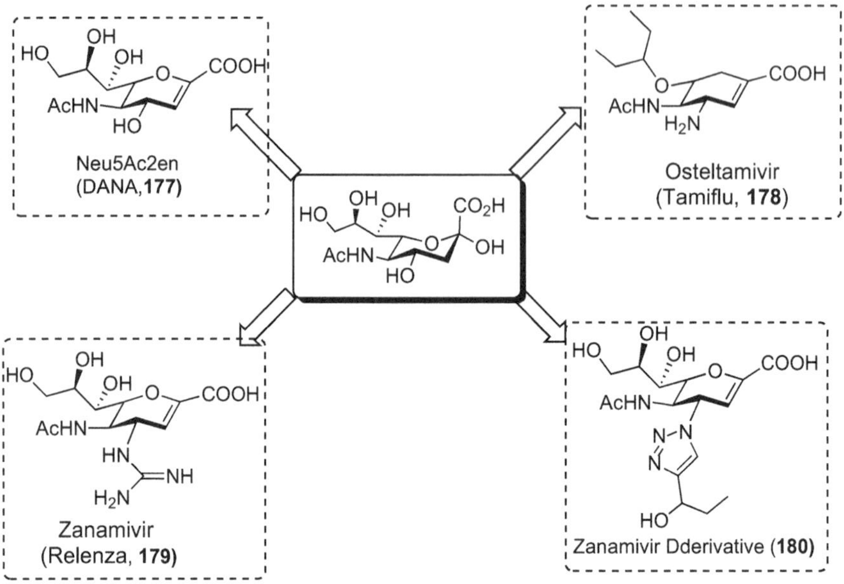

Figure 1.46 ***Structure of Zanamivir and Oseltamivir, well-known neuraminidase inhibitors.***

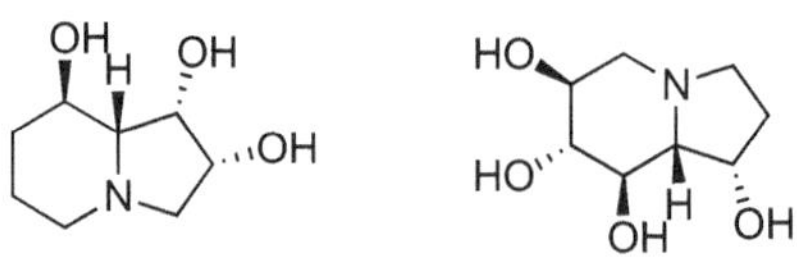

Swainsonine (**181**) Castanospermine (**182**)

Figure 1.47 ***Imino sugar alkaloids as anti-cancer glycosidase inhibitors.***

Two naturally occurring imino sugar alkaloids, Swainsonine (**181**) and Castanospermine (**182**) (Fig. 1.47) have shown significant glycosidase inhibitor activity. Swainsonine is an efficient inhibitor of Golgi α--mannosidase II, and is under clinical trial for anti-cancer activity assessment.[178] Castanospermine is isolated from the seeds of *Castanospermum australe* and a potential antiviral drug under clinical trials.[179]

3.12 Cardiac glycosides as therapeutics

Cardiac glycosides are organic compounds consisting of a steroid molecule attached to a carbohydrate unit. These glycosides act on cellular sodium-potassium ATPase pump resulting into interference in the heart functioning by an increased rate of contractions. These glycosides have shown potency

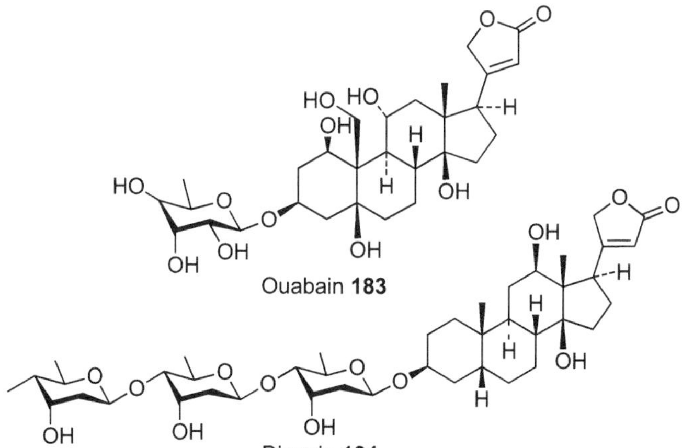

Figure 1.48 ***Structure of some commendable cardiac glycosides.***

in treatment of heart conditions. Ouabain (**183**) (Fig. 1.48) is derived from bark of *Acokanthera ouabaio* and seeds of *Strophanthus gratus*and shows inhibitory effect against Na-K (+)-exchanging ATPase.[180] Digoxin (**184**) is another cardiac glycoside which is extracted from foxglove plant *Digitalis lanata*. It is used to cure atrial flutter and atrial fibrillation.[181]

3.13 Carbohydrate-based molecules with miscellaneous activities

There are several carbohydrate-derivatives and glycoconjugates which show extra ordinary bioactivity and have significant medicinal use (Fig. 1.49). For example, Auranofin (**185**) is a sugar-based gold complex is used as orally administered anti-rheumatic agent. Recent researches revealed its anti-HIV properties as it shows reducing effect over HIV viral reservoir.[182] Topiramate (**186**), is a FDA approved sulfamate substituted monosaccharide which is used to treat epilepsy, for prevention of migraines and for Linnox-Gastaut Syndrome.[183] Prumycin (**187**) shows anti-tumor activity. Vidarabine (**188**), an arabinosyl nucleoside which was initially developed as an anti-cancer drug but succeeding researches revealed its anti-viral properties. It is used to treat infections caused by varicella zoster and herpes simplex viruses.[184] Streptozotocin (**189**), is a well-known FDA approved drug, used for treatment of cancer of the Islets of Langerhans. It is a naturally occurring alkylating antineoplastic agent toxic for β-cells of pancreas accountable for secretion of insulin.[185] Lactulose (**190**) is another carbohydrate-based drug which is used to treat chronic constipation and hepatic encephalopathy. This

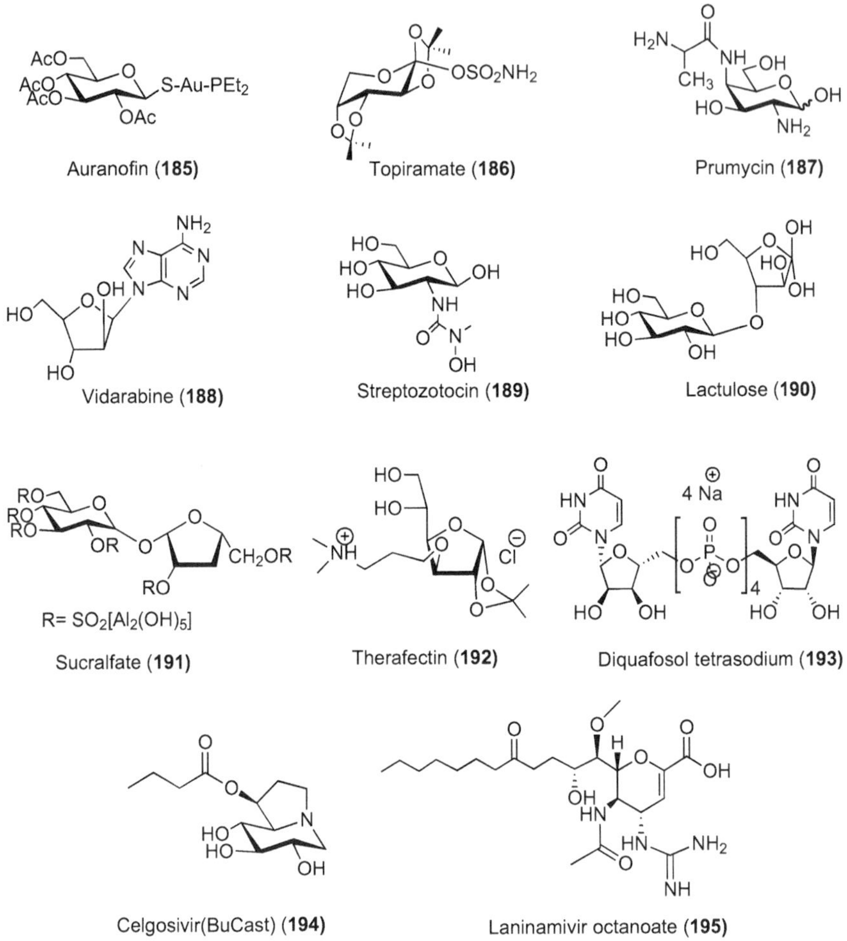

Figure 1.49 ***Structure of some carbohydrate-based drugs of miscellaneous profiles.***

synthetic disaccharide (fructose and Galactose) is labelled as an osmotic laxative by W.H.O. norms.[186]

Sucralfate (**191**), a complex of sucrose sulfate with aluminum hydroxide, is known to be the duodenal ulcers therapy. Amilprilose and its hydrochloride salt Therafectin (**192**) show anti-inflammatory and anti-proliferative activities. Amilprilose is also a successful immunomodulator effective against conditions related to autoimmune disorders, such as, arthritis, eczema and systemic lupus erythematosus with advantages of reduced side effects and lesser toxicity.[187] A carbohydrate-based neuraminidase inhibitor Laninamivir Octanaotate (**193**, CS-8958**)**, was developed in 2010 as potential anti-viral agent against influenza virus A and B and at present it is under Phase III clinical trials. A 6-*O*-butanoylcastanospermin, Celgosivir (BuCast, **194**), found effective to inhibit biosynthesis of carbohydrates and is useful for HIV positive patients. Diquafosol tetra sodium (**195,** Fig. 1.49) another carbohydrate-based drug, has shown efficacy for treatment of dry eye disease. It is now an approved drug in Japan, whereas, in United States, it is under Phase III clinical trials.[188]

4 Carbohydrate-based metallo drugs

The therapeutic value of carbohydrate-containing metal-complexes is result of amalgamation of two activities, first comes from metal ion and the other because of carbohydrate ligand. These complexes show low toxicity, biocompatibility, water solubility, and uptake enhancement. Platinum complexes of carbohydrate-derived ligands (Fig. 1.50) have been largely explored in tumor therapy. The uptake of glucose and derivatives by growing tumor cells increases due to over expression of GLUT 1–4 on cell membrane.[189] Mikata et al. reported the antitumor activity of carbohydrate-platinum complexes **196** and **197**. The activities of these complexes in terms of mice lifetime (*T/C* value) significantly varied on mere anomeric (α/β) change in sugar ligand.[190] Patra et al. have recently designed glucoconjugated platinum complexes in which a malonate tethered glucose-derivative was used as ligand. Among all the developed complexes,

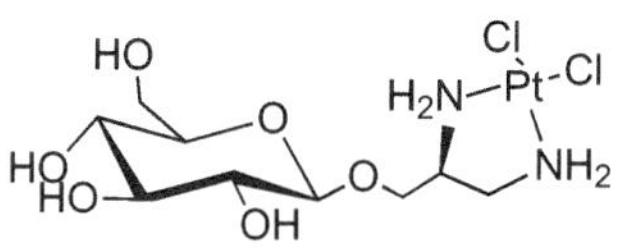

196

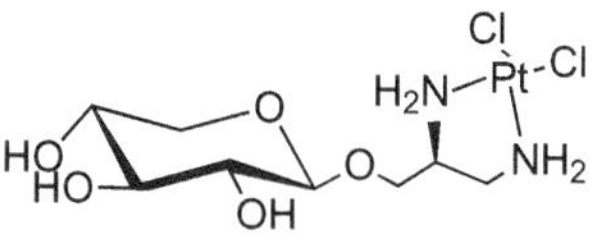

197

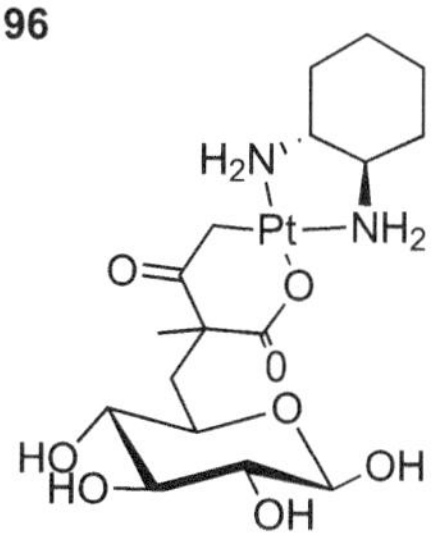

198

Figure 1.50 *Pt-complexes of carbohydrate derivatives with antitumor/anticancer activity.*

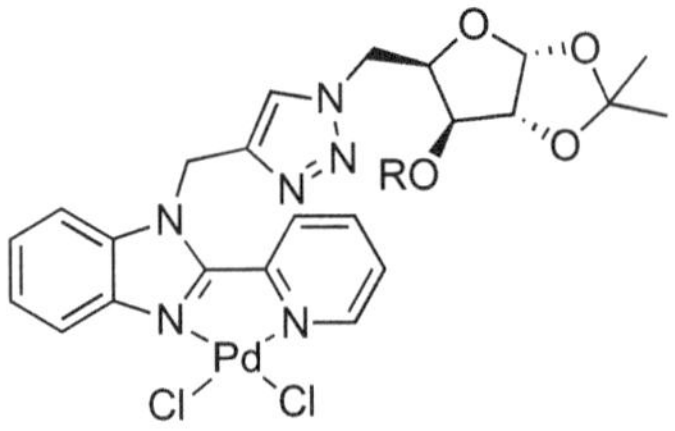

199

R = H, Ac, Bn

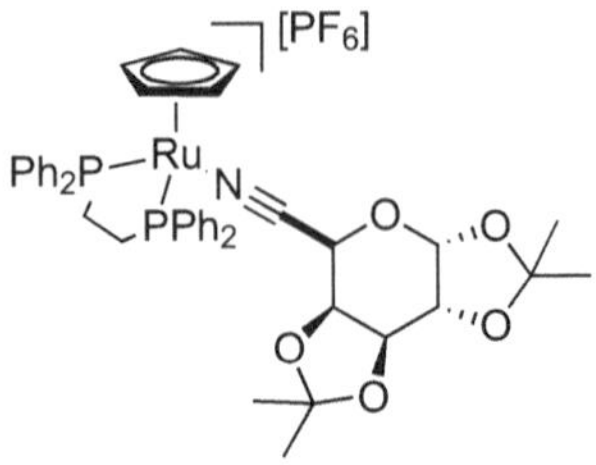

200

Figure 1.51 ***Non-platinum complexes of carbohydrate-derivatives in therapy.***

198 exhibited good anticancer activity in terms of uptake and cytotoxicity in GLUT1-expressingA2780 cells in an 8 hours incubation time.[191]

Complexes of carbohydrate-derived ligands with non-platinum metals have also shown biological significance. Palladium complexes generally show low toxicity in comparison to platinum complexes. In a recent report, Deepthi et al. explored the antimicrobial properties of Pd-complex of 2-pyridyl-benzimidazole glycoconjugate ligand (**199**) and found some promising results.[192] Florindo and co-workers have recently reported the preparation and biological properties of Ru-complexes of carbohydrate derivatives. Complex **200** (Fig. 1.51) exhibited high chemotherapeutic value against colon cancer which was comparable to the established drug oxaliplatin. Also, this complex showed decreased cell viability, high cell death induction, increased caspase-3 and caspase-7 activity and apoptosis induction as compared to oxaliplatin.[193]

Alzheimer's disease is believed to be a result of amyloid-β peptide aggregation due to the increased oxidative stress. Metal ions, such as, Cu(II), Zn(II), and Fe(II) are found in high concentration in the amyloid-β plaque.[194] The Orvig research group reported two glucose-derived ligands **201** and **202** (Fig. 1.52) as promising drug candidates able to bind with Zn(II) and Cu(II) ions effectively.[195]

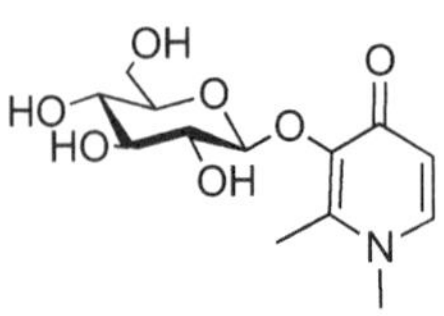

201

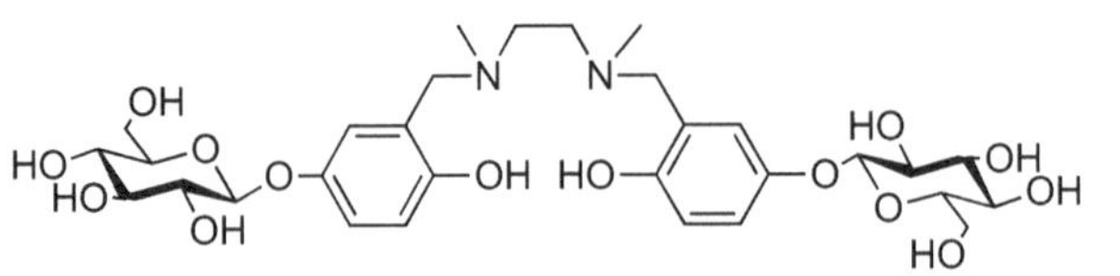

202

Figure 1.52 ***Glucose-derivatives as promising metallo-drug candidates in therapy of Alzheimer's disease.***

5 Carbohydrate-based vaccines

Vaccines are preventive measures which provide active acquired immunity toward a particular disease. The therapy based on vaccines health involves immunotherapy and hence is sometimes considered better than drug-based therapy because of related drug resistance issues. The presence of carbohydrates on cell surface of host, as well as, pathogen in a specific way, offers them as suitable candidates based on which vaccines against a particular disease can be developed. The first carbohydrate-based vaccine was developed in 1983 from *Pneumovax*™, a capsular polysaccharide as *Pneumovax*™ *23* and used for prevention of pneumococcal disease.[196] Typhim Vi, a polysaccharide vaccine, is one of the two typhoid vaccines recommended by WHO.[197] GMK is a ganglioside conjugate vaccine developed by Progenics Pharmaceuticals for the prevention of melanoma, a type of cancer involving pigment cells.[198] MGV is a bivalent GM2 and GD2 ganglioside conjugate which is used as vaccination against various cancers like gastric, colorectal and small-cell lung cancer.[199] Prevnar is a Pneumococcal conjugate vaccine used against pneumonia caused by *Streptococcus pneumonia*.[196] Theratope (**203,** Fig. 1.53), a (Sialyl Tn Ag) conjugate vaccine has been clinically developed as an immunotherapy against metastatic colorectal cancer and breast cancer.[200]

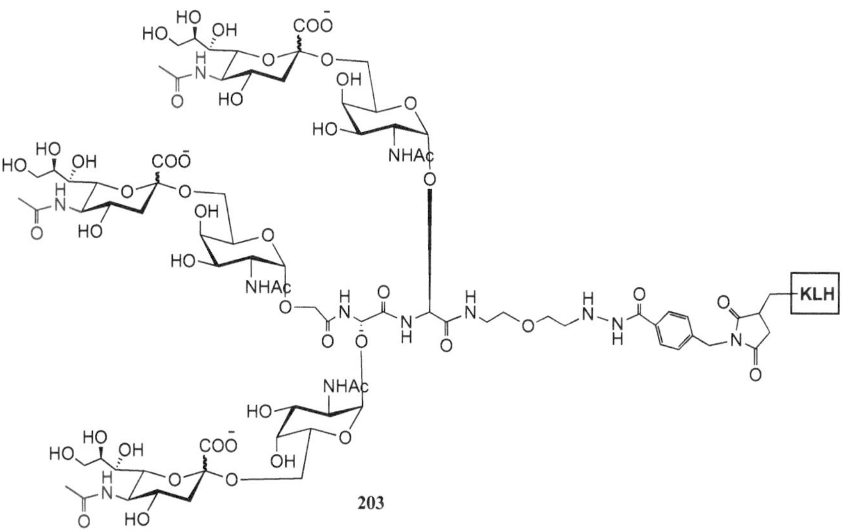

203

Figure 1.53 ***Structure of Theratope, A Sialyl-Tn Ag conjugate vaccine.***

Table 1.1 Examples of carbohydrate based vaccines.

Vaccine	Target disease
Pneumovax™ 23	Pneumococcal disease
Typhim Vi	Typhoid
GMK	Malignant melanoma
MGV	Colorectal, gastric, small-cell lung cancer
Prevnar	Pneumonia caused by *Streptococcus pneumonia.*
Theratope	Metastatic colorectal, breast cancer
IGN 301	Cancer or tumors
ActHiB, OmniHiB	Influenza type b

IGN 301 is an anti- idiotypic antibody clinically developed to be used as vaccine in cancer immunotherapy. This vaccine is under Phase Ib clinical trial.[201] Theimmunogenicity of *Haemophilus* b (Hib) conjugate vaccines is governed by many the factors like the nature of proteins and polysaccharides carriers used in the construction of vaccine and the conjugation method. Construction of Hib conjugate vaccine via chemical synthesis of a Hib saccharide antigen has opened doors towards facile development of such vaccines. ActHiB and OmniHiB are two Hib conjugate vaccines used against influenza type b infection (Table 1.1).[202] Quimi-Hib (**204**, Fig. 1.54), a synthetic Hib saccharide conjugate has recently got approval as a vaccine against *Neisseria* species causing meningitis.[203–205]

Over expression of extracellular tumor-specific carbohydrate antigens (TACAs) is observed on the surface of cancer cells and is considered as a

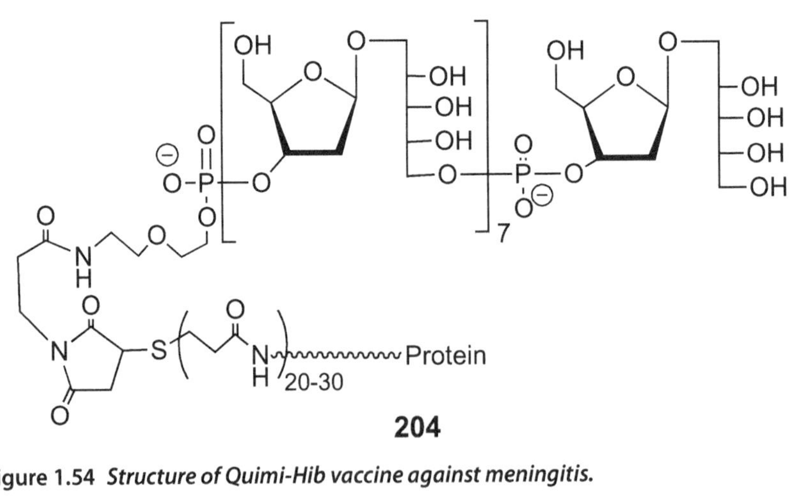

204

Figure 1.54 ***Structure of Quimi-Hib vaccine against meningitis.***

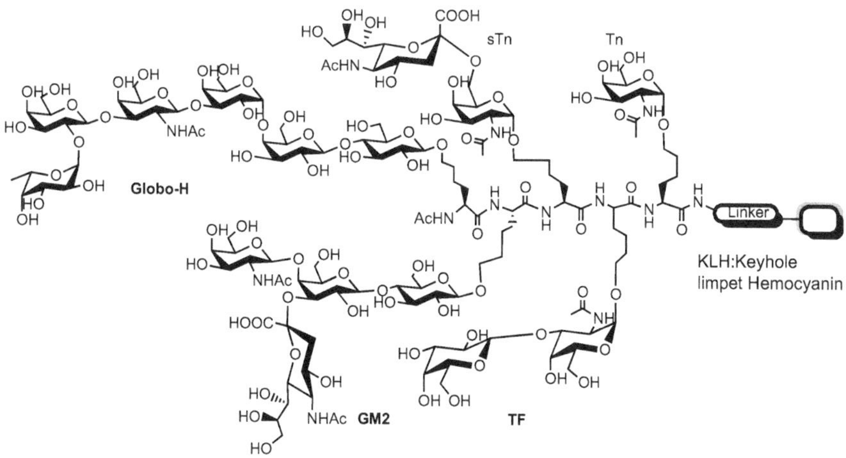

KLH conjugate of Globo-H, sTn, Tn, GM2 and TF (**207**)

Figure 1.55 ***Representative example of some commendable carbohydrate-based vaccines.***

biomarker of cancer. The aberrant glycosylation on the surface of cancer cells is the main inspiration towards development of carbohydrate-based anti-tumor and anti-cancer vaccines.[206,207] KLH conjugate of GM2 (**205**), a mono-epitotic conjugate vaccine consisting of single type of TACAs is under Phase III clinical trials. Tn (c)-KLH (**206**), a mono-epitotic conjugate vaccine consisting of clusters of TACAs, is under Phase I clinical trials against prostate cancer. The clinical assessment of Tn (c)-KLH showed marked declination of prostate specific antigen (PSA) in prostate cancer patients along with saponin adjuvant QS-21. Multi-epitotic conjugate vaccine consisting of several types of TACAs, such as, KLH conjugate of Tn, sTn, GM2, Globo-H, and TF (**207**) (Fig. 1.55) are under Phase I clinical trials.[208]

Despite being a promising lead towards immunotherapy development, semi-synthetic vaccines because of their composition limits and ambivalent structure and efficacy the need of fully synthetic glycoconjugate vaccines with sturdy immune responses has enhanced. The modification with immunological epitopes and adjuvants make these vaccines further more efficient.[209] Boons et al. developed a vaccine with tumor- associated MUC1 glycopeptide, PV Th as epitope and Pam2CysSK4 or Pam3CysSK4 as an adjuvant (three component vaccine (**208,** Fig. 1.56) which exhibited unusually high IgG antibody titer.[210]

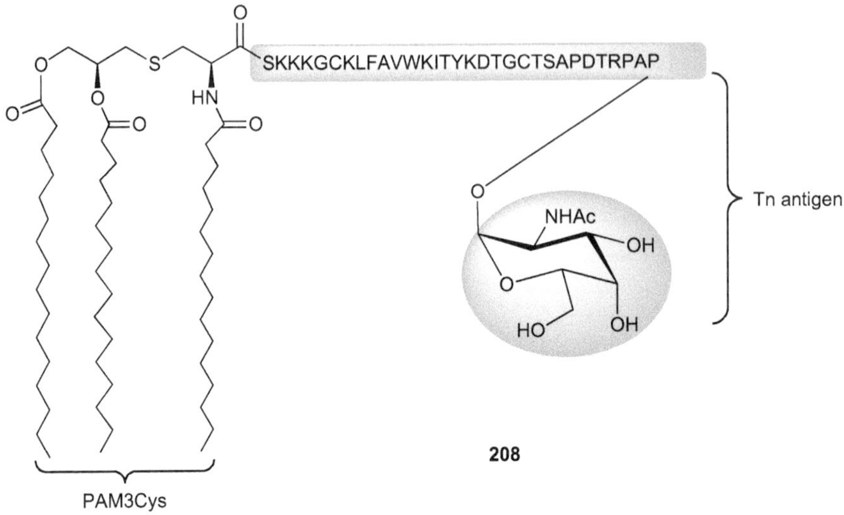

208

Figure 1.56 ***Carbohydrate-based antitumor and anticancer vaccine leads.***

Kunz and co-workers have extensively contributed in discovery and development of MUC1glycopeptide containing synthetic anti-tumor vaccines.Vaccine lead **209** (Fig. 1.57) was reported by Kunz et al. as T_N-MUC1 antigen vaccine which showed strong binding affinity with tumor associated MUC1–glycopeptides.[211]

In another report by the same group, T-antigen-MUC1-tetanus-toxoid vaccine leads **210** and **211** found efficient to induce strong immune response against tumor cells directed to various aberrant glycopeptides structures which are found on a tumor cells.[212] Antitumor vaccine leads **212** and **213** (Fig. 1.58) consisting of MUC1 glycopeptides which contain two immunodominant domains were reported by Kunz et al. In 2011 with high efficiency to induce tumor related antibodies in mice.[213] A number of anti-tumor and anti-cancer synthetic vaccines have been reported in recent years. **Prof. Balram Mukhopadhyay Chapter** of this book gives a detailed account of significance of carbohydrate moiety in development of anti-cancer and anti-bacterial vaccines.

The biological complexity of parasites makes the development of anti-parasite vaccines a difficult task.[214] Glycosylphosphatidylinositols (GPIs) have been detected as a target for development of vaccines against malarial parasite *P. falciparum*. Hexasaccharide GPIs were conjugated with ovalbumin or keyhole limpet hemocyanin (KLH) protein through a spacer and the glycoconjugate thus formed was assessed for anti-malarial activity as vaccine (**209**, Fig. 1.59).[215]

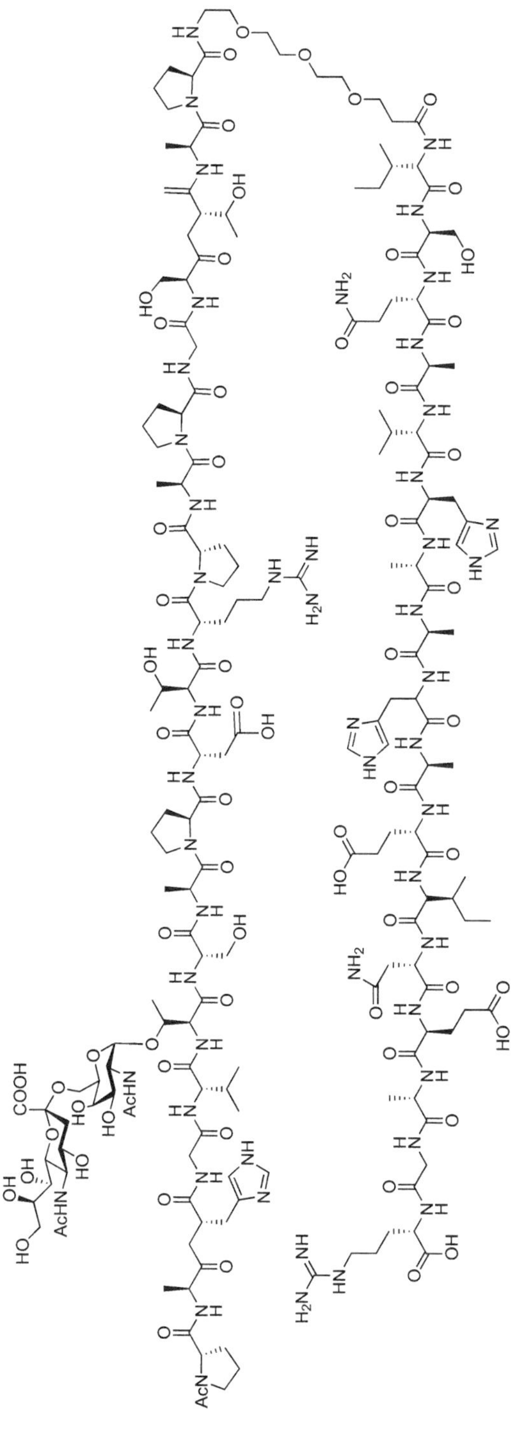

209

Figure 1.57 *Carbohydrate-based anti-tumor and anti-cancer vaccine leads.*

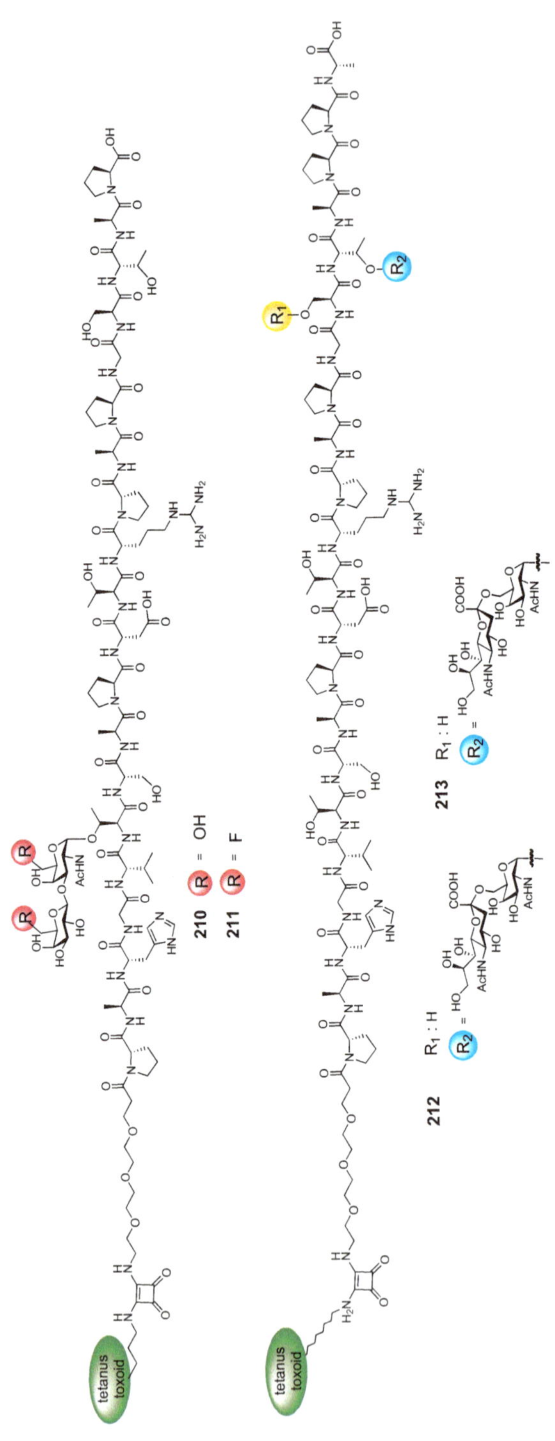

Figure 1.58 *Carbohydrate-based antitumor and anticancer vaccine leads.*

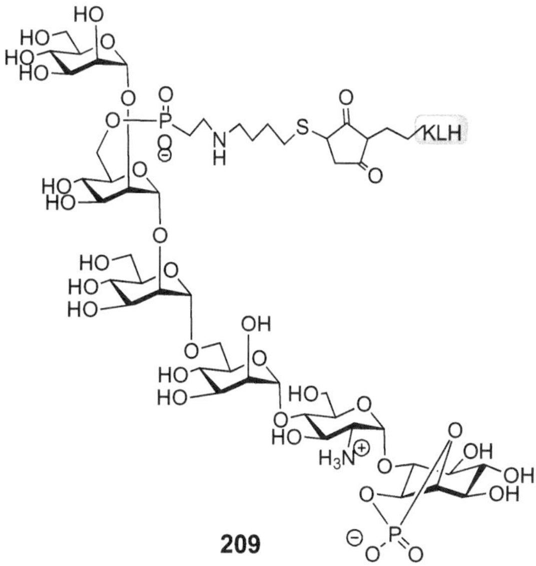

209

Figure 1.59 ***Chemical structure of carbohydrate-based anti-malarial vaccine lead 209.***

6 Conclusions and future perspective

The role of carbohydrates in biological functions makes it a fascinating moiety for biochemistry and medicinal chemistry. Abundance in nature, easy modifiable structures, promising pharmacokinetics and non-toxic nature catch the attention of the pharma world to think of carbohydrate as a favorable moiety for development of pharmaceuticals. Many naturally occurring carbohydrate and derivatives have shown potency as drugs and are being used worldwide as established medicines for various ailments. Many of these drug molecules have been further modified to achieve better selectivity and high activity as compared to the parent drug molecule. Also, a number of synthetic derivatives of carbohydrates have emerged as novel drug leads and many of them have now been approved as efficient drugs. Hundreds of carbohydrate derivatives have exhibited astonishing biological activities in *in vitro* and *in vivo* analyses. Carbohydrate-based therapeutics possess a broad spectrum of activities ranging from anti-tubercular to glycosidase-inhibitor, anti-parasite to anti-bacterial, anti-diabetic to anti cancer and anti-HIV. Carbohydrate, being the pool of functionality, is also an appealing moiety for development of ligands making complex with metal ions. This advantage of carbohydrate moiety assists the development of carbohydrate-based metallodrugs. The versatility of carbohydrate functioning has led the development of carbohydrate-based immunotherapy in form of highly effectual vaccines. Despite the success of carbohydrate-based therapeutic agents, the fraction of these drugs and vaccines is very low as compared to amazing pharmacological traits of carbohydrate moiety. Hence, there is still need to explore the possibilities and potential of glycochemistry in drug development so that the carbohydrate skeleton can be mould in a manner to fulfil the requirements of target-based drug development in order to display amazing pharmaceutical properties. Also, some naturally occurring important carbohydrates with valuable pharmacokinetic significance (for example, sialic acid) need to be explored meticulously for development of new drug leads. Moreover, the drug discovery dealing with the carbohydrate moiety should be more focused on the modification in carbohydrate moiety, recognition of carbohydrate specific targets and accordingly modification in drug delivery, exploitation of naturally occurring glycans for drug development and carbohydrate-based immunotherapy systems. The continued development in the field of glycochemistry and carbohydrate-therapeutics, new dimensions in the drug discovery and development may be added.

Acknowledgments

VKT sincerely thanks Dr. R. P. Tripathi, former senior scientist at CSIR-Central Drug Research Institute, Lucknow for his throughout help and also gratefully acknowledge Council of Scientific and Industrial Research (CSIR), New Delhi (Grant No.: P-25/370) for the funding.

References

1. (a) Barreiro, E. J.; Privileged Scaffolds in Medicinal Chemistry: An Introduction, in S., Brase, (Ed.), Privileged Scaffolds in Medicinal Chemistry: Design, Synthesis, Evaluation, RSC: *London*, **2015**, 1–15; (b) Varin, T.; Schuffenhauer, A.; Ertl, P.; Renner, S. *J. Chem. Inf. Model.*, **2011**, *51*, 1528–1538; (c) Nakagawa, T.; Miyao, T.; Funatsu, *K. Mol. Inform.* **2018**, *37*, 1–2.
2. Boysen, M. M. K. *Chem. Eur. J.* **2007**, *13*, 8648–8659.
3. Varki, A. *Glycobiology* **1993**, *3*, 97–130.
4. Fundamentals of Glycosciences, Ed. Gabius, H.-J. Wiley-VCH, Weinheim, **2009**.
5. (a) Varki, A.; Cummings, R. D.; Esko, J. D.; Freeze, Stanley, H. H. P.; Bertozzi,; Hart, C. R. G. W.; Etzler, M. E. (Eds.), Essentials of Glycobiology, second Edition, Cold Spring Harbor Laboratory Press, *New York*, **2009**; (b) Dwek, R. A.; *Chem. Rev.* **1996**, *96*, 683–720.
6. Bertozzi, C. R.; Kiessling, L. L. *Science* **2001**, *291*, 2357–2364.
7. Okajima, T.; Irvine, K. D. *Cell* **2002**, *111*, 893–904.
8. (a) Current R&D Highlights, Carbohydrate based Drugs, Drugs & Pharma., 2010, 33, 16–19; (b) Mishra, S.; Upadhayay, K.; Mishra, K. B.; Tripathi, R. P.; Tiwari, V. K.; *Studies in Natural Product Chemistry*, **2016**, *49*, 307–361.
9. Feizi, T. *Curr. Opin. Struct. Biol.* **1993**, *3*, 701–710.
10. Fukuda, M. *Biorg. Med. Chem.* **1995**, *3*, 207–215.

11. Ikehara, Y.; Niva, T.; Biao, L.; Ikehara, S. K.; Ohashi, N.; Kobayashi, T.; Shimizu, Y.; Kojima, N.; Nakanishi, H. *Cancer Res.* **2006**, *66*, 8740–8748.
12. Lowe, J. B. *Cell* **2001**, *104*, 809–812.
13. Pouysségur, J.; Willingham, M.; Pastan, I. *Proc. Natl. Acad. Sci. U. S. A.* **1977**, *74*, 243–247.
14. Nelson, David L. *Principles of Biochemistry*; 4th Ed.; W. H. Freeman, 2004.
15. Angata, T.; Varki, A. *Chem. Rev.* **2002**, *102*, 439–469.
16. (a) Schauer, R. Glycoconj. J., **2000**, *17*, 485–499; (b) Ilver, D. et al., *Methods Enzymol.*, **2003**, *363*, 134–157; (c) Lehmann, F. et al., *Cell. Mol. Life Sci.*, **2006**, *63*, 1331–1354.
17. Wang, B.; Brand-Miller, J. *Eur. J. Clinical Nutrition* **2003**, *57*, 1351–1369.
18. (a) Ley, K. Trends *Mol. Med.*, **2003**, 9, 263–268; (b) Mc Ever, R. P. Curr. Opin. *Cell Biol.*, **2002**, *14*, 581–586. (c) Dall'Olio, F.; Chiricolo, M. Glycoconj. J., **2001**, *18*, 841–850.
19. Varki, A.; Schauer, R. In *Essentials of Glycobiology*, 2nd ed. Varki, A.; Cummings, R. D.; Esko, J. D.; Freeze, H. H.; Stanley, P.; Bertozzi, C. R.; Hart, G. W.; Etzler, M. E. Eds. Cold Spring Harbor (NY) chapter 14: Sialic Acids, **2009**.
20. Sibylle, A.; Gruner, W.; Locardi, E.; Lohof, E.; Kessler, H. *Chem. Rev.* **2002**, *102*, 491–514.
21. (a) Sperandio, M.; Gleissner, C. A.; Ley, *K. Immunol. Rev.*, **2009**, *230*, 97–113; (b) van Die, I.; Cummings, R. *D. Chem. Immunol. Allergy*, **2006**, *90*, 91–112; (c) McEver, R. P.; *Glycoconj. J.* **1997**, *14*, 585–591.
22. (a) Gabius, H. J.; Andr, S.; Jimnez-Barbero, J.; Romero, A.; Sols, *D. Trends Biochem. Sci.* **2011**, *36*, 298–313. (b) Sols, D.; Bovin, N.V.; Davis, A. P.; Jimnez-Barbero, J.; Romero, A.; Roy, R.; Smetana Jr, K.; Gabius, H.-J. *Biochim. Biophys. Acta.* **2015**, *1850*, 186–235.
23. Magnani, J. L.; Ernst, B. *Discov. Med.* **2009**, *8*, 247–252.
24. Dube, D. H.; Bertozzi, C. R. *Nat. Rev. Drug Disc.* **2005**, *4*, 477–488.
25. (a) Adamczyk, B.; Tharmalingam, T.; Rudd, P. M. *Biochim. Biophys. Acta*, **2012**, 1820, 1347–1353; (b) Shigeoka, A. O.; Pincus, S. H.; Rote, N. S.; Hill, H. R. *J. Infec. Dis.*, **1884**, *149*, 363–372.
26. Murrey, H. E.; Wilson, L. C. H. *Chem. Rev.* **2008**, *108*, 1708–1731.
27. Roedern, E. G.V.; Lohrf, E.; Hesseler, G.; Hoffmann, M.; Kesseler, H. *J. Am. Chem. Soc.* **1996**, *118*, 10156–10167.
28. Bemis, G. W.; Murcko, M. A. *J. Med. Chem.* **1996**, *39*, 2887–2893.
29. Driguez, P. A.; Potierb, P.; Trouilleuxc, P. *Nat. Prod. Rep.* **2014**, *31*, 980–989.
30. Jorpes, J. E. *Am. J. Med.* **1962**, *33*, 692–702.
31. Scheen, A. *J. Diabetes Metab.* **1998**, *24*, 311–320.
32. Okutomi, T.; Kato, Y.; Ichihara, H.; Hyodo, I.; Fujitsuka, H.; Yasuda, S.; Tatematsu, N. *Gan Kagaku Ryoho* **2000**, *27*, 65–71.
33. Boucher, W. S.; Letourneau, R.; Huang, M.; Kempuraj, D.; Green, M.; Sant, G. R.; Theoharides, T. C. *J. Urology.* **2002**, *167*, 380–384.
34. (a) Harris, C.; Pierce, K.; King, G.; Yates, K. M.; Hall, J.; Tizard, I., *Mol. Biother.*, **1991**, *3*, 207–213; (b) Femenia, A.; Sanchez, E. S.; Simal, S.; Rossello, C. *Carbohydr. Polym.*, **1999**, *39*, 109–117; (c) Reynolds, T.; Dweck, A. C. *J. Ethnopharmacol.*, **1999**, *68*, 3–37.
35. Mokhtari, R. B.; Baluch, N.; Homayouni, T. S.; Morgatskaya, E.; Kumar, S.; Kazemi, P; Yeger, H. *J. Cell Commun. Signal* **2018**, *12*, 91–101.
36. Pachamuthu, K.; Schmidt, R. R. *Chem. Rev.* **2006**, *106*, 160–187.
37. Zhao, Q.; Chen, X.-Y.; Martin, C. *Sci. Bull.* **2016**, *61*, 1391–1398.
38. Nayak, M. K.; Agrawal, A. S.; Bose, S.; Naskar, S.; Bhowmick, R.; Chakrabarti, S., et al. *J. Antimicrob. Chemother.* **2014**, *69*, 1298–1310.
39. (a) Wu, J.; Li, H.; Li, M. **2015**, Int. J. Clin. Exp. Med., *8*, 2128–2137. (b) Lee, W., Ku, S. K., *Bae, J. S.*, **2015**, *38*, 110–125.
40. Wang, S. C.; Chen, S. F.; Lee, Y. M.; Chuang, C. L.; Bau, D.T.; Lin, S. S. **2013**, *27*, 707–714.
41. Zheng, W. X.; Wang, F.; Cao, X. L.; Pan, H.Y.; Liu, X.Y.; Hu, X. M., et al. *Brain Injuries* **2014**, *28*, 227–234.

42. Peng-Fei, L.; Fu-Gen, H.; Bin-Bin, D.; Tian-Sheng, D.; Xiang-Lin, H.; Ming-Qin, Z. *J. Food Sci. Technol.* **2013**, *50*, 615–619.
43. (a) Waisundara, V. Y.; Hsu, A.; Huang, D. J.; Tan, B. K. *H. Am. J. Chinese Med.*, **2008**, *36*, 517–540; (b) Banerjee, K.; Oulkar, D. P.; Patil, S. B.; Jadhav, M. R.; Dasgupta, S.; Patil, S. H.; Bal, S.; Adsule, P. G., *J. Agric. Food Chem.*, **2009**, *57*, 4068–4078.
44. Pereira, L. G. B.; Petacci, F.; Fernandes, J.; Correa, A.; Vieira, P. C.; De Silva, M. F. G. F.; Malaspina, O. *Pest Manag. Sci.* **2002**, *58*, 503–507.
45. (a) Yu, J.; Xiao, Z.; Zhao, R.; Lu, C.; Zhang, Y. *Mol. Med. Rep.*, **2017**, *17*, 3789–3796; (b) Meng, Q. F.; Zhang, Z.; Wang, Y. J.; Chen, W.; Li, F. F.; Yue, L. T.; et al. *J. Neuroimmunol.*, **2016**, *298*, 138–145.
46. Moulari, B.; Pellequer, Y.; Lboutounne, H.; Girard, C.; Chaumont, J. P.; Millet, J.; Muyard, F. *J. Ethnopharmacol.* **2006**, *106*, 272–278.
47. Fan, B.-Y.; Gu, Y.-C; He, Y.; Li, Z.-R.; Luo, J. G.; Kong, L. Y. *J. Nat. Prod.* **2014**, *77*, 2264–2272.
48. (a) Cao, S.; Norris, A.; Wisse, J. H.; Miller, J. S.; Evans, R.; Kingston, D.G. *J. Nat. Prod.*, **2005**, *68*, 487–492; (b) Kingston, D. G.; Cao, S. US2006264383 (A1) **2006**; (c) Cao, S.; Norris, A.; Wisse, J. H.; Miller, J. S.; Evans, R.; Kingston, D. G. *Nat. Prod. Res.*, **2007**, *21*, 872–876.
49. (a) Fürstner, A.; Nagano, T. *J. Am. Chem. Soc.*, **2007**, *129*, 1906–1907; (b) Nagano, T.; Pospí-sil, J.; Chollet, G.; Schulthoff, S.; Hickmann, V.; Moulin, E.; Herrmann, J.; Müller, R.; Fürstner, *A. Chem. Eur. J.*, **2009**, *15*, 9697–9706; (c) Postema, M. H. D.; TenDyke, K.; Cutter, J.; Kuznetsov, G.; Xu, Q. *Org. Lett.*, **2009**, *11*, 1417–1420; (d) Zong, G.; Barber, E.; Aljewari, H.; Zhou, J.; Hu, Z.; Du, Y.; Shi, W. Q. *J. Org. Chem.*, **2015**, *80*, 9279–9291.
50. Zong, G.; Aljewari, H.; Hu, Z.; Shi, W. Q. *Org. Lett.* **2016**, *18*, 1674–1677.
51. Li, H.; Guo, W.; Ma, X. J.; Li, J. S.; Song, X. *Appl. Biochem. Biotechnol.* **2017**, *181*, 1372–1387.
52. Ribeiro, I. A.; Faustino, C. M.; Guerreiro, P. S.; Frade, R. F.; Bronze, M. R.; Castro, M. F.; Ribeiro, M. H. *J. Mol. Recognit.* **2015**, *28*, 155–165.
53. Thompson, A. M. G.; Iancu, C.V.; Nguyen, T. T. H.; Kim, D.; Choe, J.-Y. *Sci. Rep.* **2015**, *5*, 12804.
54. Velter, I.; Ferla, B. L.; Nicotra, F. *J. Carbohydr. Chem.* **2006**, *25*, 97–138.
55. Kohm, A. P.; Fuller, K. G.; Miller, S. D. *Trends Microbiol.* **2003**, *11*, 101–105.
56. Papageorgiou, C.; Haltiner, R.; Bruns, C.; Petcher, T. J. *Biorg. Med. Chem. Lett.* **1992**, *2*, 135–140.
57. Hirschmann, R.; Nicolaou, K. C.; Pietranico, S.; Leahy, E. M.; Salvino, J.; Arison, B.; Cichy, M. A.; Spoors, P. G.; Shakespeare, W. C.; Sprengeler, P. A.; Hamley, P.; Smith, A. B.; Reisine, T.; Raynor, K.; Maechler, L.; Donaldson, C.; Vale, W.; Freidinger, R. M.; Cascieri, M. R.; Strader, C. D. *J. Am. Chem. Soc.* **1993**, *115*, 12550–12568.
58. Singh, G. S.; Pandeya, S. N. Natural Products in Discovery of Potential and safer antibacterial agents, in Opportunity Challenge and Scope of Natural Products in Medicinal Chemistry (V K Tiwari, Ed.), *Research Signpost, India*, **2011**, pp. 63–102.
59. Ritter, T. K.; Wong, C. H. *Angew. Chem. Int. Ed.* **2001**, *40*, 3508–3533.
60. (a) Umezawa, S. Adv. Carbohydr. *Chem. Biochem.*, **1994**, *30*, 111–182; (b) Umezawa, H.; Hooper I. R. (Eds.) *Aminoglycoside Antibiotics, Springer Verlag*, **1982**, *62*, 1–35.
61. Hellman, L. M. *Am. J. Obstetr. Gynecolog.* **1949**, *58*, 1224.
62. Weinstein, M. J. *J. Bacteriol.* **1967**, *94*, 789–790.
63. Savage, G. M. *Infection* **1974**, *2*, 152–159.
64. (a) Chen, L.; Walker, D;. Sun, B.; Hu, Y.; Walker, S.; Kahne, *D. Proc. Natl. Acad. Sci. U.S.A.*, **2003**, *100*, 5658–5663; (b) Halliday, J.; Keveney, M.; Muldon, C.; Rajarathan, P.; Meutermans, *W. Biochem. Pharmacol.*, **2006**, *71*, 957–967; (c) Ostasha, B.; Walkera, *S. Nat. Prod. Rep.*, **2013**, *30*, 161–174.
65. Ostash, B.; Walker, S. *Nat. Pro. Rep.* **2010**, *11*, 1594–1617.
66. Revill, P.; Serradell, N.; Bolos, J. *Drugs Fut.* **2006**, *31*, 494–497.
67. Brandish, P. E.; Kimura, K. I.; Inukai, M.; Southgate, R.; Lonsdale, J. T.; Bugg, T. D. *Antimicrob. Agents Chemother.* **1996**, *40*, 1640–1644.

68. Somner, E. A.; Reynolds, P. E. *Antimicro. Agents Chemother.* **1990**, *34*, 413–419.
69. Farver, D. K.; Hedge, D. D.; Lee, S. C. *Ann. Pharmacother.* **2005**, *39*, 863–868.
70. Fulco, P.; Wenzel, R. P. *Expert Rev. Anti-Infect. Ther.* **2006**, *4*, 939–945.
71. Woodward, R. B. *Angew. Chem.* **1957**, *69*, 50–58.
72. Hammerschlag, M. R.; Sharma, R. *Exp. Opin. Inv. Drugs.* **2008**, *17*, 387–400.
73. Stucki, A.; Gerber, P.; Acosta, F.; Cottagnoud, M.; Cottagnoud, P.; Jiang, L.; Nguyen, P.; Wachtel, D.; Wang, G.; Phan, L. T. *J. Antimicrob. Chemother.* **2008**, *61*, 665–669.
74. Dreier, J.; Amantea, E.; Kellenberger, L. *Chemother* **2007**, *51*, 4361–4365.
75. Blondeau, J. M. *Exp. Opin. Pharma* **2002**, *3*, 1131–1151.
76. Kagami, S.; Esumi, Y.; Nakakoshi, M.; Yoshihama, M.; Kumura, K.-I. *J. Antibiotics* **2003**, *56*, 552–556.
77. Isono, F.; Inukai, M. *Antimicrob. Agents Chemother.* **1991**, *35*, 234–236.
78. Wyszynski, F.; Hesketh, A.; Bibb, M.; Davis, B. *Chem. Sc.* **2010**, *1*, 581–589.
79. Higgins, D. L.; Chang, R.; Debabov, D.V. *Chemother* **2005**, *49*, 1127–1134.
80. Williams, D. H.; Bardsley, B. *Angew. Chem. Int. Edn* **1999**, *38*, 1172–1193.
81. Taylor, J. G.; Li, X.; Oberthur, M.; Zhu, W.; Kahne, D. E. *J. Am. Chem. Soc.* **2006**, *128*, 15084–15085.
82. Billeter, M.; Zervos, M. J.; Chen, A. Y.; Dalovisio, J. R.; Kurukularatne, C. *Clin. Infect. Dis.* **2008**, *46*, 577–583.
83. Martin, E.; Peter, S.; Potgieter, D.; Yu-Ping, Li; Steven, L.; Barriere, A. C.; Kingsley, J.; Corey, G. R. *Antimicrob. Agents Chemother.* **2012**, *56*, 5476–5483.
84. (a) Nicolaou, K. C.; Mitchell, H. J.; Jain, N. F.; Winssinger, N.; Hughes, R.; Bando, T. *Angew Chem., Int. Edn.,* **1999**, *38*, 240–244; (b) Nicolaou, K. C.; Rigol, S. *J. Antibiot.,* **2018**, *71*, 153–184.
85. Tiwari, V. K.; Mishra, B. B.; Mishra, K. B.; Mishra, N.; Singh, A. S.; Chen, X. *Chem. Rev.* **2016**, *116*, 3086–3240.
86. Thirumurugan, D.; Matosiuk, K. J. *Chem. Rev.* **2013**, *33*, 4905–4979.
87. Lin, H.; Walsh, C. T. *J. Am. Chem. Soc.* **2004**, *126*, 11398–14003.
88. Warburg, O. *Science* **1956**, *123*, 309–314.
89. Singh, S.; Lomelino, C. L.; Mboge, M. Y.; Frost, S. C.; McKenna, R. *Molecules* **2018**, *30*, E1045.
90. Okusaka, T.; Okada, S.; Ishii, H.; Ikeda, M.; Nakasuka, H.; Nagahama, H.; Iwata, R.; Furukawa, H.; Takayasu, K.; Nakanishi, Y.; Sakamoto, M.; Hirohashi, S.; Yoshimori, M. *Oncology* **1998**, *55*, 276–283.
91. Rowe, J. M.; Löwenberg, B. *Blood* **2013**, *121*, 4838–4841.
92. Mishra, K. B.; Tiwari, N.; Bose, P.; Singh, R.; Rawat, A. K.; Mishra, R. C.; Singh, R. K.; Tiwari, V. K. *Chemistry Select* **2019**, *4*, 2644–2648.
93. DiJoseph, J. F.; Dougher, M. M.; Armellino, D. C.; Evans, D. Y.; Damle, N. K. *Leukemia* **2007**, *21*, 2240–2245.
94. Erickson, H. K.; Park, P. U.; Widdison, W. C.; Kovtun, Y. V.; Garrett, L. M.; Hoffman, K.; Lutz, R. J.; Goldmacher, V. S.; Blattler, W. A. *Cancer Res.* **2006**, *66*, 4426–4433.
95. Flick, A. C.; Leverett, C. A.; Ding, H. X.; McInturff, E. L.; Fink, S. J.; Helal, C. J.; O'Donnell, C. J. *J. Med. Chem.* **2019**, *62*, 7340–7382.
96. Kantarjian, H. M.; DeAngelo, D. J.; Stelljes, M.; Liedtke, M.; Stock, W.; Gökbuget, N.; O'Brien, S. M.; Jabbour, E.; Wang, T.; White, J. L.; Sleight, B.; Vandendries, E.; Advani, A. S. *Cancer* **2019**, *125*, 2474–2487.
97. Arcamone, F.; Cassinelli, G.; Fantini, G.; Grein, A.; Orezzi, P.; Pol, C.; Spalla, C. *Biotechnol. Bioeng.* **1969**, *11*, 1101–1110.
98. Tacar, O.; Sriamomsak, P.; Dass, C. R. *J. Pharmacy Pharmacol.* **2013**, *65*, 157–170.
99. Pommier, Y.; Leo, Y.; Zhang, H.; Marchand, C. *Chem. Biol.* **2010**, *17*, 421–433.
100. Freiderick, C. A.; Williams, L. D.; Ughetto, G. *Biochemistry* **1990**, *29*, 2538–2549.
101. Dubost, M.; Ganter, P.; Maral, R.; Ninet, L.; Pinnert, S.; Preudhomme, J. *Cancer Chemother. Rep.* **1964**, *41*, 35–36.

102. Arcamone, F.; Bernardi, L.; Giardino, P.; Patelli, B.; Marco, A.; Casazza, A. M.; Pratesi, G.; Reggiani, P. *Cancer Treat. Rep.* **1976**, *60*, 829–834.
103. Arcamone, F.; Penco, S.; Vigevani, A. *Cancer Chemother. Rep.* **1975**, *6*, 123–129.
104. Bottley, A.; Hayes, C.; Seymour, G. 2015, WO 2015150839(A1).
105. Srikanth, V.; Prasad, R. B. N.; Poornachandra, Y.; Babu, V. S. P.; Kumar, C. G.; Jagadeesh, B.; Jala, R. C. R. *Eur. J. Med. Chem.* **2016**, *109*, 134–145.
106. Vetro, M.; Costa, B.; Donvito, G.; Arrighetti, N.; Cipolla, L.; Perego, P.; Compostella, F.; Ronchetti, F.; Colombo, D. *Org. Biomol. Chem.* **2015**, *13*, 1091–1099.
107. (a) Wu, X.; Kassie, F.; Mersch-Sundermann, *V. Mutat. Res.*, **2005**, *589*, 81–102; (b) Santelli-Rouvier, C.; Barret, J. M.; Farrell, C. M.; Sharples, D.; Hill, B. T.; Barbe, J. Eur. *J. Med. Chem.*, **2004**, *39*, 1029–1038. (c) Oommen, S.; Anto, R. J.; Srinivas, G.; Karunagaran, D. Eur. *J. Pharmacol.*, **2004**, *485*, 97–103.
108. Sawada, Y.; Omoto, K.; Kohei, N.; Sakaguchi, K.; Miura, M.; Tanabe, K. *Int. J. Urol.* **2015**, *22*, 590–595.
109. Yagisawa, T.; Okumi, M.; Omoto, K.; Sawada, Y.; Morikawa, S.; Tanabe, K. *Int. J. Urol.* **2016**, *23*, 270–272.
110. Carbonell, J. I.; Kawakubo, H.; Murata, H.; Tanabe, A.; Takeuchi, T.; Kusayanagi, T.; Tsukuda, S.; Hirakawa, T.; Iwabata, K.; Kanai, Y.; Ohta, K.; Miura, M.; Sakaguchi, K.; Matsunaga, S.; Sahara, H.; Kamisuki, S.; Sugawara, F. *Sci. Rep.* **2015**, *5*, 15136.
111. Costa, B.; Dangate, M.; Vetro, M.; Donvito, G.; Gabrielli, L.; Amigoni, L.; Cassinelli, G.; Lanzi, C.; Ceriani, M.; De Gioia, L.; Filippi, G.; Cipolla, L.; Zaffaroni, P.; Perego, N.; Colombo, D. *Bioorg. Med. Chem.* **2016**, *24*, 3396–3405.
112. Vudhgiri, S.; Koude, D.; Veeragoni, D. K.; Misra, S.; Prasad, R. B. N.; Jala, R. C. R. *Bioorg. Med. Chem. Lett.* **2017**, *27*, 3370–3373.
113. Zhu, J.; Zhou, Y.; Wang, G. N.; Tai, G.; Ye, X. S. *Eur. J. Pharmacol.* **2014**, *731*, 65–72.
114. Elías-Rodríguez, P.; Moreno-Clavijo, E.; Carri on-Jim-enez, S.; Carmona, A. T.; Moreno-Vargas, A. J.; Caffa, I.; Montecucco, F.; Cea, M.; Nencioni, A.; Robina, I. *ARKIVOC* **2014**, *3*, 197–214.
115. Kushwaha, D.; Tiwari, V. K. *J. Org. Chem.* **2013**, *78*, 8184–8190.
116. Mishra, S.; Singh, A. S.; Mishra N; and Tiwari, V. K. (2016): Impact of Carbohydrate in Drug Development against Diabetes, in 'Natural Product in Anti-diabetes Agents', Elsevier Publication (ed. , G., Brahmachari,), **2016**.
117. Scheen, A. J. *Diabetes Metab.* **1998**, *24*, 311–320.
118. Scott, L. J.; Spencer, C. M. *Drugs* **2000**, *59*, 521–549.
119. Lembcke, B.; Fölsch, U. R.; Gatzemeier, W.; Lücke, B.; Ebert, R.; Siegel, E.; Creutzfeldt, W. *Eur. J. Clin. Pharmacol.* **1991**, *41*, 561–567.
120. Dabhi, A. S.; Bhatt, N. R.; Shah, M. J. *J. Clin. Diagn. Res.* **2013**, 7, 3023–3027.
121. Rossetti, L.; Smith, D.; Shulman, G. I.; Papachristou, D.; DeFronzo, R. A. *J. Clin. Invest.* **1987**, *79*, 1510–1515.
122. Hanefeld, M.; Forst, T. *Lancet* **2010**, *375*, 2196–2198.
123. Bailey, C. J.; Gross, J. L.; Pieters, A.; Bastien, A.; List, J. F. *Lancet* **2010**, *375*, 2223–2233.
124. Davies, M. J.; D'Alessio, D. A.; Fradkin, J.; Kernan, W. N.; Mathieu, C.; Mingrone, G.; Rossing, P.; Tsapas, A.; Wexler, D. J.; Buse, J. B. *Diabetologia* **2018**, *61*, 2461–2498.
125. Ding, H. X.; Leverett, C. A.; Kyne, R. E.; Liu, K. K.-C.; Fink, S. J.; Flick, A. C.; O'Donnell, C. J. *Bioorg. Med. Chem.* **2015**, *23*, 1895–1922.
126. Porte, J. *D. Diabetes Metab. Res. Rev.*, **2001**, **17**, 181–188; (b) Kikkawa, R.; Br. **J. Nutr.**, **2000**, *84*, 183–185; (c) Edelman, S. *V. Med. Clin. North Am.*, **1998**, *82*, 665–687.
127. Neumiller, J. J.; White, J. R. J.; Campbell, R. K. *Drugs* **2010**, *70*, 377–385.
128. Meng, W.; Ellsworth, B. A.; Nirschl, A. A.; McCann, P. J.; Patel, M.; Girotra, R. N.; Wu, G.; Sher, P. M.; Morrison, E. P.; Biller, S. A.; Zahler, R.; Deshpande, P. P.; Pullockaran, A.; Hagan, D. L.; Morgan, N.; Taylor, J. R.; Obermeier, M. T.; Humphreys, W. G.; Khanna, A.; Discenza, L.; Robertson, J. G.; Wang, A.; Han, S.; Wetterau, J. R.;

Janovitz, E. B.; Flint, O. P.; Whaley, J. M.; Washburn, W. N. *J. Med. Chem.* **2008**, *51*, 1145–1149.
129. Mehta, R.; Patel, H. P.; Patel, K.; Tikson, P. M. *Int. Res. J. Pharma.* **2011**, *2*, 1–9.
130. Chen, Z. H.; Wanga, R. W.; Qing, F. L. *Tetrahedron Let.* **2012**, *53*, 2171–2176.
131. Kim, M. J.; Lee, S. H.; Park, S. O.; Kang, H.; Lee, J. S.; Lee, K. N.; Jung, M. E.; Kim, J.; Lee, J. *Bioorg. Med. Chem.* **2011**, *19*, 5468–5479.
132. Lv, B.; Xu, B.; Feng, Y.; Peng, K.; Xu, G.; Du, J.; Zhang, L.; Zhang, W.; Zhang, T.; Zhu, L.; Ding, H.; Sheng, Z.; Welihinda, A.; Seed, B.; Chen, Y. *Bioorg. Med. Chem. Lett.* **2009**, *19*, 6877–6881.
133. Markham, A. *First global approved Drugs* **2018**, *78*, 513–519.
134. Gagneux, S. *Nat. Rev. Microbiol.* **2018**, *16*, 202–213.
135. Wu, Y.; Xiong, D-C.; Chen, S-C.; Wang, Y.-S.; Ye, X.-S. *Nat. Commun.* **2017**, *8*, 14851.
136. Tripathi, R. P.; Tewari, N.; Dwivedi, N.; Tiwari, V. K. *Med. Res. Rev.* **2005**, *25*, 93–131.
137. Trias, J.; Benz, R. *Mol. Microbiol.* **1994**, *19*, 283–295.
138. (a) Rastogi, N. K.; Goh, S.; David, H. L. Antimicrob. *Agents Chemother.,* **1990**, *34*, 750; (b) Russell, A. *D. Prog. Med. Chem.,* **1998**, *35*, 133–197; (c) Brennan, P. J.; Nikaido, *H. Ann. Rev. Biochem.,* **1995**, *64*, 29–63.
139. Tripathi, R. P.; Tripathi, R.; Tiwari, V. K.; Bala, L.; Sinha, S.; Srivastava, A.; Srivastava, R.; Srivastava, B. S. *Eur. J. Med. Chem.* **2002**, *37*, 773–781.
140. (a) Van Heijennoort, J.; *Glycobiology,* **2001**, *11*, 25–26; (b) Scherman, M.; Weston, A.; Duncan, K.; Whittington, A.; Upton, R.; Deng, L.; Comber, R.; Friedrich, J. D.; Neil, M. R. *J. Bacteriol.,***1995**, *177*, 7125–7130.
141. Katiyar, D.; Tiwari, V. K.; Tewari, N.; Verma, S. S.; Sinha, S.; Gaikwad, A.; Srivastava, A. K.; Chaturvedi, V.; Srivastava, R.; Srivastava, B. S.; Tripathi, R. P. Eur. *J. Med. Chem.,* **2005**, *40*, 351–360; (b) Tripathi, R. P.; Tiwari, V. K.; Tewari, N.; Katiyar, D.; Saxena, N.; Sinha, S.; Gaikwad, A.; Srivastava, A. K;. Chaturvedi, V.; Manju, Y. K.; Srivastava, R.; Srivastava, B. S. Bioorg. *Med. Chem.,* **2005**, *13*, 5668–5679.
142. Tewari, N.; Tiwari, V. K.; Tripathi, R. P.; Gaikwad, A.; Sinha, S.; Shukla, P. K.; Srivastava, R.; Srivastava, B. S. *Bioorg. Med. Chem. Lett.* **2004**, *12*, 329–332.
143. Srivastava, S. K.; Tripathi, R. P.; Ramachandran, R. *J. Biol. Chem.* **2005**, *280*, 30273–30281.
144. Singh, N.; Mishra, B. B.; Bajpai, S.; Singh, R. K.; Tiwari, V. K. *Bioorg. Med. Chem* **2014**, *22*, 18–45.
145. Springer Science & Business Media, **2012**, pp. 109. ISBN 9789400926592.
146. Dwivedi, P.; Mishra, K. B.; Mishra, B. B.; Singh, N.; Singh, R. K.; Tiwari, V. K. *Glycoconj. J.* **2017**, *34*, 61–70.
147. Peet, V. D. P.; Gannon, C. T.; Walker, I.; Dinev, Z.; Angelin, M.; Tam, S.; Ralton, J. E.; McConville, M. J.; Williams, S. J. *Chem. Bio. Chem.* **2006**, 7, 1384–1391.
148. Mishra, R. C.; Tripathi, R.; Katiyar, D.; Tewari, N.; Singh, D.; Tripathi, R. P. *Bioorg. Med. Chem.* **2003**, *11*, 5363–5374.
149. Kinfe, H. H.; Moshapo, P. T.; Makolo, F. L.; Gammon, D. W.; Ehlers, M.; Schmuck, C. *Eur. J. Med. Chem.* **2014**, *87*, 197–202.
150. Tiwari, V. K.; Tewari, N.; Katiyar, D.; Tripathi, R. P.; Arora, K.; Gupta, S.; Ahmad, R.; Srivastava, A. K.; Khan, M. A.; Murthy, P. K.; Walter, R. D. *Bioorg. Med. Chem.* **2003**, *11*, 1789–1800.
151. Clercq, E. D. *Int. J. Antimicro. Agents.* **2009**, *33*, 307–320.
152. Overington, J. P.; Lzikani, B. A.; Hopkins, A. L. *Nature Rev. Drug Disc.* **2006**, *5*, 993–996.
153. Haynes, B. F.; Pantaleo, G.; Fauci, A. S. *Science* **1996**, *271*, 324–328.
154. Silva, F. C.; Souza, M. C. B.V.; Frugulhetti, I. I. P.; Castro, H. C.; Souza, S. L.; Souza, T. M.; Rodrigues, D. Q.; Souza, A. M. T.; Abreu, A. P.; Passamani, F.; Rodrigues, C. R.; Ferreira, V. F. *Eur. J. Med. Chem.* **2009**, *44*, 373–383.
155. Pawar, S. A.; Jabgunde, A. M.; Govender, P.; Maguire, G. E. M.; Kruger, H. G.; Parboosing, R.; Soliman, M. E. S.; Sayed, Y.; Dhavale, D. D.; Govender, T. *Eur. J. Med. Chem.* **2012**, *53*, 13–21.

156. Bemis, G.W.; Murcko, M.A. *J. Med. Chem.* **1996**, *39*, 2887–2893.
157. (a) Rosenberg, R. D.; Damus, P. S. *J. Biol. Chem.*, **1973**, *248*, 6490–6505; (b) Warkentin, T. E.; Chong, B. H.; Greinacher, *A. Thromb. Haemost.*, **1998**, *79*, 1–7; (c) Viskov, C.; Just, M.; Laux, V.; Mourier, P.; Lorenz, M. *J. Thromb. Haemostasis,* **2009**, 7, 1143–1151.
158. Driguez, P. A.; Potierb, P.; Trouilleuxc, P. *Nat. Prod. Rep.* **2014**, *31*, 980.
159. (a) Harduin Lepers, A.; Krzewinski-Recchi, M. A.; Hebbar, M.; Samyn-Petit, B.; Vallejo-Ruiz, V.; Julien, S.; Peyrat, J. P.; Delannoy, *P. Recent Res. Dev. Cancer,* **2001**, *3*, 111–126; (b) Burchell, J.; Poulsom, R.; Hanby, A.; Whitehouse, C.; Cooper, L.; Clausen, H.; Miles, D.; Taylor-Papadimitriou, *J. Glycobiology,* **1999**, *9*, 1307–1311.
160. Kajihara, Y.; Kodama, H.; Wakabayashi, T.; Sato, K.; Hashimoto, H. *Carbohydr. Res.* **1993**, *247*, 179–193.
161. Okazaki, K.; Nishigaki, S.; Ishizuka, F.; Kajihara, Y.; Ogawa, S. *Org. Biomol. Chem.* **2003**, *1*, 2229–2230.
162. Schaub, C.; Muller, B.; Schmidt, R. R. *Glycoconj. J.* **1998**, *15*, 345–355.
163. Schaub, C.; Muller, B.; Schmidt, R. R. *Eur. J. Org. Chem.* **2000**, 1745–1758.
164. Amann, F.; Schaub, C.; Muller, B.; Schmidt, R. R. *Chem. Eur. J.* **1998**, *4*, 1106–1115.
165. Muller, B.; Schaub, C.; Schmidt, R. R. *Angew Chem. Int. Ed.* **1998**, *37*, 2893–2897.
166. Schworer, R.; Schmidt, R. R. *J. Am. Chem. Soc.* **2002**, *124*, 1632–1637.
167. Gerber-Lemaire, S.; Juillerate-Jeanneret, L. *Mini-Rev. Med. Chem.* **2006**, *6*, 1043–1052.
168. D'Alonzo, D.; Guaragna, A.; Palumbo, G. *Curr. Med. Chem.* **2009**, *16*, 473–505.
169. Cox, T. M., et al. *J. Inherit. Metab. Dis.* **2003**, *26*, 513–526.
170. Asano, N. *Glycobiology* **2003**, *13*, 93–104.
171. Truscheit, E.; Frommer, W.; Junge, B.; Müller, L.; Schmidt, D. D.; Wingender, W. *Angew. Chem. Int. Edn.* **1981**, *20*, 744–761.
172. Neufeld, E. F. *Ann. Rev. Biochem.* **1991**, *60*, 257–280.
173. Tifft, C. J.; Proja, R. L. *Glycobiology* **2000**, *10*, 1249–1258.
174. Platt, F. M.; Neises, G. R.; Dwek, R. A.; Butters, T. D. *J. Biol. Chem.* **1994**, *269*, 8362–8365.
175. Magano, J. *Chem. Rev.* **2009**, *109*, 4398–4438.
176. Michiels, B.; Van Puyenbroeck, K.; Verhoeven, V.; Vermeire, E.; Coenen, S. *PLoS ONE* **2013**, *4*, e60348.
177. Itzstein, M.V.; Wu, W.Y.; Kok, G. B.; Pegg, M. S.; Dyason, J. C.; Jin, B.; Phan, T.V.; Smythe, M. L.; White, H. F.; Oliver, S.W.; Coleman, P. M.; Verghese, J. N.; Ryan, D. M.; Woods, J. M.; Bethell, R. C.; Hotham, V. J.; Cameron, J. M.; Penn, C. R. *Nature* **1993**, *363*, 418–423.
178. Sun, J.Y.; Yang, H.; Miao, S.; Li, J. P.; Wang, S.W.; Zhu, M. Z.; Xie, Y. H.; Wang, J. B.; Liu, Z.; Yang, Q. *Phytomedicine* **2009**, *16*, 1070–1074.
179. Whitby, K.; Pierson, T. C.; Geiss, B.; Lane, K.; Engle, M.; Zhou, Y.; Doms, R. W.; Diamond, M. S. *J. Virol.* **2005**, *79*, 8698–8706.
180. Ferrari, P.; Ferrandi, M.; Valentini, G.; Manunta, P.; Bianchi, G. *Med. Hypotheses* **2007**, *68*, 1307–1314.
181. (a) Hollman, A.; British *Med. J.*, **1996**, *312*, 911–912; (b) Master, J.; Schweitzer, P.; Cardiol. J., 2009, 16, 483–486; (c) The Digitalis Investigation Group, New Eng. *J. Med.*, **1997**, *336*, 525–533.
182. Jeon, K.; Byun, M.; Jue, D. *Exp. Mol. Med.* **2003**, *35*, 61–66.
183. Maryanoff, B. E.; Costanzo, M. J.; Nortey, S. O.; Greco, M. N.; Shank, R. P.; Schupsky, J. J.; Ortegon, M. P.; Vaught, J. L. *J. Med. Chem.* **1998**, *41*, 1315–1343.
184. Brentjens, R.; Saltz, L. *Surg. Clinics North Am.* **2001**, *81*, 527–542.
185. Wang, Z.; Gleichmann, H. *Diabetes* **1998**, *47*, 50–56.
186. Clausen, M. R.; Mortensen, P. B. *Drugs* **1997**, *53*, 930–942.
187. Witczak, Z. J.; Nieforth, K. A. *Carbohydrat. Drug Design.* **1997**, 23–24.
188. Liu, K. K. C.; Sakya, S. M.; O'Donnell, C. J.; Flick, A. C.; Ding, H. X. *Bioorg. Med. Chem.* **2012**, *20*, 1155–1174.

189. Warburg, O. *Science* **1956**, *123*, 309–314.
190. Mikata, Y.; Shinohara, Y.; Yoneda, K.; Nakamura, Y.; Brudzinska, I.; Tanase, T.; Kitayama, T.; Takagi, R.; Okamoto, T.; Kinoshita, I.; Doe, M.; Orvig, C.; Yano, S. *Bioorg. Med. Chem. Lett.* **2001**, *11*, 3045–3047.
191. Patra, M.; Johnstone, T. C.; Suntharalingam, K.; Lippard, S. J. *Angew. Chem. Int. Ed.* **2016**, *55*, 2550–2554.
192. Deepthi, S. B.; Ramesh, P.; Trivedi, R.; Buddana, S. K.; Prakasham, R. S. *Inorg. Chim. Acta* **2015**, *435*, 200–205.
193. Florindo, P. R.; Pereira, D. M.; Borralho, P. M.; Rodrigues, C. M. P.; Piedade, M. F. M.; Fernandes, A. C. *J. Med. Chem.* **2015**, *58*, 4339–4347.
194. Sastre, M.; Ritchie, C.W.; Hajji, N. *JSM Alzheimer's Dis Related Dementia* **2015**, *2*, 1014.
195. (a) Schugar, H.; Green, D. E.; Bowen, M. L.; Scott, L. E.; Storr, T.; Bohmerle, K.; Thomas, F.; Allen, D. D.; Lockman, P. R.; Merkel, M.; Thompson, K. H., Orvig, C. Angew. *Chem. Int. Ed.,* **2007**, *46*, 1716–1718; (b) Storr, T.; Merkel, M.; Song-Zhao, G. X.; Scott, L. E.; Green, D. E.; Bowen, M. L.; Thompson, K. H.; Patrick, B. O.; Schugar, H. J.; Orvig, C. *J. Am. Chem. Soc.,* **2007**, *129*, 7453–7463.
196. Daniels, C. C.; Rogers, P. D.; Shelton, C. M. *J. Ped. Pharmacol. Ther.* **2016**, *21*, 27–35.
197. Lin, F. Y.; Ho, V. A.; Khiem, H. B.; Trach, D. D.; Bay, P. V.; Thanh, T. C.; Kossaczka, Z.; Bryla, D. A.; Shiloach, J.; Robbins, J. B.; Schneerson, R.; Szu, S. C. *N. Eng. J. Med.* **2001**, *344*, 1263–1269.
198. Knutson, K. L. *Curr. Opin. Investig. Drugs* **2002**, *3*, 159–164.
199. Chapman, P. B.; Morrisey, D.; Panageas, K. S.; Williams, L.; Lewis, J. J.; Israel, R. J.; Hamilton, W. B.; Livingston, P. O. *Clinic. Cancer Res.* **2000**, *6*, 4658–4662.
200. Ibrahim, N. K.; Murray, J. L. *Clinic. Breast Cancer* **2003**, *3*, S139–S143.
201. Thakkar, S. G.; Heeger, P.; Wacker, B.; Roddy, M.; Waxenecker, G.; Himmler, G.; Loibner, H.; Elson, P.; Bukowski, R. *J. Clinic. Oncolog.* **2004**, *22*, 2586–12586.
202. Zarei, A. E.; Almehdar, H. A.; Redwan, E. M. *J. Immunol. Res.* **2016**, 7203587.
203. Johannes, F. G.; Vliegenthart, F. *FEBS Lett.* **2006**, *580*, 2945–2950.
204. Robbins, J. B.; Schneerson, R.; Szu, S. C.; Pozsgay, V. *Pure Appl. Chem.* **1999**, *71*, 745–754.
205. Bencomo, V.V.; Santana, V. F.; Hardy, E.; Toledo, M. E.; Rodriguez, M. C.; Heynngnezz, L.; Rodriguez, A.; Baly, A.; Herrera, L.; Izquierdo, M.; Villar, A.; Valdes, Y.; Cosme, K.; Deler, M. L.; Montane, M.; Garcia, E.; Ramos, A.; Aguilar, A.; Medina, E.; Torano, G.; Sosa, I.; Hernandez, I.; Martinez, R.; Muzachio, A.; Carmenates, A.; Costa, L.; Cardoso, F.; Campa, C.; Diaz, M.; Roy, R. *Science* **2004**, *305*, 522–525.
206. Fuster, M. M.; Esko, J. D. *Nat. Rev. Cancer.* **2005**, *5*, 526–542.
207. (a) Kozbor, D.; *Immunologic Res.,* **2010**, *46*, 23–31; (b) Borman, S. Chem. Eng. News, **2011**, *89*, 53–57.
208. Guo, Z.; Wang, Q. *Current Opinion Chem. Biol.* **2009**, *13*, 608–617.
209. Roy, R. In Drug Disc. Today: Techn. Owens J. Elsevier Sc.(Ed.), 2004, 327–336.
210. Buskas, T.; Ingale, S.; Boons, G. J. *Angew. Chem. Int. Ed* **2005**, *44*, 5985–5988.
211. Westerlind, U.; Hobel, A.; Gaidzik, N.; Schmitt, E.; Kunz, H. *Angew. Chem. Int. Ed.* **2008**, *47*, 7551–7556.
212. Hoffmann-Roder, A.; Kaiser, A.; Wagner, S.; Gaidzik, N.; Kowalczyk, D.; Westerlind, U.; Gerlitzki, B.; Schmitt, E.; Kunz, H. *Angew. Chem. Int. Ed.* **2010**, *49*, 8498–8503.
213. Gaidzik, N.; Kaiser, A.; Kowalczyk, D.; Westerlind, U.; Gerlitzki, B.; Sinn, H. P.; Schmitt, E.; Kunz, H. *Angew. Chem. Int. Ed.* **2011**, *50*, 9977–9981.
214. Tarlton, R. L. *Cell. Microbiol.* **2005**, 7, 1379–1386.
215. Hewitt, M. C.; Snyder, D.A.; Seeberger, P. H. *J. Am. Chem. Soc.* **2002**, *124*, 13434–13436.

CHAPTER TWO

Heparin mimetics as tools for modulation of biology and therapy

Jicheng Zhang[a], Xuefei Huang[a,b,c]
[a]Department of Chemistry, Michigan State University, East Lansing, MI, United States
[b]Department of Biomedical Engineering, Michigan State University, East Lansing, MI, United States
[c]Institute for Quantitative Health Science and Engineering, Michigan State University, East Lansing, MI, United States

1 Introduction

Glycosaminoglycans (GAGs) are highly negatively charged, linear polysaccharides. As members of the GAG family, heparin and heparan sulfate (HS) have highest densities of negative charges among all biological macromolecules. The structures of heparin and HS are remarkably heterogeneous, which contains complex linear disaccharide repeating units of D-glucosamine (GlcN) α-(1-4)-linked to a uronic acid (90% L-iduronic acid (IdoA) and 10% D-glucuronic acid (GlcA)). 2—OH of GlcA and IdoA residues can be sulfated. The GlcN monosaccharide can be either *N*-acetylated (GlcNAc) or *N*-sulfated (GlcNS), while *O*-sulfation could be at 6—OH and 3—OH (Fig. 2.1).[1]

Heparin has been used as an anti-coagulant drug for over 80 years. Doyon and Howell firstly reported heparin as an anticoagulant in early 1910s.[2] In 1918–1920, Howell improved the isolation method and named the polysaccharide anticoagulant "heparin". It was not until the 1930s that preparations of heparin were ready for clinical trials. One of the most famous landmarks in heparin analysis was the identification of *L*-iduronic acid in an acid hydrolysate of heparin by Cifonelli and Dorfman in 1962, as early on, it was mistakenly considered that *D*-glucuronic acid was the only uronic acid in the heparin structure. The configuration of glucosaminidic and uronidic linkages was established by Roden in 1989.[3-6] The description of the minimal binding requirement for anti-thrombin is another important landmark for heparin structure research. Lindahl's group was able to identify a unique pentasaccharide sequence from an isolated octasaccharide, which was thought to be the smallest fragment required for anti-thrombin binding.[7]

Carbohydrates in Drug Discovery and Development. http://dx.doi.org/10.1016/B978-0-12-816675-8.00002-6

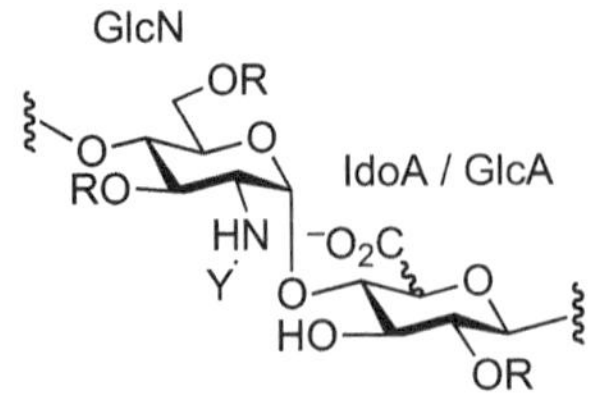

Figure 2.1 ***Structure of the disaccharide repeating unit of heparin and HS.***

There are three forms of heparin approved by US Food and Drug Administration (FDA) as anti-coagulant drugs: unfractionated heparin (UFH, average molecular weight (MW) 14,000 Da), low-molecular-weight heparin (LMWH, MW 3500–6500 Da) (12–20 saccharide units) and the synthetic pentasaccharide, fondaparinux (MW 1508.3 Da).[1] As a well-known anticoagulant drug, UFH is a mixture of polysaccharides in different lengths with various sulfation patterns.[8,9] As it is sourced from animals (primarily from porcine intestinal mucosa with some from bovine), with the heterogeneity of compounds from nature, the reliability and safety of heparin preparations have received increasing attention. This is especially the case after the incident caused by the worldwide distribution of over sulfated chondroitin sulfate contaminated heparin in 2007–08, with adverse reactions identified in over 100 patients in 13 states alone in less than three months.[10,11] As a result, less heterogeneous or even homogenous heparin derivatives are highly desired. Bemiparin and M-Enoxaparin are currently marketed in Europe and the U.S. as LMWH anticoagulant drugs.[12-14] Ultralow molecular weight (ULMW) heparin pentasaccharide fondaparinux as a chemically pure anticoagulant is the most expensive drug among heparins as results of its challenging chemical synthesis (~50 steps) with an overall low yield (~0.1%).[15,16]

Besides the anticoagulant activities, heparin can mediate many other protein functions. It is involved in tumor growth and metastasis in part as a result of its inhibition of heparanase, a β-endoglycosidase capable of cleaving HS side chains of heparan sulfate proteoglycans (HSPGs) on the cell surface and in the extracellular matrix (ECM). Furthermore, it can interact with angiogenesis mediators such as fibroblast growth factors (FGFs) and vascular endothelial growth factor (VEGF) in the ECM to promote cell proliferation.[17-20]

Synthesis of homogeneous heparin polysaccharides is a formidable challenge due to their length and structural complexity. Heparin's polypharmacy and anticoagulant properties suggest that they need to be tailored for clinical use in applications other than anticoagulation. Synthetic or semi-synthetic heparin mimetics could be designed to address these challenges associated with heparin. In this review, we focus on compounds that can mimic the structural characters and activities of heparin in three common functions, i.e., anti-coagulant, growth factor binding, and heparanase inhibition. Due to space limitation, more native-like heparin oligo and longer saccharide sequences prepared through innovative chemical[21-26] as well as chemoenzymatic syntheses,[27-34] are not included. Interested readers are referred to several excellent reviews covering those compounds.[33,35-37]

2 Heparin mimetics as anti-coagulants

According to the World Health Organization (WHO) 2017 fact sheets, cardiovascular diseases (CVDs) are the number one cause of death (31%) globally, accounting for 17.9 million deaths in 2016. Thrombosis and inflammation are strongly related to different types of CVDs, including heart attack, stroke, arterial thrombosis, and venous thromboemlism.[38-40] UFH, as a commonly used anticoagulant with low costs, is rapid-onset, and its effects can be reversed with the administration of protamine.[41] However, heparin-induced thrombocytopenia (HIT) is a significant and life-threatening side effect caused by the interaction of UFH with platelet factor 4 (PF4), a positively charged chemokine.[42-45] Besides the side effects, short half-life (< 1 h) and low dose-response relationships encourage scientists to discover improved anticoagulants.[46] LMWH and ULMH are currently approved anticoagulant drugs. LMWH was introduced in the 1940s with its subcutaneous administration route, improved bioavailability, and longer half-life.[1,47] With those advantages, LMWH has been the most prescribed heparin in the United States.[48] However, LMWH preparation still depends on the source of UFH, and incomplete neutralization with protamine occurs with risks of bleeding.[32] On the other hand, fondaparinux, the chemically synthetic pentasaccharide, reduces the risk of contamination and limits the interactions with other plasma proteins. Nevertheless, the laborious chemical synthesis and high cost impede it being widely prescribed. Thus, heparin mimetics are developed in recent years to overcome these limitations of heparin and heparin derivatives.

"Non-glycosamino" glycan analogs of heparins were reported in 1992 to reduce the number of synthetic steps of antithrombin III binding pentasaccharide.[49] Glucose was used to substitute the glucosamine unit (Compounds **1–6**, Fig. 2.2), which simplified building block preparations, and avoided the need for selective *N*-sulfation. Furthermore, only acyl and benzyl esters were applied as temporary protecting groups for building blocks used in synthesis, which reduced the number of deprotection steps needed. Overall, the synthetic route was shortened to approximate 25 steps in comparison to ~50 steps for fondaparinux preparation. Idraparinux **4** was found as a potent pentasaccharide derivative, which showed inhibition of thrombin generation *via* both the extrinsic and intrinsic pathways. 1611 anti-Xa units/mg and a long *in vivo* half-life (9.2 hours in rats and 61.9 hours in baboons) were observed after intravenous (IV) and subcutaneous (SC) administration with a linear pharmacokinetic profile. Idraparinux had been tested to prevent stroke in atrial fibrillation and venous thromboembolism (VTE). However, its antithrombotic activity cannot be reversed due to a lack of neutralizing agents. Idrabiotaparinux **7** (Fig. 2.3) is a biotinylated Idraparinux derivative to bestow the reversibility with avidin, which has no

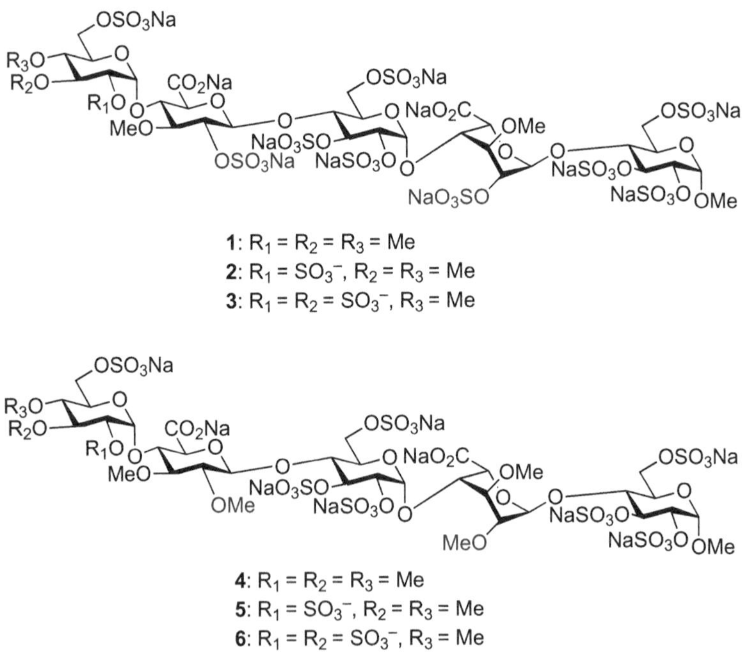

Figure 2.2 *"Non-glycosamino" glycan analogs.*

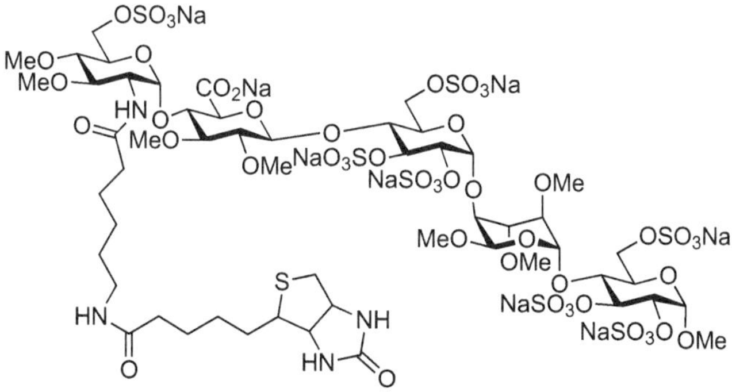

Figure 2.3 ***Structure of Idrabiotaparinux.***

reported toxicity and a very short pharmacokinetic half-life.[50-53] The pharmacodynamic results from the Phase I study confirmed that by forming an avidin-idrabiotaparinux complex, avidin effectively reversed the anti-FXa activity of idrabiotaparinux.[54-55] Phase III clinical trial study of idrabiotaparinux for arterial fibrillation was halted because it did not show better results than standard long-term warfarin treatment for VTE.[56]

To obtain more potent anti-thrombic drugs, Petitou et al. synthesized a series of "non-glycosamino" glycan heparin mimetics to reach full anticoagulant activity, including thrombin inhibitory properties.[57] As discussed above, UFH binds with many plasma proteins causing significant side effects, particularly PF4. In this study, various lengths of oligosaccharides were designed to discriminate between thrombin and PF4. 6-mer to 14-mer **8–11** (Fig. 2.4) did not display high anti-thrombin activities. The size-dependent increases in activities were observed for 16-mer (**12**) to 20-mer (**14**), with the 20-mer (**14**) being half as potent as UFH. To mimic the full anticoagulant activity of heparin, the structure of an oligosaccharide or even longer saccharide sequence would include an anti-thrombin-binding domain (A-domain) coupled to a thrombin-binding domain (T-domain). Another important issue is how to link these two domains. Repeating α and β-linked 3-*O*-methyl-2,6-di-*O*-sulfo-D-glucose units were attached to both the reducing and non-reducing ends as a thrombin-binding domain. Compounds **15–17** with T-domain at the non-reducing end were inhibitors of both factor Xa and thrombin. Thrombin inhibition was also size-dependent, and compound **17** was about twice more potent than heparin according to their IC_{50} values. With the A-domain at the non-reducing end and T-domain at the reducing end, 18-mer (**18**) showed 30 to 100 times weaker activities for thrombin inhibition than the corresponding 17/19-mer saccharides **16**

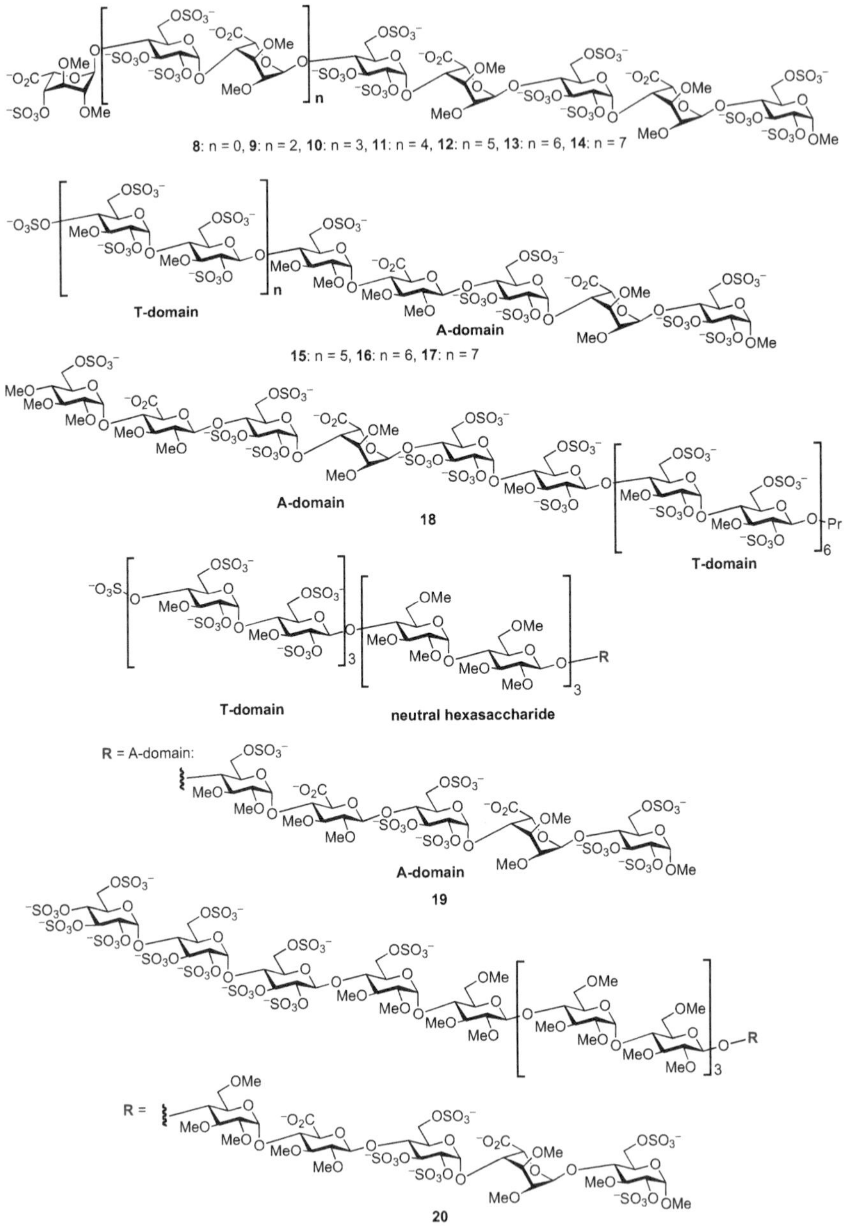

Figure 2.4 ***A series of "non-glycosamino" glycan heparin mimetics.***

and **17** bearing the T-domain at the non-reducing end instead. All compounds **8–18** showed effective *in vivo* anti-FXa activities, and dose-dependent thrombin inhibition activities were observed for compounds **12–17.** Even though compounds **15–17** inhibited both factor Xa and thrombin, the cross-reactivity with PF4 was hardly overcome, with the compounds failing to abolish the interactions with PF4.

To reduce PF4 binding, the charge of a molecule is an essential parameter for consideration to avoid the nonspecific binding. Compound **19** was designed with a pentasaccharide as A-binding domain and a hexasaccharide at the non-reducing end as the T-domain, which was connected by a neutral methylated hexasaccharide. Both domains contain oligosaccharides that are shorter than 8-mer, the minimum unit required for PF4 binding.[58] Compound **19** showed no significant interactions with PF4 even at high concentrations (100 ug/mL) and exhibited higher antithrombotic activity compared to heparin.[57] SA123781A, compound **20**, is another "non-glycosaminoglycan" synthetic heparin mimetic based on the strategy of charge reduction for polyanionic compounds, which is a structural variant of compound **19**.[59] SR123781A exhibited high affinity for human anti-thrombin and potent factor Xa and thrombin inhibition in multiple animal models for arterial-venous thrombosis. However, its development was halted during phase IIb clinical study due to the success of the LMWH.[60]

To display full anticoagulant activities, both anti-thrombin and thrombin binding domains are required to be present on the same polysaccharide chain. Glycoconjugate was designed to link both domains to facilitate the anticoagulant activity of the compound. Glycoconjugate **24** (Fig. 2.5) comprises the antithrombin binding pentasaccharide, a linear spacer, and a persulfated maltotrioside thrombin binding region.[61] Polysulfated maltotrisaccharide 1-amino-hexaethylene glycol derivative (**21**) was prepared for thrombin binding, and a thioacetyl-substituted spacer was linked to a pentasaccharide as the antithrombin binding domain **22**. Sulfosuccinimidyl (4-iodoacetylamino) benzoate (sulfo-SIAB) (**23**) was applied to connect **21** with **22** to form glycoconjugate **24**, which has the similar length as a heparin 18-mer oligosaccharide. In order to study the binding requirement for thrombin, cellobiose heptasulfate and maltopentaose hexadecasulfate were attached to the antithrombin binding pentasaccharide (**25, 26**). Symmetric pentasaccharide glycoconjugates were also prepared with various sulfation patterns. All the compounds showed good to potent *in vitro* anti-FXa and anti-thrombin inhibitory activity, and higher anti-thrombin activities were observed with the glycoconjugate **26** bearing higher charges. However, the binding with PF4 was not completely avoided.

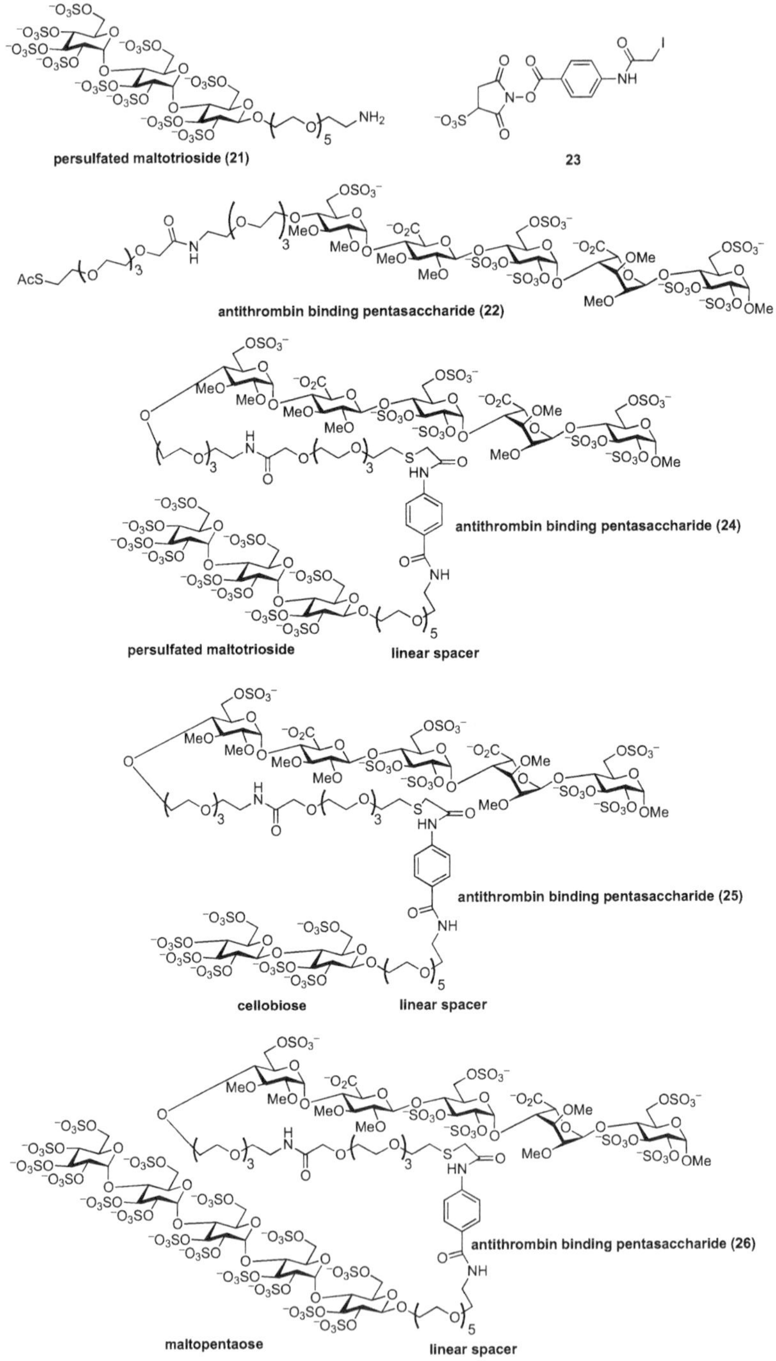

Figure 2.5 ***Maltotrioside* 21*, antithrombin binding domain* 22*, sulfo-SIAB* 23*, and glycoconjugates* 24, 25 *and* 26.**

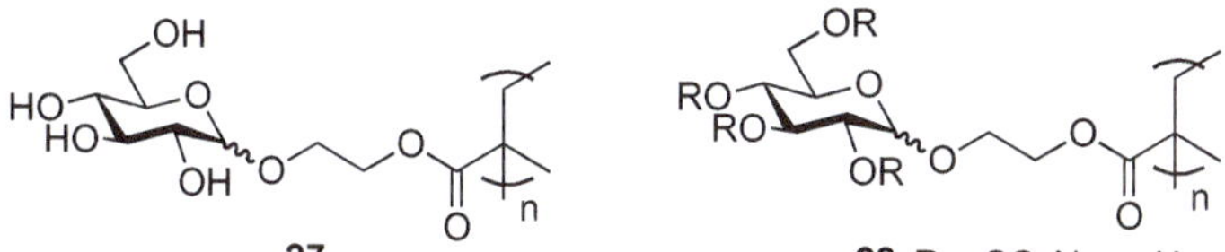

Figure 2.6 ***Structures of poly(GEMA)* 27 *and poly(GEMA) sulfate* 28.**

Glycopolymers with well-defined pendants have been developed in recent years. Free radical polymerization and copolymerization of glucosyloxyethyl methacrylate (GEMA) were reported for the preparation of highly water-soluble polymers.[62,63] Akashi and colleagues utilized the poly(GEMA) **27** with pendant glucoside residue to synthesize poly(GEMA) sulfate **28** (Fig. 2.6). The degree of sulfation in **28** varied from 1.91 to 3.75 per sugar according to the amount of DMF/sulfur trioxide complex added and the reaction time.[64] The anticoagulant activity of poly(GEMA) sulfate polymer **28** was tested, which showed anticoagulant activity with prolonged coagulation time. However, the anticoagulant effect was about 180 times weaker than that of heparin.[65] The Hsieh-Wilson group utilized ring-opening metathesis polymerization (ROMP) to generate heparin glycopolymers **29** and **30** carrying disaccharide units (Fig. 2.7).[66] Anti-FXa and anti-FIIa activities of these glycopolymers were measured via chromogenic substrate assays.[67] Impressively, glycopolymer **29–45** (45-mer) showed high potency against FIIa and FXa, with activities 100 fold higher than those of UFH or LMWH. The lack of 3-*O*-sulfation on **30–155** (155-mer) led to significantly attenuated FXa and FIIa inhibition reaffirming the importance of 3-*O*-sulfation. However, the anticoagulant activity of **29–45** was neutralized by the addition of PF4.[66] Interestingly, the anti-FIIa activity of glycopolymer **29–30** (polymer with an average length of 30 monomers) was only partially neutralized by PF4, suggesting that the PF4 reactivity can be reduced by modulating the length of the glycopolymer.

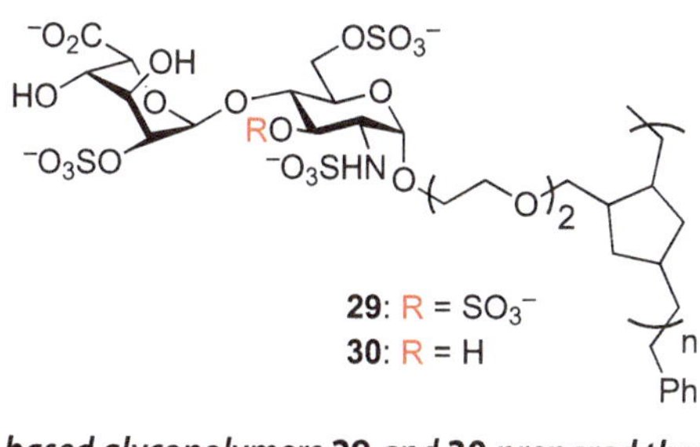

Figure 2.7 ***Heparin-based glycopolymers* 29 *and* 30 *prepared through ROMP.***

3 Heparin mimetics as growth factor binders

FGFs and VEGF initiate cell signaling pathways, and regulate cell proliferation and angiogenesis by forming ternary complexes with heparin/HS and cell surface receptors.[68] These growth factors were also found sequestered by HS in the ECM. Therefore, reducing heparin and HS binding with these growth factors by heparin mimetics has been investigated in recent years as a novel approach to block cell proliferation.

Parish and coworkers reported structurally well-defined cyclitol-based heparin mimetics with a wide variety of linker chain lengths and extent of sulfation.[69] The sulfated tetrameric cyclitols mimicked heparin disaccharide units linked through a 2 to 8 carbon atom spacer (Fig. 2.8). Growth factors exhibited distinct binding patterns with various spacer lengths. Cyclitols with a short spacer of 2 carbon atoms (**31**) effectively bound with FGF-1 but poorly inhibited FGF-2 and VEGF. Compound **32** with a longer linker possessed highest inhibitory activity for the latter two growth factors. This study indicated subtle changes in the spacer could affect the binding ability toward proteins and supported the idea that different domains of heparin account for the binding of various proteins.

A 6-azido-6-deoxy-α-D-mannopyranoside was employed as a heparin mimetic core structure template to investigate the binding affinity and selectivity against growth factors FGF-1, FGF-2, and VEGF. Preliminary binding studies showed 2,3-disulfated mannoside has a similar binding ability with FGF-1 and VEGF as a trisulfated monosaccharide.[70] Therefore, [3 + 2] copper-catalyzed alkyne azide cycloaddition and Swern oxidation-Wittig reactions were applied to further decorate the 2,3-disulfate template by selected hydrophobic or polar groups. An 18-membered library, including the lead compound **33**, was successfully prepared (Fig. 2.9).[71] All the mimetics

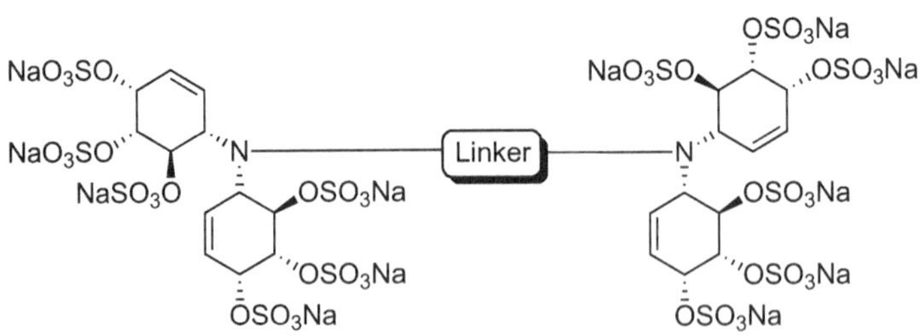

31: linker = $-(CH_2)_2-$
32: linker = $-(CH_2)_8-$

Figure 2.8 *Cyclitol-based heparin mimetics* **31** *and* **32**.

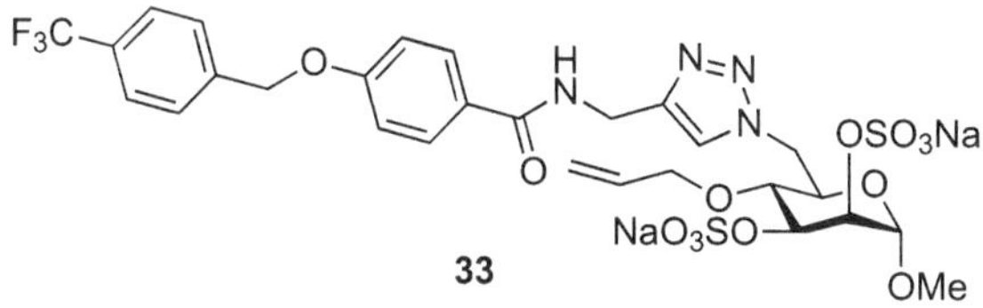

33

Figure 2.9 ***Compound* 33 *synthesized via the click reaction.***

had good binding affinity toward FGF-1 and VEGF and selectivity over FGF-2. The attachment of phenyl groups improved the binding to FGF-1 and VEGF, and an extra anionic group such as carboxylate could enhance binding. Compound **33** with a trifluoromethyl substituted aromatic linker exhibited the best binding affinity with FGF-1 and VEGF (K_D = 84 and 49 uM, respectively) and good selectivity over FGF-2 (29- and 51-fold, respectively). Ugi four-component condensation reaction was used for a monosaccharide library preparation to introduce additional hydrophobic groups (Fig. 2.10).[72] The library was used to examine the effect of the hydrophobic groups on binding with growth factors (FGF-1, FGF-2, and VEGF). Surface plasmon resonance (SPR) binding assay showed the compounds with either an aromatic group or an extra negative charge is preferred over an aliphatic group for increased affinity with growth factors, especially FGF-2 and VEGF.

Non-toxic polyanionic compounds were developed to inhibit FGF-2 induced biological activities.[73] RG-13577 is a synthetic poly-4-hydroxyphenyl acetic acid aromatic compound with an average of 5000 molecular weight. Effect of RG-13577 on FGF-2 binding was tested in competition with the binding of heparin to FGF-2, and 50% competition was observed at 3 ug/mL. Heparin-induced FGF receptor dimerization and formation of a ternary FGF-2-FGFR-heparin complex are essential steps in FGF-2 signaling.[74] Compound RG-13577 alone did not promote the binding between FGF-2 and soluble or cell-surface FGFR1 and abrogated

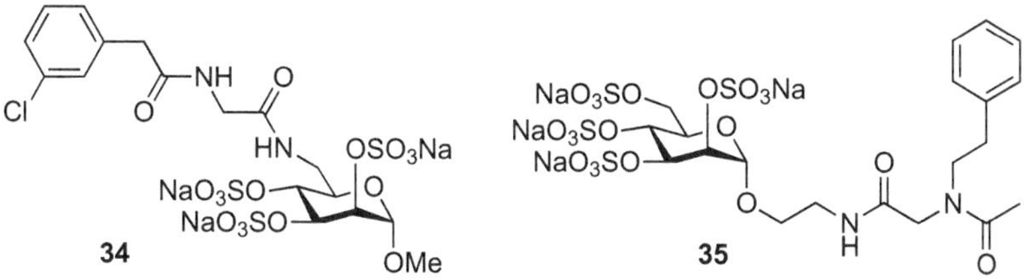

34 35

Figure 2.10 ***Representative heparin mimetic compounds synthesized via the Ugi reaction for growth factor binding.***

heparin-mediated dimerization of FGF-2 and FGFR1 at 10 ug/mL. In the presence of 5–10 ug/mL compound RG-13577, vascular endothelial cell proliferation was dramatically inhibited.

Sulfated peptide combinatorial libraries are another novel polyanionic structures that mimic heparin, and its biologic activity toward FGFs.[75] Peptides were chosen as templates for heparin mimetics due to the ease in accessing a range of peptide structures via automated solid-phase peptide synthesis. The structures can be readily modified by split–pool synthesis to create combinatorial peptide libraries. *O*-Sulfation of heparin was mimicked by sulfated serine, threonine, and hydroxyproline residues, and *N*-sulfated amino acid resembled *N*-sulfation of heparin. The sulfated peptide library (240,000 members) was screened with fluorescence-labeled FGF-1 to pinpoint heparin mimetic candidates. Eight peptides were identified and further investigated by SPR. However, only 2 out of 8 peptides showed low micromolar IC_{50} value of FGF-1 binding in competition assays against heparin immobilized on the sensor chip.

To improve binding with growth factors, multivalent constructs of heparin oligosaccharide such as dendrimers have been investigated. Hexasaccharide **36** could bind strongly with FGF-2, as demonstrated by microarray analysis (Fig. 2.11).[23] Disaccharide **37** and monosaccharide **38** also exhibited binding with FGF-2. The Seeberger group synthesized glycodendrimers incorporating synthetic heparin analogs (**36-39**) to improve the interactions with FGFs. FGF-2 was incubated with heparin glycodendrimers for competitive microarray assay against immobilized heparin. Dendrimers **41** and **42** (IC_{50} = 43 and 165 uM respectively) showed weaker inhibitory activities than **40** (IC_{50} = 1.4 uM), but remarkably better than the monovalent counterpart **37** and **38**, indicating the increased avidity associated with the multivalent dendrimers. The binding results were confirmed by SPR experiments. In order to validate the performance of heparin dendrimers in the heparin-mediated biological process, dendrimer **40** was tested for the ability to mediate FGF-FGFR complex formation, which successfully activated FGF-2-mediated signaling in L6 myoblasts.[76]

Besides dendrimers, glycopolymers have been used to enhance FGF binding. The Godula group reported synthetic neoproteoglycans (neoPGs) utilizing the poly(acrylamide) scaffold and disaccharides prepared by HS depolymerization.[77] A library of tetramethylrhodamine (TAMRA)-labeled 12 disaccharides and 5 monosaccharide glycopolymers were successfully generated. The trends of structure dependence of FGF-2 binding obtained from microarray screening were consistent with those previously reported

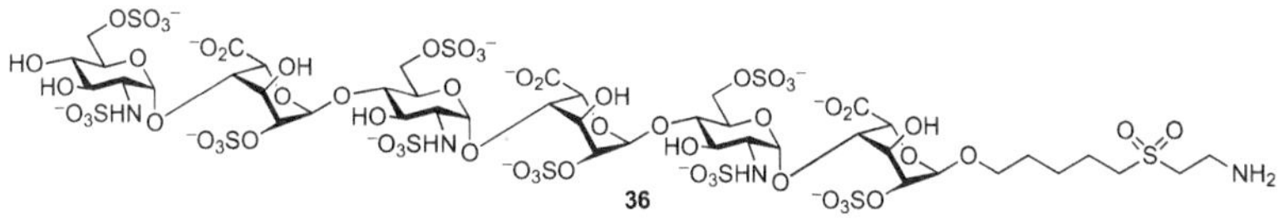

36

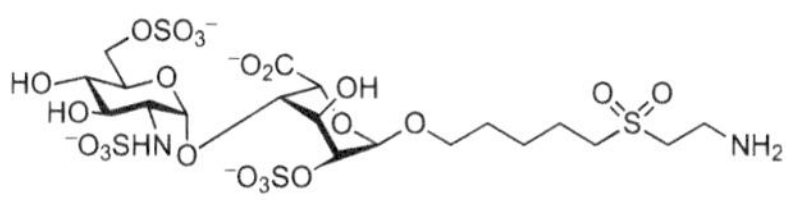

37

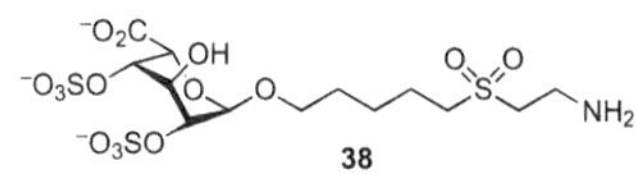

38

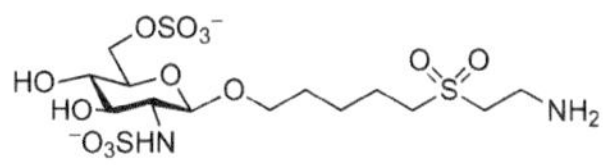

39

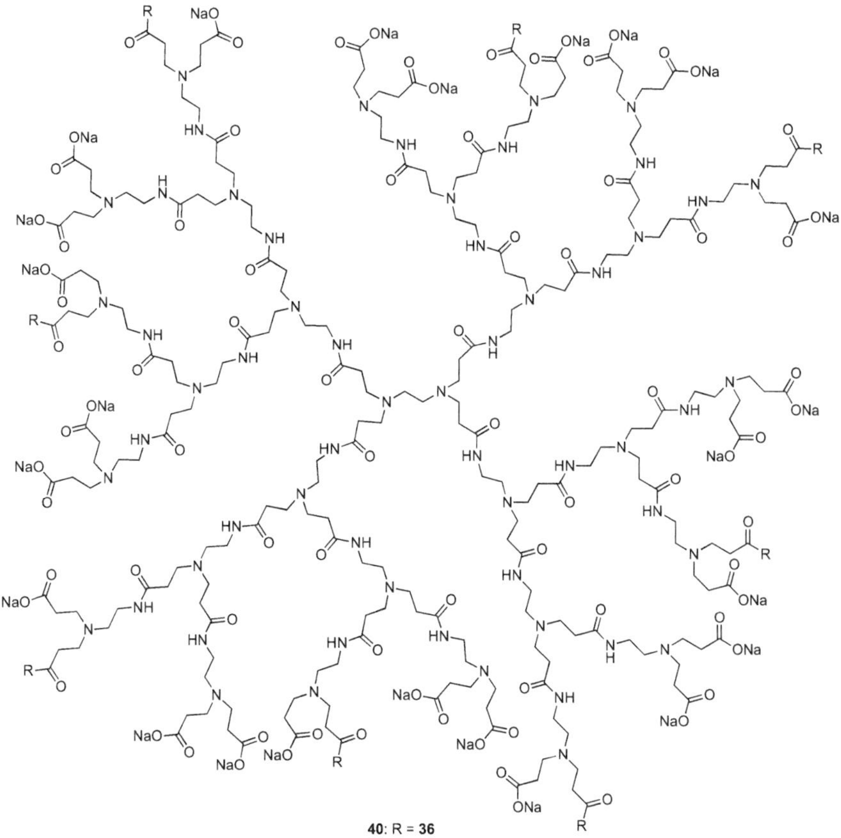

40: R = **36**
41: R = **37**
42: R = **38**
43: R = **39**
44: R = ONa

Figure 2.11 ***Structures of heparin oligosaccharides and the corresponding glycodendrimers.***

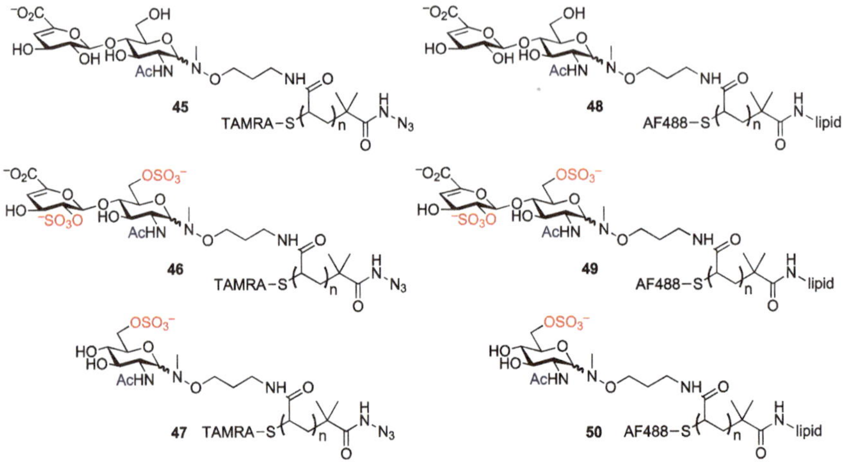

Figure 2.12 ***A library of neoPGs* 45–50.**

for HS.[21] 2-*O*-sulfations on disaccharide motifs were required for FGF-2 binding, and neoPGs **46** was the best FGF-2 binder within mimetics **45–50** (Fig. 2.12). In order to introduce neoPGs to the mouse embryonic stem cells (mESCs), phospholipid tail and Alex Fluor 488 (AF488) tag were installed for membrane insertion and imaging. With neoPGs inserted on cell membrane, increased FGF-2 binding was observed with mESCs remodeled with **49**, while NeoPGs **48** and **50** failed to bind FGF-2 on the cell surface. To confirm FGF-2 binding, a growth factor stimulation assay was performed. The cells remodeled with **47** successfully formed neural rosettes in neural monolayer differentiation experiments.[78] Furthermore, several kinases isolated from these mESCs showed significantly increased degree of phosphorylation by Western blot suggesting FGF-2 indeed facilitated signaling in cells.

While current glycopolymers with well-defined saccharide moieties and glycomimetic clusters of heparin provided multivalency in receptor binding, these compounds do not mimic the natural linear connection in heparin. A new generation of linear mimetics was designed using modified GlcN monosaccharide (compounds **51** – **57**) (Fig. 2.13).[79,80] A serine was added to GlcN to mimic the uronic acid residue. The oligomers without sulfate groups and those with 6—*O* and 3—*O* sulfate groups on the GlcN moiety were synthesized through amide bond coupling. The interactions of these compounds with FGF-2 were performed by SPR, and results showed higher binding affinity with increased chain length and 3-*O*-sulfation.

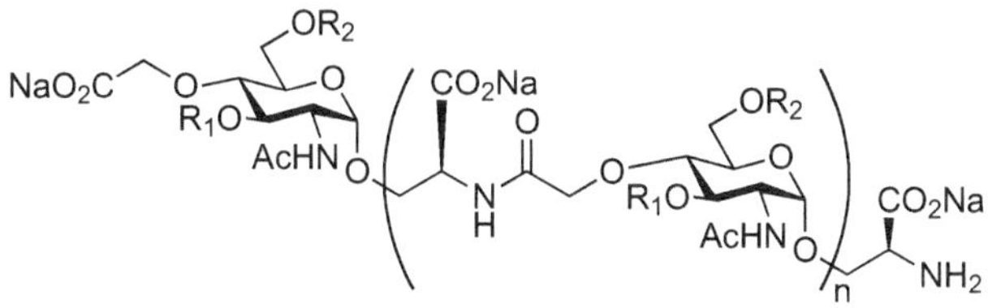

51: $R_1 = R_2 = H$; n = 1
52: $R_1 = R_2 = H$; n = 3
53: $R_1 = H$, $R_2 = SO_3Na$; n = 1
54: $R_1 = H$, $R_2 = SO_3Na$; n = 3
55: $R_1 = SO_3Na$, $R_2 = H$; n = 1
56: $R_1 = SO_3Na$, $R_2 = H$; n = 2
57: $R_1 = SO_3Na$, $R_2 = H$; n = 3

Figure 2.13 ***Linear heparin mimetics with glyco-amino acid oligomers* 51–57.**

Compounds without sulfate groups and short 6-*O*-sulfated oligomer exhibited little interactions with FGF-2, and stronger binding was observed with 3-*O*-sulfation and longer chain length. Overall, the binding affinities were still modest with the most potent compound **57** having a K_D value of 448 uM.

4 Heparin mimetics as heparanase inhibitors

With their presence on cell surface and in the ECM, heparin and HS can function as a reservoir of growth factors, chemokines, receptors and lipoproteins.[17,81] Moreover, by binding with ECM components such as laminin, fibronectin, and collagens I and IV, heparan sulfate linked to proteins, also known as heparan sulfate proteoglycans, contribute to the structural integrity of the ECM and basement membrane, and are involved in cell survival, proliferation, and migration.[82] Heparanase is the sole endoglycosidase degrading HS side chains of heparan sulfate proteoglycans. Due to its impact on cancer, inflammation, and other disease processes, heparanase has been a potential target for anti-cancer therapeutics. Heparin shares the same polysaccharide backbone structure as HS, which makes it a potential heparanase inhibitor candidate. While heparin exhibited potent anti-heparanase activity, its anticoagulant property restrains its clinical use as heparanase inhibitors. Heparin mimetics without any anticoagulant activities are considered as potential therapeutic drugs for cancers.

PI-88 (**58**), a mixture of highly sulfated phosphosulfomannan, was isolated from the yeast *Pichia holstii* NRRL Y-2448 (Fig. 2.14). PI-88 contains heterogeneous oligosaccharides ranging from di- to hexa-saccharides, with major components being tetra- (~30%) and penta-saccharides (~60%).[83]

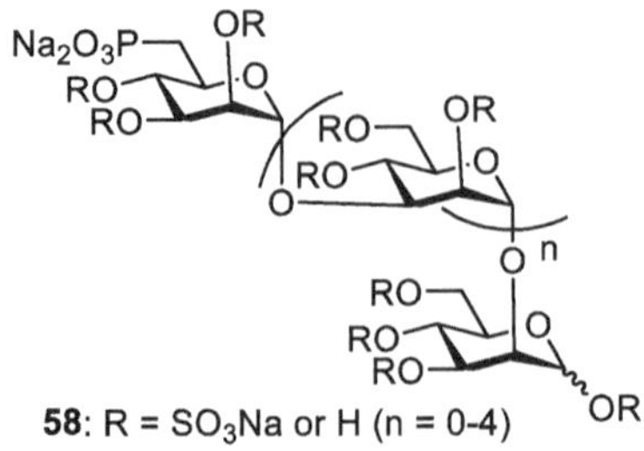

58: R = SO_3Na or H (n = 0-4)

Figure 2.14 ***PI-88, a heparin mimetic with heparanase inhibitory activities.***

PI-88 showed 2 ug/mL heparanase inhibitory IC_{50} value, which is comparable to that of heparin (1 ug/mL). Besides its inhibition of heparanase, PI-88 inhibited angiogenesis by abolishing the interactions between growth factors and their receptors.[84-86] *In vivo* study of PI-88 showed that it was effective in inhibiting tumor growth, metastasis, and angiogenesis.[84] PI-88 also minimized the malignant cell load in rodent models of human myeloid leukemia.[87] In addition to heparanase, Hsulf-1 and Hsulf-2, two human extracellular endoglucosamine 6-sulfatases, which are upregulated in several types of cancers, are inhibited by PI-88 in a concentration-dependent manner.[88] However, a phase III study for hepatocellular carcinoma has been terminated in 2015 due to a lack of significant improvements in disease-free survival compared to other treatment methods suggesting further development of PI-88 like mimetics is needed.

PG 500 is a library of synthetic, fully sulfated, and anomerically pure oligosaccharides with the reducing ends modified by various lipophilic aglycones.[89] PG500 compound series were reported as dual inhibitors of heparanase and angiogenesis via inhibition of the enzyme and growth factors. They exhibited low nanomolar binding with FGF-1, FGF-2, and VEGF through the BIA core binding assay. Both tube formation and rat aortic *in vivo* assays indicated the antiangiogenic activity of PG500 compounds. B16 solid tumor and metastatic model studies also showed the inhibition of tumor development by PG500. Based on the biological results and accessibility of the compounds, a member of the PG500 series, PG545 (**59**) (Fig. 2.15), had been selected as the lead clinical candidate for further research.[89] The Hammond group reported the inhibition of tumor growth and metastasis in the orthotopic 4T1 breast cancer model by PG545.[90] Recently, PG545 was reported to attenuate the growth of patient-derived lung cancer xenografts from fifteen patients.[91] Phase I clinical trial (NCT02042781) demonstrated the safety and tolerability of PG545 in patients with advanced solid tumors.[92]

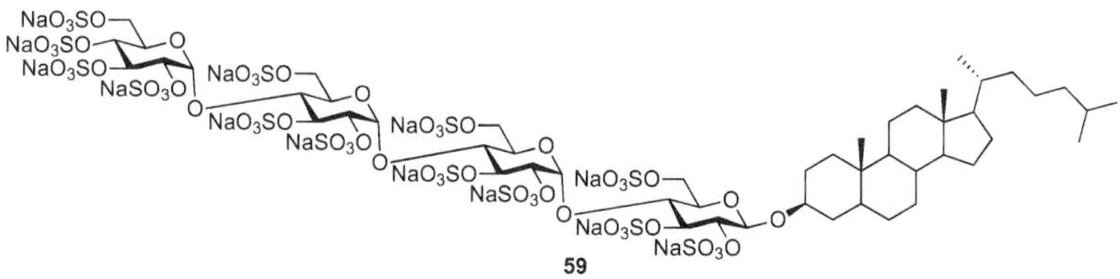
59

Figure 2.15 ***Structure of PG545.***

Oligomannurarate sulfate (JG3) (**60**) are derived from sodium alginate, ranging from tetra- to deca-saccharides (Fig. 2.16).[93] JG3 inhibited heparanase activity by binding to the KKDC and QPLK domains of the heparanase molecule. *In vitro* studies showed the inhibition of heparanase derived from NIH 3T3 and MDA-MB-435 human breast cancer cells by JG3. In an *ex vivo* aorta sprout outgrowth assay, 100 ug/mL JG3 inhibited 77% microvessel growth, and the antiangiogenic property of JG3 was confirmed in an *in vivo* chorioallantoic membrane model and the direct binding with FGF-2. JG3 modulated tumor angiogenesis and metastasis.[93] Moreover, it inhibited nuclear factor kappa B (NF-$_{\kappa}$B) transcription factor activation, which is aberrantly regulated in cancer development and involved in chemotherapy resistance. The inhibition mechanism is attributed to antagonizing the doxorubicin triggered Ataxia-telangiectasia-mutated kinase (ATM) and the successive MEK/ERK/p90Rsk/IKK signaling pathway. Thus, JG3 could be a potential chemotherapy sensitizer and NF-$_{\kappa}$B inhibitor.[94]

Suramin (**61**), a polysulfated naphthylurea, has been tested to treat acquired immunodeficiency syndrome (Fig. 2.17). Suramin was investigated as a heparanase inhibitor due to its potential inhibitive activity of

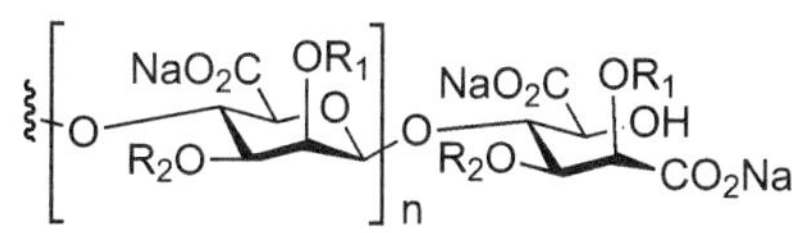
$R_1 = SO_3Na$, $R_2 = H$ or SO_3Na, n = 3-9

60

Figure 2.16 ***Structure of JG3.***

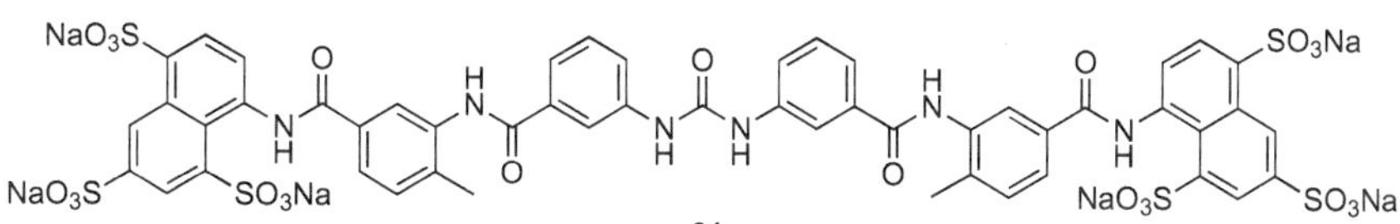
61

Figure 2.17 ***Structure of suramin.***

glycosaminoglycan catabolism. Melanoma-derived heparanase was completely inhibited by suramin at ~100 uM (IC_{50} = 46 uM), which was more potent than similar-sized oversulfated heparin tetrasaccharide. B16 melanoma cell invasion was effectively inhibited by suramin as well.[95] Suramin inhibited the human ovarian and cervical cancer cell growth with modest potencies (IC_{50} values around 350 ug/mL) and significantly downregulated heparanase expression.[96] However, suramin had been associated with multitoxicity at the therapeutic concentration in patients,[97-99] which led to the disapproval of its use by the FDA. Several suramin analogs have been reported to address the toxicity issue.[100,101]

Glycopolymers that could mimic the multivalent property of HS for heparanase inhibition have been reported by the Nguyen group.[102] [GlcNS(6S)α(1,4)GlcA] disaccharide unit was applied on diantennary monomer **62** (Fig. 2.18) with a carboxylate group on the scaffold. The monomer **63** served as a control to shine light on the effect of the carboxylate group on heparanase inhibitory activity. ROMP was applied to form glycopolymers from monomers **62** and **63**. Degrees of polymerization (DP = 5–12) were controlled by fine-tuning the amount of Grubbs third-generation catalyst (9-20 mol%). The heparanase inhibitory activities were measured by TR-FRET assay against fluorescently tagged HS.[103] Monomer **62** exhibited ~4.3-fold higher inhibition than monomer **63,** suggesting the additional carboxylate moiety can enhance inhibition. While the polymer backbone did not show any heparanase inhibitory activity, the polymer of monomer **62** (n = 12) had stronger heparanase inhibition (IC_{50} = 0.10 ± 0.036 nM) than heparin (IC_{50} = 0.54 ± 0.028 nM). To investigate the effect of local saccharide density, an additional *N*- and *O*- sulfated GlcN unit was placed generating the diantennary monomer **64**. ROMP of diantennary monomer **64** afforded its corresponding glycopolymer

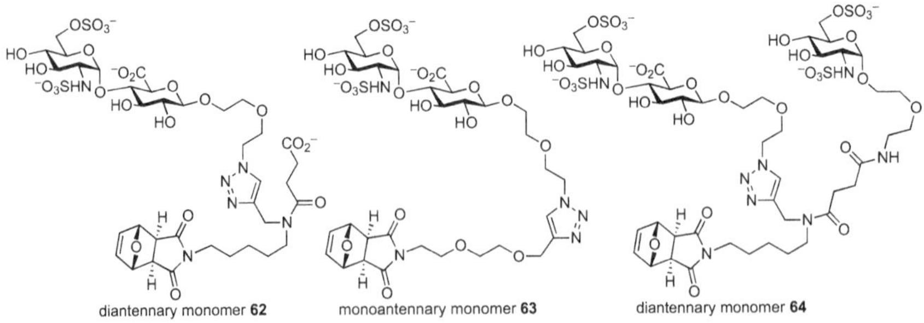

Figure 2.18 ***Diantennary and monoantennary monomers* 62, 63, *and* 64.**

(DP =8). The diantennary polymer showed similar activity against heparanase as the polymer from monoantennary monomer (DP = 9). The lack of further enhancement with the diantennary polymer could be a result of the steric crash and electronic repulsion of the additional GlcNS moiety based on docking study results.[104] The glycopolymer with 12 repeating monomer **62** has the most potent heparanase inhibitory activity. Next, the binding of this glycopolymer was investigated with a variety of other potential targets for heparin interactions through biolayer interferometry (BLI). The glycopolymers showed very low binding affinity to growth factors (FGF-1, FGF-2, and VEGF) and PF4, suggesting the high selectivity of the glycopolymer as heparanase inhibitors. The glycopolymer was effective in reducing cancer metastasis when evaluated in a mouse 4T1 breast cancer model, demonstrating its high translational potential.

In order to investigate the relationship between the sulfation pattern on the pedant disaccharides and their biological functions, a systematic study on sulfation patterns of glycopolymers was reported.[105] Structurally, the GlcN unit in the disaccharide module may be the key for heparanase recognition. Disaccharide motifs bearing different sulfation patterns on GlcN were designed (**65–70**) (Fig. 2.19). The correlation between sulfation patterns of the GlcN and heparanase inhibition had been investigated. 6-*O*-sulfation was found crucial for heparanase inhibition, as removal of 6—*O* sulfate (**65** vs **66**) drastically reduced the inhibition. The addition of 3-*O*-sulfate was detrimental to heparanase inhibition. Replacement of *N*-sulfate with acetamide or ammonium (compounds **68** and **69**) did not significantly affect the inhibitory activities.

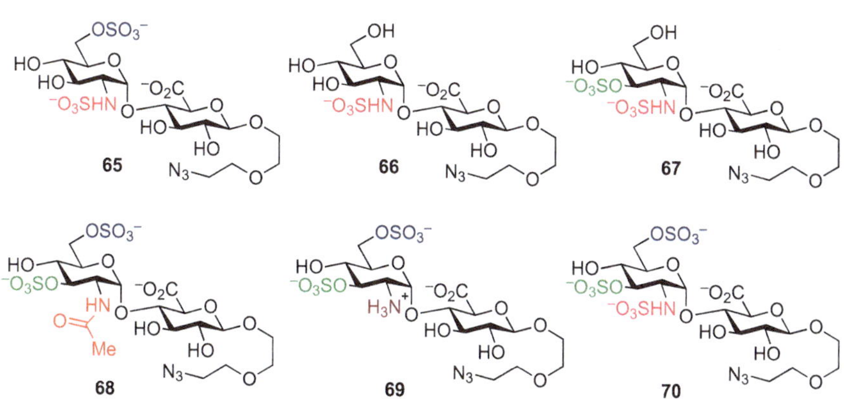

Figure 2.19 ***Disaccharide motifs 65–70 with various sulfation patterns used for glycopolymer construction.***

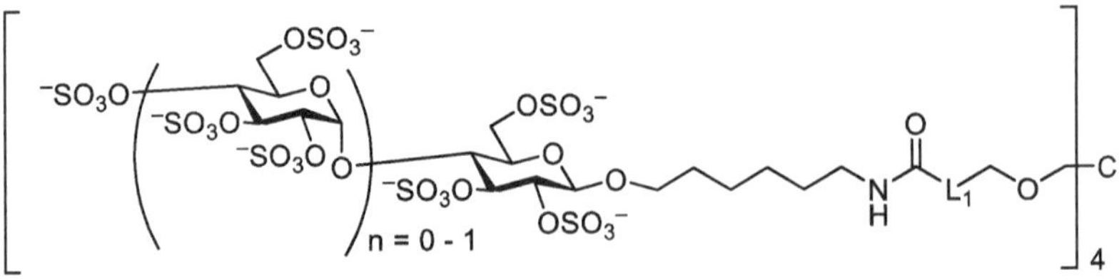

Figure 2.20 ***Dendrimer glycomimetics* 71–76 *for heparanase inhibition.***

Recently, Turnbull and coworkers employed multivalent single-entity heparin sulfate dendrimers to enhance their heparanase inhibitory activities (Fig. 2.20).[106] Since *N*-sulfated heparin oligosaccharide dendrimer did not exhibit high potency in heparanase inhibition (IC_{50} ~ 1 uM), glucose and maltose with higher sulfation levels were chosen as dendritic pendants with various dendritic core length. Glucose tetramers moderately inhibited heparanase with IC_{50} values over 1 uM. In comparison, maltose tetramers, especially "short-armed" ones such as **74** and **75**, showed similar potencies (IC_{50} = 11 and 23 nM respectively) as PG 545 (IC_{50} = 8 nM) as the positive control. The glycodendrimer **75** significantly reduced tumor growth in a xenograft mouse model with human myeloma cells (88.5% inhibition at 4 weeks). The dendrimers **74** and **75** also showed potential inhibitory activity against angiogenesis. Importantly, unlike PG 545, the glycodendrimers exhibited little anticoagulant capabilities, reducing the potential concerns of anti-coagulation side effects, a hurdle for clinical applications as heparanase inhibitors.

5 Conclusion and future perspectives

Heparin, a well-known anticoagulant drug, has been widely prescribed. However, in some patients, heparin can induce life-threatening side effects such as HIT. The heparin contamination crisis in 2007 and 2008 has further stimulated the interests in discovery of more structurally defined heparin compounds for biomedical applications. LMWH has improved pharmacokinetic profiles, but its heterogeneity and the reliance on animal source for heparin are still concerns for wide applications. Fondaparinux, a synthetic heparin oligosaccharide of defined structures, demonstrated the feasibility of obtaining pure heparin structures through organic synthesis. However, the long synthesis needed (around 50 steps) to prepare fondaparinux hinders its wide applications.

To expedite biomedical applications, heparin mimetics have been investigated as attractive substitutes of native heparin, with some of the innovative design and applications of heparin mimetics summarized in this chapter. Three major types of biological functions have been discussed, which include anti-coagulants, growth factor binding, and heparanase inhibition.

For anti-coagulation applications, "non-glycosamino" glycan analogs exemplified by idraparinux **4** have been designed, which contains glucose moiety in lieu of the glucosamine unit. As a result, the synthesis was significantly shortened with idraparinux taking about 25 steps to prepare, half of those required for fondaparinux. At the same time, these synthetic analogs can maintain potent anti-coagulant activities. Furthermore, structural features can be built into these mimetics to recruit further factors involved in anticoagulation and to modulate neutralization of the compounds if necessary.

Growth factors play important roles in regulating cell proliferation and angiogenesis. A variety of sulfated glycans and cyclitols have been investigated for growth factor binding. To expand structural diversity, multicomponent Ugi reaction has been utilized to introduce additional structural features such as hydrophobic groups and anions into the mimetics, which can help enhance growth factor binding. Furthermore, with the ease in automated solid-phase synthesis, sulfated peptide libraries were screened for growth factor binding leading to binders with modest affinity to FGF-1. To enhance the avidity towards growth factors, glycopolymers and glycodendrimers bearing sulfated mono- or oligo-saccharides have been prepared. These mimetics can present sugar moieties in a multivalent manner, which can be used to modulate heparin-protein interactions, including the regulation of embryonic stem cell differentiation.

Heparanase plays important roles in cell survival, proliferation and migration. Heparin is a potent heparanase inhibitor, which can potentially inhibit tumor growth and metastasis. Nevertheless, the anticoagulant activities of heparin impede its clinical use as heparanase inhibitors. A variety of sulfated glycans including phophosulfomannan PI-88, hydrophobic aglycon bearing PG500 series, oligomannurarate sulfate JG3, as well as polysulfated naphthylurea suramin have been investigated. However, they have relatively modest efficacy and/or potential toxicities, which need to be improved to enable further clinical evaluations.

Glycodendrimers and glycopolymers bearing sulfated mono- and disaccharides have shown promising heparanase inhibitory activities with potency (IC_{50} < 1 nM) of some compounds higher than that by native heparin. Furthermore, the structures of these mimetics can be tuned to avoid the interactions with other heparin-binding proteins including growth factors, and blood coagulating factors. When evaluated in mouse tumor models, these novel mimetics could significantly reduce tumor development, thus providing promising leads for further development.

Compared to native heparin, heparin mimetics can be synthesized with more homogeneous structures. While the field has witnessed great progress, there are still some limitations with current heparin mimetics. Native heparin contains sulfated and non-sulfated domains connected linearly. Current mimetics have not been able to approach the structural complexity and diversity present in heparin yet. For example, glycopolymers and glycodendrimers are only constructed in a branched fashion. Linearly linking the sulfated and non-sulfated modules may more closely mimic the native heparin structures. With the great advances achieved in synthesis, especially chemoenzymatic synthesis of heparin like oligosaccharides, more sophisticated heparin mimetics will be sure to be prepared in the near future to aid in more thorough understanding of the structure-activity relationship of heparin and the development of advanced agents for biomedical applications.

Acknowledgments

We are grateful for the financial support from the National Institute of General Medical Sciences, NIH (R01GM072667).

References

1. Linhardt, R. J.; Liu, J. *Curr. Opin. Pharmacol.* **2012**, *12*, 217–219.
2. Jorpes, J. E. *Ann. N.Y. Acad. Sci.* **1964**, *115*, 392–398.
3. Cifonelli, J. A.; Dorfman, A. *Biochem. Biophys. Res. Commun.* **1962**, 7, 41–45.
4. Wardrop, D.; Keeling, D. *Br. J. Haematol.* **2008**, *141*, 757–763.
5. Barrowcliffe, T. W. *Handb. Exp. Pharmacol.* **2012**, *207*, 3–22.
6. Marcum, J. A. *J. Hist. Med. Allied Sci.* **2000**, *55*, 37–66.
7. Thunberg, L.; Bäckström, G.; Lindahl, U. *Carbohydr. Res.* **1982**, *100*, 393–410.
8. Alquwaizani, M.; Buckley, L.; Adams, C.; Fanikos, J. *Curr. Emerg. Hosp. Med. Rep.* **2013**, *1*, 83–97.
9. Oduah, E. I.; Linhardt, R. J.; Sharfstein, S. T. *Pharmaceuticals* **2016**, *9*, 38.
10. Blossom, D. B.; Kallen, A. J.; Patel, P. R.; Elward, A.; Robinson, L.; Gao, G.; Langer, R.; Perkins, K. M.; Jaeger, J. L.; Kurkjian, K. M.; Jones, M.; Schillie, S. F.; Shehab, N.; Ketterer, D.; Venkataraman, G.; Kishimoto, T. K.; Shriver, Z.; McMahon, A. W.; Austen, K. F.; Kozlowski, S.; Srinivasan, A.; Turabelidze, G.; Gould, C. V.; Arduino, M. J.; Sasisekharan, R. *N. Engl. J. Med.* **2008**, *359*, 2674–2684.

11. Chess, E. K.; Bairstow, S.; Donovan, S.; Havel, K.; Hu, P.; Johnson, R. J.; Lee, S.; McKee, J.; Miller, R.; Moore, E.; Nordhaus, M.; Ray, J.; Szabo, C.; Wielgos, T. *Handb. Exp. Pharmacol.* **2012**, *207*, 99–125.
12. Martínez-González, J.; Rodríguez, C. *Expert Rev. Cardiovasc. Ther.* **2010**, *8*, 625–634.
13. Rostoker, G.; Durand-Zaleski, I.; Petit-Phar, M.; Ben Maadi, A.; Jazaerli, N.; Radier, C.; Rahmouni, A.; Mathieu, D.; Vasile, N.; Rosso, J.; Meignan, M.; Remy, P.; Lang, P.; weil, B. *Nephron* **1995**, *69*, 20–28.
14. Joannidis, M.; Kountchev, J.; Rauchenzauner, M.; Schusterschitz, N.; Ulmer, H.; Mayr, A.; Bellmann, R. *Intensive Care Med.* **2007**, *33*, 1571–1579.
15. Sinaÿ, P.; Jacquinet, J. C.; Petitou, M.; Duchaussoy, P.; Lederman, I.; Choay, J. *Carbohydr. Res.* **1984**, *132*, C5–C9.
16. Li, T.; Ye, H.; Cao, X.; Wang, J.; Liu, Y.; Zhou, L.; Liu, Q.; Wang, W.; Shen, J.; Zhao, W.; Wang, P. *Chem Med Chem* **2014**, *9*, 1071–1080.
17. Bishop, J. R.; Schuksz, M.; Esko, J. D. *Nature* **2007**, *446*, 1030–1037.
18. Vlodavsky, I.; Ilan, N.; Naggi, A.; Casu, B. *Curr. Pharm. Des.* **2007**, *13*, 2057–2073.
19. Miao, H. Q.; Liu, H.; Navarro, E.; Kussie, P.; Zhu, Z. *Curr. Med. Chem.* **2006**, *13*, 2101–2111.
20. Kilarski, W. W.; Bikfalvi, A. *Curr. Pharm. Biotechnol.* **2007**, *8*, 3–9.
21. Hu, Y. P.; Zhong, Y. Q.; Chen, Z. G.; Chen, C. Y.; Shi, Z.; Zulueta, M. M.; Ku, C. C. Y. L. P.; Wang, C. C.; Hung, S. C. *J. Am. Chem. Soc.* **2012**, *134*, 20722–20727.
22. Zong, C.; Venot, A.; Li, X.; Lu, W.; Xiao, W.; Wilkes, J. L.; Salanga, C. L.; Handel, T. M.; Wang, L.; Wolfert, M. A. *J. Am. Chem. Soc.* **2017**, *139*, 9534–9543.
23. Noti, C.; de Paz, J. L.; Polito, L.; Seeberger, P. H. *Chem. Eur. J.* **2006**, *12*, 8664–8686.
24. Miller, G. J.; Hansen, S. U.; Avizienyte, E.; Rushton, G.; Cole, C.; Jayson, G. C.; Gardiner, J. M. *Chem. Sci.* **2013**, *4*, 3218–3222.
25. Hansen, S. U.; Miller, G. J.; Jayson, G. C.; Gardiner, J. M. *Org. Lett.* **2012**, *15*, 88–91.
26. Hansen, S. U.; Miller, G. J.; Cliff, M. J.; Jayson, G. C.; Gardiner, J. M. *Chem. Sci.* **2015**, *6*, 6158–6164.
27. Kreuger, J.; Kjellén, L. *J. Histochem. Cytochem.* **2012**, *60*, 898–907.
28. Xu, Y.; Masuko, S.; Takieddin, M.; Xu, H.; Liu, R.; Jing, J.; Mousa, S. A.; Linhardt, R. J.; Liu, J. *Science* **2011**, *334*, 498–501.
29. Oosta, G. M.; Gardner, W. T.; Beeler, D. L.; Rosenberg, R. D. *Proc. Natl. Acad. Sci. U.S.A.* **1981**, *78*, 829–833.
30. Xu, Y.; Pempe, E. H.; Liu, J. *J. Biol. Chem.* **2012**, *287*, 29054–29061.
31. Chandarajoti, K.; Xu, Y.; Sparkenbaugh, E.; Key, N. S.; Pawlinski, R.; Liu, J. *Glycobiology* **2014**, *24*, 476–486.
32. Xu, Y.; Cai, C.; Chandarajoti, K.; Hsieh, P. H.; Li, L.; Pham, T. Q.; Sparkenbaugh, E. M.; Sheng, J.; Key, N. S.; Pawlinski, R.; Harris, E. N.; Linhardt, R. J.; Liu, J. *Nat. Chem. Biol.* **2014**, *10*, 248–250.
33. Liu, J.; Linhardt, R. J. *Nat. Prod. Rep.* **2014**, *31*, 1676–1685.
34. Wang, T.; Liu, L.; Voglmeir, J. *BBA - Proteins and Proteomi* **2020**, *1868*, 140301.
35. Mohamed, S.; Coombe, D. R. *Pharmaceuticals* **2017**, *10*, 78.
36. Mende, M.; Bednarek, C.; Wawryszyn, M.; Sauter, P.; Biskup, M. B.; Schepers, U.; Brase, S. *Chem. Rev.* **2016**, *116*, 8193–8255.
37. Dulaney, S. B.; Huang, X. *Adv. Carbohydr. Chem. Biochem.* **2012**, *67*, 95–136.
38. Gustafsson, D.; Bylund, R.; Antonsson, T.; Nilsson, I.; Nyström, J. E.; Eriksson, U.; Bredberg, U.; Teger-Nilsson, A. C. *Nat. Rev. Drug Discovery* **2004**, *3*, 649–659.
39. Furie, B.; Furie, B. C. *N. Engl. J. Med.* **2008**, *359*, 938–949.
40. Raskob, G. E.; Angchaisuksiri, P.; Blanco, A. N.; Buller, H.; Gallus, A.; Hunt, B. J.; Hylek, E. M.; Kakkar, A.; Konstantinides, S. V.; McCumber, M.; Ozaki, Y.; Wendelboe, A.; Weitz, J. I. *Arterioscler., Thromb., Vasc. Biol.* **2014**, *34*, 2363–2371.
41. Rabenstein, D. L. *Nat. Prod. Rep.* **2002**, *19*, 312–331.

42. Warkentin, T. E.; Chong, B. H.; Greinacher, A. *J. Thromb. Haemostasis* **1998**, *79*, 1–7.
43. Thomas, D. P. *J. Thromb. Haemostasis* **1997**, *78*, 1422–1425.
44. Shen, J. I.; Winkelmayer, W. C. *Am. J. Kidney Dis.* **2012**, *60*, 473–486.
45. Arepally, G. M.; Ortel, T. L. *N. Engl. J. Med.* **2006**, *355*, 809–817.
46. Diuguid, D. L. *N. Engl. J. Med.* **2001**, *345*, 1340–1342.
47. Pineo, G. F.; Hull, R. D. *Annu. Rev. Med.* **1997**, *48*, 79–91.
48. Melnikova, I. *Nat. Rev. Drug Discovery* **2009**, *8*, 353–354.
49. Basten, J.; Jaurand, G.; Olde-Hanter, B.; Duchaussoy, P.; Petitoub, M.; van Boeckel, C. A. A. *Bioorg. Med. Chem. Lett.* **1992**, *2*, 905–910.
50. Bousser, M. G.; Bouthier, J.; Büller, H. R.; Cohen, A. T.; Crijns, H.; Davidson, B. L.; Halperin, J.; Hankey, G.; Levy, S.; Pengo, V.; Prandoni, P.; Prins, M. H.; Tomkowski, W.; Torp-Pedersen, C.; Wyse, D. G. *Lancet* **2008**, *371*, 315–321.
51. Buller, H. R.; Cohen, A. T.; Davidson, B.; Decousus, H.; Gallus, A. S.; Gent, M.; Pillion, G.; Piovella, F.; Prins, M. H.; Raskob, G. E. *N. Engl. J. Med.* **2007**, *357*, 1094–1104.
52. Herbert, J. M.; Hérault, J. P.; Bernat, A.; van Amsterdam, R. G. M.; Lormeau, J. C.; Petitou, M.; van Boeckel, C.; Hoffmann, P.; Meuleman, D. G. *Blood* **1998**, *91*, 4197–4205.
53. Investigators *J. Thromb. Haemostasis* **2004**, *2*, 47–53.
54. Savi, P.; Herault, J. P.; Duchaussoy, P.; Millet, L.; Schaeffer, P.; Petitou, M.; Bono, F.; Herbert, J. M. *J. Thromb. Haemostasis* **2008**, *6*, 1697–1706.
55. Paty, I.; Trellu, M.; Destors, J. M.; Cortez, P.; Boëlle, E.; Sanderink, G. *J. Thromb. Haemostasis* **2010**, *8*, 722–729.
56. Song, Y.; Li, X.; Pavithra, S.; Li, D. *PLoS One* **2013**, *8*, e78972.
57. Petitou, M.; Hérault, J. P.; Bernat, A.; Driguez, P. A.; Duchaussoy, P.; Lormeau, J. C.; Herbert, J. M. *Nature* **1999**, *398*, 417–422.
58. Maccarana, M.; Lindahl, U. *Glycobiology* **1993**, *3*, 271–277.
59. Herbert, J. M.; Hérault, J. P.; Bernat, A.; Savi, P.; Schaeffer, P.; Driguez, P. A.; Duchaussoy, P.; Petitou, M. *J. Thromb. Haemostasis* **2001**, *85*, 852–860.
60. Hoppensteadt, D. A.; Cunanan, J.; Geniaux, E.; Lorenz, M.; Viskov, C.; Ythier-Moury, P.; Brin, J. F.; Jeske, W. P. *Blood* **2007**, *110*, 1881.
61. Westerduin, P.; Basten, J. E. M.; Broekhoven, M. A.; de Kimpe, V.; Will, I.; Kuijpers, H. A.; van Boeckel, C. A. A. *Angew. Chem. Int. Ed.* **1996**, *35*, 331–333.
62. Kitazawa, S.; Okumura, M.; Kinomura, K.; Sakakibara, T. *Chem. Lett.* **1990**, *19*, 1733–1736.
63. Fukudome, N.; Suzuki, K.; Yashima, E.; Akashi, M. *J. Appl. Polym. Sci.* **1994**, *52*, 1759–1763.
64. Nakamae, K.; Miyata, T.; Jikihara, A.; Hoffman, A. S. *J. Biomater. Sci., Polym. Ed.* **1994**, *6*, 79–90.
65. Akashi, M.; Sakamoto, N.; Suzuki, K.; Kishida, A. *Bioconjugate Chem.* **1996**, 7, 393–395.
66. Oh, Y. I.; Sheng, G. J.; Chang, S. K.; Hsieh-Wilson, L. C. *Angew. Chem., Int. Ed. Engl.* **2013**, *52*, 11796–11799.
67. Harder, S.; Parisius, J.; Picard-Willems, B. *Thromb. Res.* **2008**, *123*, 396–403.
68. Yun, Y. R.; Won, J. E.; Jeon, E.; Lee, S.; Kang, W.; Jo, H.; Jang, J. H.; Shin, U. S.; Kim, H. W. *J. Tissue Eng.* **2010**, 218142.
69. Freeman, C.; Liu, L.; Banwell, M. G.; Brown, K. J.; Bezos, A.; Ferro, V.; Parish, C. R. *J. Biol. Chem.* **2005**, *280*, 8842–8849.
70. Liu, L.; Li, C. P.; Cochran, S.; Ferro, V. *Bioorg. Med. Chem. Lett.* **2004**, *14*, 2221–2226.
71. Liu, L.; Li, C.; Cochran, S.; Jimmink, S.; Ferro, V. *Chem Med Chem* **2012**, *7*, 1267–1275.
72. Liu, L.; Li, C.; Cochran, S.; Feder, D.; Guddat, L. W.; Ferro, V. *Bioorg. Med. Chem. Lett.* **2012**, *22*, 6190–6194.
73. Miao, H. Q.; Ornitz, D. M.; Aingorn, E.; Ben-Sasson, S. A.; Vlodavsky, I. *J. Clin. Invest.* **1997**, *99*, 1565–1575.

74. Schlessinger, J.; Plotnikov, A. N.; Ibrahimi, O. A.; Eliseenkova, A. V.; Yeh, B. K.; Yayon, A.; Linhardt, R. J.; Mohammadi, M. *Mol. Cell.* **2000**, *6*, 743–750.
75. Vázquez-Campos, S.; St. Hilaire, P. M.; Damgaard, D.; Meldal, M. *QSAR Comb. Sci.* **2005**, *24*, 923–942.
76. de Paz, J. L.; Noti, C.; Böhm, F.; Werner, S.; Seeberger, P. H. *Chem. Biol.* **2007**, *14*, 879–887.
77. Huang, M. L.; Smith, R. A.; Trieger, G. W.; Godula, K. *J. Am. Chem. Soc.* **2014**, *136*, 10565–10568.
78. Ying, Q. L.; Stavridis, M.; Griffiths, D.; Li, M.; Smith, A. *Nat. Biotechnol.* **2003**, *21*, 183–186.
79. Jiménez Blanco, J. L.; Ortega-Caballero, F.; Ortiz Mellet, C.; García Fernández, J. M. *Beilstein J. Org. Chem.* **2010**, *6*, 20.
80. Revuelta, J.; Fuentes, R.; Lagartera, L.; José Hernáiz, M.; Bastida, A.; García-Junceda, E.; Fernández-Mayoralas, A. *Chem. Commun.* **2018**, *54*, 13455–13458.
81. Xu, D.; Esko, J. D. *Annu. Rev. Biochem.* **2014**, *83*, 129–157.
82. Reiland, J.; Sanderson, R. D.; Waguespack, M.; Barker, S. A.; Long, R.; Carson, D. D.; Marchetti, D. *J. Biol. Chem.* **2004**, *279*, 8047–8055.
83. Karoli, T.; Liu, L.; Fairweather, J. K.; Hammond, E.; Li, C. P.; Cochran, S.; Bergefall, K.; Trybala, E.; Addison, R. S.; Ferro, V. *J. Med. Chem.* **2005**, *48*, 8229–8236.
84. Parish, C. R.; Freeman, C.; Brown, K. J.; Francis, D. J.; Cowden, W. B. *Cancer Res.* **1999**, *59*, 3433–3441.
85. Cochran, S.; Li, C.; Fairweather, J. K.; Kett, W. C.; Coombe, D. R.; Ferro, V. *J. Med. Chem.* **2003**, *46*, 4601–4608.
86. Liang, X. J.; Yuan, L.; Hu, J.; Yu, H. H.; Li, T.; Lin, S. F.; Tang, S. B. *Mol. Vision* **2012**, *18*, 1649–1657.
87. Iversen, P. O.; Sorensen, D. R.; Benestad, H. B. *Leukemia* **2002**, *16*, 376–381.
88. Hossain, M. M.; Hosono-Fukao, T.; Tang, R.; Sugaya, N.; van Kuppevelt, T. H.; Jenniskens, G. J.; Kimata, K.; Rosen, S. D.; Uchimura, K. *Glycobiology* **2010**, *20*, 175–186.
89. Dredge, K.; Hammond, E.; Davis, K.; Li, C. P.; Liu, L.; Johnstone, K.; Handley, P.; Wimmer, N.; Gonda, T. J.; Gautam, A.; Ferro, V.; Bytheway, I. *Invest. New Drugs* **2010**, *28*, 276–283.
90. Ferro, V.; Liu, L.; Johnstone, K. D.; Wimmer, N.; Karoli, T.; Handley, P.; Rowley, J.; Dredge, K.; Li, C. P.; Hammond, E.; Davis, K.; Sarimaa, L.; Harenberg, J.; Bytheway, I. *J. Med. Chem.* **2012**, *55*, 3804–3813.
91. Katz, A.; Barash, U.; Boyango, I.; Feld, S.; Zohar, Y.; Hammond, E.; Ilan, N.; Kremer, R.; Vlodavsky, I. *Oncotarget* **2018**, *9*, 19294–19306.
92. Dredge, K.; Brennan, T. V.; Hammond, E.; Lickliter, J. D.; Lin, L.; Bampton, D.; Handley, P.; Lankesheer, F.; Morrish, G.; Yang, Y.; Brown, M. P.; Millward, M. *Br. J. Cancer* **2018**, *118*, 1035–1041.
93. Zhao, H.; Liu, H.; Chen, Y.; Xin, X.; Li, J.; Hou, Y.; Zhang, Z.; Zhang, X.; Xie, C.; Geng, M.; Ding, J. *Cancer Res.* **2006**, *66*, 8779–8787.
94. Zhang, J.; Xin, X.; Chen, Q.; Xie, Z.; Gui, M.; Chen, Y.; Lin, L.; Feng, J.; Li, Q.; Ding, J.; Geng, M. *Int. J. Cancer* **2012**, *130*, 467–477.
95. Nakajima, M.; DeChavigny, A.; Johnson, C. E.; Hamada, J.; Stein, C. A.; Nicolson, G. L. *J. Biol. Chem.* **1991**, *266*, 9661–9666.
96. Li, H.; Li, H.; Qu, H.; Zhao, M.; Yuan, B.; Cao, M.; Cui, J. *Cancer Cell Int.* **2015**, *15*, 52.
97. Kobayashi, K.; Weiss, R. E.; Vogelzang, N. J.; Vokes, E. E.; Janisch, L.; Ratain, M. J. *Cancer* **1996**, *78*, 2411–2420.
98. La Rocca, R. V.; Meer, J.; Gilliatt, R. W.; Stein, C. A.; Cassidy, J.; Myers, S. E.; Dalakas, M. C. *Neurology* **1990**, *40*, 954–960.
99. Figg, W. D.; Cooper, M. R.; Thibault, A.; Headlee, D.; Humphrey, J.; Bergan, R. C.; Reed, E.; Sartor, O. *Cancer* **1994**, *74*, 1612–1614.

100. Firsching, A.; Nickel, P.; Mora, P.; Allolio, B. *Cancer Res.* **1995**, *55*, 4957–4961.
101. Marchetti, D.; Reiland, J.; Erwin, B.; Roy, M. *Int. J. Cancer* **2003**, *104*, 167–174.
102. Loka, R. S.; Yu, F.; Sletten, E. T.; Nguyen, H. M. *Chem. Commun.* **2017**, *53*, 9163–9166.
103. Roy, S.; El Hadri, A.; Richard, S.; Denis, F.; Holte, K.; Duffner, J.; Yu, F.; Galcheva-Gargova, Z.; Capila, I.; Schultes, B.; Petitou, M.; Kaundinya, G. V. *J. Med. Chem.* **2014**, *57*, 4511–4520.
104. Sletten, E. T.; Loka, R. S.; Yu, F.; Nguyen, H. M. *Biomacromolecules* **2017**, *18*, 3387–3399.
105. Loka, R. S.; Sletten, E. T.; Barash, U.; Vlodavsky, I.; Nguyen, H. M. *ACS Appl. Mater. Interfaces* **2019**, *11*, 244–254.
106. Zubkova, O. V.; Ahmed, Y. A.; Guimond, S. E.; Noble, S. L.; Miller, J. H.; Smith, R. A. A.; Nurcombe, V.; Tyler, P. C.; Weissmann, M.; Vlodavsky, I.; Turnbull, J. E. *ACS Chem. Biol.* **2018**, *13*, 3236–3242.

CHAPTER THREE

Bioactive C-glycosides inspired from natural products towards therapeutics

Rajeswara Reddy Mannem, Naveenkumar Thoti, Indrapal Singh Aidhen
Department of Chemistry, Indian Institute of Technology Madras, Chennai, India

1 Introduction

Nature in its repertoire has offered to mankind several natural products. It could be of terrestrial or marine origin and either from plants or microbial world. Inspired from the structural diversity and biological activity associated with these natural products, the scientific community was able to provide some of the most epoch-making drugs during the last century, as remedies for several ailments (Fig. 3.1). Unfortunately, around mid-nineties, either attracted or distracted by new directions, the pharmaceutical industry interest in natural products (NPs) based drug discovery started declining.[1] Presumably, the disinterest was either driven by (1) low success rate in discovery of natural product based drugs and increasing expenditure, (2) difficulties associated with producing and scaling up NPs in quantities sufficient for development and coupled with other market forces. However, renewed interest in the use of NPs along with synthetic modifications is re-emerging.[2] The award of Nobel Prize in Physiology and Medicine, in 2015, to William C. Campbell and Satoshi Moura, and Youyou Tu for the discovery of avermectins and artemisinin respectively stands witness to the importance and significance of natural products.[3]

2 Glycosides in nature

The natural products, belonging to any class of compounds, when attached to glycose unit (glycone part), be it monosaccharide or oligosaccharide are referred as *O*-glycosides when appended through an oxygen atom and as *C*-glycosides when attached through C—C bond at the anomeric carbon (Fig. 3.2). The biological activity associated with these glycosides

Carbohydrates in Drug Discovery and Development. http://dx.doi.org/10.1016/B978-0-12-816675-8.00003-8

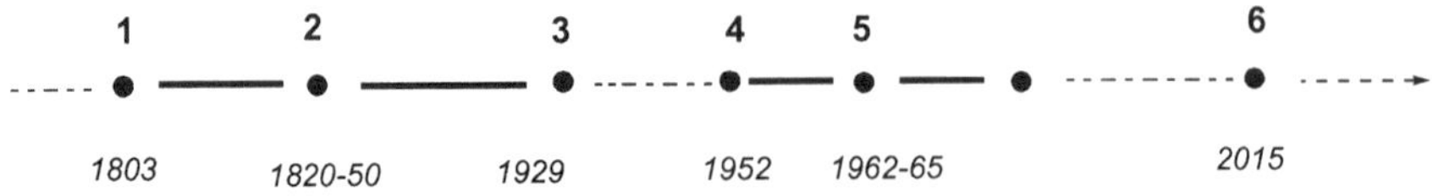

1: Isolation of Morphine from Opium and later used as nature derived medicine
2: Discovery of other natural products, such as Quinine, Colchicine, Codeine, Salicin etc
3: Discovery of Penicillin by Fleming (Nobel Prize in Physiology and Medicine, 1945)
4: Nobel Prize in Physiology and Medicine, 1952, for Discovery of Streptomycin
5: Isolation of Camptothecin and Taxol, later to evolve as clinically used anti-cancer drugs
6: Nobel Prize in Physiology and Medicine, 2015 for the Discovery of avermectins and artemisinin

Figure 3.1 ***Illustrative historical milestones of drug discovery from natural products.***

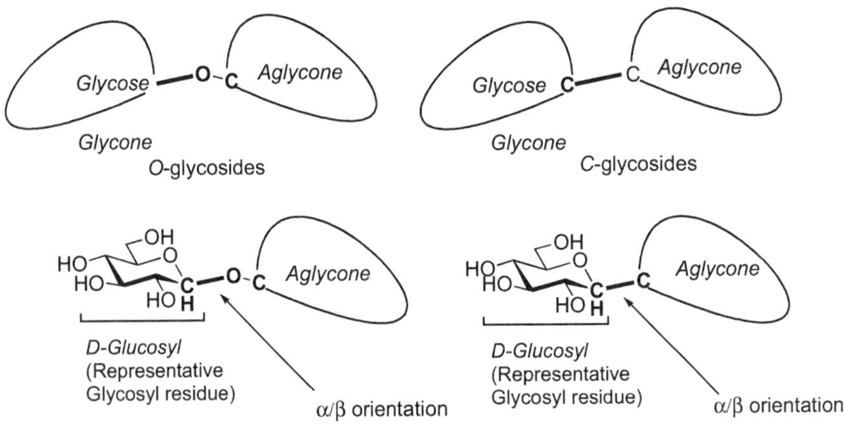

Figure 3.2 ***Definition of O- and C-Glycosides.***

has been the source of motivation for their isolation and synthesis. The diversity of the aglycone part present in the *O*-glycosides is fairly large. It can be macrolide, lignan, phytosterols, flavonoids, etc. A brief survey covering representative examples of important *O*-glycosides from nature obtained for their valuable biological activity is described herein.

Erythromycin A (**1**) and Erythromycin B (**2**) are *O*-glycosides isolated from the fungus *Saccharopolyspora erythraea*[4] and became the first clinically used macrolide antibiotics, where in the aglycone part is a macrolide (Fig. 3.3). The desosamine residue on the hydroxyl group at C-5 and a cladinose group on the hydroxyl function at C-3 represent the glycosyl portion of these *O*-glycosides. The desosamine, glycosyl residue is critical for the observed activity as against, cladinose. The ready availability of this *O*-glycoside through fermentation enabled its pharmaceutical application and access to equally important semi-synthetic antibiotic, azithromycin **3** through four step synthetic operation.[5] Another 24-membered polyketide

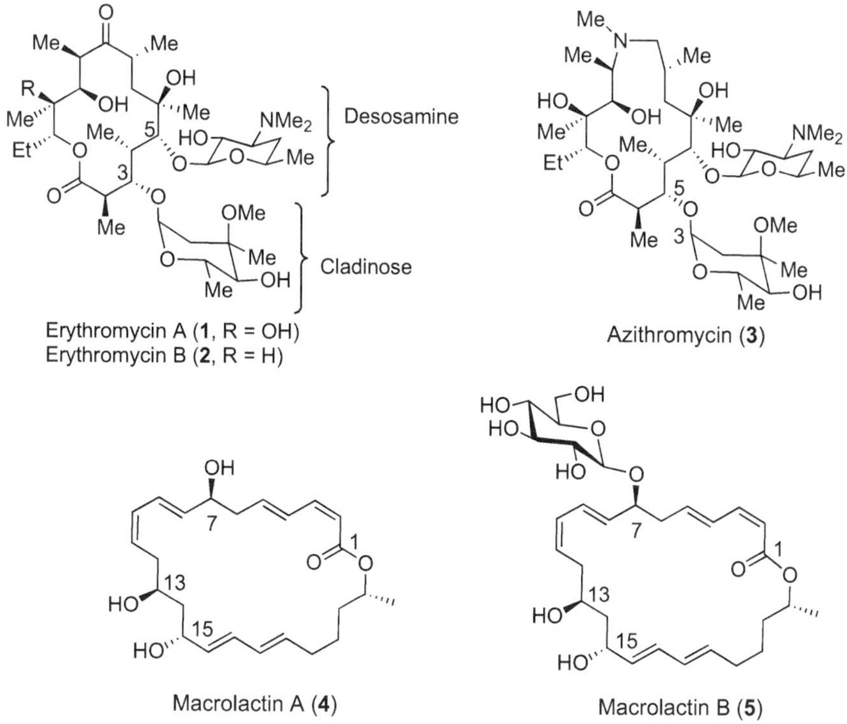

Figure 3.3 ***Representative O-Glycosides with macrolides as aglycone part.***

derived lactone **4**, Macrolactin A and derivatives isolated from *B. subtilis* drew attention of chemists and pharmacologists, due to its antibacterial, antiviral and cytotoxic activities.[6] These were found to inhibit *S. aureus*, *Herpes simplex* type I and II, murine tumor cells and several fungi. Macrolactin B (**5**), an *O*-glycoside, is 7-*O*-*β*-D-glucosylated derivative and was first time isolated by Gustafson and colleagues,[7] in quantities not sufficient for evaluation of bioactivity. This has been re-isolated[8] from marine *B. subtilis*, thoroughly characterized and found to exhibit significant broad range antifungal activity.

Among the lignans as aglycones, podophyllotoxin **6** and (epi)-podophyllotoxin **7** have been among the most prominent examples (Fig. 3.4). In search of anti-cancer drugs, efforts on this natural product **6** and its epimer **7**, were rewarded with the discovery of etoposide **8** and teniposide **9**,[9] two clinically used anti-tumor agents, specifically for the treatment of testicular and small-cell lung cancers, lymphoma and leukemia. Both these drugs are *O*-glycosylated at the C-4 hydroxy group present in the lignan

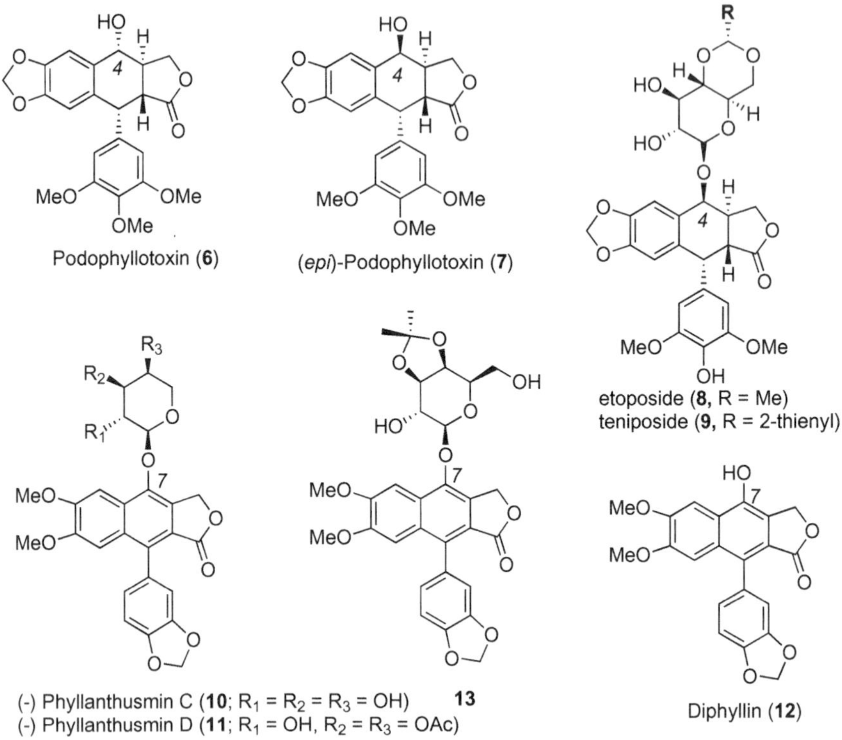

Figure 3.4 ***Representative O-Glycosides with lignan as aglycone part.***

frame-work. The side effects of etoposide which includes myelosuppression and the development of secondary leukemia linked to topo II inhibitory activity have spurred search of other lignans displaying anti-cancer activity through other mechanisms. In this context, the *O*-glycosides phyllanthusmin C (**10**) and phyllanthusmin D (**11**) of lignan, diphyllin (**12**) isolated[10] from different plant parts of *Phyllanthus poilanei* Beille (Phyllanthaceae) have surfaced as promising alternatives. The natural product, phyllanthusmin C and D has a α-L-arabinose unit as glycosyl residue appended onto C-7 phenolic hydroxyl present in lignan **12**. The *O*-glycoside, **13**, diphyllin-β-galactoside, is a synthetic analog,[11] with *D*-galactose as glycosyl unit. These compounds **10**, **11** and **13** have exhibited their anti-tumor activity through stimulation of human natural killer (NK) cell activity,[12] induction of cell apoptosis by activation of caspase-3[10] and inhibition of autophagy,[13] respectively.

The *O*-glycosides, **14**[14] isolated and identified from the aerial parts of *Phyllanthus songboiensis* represents a case, wherein a phytosterol, β-sitosterol (**15**), is the aglycone part. Similarly, **16** is another *O*-glycoside of β-sitosterol

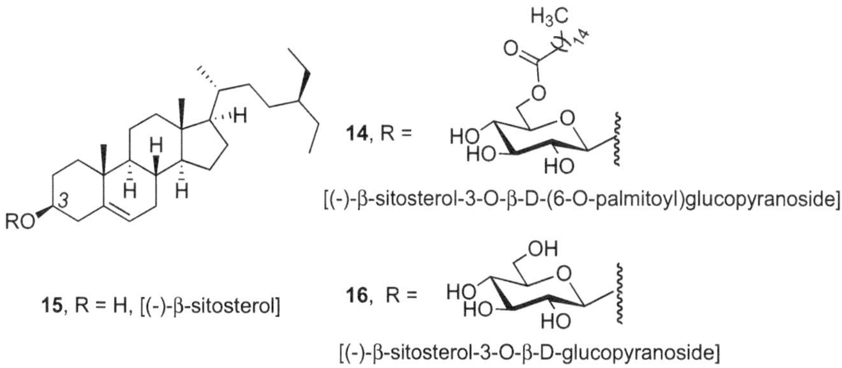

Figure 3.5 ***Representative O-Glycosides with phytosterol as aglycone.***

(**15**) isolated from the stem of *Cratoxylum cochinchinense* (Fig. 3.5). Since the adaptive and innate immune NK cells play a vital role in the host cancer immune surveillance and immune editing networks,[15] novel use of these cells through its stimulation, proliferation or enhancement of its activity, is emerging as a new promise for combating cancer. The *O*-glycoside **14** has displayed NK stimulatory effects when treated with human NK cells.[16] Although other natural products, such as curcumin, genistein have shown NK cell stimulatory activity,[17] the activity observed with *O*-glycosides **14** and **16** will be a new addition towards design of effective cancer chemotherapy targeting NK cells.

The *O*-glycosides of plant phytosterols in general[16] and those present in pulses[18a] are emerging as promising candidates for health-promoting uses. The ability of saponins to disrupt the lipid bilayers through formation of pores and enabling uptake of molecules is being viewed as novel antifungal agents for clinical use either by themselves or in conjunction with other antifungal agents in use. In the context that, *Candida* spp. has been seen as a leading cause of bloodstream infections posing serious challenge and concern, this ability of saponins may be useful.[18b]

The nitrogen-containing steroidal glycosides prominently found in plants of the Solanum genus, represented by potato (*Solanum tuberosum*), tomato (*Solanum lycopersicum*), and eggplant (*Solanum melongena*), as secondary metabolites, are *O*-glycosides, drawing great attention and offering new direction towards inhibition of cancer.[19] The aglycone part present in these *O*-glycosides are essentially nitrogen analogs of steroidal sapogenins,[20] such as diosgenin **17** (Fig. 3.6). The major glycoside (comprising >95%) found in commercial potatoes are α-chaconine (**21a**) and α-solanine (**21b**) with solanidine **18** as aglycone. The *O*-glycosides, α-solasonine (**23a)** and

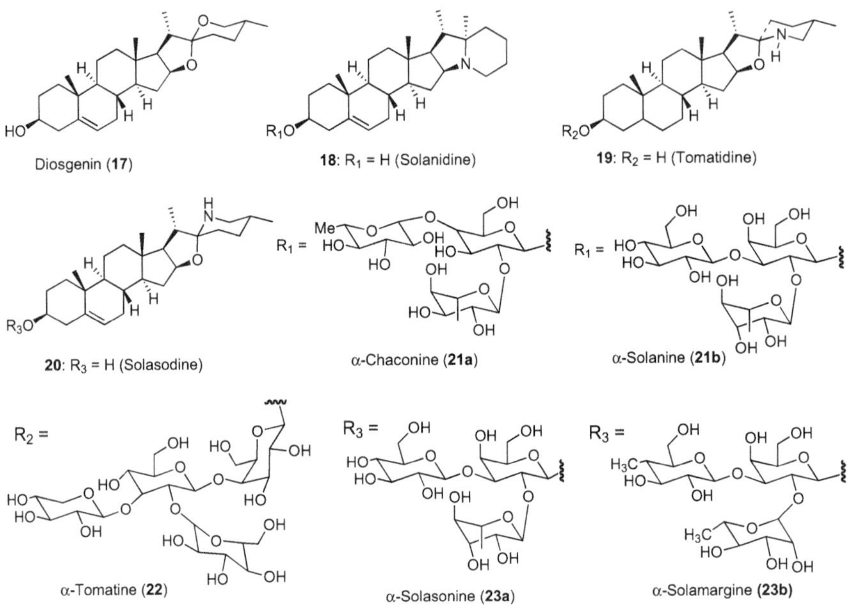

Figure 3.6 ***Representative O-Glycosides with steroidal alkaloid as aglycone.***

α-solamargine (**23b**) are present in eggplant (brinjal or aubergines), while α-tomatine **22** is present in tomatoes. Tomatidine (**19**) and Solasodine **20** are the corresponding aglycones, respectively.

The *O*-glycosides, digoxin (**24**), ouabain (**25**) and digitoxin (**26**) popularly known as cardiac glycosides and known for their efficacy in treatment of congestive heart failure and as antiarrhythmic agents,[21] are now being repurposed for cancer therapy (Fig. 3.7).[22]

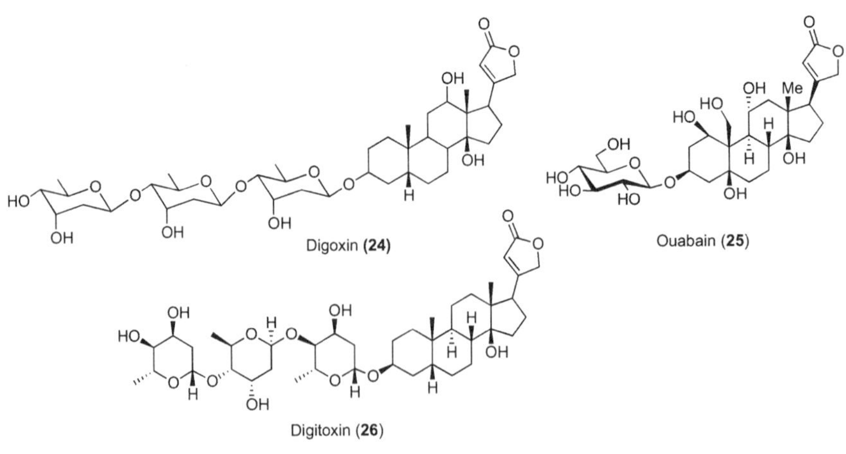

Figure 3.7 ***Structures of some cardiac O-Glycosides.***

The secondary metabolism in plants operating through two fundamental metabolic pathways, the shikimate and the acetate pathway[23,24] have resulted in the formation of a large number of compounds, commonly referred as polyphenolics. Structurally they are characterized by presence of hydroxyl groups on the aromatic ring as a common feature. The phenolic compounds commonly found in foods and natural health products have attracted great attention due to their medicinal properties.[25,26] Structurally these are broadly classified into simple phenols, hydroxybenzoic acids and hydroxycinnamic acid derivatives, flavonoids, stilbenes and lignans, as well as condensed tannins (Fig. 3.8).

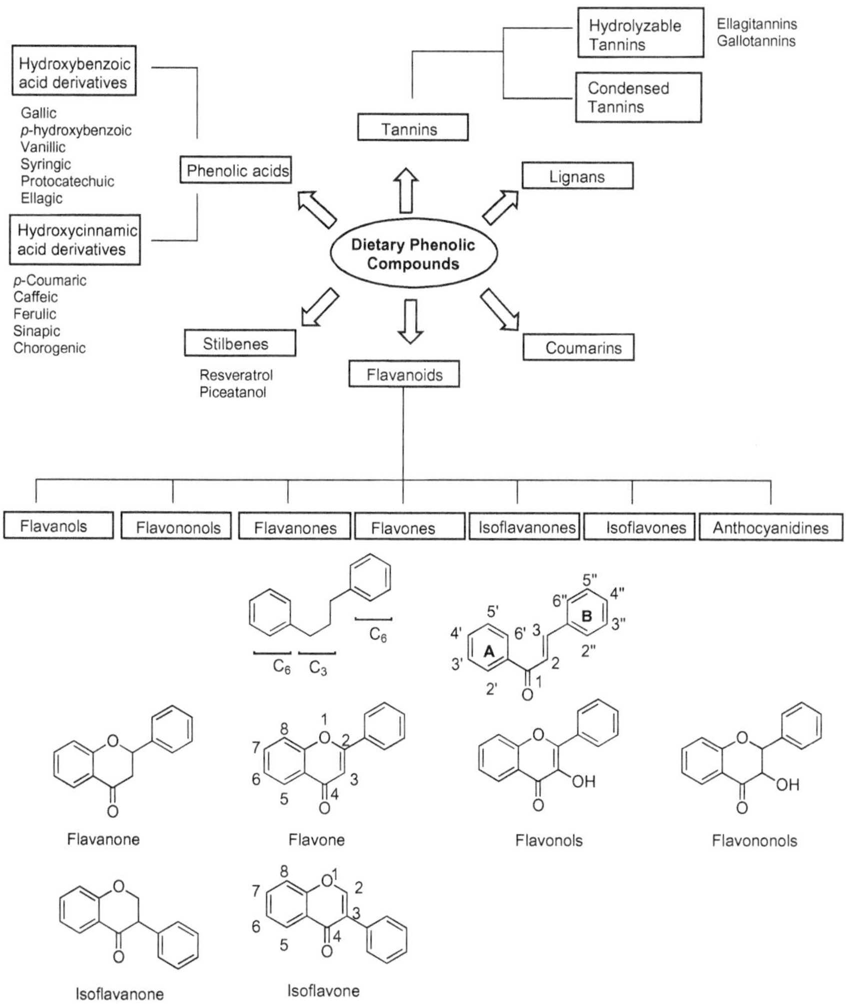

Figure 3.8 ***Classification of major dietary phenolic compounds.***

Among these polyphenolic phytochemicals, the flavonoid class either by themselves[27] or as *O*/*C*-glycosylated derivatives has attracted lot of attention, due to their beneficial health and healing potentials, and because of their presence in our common dietary cereals and legumes. The basic flavonoid structure contains the flavan nucleus, which consists of 15 carbon atoms derived from a C6—C3-C6 skeleton. Depending upon the presence or absence of additional oxygen-containing heterocyclic ring and by positional differences of B-ring, they are classified in different categories. They are usually present as glycosides in vacuoles of flowers, leaves, seeds, stems and roots.

For the *in vivo* bioactivity of flavonoid class of compounds, the absorption and bioavailability are crucial and it is in this context that flavonoid glycosides are more relevant and important.[28] Besides imparting water solubility, the glycosyl residue also improves stability. Few prominent and important *O*-glycosides among the polyphenolics displaying the diversity in the nature of aglycones are described herein (Fig. 3.9). The cinnamic acid ester glycosides **27a** and **27b** isolated from soybean molasses represent an example of *O*-glycosides, wherein hydroxy-substituted cinnamic acid is aglycone and a trisaccharide, α-L-arabinofuranosyl (1→3)] [glucopyranosyl

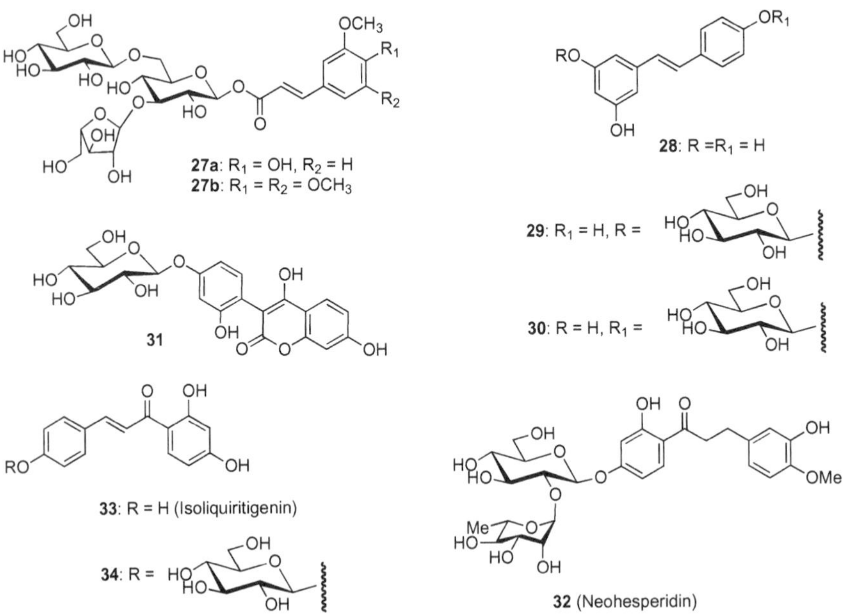

Figure 3.9 ***O-Glycosides of hydroxy-cinnamic acid, stilbene, coumarin and chalcone as aglycones.***

(1→6)-β-D-glucopyranose, is the glycone part. Resveratrol **28**, and its *O*-glucosides **29** and **30**, are polyphenolic compounds, classified under stilbene category. Resveratrol is abundantly found in grapes and gained prominence in early nineties, when cardio-protection was attributed to consumption of red wine containing rich amounts of resveratrol.[29] Further upsurge in the interest of resveratrol occurred due to its anti-neoplastic activity and other pharmacological activities.[30] Guided by an assay for inhibition of microbial activity, the first 3-aryl-coumarin based *O*-glycoside **31** was isolated from a plant in the family liliaceae.[31] Neohesperidin (**32**), a natural *O*-glycoside of dihydrochalcone is an approved sweetener by the European Union,[32] whereas isoliquiritine **34**[33a] an *O*-glucoside of a chalcone **33**, obtained from *Glycyrrhiza Radix*, is a well known aldose reductase inhibitor.[33b]

Among the flavonoids, flavone class is prominently represented by luteolin **35** and apigenin **36** as aglycones (Fig. 3.10). Luteolin, one of the most common flavone present in edible plants such as carrots, peppers, olive oil and peppermint, along with its glycosides are emerging as promising agents towards neuroprotection through their anti-inflammatory activities.[34,35] The 7-*O*-glucosides **37, 38** and 8-*C*-glucoside **39** in particular have been investigated as therapeutics for treating hyperuricemia and gouty arthritis.[36]

Apigenin **36,** found in wide variety of plants, vegetables, herbs and fruits, has been implicated for use in therapeutics.[37-39] Mung beans (*Vigna radiata* L; Family Fabaceae), also popularly referred as green gram has been consumed worldwide, for more than 3,500 years, under different names. The extracts of mung bean seeds and leaves, besides other phenolic compounds, contains

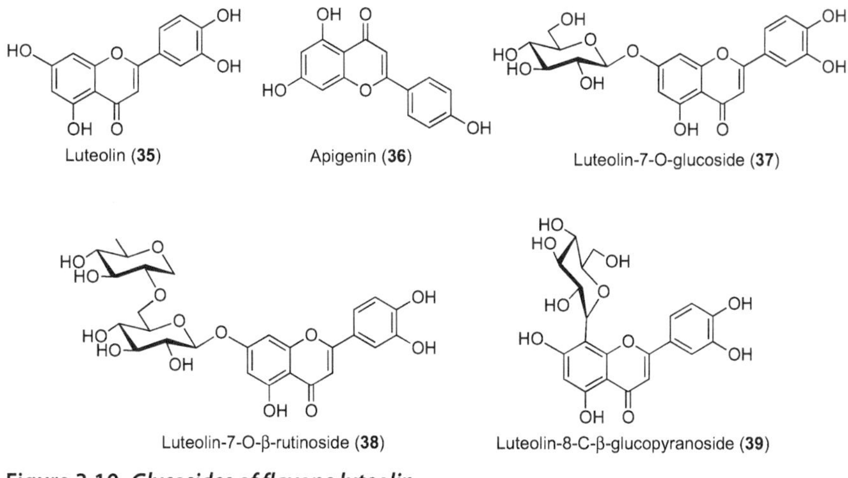

Figure 3.10 ***Glycosides of flavone luteolin.***

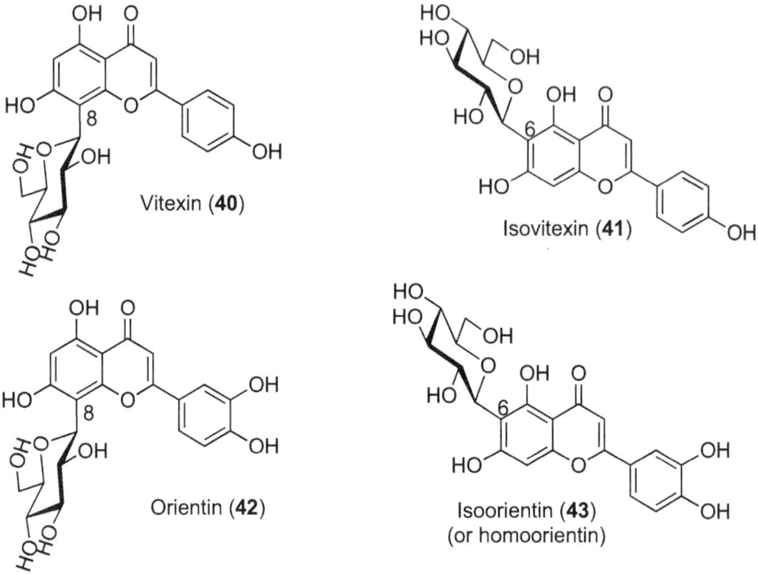

Figure 3.11 ***Glycosides of flavone Apigenin and Luteolin.***

two important *C*-glucosides of apigenin **36**, vitexin **40**, isovitexin **41** and other di-*C*-glucosides with promising pharmacological effects (Fig. 3.11).[40] They are predominantly found in the seed coat of mung beans and amount of these glucosides have been found to be several fold higher in mung bean sprouts, following germination. Both these *C*-glucosides have shown wide range of biological activities, including antioxidant, anti-inflammatory, anti-diabetic, anti-viral, anti-cancer, anti-bacterial, anti-fungal and hepato-protective activity. In this context, consumption of mung bean sprouts is viewed as health promoting,[41] as it contains several polyphenolics, with proven therapeutic potential against several human ailments. Leaves of the medicinal herb *Microcos paniculata* have been traditionally used for treating upper airway infections, by virtue of its content of flavonoids such as apigenin *C*-glycosides (ACGs).[42] Orientin **42** and homoorientin **43** are the corresponding *C*-8 and *C*-6 glucosides of flavone luteonin as the aglycone. These *C*-glucosides have been isolated from bamboo leaves and purified using high-performance liquid chromatography (HPLC).[43]

The α-glucosidase inhibitory effects of vitexin **40**, isovitexin **41** and isorhamnetin-*O*-glycoside **44**, carried out using acarbose **45** (IC_{50} = 1007 μM) as positive control, showed a dose-dependent inhibition of α-glucosidase activity (Fig. 3.12). The IC_{50} values for compounds vitexin, isovitexin and

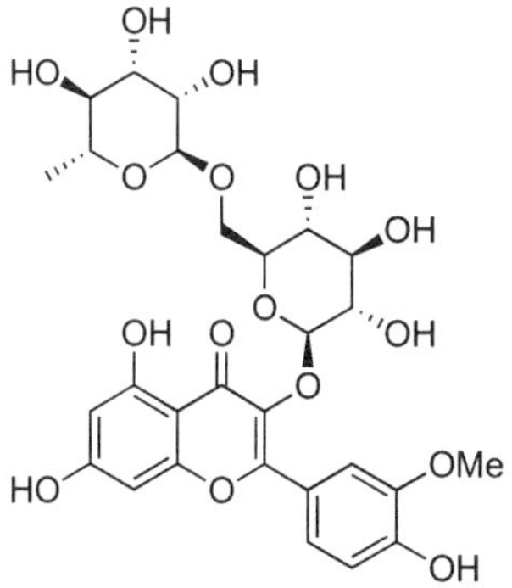

Isorhamnetin 3-O-β-D-rutinoside **(44)**

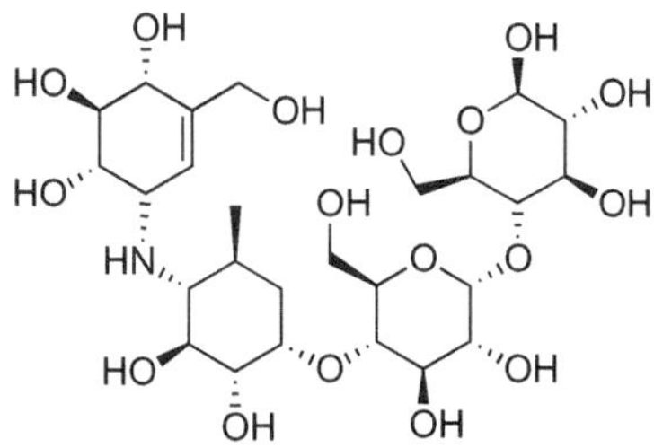

Acarbose (**45**)

Figure 3.12 ***O-Glycoside of flavonol isorhamnetin.***

isorhamnetin-*O*-glycoside were determined to be 244.0 μM (96.9 μg/mL), 266.2 μM (115.1 μg/mL) and 275.4 μM (172.0 μg/mL), respectively. Thus, vitexin and isovitexin are promising α-glucosidase inhibitory candidates.[44]

Doxorubicin **46** (DOX), an effective chemotherapeutic agent, frequently used to treat various malignancies has certain amount of cardio-toxicity originating through mechanisms other than those mediating its antitumor effect.[45] Studies on the rats have suggested that premedication with vitexin prior to treatment with DOX for 4 weeks improved the cardiac function of rats as reflected through decrease of LDH and CK–MB levels in serum. This has been attributed to decrease in oxidative stress and inflammatory cytokine levels and in this context, vitexin may be an effective therapeutic agent against DOX-induced cardio-toxicity.[46]

The sulfated flavone glycosides **47a-c**, known as Thalassiolins A–C (Fig. 3.13), isolated from the Caribbean Sea grass *Thalassia testudinum*[47] have been found to inhibit HIV replication at IC_{50} 30 μM, by targeting the

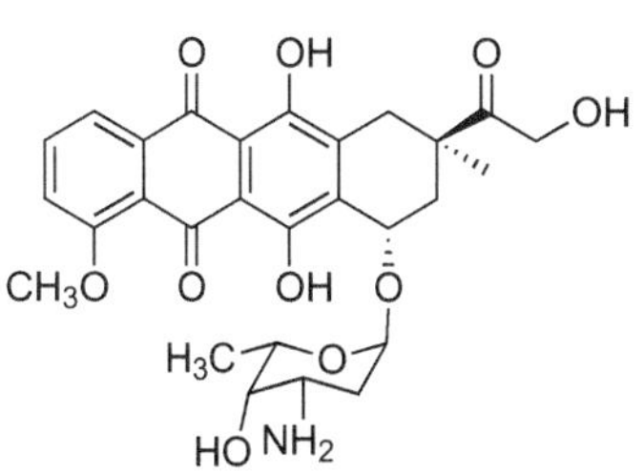

Doxorubicin (**46**)

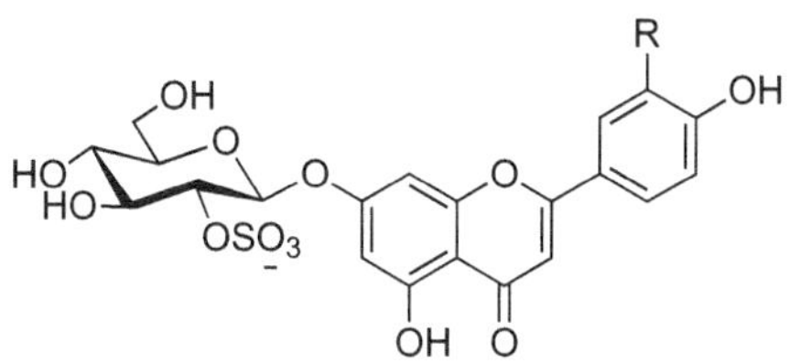

47a: Thalassiolin A, R = OH
47b: Thalassiolin B, R = OMe
47c: Thalassiolin C, R = H

Figure 3.13 ***Structure of Doxorubicin and sulfated O-Glycoside with flavone as aglycone.***

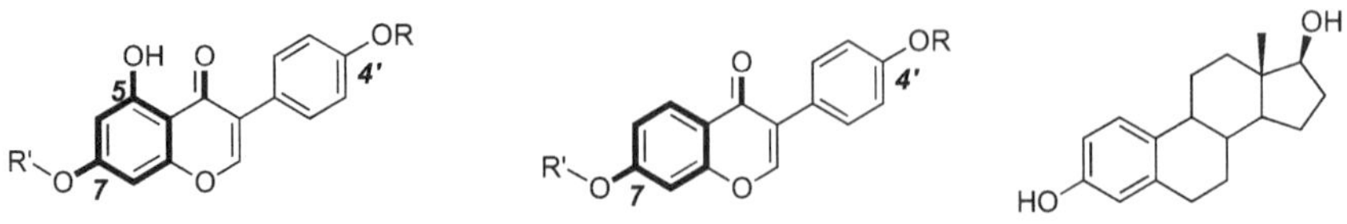

48a: R = H: Genistein
48b: R = Me, Biochanin
48c: R = H, R' = 1-β-D-glucosyl (Genistin)
49a: R, R' = H, Daidzein
49b: R = Me,R' = H, Formonenetin
49c: R = H, R' = 1-β-D-glucosyl (Diadzin)
50: β-estradiol

Figure 3.14 ***Structures of some important isoflavones.***

integrase catalyzed strand transfer at IC_{50} 0.4 μM. The observed activity against HIV integrase is due to the presence of sulfated glucose functionality attached through *O*-glycosidic linkage.

Among the flavonoids, isoflavones are predominantly and commonly found in members of the Leguminosae plant family.[48,49] Genistein **48a** and daidzein **49a** are the most abundant aglycones found in soybeans,[50] where they occur as their 7-(6-*O*-acetyl)-glucosides known as genistin **48c** and daidzin **49c**, respectively (Fig. 3.14). Isoflavones have been viewed as phytoestrogens, structurally similar to estradiol **50** and for mimicking it effects. Although the *O*-glucosides of genistein and daidzein are biologically inactive, the aglycones by themselves have aroused great interest in conjunction to glucose and lipid metabolism and for understanding their health benefits.[51]

The mango peels of different cultivars originating from Africa and Asia has yielded mangiferin **51** and isomangiferin **52**, two *C*-glucosides,[52] wherein xanthone **53** represents the aglycone part (Fig. 3.15). Due to plethora of diverse and specific bioactivities associated with xanthones, this class of natural products is resurfacing as attractive targets for total synthesis, in the recent times.[53] Besides mangiferin, the extract of leaves of the mango tree (*Mangifera indica* L.) has additionally furnished seven new *C*-glycosides, wherein benzophenone is the agylcone part.[54] These glycosides have been

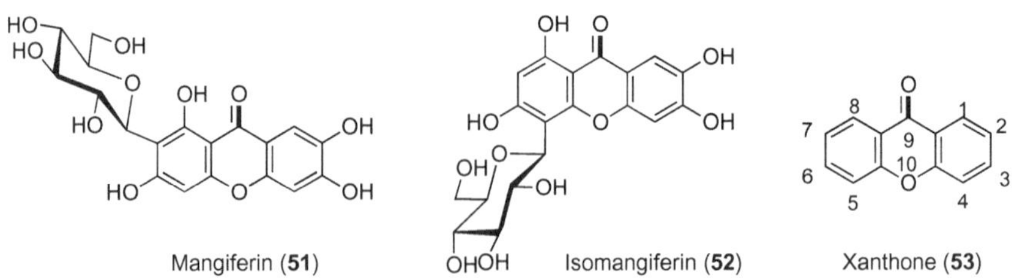

Mangiferin (**51**) Isomangiferin (**52**) Xanthone (**53**)

Figure 3.15 ***C-glucosides of Xanthone as aglycone.***

evaluated for inhibitory effect on triglyceride accumulation in 3T3-L1 adipocytes.

3 C-Glycosides: an introduction

Although, presence of *O*-glycosides is dominant in nature, the *C*-glycosides are also remarkable in number. The plant derived *C*-glycosides with aromatic aglycones such as flavones **54**, chromone **55**, xanthone **56**, anthrones **57** are fairly common (Fig. 3.16). The aglycone part arises through series of enzyme mediated biosynthetic processes, such as acetate-malonate pathway, shikimic acid pathway and mixed acetate-malonate-shikimic acid pathway.[55] Although natural dietary flavonoids exist either as *O*-or *C*-glycosides, the *C*-glycosides have started receiving attention from health perspectives, only recently.[56]

The review by Veitch and Grayer[57] summarizes the new flavonoid aglycones and glycosides reported during the period 2007 to 2009 and it is evident that there is a significant increase in obtainment of new *O*-glycosides compared to those reported during the preceding three years (2004-2006). Although *C*-glycosides remain less in number compared to *O*-glycosides, it has been observed, in most cases, that *C*-glycosyl flavonoids demonstrated

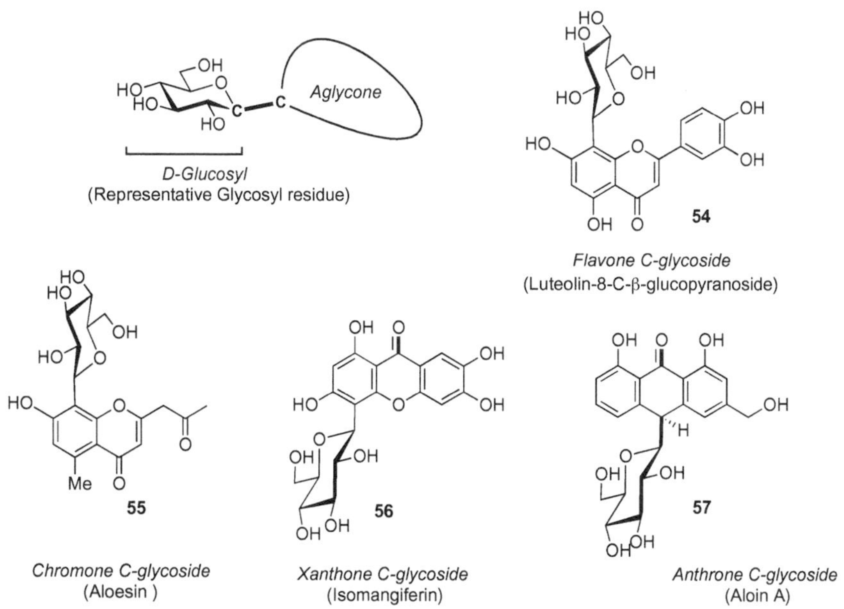

Figure 3.16 ***Representative plant natural C-Glycosides.***

higher antioxidant and anti-diabetes activity compared to their corresponding *O*-glycosyl flavonoids. Additionally, the *C*-multiglycosides are absorbed unchanged in the intestine and distributed to other tissues, whereas *C*-monoglycosides are poorly absorbed. In the context that dietary *C*-glycosides are gaining prominence and value for their wide-ranging health benefits, the advances in biotechnological methods for *C*-glycosylation,[58] would be relevant and necessary for their convenient production.

The prominent red dye, carminic acid **58** known since centuries, is insect derived and obtained from the cochineal insects (*Dactylopius* sp.).[59] The secondary metabolism in bacteria, particularly *Streptomyces* have also been the source of important *C*-glycosides (Fig. 3.17) with significant biological activities.[60] Urdamycin A **59** among the angucyclines group of antibiotics is an example of *C*-glycoside displaying multitude of activities such as antibacterial and antiviral. The modification of the glycone part in *C*-glycosides, aquayamycin **60** and chrysomycin A **(61)**, have demonstrated the significance of glycosyl residues towards the biological activity.[61]

Kidamycin **62** is another *C*-glycoside, carrying two different amino-sugar residues, equatorially linked angolosamine at C8 and axially linked vancosamine at C10. Kidamycin **62** was isolated from soil bacteria *Streptomyces* in early 1970s.[62] In an attempt toward total synthesis of kidamycin, regioselective introduction of the amino-sugars angolosamine and vancosamine has been realized[63] on the hydroxylated anthrapyran aglycone.

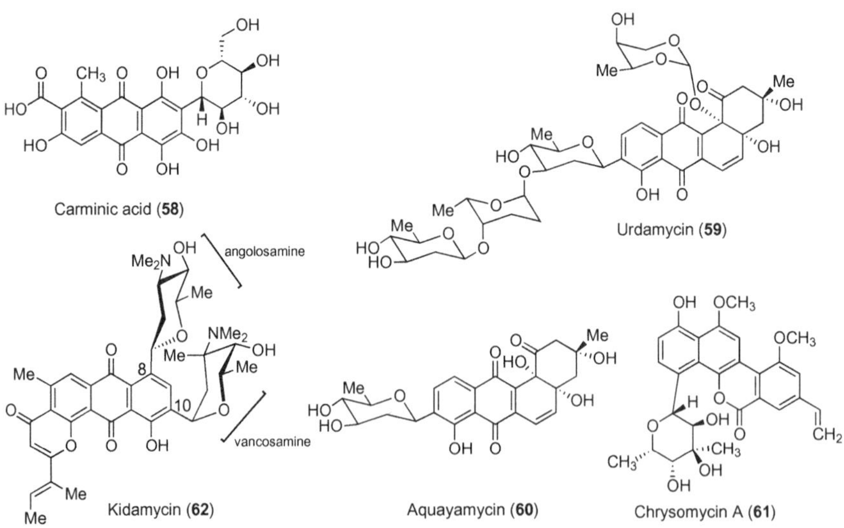

Figure 3.17 ***Representative natural C-Glycosides from bacteria and insect.***

4 Synthesis of bioactive C-glycosides

The well-known rationale for the synthesis and study of *C*-glycosides, as against *O*-glycosides, based on susceptibility toward enzymatic degradation has advanced and reached a center stage during the last three decades. The urge to obviate the hydrolytic instability of *O*-glycosides, while retaining or improving the biological activity has provided necessary motivation for synthetic organic chemists to engage in devising new strategies for constructing the C—C bond at the anomeric position or optimizing the well-known routes already existing in the literature. Literature has witnessed, periodic and timely appearance of comprehensive reviews covering the synthetic approaches.[64–78] These reviews, paralleled growing interest in biological activity and associated synthetic challenges for constructing the C—C bond at the anomeric position, efficiently. In presence of excellent reviews, covering comprehensively, various approaches for the formation of glycosidic C—C linkage, no effort is being made to delineate the same once again here. The four main approaches used for the synthesis of *C*-glycosides are graphically presented in Fig. 3.18. The most common and major among these approaches, banks on incorporating the

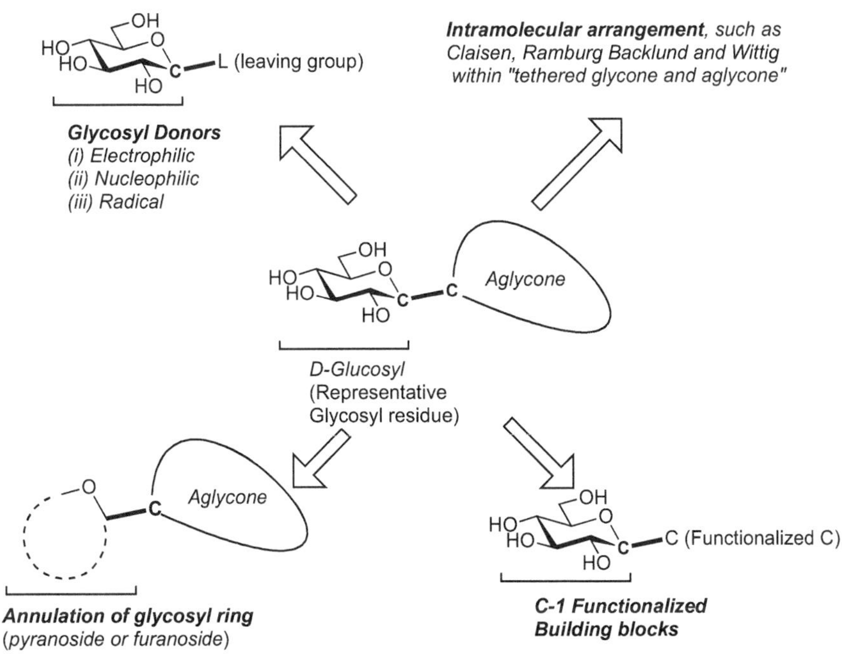

Figure 3.18 ***Prominent approaches for the synthesis of C-Glycosides.***

already built-in glycosyl residues through the use of appropriate glycosyl-donors, whereby the glycosyl-unit is brought-in either as electrophilic or anionic or radical species or transition-metal complex, during the C—C bond formation reaction at the anomeric position. The use of intramolecular reorganization invoking the use of Claisen, Ramburg-Backlund and 1,2-Wittig rearrangement is another valuable methods for formation of the key C—C bond at the anomeric position for synthesis of *C*-glycosides. Annulation of the glycosyl ring (pyranoside/furanoside) after construction of the predetermined glycosidic C—C bond, is also appearing among new methods for synthesis of *C*-glycosides. The concept of C-1 functionalized building blocks with carbon containing functionality, wherein the C—C bond has been already established in the making of building block, is also a promising approach, due to the simplicity associated with making of these blocks.

Few illustrative synthetic endeavors under the realm of *C*-glycosides, driven by the objective of exploiting promising biological activity and possible development of a drug for a defined therapeutic end, is described in the rest of the chapter. The chosen examples are without any bias and are incidental while compilation of literature material for this chapter.

5 *C*-Glycoside: Dapagliflozin, a novel drug in the market for diabetes

Inspiration drawn from the natural product, Phlorizin **63**,[79] an *O*-glucoside isolated from the bark of an apple tree and known, more than century ago has paved the way for the discovery of a novel anti-diabetic drug, Dapagliflozin **64** (Fig. 3.19). The inhibition of sodium-glucose linked transporter 2 (SGLT2) protein present in the kidney and responsible for re-absorption of glucose, has emerged as a successful and novel strategy for the treatment of type 2 diabetes,[80] which is insulin-independent. The SGLT2

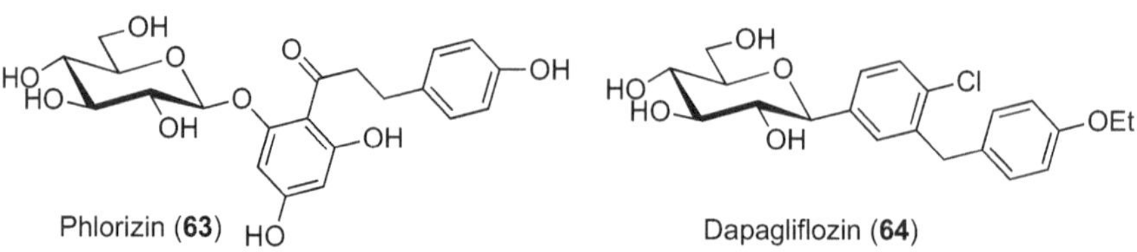

Figure 3.19 ***C*-glucoside, anti-diabetic drug, Dapagliflozin 64.**

inhibition enables urinary excretion of glucose and thereby reduction of elevated blood glucose levels during hyperglycemic situation. Although non selective, the natural product phlorizin **63**, was found to be the first inhibitor of SGLT1 and SGLT2. The journey undertaken by synthetic and medicinal chemists together, following inspiration from natural product **63**, has paid rich dividends to the community engaged in drug discovery through the novel strategy involving SGLT2 inhibition.

Following the initial disclosure of orally active phenol-*O*-glucoside SGLT2 inhibitor **65b** (T-1095),[81] with a 6'-*O*-alkyl carbonate mask, few other *O*-glucosides including Sergliflozin (**66b**)[82] and Remogliflozin (**67**)[83] were developed for SGLT2 inhibition (Fig. 3.20). The *O*-glucosides in general have poor oral bioavailability due to liability of glycosyl bonds, thereby meriting masking of glucose moiety as a prodrug. In contrast to *O*-glucosides the *N*-glucosides and *C*-glucosides are metabolically more stable, display higher oral bioavailability and plasma exposure demanding no need for conversion to a prodrug. The improved pharmacokinetic profiles of *C*-glucosides paved their way for successfully emerging as drugs for type-2 diabetes, based on SGLT2 inhibition.

Under the *C*-glucosides class, Dapagliflozin (**64**),[84] Canagliflozin (**68**),[85] Empagliflozin (**69**),[86] Ipragliflozin (**70**),[87] and Ertugliflozin (**71**)[88] have been approved by the FDA for treatment of type 2 diabetes. Many other SGLT2 inhibitors are under different phases of clinical trials.[89] The other analogs resulting from variation in the glycone part are represented by Tofogliflozin (**72**),[90] Ertugliflozin (**71**), and Luseogliflozin (**73**)[91] (Fig. 3.21).

Nomura, in 2013, probably inspired by the partial success of *O*-glucosides and the significance of the aglycone part present in Dapagliflozin,

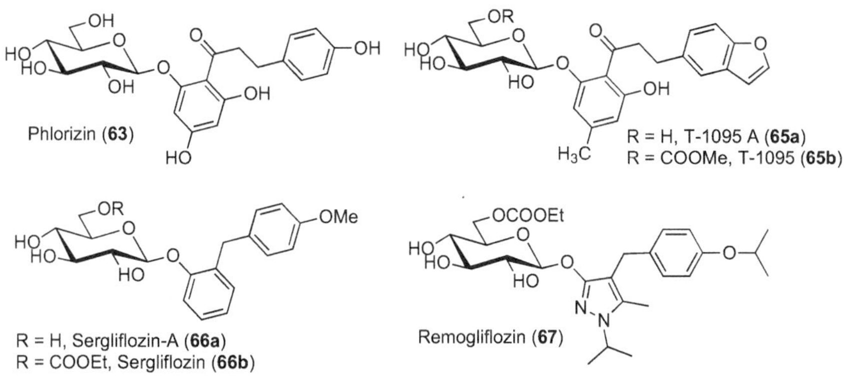

Figure 3.20 ***Examples of some O-Glucosides as SGLT inhibitors with masked glucose.***

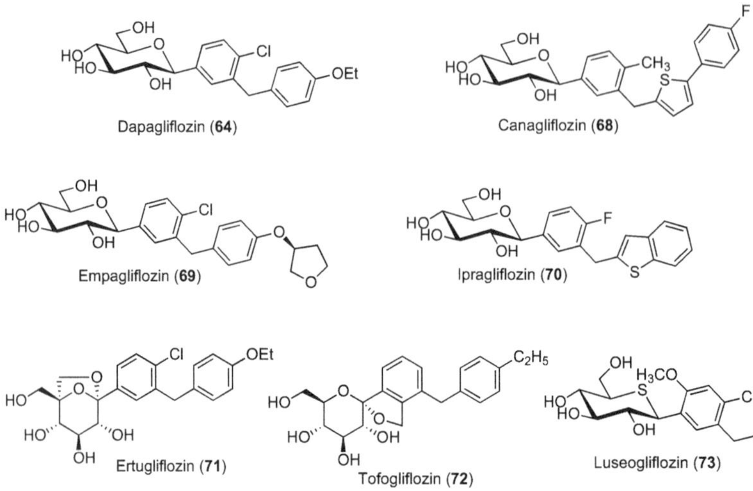

Figure 3.21 ***Important C-glycosides as SGLT2 inhibitors.***

synthesized and evaluated the *N*-glucoside **75**[92] (Fig. 3.22) for SGLT2 inhibition. Although **75** exhibited strong hSGLT2 inhibitory activity (IC_{50}= 3.9 nM) comparable to **74** (IC_{50}= 5.1 nM) its hydrolytic stability remained a challenge.

Pre- and post-discovery period of Dapagliflozin, has witnessed several structural modifications in the glycone or aglycone part and efforts continue to remain unabated in the synthetic community. We have achieved

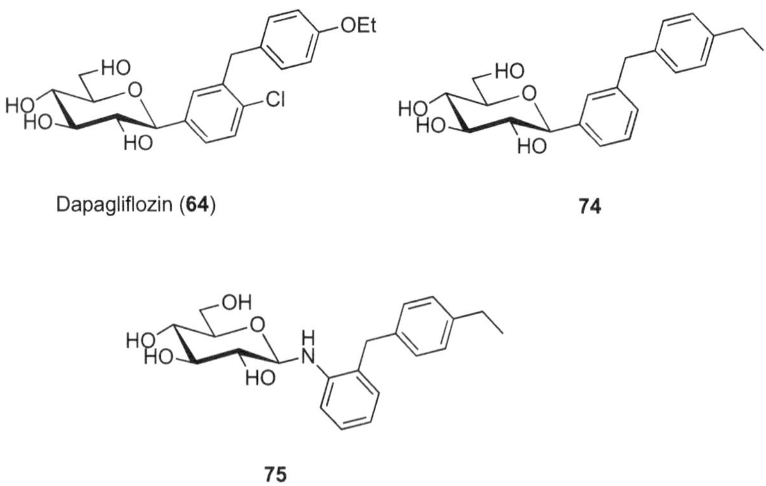

Figure 3.22 ***N-Glycoside as SGLT2 inhibitor.***

synthesis of hitherto unreported and challenging *C*-benzyl-glucoside **76** in particular and analogs **77** resulting from variation in the distal aryl ring of biarylmethane-aglycone. The target **76** was concieved based on the rationale that the original lead molecule from nature, phlorizin **63,** had one atom spacer (oxygen) between glucosyl residue and the aglycone part, hence the proposed targets **76/77** paralleled the concept by replacing the oxygen atom with an isosteric methylene-unit between the glucosyl residue and the successful biarylmethane-aglycone of Dapagliflozin. The synthesis of the targeted compounds **76** and **77** have been achieved[93] using the key building block **78** (Scheme 3.1).

Paul Knochel's elegant contribution towards synthesis of functionalized aryl magnesium reagents[94] in presence of electrophilic functional groups such as cyano, nitro, amide and ester enabled generation of organomagnesium bromide **80** from the iodide **79**. This was the first instance wherein Knochel's protocol has been used for preparation of functionalized aryl magnesium halides containing Weinreb-amide (WA) functionality. The successful formation of **78** has further strengthened our long interest in the

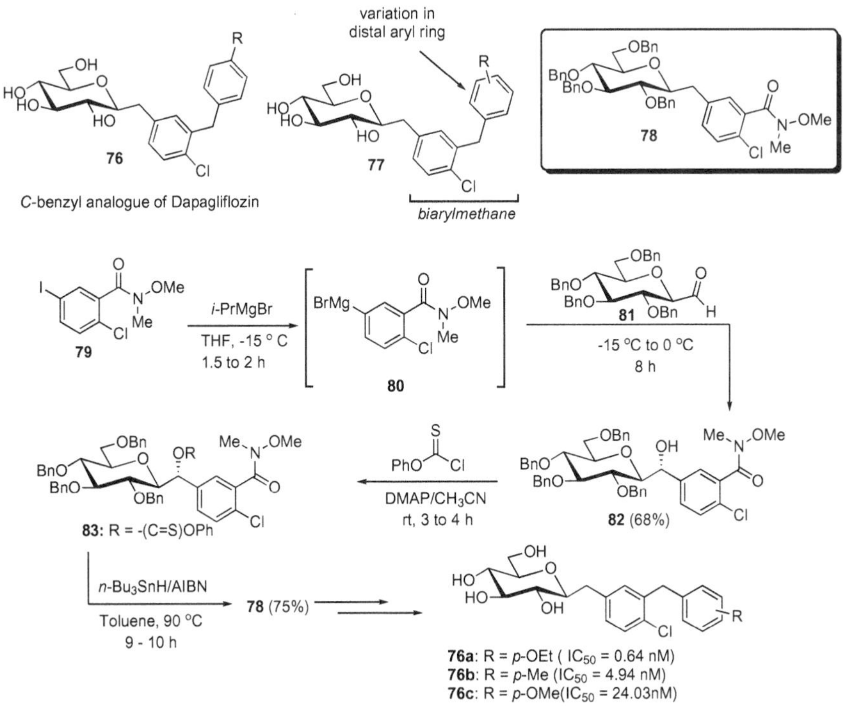

Scheme 3.1 ***New C-benzyl analogs of Dapagliflozin for SGLT2 inhibition.***

use of WA functionality for synthetic endeavours.[95] Among the synthesized compounds two benzyl analogs of dapagliflozin **76a** and **76b** have demonstrated promising SGLT2 inbitory activity with an IC_{50} values of 0.64 nM and 4.94 nM, respectively.[93]

Modification in the distal ring of biarylmethane-aglycone of dapagliflozin has paved the way for the discovery of orally active SGLT1/SGLT2 dual inhibitors. Compound **84** with benzocyclobutane residue as the distal ring (Fig. 3.23) showed excellent inhibitory potency of IC_{50} = 45 nM at SGLT1 and of IC_{50} = 1nM at SGLT2.[96] Treatment of bromide **86** with *n*-BuLi at low temperature followed by condensation with tetra-*O*-benzyl-D-gluconolactone **85** and subsequent deoxygenation at C1 in the lactol **87** with $BF_3 \cdot Et_2O/Et_3SiH$ afforded the desired compound **84** in the most convenient way.

5.1 Proximal ring modifications in Dapagliflozin

Proximal ring modification in dapagliflozin is one of the directions along which further efforts have been expended for possible improvement in activity. In this context our research activity, included synthesis of 2-*β*-D-glucopyranosylpyridines **92** as new structural variants of dapagliflozin. The synthesis banked on the use of Bohlmann–Rahtz reaction for annulation of pyridine ring (Scheme 3.2)[97] and essentially required good access to glucosyl alkynone **88** for condensation with β-amino crotonate for construction of the pyridine ring. The alkynone **88** was synthesized in three steps using literature known glucosyl aldehyde **81** as C1-functionalized

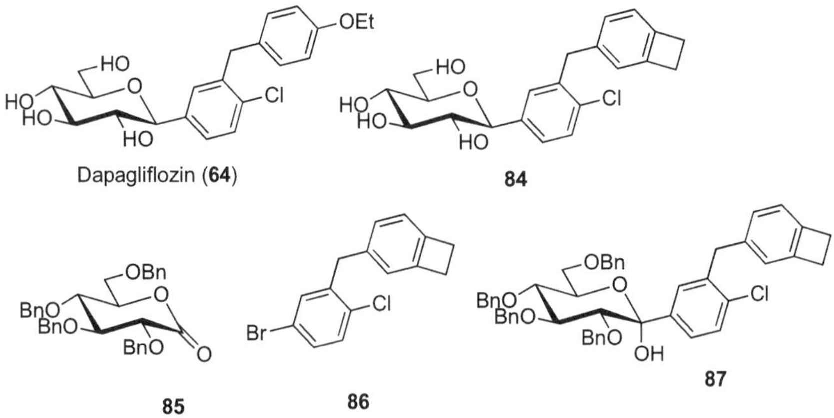

Figure 3.23 *New C-glycoside analogs of Dapagliflozin.*

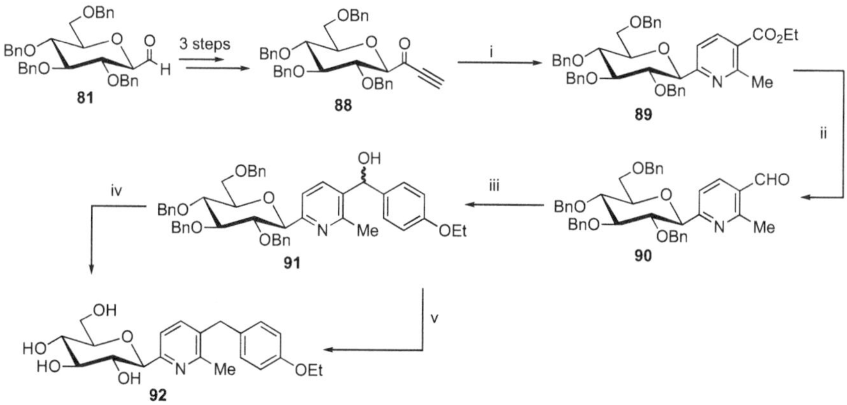

(i) β-aminocrotanate, toluene: AcOH, 70 °C, 6 h, 90%; (ii) a. $LiAlH_4$, Dry THF, rt, 2 h, 89%, b. IBX, DMSO,70 °C, 3 h, 95%; (iii) ArMgBr,Dry THF, 0 °C-rt, 3 h, 72-90%; (iv) Pd/C (10 %) or $Pd(OH)_2$ (20 %), THF: AcOH, rt, 36-72 h, 45-90%

Scheme 3.2 ***Synthesis of glucopyranosylpyridine* 92 *as analogs of Dapagliflozin.***

building block. The Bohlmann-Rahtz reaction was performed between alkynone **88** and preformed ethyl β-aminocrotonate in toluene and acetic acid mixture, affording the anticipated glucosylated pyridine **89** in good yields. The addition of various arylmagnesium bromides onto **90**, resulted in the formation of the corresponding addition products **91,** as diastereomeric mixtures. Direct hydrogenation in **91** using Pd/C or $Pd(OH)_2$ in THF and acetic acid mixture (1:1) or a two-step reaction sequence involving Barton-McCombie deoxygenation followed by Lewis acid mediated debenzylation paved the way for obtainment of Dapagliflozin analogs **92**, with pyridine ring as the proximal ring (Scheme 3.2). SGLT2 inhibition studies are underway.

Few other analogs resulting from modification or replacement of proximal aryl ring of dapagliflozin are depicted in Fig. 3.24. The ortho sustituted *C*-aryl glucosides **93** showed SGLT2 inhibition at 0.9 nM concentration and with increased selectivity.[98] The benzoisothiazole-β-glucopyranoside **94** was found to be an inhibitor of SGLT2 with an IC_{50} of 10 nM.[99] Probably in a bid to understand significance of the proximal aryl-ring of dapagliflozin, Lee et.al. in 2010, synthesized and evaluated the biological activity of pyridazine (**95**) and thiazole-*C*-glucosides (**96**) as SGLT2 inhibitors.[100] However, neither have shown potent inhibitory activity probably due to the unfavorable electronic environments around proximal ring of those compounds.

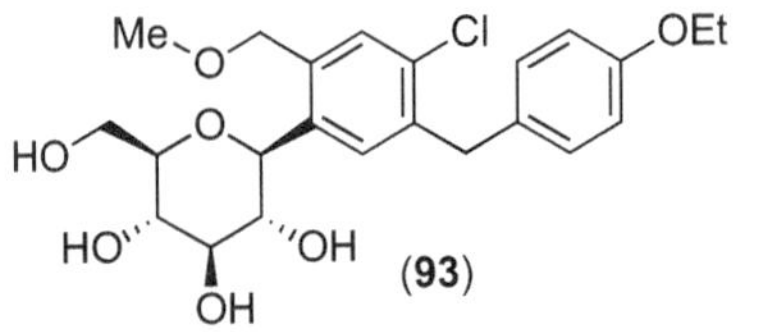

Ortho-substituted *C*-aryl glucosides

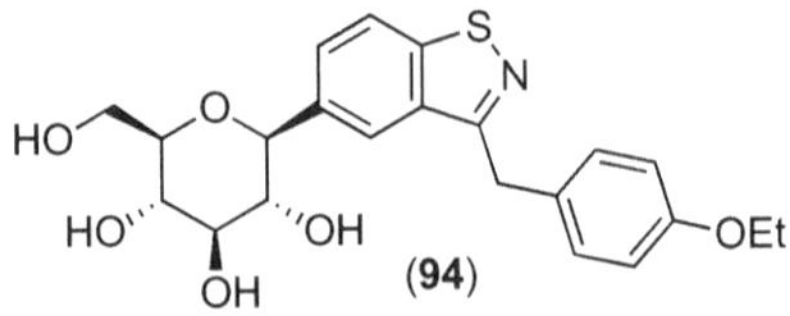

Benzisothiazole-β-glucopyranoside

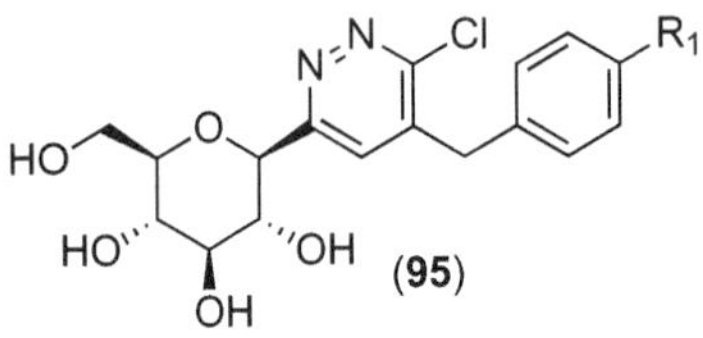

Pyridazine analogs as SGLT2 inhibitors

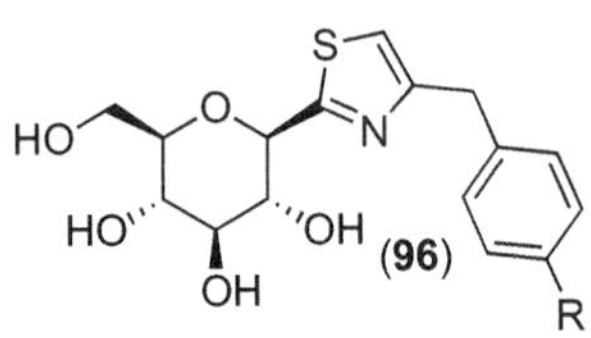

Thiazole analogs as SGLT2 inhibitors

Figure 3.24 ***Proximal ring modifications on Dapagliflozin.***

5.2 Modification in the glycone part of Dapagliflozin

The glycone part of Dapagliflozin, has also provided opportunity for structural variations, both for understanding its role and for search of better candidates. The inspiration for sugar ring modification came from SGLT inhibition displayed by **97**,[101] an *O*-glucoside with thio-sugar as the glycone. Despite promising activity, requirement of larger doses because of moderate stability in the biological system, marred its promise as a successful drug. Probably to address this issue, Kakinuma et al in 2010 proposed and synthesised *C*-glycoside, Luseogliflozin **73**[91] with thio-sugar as the glycone (Scheme 3.3).

Using standard reaction strategy involving the addition of organomagnesiumbromide obtained from bromide **99**, on to thio-lactone **98** followed

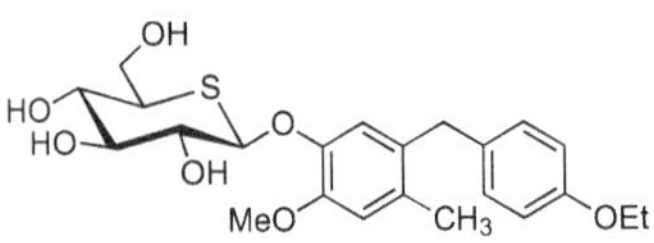

O-aryl 5-thio-β-glucoside (**97**)

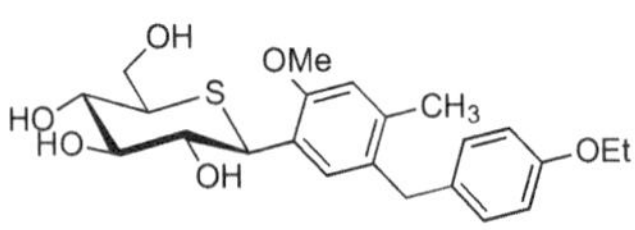

Luseogliflozin (**73**)

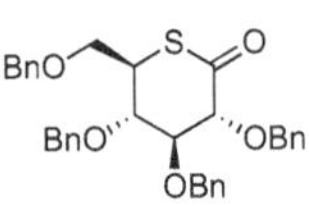

98

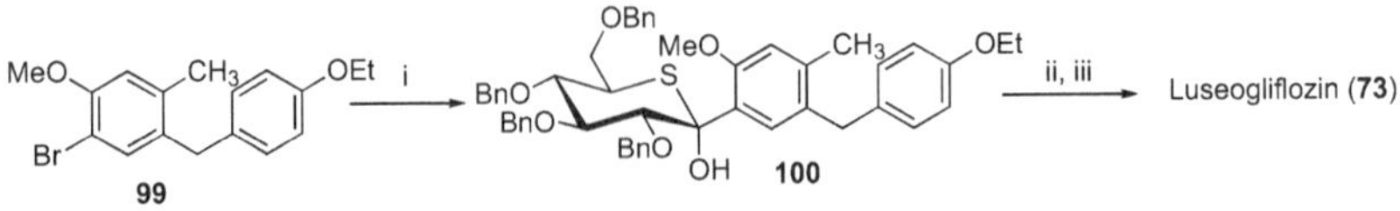

(i) **98**, Mg, $BrCH_2CH_2Br$, dry THF, reflux, 2 h; (ii) Et_3SiH, BF_3OEt_2, CH_3CN; (iii) $Pd(OH)_2/H_2$

Scheme 3.3 ***Synthesis of Luseogliflozin*** **73.**

by anomeric de-oxygenation in the corresponding lactol **100** and subsequent de-protection of benzyl ethers afforded the targeted compound **73**. The biological activity of synthesized compound **73** indeed exhibited good SGLT inhibition (IC_{50} of 2.26 nM) and high selectivity toward SGLT2 over SGLT1.

Y. Ohtake et al. (2012), designed and synthesized the Tofogliflozin **72**, as promising SGLT2 inhibitor containing an *O*-spiroketal ring system in the glycone part (Scheme 3.4).[90] Exposure of lactol **102** with $BF_3{\cdot}OEt_2$ in presence of triethylsilane furnished the spiro compound **103** in 56% yield. Apparently triethylsilane is not essential for spiro-cyclization, however its beneficial role has been rationalized. Presumably participation of benzylic oxygen atom precedes the hydride delivery from triethylsilane occurs on the trityl-quaternary carbon and its removal under reductive condition. Concomitant formation of triphenylmethane is indeed observed.

The targeted compound **72** was accessed through two approaches; either by oxidation of **103**, followed by addition of organometallic reagent on **104** and subsequent reduction and deprotection or debenzylation of **103** followed by selective chlorination of benzylic alcohol with chloro trimethylsilane in DMSO. The later approach furnished **106** ready for Suzuki

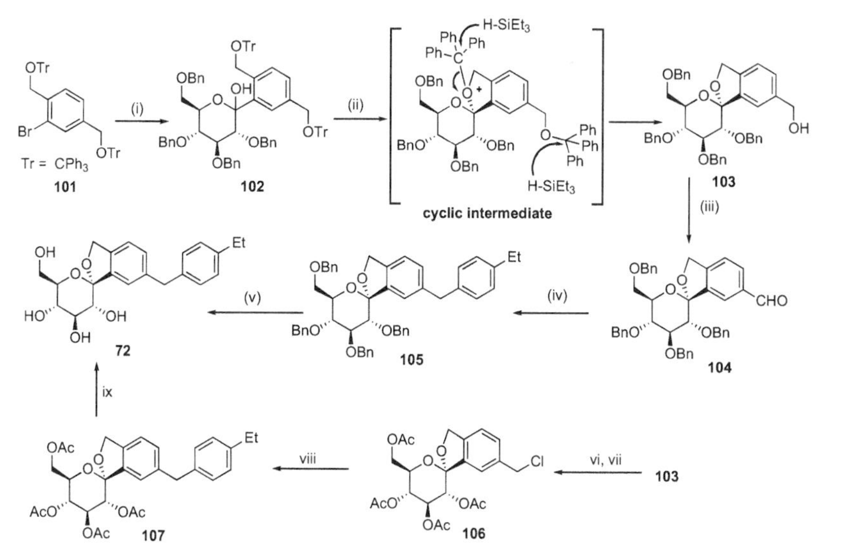

i) BuLi, toluene, -78 °C - rt, 30 min, 80%,**85**; (ii) Et_3SiH, $BF_3{\cdot}OEt_2$, CH_3CN, -40 to 0 °C,1 h, 56%; (iii) DMP, CH_2Cl_2, rt, 30 min, 33%; (iv) a. 4-(Et)-C_6H_4MgBr, THF: b. Et_3SiH, $BF_3{\cdot}OEt_2$, CH_3CN, -40 to 0 °C , 1 h, 56%; (v) H_2, $Pd(OH)_2/C$, MeOH-AcOEt, 2N HCl aq, rt,1h,(80-99%);(vi) BCl_3,Me_5-benzene,CH_2Cl_2,-78 °C,2h, 67%; (vii) TMSCl, DMSO, rt, 1.5 h, Ac_2O, NMM, DMAP, 0 °C, 1 h, 76%, (viii) 4-(Et)-$C_6H_4B(OH)_2$, $Pd(OAc)_2$, Ph_3P, K_3PO_4; toluene,80°C, 15 h, 93%; (ix) K_2CO_3, MeOH, rt, 12 h.

Scheme 3.4 ***Synthesis of Tofogliflozin* 72.**

coupling with various 4-substituted arylboronic acids for incorporation of the distal aryl ring. Compound **72** is currently undergoing phase III clinical trials for the treatment of type 2 diabetes through the SGLT2 Inhibition with an IC_{50} of 2.9 nM.

Goodwin et al. in 2009 proposed to replace the *D*-glucose residue in Dapagliflozin with *L*-xylose.[102] This was with the assumption that additional stability may be secured against glucosidases and also avoid other cross interactions with glucose binding enzymes. The proposed analogs **113** were synthesized from *L*-xylose derived aldehyde **108** (Scheme 3.5). A small library of compounds having different groups at the anomeric position was screened. It was found that compounds with substituents smaller in size and with β–configuration at anomeric position displayed more activity against mSGLT2 and hSGLT2. The compound **113** in specific showed promising SGLT2 inhibitory activity (IC_{50} = 7.1 nM).

The original route to SGLT2 inhibitor **71**, presenting sugar ring modification by Mascitti et al. in 2010[103] was found to be less suitable for access to specific compounds in the series, on a larger scale. New stereoselective route developed by the same group,[104a] for efficient and convenient access to **71** banked on readily available diacetone-α-D-mannofuranose **114** as starting material (Scheme 3.6). This was converted to the corresponding aldehyde **115** using literature procedure,[104b] and subjected to nucleophilic addition from the carbanion derived from 2-aryl-dithianes. The obtained product **116** on desilylation afforded the corresponding lactol **117** in 75% yields. This lactol on treatment with excess formaldehyde afforded the key

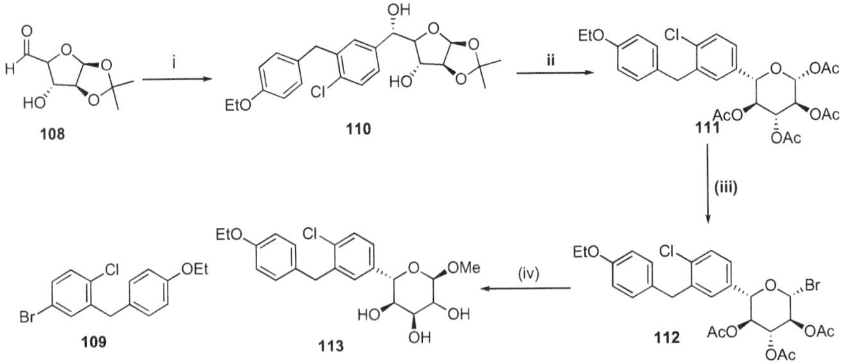

(i) **109**, nBuLi, THF, -78 °C, 45 %; (ii) a) HOAc, H_2O,100 °C, b) Ac_2O, pyridine, 87 %; (iii) ZnO, MeOH, 60 °C, 78% (iv) K_2CO_3, MeOH

Scheme 3.5 ***Synthesis of L-xylose derived Dapagliflozin analogs.***

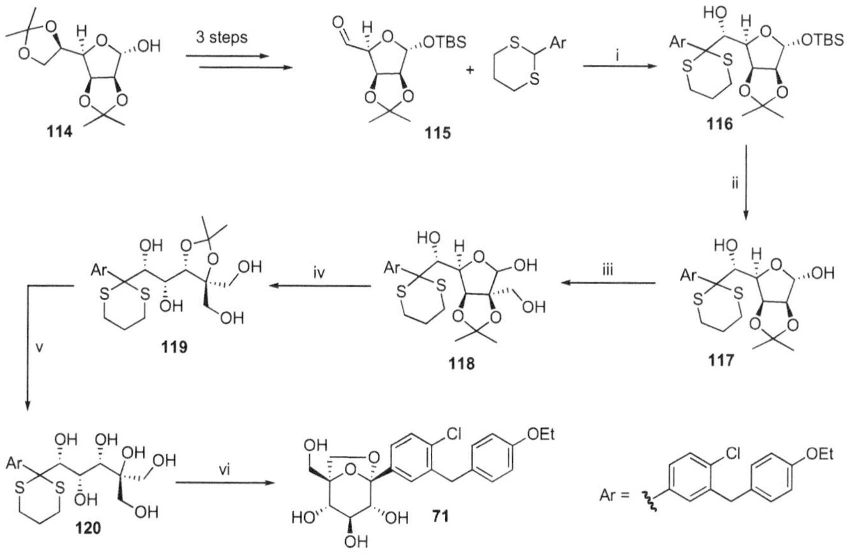

(i) n-BuLi, THF, -78 to -20 °C, 53%; (ii) TBAF, THF, 0 °C, 75%; (iii) K_2CO_3, H_2CO; $MeOH/H_2O$, 85 °C, 65%; (iv) $NaBH_4$, MeOH, 23 °C, 33%; (v) TFA/H_2O, air, 23 °C, 77%;

Scheme 3.6 ***Synthesis of Ertugliflozin* 71.**

intermediate **118**. Reduction of compound **118** with sodium borohydride followed by treatment of diol **119** with TFA/H_2O (9/1) afforded the desired compound **71**. The process includes removal of acetonide protection and formation of intermediate **120**. Internal ketalization on to unmasked carbonyl, presumably occurs through a putative thionium ion leading to the formation of spiro-ketal **71** under thermodynamic control. The compound **71**, known as Ertugliflozin, has been approved by the US-FDA in 2017 for treatment of type 2 diabetes.

Mascitti et al. in 2010 also synthesized compound **126** with spirocyclic ring at C-5 position of pyranose ring (Scheme 3.7).[105] The C-6 and C-4 hydroxyl group in Dapagliflozin **64** was regioselectively protected as benzylidene acetal. Subsequent protection of C-2, C 3 hydroxyl as benzyl ether, followed by regioselective opening of benzylidene acetal with $AlCl_3$/$LiAlH_4$ provided the compound **122**. Swern oxidation of the primary hydroxyl group and treatment with formaldehyde gave the corresponding diol **123**. The diol **123** was sequentially treated with *n*-BuLi-tosyl chloride- *n*-BuLi (1:1:1) which furnished the spiro compound **125**, containing an oxetane ring at C-5 position, *via* the intermediate **124**. Finally, hydrogenolysis of **125** with Pd/C, HCOOH, EtOH/THF mixture provided the targeted

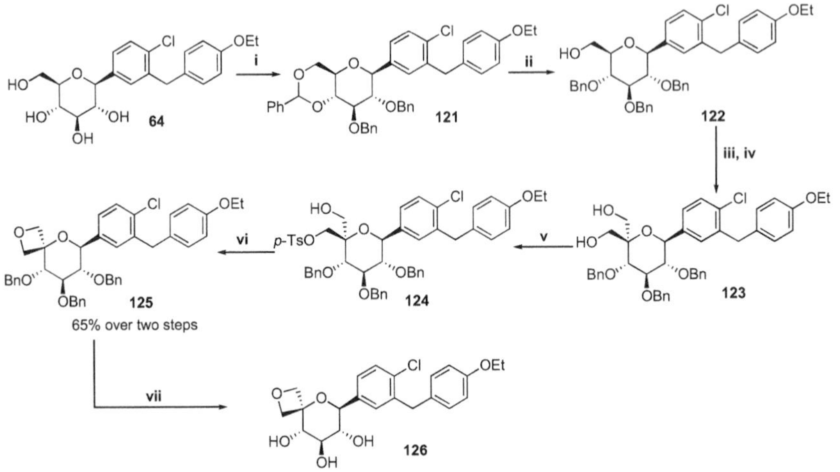

(i) $PhCH(OMe)_2$ (1.3 eq),*p*-TsOH (Cat), DMF, 60 °C; (ii) BnBr, NaH, DMF, 0-23 °C, 82%; (ii) $LiAlH_4$, $AlCl_3$; Et_2O/CH_2Cl_2, reflux, 90%; (iii) Swern Oxidation; (iv) (a) NaOH, H_2CO (excess). (b) H_2O/ 1,4-dioxane, 23 °C, 45%; (v) *n*-BuLi (1 eq), *p*-TsCl (vi) n-BuLi, THF, 0-65 °C; (vii) Pd/C, HCO_2H, EtOH/THF, 23 °C.

Scheme 3.7 ***Synthesis of C5-spirocyclic C-aryl Glycosides* 126.**

compound **126**. The target exhibited moderate SGLT2 inhibitory activity (IC_{50} = 32 nM).

To investigate the significance of hydroxyl functionality and their stereochemistry in *C*-aryl glucosides, displaying SGLT2 inhibition, such as **127** and **135**, R. P. Robinson et al. in 2010 synthesized a series of molecules by systematically modifying the sugar moiety with deletion of the hydroxyl functionality or inverting their stereochemistry by using simple functional group transformations. Synthesis of most promising *C*-aryl glycosides, among all other compounds, are presented in Scheme 3.8 and Scheme 3.9.[106]

Biological studies with the synthesized compounds, which included modification at C-2, C-3 either by deletion of hydroxyl functionality or their conversion to methyl ether (OH → OMe) or inversion of stereochemistry, revealed only lowering of activity and selectivity towards SGLT2 over SGLT1.

In 2012, Chen et al., designed and synthesized Dapagliflozin analog **147**, wherein the hydrogen atom and the hydroxyl group at C-4-position in pyranose ring was substituted with two fluorine atoms.[107] This proposal seems to be in the context that structural modifications at C-2 and C-3 positions resulted in significant loss of inhibitory activity, while changes at positions C-4 and C-5 have less effect on activity.[106] Moreover, the fluorine atom is normally expected to be able to mimic some properties of hydroxyl group

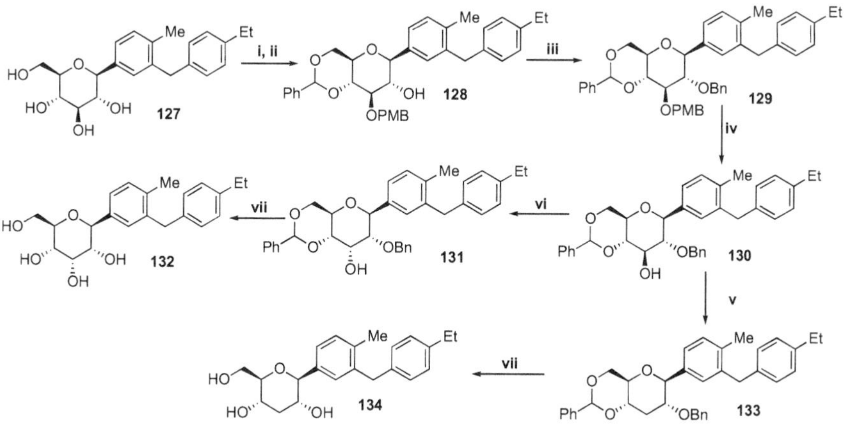

(i) $PhCH(OMe)_2$, $HBF_4{\cdot}OEt_2$, DMF, rt, 12h; (ii) (a). Bu_2SnO, toluene, reflux, 4h. (b). PMBBr, TBAI, reflux, 12h; (iii) BnBr, TBAI, reflux, 12 h, (iv)DDQ, CH_2Cl_2, rt, 3 h; (v) (a). NaH, imidazole, THF, CS_2, reflux, 1 h, MeI, reflux, 0.5 h, (b) Bu_3SnH, AIBN, toluene, reflux, 1.5 h; (vi) (a). Oxidation, (b) $NaBH_4$, MeOH, 0-10 °C, 1 h.(vi) H_2, $Pd(OH)_2$, HCl, EtOH;

Scheme 3.8 ***Modification at C3-position of sugar ring.***

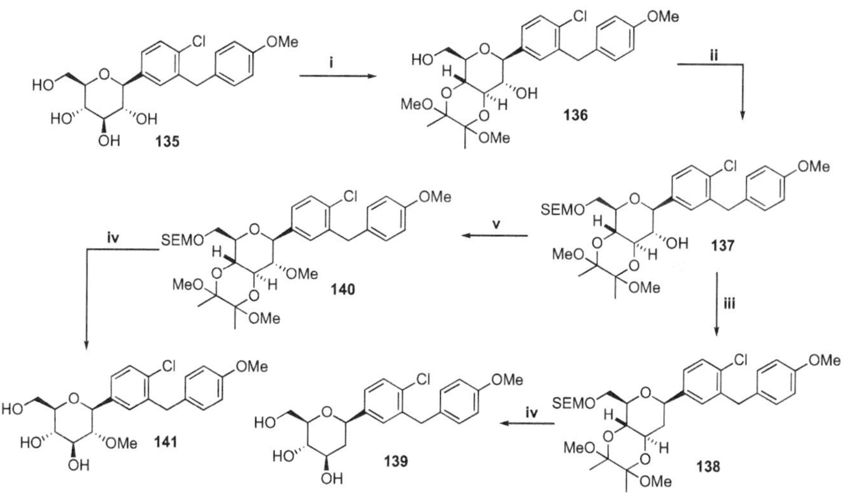

(i) 2,3-butadione, $HC(OMe)_3$, $BF_3{\cdot}OEt_2$, 60 °C,3 h; (ii) SEMCl, $^iPr_2NEt$, CH_2Cl_2, rt, 12 h; (iii) (a)TCDI, toluene, reflux 18 h, AIBN, (b) $(TMS)_3SiH$, 120 °C, 12 h; (iv) TFA, H_2O,rt, 1h; (iv) MeI, NaH, THF, rt, 12 h

Scheme 3.9 ***Modification at C2-position of sugar ring.***

and the gem-difluoromethylene group is strongly electron withdrawing too.[108] The negative charge density on the vicinal ring oxygen will be decreased dramatically and may thereby influence the binding affinity. Synthesis of compounds with difluoro substitution at C-4 in the pyranose ring and represented by structure **147** banked on the lactone **142**.This was made

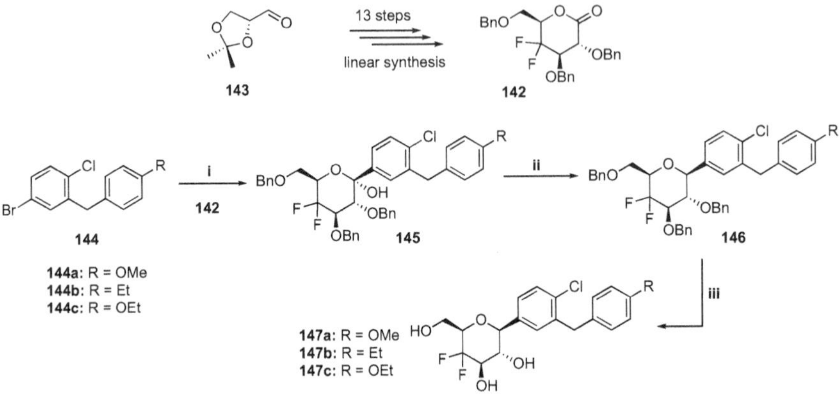

(i) n-BuLi, -78 °C (ii) Et_3SiH, $BF_3{\cdot}OEt_2$, 70- 80% in two steps; (iii) Pd/ C, H_2, 1,2 - dichloro benzene (20 eq.) EtOAc / MeOH = 2:3, 83-90%.

Scheme 3.10 ***C-4 modification of Dapagliflozin, synthesis of compound* 147.**

through a 13-step synthetic scheme using (*R*)-glyceraldehyde acetonide **143** as a readily available starting material (Scheme 3.10). The organolithiums generated from varied bromodiphenylmethane substrates **144a-c** were added onto lactone **142** and furnished the corresponding hemiketals **145**.

Stereoselective deoxygenation at anomeric position followed by debenzylation afforded the targeted compounds **147a-c** in good to excellent yields. The *in-vitro* inhibitory activity of synthesized compound **147a-c** against human SGLT2 showed comparable inhibitory activity (IC_{50} = 0.35, 0.42 and 5.54 nM respectively) with Dapagliflozin.

6 C-Glycosides of flavones

The flavone *C*-glycoside Chafuroside A **148a** and Chafuroside B **148b**, isolated as minor constituents of oolong tea extract has aroused tremendous interest because chafuroside A has been found to be 1000 times more potent than conventionally used dexamethasone. It has a *manno*-configured sugar moiety fused onto C-6 carbon and C-7 hydroxyl in the flavone ring (Fig. 3.25). The low amount present in natural sources has warranted efficient synthesis of Chafuroside A.[109a–c]

The glycosylation of phloroacetophenone derivative **149** with *O*-(2,3,4,6-tetra-*O*-benzyl-α-D-glucopyranosyl)trichloroacetimidate **150** in the presence of catalytic amounts of TMSOTf accomplishes the appending of sugar unit on to the aromatic ring via *O*-to-*C* rearrangement in 69% yield (Scheme 3.11). The obtained *C*-glucoside product **151** is β-configured.

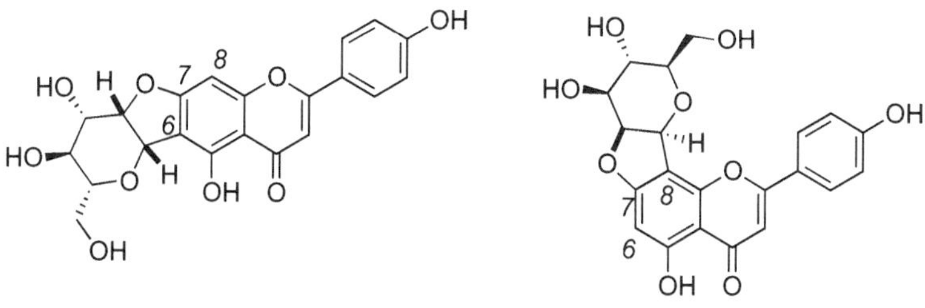

Figure 3.25 ***Flavone C-Glycosides from Oolong tea.***

Subsequent condensation of **151** with 4-*O*-MOM-protected benzoic acid and refluxing of the formed mono-*O*-acylated derivative **152** in pyridine solution in the presence of K_2CO_3 furnished **153** (17%) and **154** (15%) as a separable mixture. The formation of two isomeric flavones **153** and **154**

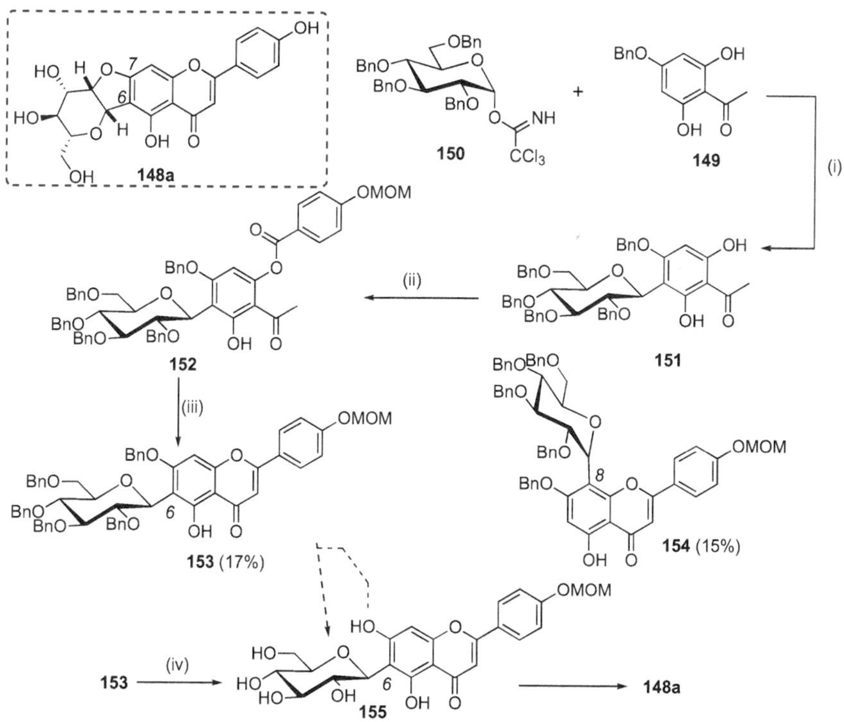

(i) TMSOTf, CH_2Cl_2, 0 °C to rt, 69%, (ii) 4-OMOM-benzoic acid, DCC, DMAP, CH_2Cl_2, r.t., 82%
(iii) K_2CO_3, pyridine reflux, 1 h, **153** (17%); **154** (15%), (iv) $Pd(OH)_2$, H_2, EtOH, 95%
(v) (a) 1,10-azobis(*N,N*-dimethylformamide), Bu_3P, THF, 50 °C, (b) HCl–Dioxane, MeOH, rt, 32% (two steps)

Scheme 3.11 ***Synthesis of Chafuroside.***

are indicative of C ring construction occurring via Baker–Venkataraman rearrangement and successive cyclization under the reaction conditions. Intramolecular cyclization under modified Mitsunobu conditions in THF followed by MOM group removal under acidic conditions afforded the flavone **148a** in 32%, over two steps. The stereochemistry at C-2 position in the glucosyl residue gets inverted during *Mitsunobu* cyclization, leading to the desired *manno*-configuration.[109b]

The 8-β-D-glucopyranosylgenistein **156**, isoflavone *C*-glucoside isolated from *Genista tenera* has emerged as a promising lead for intervention in amyloid events of both diabetes and Alzheimer's disease.[110] Further studies are necessary, as its use as a functional food ingredient may provide specific health benefits, such as prevention of functional and cognitive disorder, during aging and for promoting a healthy life. Efficient synthesis of **156** has been achieved, wherein construction of the *C*-glucosidic bond is realized through condensation of perbenzylated glucosyl acetate **157** as electrophilic glucosyl donor and electron rich protected acetophloroglucinol **158**, under Lewis acid condensation (Scheme 3.12). The *C*-glucosylated acetophenone derivative **159** enabled Aldol condensation with 4-benzyloxybenzaldehyde

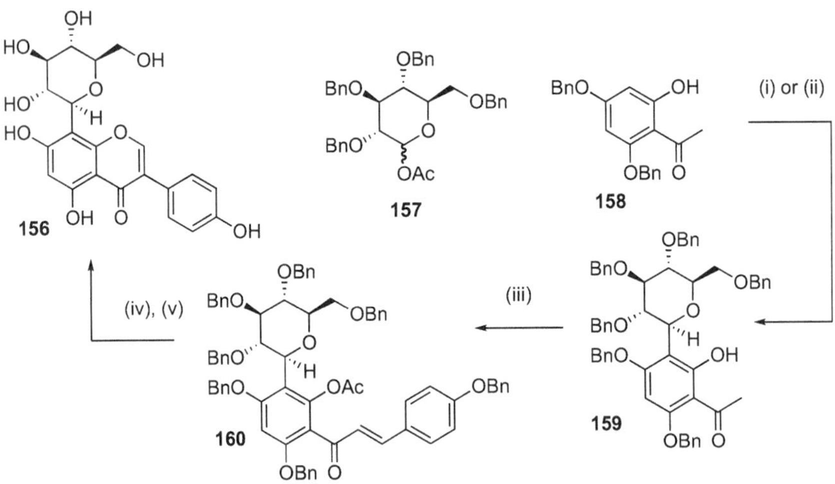

(i) Sc(OTF)$_3$, DCE, Drierite, -30°C, 30 min, then RT 5h, 49%
(ii) TMSOTf, CH_2Cl_2 -CH_3CN (1:1), Drierite -40°C, 30 min, then RT 3h, 56%
(iii) (a) 50% (w/v) aq. NaOH, 1,4-dioxane, 4-benzyloxybenzaldehyde (b) Ac_2O, Py, DMAP, 1h 60%
(iv)(a) Tl(NO_3)$_3$, (MeO)$_3$CH, MeOH, 24 h, 40 °C (b) THF, MeOH, aq. NaOH 50% (w/v), 4h rt, 63%
(v) MeOH, EtOAc, Pd/C, H_2, 6h, rt, 96%

Scheme 3.12 ***Synthesis of 8-β-D-glucopyranosylgenistein*** **156.**

to afford chalcone **160**. Oxidative rearrangement with thallium (III) nitrate under basic conditions furnished the desired isoflavone ring and final debenzylation under hydrogenolytic condition gave the desired 8-β-D-glucopyranosylgenistein **156**.[111]

Although, *C*-glycosylated flavonoids, widely distributed in plants, represented by vitexin **40**, isovitexin **41**, orientin **42** and isoorientin **43** (Fig. 3.11) had aroused synthetic interest, since last three decades,[112–116] the diverse pharmacological activities particularly associated with vitexin and isovitexin has spurred renewed interest in the recent times.[117–119] This is because of their rich presence in sprouted Mung beans (*Vigna radiate* L.), considered to be part of a healthy diet. Vitexin has displayed promising analgesic effect in a variety of inflammatory pain models through modulation of cytokine production.[120] Since it has been found to be safe with regard to liver damage, gastric mucosa injuries, and also showed no signs of toxicity over a period of 14 days in rats and mice at 2000 mg/kg dose, it does merit further preclinical and clinical investigations.

Passiflora edulisis, commonly referred to as passion fruit is cultivated in many countries for commercial value as edible fruit and also for its health benefits. The leaf extracts have been used in Europe and America as a remedy for many neurogenic diseases. Besides five known compounds, the *n*-butanol extracted fraction of air-dried stems and leaves has offered for the first time, isolation of four new *C*-glycosyl flavones **161–164** (Fig. 3.26) wherein 2,6-dideoxyhexoses, constitute the glycone part.[121] In the context, this plant is used both for medicine and for its juice material, unraveling the significance of these constituents will be pertinent and relevant.

In the Ayurvedic system of medicine, water stored in a wooden glass made up of heartwood *Pterocarpus marsupium* (Leguminaceae) has been

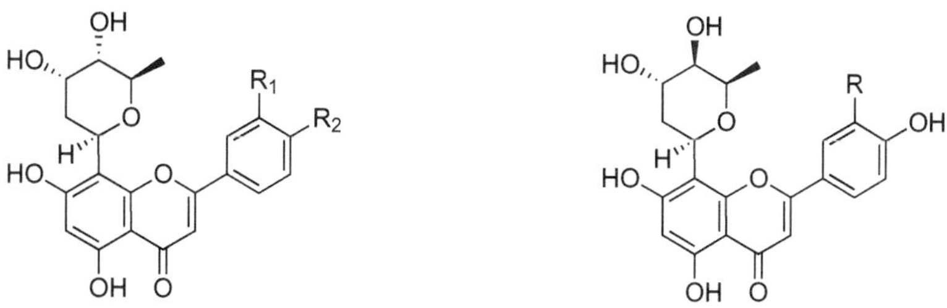

161: R_1 = OH, R_2 = O-β-D-glucopyranosyl
162: R_1 = OH, R_2 = OH

163: R = H
164: R = OH

Figure 3.26 ***C-Glycosyl Flavones.***

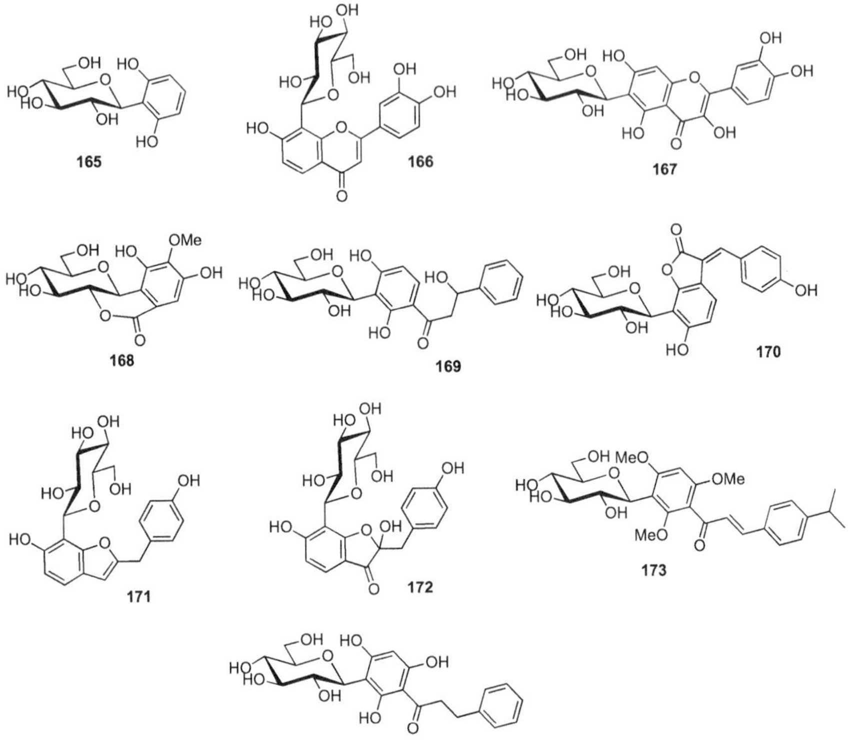

Figure 3.27 ***C-Glycosides* 165–172 *extracted from heartwood of Pterocarpus marsupium.***

found to be effective for non-insulin dependent diabetes mellitus patients.[122] The *C*-glycosides **165–172** (Fig. 3.27) have been isolated from the aqueous extract of heartwood of *Pterocarpus marsupium* (Leguminaceae). Inspired from this observation, Maurya's group has ventured into synthesis of several phenolic *C*-glycosides and evaluated their anti-hyperglycemic activity.[123] Chalcone *C*-glucoside **173** and dihydrochalcone *C*-glucoside **174** lowered the blood glucose levels to 33.6% and 26.5% respectively after 24 hours on STZ model, which was comparable to the standard clinically used drug metformin.

7 C-Glycosides of chalcones

Aspalathin **176**, a dihydrochalcone *C*-glucopyranoside, first characterized by Koeppen and co-workers[124] in 1965, is showing renewed interest as a nutraceutical ingredient in foods and beverages[125] due to its beneficial effects on glucose homeostasis in type 2 diabetes by stimulating glucose-uptake in muscle cells and insulin secretion. Limited supply of the same from natural sources has prompted synthetic efforts.[125,126] Convenient access to multi-gram scale quantities of aspalathin **176** has been realized through the *C*-glucoside **182** (Scheme 3.13).[126] The *O*-glucoside **181** obtained through BF_3 catalyzed coupling of phloroacetophenone **180** with **177, 178** and **179** as glucosyl donors underwent O→C rearrangement affording *C*-glucoside **182** in good yields. The *O*-glucosylation reaction was found to be high yielding (84%), using **179** as donor. Condensation of **182** with aldehyde **183** afforded chalcone **184** in 96% yield. Catalytic hydrogenation of **184** furnished Aspalathin **176** in 99% yield.

Our group in 2019 has synthesized *C*-glucosylated chalcone **185** and dihydrochalcone **186** inspired from the natural product isoliquiritigenin (ISL) **187**, a chalcone isolated form *Glycyrrhiza radix* (Licorice)[127] and davidigenin **188**, the dihydro-analog, independently isolated from Artemisia plant.[128] Although both the natural products displayed potent aldose reductase inhibiting activity, poor solubility of these compounds in aqueous

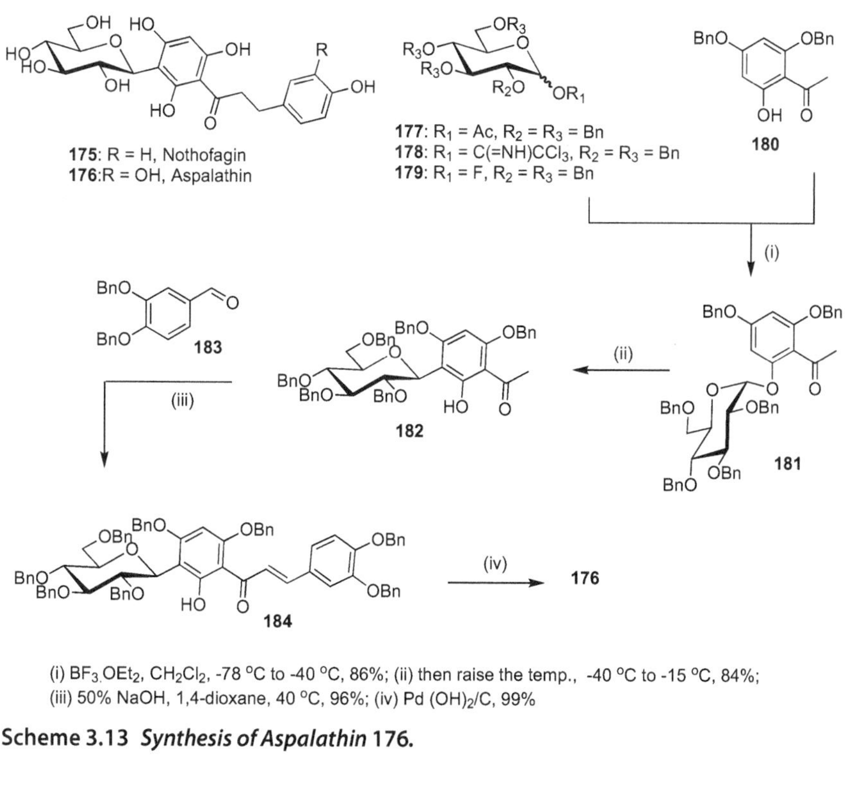

(i) $BF_3.OEt_2$, CH_2Cl_2, -78 °C to -40 °C, 86%; (ii) then raise the temp., -40 °C to -15 °C, 84%; (iii) 50% NaOH, 1,4-dioxane, 40 °C, 96%; (iv) Pd $(OH)_2/C$, 99%

Scheme 3.13 ***Synthesis of Aspalathin*** **176.**

medium, has limited their further exploration. With the view to increase the water solubility and bioavailability, 4-*C*-glucosylated targets **185** and **186** were synthesized (Scheme 3.14).[129] The key building block **192** was readily made as described in synthetic Scheme 3.14. It involved addition of functionalized aryl magnesium bromide from ethyl-4-iodobenzoate **189** onto protected *D*-gluconolactone **85** for desired C—C bond formation at the anomeric position. Reductive anomeric deoxygenation and subsequent functional group interconversions on **190** afforded the compound **192**.

The aldehyde **192** underwent Claisen-Schmidt reaction with various acetophenones and furnished, the targeted *C*-glucosylated chalcones **193** in general and **185** in particular. Hydrogenation of chalcones **193** conveniently afforded the corresponding dihydrochalcones **194** in general and **186** in particular. All *C*-glucosylated chalcones and dihydrochalcones were evaluated for aldose reductase inhibition activity. *C*-glucosylated ISL **185** with improved aqueous solubility, indeed exhibited 100% inhibition at 120 μM concentration with an IC_{50} of 21 μM. The natural product ISL exhibits an IC_{50} of 19 μM.[129]

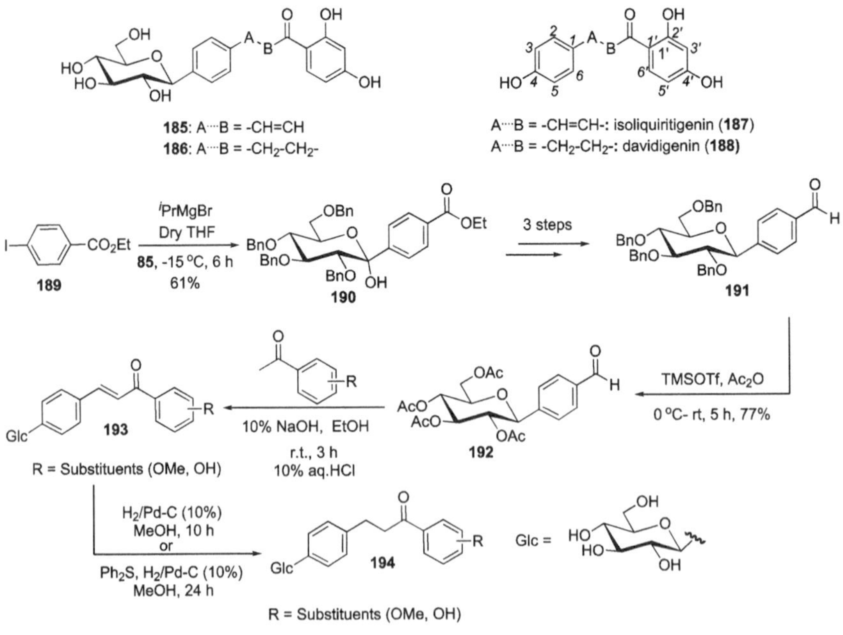

Scheme 3.14 ***Proposed glucosylated chalcones* 185, 186 *and their synthesis.***

8 C-glycosides of xanthones

Mangiferin (**51**) and neo-mangiferin (**195**) are the major active constituents of *Anemarrhena asphodeloides* (Liliaceae family) and *Mangifera indica L.* (Anacardiaceae family). Plants from these families have been often used as a part of herbal medicines and functional foods for treating inflammation, asthma and pain.[130,131] Neo-mangiferin (mangiferin-7-*O*-β-D-glucoside) is more hydrophilic than mangiferin and exhibits anti-osteoporotic and anti-lipidemic properties,[131,132] anti-inflammatory effect[133] and lowering of blood glucose levels.[134] A 70% ethanol-water extract from the leaves of *Mangifera indica L.* was found to inhibit triglyceride accumulation in 3T3-L1 adipocytes and the active fraction afforded seven new benzophenone *C*-glycosides **196a-196g**, besides mangiferin (Fig. 3.28). [135]

Given the fact that mangiferin is abundantly and easily available from many other plants such as *Mangifera indica, Belamcanda chinensis, Foliaum pyrrosiae*, and *Coffea pseudozanguebariae*, efficient semi-synthetic scheme has been realized for convenient access to neomangiferin (Scheme 3.15).[136]

After carefully deciphering the reactivity pattern of the four phenolic hydroxyl groups present in mangiferin, appropriate reaction conditions were developed which allowed exclusive formation of 3,6,7-tri-acetyl ester derivative **197** of mangiferin, instead of the expected per-acetylated product **198**. Banking on literature known fact that acetyl-ester protected

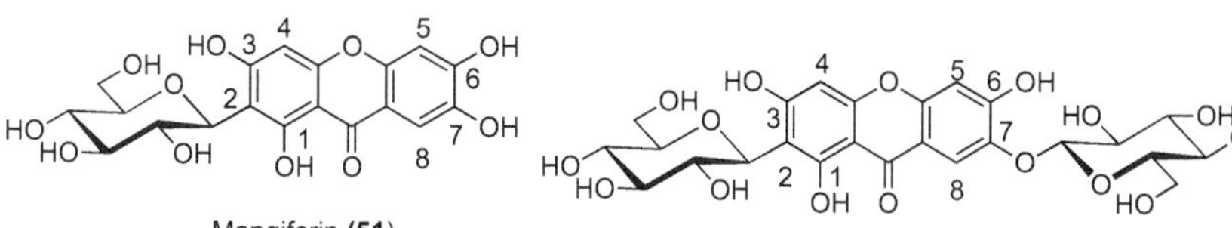

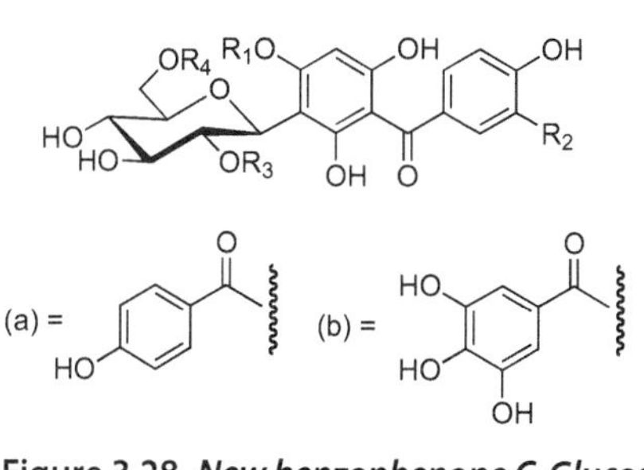

	R_1	R_2	R_3	R_4
Foliamangiferoside A **196a**:	CH_3	H	H	H
Foliamangiferoside A_1 **196b**:	CH_3	H	(a)	H
Foliamangiferoside A_2 **196c**:	CH_3	H	H	(a)
Foliamangiferoside B **196d**:	CH_3	OMe	H	H
Foliamangiferoside C_1 **196e**:	H	H	H	H
Foliamangiferoside C_2 **196f**:	H	H	(a)	(a)
Foliamangiferoside C_3 **196g**:	H	H	H	(b)

Figure 3.28 ***New benzophenone C-Glucosides*** **196a-196g.**

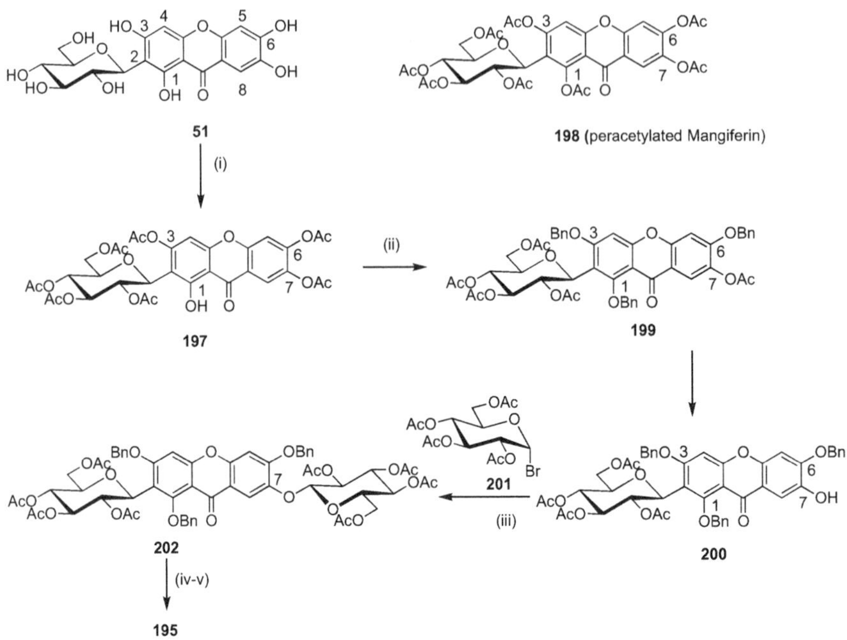

(i) $Ac_2O/AcOH$ (v/v = 1/2), NaOAc, 120 °C, 3 h, 94%; (ii) BnBr, KI, K_2CO_3, acetone, 60 °C, 13 h, 85%
(iii) 3.0 equiv. of **201**, TBAB, 5% aq. NaOH (7 equiv.), $CHCl_3$, 78%
(iv) $Pd(OH)_2$, H_2, $CH_2Cl_2/MeOH$, r.t. 6 h, 94%; (v) NaOMe/MeOH, r.t. 2 h, 96%

Scheme 3.15 ***Synthesis of Neomangiferin*** **195.**

phenolic-hydroxy group, positioned para to carbonyl group in flavones can be transformed to benzyl ether protection, **197** was converted to **199** in 85% yield. As expected, the reaction conditions, not only offered benzyl-ether protection at C3 and C6 position, but also enabled benzyl-ether protection to the free phenolic hydroxyl at C1 in **197**. The intact acetyl-ester protection at C7 position in **197** was now removed for *O*-glucoside formation which occurred readily between **200** and **201** as the glucosyl donor under phase transfer conditions and afforded the desired *O*-glucoside **202** in 78% yield. Final deprotection completed the synthesis of neomangiferin, **195** in 47% over all yield.

9 C-Glycosides inspired from Adenophostin A

The D-myo-inositol 1,4,5-trisphosphate (IP_3) **203** released from the hydrolysis of phosphatidylinositol 4,5-bisphosphate by enzyme phospholipase C is shown to mobilize Ca^{2+} from the intracellular stores of most

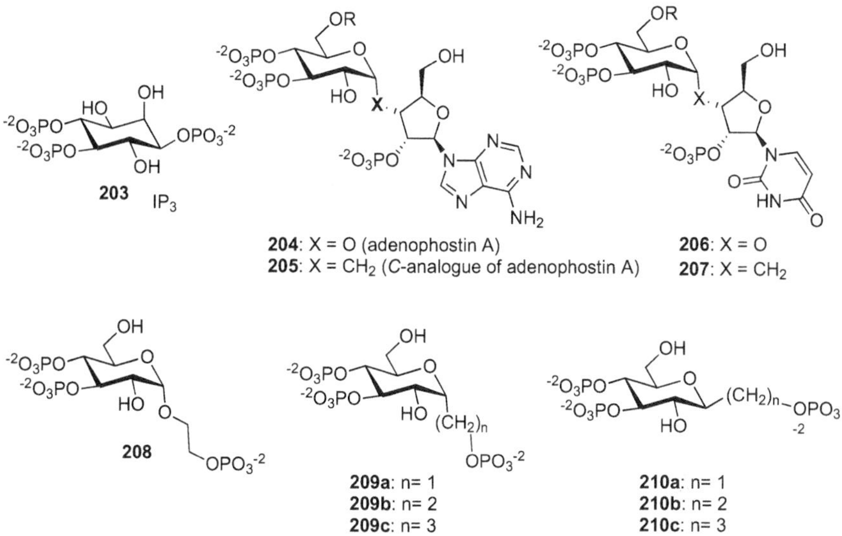

Figure 3.29 ***C-glucoside analogs of D-myo-inositol 1,4,5-trisphosphate (IP_3).***

mammalian cells. Diverse array of cell-surface receptors on stimulation, trigger IP_3 mediated pathways. The high-affinity agonist activity of natural product adenophostin A **204** (Fig. 3.29), isolated from *Penicillium brevicompactum*, being several folds higher than IP_3 (10 to 100),[137] motivated synthesis of stable analogs of IP_3 possibly for investigating mechanisms of IP_3-mediated Ca^{2+} signaling and for medicinal applications. Shuto's group deciphered *D*-glucopyranose ring as a good bioisostere of the myo-inositol ring present in IP_3 and the adenine moiety can be replaced by other aromatic rings too.

Synthesis of *C*-glycosidic analog **205** of adenophostin A and its uracil congener **207** has been achieved.[138] The *O*-glucoside trisphosphate **208** is an agonist of the IP_3 receptor, however its affinity is more than 10-fold lower than that of IP_3. In this context, a series of α-*C*-glucosides **209a-c** trisphosphates and corresponding *β*-*C*-glucoside trisphosphates **210a-c** with different lengths were proposed as stable mimics of the *O*-glucoside **208** (Fig. 3.29).[139] The *C*-analog of adenophostin A **205** was found to be a potent agonist with EC_{50} value of 4.0 ± 0.2 nM, 6-fold more potent than the natural ligand IP_3 (EC_{50}= 24.8 ± 2.1nM) and only 2-fold less potent than adenophostin A (EC_{50} = 2.1 ± 0.2 nM). The *C*-analog with uracil unit **207** was also active (EC_{50} = 11.3 ± 2.7 nM), about 2-fold more potent than IP_3 and only 5-fold less potent than adenophostin A.[140]

10 C-Glycosides of KRN 7000

The natural *O*-glycoside **211** in particular isolated along with other *O*-glycosides, from the Okinawan marinesponge *Agelas mauritianus*, way back in 1993[141] and termed as agelasphins (AGLs) spurred intense interest, that the research associated with this lead, both from the chemistry and biological perspectives has not only spanned over period more than two decades, but thrived and advanced, with each passing year. The *O*-glycoside **211** presenting α-galactosylceramide (α-GalCer) structure, displayed potent anti-tumor activities against *in vivo* models of several murine tumor cells. However, right at the beginning, the scarcity of this *O*-glycoside from the natural sources, compelled development of synthetic methods for enabling access not only to **211** but also to several other analogs. From the synthesized analogs, the *O*-glycoside **212**, referred as KRN 7000 (Fig. 3.30), surfaced as the most promising compound towards possible clinical applications.[142]

Further structural modifications on KRN 7000 centering on (1) sugar moiety, (2) configuration and the nature of the glycoside bond, (3) polar portion of the ceramide, and (4) variation on the lipid chains, have paid rich dividends with regard to obtainment of new and potent analogs.[143] The synthesis of *C*-analog of KRN 7000, namely **213** drew attention from several groups. The first report of synthesis of *C*-analog, termed as α-*C*-GalCer, appeared in the patent assigned to Kotobuki Pharma in 2002.[144] A brief summary of various approaches toward synthesis of α-C-GalCer has been covered by Franck.[145] Franck's group, also popularly referred as "*Hunter group*,"[146] has provided four different approaches of this important target and discovered that it was 100 times more effective in comparison to *O*-glycoside at 10 ng dosage level in a melanoma challenge assay on C57BL/6 mice.[147] The first synthetic strategy from Franck's group in 2004 banked

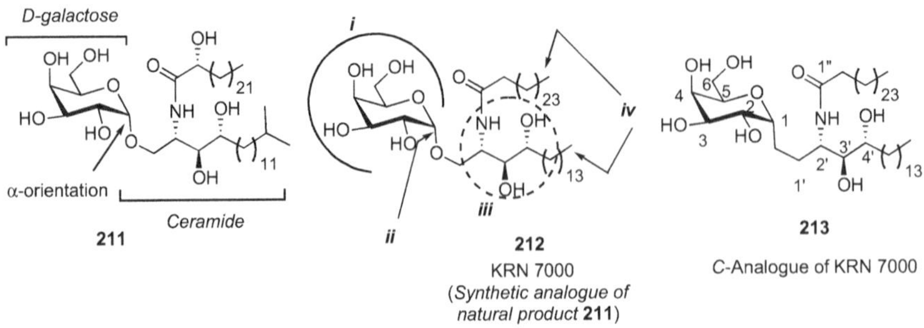

Figure 3.30 ***Structure of KRN 7000 and site of structural modifications.***

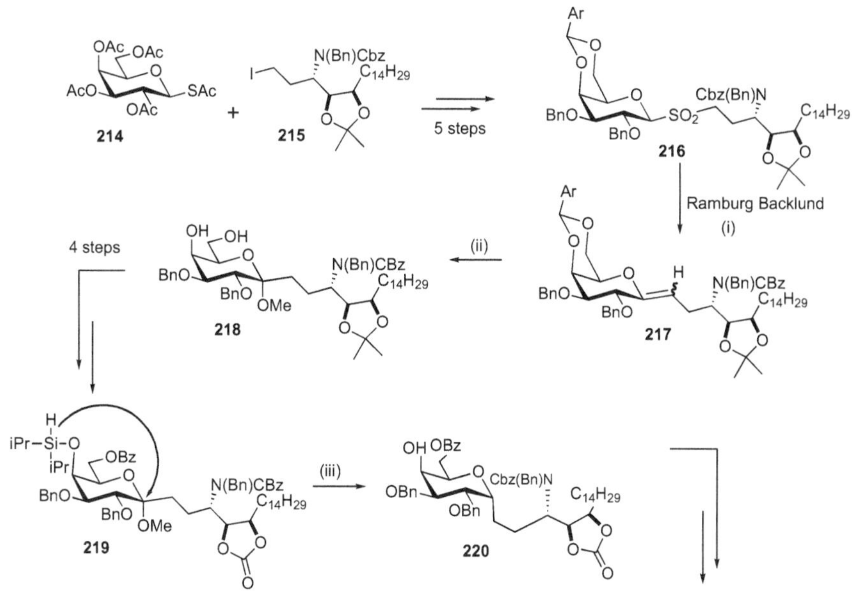

Scheme 3.16 ***Synthesis of C-analog of KRN 7000.***

on the use of Ramberg-Backlund reaction[147e] on the highly functionalized sulfone **216** for the construction of *C*-glycoside bond (Scheme 3.16).[147a]

Internal hydride transfer in the advanced precursor **219** affording compound **220** ensured α-configuration at the anomeric center in the target **213** (Scheme 3.16).[147a] The synthetic scheme, although long (26 steps) did provide material sufficient enough for biological assays. Other synthetic routes from Franck's group entailed use of C1-functionalized *D*-galactose **221**[148] and **222**[149] wherein the α-oriented C—C bond is built-in at the anomeric position.The α-C1-formyl derivative enabled requisite C—C bond formation between the *D*-galactose unit and latent ceramide part using modified Julia olefination reaction between **221** and sulfone **225**[150] while olefin metathesis reaction enabled C—C formation between alkene building block **222** and **226** (Fig. 3.31).[149] The developed synthetic scheme using olefin metathesis was convergent, short (11 steps) and more efficient (30% yield). Use of other building block **223** banked on Sharpless asymmetric epoxidation (SAE) for attending to the stereochemistry and the functionalities desired in the ceramide part through the advanced epoxide **227**.[151] Poor regioselectivity in opening of the epoxide ring with sodium azide for incorporation of the amine functionality and subsequent difficulties in

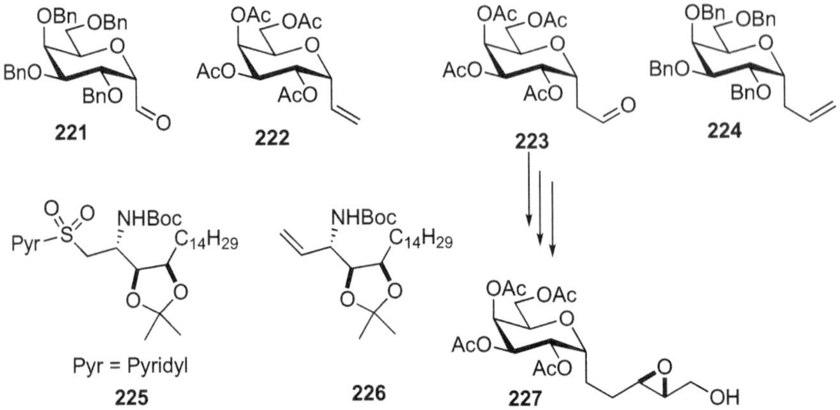

Figure 3.31 ***Important building blocks for the synthesis of KRN 7000 and analogs.***

carbon chain extension using Wittig reaction undermined the efficiency of the synthetic scheme through epoxide **227**.

10.1 Sugar ring annulation for the access to C-glycosides

Synthetic approaches, for the synthesis of *C*-glycosides, wherein sugar ring (pyranoside/furanoside) is annulated after the construction of the predetermined glycosidic C-C bond, is aptly demonstrated by two examples described below. The β-isomer of α-*C*-GalCer synthesized in the same year as α-*C*-GalCer, by Postema,[152a] used ring closing metathesis for construction of the *D*-galactose ring, it also exhibited anti solid-tumor activity. Ester formation between **228** and functionalized carboxylic acid **229**, followed by ester carbonyl methylenation using Takai methylenenation[152b] protocol afforded the much desired intermediate **231** (Scheme 3.17) for implementing ring closing metathesis using Grubb's second-generation catalyst. Glycal **232** so obtained from ring closing metathesis reaction was subjected to hydroboration reaction, followed by benzyl-ether protection to furnish the β-*C*-glycoside **233** as an advanced intermediate for further functional group manipulations toward synthesis of β-*C*-GalCer.

Lankalapalli accomplished the synthesis of new aza-variant analog **236** of β-*C*-GalCer using the concept of double reductive amination on the open chain dicarbonyl precursor **237**.[153] This double reductive amination enabled the aza-sugar ring construction and formation of **238**. The open chain dicarbonyl precursor required for this approach was obtained through Horner-Wadsworth-Emmons reaction between β-keto-phosphonate **239** and aldehyde **240**, followed by removal of *p*-methoxybenzyl ether protection on C5—OH and oxidation (Scheme 3.18).

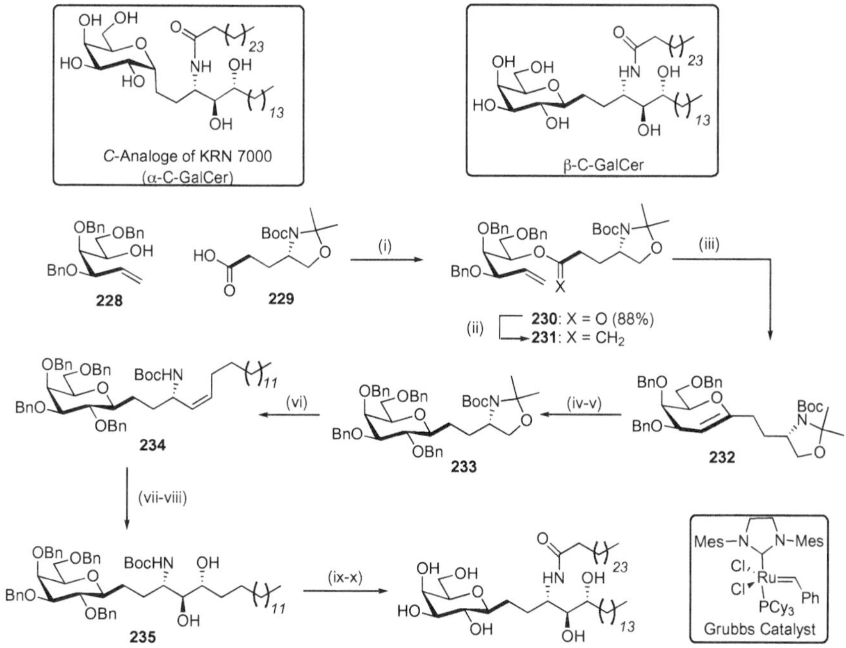

(i) DCC, 4-DMAP; (ii) $TiCl_4$, CH_2Br_2, TMEDA, $PbCl_2$; (iii) Grubbs II-Gen Catalyst, 60 °C; (iv) B_2H_6, THF/ NaOH, H_2O_2; (v) NaH/BnBr; (vi) (a) p-TsOH, MeOH, (b) Swern oxidn, (c) $Ph_3P{=}CH(CH_2)_{13}CH_3$; (vii) OsO_4, NMO; (viii) AcCl, MeOH; (ix) $CH_3(CH_2)_{24}CO_2OC_6H_4$p-$NO_2$; (x) H_2, Pd/C

Scheme 3.17 ***Synthesis of β-C-GalCer.***

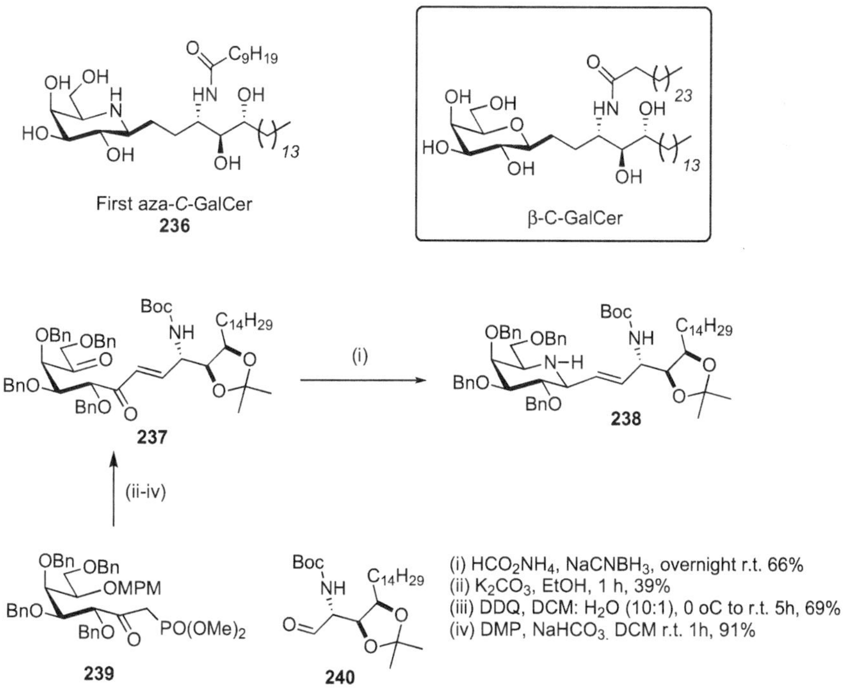

(i) HCO_2NH_4, $NaCNBH_3$, overnight r.t. 66%
(ii) K_2CO_3, EtOH, 1 h, 39%
(iii) DDQ, DCM: H_2O (10:1), 0 oC to r.t. 5h, 69%
(iv) DMP, $NaHCO_3$. DCM r.t. 1h, 91%

Scheme 3.18 ***Synthesis of aza analogs of β-C-GalCer.***

11 C-Glycosides: annulating glycone and aglycone part, an illustration

Synthesis of *C*-glycosides, wherein both sugar ring and aryl and hetero-aryl residue (representing aglycone part) are formed as a result of ring annulation is illustrated by the synthesis of 2-deoxy-*C*-glycoside **246** and **251** (Scheme 3.19).[154] An ene reaction between **241** and ethyl glyoxylate **242** furnished the homoallylic alcohol **243** with *Z*-geometry in low yield (37%). Intramolecular Sakurai cyclization followed by aminolysis of C1-ester group enabled construction of tetrahydropyran ring of the sugar residue **244**. Further functional group interconversions allowed construction of the thiazole and synthesis of *C*-glycoside **246**. A similar reaction protocol between aldehyde **247** and allylsilane **241** enabled construction of the tetrahydropyran ring **249** with alkynyl residue at C1 carbon. [2 + 2 + 2] cyclotrimerization of alkyne **249** with propargyl ether catalyzed

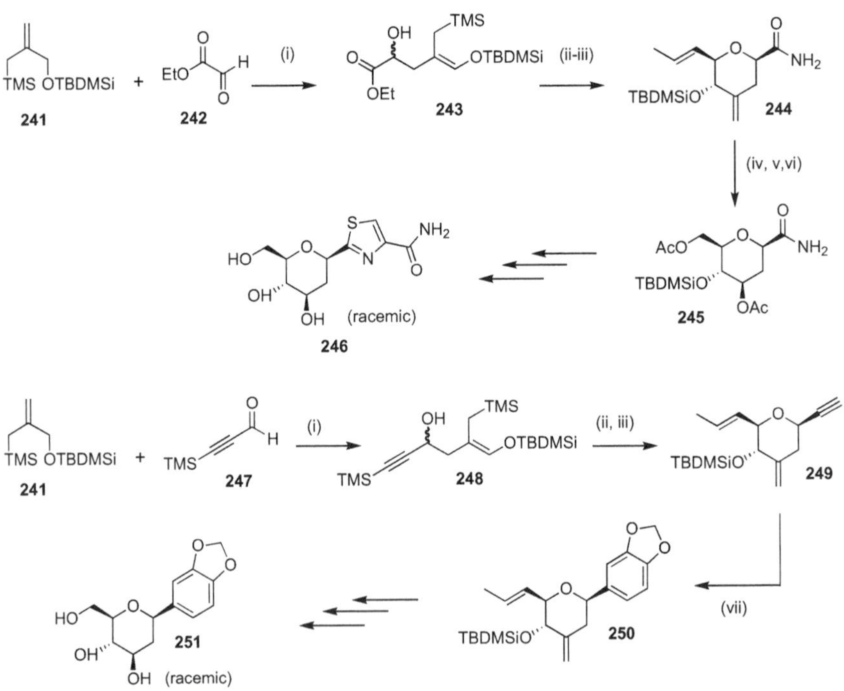

(i) Et_2AlCl, CH_2Cl_2, 37%; (ii) crotonaldehyde, $BF_3 \cdot Et_2O$, CH_2Cl_2; (iii) NH_3 (g), MeOH; (iv) O_3, $MeOH/CH_2Cl_2$, then Me_2S; (v) $NaBH_4$, MeOH; (vi) Ac_2O, pyridine; (vii) propargyl ether, $RhCl(PPh_3)$

Scheme 3.19 ***Synthesis of C-glycoside 246 and 251.***

by Wilkinson's catalyst afforded the *C*-aryl-advanced precursor **250**. Further functionalization on the tetrahydropyran ring afforded the targeted 2-deoxy-*C*-aryl glycoside **251** as racemic material.

12 C-1 functionalized building blocks for synthesis of C-Glycosides

The commonly found C1-functionalized building blocks (BB) in literature are presented in Fig. 3.32. Use of these building blocks has been made more often for access to β-configured *C*-glucosides. Knoevenagel condensation reaction between unprotected *D*-glucose and pentane-2,4-dione in aqueous medium and under basic conditions has paved the way for convenient access to β-C-glucosyl propanone **252** as a valuable BB.[155a] Simplicity of this condensation and convenient accessibility of **252**, containing functionalized unit at C1 has attracted the attention of many, for further chemistry and applications (Fig. 3.32). Peracetylation or perbenzylation affords building blocks **253**,[155b] and **254**[155c] respectively. Prior to use of Knoevenagel condensation for introduction of the acetonyl unit at C1 position, use of allyl *C*-glucosides or Wittig-type reaction on reducing sugars has been made for the same objective.[156]

The BB **255**[157] and **256**[158] are easily accessible in high yields through addition of allylmagnesium chloride and TMS protected lithium acetylide,

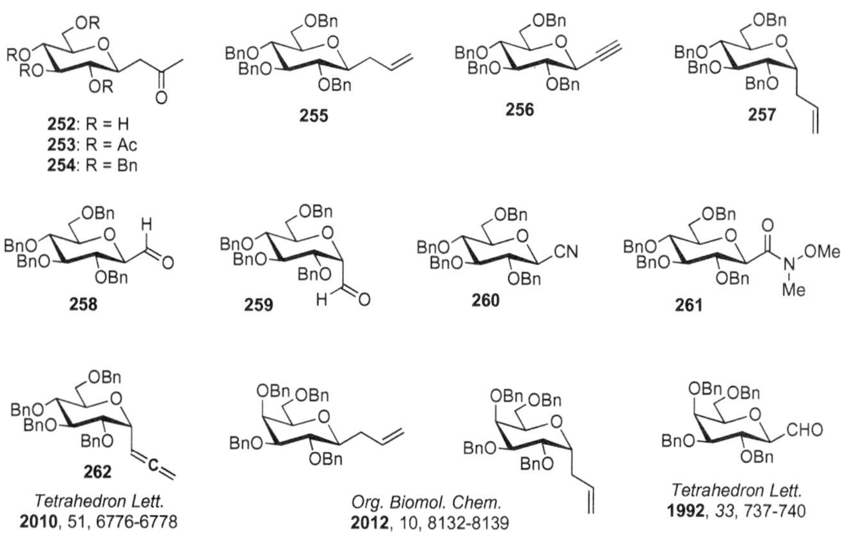

Figure 3.32 ***Structures of C1-Functionalized Building Blocks.***

respectively on tetra-*O*-benzyl-ether protected *D*-glucono lactone followed by stereoselective deoxygenation with Et_3SiH. Stereoselective and efficient synthesis of the α-C-allyl BB **257** was first reported by Kishi et al. which involves reaction of 2,3,4,6-tetrabenzyl-α-(p-nitro-benzoyl)glucopyranose with allylsilane in presence of $BF_3{\cdot}OEt_2$ in acetonitrile.[157b] The method replacing $BF_3{\cdot}OEt_2$ with trimethylsilyl triflate (TMSOTf), has provided a equally good route for **257** with excellent stereoselectivity.[159a] Recently, Nicolas Petry et al. reported that allylation of glucopyranosides, either using $BF_3{\cdot}OEt_2$ or TMSOTf, although predominantly yields α-anomer, formation of the β-anomer is also observed and therefore needs separation and careful characterization.[159b] An efficient synthetic scheme for the synthesis of α-*C*-allyl BB **257** is reported by Schmidtmann et al. in 2008[159c] involves reaction of protected *D*-glucosylsulfoxides as donor, with allyl trimethylsilane in presence of triflic anhydride.

The BB **258** and **259** with a formyl unit at the anomeric position are valuable starting materials for synthesis of *C*-glucosides. Dondoni has synthesized these BB using benzothiazole as formyl group equivalent.[160a-c] Addition of 2-lithiobenzothiazole onto tetra-*O*-benzyl-ether protected *D*-glucono lactone followed by deoxygenation with Et_3SiH/TMSOTf affords a mixture of benzothiazolyl α- and β-D-C-glucosides (4:6), which on separation and unmasking afforded both **258** and **259**.[160a] For the exclusive formation of **258** alone, Labeguere's approach is noteworthy, wherein bis(methylthio)methane has been used as an equivalent of formyl group. The addition of bis(methylthio)methyl carbanion onto tetra-*O*-benzyl-ether protected *D*-glucono lactone, followed by reductive deoxygenation at anomeric position afforded only β-isomer.[160d] Exclusive obtainment of **259** has been realized by ozonolysis of α-linked *C*-glucosyl allene **262**.[160e] The nitrile based BB **260** is easily accessible through nucleophilic substitution at anomeric position of *D*-glucosyl acetate with trimethylsilyl cyanide (TMSCN).[161] In the recent times, simple oxidation of BB **258** and amide formation with *N*-methoxy-*N*-methyl amine has afforded Weinreb amide (WA) based building block **261** from our group.[162] Although BB **258** and **260** allow C—C bond formation through Grignard addition, formation of glycal as a side product is the major limitation. However, the reactivity of WA functionality has been observed to be intermediary, while comparing with more electrophilc aldehyde and less electrophilic nitrile group. This has enabled exclusive addition of organometallic onto carbonyl group, with no glycal formation.

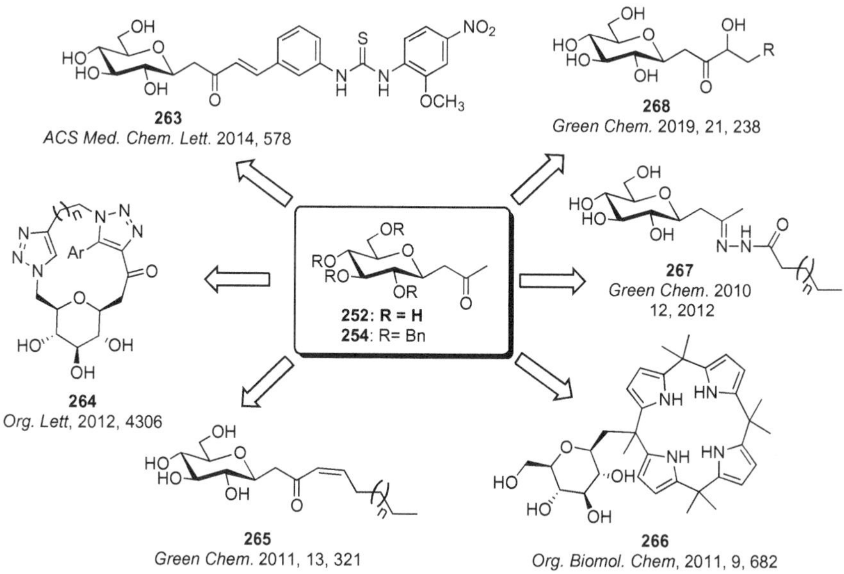

Figure 3.33 ***Representative uses of BB*** **252.**

12.1 Applications of C-1 functionalized building blocks

The building block, β-D-Glucosyl-propan-2-one **252**, has been used for synthesis of biologically active compounds.[163]Few illustrative examples are presented in Fig. 3.33.Tripathi R.P. et al. in 2014, has prepared a new series of *C*-linked phenyl butenonylglycoside bearing thioureidyl group, represented by compound **263** using building block **252** and evaluated their *in vitro* antimalarial activities against *plasmodium falciparum* 3D7 (CQ sensitive) and K1 (CQ resistant) strains. Antimalarial activities against both 3D7and K1 were observed and IC_{50} values were found in micromolar range.[163a] Same group in 2012, using BB **252** synthesized a homologous series of 14-, 15-, and 16-membered drug-like, macrocyclic glycoconjugates **264** involving TBAHS (tetrabutylammonium hydrogen sulfate) promoted azide-propenone intramolecular cycloaddition.[163b] Pavel D. et al. (2011) have prepared first calix[4]pyrrole **266** containing unprotected carbohydrate moiety directly linked to *meso*-position of oligopyrrole through stable "*C*-glycosidic" bond.The calix[4]pyrrole represented by **266** is being aimed as synthon for build-up of supramolecular devices, sensors and ionophores.[163c]

Other BB very often used are the allyl *C*-glycopyranosides with glycose unit being *D*-glucose or *D*-galactose. Few examples illustrating their applications[164] using D-*gluco*-configured BB **255** and **257** as BB are graphically

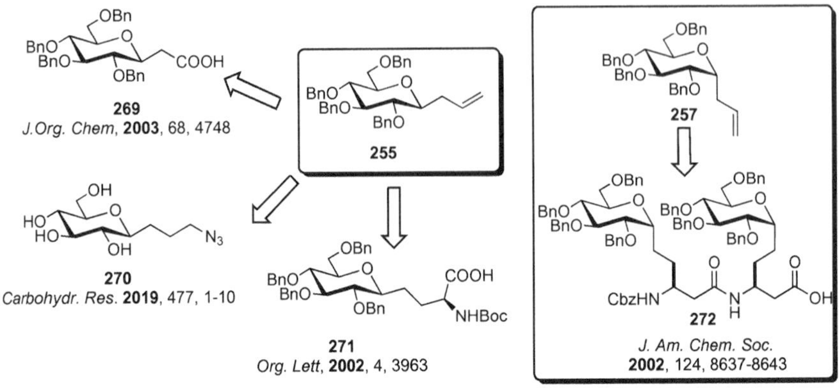

Figure 3.34 ***Representative uses of BB 255 and 257.***

presented in Fig. 3.34. The carboxylic acid **269** prepared from allyl *C*-glucopyranoside **255**, paved the way for synthesis of *C*-disaccharides which were screened for their efficacy as inhibitors of β-glycosidase and against several solid-tumor cell lines for *in vitro* differential cytotoxicity.[164a] Few value-added *C*-glycosyl BB represented by **270** have been developed with the use of BB **255** for bioactive glycoconjugates containing *C*-glycosides.[164b] Given the importance of serine-*O*-glycoconjugates, the *C*-linked glycopyranosyl serines in general and **271** in particular have been synthesized using BB **255**.[164c] The α-allyl-glucopyranoside BB **257** has served as a valuable starting material for the total synthesis of sugar-peptide hybrids represented by **272**.[164d]

The alkynyl *C*-glucoside BB **256** has been used in many places for synthesis of complex carbohydrate molecules (Fig. 3.35).[165] Vinyl-*C*-glucoside represented by compound **273**, by itself is a valuable BB for entry into more functionalized *C*-glucosides and has been synthesized on gram-scale quantities using alkynyl *C*-glucoside BB **256**, through controlled reduction.[165a] Triazole-linked *C*-heteroaryl glucoside **274**[165b] and glycopeptides **275**[165c] have been prepared from alkyne **256** using click reaction chemistry. Similarly, Toshio Nishikawa et al. (2006) developed an efficient synthetic route for the synthesis of glucose analog of *C*-Mannosyltryptophan (α-*C*-Man-Trp) **276**, for novel post-translational modification of tryptophan found in biologically important glycoproteins.[158b] Addition of lithiated alkyne **256** on to 2,3,5-tri-*O*-benzyl-arabinofuranose followed by gold catalyzed cycloisomerization afforded the synthesis of *C*-disaccharide **277**.

Among the C1-functonalized building blocks, 1-formyl *C*-glucopyranoside **258** appears to be the most used in synthetic endeavours.[166] Few

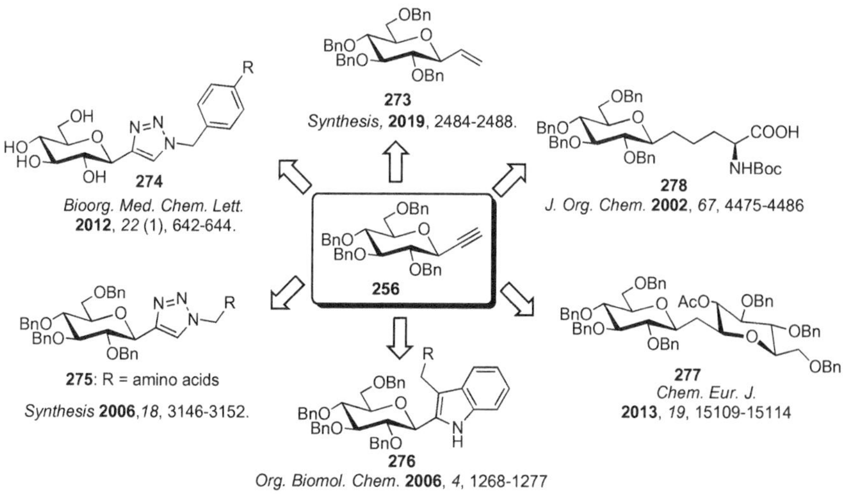

Figure 3.35 ***Representative uses of BB*** **256.**

illustrative uses of this building block are presented in Fig. 3.36. Our research group in 2017, reported the first synthesis of 3-glycosylated isocoumarins in general and 3-glucosylated isocoumarins **279** in particular.[166a] In 2018, using the same BB, we have also synthesized glucopyranosyl pyridines **280** as BB for synthesis of pyridyl-analogs of Dapagliflozin, as sodium glucose co-transport SGLT 2 inhibitors.[97] The same building block enabled access

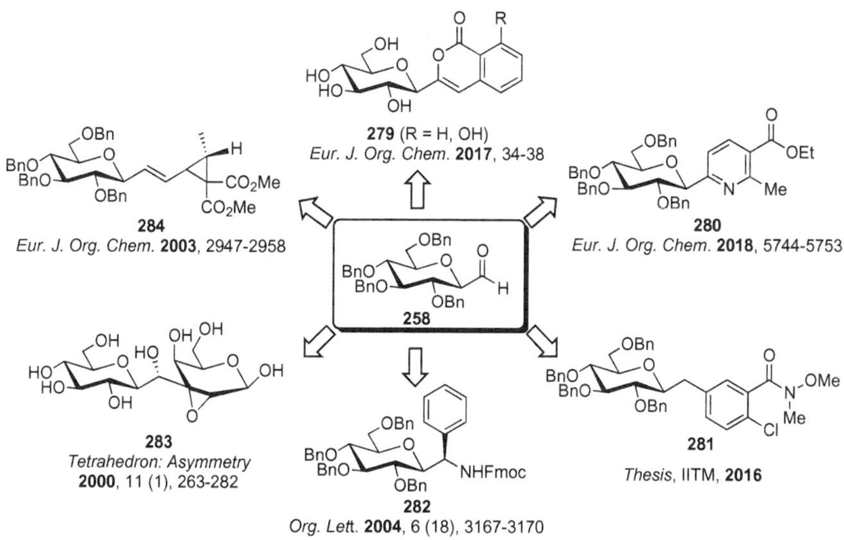

Figure 3.36 ***Representative uses of BB*** **258.**

to another Weinreb-amide BB **281**, for synthesis of benzyl-C-analogs of Dapagliflozin.[93] Overhand. M. et al. (2004), synthesized a novel and highly functionalized cyclic tetramer via compound **282**.[166b] Diastereoselective addition of aryl magnesium bromide onto sulfinimine derived from BB **258** enabled incorporation of aryl and amino-functionality for the obtainment of **282**. Genet J.P. et al. in 2003, using BB **258**, synthesized the D-*gluco*-analog, **284**, representing the "western" part of natural product ambruticin, known for antifungal antibiotic activity.[166c] Stereoselective addition of carbanion derived from 3-(tert-butyldimethylsilyloxy)-but-1-yne added onto BB **258**, constitutes the first step in their efforts towards **284**.

The easy availability of glucopyranosyl cyanides **260**[161] enabled its utilization in many synthetic schemes (Fig. 3.37).[167] Synthesis of aryl-2-(β-D-glucopyranosyl)-imidazoles **285**, currently the most efficient glucose derived inhibitors of glycogen phosphorylase enzymes were synthesized using this BB as the starting material. The *D*-glucosylated imidazole derivatives **285b** and **285c** proved less efficient inhibitors (Ki values of 1141 and 411 nM, respectively) than their unsubstituted counterpart **285a** (Ki ¼ 280 nM).[167a] Guillarme S. et al. have recently reported the reactivity of glycopyranosylcyanide towards organomagnesium and organolithium reagents leading to the formation of acyl-*C*-glycosides **287**.[167b]

Formation of glycal as a side product has been obviated by using glycopyranosyl cyanide, such as **286** bearing a unprotected hydroxyl group at C-2 position.[167c] Recently (2019), to overcome the same difficulty of glycal formation, we have used the WA-based BB **261** (Fig. 3.37). The acyl-*C*-β-glucosides **288** and benzyl-*C*-β-glucosides **289** were obtained in good yields. Use of Weinreb-amide (WA) functionality was crucial for this accomplishment as it is less electrophilic, compared to the C-1 formyl group of BB **258**. In the intial studies, benzyl-*C*-glucosides **289** were found to be

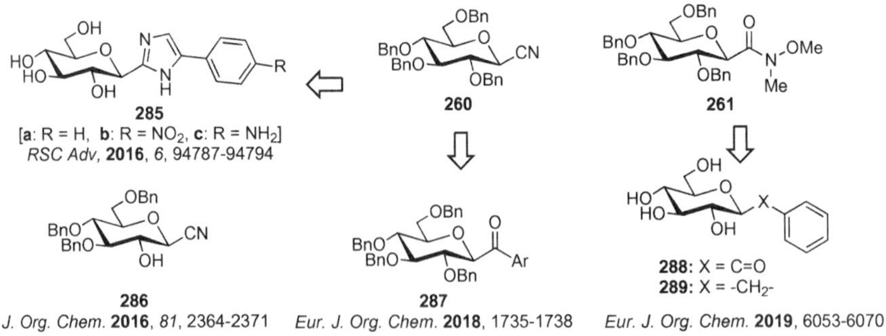

Figure 3.37 ***Representative uses of BB* 260.**

far superior in promoting glucose uptake in C2C12 mouse skeletal muscle cell line, at 10 μM, when compared to the corresponding acyl-*C*-glucosides.[162] The efficacy of benzyl-*C*-glucosides needs to be further ascertained using *in vivo* animal models, for their possible development as new antidiabetic lead substances.

13 Conclusions and future perspectives

Search for bioactive *O*- or *C*-glycosides from natural products towards various therapeutic ends will remain unabated. In the context that botanical dietary supplements, traditionally used for health benefits and disease prevention, are emerging with a promise,[168] a closer and careful look into their *O*- or *C*-glycosidic components will be important and crucial for the envisaged therapeutic use. Since dietary supplements are regulated differently, compared to drugs, their application may be quicker after proper validation, the information obtained during the process, at the molecular level will continue to inspire synthetic chemists with new leads. Mangiferin, a natural xanthone based *C*-glycoside, endowed with variety of pharmacological activities, provides a good illustration of lead availed by synthetic chemists for an important therapeutic application, as heparinoid mimetic.[169] Simple one step polysulfation of mangiferin at phenolic and glycone hydroxyl groups has yielded a well-defined heptasulfated product, possessing both anticoagulant and antiplatelet effects and hence a promise toward possible alternative to heparin. A revisit to all O/C-glycosides present in botanical supplements, traditionally used as therapeutics, is likely to provide valuable direction to drug discovery processes. Limited solubility of low molecular weight and pharmacologically active aglycones derived from these dietary supplements will merit, glycosylation for improving bioavailability. This will offer great opportunities for enzymatic glycosylation routes[170] for large scale synthesis of therapeutically relevant *O*/*C*-glycosides. Synthetic schemes based on use of unprotected sugars[171] for access to O/C/N-glycoconjugates will hold equally good promise, for synthesis of analogs of important lead compounds.

References

1. (a) Lam, K. S. *Trends Microbiol.* **2007**, *15*, 279–289. (b) Harvey, A.L. *Drug Discovery Today* **2008**, *13*, 894–901.
2. (a) Li, F.; Wang, Y.; Li, D. Chen, Y.; Dou, Q. P., *Expert Opin. Drug Discov.*, **2019**, *14*, 5, 417–420. (b) Newman, D.; Cragg, G., *J. Nat. Prod.* **2016**, *79*, 629. (c) Harvey, A. L.; Edrada-Ebel, R.; Quinn, R. *J. Nat. Rev. Drug Discovery* **2015**, *14* (2), 111–129. (d) Szychowski, J.; Truchon, J.-F.; Bennani, Y. L. *J. Med. Chem.* **2014**, *57*, 9292–9308. (e) Rodrigues, T.; Reker, D.; Schneider, P.; Schneider, G. *Nat. Chem.* **2016**, *8* (6), 531−541.

3. (a) Chen, W. J. *Biomed. J.* **2016**, *39* (2), 93–97. (b) Shen, B. A. **2015**, *163* (6), 1297–1300. (c) Molyneux, D. H.; Ward, S. A. *Trends. Parasitol.* **2015**, *31* (12), 605-607.(d) Efferth, T.; Zacchino, S.; Georgiev, M. I.; Liu, L.; Wagner, H. *Phytomedicine* **2015**, *22* (13), A1–A3.
4. McGuire, J. M.; Bunch, R. L.; Anderson, R. C.; Boaz, H. E.; Flynn, E. H.; Powell, H. M.; Smith, J. W. Ilotycin. *Antibiot. Chemother.* **1952**, *2*, 281–283.
5. Jelic, D.; Antolovic, R. *Antibiotics* **2016**, *5*, 29.
6. Rychnovsky, S. D.; Skalitzky, D. J.; Pathirana, C.; Jensen, P. R.; Fenical, W. *J. Am. Chem. Soc.* **1992**, *114*, 671–677.
7. Gustafson, K.; Roman, M.; Fenical, W. *J. Am. Chem. Soc.* **1989**, *111*, 7519–7524.
8. Tareq, F. S.; Kim, J. H.; Lee, H.-S.; Lee, J.-S.; Lee, Y.-J.; Shin, H. J. *J. Agric. Food Chem.* **2013**, *61*, 3428–3434.
9. (a) Botta, B.; Monache, G. D.; Misiti, D.; Vitali, A.; Zappia, G. *Curr. Med. Chem.* **2001**, *8*, 1363. (b) Xu, H.; Lv, M.; Tian, X. *Curr. Med. Chem.* **2009**, *16*, 327–349.
10. Ren, Y.; Lantvit, D. D.; Deng, Y.; Kanagasabai, R.; Gallucci, J. C.; Ninh, T. N.; Chai, H.-B.; Soejarto, D. D.; Fuchs, J. R.; Yalowich, J. C.; Yu, J.; Swanson, S. M.; Kinghorn, A. D. *J. Nat. Prod.* **2014**, 77, 1494–1504.
11. Shi, D.-K.; Zhang, W.; Ding, N.; Li, M.; Li, Y.-X. *Eur. J. Med. Chem.* **2012**, *47*, 424–431.
12. Deng, Y.; Chu, J.; Ren, Y.; Fan, Z.; Ji, X.; Mundy-Bosse, B.; Yuan, S.; Hughes, T.; Zhang, J.; Cheema, B.; Camardo, A. T.; Xia, Y.; Wu, L.-C.; Wang, L.-S.; He, X.; Kinghorn, A. D.; Li, X.; Caligiuri, M. A.; Yu, J. *J. Immunol.* **2014**, *193*, 2994–3002.
13. Young, A. N.; Herrera, D.; Huntsman, A. C.; Korkmaz, M. A.; Lantvit, D. D.; Mazumder, S.; Kolli, S.; Coss, C. C.; King, S.; Wang, H.; Swanson, S. M.; Kinghorn, A. D.; Zhang, X.; Phelps, M. A.; Aldrich, L. N.; Fuchs, J. R.; Burdette, J. E. *Mol. Cancer Ther.* **2018**, *17*, 2123–2135.
14. Ren, Y.; Yuan, C.; Deng, Y.; Kanagasabai, R.; Ninh, T. N.; Tu, V. T.; Chai, H.-B.; Soejarto, D. D.; Fuchs, J. R.; Yalowich, J. C.; Yu, J.; Kinghorn, A. D. *Phytochemistry* **2015**, *111*, 132–140.
15. Mehta, R. S.; Randolph, B.; Daher, M.; Rezvani, K. *Int. J. Hematol.* **2018**, *107*, 262–270.
16. Thornthwaite, J. T.; Shah, H.; Shah, P.; Respess, H. *J. Immune Based Ther., Vaccines Antimicrob.* **2012**, *1*, 21–52.
17. Moreau, R. A.; Nystrom, L.; Whitaker, B. D.; Winklier-Moser, J. K.; Baer, D. J.; Gebauer, S. K.; Hicks, K. B. *J. Lipid Res.* **2018**, *70*, 35–61.
18. (a) Singh, B.; Singh, J.P.; Singh, N.; Kaur, A. Food Chem. 2017, 233, 540–549. (b) Coleman, J. F.; *ACS Chem. Biol.* **2010**, *5*, 3, 321–332.
19. Friedman, M. *J. Agric. Food Chem.* **2015**, *63*, 3323–3337.
20. For comprehensive review on saponins as *O*-glycosides and sapogenins as aglycones see: *Chemistry & Pharmacology of Natural Products: Saponins.* K. Hostettmann and A. Marston. Cambridge University Press, New York, NY. **1995**, 548 pp.
21. Gheorghiade, M.; van Veldhuisen, D. J.; Colucci, W. S. *Circulation* **2006**, *113*, 2556–2564.
22. Elbaz, H. A.; Stueckle, T. A.; Tse, W.; Rojanasakul, Y.; Dinu, C. Z. *Exp. Hematol. Oncol.* **2012**, *1* (4), 1–10.
23. Bravo, L. *Nutr. Rev.* **1998**, *56*, 317–333.
24. Manach, C.; Scalbert, A.; Morand, C.; Rémésy, C.; Jiménez, L. *Am. J. Clin. Nutr.* **2004**, *79*, 727–747.
25. Shahidi, F.; Ambigaipalan, P. *J. Funct. Foods* **2015**, *18*, 820–897.
26. Santhakumar, A. B.; Battino, M.; Alvarez-Suarez, J. M. *Food Chem. Toxicol.* **2018**, *113*, 49–65.
27. Singh, M.; Kaur, M.; Silakari, O. *Eur. J. Med. Chem.* **2014**, *84*, 206–239.
28. Yang, B.; Liu, H.; Yang, J.; Gupta, V. K.; Jiang, Y. *Trends Food Sci. Technol.* **2018**, *79*, 116–124.
29. Kopp, P. *Eur. J. Endocrinol.* **1998**, *138*, 619–620.
30. Goswami, S. K.; Das, D. K. *Cancer Lett.* **2009**, *284*, 1–6.

31. El-Seedi, H. R. *J. Nat. Prod.* **2007**, *70*, 118–1120.
32. Winnig, M.; Bufe, B.; Kratochwil, N. A.; Slack, J. P.; Meyerhof, W. *BMC Struct. Biol.* **2007**, 7, 1–12.
33. Nakatani, Y.; Kobe, A.; Kuriya, M.; Hiroki, Y.; Yahagi, T.; Sakakibara, I.; Matsuzaki, K.; Amano, T. *Eur. J. Pharmacol.* **2017**, *815*, 381–390.
34. Nabavi, S. F.; Braidy, N.; Gortzi, O.; Sobarzo-Sanchez, E.; Daglia, M.; Skalicka-Wozniak, K.; Nabavi, S. M. *Brain Res. Bull.* **2015**, *119*, 1–11.
35. Bakoyiannis, I.; Daskalopoulou, F.; Pergialiotis, V.; Perrea, D. *Biomed. Pharmacother.* **2019**, *109*, 1488–1497.
36. Lin, Y.; Liu, P.-G.; Liang, W.-Q.; Hu, Y.-J.; Xu, P.; Zhou, J.; Pu, J.-B. *Phytomedicine* **2018**, *41*, 54–61.
37. Kashyap, D.; Sharma, A.; Tuli, H. S.; Sak, K.; Garg, V. K.; Buttar, H. S.; Setzer, W. N.; Sethi, G. *J. Funct. Foods* **2018**, *48*, 457–471.
38. Mudnic, J.; Mudnic, I. V.; Gajski, G.; Popic, J.; Garaj-Vrhovac, V. *Cancer Lett.* **2018**, *413*, 11–22.
39. Nabavi, S. F.; Khan, H.; D'onofrio, G.; Samec, D.; Shirooie, S.; Dehpour, A. R.; Arguelles, S.; Habtemariam, S.; Sobarzo-Sanchez, E. *Pharmacol. Res.* **2018**, *128*, 359–365.
40. He, Miao.; Min, J.-W.; Kong, W.-L.; He, X.-H.; Li, J.-X.; Peng, B.-W. *Fitoterapia* **2016**, *115*, 74–85.
41. Ganesan, K.; Xu, B. *Food Science and Human Wellness* **2018**, 7, 11–33.
42. Li, K.; He, Z.; Wang, X.; Pineda, M.; Chen, R.; Liu, H.; Ma, K.; Shen, H.; Wu, C.; Huang, N., et al. *Free Radic. Biol. Med.* **2018**, *124*, 163–175.
43. Zhang, Y.; Jiao, J.; Liu, C.; Wu, X.; Zhang, Y. *Food Chem.* **2008**, *107*, 1326–1336.
44. Chen, Y. G.; Li, P.; Li, P., et al. *Molecules* **2013**, *18*, 4221–4232.
45. El-Sayed, E. M.; Mansour, A. M.; Abdul-Hameed, M. S. *J. Biochem. Mol. Toxicol.* **2016**, *30*, 37–44.
46. Sun, Z.; Yan, B.; Yu, W. Y.; Yao, X.; Ma, X.; Sheng, G.; Ma, Q. *Exp. Ther. Med.* **2016**, *12*, 1879–1884.
47. Rowley, D. C.; Hansen, M. S. T.; Rhodes, D.; Sotriffer, C. A.; Ni, H.; McCammon, J. A.; Bushman, F. D.; Fenical, W. *Bioorg. Med. Chem.* **2002**, *10*, 3619–3625.
48. Dewick, P. M., Harborne, J. B. Ed.; Chapman and Hall: *New York*, **1988**, Chapter 5, pp 187–192.
49. Dewick, P. M. *Medicinal Natural Products: A Biosynthetic Approach.;* John Wiley & Sons: Chichester, U.K, 2009.
50. Hosny, M.; Rosazza, J. P. N. *J. Nat. Prod.* **1999**, *62*, 853–858.
51. Caderroth, C. R.; Nef, S. *Mol. Cell. Endocrinol.* **2009**, *304*, 30–42.
52. Beradini, N.; Fezer, R.; Conrad, J.; Beifuss, U.; Carle, R.; Schieber, A. *J. Agric. Food Chem.* **2005**, *53*, 1563–1570.
53. Masters, K.-S.; Brase, S. *Chem. Rev.* **2012**, *112*, 3717–3776.
54. Zhang, Y.; Qian, Q.; Ge, D.; Li, Y.; Wang, X.; Chen, Q.; Gao, X.; Wang, T. *J. Agric. Food Chem.* **2011**, *59*, 11526–11533.
55. (a) Bililign, T.; Griffith, B. R.; Thorson, J. S. *Nat. Prod. Rep.* **2005**, *22*, 742–760. (b) Franz, G.; Grun, M. *Planta Med.* **1983**, *47*, 131–140. (c) He, X.; Blount, J. W.; Ge, S.; Tang, Y.; Dixon, R. A. *Planta* **2011**, *233*, 843–855.
56. (a) Xiao, J.; Capanoglu, E.; Jassbi, A. R.; Miron, A. *Crit. Rev. Food Sci. Nutr.* **2016**, *56*, S29-S45. (b) Yang, B.; Liu, H.; Yang, J.; Gupta, V. K.; Jiang, Y. Trends. *Food Sci. Technol.* **2018**, *79*, 116–124. (c) Zeng, P.; Zhang, Y.; Pan, C.; Jia, Q.; Gua, F.; Li, Y.; Zhu, W.; Chen, K. *Acta Pharmaceutica Sinica B* **2013**, *3*(3), 154–162.
57. Veitch, N. C.; Grayer, R. J. *Nat. Prod. Rep.* **2011**, *28*, 1626–1695.
58. Xiao, J. B.; Muzashvili, T. S.; Georgiev, M. I. *Biotechnol. Adv.* **2014**, *32*, 1145–1156.
59. Eisner, T.; Nowicki, S.; Goetz, M.; Meinwald, J. *Science* **1980**, *208*, 1039–1042.
60. Hultin, P. G. *Curr. Top. Med. Chem.* **2005**, *5*, 1299–1331.

61. Kirschning, A.; Chen, G.-W.; Drager, G.; Schuberth, I.; Tietze, L. F. *Bioorg. Med. Chem.* **2000**, *8*, 2347–2354.
62. Kanda, N. *J. Antibiot.* **1971**, *24*, 599.
63. Fei, Z.; McDonald, F. E. *Org. Lett.* **2007**, *9*, 3547–3550.
64. Kitamura, K.; Ando, Y.; Matsumoto, T.; Suzuki, K. *Chem. Rev.* **2018**, *118* (4), 1495–1598.
65. Bennett, C. S.; Galan, M. C. *Chem. Rev.* **2018**, *118* (17), 7931–7985.
66. Nielsen, M. M.; Pedersen, C. M. *Chem. Rev.* **2018**, *118* (17), 8285–8358.
67. Yang, Y.; Yu, B. *Chem. Rev.* **2017**, *117* (19), 12281–12356.
68. Bokor, E.; Goyard, D.; To, M. C. *Chem. Rev.* **2017**, *117*, 1687–1764.
69. Elshahawi, S. I.; Shaaban, K. A.; Kharel, M. K.; Thorson, J. S. *Chem. Soc. Rev.* **2015**, *44* (21), 7591–7697.
70. Lalitha, K.; Muthusamy, K.; Prasad, Y. S.; Vemula, P. K.; Nagarajan, S. *Carbohdr. Res.* **2015**, *402*, 158–171.
71. Lee, D.; He, M. *Curr. Top. Med. Chem.* **2005**, *5* (14), 1333–1350.
72. Bililign, T.; Griffith, B. R.; Thorson, J. S. *Nat. Prod. Res.* **2005**, (22), 742–760.
73. Du, Y.; Linhardt, R. J.; Vlahov, I. R. *Tetrahedron* **1998**, *54* (462), 9913–9959.
74. Carlos Jarmillo, S. K. *Synthesis* **1994**, 1–20.
75. Carlos Jarmillo, S. K. *Synthesis* **1993**, 1–20.
76. Postema, M. H. D. *Tetrahedron* **1992**, *48*, 8545–8599.
77. Hackell, U. L. I.; Daves, G. D. *Pro. Med. Chem.* **1985**, *22*, 1–65.
78. "Synthesis of C-Glycosides: Stable Mimics of O-GlycosidicLinkages": (a) C. R. Bertozzi, M. D. Bednarski in Modern Methods in Carbohydrate Synthesis, Harwood Academic Publishers, Reading, **1996**, pp. 316–351; (b) "C-Oligosaccharide Synthesis": T. Skrydstrup, B. Vauzailles, J.-M. Beau in Glycoscience: Chemistry and Chemical Biology, Vol. 3(Eds.: B. Fraser Reid, K. Tatsuta, J. Thiem), *Springer, Heidelberg,* **2001**,p. 2679.
79. (a) McKee, F. W, Hawkins, W. B. *Physiol. Rev.* **1945**, *25*, 255–280; (b) Stiles, P. G. Lusk, G. Amer. *J. Physiol.* **1903**, *10*, 67–79; (c) Chasis, H. Jolliffe, N. Smith, H. W. *J. Clin. Invest.* **1933**, *12*, 1083–90; (d) Ehrenkranz, J. R. L.; Lewis, N. G.; Kahn, C. R.; Roth, *J. Diabetes Metab. Res. Rev.* **2005**, *21* (1), 31-38.
80. Chao, E. C.; Henry, R. R. *Nat. Rev. Drug Discov.* **2010**, *53*, 551–559.
81. (a) Oku, A.; Ueta, K.; Arakawa, K.; Ishihara, T.; Nawano, M.; Kuronuma, Y.; Matsumoto, M.; Saito, A.; Tsujihara, K.; Anai, M.; et al. *Diabetes* **1999**, *48* (9), 1794–1800. (b) Tsujihara, K.; Hongu, M.; Saito, K.; Kawanishi, H.; Kuriyama, K.; Matsumoto, M.; Oku, A.; Ueta, K.; Tsuda, M.; Saito, A. *J. Med. Chem.* **1999**, *42* (26), 5311–5324. (c) Oku, A.; Ueta, K.; Nawano, M.; Arakawa, K.; Kano-Ishihara, T.; Matsumoto, M.; Saito, A.; Tsujihara, K.; Anai, M.; Asano, T. *Eur. J. Pharmacol.* **2000**, *391* (1–2), 183–192.
82. Katsuno, K.; Fujimori, Y.; Takemura, Y.; Hiratochi, M.; Itoh, F.; Komatsu, Y.; Fujikura, H.; Isaji, M. *J. Pharmacol. Exp. Ther.* **2007**, *320*, 323–330.
83. Katsuno, K.; Fujimori, Y.; Takemura, Y.; Hiratochi, M.; Itoh, F.; Komatsu, Y.; Fujikura, H.; Isaji, M. *J. Pharmacol. Exp. Ther.* **2007**, *320* (1), 323–330.
84. Meng, W.; Ellsworth, B.; Nirschl, A.; McCann, P.; Patel, M.; Girotra, R.; Wu, G.; Sher, P.; Morrison, E.; Biller, S., et al. *J. Med. Chem.* **2008**, *51* (5), 1145–1149.
85. Nomura, S.; Sakamaki, S.; Hongu, M.; Kawanishi, E.; Koga, Y.; Sakamoto, T.; Yamamoto, Y.; Ueta, K.; Kimata, H.; Nakayama, K., et al. *J. Med. Chem.* **2010**, *53* (17), 6355–6360.
86. Grempler, R.; Thomas, L.; Eckhardt, M.; Himmelsbach, F.; Sauer, A.; Sharp, D. E.; Bakker, R. A.; Mark, M.; Klein, T.; Eickelmann, P. Empagliflozin. *Diabetes Obes. Metab.* **2012**, *14* (1), 83–90.
87. Imamura, M.; Nakanishi, K.; Suzuki, T.; Ikegai, K.; Shiraki, R.; Ogiyama, T.; Murakami, T.; Kurosaki, E.; Noda, A.; Kobayashi, Y., et al. *Bioorg. Med. Chem.* **2012**, *20* (10), 3263–3279.

88. (a) Mascitti, V.; Maurer, T. S.; Robinson, R. P.; Bian, J.; Boustany-Kari, C. M.; Brandt, T.; Collman, B. M.; Kalgutkar, A. S.; Klenotic, M. K.; Leininger, M. T.; et al. *J. Med. Chem.* **2011**, *54* (8), 2952–2960. (b) Cinti, F.; Moffa, S.; Impronta, F.; Cefalo, C. M. A.; Sun, V. A.; Sorice, G.; Mezza, T.; Giaccari, A: Evidence to Date. *Drug. Des. Devel. Ther.* **2017**, *11*, 2905–2919.
89. (a) Jesus, A. R.; Vila-Viçosa, D.; Machuqueiro, M.; Marques, A. P.; Dore, T. M.; Rauter, A. P. *J. Med. Chem.* **2017**, *60* (2), 568–579. (b) Goodwin, N. C.; Ding, Z.-M.; Harrison, B. A.; Strobel, E. D.; Harris, A. L.; Smith, M.; Thompson, A. Y.; Xiong, W.; Mseeh, F.; Bruce, D. J.; et al. *J. Med. Chem.* **2017**, *60* (2), 710–721.
90. Ohtake, Y.; Sato, T.; Kobayashi, T.; Nishimoto, M.; Taka, N.; Takano, K.; Yamamoto, K.; Ohmori, M.; Yamaguchi, M.; Takami, K., et al. Discovery of Tofogliflozin. *J. Med. Chem.* **2012**, *55* (17), 7828–7840.
91. Kakinuma, H.; Oi, T.; Hashimoto-Tsuchiya, Y.; Arai, M.; Kawakita, Y.; Fukasawa, Y.; Iida, I.; Hagima, N.; Takeuchi, H.; Chino, Y., et al. *J. Med. Chem.* **2010**, *53* (8), 3247–3261.
92. Yamamoto, Y.; Kawanishi, E.; Koga, Y.; Sakamaki, S.; Sakamoto, T.; Ueta, K.; Matsushita, Y.; Kuriyama, C.; Tsuda-Tsukimoto, M. Nomura. *Bioorg. Med. Chem. Lett.* **2013**, *23* (20), 5641–5645.
93. (a) Ramesh, M., Thesis title: Development of Functionalized Building Blocks and Syntheses of Biologically Important Targets, Department of Chemistry, Indian Institute of Technology, Madras, **2016**. (b) Mukkamala R., Kumar R., Banerjee S. K., I. S. Aidhen, *Eur. J. Org. Chem.* **2020**, 1828–1839.
94. Knochel, P.; Dohle, W.; Gommermann, N.; Kneisel, F. F.; Kopp, F.; Korn, T.; Sapountzis, I.; Vu, V. A. *Angew. Chem. Int. Ed.* **2003**, *42* (36), 4302–4320.
95. (a) Balasubramaniam S., Aidhen I. S., *Synthesis* **2008**, 3707–3738; (b) Sivaraman B., Aidhen I. S., *Eur. J. Org. Chem.* **2010**, 4991–5003.
96. Kuo, G.-H.; Gaul, M. D.; Liang, Y.; Xu, J. Z.; Du, F.; Hornby, P.; Xu, G.; Qi, J.; Wallace, N.; Lee, S., et al. *Bioorg. Med. Chem. Lett.* **2018**, *28* (7), 1182–1187.
97. Reddy, M. R.; Aidhen, I. S. *Eur. J. Org. Chem.* **2018**, 5744–5753.
98. Xu, B.; Feng, Y.; Lv, B.; Xu, G.; Zhang, L.; Du, J.; Peng, K.; Xu, M.; Dong, J.; Zhang, W., et al. *Bioorg. Med. Chem.* **2010**, *18* (12), 4422–4432.
99. Zhou, H.; Danger, D. P.; Dock, S. T.; Hawley, L.; Roller, S. G.; Smith, C. D.; Handlon, A. L. *ACS. Med. Chem. Lett.* **2010**, *1* (1), 19–23.
100. Lee, J.; Kang, S. Y.; Song, K. S.; Lee, J.; Lee, S. H. *Bioorg. Med. Chem.* **2010**, *18* (16), 6069–6079.
101. Sato, M.; Kakinuma, H.; Asanuma, H. PCT Int Appl. WO2004014931, 2004. *Chem. Abstr.* **2004**, *140*, 199631.
102. Goodwin, N. C.; Mabon, R.; Harrison, B. A.; Shadoan, M. K.; Almstead, Z. Y.; Xie, Y.; Healy, J.; Buhring, L. M.; DaCosta, C. M.; Bardenhagen, J., et al. *J. Med. Chem.* **2009**, *52* (20), 6201–6204.
103. Mascitti, V.; Collman, B. M. WO10023594, 2010.
104. (a) Brewster, K.; Harrison, J. M.; Inch, T. D.; Williams, N. *J. Chem. Soc. Perkin. Trans 1.* **1987**, 21–26. (b) Preville, V. M.; Pfizer, C.; *Org. Lett.* **2010**, No. *12*, 2940–2943.
105. Mascitti, V.; Robinson, R. P.; Préville, C.; Thuma, B. A.; Carr, C. L.; Reese, M. R.; Maguire, R. J.; Leininger, M. T.; Lowe, A.; Steppan, C. M. *Tetrahedron Lett.* **2010**, *51* (14), 1880–1883.
106. Robinson, R. P.; Mascitti, V.; Boustany-Kari, C. M.; Carr, C. L.; Foley, P. M.; Kimoto, E.; Leininger, M. T.; Lowe, A.; Klenotic, M. K.; MacDonald, J. I., et al. *Bioorg. Med. Chem. Lett.* **2010**, *20* (5), 1569–1572.
107. Chen, Z. H.; Wang, R. W.; Qing, F. L. *Tetrahedron Lett.* **2012**, *53* (17), 2171–2176.
108. Goldman, P. *Science* **1969**, *164*, 1123–1130.
109. (a) Nakatsuka, T.; Tomimori, Y.; Fukuda, Y.; Nukaya, H. *Bioorg. Med. Chem. Lett.* **2004**, *14* (12), 3201–3203. (b) Furuta, T.; Kimura, T.; Kondo, S.; Mihara, H.; Wakimoto,

T.; Nukaya, H.; Tsuji, K.; Tanaka, K. *Tetrahedron* **2004**, *60*, 9375–9379. (c) Furuta, T.; Nakayama, M.; Suzuki, H.; Tajimi, H.; Inai, M.; Nukaya, H.; Wakimoto, T.; Kan, T. *Org. Lett.* **2009**, *11* (11), 2233–2236.

110. Rauter, A. P.; Martins, A.; Borges, C.; Ferreira, J.; Justino, J.; Bronze, M.-R.; Coelho, A.V.; Choi, Y. H.; Verpoorte, R. *J. Chromatogr. A* **2005**, *1089* (1–2), 59–64.
111. Jesus, A. R.; Dias, C.; Matos, A. M.; de Almeida, R. F. M.; Viana, A. S.; Marcelo, F.; Ribeiro, R.T.; Macedo, M. P.; Airoldi, C.; Nicotra, F., et al. *J. Med. Chem.* **2014**, *57* (22), 9463–9472.
112. Eade, R. A.; Pham, H.-P. *Aust. J. Chem.* **1979**, *32*, 2483–2493.
113. Frick, W.; Schmidt, R. R. *Liebigs Ann. Chem.* **1989**, 565–570.
114. Mahling, J.-A.; Jung, K.-H.; Schmidt, R. R. *Liebigs Ann.* **1995**, 461–466.
115. Kumazawa, T.; Minatogawa, T.; Matsuba, S.; Sato, S. Onodera. *J. Carbohydr. Res.* **2000**, *329*, 507–513.
116. Kumazawa, T.; Kimura, T.; Matsuba, S.; Sato, S. Onodera. *J. Carbohydr. Res.* **2001**, *334*, 183–193.
117. Bhardwaj, M.; Cho, H. J.; Paul, S.; Jakhar, R.; Khan, I.; Lee, S. J.; Kim, B.Y.; Krishnan, M.; Khaket, T. P.; Lee, H. G.; Kang, S. C. *Oncotarget* **2018**, *9*, 3278–3291.
118. Bai, Y.; Chang, J.; Xu, Y.; Cheng, D.; Liu, H.; Zhao, Y.; Yu, Z. *J. Agric. Food Chem.* **2016**, *64*, 4648–4655.
119. (a) Choudhury, H.; Pandey, M.; Hua, C. K.; Mun, C. S.; Jing, J. K.; Kong, L.; Ern, L.Y.; Ashraf, N. A.; Kit, S.W.; Yee, T. S.; Pichika, M. R.; Gorain, B.; Kesharwani, P. *J. Tradit. Complement. Med.* **2018**, *8*, 361–376. See also references 37, 40, 41 and 56a above.
120. Borghi, S. M.; Carvalho, T. T.; Staurengo-Ferrari, L.; Hohmann, M. S. N.; Pinge-Filho, P.; Casagrande, R.; Verri, W. A., Jr. *J. Nat. Prod.* **2013**, *76*, 1141–1149.
121. Xu, F.; Wang, C.; Yang, L.; Luo, H.; Fan, W.; Zi, C.; Dong, F.; Hu, J.; Zhou, J. *Food Chem.* **2013**, *136*, 94–99.
122. Indian Council of Medical Research (ICMR), *Indian, J. Med. Res.* **1998**, *108*, 24.
123. Rawat, P.; Kumar, M.; Rahuja, N.; Srivastava, D. S. L.; Srivastava, A. K.; Maurya, R. *Bioorg. Med. Chem. Lett.* **2011**, *21*, 228–233.
124. Koeppen, B. H.; Roux, D. G. *Biochem. J.* **1966**, *99*, 604–609.
125. Yepremyan, A.; Salehani, B.; Minehan, T. G. *Org. Lett.* **2010**, *12*, 1580–1583.
126. Han, Z.; Achilonu, M. C.; Kendrekar, P. S.; Joubert, E.; Ferreira, D.; Bonnet, S. L.; van der Westhuizen, J. H. *J. Nat. Prod.* **2014**, *77*, 583–588.
127. Aida, K.; Tawata, M.; Shindo, H.; Onaya, T.; Sasaki, H.; Yamaguchi, T.; Chin, M.; Mitsuhasi, H. *Planta Medica* **1990**, *56* (3), 254–258.
128. Logendra, S.; Ribnicky, D. M.; Yang, H.; Poulev, A.; Ma, J.; Kennelly, E. J.; Raskin, I. *Phytochemistry* **2006**, *67* (14), 1539–1546.
129. Reddy, M. R.; Aidhen, S.; Reddy, U. A. *Eur. J. Org. Chem.* **2019**, 3937–3948.
130. Wang, Y.; Dan, Y.; Yang, D.; Hu, Y.; Zhang, C.; Zhu, H.; Cui, Z.; Li, M.; Liu, Y. *J. Ethnopharmacol.* **2014**, *153*, 42–60.
131. Zhou, C.; Zhou, J.; Han, N.; Liu, Z.; Xiao, B.; Yin, J. *Int. Immunopharmacol.* **2015**, *25*, 218–228.
132. Qin, L.; Han, T.; Zhang, Q.; Cao, D.; Nian, H.; Rahman, K.; Zheng, H. *J. Ethnopharmacol.* **2008**, *118*, 271–279.
133. Lim, S.-M.; Kang, G.-D.; Jeong, J.-J.; Choi, H. S.; Kim, D.-H. *Phytomedicine* **2016**, *23*, 131–140.
134. (a) He, B. F.; Wu, X. M.; Chu, J. I. *Chin. Pat.* CN102863484A, **2011**. (b) Miura, T.; Ichiki, H.; Iwamoto, N.; Kato, M.; Kubo, M.; Sasaki, H.; Okada, M.; Ishida, T.; Seino, Y.; Tanigawa, K. *Biol. Pharm. Bull.* **2001**, *24*, 1009–1011. (c) Miura, T.; Iwamoto, N.; Kato, M.; Ichiki, H.; Kubo, M.; Komatasu, Y.; Ishida, T.; Okada, M.; Seino, Y.; Tanigawa, K. *Biol. Pharm. Bull.* **2001**, *24*, 1091–1092. (d) Ichiki, H.; Miura, T.; Kubo, M.; Ishihara, E.; Komatasu, Y.; Tanigawa, K.; Okada, M. *Biol. Pharm. Bull.* **1998**, *21*, 1389.

135. Zhang, Y.; Qian, Q.; Ge, D.; Li, Y.; Wang, X.; Chen, Q.; Gao, X.; Wang, T. *J. Agric. Food Chem.* **2011**, *59*, 11526–11533.
136. Wei, X.; Liang, D.; Ning, M.; Wang, Q.; Meng, X.; Li, Z. *Tetrahedron Lett.* **2014**, *55*, 3083–3086.
137. Takahashis, M.; Tanzawa, K.; Takahashi, S. *J. Bio. Chem.* **1994**, *269* (1), 369–372.
138. Abe, H.; Shuto, S.; Matsuda, A. *Tetrahedron Lett.* **2000**, *41* (14), 2391–2394.
139. (a) Shuto, S.; Yahiro, Y.; Ichikawa, S.; Matsuda. A. *J. Org. Chem.* **2000**, *65* (18), 5547–5557. (b) Rosenberg, H. J.; Riley, A. M.; Correa, V.; Taylor, C. W.; Potter, B.V. L. *Carbohydr. Res.* **2000**, *329* (1), 7–16.
140. Terauchi, M.; Abe, H.; Tovey, S. C.; Dedos, S. G.; Taylor, C. W.; Paul, M.; Trusselle, M.; Potter, B.V. L.; Matsuda, A.; Shuto, S. A. *J. Med. Chem.* **2006**, *49* (6), 1900–1909.
141. (a) Natori, T.; Koezuka, Y.; Higa, T. Agelasphins, *Tetrahedron Lett.* **1993**, *34*, 5591–5592. (b) Natori, T.; Morita, M.; Akimoto, K.; Koezuka, Y. *Tetrahedron* **1994**, *50*, 2771–2784.
142. Morita, M.; Motoki, K.; Akimoto, K.; Natori, T.; Sakai, T.; Sawa, E.; Yamaji, K.; Koezuka, Y.; Kobayashi, E.; Fukushima, H. *J. Med. Chem.* **1995**, *38*, 2176–2187.
143. For reviews, see: (a) Wu, D.; Fujio, M.; Wong, C.-H. *Bioorg. Med. Chem.* **2008**, *16*, 1073–1083. (b) Banchet-Cadeddu, A.; Hénon, E.; Dauchez, M.; Renault, J.-H.; Monneaux, F.; Haudrechy, A. *Org. Biomol. Chem.* **2011**, *9*, 3080–3104.
144. Tomiyama, H.; Yanagisawa, T.; Nimura, M.; Noda, A.; Tomiyama, T. U.S. Patent 6635622.
145. Franck, R. W. *Comptes. Rendus. Chemie* **2012**, *15*, 45–56.
146. Due to their affiliation with Hunter College of CUNY, New York.
147. (a) Yang, G.; Schmieg, J.; Tsuji, M.; Franck, R. W. *Angew. Chem., Int. Ed.*, **2004**, *43*, 3818–3822; (b) Tsuji, M.; Franck, R. W.; Chen, G. US Pat. WO 2005/102049 A1, 2005; (c) Franck, R. W.; Tsuji, M. *Acc. Chem. Res.*, **2006**, *39*, 692–701; (d) Schmieg, J.; Yang, G.; Franck, R. W.; Tsuji, M. *J. Exp. Med.*, **2003**, *198*, 1631–1641. (e) Taylor, R. J. K.; McAllister, G. D.; Franck, R. W. *Carbohydr. Res.* **2006**, *341* (10), 1298–1311.
148. Kobertz, W. R.; Bertozzi, C. R.; Bednarski, M. D. *Tetrahedron Lett.* **1992**, *33* (6), 737–740.
149. Chen, G.; Schmieg, J.; Tsuji, M.; Franck, R. W. *Org. Lett.* **2004**, *6* (22), 4077–4080.
150. Chen, G.; Chien, M.; Tsuji, M.; Franck, R. W. E. *Chem. Bio. Chem.* **2006**, 7 (7), 1017–1022.
151. Franck, R. W.; Tsuji, M. *Acc. Chem. Res.* **2006**, *39* (10), 692–701.
152. (a) Chaulagain, M. R.; Postema, M. H. D.; Valeriote, F.; Pietraszkewicz, H. *Tetrahedron Lett.* **2004**, *45* (41), 7791–7794. (b) Takai, K.; Nitta, K.; Utimoto, K. *J. Am. Chem. Soc.* **2002**, *108* (23), 7408–7410.
153. Gorantla, J. N.; Lankalapalli, R. S. *J. Org. Chem.* **2014**, *79* (11), 5193–5200.
154. Redpath, P.; Ness, K. A.; Rousseau, J.; Macdonald, S. J. F.; Migaud, M. E. *Carbohydr. Res.* **2015**, *402*, 25–34.
155. (a) Rodrigues, F.; Canac, Y.; Lubineau, A.; Chimie, L. De; Multifonctionnelle, O.; Cnrs, A.; Chimie, I. De; Sud, P.; *E*-mail, F. *Chem. Commun.* **2000**, *6*, 2049–2050. (b) Misra A. K., TiwariP., Madhusudan S. K., *Carbohydr. Res.* **2005**, *340*, 325–329. (c) Norsikian, S.; Zeitouni, J.; Rat, S.; Gérard, S.; Lubineau, A. *Carbohydrate Res.* **2007**, *342* (18), 2716–2728.
156. (a) Allevi, P.; Anastasia, M.; Ciuff, P.; Fiecchi, A.; Scala, A. *J. Chem. Soc., Chem. Commu.* **1987**, 101–102. (b) Allevi, P.; Anastasia, M.; Ciuff, P.; Fiecchi, A.; Scala, A. *Chem. Commun.* **1988**, 57–58. (c) Aucagne, V.; Gueyrard, D.; Tatibouët, A.; Quinsac, A.; Rollin, P. *Tetrahedron* **2000**, *56* (17), 2647–2654.
157. (a) Brenna, E.; Fuganti, C.; Grasselli, P.; Serra, S.; Zambotti, S. *Chem. Eur. J.* **2002**, *8* (8), 1872–1878. (b) D. Lewis, M.; Kun Cha, J.; Kishi, Y. *J. Am. Chem. Soc.* **1982**, *104* (18), 4976–4978.
158. (a) Lowary, T.; Meldal, M.; Helmboldt, A.; Vasella, A.; Bock, K. *J. Org. Chem.* **1998**, *63* (26), 9657–9668. (b) Nishikawa, T.; Koide, Y.; Kanakubo, A.; Yoshimura, H.; Isobe, M. *Org. Biomol. Chem.* **2006**, *4* (7), 1268–1277.

159. (a) Gaertzen, O.; Misske, A. M.; Wolbers, P.; Hoffmann, H. M. R. *Tetrahedron Lett.* **1999**, *40* (35), 6359–6363. (b) Petry, N.; Vucko, T.; Collet, C.; Lamandé-Langle, S.; Pellegrini-Moïse, N.; Chrétien, F. *Carbohydr. Res.* **2017**, *445*, 61–64. (c) McGarvey J., G.; LeClair A., C.; Schmidtmann A., B. *Org. Lett.* **2008**, *10* (21), 4727–4730.
160. (a) Dondoni, A.; Marra, A. *Tetrahedron Lett.* **2003**, *44* (1), 13–16.(b) Dondoni, A.; Scher-rmann, M. C. *Tetrahedron Lett.* **1993**, *34* (45), 7319–7322. (c) Dondoni, A.; Scher-rmann, M. C. *J. Org. Chem.* **1994**, *59* (21), 6404–6412; see also, Genet, J. P. et al., *Synlett* **1994**, 705–708. (d) Labéguère, F.; Lavergne, J.-P.; Martinez, J. *Tetrahedron Lett.* **2002**, *43* (40), 7271–7272. (e) Geng, Y.; Kumar, A.; Faidallah, H. M.; Albar, H. A.; Mhkalid, I. A.; Schmidt, R. R. *Bioorg. Med. Chem.* **2013**, *21* (16), 4793–4802.
161. (a) Eszter Szennyes, a Eva Bokor, a Gyula Batta, a Tibor Docsa, b P. G. and L. S. *RSC Adv.* **2016**, *6*, 94787–94794. (b) María-Teresa García López a, F. G. D. las H. a & A. S.; A, F. J. *Carbohydr. Chem.* **1987**, *6* (2), 37–41.
162. Reddy, M. R.; Hemaiswarya, S.; Kommidi, H.; Aidhen, I. S.; Doble, M. *Eur. J. Org. Chem.* **2019**, 6053–6070.
163. (a) Ramakrishna, K. K. G.; Gunjan, S.; Shukla, A. K.; Pasam, V. R.; Balaramnavar, V. M.; Sharma, A.; Jaiswal, S.; Lal, J.; Tripathi, R.; Anubhooti; et al. *ACS Med. Chem. Lett.* **2014**, *5* (8), 878–883. (b) Ajay, A.; Sharma, S.; Prasad Gupt, M.; Bajpai, V.; Hamidullah; Kumar, B.; Prasad Kaushik, M.; Konwar, R.; Sankar Ampapathi, R.; Pati Tripathi, R. *Org. Lett.* **2012**, *14* (17), 4306–4309. (c) Tpánek, P.; Imák, O.; Nováková, Z.; Wimmer, Z.; Draar, P. *Org. Biomol. Chem.* **2011**, *9* (3), 682–683. (d) Norsikian, S.; Zeitouni, J.; Rat, S.; Gérard, S.; Lubineau, *Carbohydr. Res.* **2007**. *342*, 2716. (e) Carpenter, C. A.; Kenar, J. A.; Price, N. P. J. *Green Chem.* **2010**, *12* (11), 2012–2018. (f) Foley, P. M.; Phimphachanh, A.; Beach, E. S.; Zimmerman, J. B.; Anastas, P. T. *Green Chem.* **2011**, *13* (2), 321–325. (g) Riafrecha, L. E.; Bua, S.; Supuran, C. T.; Colinas, P. A. *Bioorg. Chem.* **2018**, *76*, 61–66. (*h*) Areti, S.; Bandaru, S.; Kandi, R.; Rao, C. P. *ACS Omega* **2019**, *4* (1), 1167–1177. (i) Petitjean, L.; De Winter, T. M.; Petrovic, P. V.; Coish, P.; Hitce, J.; Moreau, M.; Bordier, T.; Erythropel, H.C.; Anastas, P. T. *Green Chem.* **2019**, *21* (2), 238–244.
164. (a) Postema H. D., M.; Piper L., J.; Liu, L.; Shen, J.; Faust, M.; Andreana, P. *J. Org. Chem.* **2003**, *68* (12), 4748–4754. (b) Vucko, T.; Pellegrini Moïse, N.; Lamandé-Langle, S. *Carbohydr. Res.* **2019**, *477*, 1–10. (c) Nolen G., E.; Watts M., M.; Fowler J., D. *Org. Lett.* **2002**, *4* (22), 3963-3965.(d) Palomo, C.; Oiarbide, M.; Landa, A.; Concepción González-Rego, M.; García M., J.; González, A.; Odriozola M., J.; Martín-Pastor, M.; Linden, A. *J. Am. Chem. Soc.* **2002**, *124* (29), 8637–8643. (e) Lazzaroni, R.; Rocchiccioli, S.; Iuliano, A.; Cipolla, L.; Nicotra, F. *Tetrahedron Asymmetry*. **2005**, *16* (22), 3661–3666.
165. (a) Rouzier, F.; Sillé, R.; Nourry, A.; Tessier, A.; Pipelier, M.; Guillarme, S. *Synthesis* **2019**, *51* (12), 2484–2488. (b) Li, L.-T.; Zhou, L.-F.; Li, Y.-J.; Huang, J.; Liu, R.-H.; Wang, B.; Wang, P. *Bioorg. Med. Chem. Lett.* **2012**, *22* (1), 642–644. (c) Groothuys, S.; Kuijpers, B. H. M.; Quaedflieg, P. J. L. M.; Roelen, H. C. P. F.; Wiertz, R. W.; Blaauw, R. H.; Van Delft, F. L.; Rutjes, F. P. J. T. *Synthesis* **2006**, *18*, 3146–3152. (d) Narute, S. B.; Rout, J. K.; Ramana, C. V. *Chem. Eur. J.* **2013**, *19* (45), 15109–15114. (e) Kantsadi, A. L.; Bokor, É.; Kun, S.; Stravodimos, G. A.; Chatzileontiadou, D. S. M.; Leonidas, D. D.; Juhász-Tóth, É.; Szakács, A.; Batta, G.; Docsa, T.; et al. *Eur. J. Med. Chem* **2016**, *123*, 737–745. (f) Dondoni, A.; Marra, A. *J. Org. Chem.* **2006**, *71* (20), 7546–7557. (g) Kuijpers H. M., B.; Groothuys, S.; Keereweer, A.; J. L. Quaedflieg M., P.; Blaauw H., R.; van Delft L. F.; Rutjes P. J. T., F. *Org. Lett.* **2004**, *6* (18), 3123–3126. (h) Dondoni, A.; Mariotti, G.; Marra, A. *J. Org. Chem.* **2002**, *67* (13), 4475–4486. (i) Murty, K. V. S. N.; Vasella, A. *Helv. Chim. Acta.* **2001**, *84* (4), 939–963. (j) Dondoni, A.; Mariotti, G.; Marra, A. *Tetrahedron Lett.* **2000**, *41* (18), 3483–3487.

166. (a) Sudarshan, K.; Aidhen, I. S. *Eur. J. Org. Chem.* **2017**, (1), 34–38. (b) El Oualid, F.; van der Marel A., G.; Overkleeft S., H.; Overhand, M. *Org. Lett.* **2004**, *6* (18), 3167–3170. (c) Michelet, V.; Adiey, K.; Tanier, S.; Dujardin, G.; Genêt, J.-P. *Eur. J. Org. Chem.* **2003**, 2947–2958. (d) Guerrini, A.; Varchi, G.; Battaglia, A. *J. Org. Chem.* **2006**, *71* (18), 6785–6795. (e) Boutard, N.; Labéguère, F.; Vidal, Y.; Lavergne, J.-P.; Martinez, J. *Synth. Commun.* **2012**, *42* (10), 1461–1471. (f) Zeitouni, J.; Norsikian, S.; Merlet, D.; Lubineau, A. *Adv. Synth. Catal.* **2006**, 348 (12-13), 1662–1670. (g) Risseeuw, M. D. P.; Mazurek, J.; van Langenvelde, A.; van der Marel, G. A.; Overkleeft, H.S.; Overhand, M. *Org. Biomol. Chem.* **2007**, *5* (14), 2311–2314. (*h*) Levoirier, E.; Canac, Y.; Norsikian, S.; Lubineau, A. *Carbohydr. Res.* **2004**, *339* (17), 2737–2747. (i) Canac, Y.; Levoirier, E.; Lubineau, A. *J. Org. Chem.* **2001**, *66* (9), 3206–3210. (j) Zhu, Y.-H.; Demange, R.; Vogel, P. *Tetrahedron: Asymmetry* **2000**, *11* (1), 263–282. (k) Dondoni, A.; Marra, A.; Massi, A. *J. Org. Chem.* **1999**, *64* (3), 933–944. (m) Demange, R.; Bühlmann, C.; Vogel, P. *Helv. Chim. Acta.* **2003**, *86* (2), 361–376.
167. (a) Eszter Szennyes L. S., a Eva Bokor, a Gyula Batta, a Tibor Docsa, *RSC Adv.* **2016**, 94787–94794. (b) Dubreuil D., Saluzzo C., Guillarme S., Pipelier M., Dujardin G., Nourry A., Guisot N. E. S., Ireddy P., Ella Obame I., *Eur. J. Org. Chem.* **2018**, 1735–1738. (c) Guisot N. E. S., Ella Obame I., Ireddy P., Nourry A., Saluzzo C., Dujardin G., Dubreuil D., Pipelier M., Guillarme S., *J. Org. Chem.* **2016**, *81*, 2364–2371. (d) Barghash R. F., Massi A., Dondoni A., *Org. Biomol. Chem.* **2009**, 7, 3319–3330. (e) Dondoni A., Massi A., Minghini E., Bertolasi V., *Helv. Chim. Acta.* **2002**, *85*, 3331–3348. (f) Aldhoun M., Massi A., Dondoni A., *J. Org. Chem.* **2008**, *73*, 9565–9575. (g) Somsák L., Nagy V., *Tetrahedron: Asymmetry* **2000**, *11*, 1719–1727. (h) Mydock-McGrane L., Cusumano Z., Han Z., Binkley J., Kostakioti M., Hannan T., Pinkner J. S., Klein R., Kalas V., Crowley J., et al., *J. Med. Chem.* **2016**, *59*, 9390–9408. (i) Sipos S., Jablonkai I., *Carbohydr. Res.* **2011**, *346*, 1503–1510. (h) Singh G., Vankayalapati H., *Tetrahedron: Asymmetry* **2001**, *12*, 1727–1735. (i) Araki Y., Kobayashi N., Watanabe K., Ishido Y., *J. Carbohydr. Chem.* **1985**, *4*, 565–585.
168. van Breemen B., R. *J. Med. Chem.* **2015**, *58* (21), 8360–8372.
169. Correia-da-Silva, M.; Sousa, E.; Duarte, B.; Marques, F.; Carvalho, F. M.; Cunha-Ribeiro, L. M. M.; Pinto, M. *J. Med. Chem.* **2011**, *54* (15), 5373–5384.
170. Huang, F.-C.; Hinkelmann, J.; Schwab, W. *J. Biotechnol.* **2015**, *216*, 100–109.
171. Villadsen, K.; Martos-Maldonado, M. C.; Jensen, K. J.; Thygesen, M. B. *Chem Bio Chem.* **2017**, *18* (7), 574–612.

CHAPTER FOUR

3-Deoxy-D-*manno*-oct-2-ulosonic acid (Kdo) derivatives in antibacterial drug discovery

Maude Cloutier, Charles Gauthier
Centre Armand-Frappier Santé Biotechnologie, Institut national de la recherche scientifique (INRS), boul. des Prairies, Laval, QC, Canada

1 Introduction

First isolated in 1959,[1] 3-deoxy-D-*manno*-oct-2-ulosonic acid (Kdo, Fig. 4.1) is part of the large family of biologically relevant higher-carbon 3-deoxy-2-keto-ulosonic acids. This sugar residue is ubiquitously found in the surface polysaccharides of Gram-negative bacteria (GNB). It is indeed a highly conserved and essential element of lipopolysaccharides (LPSs)[2] found in the outer leaflet of the bacterial outer membrane (OM).

LPSs, also known as endotoxins, are amphiphilic macromolecules that can be structurally divided into three distinct regions: the lipid A region, the core region—containing the inner and outer cores—and the polysaccharide region. The hydrophobic lipid A, the main endotoxin of LPSs,[3] is a negatively charged diphosphorylated β-(1→6)-linked glucosamine disaccharide asymmetrically substituted with (*R*)-3-hydroxy fatty acid chains anchored in the outer leaflet of the OM.[4] Connection of lipid A to the inner core region is achieved through a glycosidic linkage between the *O*-6' position of the former and one Kdo unit (Kdo I). Most bacteria have a second Kdo (Kdo II) moiety attached at the *O*-4 position of Kdo I, thus forming the so-called *Re*-LPS or lipid A-Kdo_2.[5] Noteworthy, LPS composed only of lipid A and one Kdo residue substituted with a negatively charged unit stands as the smallest LPS structure known to sustain growth and viability in GNB. To complete the LPS inner core usually found in GNB, Kdo I bear two or more L-*glycero*-α-D-*manno*-heptopyranose residues at its *O*-5 position. The second heptose serves as an anchorpoint between the inner and outer cores.[5] Finally, in constrast with the inner core and lipid A, the outer core and *O*-antigenic polysaccharide regions show high variabilities between bacterial species.

Carbohydrates in Drug Discovery and Development. http://dx.doi.org/10.1016/B978-0-12-816675-8.00004-X

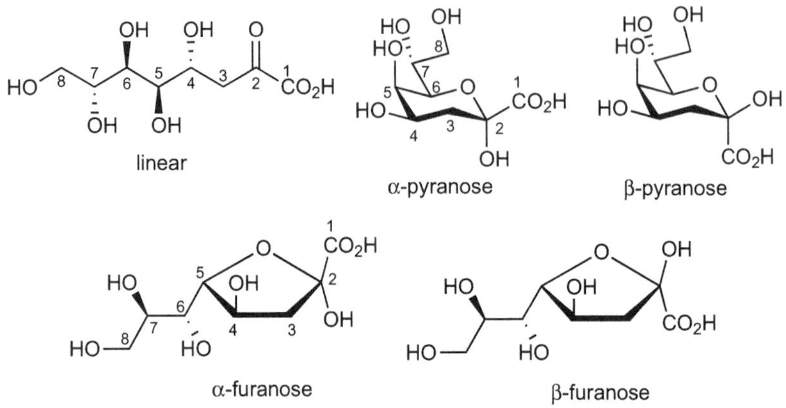

Figure 4.1 ***Chemical structure of d-Kdo in linear and cyclic forms.***

Biosynthesis of the activated form of Kdo in *E. coli*, that is, cytidine monophosphate (CMP)-Kdo, involves the sequential activity of four different enzymes, as depicted in Fig. 4.2. The first biosynthetic step involves the conversion of D-ribulose-5-phosphate into D-arabinose-5-phosphate through the catalytic activity of the enzyme arabinose-5-phosphate isomerase[6] (KdsD). Kdo-8-phosphate synthase (KdsA) then condenses the resulting D-arabinose-5-phosphate with phosphoenol pyruvate, furnishing Kdo-8-phosphate.[7-9] Removal of the phosphate group is processed by Kdo-8-phosphatase[10] (KdsC) and the resulting Kdo is activated into CMP-Kdo by CMP-Kdo synthase[7,11,12] (CKS or KdsB). Importantly, the biosynthetic intermediate CMP-Kdo is found in the β-form,[13] which makes it highly unstable, whereas Kdo is usually found in its α-form within LPS. Incorporation of CMP-Kdo into the growing lipid A, achieved through the action of Kdo transferase[14,15] (WaaA), occurs with inversion of configuration. WaaA also catalyzes the incorporation of Kdo II. In some cases, the resulting lipid A-Kdo_2 can be subjected to acyltransferases for the acylation of the fatty acid chains.[16] Finally, one heptose unit is attached to Kdo I by heptosyltransferase WaaC[17-19] and the resulting construct can be further elongated into the whole LPS structure.

The rate-determining step of this enzymatic process is the formation of key intermediate CMP-Kdo, which, as previously mentioned, is highly unstable. To gain more insights into the mechanisms involved in the instability of this biosynthetic intermediate, the Chi-Huey Wong group[20] performed a stability study on different analogues of CMP-β-Kdo (Fig. 4.3). As opposed to CMP-Kdo, which has a half-life of 34 min under specific

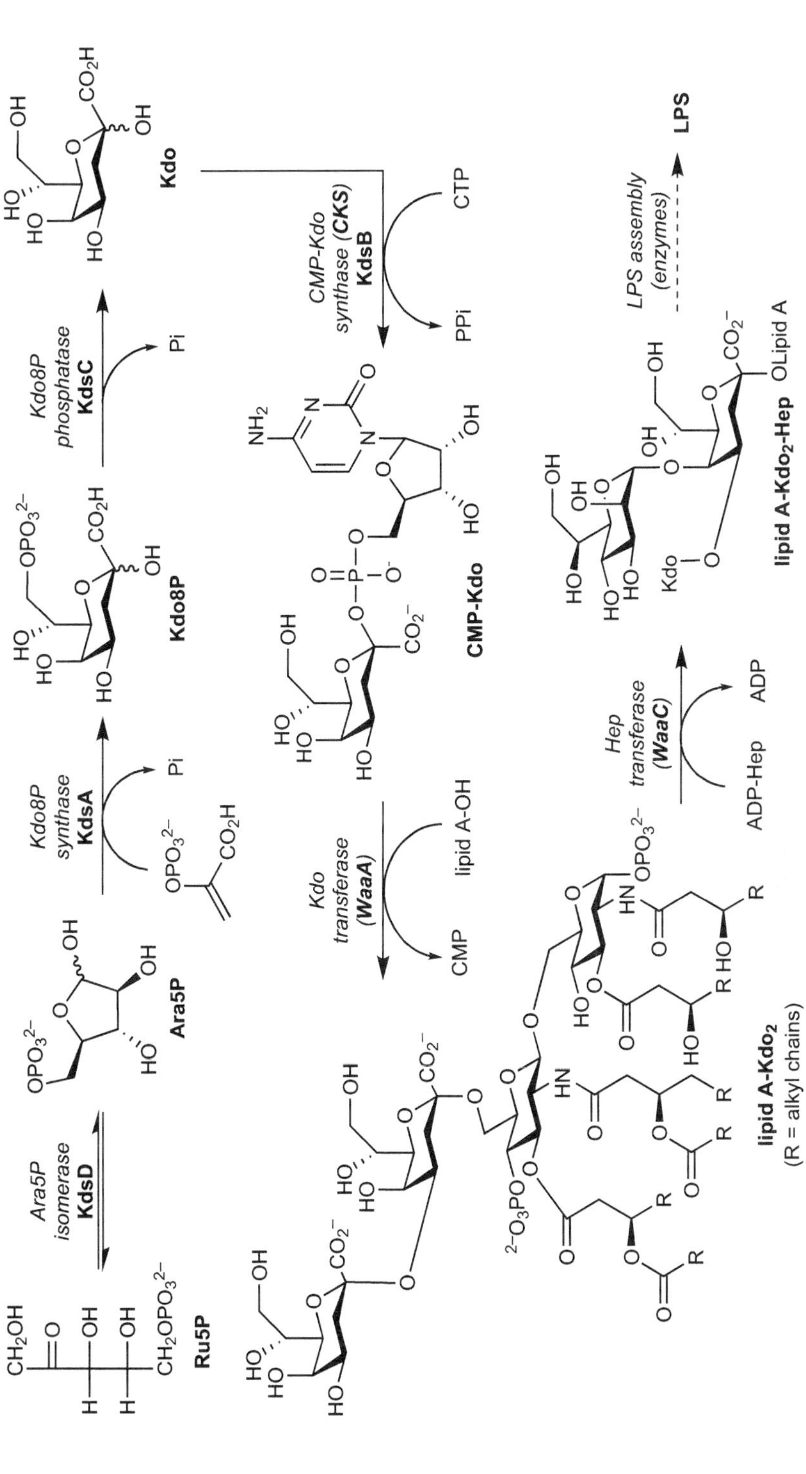

Figure 4.2 ***Biosynthetic LPS machinery towards Kdo$_2$-lipid A and beyond.***

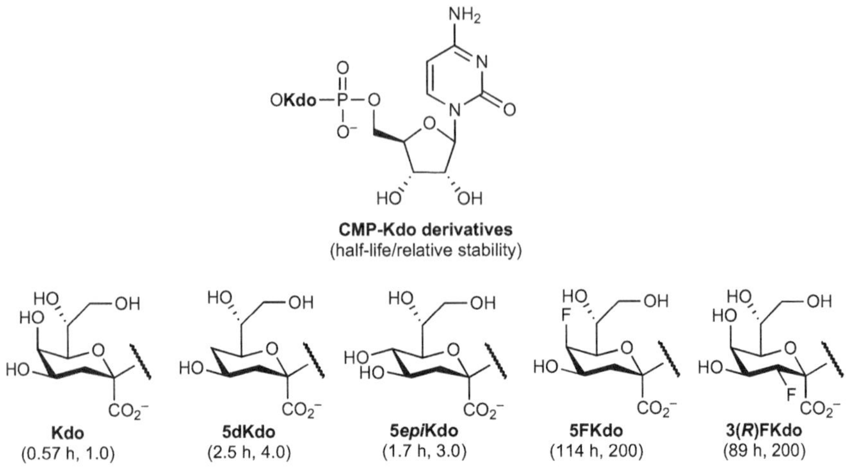

Figure 4.3 ***CMP-Kdo derivatives and their relative stability.***[20]

conditions (25 °C, 0.1 M *Tris*-HCl buffer, pH 7.5, 20 mM $MgCl_2$), CMP-5dKdo, CMP-5*epi*Kdo, CMP-5FKdo, and CMP-3(*R*)FKdo derivatives showed enhanced stability with half-life ranging from 1.7 h to 114 h. Mecanistically, it was proposed that CMP-Kdo hydrolysis proceeded through a twist-boat conformation enabled by hydrogen-bonding interactions between the different *exo*-cyclic oxygen atoms and the negatively charged phosphate moiety.

Appart from LPS, Kdo moieties are also found in capsular polysaccharides (CPSs) of GNB,[2] which allow them to evade the host immune system.[21] More specifically, CPSs of *Escherichia coli*, *Neisseria meningitidis, Campylobacter jejuni*, *Haemophilus influenza*, and *Pasteurella multocida* are bound to membrane-anchored (lyso)phosphatidylglycerol through a linker composed of five to nine β-(2→4) and β-(2→7)-linked Kdo units.[22] The Whitfield group discovered in 2013 that the incorporation of these sugar units was achieved through the catalysis of KpsS and KpsC, the two first-identified Kdo transferases retaining the β-configuration of CMP-Kdo. As shown in Fig. 4.4, KpsS adds the first Kdo residue to the lipidic chain, whereas KpsC allows chain extension to occur by transferring additional Kdo units to the growing linker.[23] Because of the ubiquitous presence of these enzymes in encapsulated GNB, KpsC and KpsS stand as exquisite targets for developing novel antibiotic adjuvants working through the "antivirulence" strategy.[24]

As opposed to α-Kdo, its β-anomer is therefore only found sporadically in pathogenic bacteria. As an example, the CPS of *Actinobacillus pleuropneumoniae* serotype 5b is composed of a trisaccharide repeating unit bearing

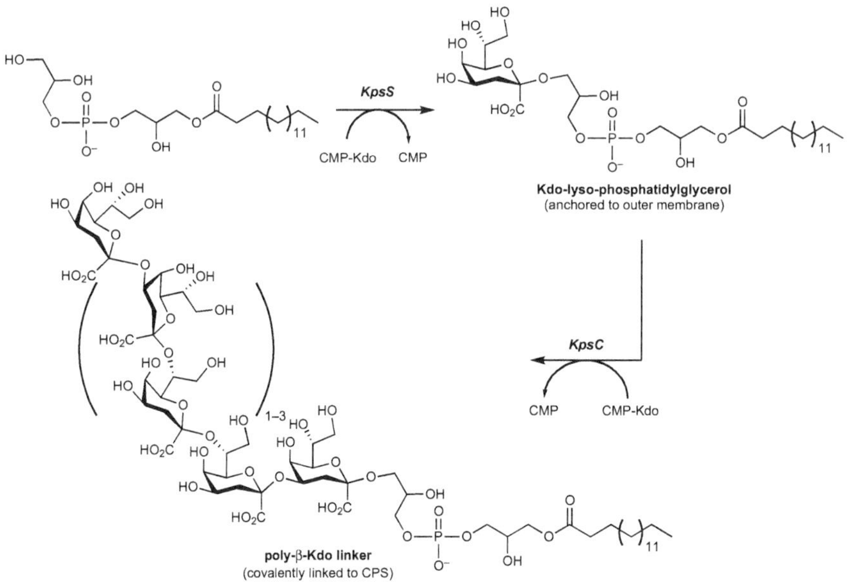

Figure 4.4 ***Biosynthesis of poly-β-Kdo linker via KpsS/KpsC catalysis.***

glucose, *N*-acetyl glucosamine, and Kdo residues (Fig. 4.5), whereas the serotype 5a lacks the glucose residue.[25,26] β-Kdo residues are also found in the CPS of *Kingella kingae*, a GNB causing septic arthritis in young children,[27] *Moraxella nonliquefaciens*,[28] a potentially pathogenic bacterium in patients suffering from respiratory tract diseases,[29] as well as *Pseudoalteromonas flavipulchra*,[30] a marine GNB. β-Kdo glycosides can also be found in exopolysaccharides (EPS) of the potential bioterrorism agent *Burkholderia pseudomallei* as well as some species of the *B. cepacia* complex. In that regard, the natural occurrence of Kdo in bacterial polysaccharides has been extensively reviewed elsewhere.[2]

The search for new therapeutic targets and antibiotics in recent years has been catalyzed by the re-emergence of pathogenic bacteria as well as their growing resistance to currently used antibiotics. As loss of functional enzymes involved in the biosynthesis of LPS has been shown to affect bacteria viability or virulence, there is a growing interest in the discovery of potential inhibitors of this enzymatic pathway, which would act as antibiotics or antibiotic adjuvants. Indeed, Bozue et al.[31] highlighted that a KdsD mutant strain of the pathogenic bacterium *Francisella tularensis* showed attenuated virulence, growth defects, and enhanced sensitivity to hydrophobic agents. More particularly, as Kdo is highly conserved, is a critical

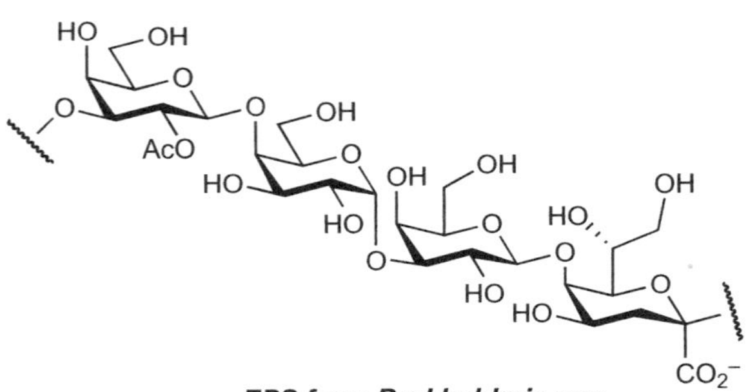

EPS from *Burkholderia* spp.
[→3)-2-*O*-Ac-β-D-Gal*p*-(1→4)-α-D-Gal*p*-(1→3)-β-D-Gal*p*-(1→5)-β-D-Kdo*p*-(2→]

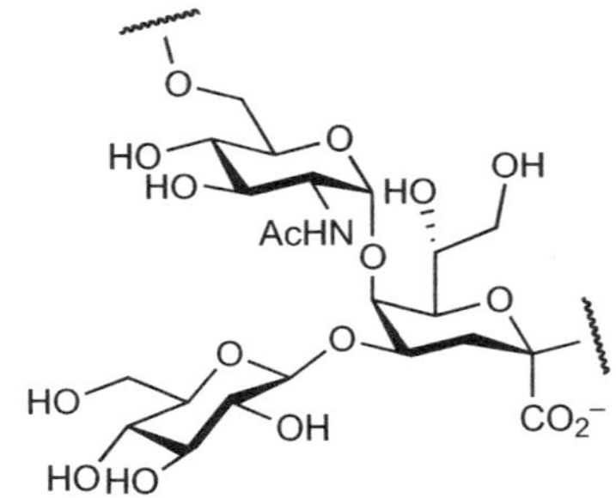

CPS from *Actinobacillus pleuropneumoniae* serotype 5b
[→5)-β-D-Kdo*p*-[→4)-β-D-Glc*p*-(1→]-(2→6)-α-D-Glc*p*NAc-(1→]

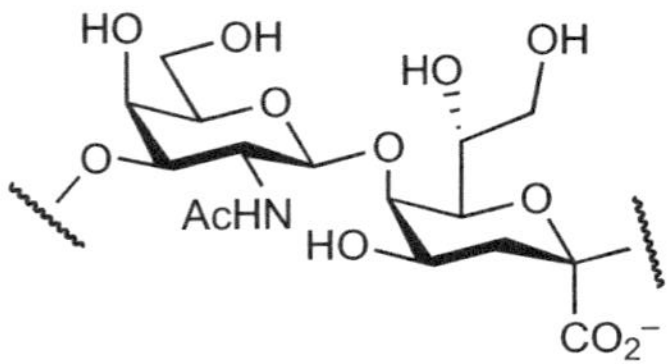

CPS from *Kingella kingae* and *Moraxella nonliquefaciens*
[→3)-β-D-Gal*p*NAc-(1→5)-β-D-Kdo*p*-(2→]

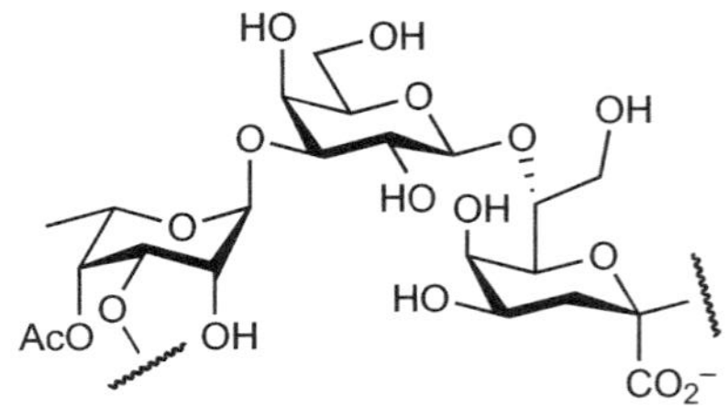

CPS from *Pseudoalteromonas flavipulchra*
[→3)-β-D-Gal*p*NAc-(1→5)-β-D-Kdo*p*-(2→]

Figure 4.5 ***Examples of β-Kdo-containing bacterial glycans.***

unit for LPS assembly, and is not found in mammals, lots of research has been focused on the inhibition of its biosynthetic pathway. Development of specific inhibitors of Kdo transferases retaining the β-configuration, such as those involved in the assembly of the poly-β-Kdo-linker found in the CPS of some GNB, could also be valuable tools in the fight against drug-resistant pathogenic bacteria.

As the enzymes involved in the biosynthesis of Kdo-containing polysaccharides stand as potential therapeutic targets, this subject has been extensively reviewed. Many groups indeed published comprehensive investigations of the biosynthesis of bacterial polysaccharides,[32] Kdo[16,33] and LPS,[34–36] including the development of related therapeutics against GNB.[37–39] Additionally, Cipolla et al. reviewed in 2011 the biosynthesis of Kdo and its analogues as potential inhibitors of the corresponding enzymes for the development of antibiotics or antibiotic adjuvants against GNB.[40] Furthermore, other research groups have been interested in the advances surrounding the chemistry of Kdo-containing oligosaccharides, as such synthetic compounds are promising candidates for the development of prophylactic measures against pathogenic bacteria or new diagnostic tools, and are inherently useful for the elucidation of biological processes in GNB.[41–43] In the present chapter, we wish to provide a comprehensive account regarding the total synthesis of Kdo and its derivatives as potential inhibitors of Kdo-processing enzymes. We also describe innovative imaging approaches involving the Kdo biosynthetic machinery, which have been recently developed for the LPS labeling of living organisms.

2 Total syntheses of Kdo

Since the first identification of Kdo from bacterial cultures and the demonstration of its critical biological role, there has been tremendous advancements surrounding its total synthesis. Although already reviewed in 2003 by Li and Wu,[44] we herein wish to offer an up-to-date account of the total synthesis of Kdo from different precursors. Generalization of the various synthetic approaches allows to divide them in three main categories based on the starting material that is used. One of the most studied approach relies on the addition of a C-3 building block to a carbohydrate-derived intermediate ([3 + X] methodology) such as *d*-arabinose diacetone (Fig. 4.6), thus mimicking the previously described biosynthetic pathway. Similarily, the second approach consists in the addition of a C-2 building block to a five-carbon sugar ([2 + X] methodology) such as *d*-mannose and

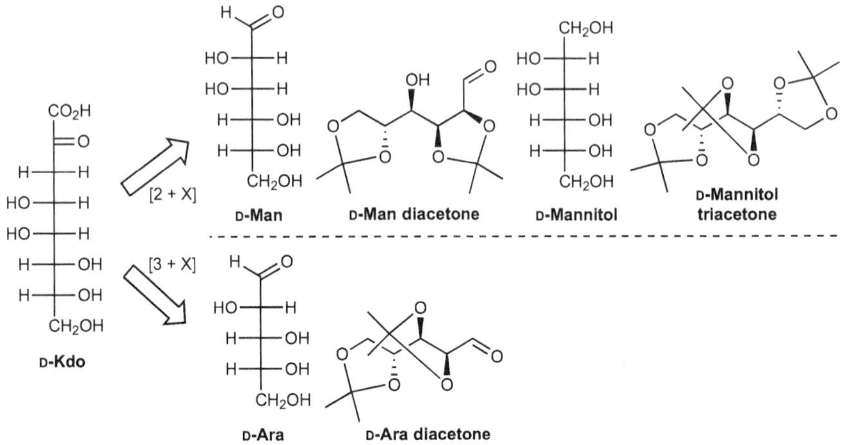

Figure 4.6 ***Most common synthetic precursors for the total synthesis of Kdo.***

d-mannitol, which allows to introduce the α-keto acid moiety characteristic of Kdo. The third category includes a variety of approaches employing non-carbohydrate starting materials and involving reactions such as Diels-Alder cycloadditions. These general types of syntheses will be discussed in the following sections, which will first focus on the use of *d*-mannose as starting material, followed by *d*-arabinose, and *d*-mannitol. Then, we will describe the miscellaneous approaches relying on the use of non-sugar intermediates.

2.1 Total synthesis of Kdo from D-arabinose

In 1958, Cornforth reported the first total synthesis of the sialic acid *N*-acetylneuraminic acid (Neu5Ac), which plays critical biological functions in glycoproteins and glycolipids,[45] through a [3 + X] methodology by condensing *N*-acetyl-D-glucosamine with oxaloacetic acid under basic conditions.[46] Since then, this method has been adapted by Ogura and co-workers for the synthesis of Kdo and its 4-*epi*-isomer by using *d*-arabinose rather than *N*-acetyl-D-glucosamine.[47] The synthesis, as depicted in Scheme 4.1, can be achieved through a two-step sequence involving an aldol condensation followed by decarboxylation of the C3 position under $NiCl_2$ catalysis.[48]

To bypass the disadvantage of the previous method, which lies in the formation of diastereoisomers hampering the purification step, Schmidt developed in 1984 a method involving the use of *C*-lithiated acrylic acid **1** as a C-3 building block (Scheme 4.2).[49] Thus, addition of the latter to

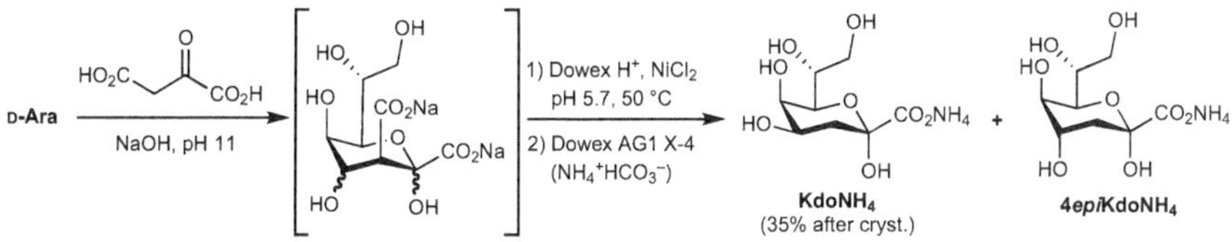

Scheme 4.1 ***Synthesis of Kdo using the modified Cornforth procedure.***[46-48]

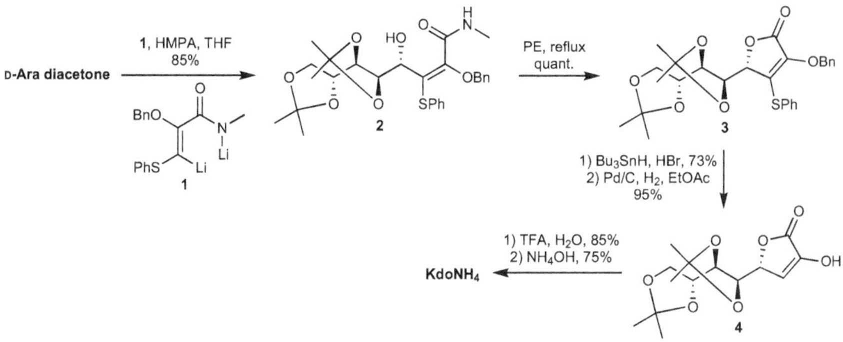

Scheme 4.2 ***Schmidt first generation synthesis of Kdo.***[49]

acetonide-protected arabinose proceeded with high stereoselectivity to furnish *manno*-configured intermediate **2**. Quantitative intramolecular cyclization of compound **2** into butanolide **3** was achieved upon heating in petroleum ether. The thiophenyl group was removed through the formation of a tributyltin intermediate, which was subsequently cleaved using HBr. Hydrogenolysis of the resulting intermediate furnished alcohol **4** while acidic cleavage of the isopropylidene groups followed by treatment with aqueous ammonia ultimately afforded the ammonium salt of Kdo.

The Dondoni group's approach towards the synthesis of Kdo relied on the use of the lithium enolate of 2-acetylthiazole **5** for the elongation of *d*-arabinose diacetone, furnishing intermediate **6** with high-stereoselectivity (Scheme 4.3).[50] Importantly, the use of such thiazole as a formyl group equivalent proved useful due to its low lability, which eased the functionalization steps, and high reactivity. Hemiketalization of compound **6** using methanolic HCl and simultaneous cleavage of the isopropylidene groups furnished Kdo derivative **7**. Following reprotection of the diols, thiazole unmasking was achieved in a three-step sequence involving: (1) formation of an *N*-methylthiazolium intermediate; (2) reduction of the thiazole ring; and (3) $HgCl_2$-promoted hydrolysis of the reduced *N*-methylthiazole group. Oxidation of resulting aldehyde **9** with Ag_2O, cleavage of the isopropylidene

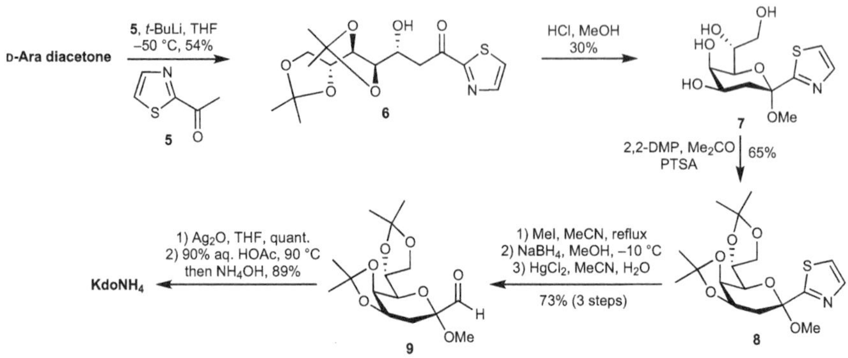

Scheme 4.3 ***Dondoni synthesis of Kdo.***[50]

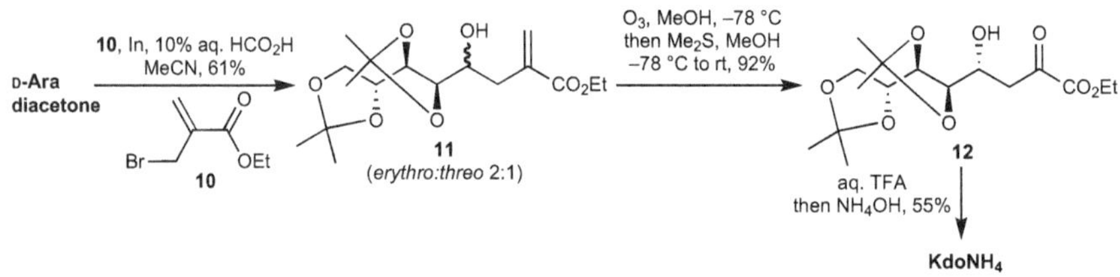

Scheme 4.4 ***Whitesides synthesis of Kdo.***[51]

groups, and treatment with aqueous ammonia finally furnished ammonium Kdo.

Three years later, a methodology involving the use of ethyl α-(bromomethyl)acrylate **10** as a latent C3 pyruvate equivalent was developed by Whitesides et al., as depicted in Scheme 4.4.[51] Thus, inspired by the work of Schmid and co-workers,[52] indium-promoted allylation of *d*-arabinose diacetone with compound **10** in aqueous acidic media yielded alcohol **11** as a 2:1 mixture of *erythro:threo* diastereoisomers. Upon isolation of the desired *erythro* isomer by column chromatography, the latter was ozonized into α-keto ester **12** in excellent yield, which was ultimately hydrolysed and treated with aqueous ammonia to provide Kdo.

Alternatively, Wu's approach towards the synthesis of Kdo was based on the propargylation of acetonide-protected *d*-arabinose (Scheme 4.5).[53] Propargylation was achieved by reacting the latter with the C3 building block 3-bromopropyne in the presence of activated zinc. The resulting alcohol was protected as a methoxymethyl (MOM) ether, yielding alkyne **13** featuring the *erythro* configuration. Bromination of the terminal alkyne was achieved using *N*-bromosuccinimide (NBS) and $AgNO_3$ and the resulting bromoalkyne **14** was oxidized into α-keto ester **15** through the action

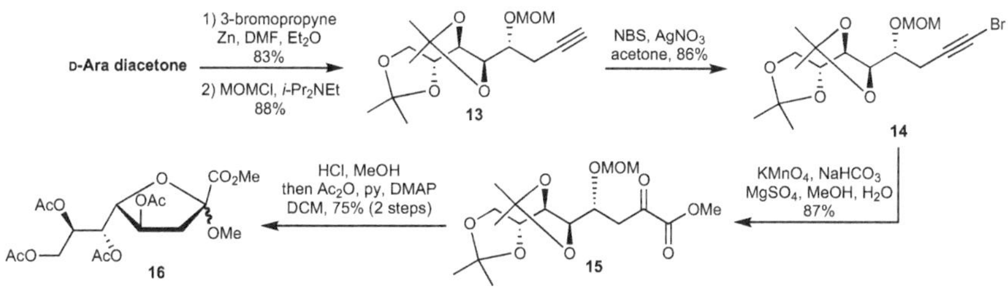

Scheme 4.5 ***Wu synthesis of Kdo in the furanose form.***[53]

of $KMnO_4$. Concentrated methanolic HCl then successfully deprotected intermediate **15** while simultaneously generating the methyl glycoside of Kdo locked in the kinetically favoured furanose form, which was subsequently peracetylated into a 4:1 anomeric mixture of Kdo derivative **16**. Interestingly, the authors hypothesized that this compound could have potent biological activities since the furanose form of Neu5Ac was shown to possess inhibitory activity against neuraminidases.[54]

In the context of finding new catalysts for asymmetric cross-aldol reactions of enolisable aldehydes with pyruvate esters, the Mlynarski group developed efficient syntheses of Kdo derivative **20** (Scheme 4.6). In 2012, they first showed that by using C3 building block 2-acetylthiazole **5** introduced by Dondoni, asymmetric aldolization of acetonide-protected *d*-arabinose could be achieved through the activity of a zinc catalyst bearing ProPh ligands.[55] As pyruvate-dependent aldolases use zinc ions to form an enolate of the donor, this developed synthetic method conceptually mimicked these enzymes.

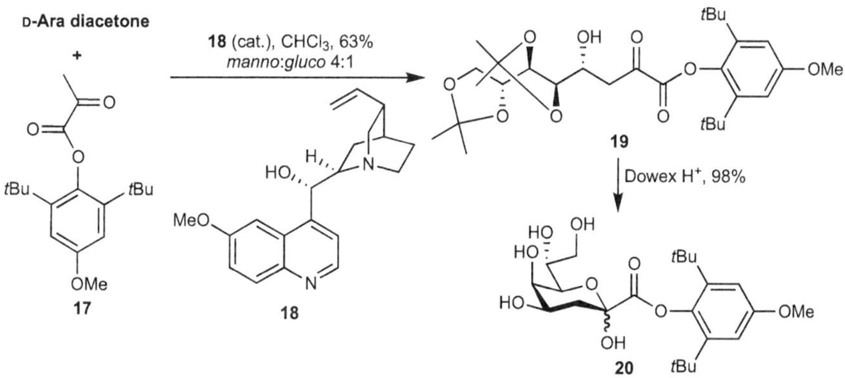

Scheme 4.6 ***Mlynarski synthesis of Kdo.***[55,56]

In order to develop more versatile synthetic approaches for asymmetric cross-aldol reactions, the authors subsequently showed that such reactions could be achieved by employing tertiairy amine-containing alkaloids naturally produced by *Cinchona*, such as quinidine **18**.[56] Indeed, *d*-arabinose diacetone could be condensed with 2,6-di-*tert*-butyl-4-methoxyphenyl pyruvate **17** through the catalytic activity of compound **18**. This reaction afforded *manno*-configured α-keto ester **19** as the main product, probably due to the formation of a network of hydrogen bonds with the catalyst, thus resulting in an asymmetric environment. Through the action of Dowex H^+, intermediate **19** was deprotected and hemiketalized into Kdo derivative **20** in an almost quantitative yield. Interestingly, this methodology also proved useful for the synthesis of other biologically relevant natural products, such as ulosonic acid 3-deoxy-α-D-*lyxo*-hept-2-ulosonic acid (KDH) and 3-deoxyglucosone (3DG).

2.2 Total synthesis of Kdo from *d*-mannose

As mentioned, another common approach towards the synthesis of Kdo is a two-carbon elongation of a chiral aldehyde such as *d*-mannose. Among the different elongation techniques, the use of a Wittig reagent has been widely reported. Thus, as depicted in Scheme 4.7, Overend's high-yielding approach first consisted of the Wittig reaction between *d*-mannose diacetone and (ethoxycarbonylmethylene)triphenylphosphorane.[57] The resulting ethyl octenoate **21** was deacetonated with aqueous trifluoroacetic acid (TFA) and hydrogenated to furnish pentaol **22**. Following lactonization and acetonation, lactone **23** was converted into 2-(dimethylaminomethylene)lactone **24**

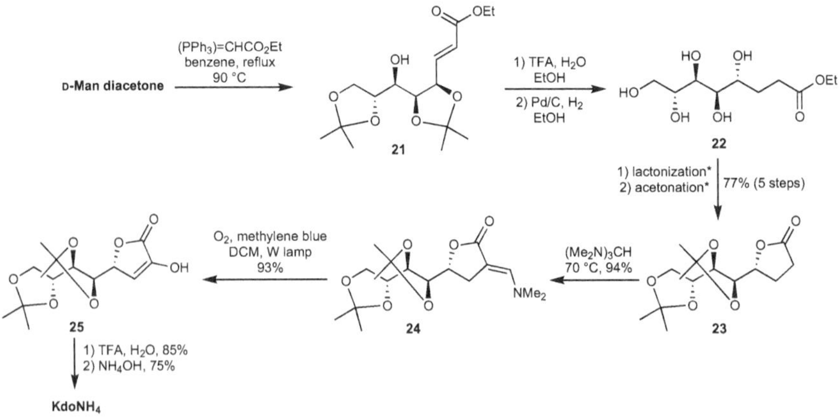

Scheme 4.7 ***Overend synthesis of Kdo.***[57]

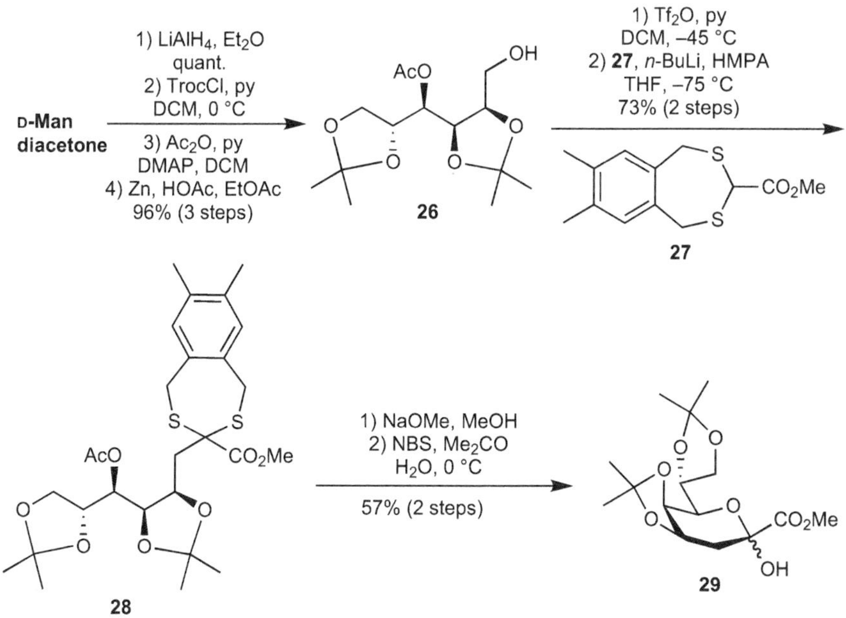

Scheme 4.8 ***Shiba synthesis of Kdo.***[58]

under the action of tris(dimethylamino)methane. Oxidation of compound **24** in oxygenated dichloromethane (DCM) with a photoflood lamp in the presence of methylene blue afforded enol **25**, which underwent standard manipulations to yield ammonium Kdo.

A few years later, Shiba completed the stereoselective synthesis of ammonium Kdo through the formation of a dithioketal (Scheme 4.8).[58] First, primary alcohol **26** was obtained in high-yield from diacetonide-protected mannose in a four-step sequence involving: (1) reduction of the aldehyde into a primary alcohol; (2) protection of the latter with a 2,2,2-trichloroethoxycarbonyl (Troc) group; (3) acetylation of the remaining OH; and (4) cleavage of the Troc protecting group using activated Zn in the presence of acetic acid. To lengthen the carbon chain, alcohol **26** was triflated and the crude material was reacted at low temperature with lithiated methyl glyoxylate dithioacetal **27**, available through condensation of glyoxylic acid with 1,2-dimethyl-4,5-bis(mercaptomethyl)benzene. Resulting compound **28** was deacetylated and the dithioacetal group was removed with NBS in an aqueous solution of acetone, thus furnishing diisopropylidene methyl ester Kdo derivative **29** as a mixture of anomers. Although not shown in Scheme 4.8, this compound was further converted into Kdo ammonium salt following cleavage of the isopropylidene groups and hydrolysis of the methyl ester.

The van Boom group reported the synthesis of Kdo as part of a program aiming at gaining more insights regarding the immunogenic potential of Kdo-containing fragments of *N. meningitidis* inner core region.[59,60] In 1989, they first revealed that Shiba's approach could be improved by forming a 1,4-cyclic sulfate of *d*-mannose diacetone, which could efficiently be opened by an anion of ethyl glyoxylate dithioketal.[60] A year later, despite the efficiency of this method, the authors chose to take advantage of the Wittig reaction for the two-carbon elongation as the ylide employed is inexpessive and the Wittig reaction is highly stereoselective.[59] Thus, similarily to Overend's work, *d*-mannose diacetone was elongated with (ethoxycarbonylmethylene)triphenylphosphorane in the presence of catalytic benzoic acid in heated toluene, resulting in the formation of ethyl octenoate **21** (Scheme 4.9). The latter, exclusively formed in the *manno*-configuration, was reduced and the free alcohol was silylated into intermediate **30** in an almost quantitative yield. Compound **30** was then reacted with the anion of methanethiosulfonate to furnish dimethyl dithioketal **31**. Tetra-*N*-butylammonium fluoride (TBAF) was employed to cleave the silyl ether and the ketone function was unmasked following treatment with NBS, allowing hemiketalisation into Kdo derivative **32**.

Following their work on the synthesis of Kdo from *d*-arabinose,[49] the Schmidt group proposed an alternative straightforward approach using *d*-mannose as starting material.[61] As shown in Scheme 4.10, acylation of

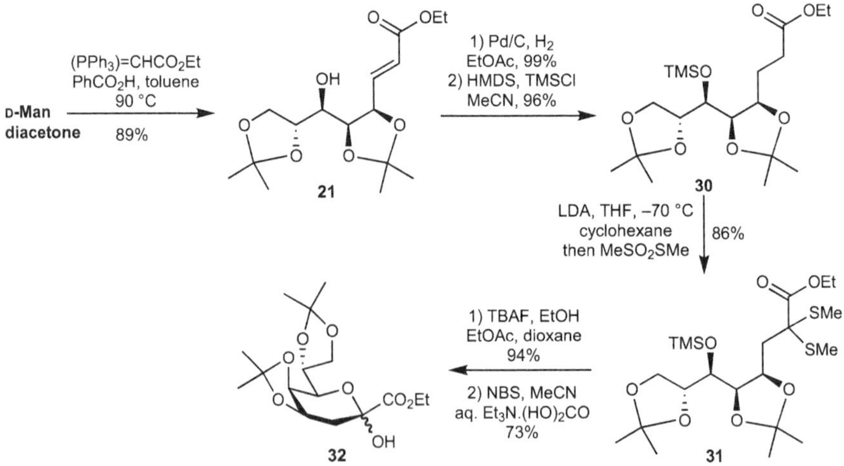

Scheme 4.9 ***van Boom synthesis of Kdo.***[59,60]

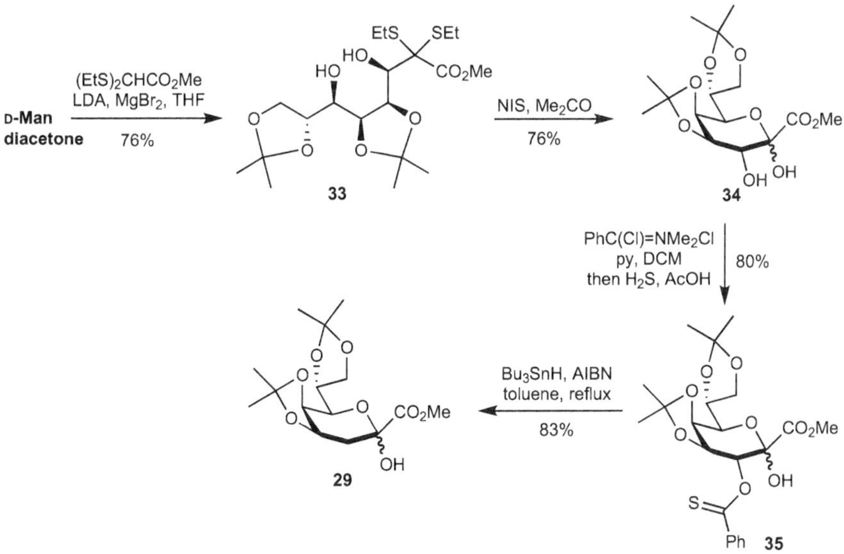

Scheme 4.10 ***Schmidt second generation synthesis of Kdo.***[61]

the latter was completed following its reaction with the anion of methyl glyoxylate diethyl mercaptal. Importantly, the reaction was conducted in the presence of $MgBr_2$, which allowed transmetallation of the less reactive lithio glyoxylate derivative. Resulting dithioketal **33** was unmasked using *N*-iodosuccinimide (NIS), yielding C3 alcohol **34**. As described by Barton,[62] dehydroxylation was accomplished by reacting intermediate **34** with *N,N*-dimethyl-α-chlorobenzimidium chloride followed by hydrogen sulfide, furnishing 2-*O*-thiobenzoyl **35**, which was converted into Kdo precursor **29** using tributyltin hydride. Interestingly, not only did this approach enable the preparation of D-*glycero*-D-*talo*-oct-2-ulosonic acid (Ko) and its C3 epimer, but the authors also highlighted the efficacy and versatility of their method by preparing Ko- and Kdo-containing disaccharides using diol **33** as starting material.

A new and versatile intermediate involved in Kdo synthesis and some analogues, *i.e.*, 1-thio-1,2-isopropylidene acetal **37**, was conceived by Mootoo group (Scheme 4.11).[9] This intermediate was prepared from diisopropylidenated *d*-mannose, which was converted into 1-*O*-acetate-1,2-*O*-isopropylidene **36** under the action of iodobenzenediacetate and I_2. Compound **36** reacted with thiophenol in the presence of a Lewis acid then it was saponified, yielding key intermediate **37**.The free hydroxyl group was esterified with furoic acid and resulting enol ether **38** was methylenated

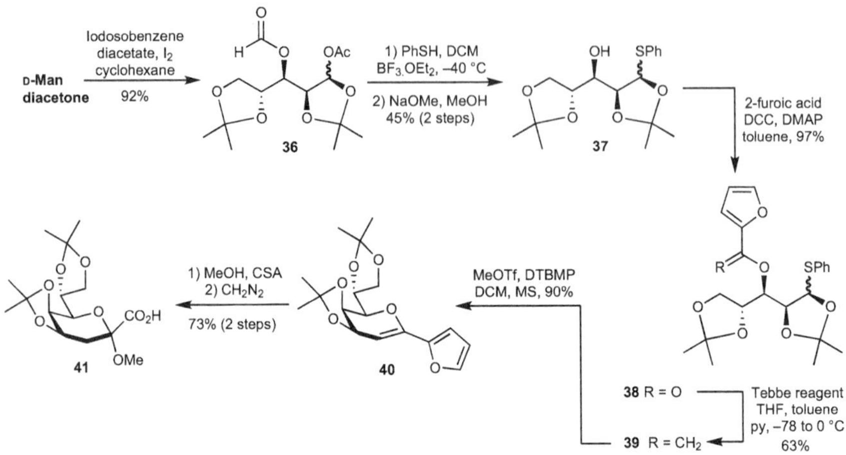

Scheme 4.11 ***Mootoo synthesis of Kdo.***[9]

under Tebbe's conditions into alkene **39**. Cyclization of the latter into glycal **40** was accomplished through the action of trifluoromethane sulfonate (MeOTf) in the presence of 2,6-di-*tert*-butyl-4-methylpyridine (DTBMP). Methanolysis followed by oxidative degradation of the furan furnished Kdo precursor **41** as a single ketal acid. Acid hydrolysis of the latter could easily be achieved to provide Kdo. Alternatively, the authors showed that glycal **40** could be used to prepare 2-deoxy-β-Kdo analogs through hydrogenation, previously described furan processing, basic isomerization, and hydrolysis of the acetals. The Mootoo group additionally synthesized β-*C*-glycosides of Kdo from alcohol **37** through a seven-step sequence, thus illustrating the versatility of their method.

Following their work on the preparation of [4.5]spiroketal glycosides from 2-oxopropyl glycosides *via* an alkylidene-carbene C—H insertion,[63] Wardrop and co-workers conceived an innovative approach to the synthesis of Kdo based on this type of intermediate.[64] First, as depicted in Scheme 4.12, olefination of *d*-mannose diacetone *via* a Wittig reaction followed by $Hg(OAc)_2$-catalyzed pyrolysis of the resulting mixture of enol ethers furnished intermediate **42**. Addition of methallyl alcohol to the latter was accomplished under acidic catalysis with complete diastereoselectivity, affording exclusively an α-glycoside, which underwent oxidative cleavage by the action of OsO_4 and $NaIO_4$. Reaction of the resulting 2-oxopropyl glycoside **43** with [(trimethylsilyl)diazomethane]lithium, as already reported, furnished spiroketal glycoside **44**. The latter was ozonolyzed to give a keto aldehyde, which was converted into α-Kdo glycoside **45**

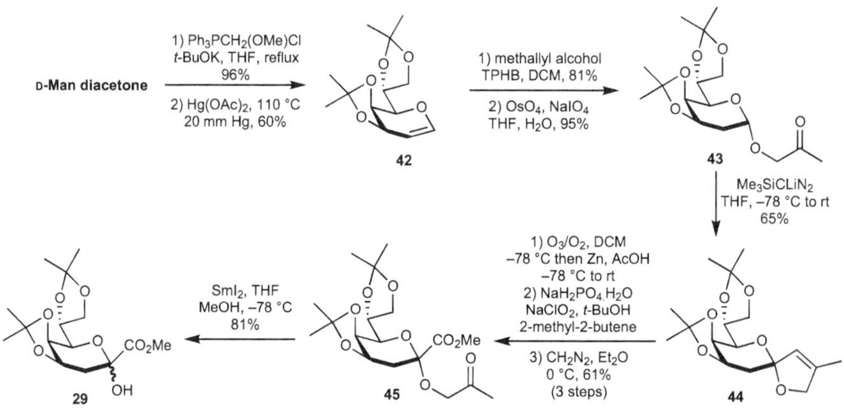

Scheme 4.12 ***Wardrop synthesis of Kdo.***[64]

through oxidation and methylation steps. Treatment of compound **45** with SmI_2 yielded hemiketal **29** as a 5:1 α/β mixture and cleavage of the isopropylidene groups followed by saponification allowed to isolate ammonium Kdo. Importantly, intermediate **45** could act as a potential Kdo glycosyl donor, as the 2-oxopropyl group can be readily converted into an acetoxymethyl group.

Due to the reliability of the Wittig olefination, Kuboki also started their Kdo synthesis through the reaction of *d*-mannose diacetone with the C2 ylide (ethoxycarbonylmethylene)triphenylphosphorane, yielding olefin **21** as a 18:1 *E*/*Z* mixture (Scheme 4.13).[65] Dihydroxylation using OsO_4 and *N*-methylmorpholine *N*-oxide (NMO) furnished a complex stereoisomeric mixture of compound **46**. A key step of the synthesis was the

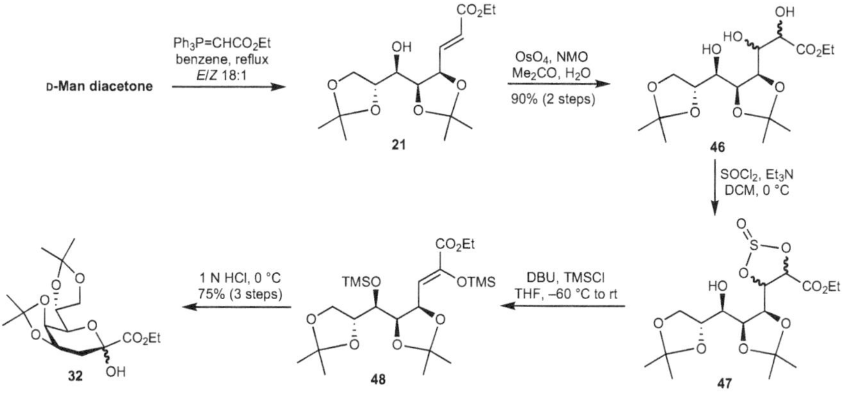

Scheme 4.13 ***Kuboki synthesis of Kdo.***[65]

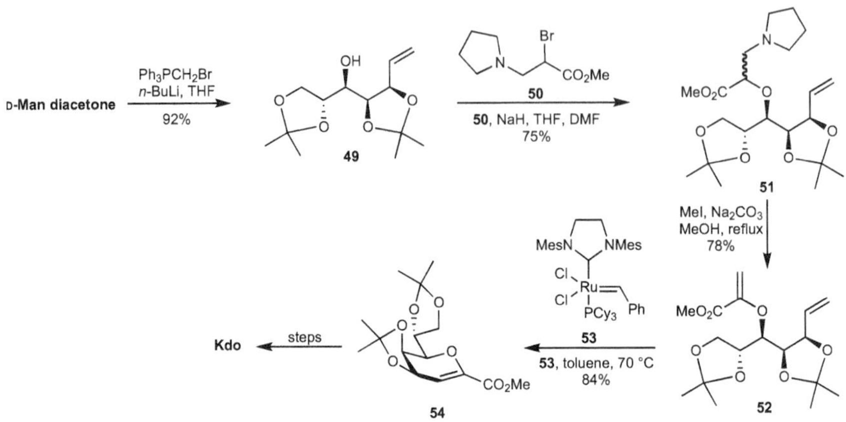

Scheme 4.14 ***Rutjes synthesis of Kdo.***[66,67]

conversion of diol **46** into five-membered cyclic sulfite **47** without affecting the C-6 hydroxyl group and subsequent β-elimination with DBU in the presence of trimethylsilyl chloride (TMSCl), affording *bis*-TMS enol ether **48**. Noteworthy, the presence of TMSCl allowed to trap the enol product, thus preventing the formation of a self-condensed butyrolactone. The labile nature of TMS protecting groups expedited their deprotection using 1 N HCl, simultaneously enabling hemiketalisation into diisopropylidene Kdo derivative **32**.

Taking advantage of their expertise on ring-closing metathesis, the Rutjes group developed a novel approach towards the synthesis of Kdo, as reported in Scheme 4.14.[66,67] Following high-yielding Wittig reaction of diacetonide-protected *d*-mannose with methyltriphenylphosphonium bromide, resulting alcohol **49** was reacted with bromide **50**, readily available from 2,3-dibromopropionic acid methyl ester and pyrrolidine, to give ester **51**. *N*-Methylation of the latter and subsequent base-mediated elimination furnished α-enol ether **52**. Ring-closing metathesis of the latter was accomplished using ruthenium catalyst **53** and produced, without any alkene isomerization, glycal **54**, which could be used as an important building block for the synthesis of Kdo-containing oligosaccharides. Final deprotection of precursor **54** was achieved in a three-step manner consisting of: (1) formation of an iodohydrin intermediate using NIS in aqueous acetonitrile; (2) iodine removal under hydrogenation conditions; and (3) acidic cleavage of the isopropylidene groups followed by saponification. As proposed by the authors, this hydroxylation and deprotection sequence is a mild method that could be employed for the synthesis of complex oligosaccharides.

To reach the Kdo carbon skeleton, Yamasaki proposed to conduct the two-carbon chain elongation through treatment of a cyclic sulfate with a glyoxylate equivalent.[68] Thus, starting from *d*-mannose, the authors prepared a 5,6-diol mannofuranose derivative in three steps, which was then reacted with thionyl chloride to afford a cyclic sulfite intermediate. Oxidation of the latter into the corresponding sulfate was followed by its alkylation with ethyl 1,3-dithiane-2-carboxylate glyoxylate equivalent and further conversion into an acetylated lactone. Cleavage of the acetyl group, reductive ring-opening, removal of the dithioacetal with NBS, and acid hydrolysis finally furnished ammonium Kdo.

With the aim of avoiding the formation of undesired furanose side-product, Mong et al. developed a highly selective approach for the preparation of the pyranose isomer of Kdo.[69] Their synthetic scheme started with the olefination of *d*-mannose diacetone through a Wittig reaction and protection of the free C5 hydroxyl group as a benzyl ether (Scheme 4.15). Hydroboration-oxidation of resulting alkene **55** with 9-borabicyclo[3.3.1] nonane (9-BBN) and H_2O_2 yielded primary alcohol **56** as the sole regioisomer. Oxidation of the latter into an aldehyde followed by Corey-Fuchs reaction[70] furnished alkyne **57**. α-Keto ester **58** was then obtained through bromination of intermediate **57** and subsequent oxidative cleavage. Upon hydrogenolysis of the benzyl ether, hemiketal Kdo precursor **29** was isolated in a 1:1 α/β ratio. Interestingly, the authors found that the stereoselectivity was highly dependent of the solvent used for the hydrogenolysis: when methanol was switched to a less polar solvent mixture such as hexanes/ethyl acetate, the α-anomer was highly favoured.

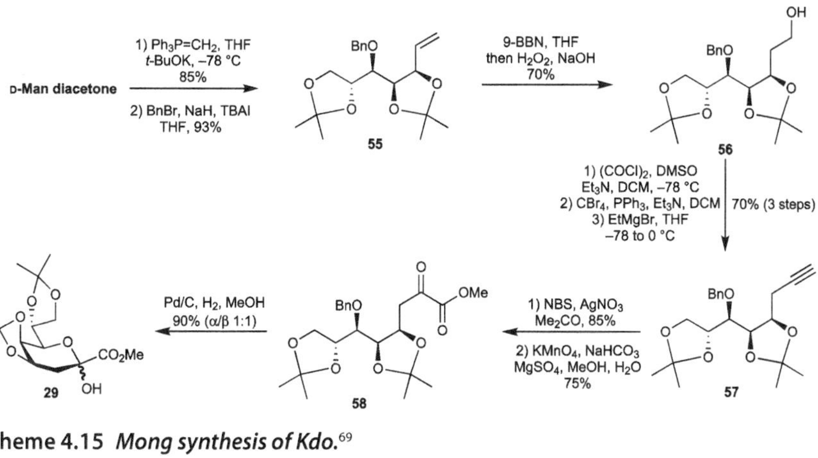

Scheme 4.15 ***Mong synthesis of Kdo.***[69]

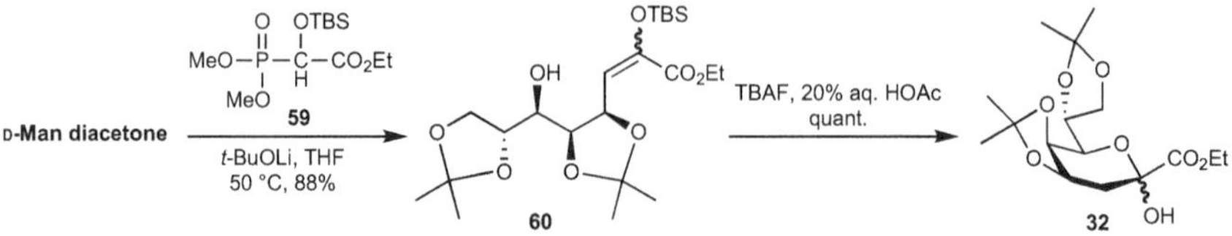

Scheme 4.16 ***Chai synthesis of Kdo.***[71]

Based on the Horner-Wadsworth-Emmons (HWE) condensation, Chai reported a high-yielding and straightforward synthesis of Kdo on a 40-gram scale (Scheme 4.16).[71] Thus, *d*-mannose diacetone underwent an HWE condensation with phosphate ester **59**. Different bases and solvents were tested by the authors and the best results were achieved using *t*-BuOLi in THF. Resulting silyl enol ether **60** was treated with TBAF and acetic acid to cleave the silyl group and allow the hemiketalization, furnishing diisopropylidene Kdo ethyl ester **32** quantitatively, which was further transformed into ammonium Kdo following previously described conditions.

2.3 Total synthesis of Kdo from *d*-mannitol

As previously stated, although not widely used, *d*-mannitol is another C-6 building block that can be employed for the [2 + 6] Kdo synthesis methodology. Among others, Tsukamoto and Takahashi employed *d*-mannitol triacetone as the starting material for their proposed synthetic scheme (Scheme 4.17).[72] Partial hydrolysis of the latter was achieved using concentrated HCl in aqueous EtOH, yielding diol **61**. The primary alcohol was selectively tosylated whereas the secondary alcohol was protected as a silyl ether with *tert*-butyldimethylsilyl chloride (TBSCl). Iodination of the

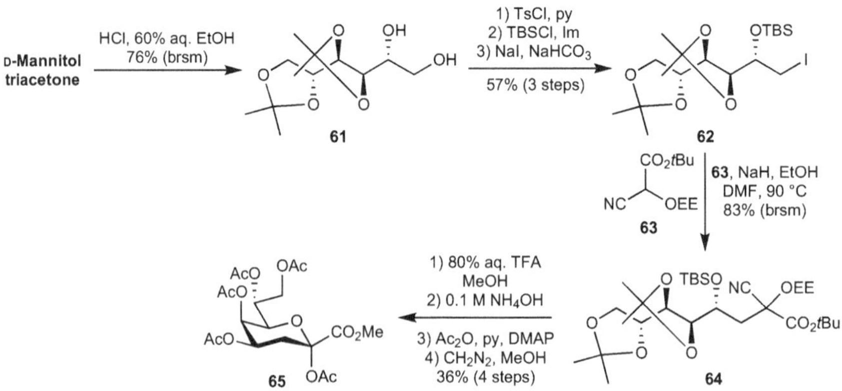

Scheme 4.17 ***Takahashi synthesis of Kdo.***[72]

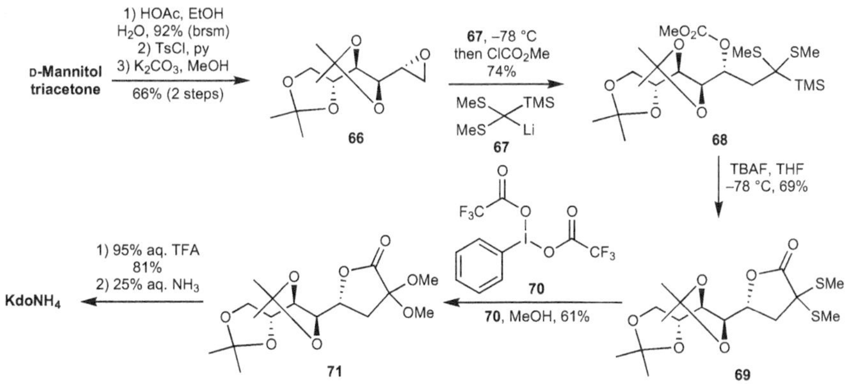

Scheme 4.18 ***Schaumann synthesis of Kdo.***[73]

tosylate yielded intermediate **62** ready to be elongated. Iodine **62** was alkylated with protected cyanohydrin **63** through the action of sodium hydride, thus furnishing compound **64**. Global deprotection was completed through reaction with aqueous TFA in methanol and treatment with aqueous ammonia, affording the ammonium Kdo. For further identification, the latter was converted into pentaacetylated methyl ester **65** *via* a two-step sequence.

As depicted in Scheme 4.18, instead of relying on the usual [2 + 6] methodology, Schaumann designed a [1 + 1 + 6] approach for the preparation of the Kdo skeleton.[73] Partial hydrolysis of *d*-mannitol triacetone furnished a 1,2-diol in which the primary alcohol was selectively tosylated. Cycloelimination of the tosyl group under the promotion of K_2CO_3 afforded epoxide **66**, which was opened with lithiated silyl dithioacetal **67**. *O*-Acylation of the resulting intermediate yielded carbonate **68** composed of a seven-carbon framework. The latter underwent a silicon-induced lactonization using TBAF to form the reactive carbanion, furnishing **69** featuring the Kdo framework. Transformation of the dithioketal into dimethylketal **71** was completed using compound **70** in methanol. Treatment with aqueous TFA enabled to release the carbonyl group and to cleave the diisopropylidene groups. Ultimately, treatment of the resulting enol with aqueous ammonia allowed the isolation of ammonium Kdo.

Rather than employing *d*-mannitol triacetone as starting material, Sametz and Burke developed an innovative elongation approach of *d*-mannitol (Scheme 4.19).[74] Hence, one-pot conversion of the latter into dibromide **72** was achieved by employing acetyl bromide and treatment with acetic anhydride. Double reductive elimination using activated zinc followed by methanolysis furnished dienediol **73**. Formation of a stannylene

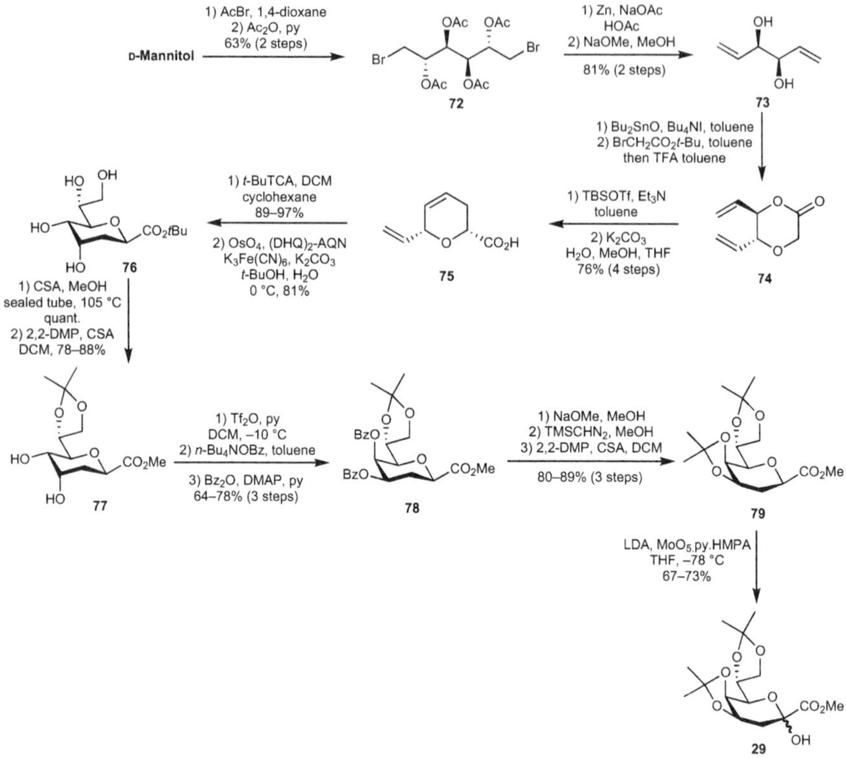

Scheme 4.19 ***Burke synthesis of Kdo.***[74]

acetal of dienediol **73** allowed its alkylation with *tert*-butyl bromoacetate with concomitant cyclization, whose completion was ensured by treatment of the crude material with TFA. Resulting dioxanone **74** was transformed into a silyl ketene acetal with TBSOTf and the resulting intermediate was subjected to Ireland-Claisen rearrangement and hydrolysis,[75,76] yielding dihydropyran **75**. The acid group was protected as a *tert*-butyl ester and double Sharpless asymmetric dihydroxylation was employed to produce tetraol **76**. *tert*-Butyl ester **76** was transesterified with methanol and the 7,8-diol was selectively protected with an isopropylidene group. This step allowed the double epimerization at positions C4 and C5 through a three-step sequence, involving the formation of a ditriflate, treatment with n-Bu_4NOBz to give a mixture of two monobenzoate alcohols with inversion at both C4 and C5 positions, and benzylation of the mixture to furnish dibenzoate **78** in the requested configuration. The authors hypothesized that this double epimerization occurred through the formation of a bridged acyloxonium species following the displacement of the second triflate by an intermediate monobenzoate. The benzoyl groups were then cleaved by saponification, the acid was reesterified with $TMSCHN_2$, and the resulting diol was protected, yielding diisopropylidene methyl ester **79**. Finally, oxidation of the lithium enolate of the latter using MoO_5·py·HMPA [hexamethylphosphoramide] furnished Kdo precursor **29**, which was eventually deprotected and isolated as its ammonium salt.

2.4 Total synthesis of Kdo from other precursors

The third type of approaches towards the preparation of Kdo are miscellaneous and thus include syntheses from non-carbohydrate precursors instead of smaller sugars. In that regard, Danishefsky proposed the total synthesis of Kdo through the cyclocondensation of oxygenated dienes with aldehydes.[77] Specifically, they achieved the cyclocondensation of a 1,3-butadiene bearing a furyl residue and an alkoxy group at C1 with α-(phenylseleno)propionaldehyde through the catalysis of BF_3 and treatment with TFA. Upon formation of the resulting glycoside precursor, ammonium Kdo was reached following ten functionalization steps.

Similarily, the Augé group designed an innovative approach for the synthesis of Kdo based on a hetero Diels-Alder cyclization using chiral diene **80** and butyl glyoxylate dienophile **81**, as depicted in Scheme 4.20.[78] The structure of compound **80** enabled a modest control of the facial selectivity,

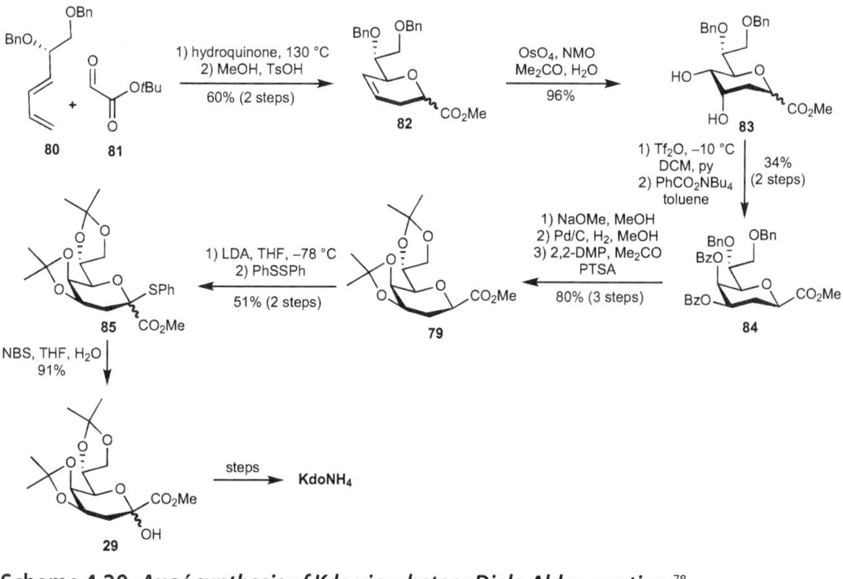

Scheme 4.20 ***Augé synthesis of Kdo via a hetero Diels-Alder reaction.***[78]

furnishing a mixture of *Si* (major) and *Re* product **82**. *Cis*-dihydroxylation of the latter was accomplished under the catalysis of OsO_4 in the presence of NMO, yielding diol **83**. C4 and C5 epimerization was achieved simultaneously by forming triflates of both alcohols followed by nucleophilic substitution with tetrabutylammonium benzoate. The major diastereoisomer **84** was isolated from the other anomer following flash chromatography. Deprotection of the benzoyl and benzyl groups was achieved, respectively, by methanolysis and hydrogenolysis. The resulting tetraol was converted into diisopropylidene 2-deoxy-α-Kdo methyl ester **79**. To achieve the complete Kdo structure, the enolate of intermediate **79** underwent sulfenylation with diphenyldisulfide, furnishing thioglycoside **85** as a mixture of anomers. The latter could either be used directly as a glycosyl donor for O- or C-glycosidation or hydrolysed into methyl ester **29** and subsequently deprotected into Kdo.

Following this work, Wu and co-workers were interested in developing a hetero Diels-Alder reaction with higher yields and selectivities for the synthesis of Kdo.[79] Their methodology was innovative in that they employed a (Salen)Co^{II} complex as a catalyst, which had never been reported for Diels-Alder type reactions. Furthermore, the authors achieved good *endo/exo* and *Si/Re* selectivities, highlighting the applicability of their method for the synthesis of Kdo as well as of 2-deoxy-Kdo.

Based on their previous work where they reported the high *syn*-selectivity of the enolate of a vinylogous urethane in aldol reactions,[80] Pettus and co-workers devised a method to prepare Kdo from vinylogous urethane **86** (Scheme 4.21).[81] Thus, the enolate of the latter was formed through the action of lithium diisopropylamide (LDA) and was reacted with acrolein equivalent (*E*)-3-(tri-*n*-butylstannyl)-2-propenol, yielding *syn*-product **87**. The authors suggested that the high diastereoselectivity of the reaction was due to cyclic reactive species produced through the formation of a lithium-nitrogen bond in the enolate of compound **86**.[82] Acidic hydrolysis of the vinylogous urethane functionality of intermediate **87** and concomitant destannylation furnished keto-lactone **88**, which underwent hydride reduction to yield single stereoisomer **89**. Ag_2CO_3 and Celite were then employed to convert compound **89** into a hydroxy lactone, the trimethylsilyl ether was removed *via* treatment with $BF_3{\cdot}OEt_2$ and the resulting diol was protected with an isopropylidene group, providing acetonide **90**. Stereospecific catalytic osmylation allowed the installation of the 7,8-diol characteristic of Kdo. Subsequent protection of the resulting diol as an acetonide furnished diisopropylidene lactone **91** as the sole product. Finally,

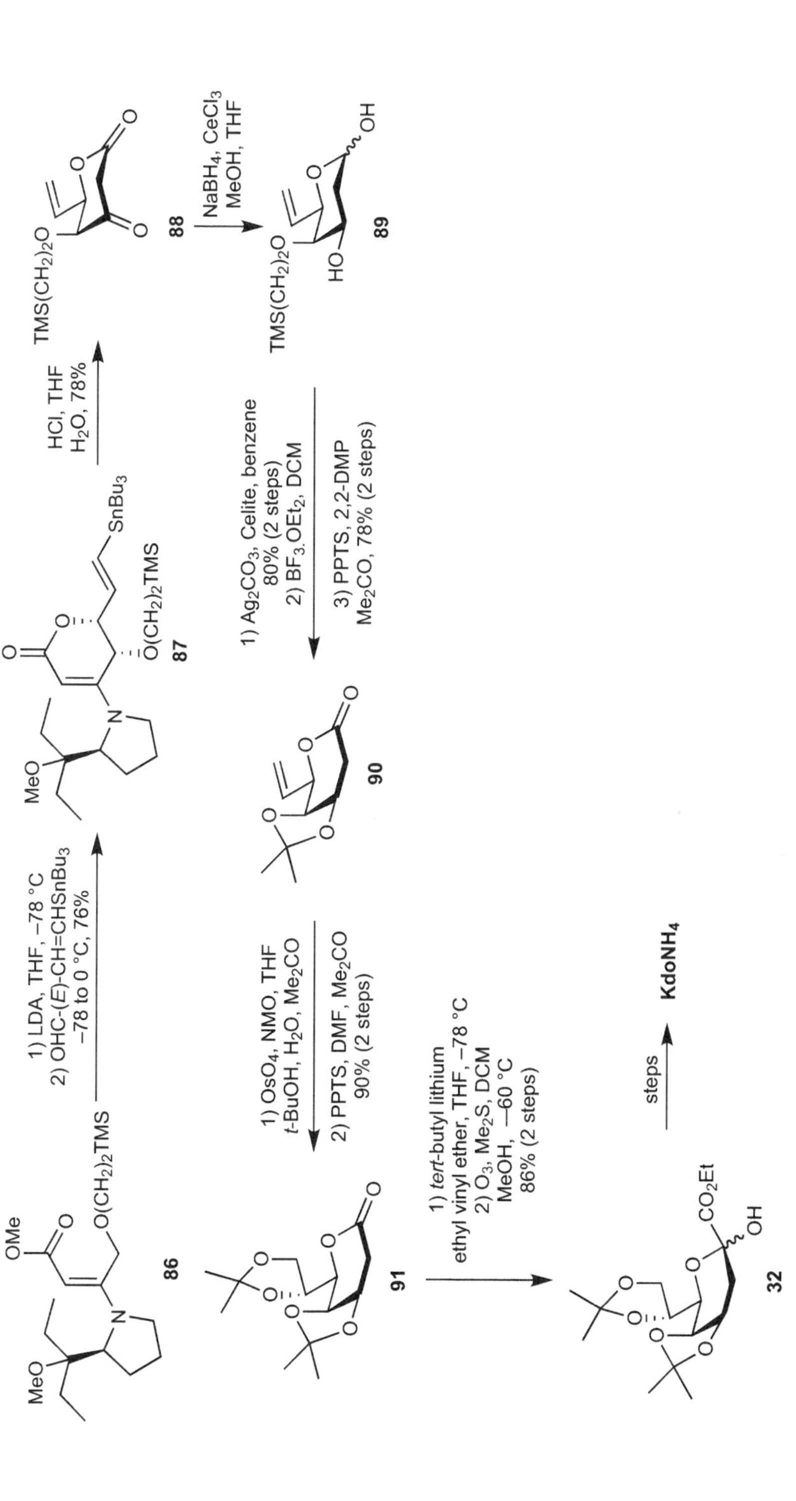

Scheme 4.21 *Pettus synthesis of Kdo.*[81]

installation of the carboxylic acid function as well as the C2 hydroxyl group to provide ethyl ester **32** was achieved in a two-step fashion: (1) addition of 1-ethoxyvinyllithium to the lactone; and (2) reductive ozonolysis with O_3 and Me_2S. Noteworthy, the usefulness of this approach for the introduction of a carboxylic acid moiety on a lactone was demonstrated by the Singh group, who employed a similar methodology for the preparation of Kdo ethyl ester **32**.[83]

As highlighted by Singh and Pettus, lactones are particularly useful for the preparation of ulosonic acids such as Kdo. Portella was therefore interested in the development of a more reliable method for the synthesis of lactone **91**.[84] Thus, they developed a high-yielding synthetic approach without any intermediate purification involving a three-step sequence consisting of: (1) treatment of *d*-mannose diacetone with lithiated bis(methylsulfanyl) trimethylsilylmethane; (2) cyclization of the resulting ketene dithioacetal in acidic conditions; and (3) treatment with $CaCO_3$, water, and I_2 to furnish lactone **91** in 82% overall yield.

Besides the chemical synthesis of Kdo, there has also been progress in its enzymatic biosynthesis. In 1992, Wong et al. prepared Kdo through an aldolase-catalyzed condensation of *d*-arabinose and pyruvate, which introduced a new *R* stereocenter at the C3 position.[85] The Kdo aldolase was isolated from a Gram-positive bacterium, that is, *Aureobacterium barkerei* strain Kdo-37-2. With this enzyme, the authors were able to synthesize Kdo on a multi-mmol scale with an overall yield of 37% following crystallization and isolation of the product as its ammonium salt. Importantly, this enzyme showed wide substrate specificity and could therefore be employed for the synthesis of lower homologs of Kdo. Alternatively, Pohl and co-workers showed that the fermentative production of Kdo could be achieved from inexpensive *d*-glucose using an engineered *E. coli* strain, which overproduces D-arabinose-5-phosphate isomerase, Kdo-8-phosphate synthase, and Kdo-8-phosphatase.[86]

3 Synthesis of Kdo derivatives as potential inhibitors of Kdo-processing enzymes

Due to the emergence of drug-resistant bacteria, there is an urgent need for developing new therapeutic agents that could act as inhibitors of antibacterial targets. In that regard, the biosynthesis of LPS is a particularly attractive target since this crucial structural component is highly characteristic of GNB. As such, synthesizing Kdo analogues that could not be converted into the normal biosynthetic product could result in the disruption of LPS biosynthesis, thus affecting cell viability. Among the most promising enzymatic targets, CKS has attracted a lot of attention as it is the rate-limiting enzyme of LPS biosynthesis. The following sections will therefore cover the different approaches developed towards the synthesis of 2-deoxy-β-Kdo and derivatives, Kdo *C*- and *O*-glycosides as well as miscellaneous Kdo analogs as potential inhibitors of Kdo-processing enzymes.

3.1 Total synthesis of 2-deoxy-β-Kdo

In 1987, a pioneering study was published by Claesson reporting the inhibitory activity of the first synthesized CKS competitive inhibitor, 2-deoxy-β-Kdo **92** (Fig. 4.7).[87] Although a potent inhibitor, this Kdo derivative lacked *in vivo* antibacterial activity due to its incapacity to cross the cytoplasmic GNB membrane where lies the enzymatic target.[88] Similarily, 8-amino derivative **93** showed high competitive inhibitory activity but also failed to kill GNB.[88-90] This derivative nonetheless proved useful for the synthesis of peptidic derivatives, which could be transported into the cytoplasm through an oligopeptide permease system, where the inhibitor could then be released by the action of intracellular aminopeptidases.[88-90]As expected, peptidic prodrugs **94**,[88-90] **95**,[88] **96**,[90] and **97**[90] showed high antibacterial activity against *Salmonella typhimurium*and *E. coli*. Although less useful *in vivo* since these peptidic bonds are degraded by mammalian peptidases[88,89] and the targeted transport mechanisms are prone to mutations and thus resistance, these results still highlight the potential of such approaches for the inhibition of Kdo-processing enzymes. Additionally, the therapeutic

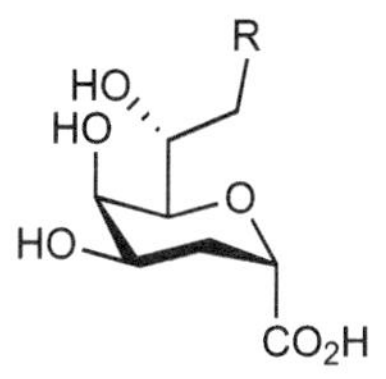

92 R = OH (**2dβKdo**)
93 R = NH_2 (**2dβKdoNH_2**)
94 R = NH-Ala-Ala
95 R = NH-Met-Pro
96 R = NH-Nva-Ala
97 R = NH-Lys-Lys

Figure 4.7 ***Structure of 2-deoxy-β-Kdo*** **(92),** ***8-amino-2,8-dideoxy-β-Kdo*** **(93)** ***and corresponding peptidic prodrugs*** **(94–97)** ***as inhibitors of CKS.***

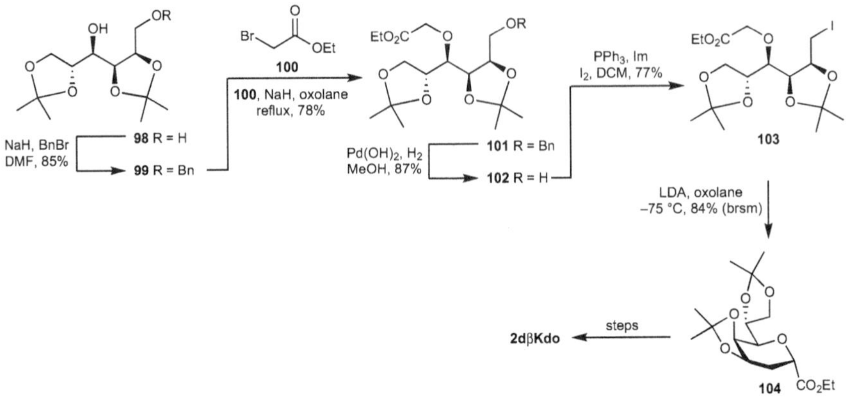

Scheme 4.22 ***Meguro synthesis of 2-deoxy-β-Kdo.***[92]

potential of such inhibitors, as demonstrated by Nishimura and co-workers, goes beyond the simple disruption of LPS biosynthesis. Indeed, peptidic prodrugs such as compounds **94** and **96** possessed synergetic activities with antibiotics like fosfomycin and kanamycin.[91]

Because of the high CKS inhibitory activity of 2-deoxy-β-Kdo, its total synthesis has been a continuous challenge for many organic chemistry groups. As shown in Scheme 4.22, the Meguro group developed a stereospecific synthesis starting from *d*-mannose derivative **98**.[92] Thus, the primary alcohol of the latter was selectively benzylated into compound **99**, which was then alkylated with ethyl bromoacetate **100**. Resulting ester **101** underwent hydrogenolysis to reveal the primary hydroxyl group. Introduction of an iodine leaving group at C1 was then accomplished following treatment of alcohol **102** with PPh_3, imidazole, and I_2. In contrast, when attempting to introduce a tosyl group at C1, the main product was an undesired tetrahydrofuran derivative. Intramolecular cyclization of intermediate **103** was achieved through the formation of its ester enolate, resulting in ethyl ester β-Kdo derivative **104**, which could be further deprotected into target 2-deoxy-β-Kdo.

Without using any catalyst, López-Herrera accomplished an aldol-type condensation of reactive diisopropylidene aldehyde **105** with diazo compound **106**, yielding a β-hydroxy-α-diazo carbonyl intermediate (Scheme 4.23).[93,94] The latter was directly acetylated as to avoid cyclization that would have resulted following *O*-6 to *O*-3 acetyl migration in the presence of silica gel. Treatment of diazo compound **107** with rhodium acetate furnished α-acetoxy-α,β-unsaturated carboxylate **108** as a 1:1 *Z*:*E*

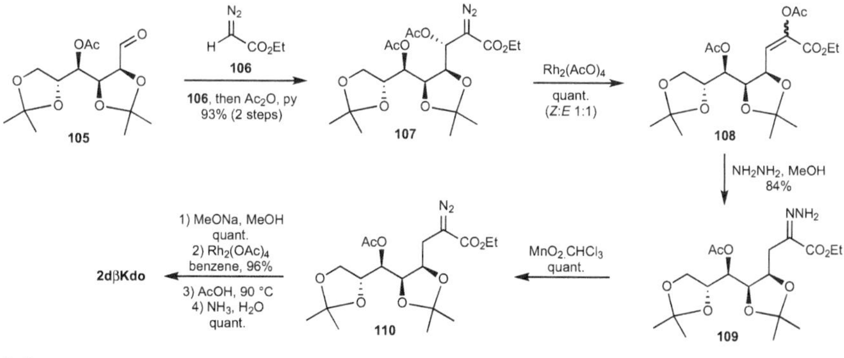

Scheme 4.23 ***Lopez-Herrera synthesis of 2-deoxy-β-Kdo.***[93,94]

mixture. As attempts at converting the latter into the corresponding ketone derivative proved unsuccessful, intermediate **108** was hydrazinolyzed into the more stable compound **109**. Oxidation of hydrazine **109** with manganese dioxide furnished diazo intermediate **110**, which was deacetylated and further treated with rhodium acetate in anhydrous benzene. Interestingly, when conducting the rhodium-mediated cyclization in anhydrous benzene, pure α-Kdo precursor was formed. Contrarywise, when non-anhydrous solvents were employed, a mixture of isomers was obtained presumably due to the solvatation of the free hydroxyl group, which would favour a conformational equilibrium. Saponification and concomitant partial epimerization followed by treatment with aqueous ammoniafinally furnished target 2-deoxy-β-Kdo in excellent yield. Alternatively, intermediate **110** could be readily converted into Kdo precursor **32** following deacetylation and oxidation with *meta*-chloroperoxybenzoic acid (*m*-CPBA).

3.2 Synthesis of 2-deoxy-β-Kdo derivatives

Owing to the inability of 2-deoxy-β-Kdo at crossing the cytoplasmic membrane, synthesis of derivatives of the latter has been tremendously investigated to circumvent this issue and to generate potentially more active compounds. Following a study revealing that carbocyclic Kdo derivatives showed no inhibitory activity,[95] it was hypothesized that the intracyclic oxygen atom forms H-bond with the enzyme that are essential for its activity. Thus, Norbeck and co-workers synthesized azacyclic 6dβKdo derivative **116** (Scheme 4.24) with the hope that it could donate strong H-bond to CKS and therefore show enhanced inhibitory activity.[96] With that aim in mind, conversion of Kdo into alcohol **111** was first achieved through a four-step sequence involving isopropylidenation to distinguish the C6 hydroxyl

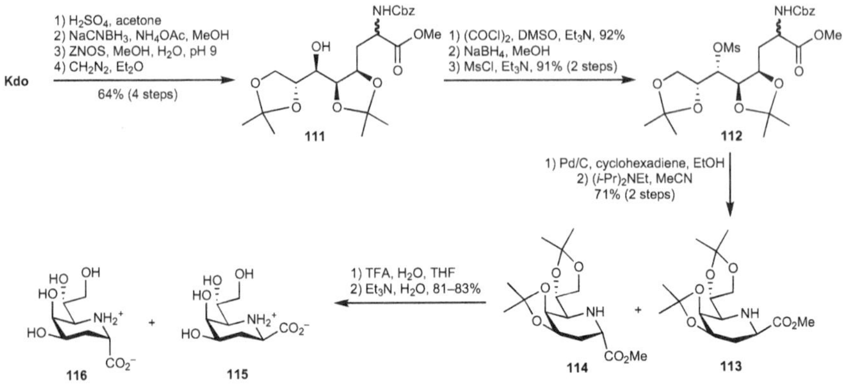

Scheme 4.24 ***Norbeck synthesis of azacyclic 2-deoxy-Kdo derivatives.***[96]

group, reductive amination, protection of the resulting amine, and esterification of the carboxylic acid moiety. Compound **111**, obtained as a 2:1 mixture of diastereoisomers, underwent Swern oxidation and stereoselective reduction to epimerize the C6 hydroxyl group, which was then mesylated. Phase-transfer conditions were used for the cleavage of the benzyl chloroformate protecting group and treatment of the resulting amine with diisopropylethylamine furnished cyclic amines **113** and **114**. Deprotection of these latter, priorly separated by column chromatography, respectively furnished anomers **115** and **116** in good and similar yields. Unfortunately, azacylic 6dβKdo **116** only showed modest inhibitory activity against CKS whereas anomer **115** showed, as expected, lower activity.

As depicted in Scheme 4.25, Pring designed 2-deoxy-β-Kdo derivatives differently modified at the C8 position.[97] As such, 8-deoxy-8-thio derivative **118** was prepared following treatment of 2-deoxy-β-Kdo methyl ester **117** with diethyl azodicarboxylate (DEAD), Ph_3P, and AcSH followed by a two-step deprotection of the resulting 8-acetylthio intermediate. In parallel, tritylation of the *O*-8 position, benzylation of the remaining hydroxyl groups, and trityl cleavage furnished key intermediate **120**. The latter was tosylated and reacted with NaI to afford iodide **121**, which underwent hydrogenation, saponification, and treatment with cation exchange resin to furnish target derivative **122**. Alcohol **120** was alternatively fluorinated into compound **123**, which provided 8-deoxy-8-fluoro-2-deoxy-β-Kdo analog **124** upon deprotection. The authors were also able to prepare amino acid **93** from the benzyl ester derivative of compound **120**. Although biological assays highlighted the potent inhibitory activity of derivatives **122** and **124**, none of the synthesized compounds showed antibacterial activities.

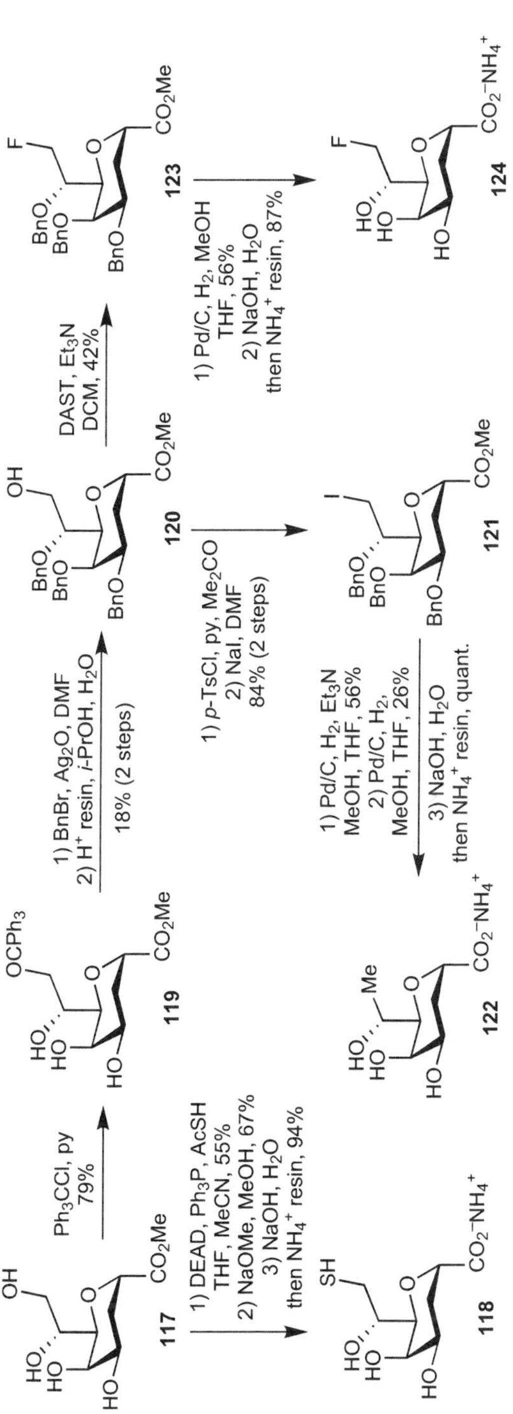

Scheme 4.25 ***Pring synthesis of 2-deoxy-β-Kdo derivatives modified at the C8 position.***[97]

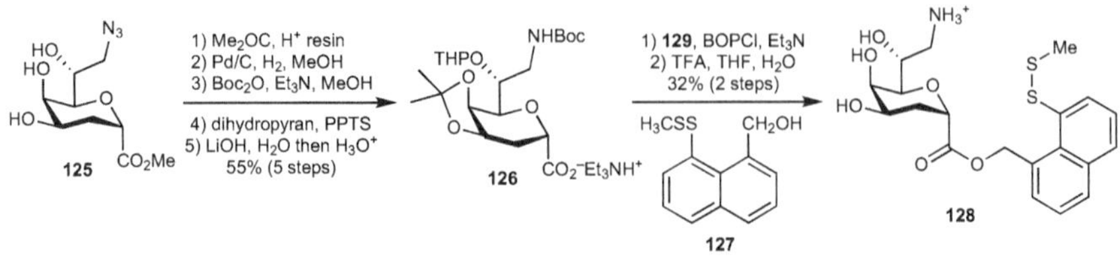

Scheme 4.26 ***Norbeck synthesis of a disulfide prodrug of 8-amino-2,8-dideoxy-β-Kdo.***[98]

Contrarywise, thiol derivative **118** exhibited lower inhibitory activtiy, which the authors rationalized by steric effects.

Based on the observations that radio-labelled 2-deoxy-β-Kdo methyl ester enters significantly into the cytoplasm but that such esters exhibit no inhibitory activity towards CKS, Norbeck proposed an innovative prodrug strategy for the uptake of 2-deoxy-8-amino-β-Kdo into the cytoplasm.[98] The authors devised derivative **128** (Scheme 4.26) bearing a disulphide ester that would allow passive diffusion through the lipid bilayer and release of the inhibitor through disulphide exchange reactions. Indeed, as GNB possess a significant amount of sulfhydryl compounds,[99] the ester would readily be cleaved into the cytoplasm. It was also expected that this type of derivative would be less prone to resistance from GNB as transport into the cytoplasm results from passive diffusion mechanisms. Synthesis of this potential antibacterial compound started from 2-deoxy-8-azido-β-Kdo methyl ester **125**. 4,5-Diol of the latter was isopropylinated, the azido group was reduced, and the resulting amine was capped with a *tert*-butyloxycarbonyl (Boc) protecting group. Dihydropyran and pyridinium *p*-toluenesulfonate (PPTS) were employed to protect the remaining hydroxyl group as a tetrahydropyran ether and the methyl ester was saponified and converted into triethylamine carboxylate **126**. Esterification of compound **126** with alcohol **127** was achieved through the action of bis(2-oxo-3-oxazolidinyl) phosphinic chloride (BOPCl) and acidic cleavage of the protecting groups furnished disulphide ester **128**. Biological assays revealed that this prodrug was more rapidly released than dipeptidic prodrug **94**, underlining the usefulness of such a prodrug approach. Although lower than that of compound **94**, disulphide ester **128** also exhibited antibacterial activity by potentially inducing the accumulation of lipid A precursors into the outer GNB membrane.

With the aim of probing the structure-activity relationship of C8 derivatives of 2-deoxy-8-amino-β-Kdo, Nishimura and co-workers synthesized derivatives **93** and **131–136** as potential inhibitors of CKS based on the

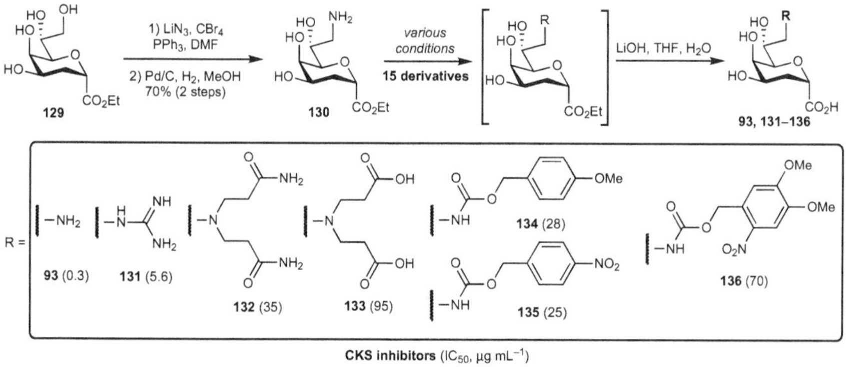

Scheme 4.27 ***Nishimura synthesis of 2-deoxy-β-Kdo derivatives modified at the C8 position.***[100]

structure of the enzyme active site (Scheme 4.27).[100] First, displacement of the C8 hydroxyl group of Kdo ethyl ester **129** was achieved using LiN_3, CBr_4, and PPh_3, then the resulting azide was reduced into amine **130**. The latter underwent different modifications to furnish a series of intermediates, which upon saponification furnished potential inhibitors **93** and **131–136.** Enzymatic inhibition studies underlined that steric effects affected the compounds inhibition ability as derivatives bearing large functional groups at C8 showed lower activity. Additionally, derivative **131** substituted with a guanidino group was less active than amino acid **93** whereas compounds **132**, **134**, and **135** showed moderate inhibition. Noteworthy, other derivatives were prepared but where excluded here as they showed no inhibitory activity.

The Nishimura group further explored the inhibitory contribution of the hydroxyl groups at C4 and C7 of 2-deoxy-8-amino-β-Kdo by synthesizing derivatives **144**, **146**, **149**, and **152** (Scheme 4.28).[101] Compound **141**, a precursor of derivatives **144**, **146**, and **149**, was therefore prepared from ethyl ester of Kdo **129**. An azide group was inserted at C8, the 4,5-*cis*-diol was protected as a benzylidene acetal, and the remaining C7 hydroxyl group was benzylated, furnishing isomers **138** and **139**. The latter was regioselectively opened under the action of titanium tetrachloride and sodium cyanoborohydride, releasing the *O*-4 position. Dess-Martin periodinane was employed to oxidize alcohol **140** into ketone **141**. From this precursor, C4 epimer **144** of 2-deoxy-8-amino-β-Kdo was obtained through the stereoselective reduction of the ketone into alcohol **142** and deprotection. Difluoro derivative **146** was prepared following fluorination of ketone **141** and standard deprotection. Alternatively, diamine **149** was synthesized. This

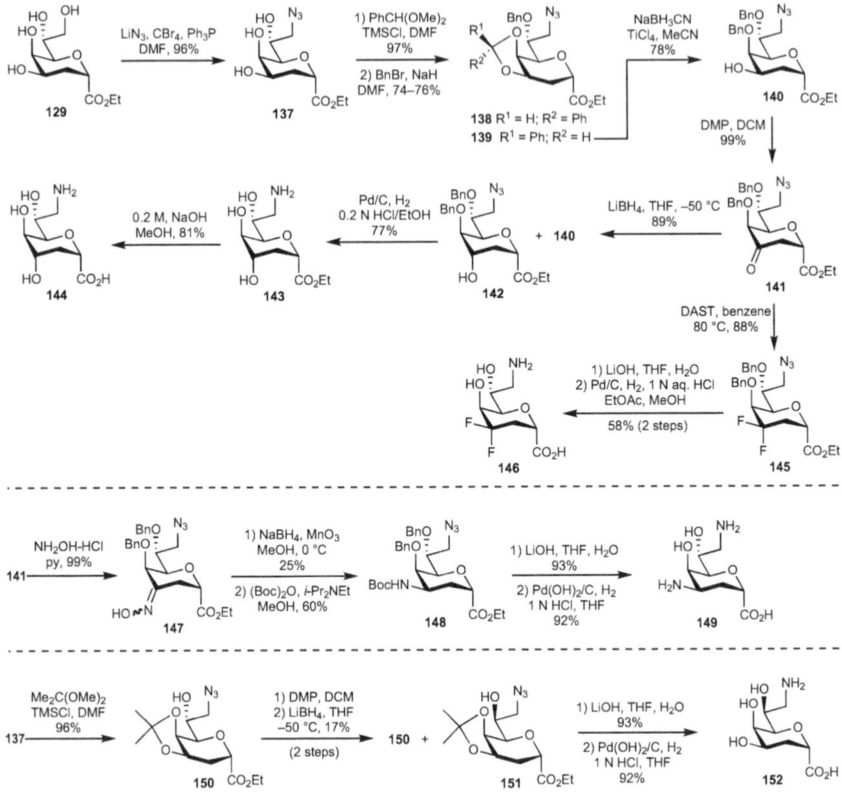

Scheme 4.28 ***Nishimura synthesis of 2-deoxy-β-Kdo derivatives.***[101]

was achieved through a five-step sequence from ketone **141** consisting of the formation of an oxime at C4, reductive amination, protection of the resulting amine with a *tert*-butoxycarbonyl group, hydrolysis, and hydrogenation. Finally, 7-*epi*-2-deoxy-8-amino-β-Kdo **152** was synthesized from intermediate **137**. Thus, isopropylidenation of the 4,5-diol furnished alcohol **150**, which was then oxidized and stereoselectively reduced into intermediate **151**. Deprotection of the latter furnished C5 epimer **152**. Inhibitory activities of these 2-deoxy-8-amino-β-Kdo derivatives against CKS were evaluated to gain more insights into the structural requirements for such activities. Results revealed the crucial role of equatorial C4 hydroxyl group as 4-*epi* and difluoro derivatives **144** and **146**, respectively, showed no inhibition. However, 4,8-diamine **149** exhibited inhibitory activity, although weaker than that of 2-deoxy-8-amino-β-Kdo **93**. Finally, it was highlighted that the C7 hydroxyl group was also required for inhibition of CKS as C7 epimer **152** showed no activity.

3.3 Synthesis of Kdo C-glycosides

Besides 2-deoxy-β-Kdo and analogs, the synthesis of Kdo glycosides has been the target of numerous research groups due to the synthetic challenge they represent. Indeed, such syntheses are challenging due to the presence of an electron-withdrawing and hindered carboxylate at C2 and the absence of a participating group at C3, often leading to the undesired formation of 2,3-glycals. These types of compounds also exhibit potential therapeutic activities. More specifically, there has been tremendous interest in the synthesis of Kdo *C*-glycosides as they stand as exquisite targets for developing potential inhibitors of Kdo-processing enzymes due to the enzymatic and chemical stability of C—C bonds.

Claesson and co-workers prepared a series of *C*-glycosides in a straightforward and β-selective manner relying on the alkylation of an enolate, as depicted in Scheme 4.29.[102] First, pentaacetate Kdo methyl ester **65** was converted into chloride **153** under the action of $TiCl_4$. The latter was hydrogenolyzed in the presence of pyridine as an acid scavenger to produce 2-deoxy-β-Kdo derivative **155** as the major epimer. Cleavage of the acetyl groups under basic conditions furnished a tetraol intermediate, which was converted into diacetonide **156**. Its lithium enolate was formed and reacted with a series of halogenated electrophiles to furnish β-*C*-glycosides **157–164** as sole products.

With the aim of mimicking CMP-Kdo, that is the substrate of rate-limiting enzyme CKS, Norbeck and co-workers reported the synthesis of *C*-glycoside CMP-Kdo phosphonate **171** (Scheme 4.30).[103] Starting from glycal **54**, Raney nickel reduction proceeded exclusively from the unhindered α-face as to provide intermediate **79**. Formation of the enolate of the

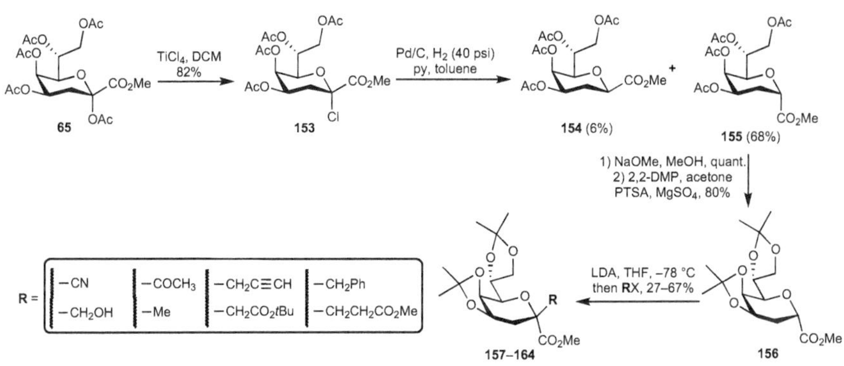

Scheme 4.29 ***Claesson first generation synthesis of Kdo C-glycosides.***[102]

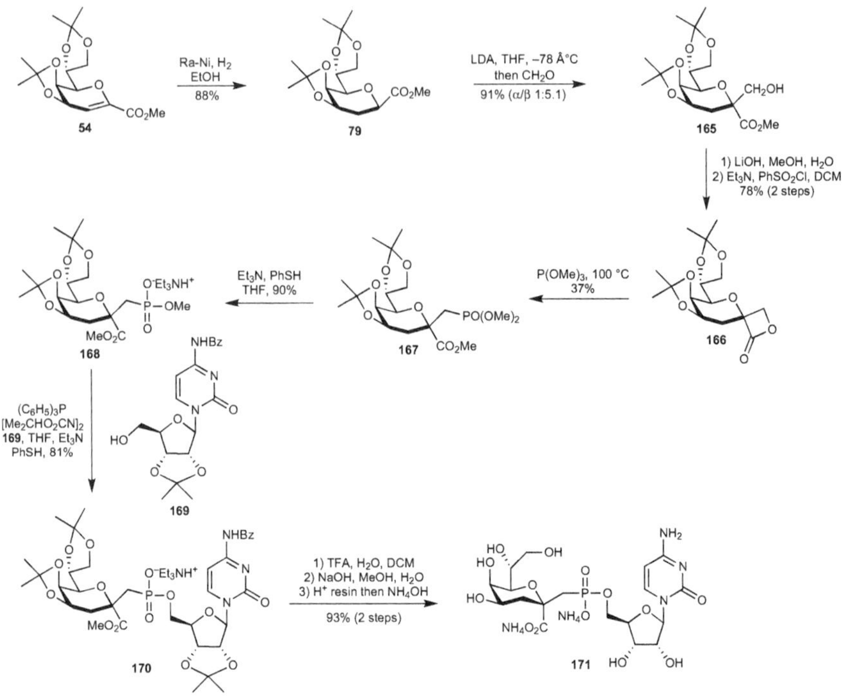

Scheme 4.30 ***Norbeck synthesis of a C-glycoside analogue of CMP-Kdo derivative.*** [103]

latter and addition of gaseous formaldehyde furnished β-*C*-glycoside **165** as the major product and its deprotection allowed to confirm the β-selectivity through NMR studies. Methyl ester **165** was then converted into lactone **166** so that the carboxyl residue could act as a leaving group for the introduction of the phosphonate moiety. Reaction of activated compound **166** with neat trimethylphosphite successfully furnished phosphonate **167**. Demethylation was accomplished with thiophenoxide and resulting intermediate **168** was esterified with cytidine derivative **169** under Mitsunobu conditions. Subsequent demethylation afforded *C*-glycoside **170**, which was readily deprotected into CMP-Kdo phosphonate **171**. The authors showed that the latter was only a modest inhibitor of CKS. Differences in bonding geometries, steric interactions with the methylene hydrogens as well as deletion of a hydrogen bond are plausible explanations for this decreased activity. Nonetheless, as it was shown that 2-deoxy-β-Kdo synergistically binds to the enzyme in presence of cytidine triphosphate (CTP), CTP-Kdo bisubstrates could be valuable tools for the development of CKS inhibitors.

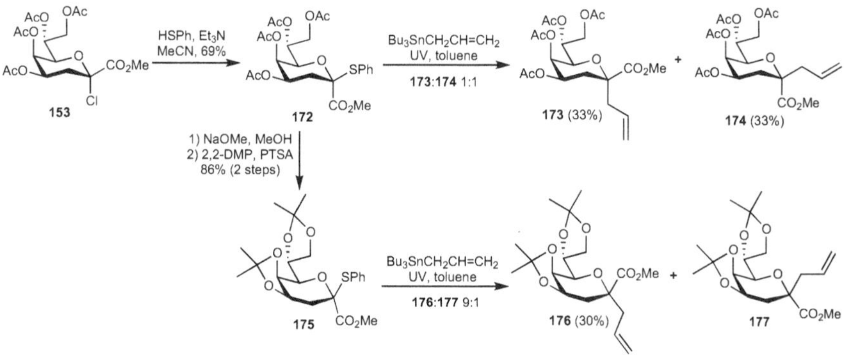

Scheme 4.31 ***Claesson second generation synthesis of Kdo C-glycosides.***[104]

Following their work on the alkylation of enolates for the formation of β-Kdo *C*-glycosides, the Claesson group investigated the impact of protecting groups on the stereoselectivity of photochemically initiated radical reactions.[104] By synthesizing tetraacetylated *C*-glycosides **173** and **174** and diisopropylinated *C*-glycosides **176** and **177** from thioglycosides **172** and **175**, respectively, the authors showed that the use of acetyl groups in such reactions enhanced the β-selectivity of the C-glycosidation (Scheme 4.31). The authors hypothesized that this could be caused by the isopropylidene groups locking the compound in a skew-boat configuration, which might be maintainted during the radical reaction.

In diazo compound **110** obtained during their zinc-catalyzed synthesis of 2-deoxy-β-Kdo,[93,94] the López-Herrera group found an interesting intermediate for the preparation of Kdo *C*-glycosides and other derivatives.[105] Thus, alkyl derivatives **178–180** were synthesized by reacting the deacetylated form of intermediate **110** with the corresponding bromoalkene in the presence of sodium hydride (Scheme 4.32). From these compounds, it was expected that reaction with catalysts such as Rh(II) and Cu(II) would lead to the formation of an oxonium ylide followed by sigmatropic rearrangement. This would result in the migration of the *O*-6-alkyl group to the diazo carbon, leading to Kdo *C*-glycosides. Other reactions could also take place, such as C—H insertion giving rise to seven-membered rings or formation of an oxonium ylide from different oxygen atoms. Different catalysts, temperature, and solvents were therefore tested to study the outcome of the reaction. It was observed that when a Rh(II) catalyst in refluxing benzene was employed, the formation of the expected six-membered rings **181–186** was favoured, although the reaction was non-stereoselective. In contrast, formation of *cis*-configured seven-membered rings **187–189** was

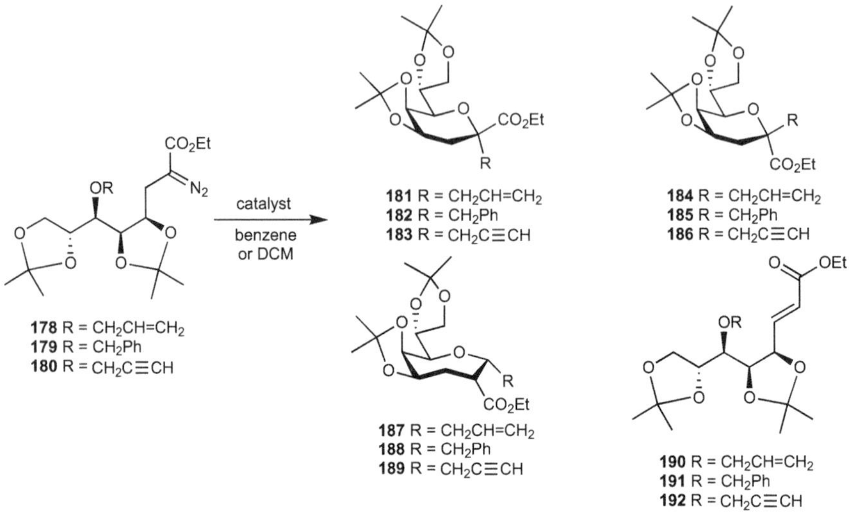

Scheme 4.32 ***Lopez-Herrera synthesis of Kdo C-glycosides.***[105]

predominant when $Cu(acac)_2$ or $Cu(hfacac)_2$ in refluxing benzene were employed. However, switching the solvent for dichloromethane led to elimination products **190-192**.

3.4 Synthesis of Kdo *O*-glycosides

The importance of Kdo-containing oligosaccharides for the development of therapeutics and prophylactics against GNB have prompted many research groups to develop new stereoselective methodologies for the preparation of such compounds. As oligosaccharides containing α-Kdo moieties are ubiquitously found in GNB, Banaszekand and co-workers were interested in the development of an efficient and stereoselective approach for the synthesis of α-linked disaccharides.[106,107] As depicted in Scheme 4.33, their methodology was based on the use of dithianyl **199** as a Kdo donor, whose synthesis was achieved starting from mannose derivative **193**. The latter was first elongated through a Grignard reaction with $Me_2PhSiCH_2MgCl$, forming a silylated intermediate as a mixture of two isomers, which underwent elimination and benzylation to afford alkene **194** as the sole product. Alcohol **195** was obtained following hydroboration-oxidation and was subsequently acetylated to allow the regioselective opening of the benzylidene group using tetraethylsilane (TES) and TFA. Cleavage of the benzylidene group was also noticed, therefore the resulting diol was converted into an *O*-8 glycosyl acceptor that was eventually used for the synthesis of disaccharide **200**.

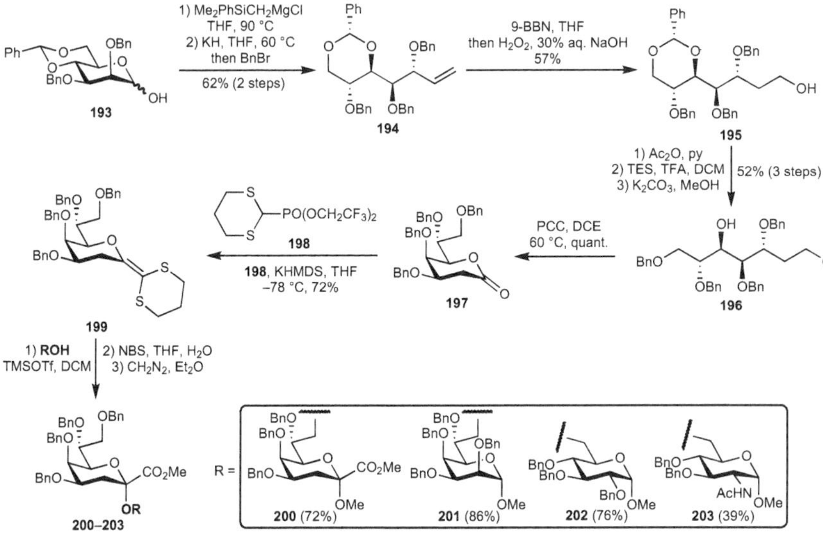

Scheme 4.33 ***Banaszek synthesis of Kdo disaccharides.***[106,107]

Then, hydrolysis of the acetyl group furnished 1,5-diol **196**, which underwent cyclization into lactone **197** through the action of pyridinium chloroformate (PCC). Horner-Emmons reaction was performed to prepare ketene dithioacetal **199** employing 2-[bis(2,2,2-trifluoroethoxy)phosphoryl]1,3-dithiane **198** and potassium bis(trimethylsilyl)amine (KHMDS) at low temperature. Glycosidation of the latter with various acceptors was achieved under the promotion ofTMSOTf, followed by NBS-mediated hydrolysis of the dithianyl group, and subsequent esterification of the resulting carboxylate with diazomethane. This three-step sequence furnished (2→8)-α-linked disaccharides **200–203** with complete stereoselectivity due to the steric hindrance of the dithianyl group.

Alternatively, Ling proposed a stereoselective approach towards the synthesis of β-Kdo glycosides relying on diisopropylinated glycal **210** bearing an aryl group at C1 (Scheme 4.34).[108] Their synthesis begun with treatment of *d*-mannose derivative **21** with NIS in MeCN, exclusively furnishing furanoside **204** rather than the desired pyranose ring. The authors attributed this phenomenon to the electron-withdrawing nature of the carboxylate moiety. This effect would unfavour the formation of a destabilized carbocation at C1, consequently leading to the attack of the C5 hydroxyl group on the more stable C2 carbocation thereby yielding a furanoside ring. The authors therefore designed compound **206** bearing an electron-donating aryl group that could stabilize the C1 carbocation and that could readily

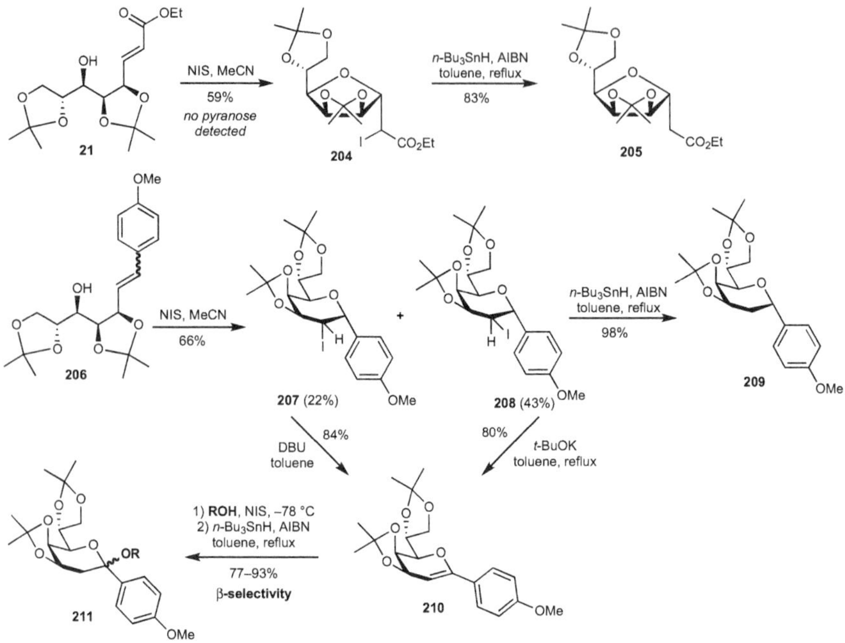

Scheme 4.34 ***Ling β-selective glycosylation of Kdo glycal*** **210.**[108]

be oxidized into a carboxylate moiety. Upon treatment with NIS, compound **206** furnished, as expected, pyranosides **207** and **208**. Importantly, deiodination with n-Bu$_3$SnH of both compounds separately resulted in the formation of *C*-glycoside **209**. Elimination of compound **207** was easily achieved using DBU in toluene, providing glycal **210**, but harsher conditions were required for the elimination of iodide **208** due to the *syn*-relationship of the 2-iodide and the C1 proton. NIS-mediated glycosidation of compound **210** and subsequent deiodination of the crude material was investigated using various alcohols (primary and secondary, including a glucose unit). Results showed that glycal **210** was highly reactive, as opposed to what was observed with peracetylated glycals bearing a carboxylic group at C1. Additionally, the reactions proved to occur with high stereoselectivity, as β-glycosides were invariably the major products. This stereoselectivity could be attributed to one of the methyl group of the endocyclic isopropylidene group, which hinders the top face of the double bond, thereby forcing the latter to attack the iodide from the bottom face. Opening of the resulting cyclic iodonium intermediate by the glycosyl acceptor would thus occur from the top face. The authors finally highlighted the usefulness of

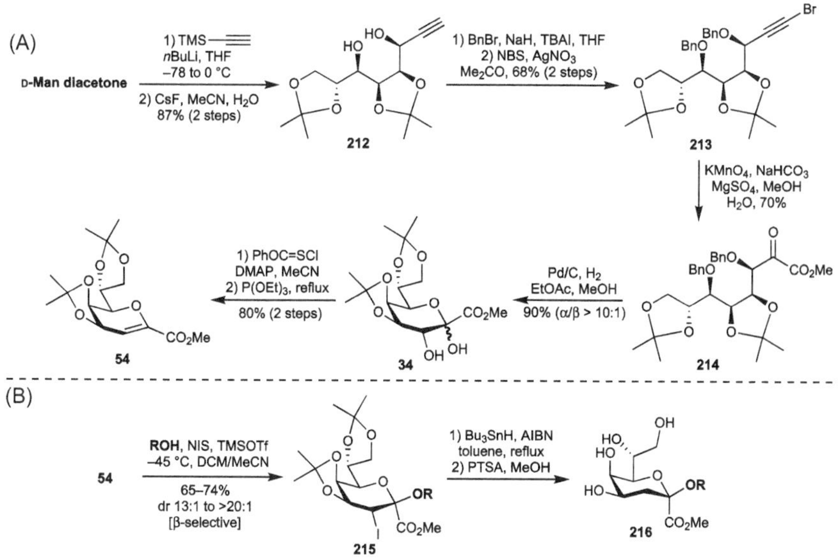

Scheme 4.35 Mong synthesis (A) and β-selective glycosylation of Kdo glycal **54** (B).[109]

their approach for the synthesis of β-Kdo glycosides by oxidizing the C1 aryl group into a carboxylate moiety.

In a similar way, the Mong group developed an efficient approach for the synthesis of a glycal that allow the stereoselective formation of β-Kdo glycosides through NIS-mediated glycosidation reactions.[109] As described in Scheme 4.35, the synthesis of glycal **54** started with the alkylation of *d*-mannose diacetone with lithiated trimethylsilylacetylide and subsequent desilylation. Dibenzylation of intermediate **212** and bromination of the terminal alkyne furnished derivative **213**. Oxidative cleavage provided methyl α-ketoester **214**, whose hydrogenolysis enabled its hemiketalization into compound **34**. The Corey-Winter procedure[110] furnished Kdo glycal **54**, which was studied as a donor for β-glycoside synthesis. Thus, various glycosylation reactions were screened under the promotion of NIS/TMSOTf and resulting glycosides were deprotected into compound **216**. Results showed that these reactions proceeded with high β-selectivity, particularly when a DCM/CH_3CN mixture was used as solvent rather than pure DCM. It was also highlighted that the isopropylidene protecting groups on Kdo glycal donor had a β-stereodirecting effect since their replacement with benzyl groups led to a significant decrease in the stereoselectivity of the coupling reactions.

Gauthier's approach towards the synthesis of β-Kdo glycosides involved the use of a long-range 4'-methoxyphenacyl participating group at C1, as shown in Scheme 4.36.[111] Benzylation of the carboxylate function of peracetylated Kdo derivative **217** followed by introduction of a thioethyl group at the anomeric position yielded thioglycoside **218**. Reduction of the benzyl ester allowed the introduction of the C1 4'-methoxyphenacyl (Phen) group, resulting in compound **219**. The Phen group was chosen as a long-range participating group because of the enhanced electronic density of the ketone favouring the formation of a probable α-spirophenacylium intermediate. For this intermediate to be formed, isomerization of *Z*-ester **219** to the less stable *E*-ester is required and would be favoured by the presence of the electron-withdrawing Phen group and by the use of polar solvents such as acetonitrile. Upon isomerization and departure of the thioethyl moiety, it was expected that the ketone would participate from the α-face of the $^{5}H_{4}$ half-chair intermediate as to minimize 1,3-diaxial interactions. The resulting α-spiro intermediate would then be attacked preferentially from the β-side. This approach enabled the synthesis of various β-Kdo glycosides using a small series of glycosyl acceptors (**220-226**). Moreover, the authors showed that the use of stereodirecting CH_3CN led to the major formation of β-glycosides, as opposed to Et_2O and DCM. The counteranion also played a role in the process, as promoters containing less coordinating anions led to a decrease in β-stereoselectivity. Finally, chemoselective deprotection of the Phen group was achieved following treatment with activated zinc powder in the presence of acetic acid.

Besides Kdo-CMP synthase, monomeric heptosyltransferase WaaC, which catalyzes the transfer of one heptosyl moiety to lipid A-Kdo_2, is an attractive target for the development of therapeutics against GNB. Vincent and co-workers showed that such enzyme could be inhibited by glucofullerenes bearing 12 heptoside residues.[112] Indeed, it was unveiled that the inhibitory activity of this construct was enhanced as compared to that of monomeric heptoside inhibitors. This enhancement was attributed to the amplification of the ligand binding affinity for its receptor due to its presentation in a multivalent fashion (multivalent effect).

This research group was then interested in the synthesis of glycofullerenes bearing α-Kdo with different chain lengths at C-2 (**236** and **238**), β-Kdo (**237**), and 4-*epi*-α-Kdo (**239**, Scheme 4.37).[113] Acyclic perbenzylated D-arabinose **226** was used as starting material and was condensed with benzyl-2-(bromomethyl)acrylate **227** under the action of chromium chloride. This reaction furnished two diastereoisomers, which were converted

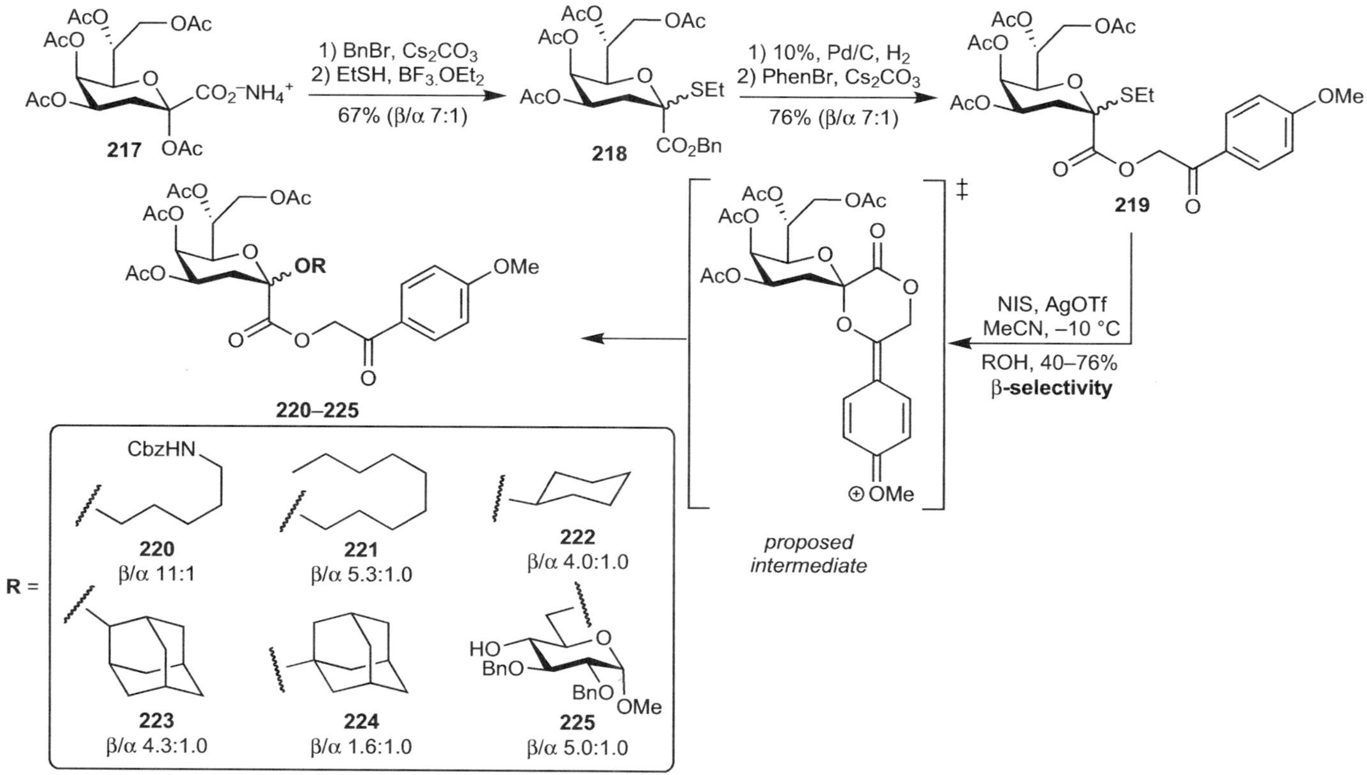

Scheme 4.36 ***Gauthier phenacyl-assisted synthesis of β-Kdo glycosides.***[111]

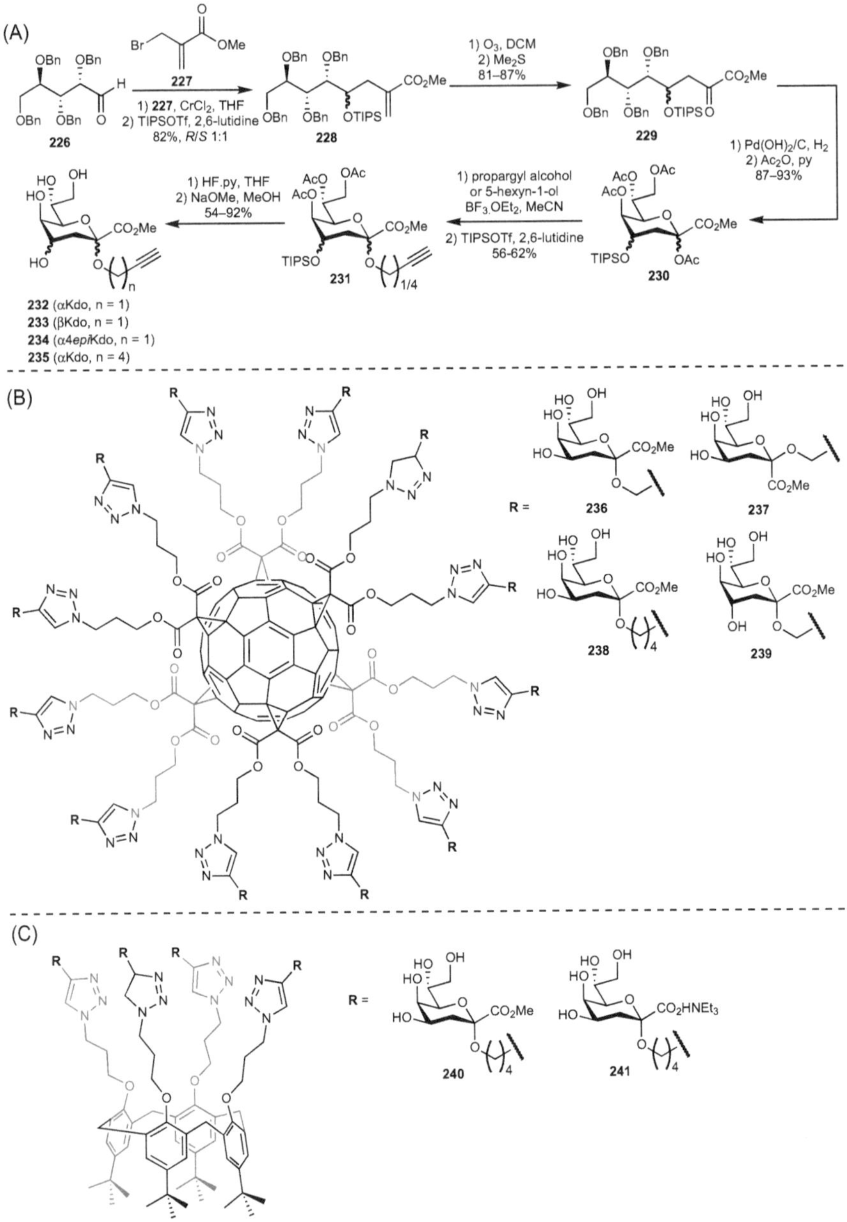

Scheme 4.37 Vincent synthesis of Kdo *C*-glycosides (A) and their covalent attachment into fullerenes (B) and callixarenes (C).[113]

into triisopropylsilyl ether **228**. Importantly, the isomers were separated by flash chromatography and the following synthetic route was applied to both compounds. Ozonolysis and subsequent reduction with dimethylsulfide furnished α-keto ester **229**. Benzyl ethers were cleaved and the resulting hydroxyl groups were acetylated into anomers **230**. Installation of spacers of different lengths was achieved by treating methyl esters **230** with either propargyl alcohol or 5-hexyn-1-ol under the promotion of $BF_3 \cdot OEt_2$. Silylation of the desilylated byproducts was accomplished and allowed the isolation of α- and β-anomers of compound **231**. Cleavage of the triisopropylsilyl (TIPS) group was achieved through treatment of intermediate **231** with HF·pyridine and was followed by Zemplén deacetylation to furnish Kdo derivatives **232–235**. These latter were coupled to fullerene units bearing 12 azido groups through copper-catalyzed azide-alkyne cycloaddition and thus furnished dodeca-Kdo fullerenes **236-239**.

Biological assays revealed the potent competitive inhibition of WaaC by the multimeric Kdo constructs and, as already highlighted by other groups, the absolute configuration of C2 and C4 was showed to significantly affect this activity.[113] Indeed, glycofullerenes **236** and **238** bearing α-Kdo residues, which are found in the enzyme substrate lipid AA-Kdo_2, exhibited the highest inhibitory activity. However, multivalent effects were not significative, as revealed through the measure of the enhancement affinity per sugar unit. Although glycofullerene **236** was prone to aggregate the enzyme, this aggregation was not considered as the main origin of inhibition. In a similar manner, tetravalent calix[4]arenes **240** and **241** were prepared to compare the role of the central unit in the inhibition process. Results showed that the fullerene scaffolds were superior to that of the calix[4]arene ones.

3.5 Synthesis of miscellaneous Kdo derivatives

Through the synthesis of various Kdo derivatives other than 2-deoxy-β-Kdo analogs and *C*- or *O*-glycosides, deep insights into the biological requirements of Kdo-processing enzyme inhibitors can be acquired. The Claesson group was involved in the synthesis of miscellaneous Kdo derivatives as potential enzyme inhibitors. In 1982, they reported the preparation of unsaturated analogue **243** from pentaacetate methyl ester **65** (Scheme 4.38).[114] Thus, elimination product **242** was obtained following treatment of compound **65** with TMSOTf, adding TMSCN to act as an acid scavenger, and standard deprotection procedures furnished target compound **243**.

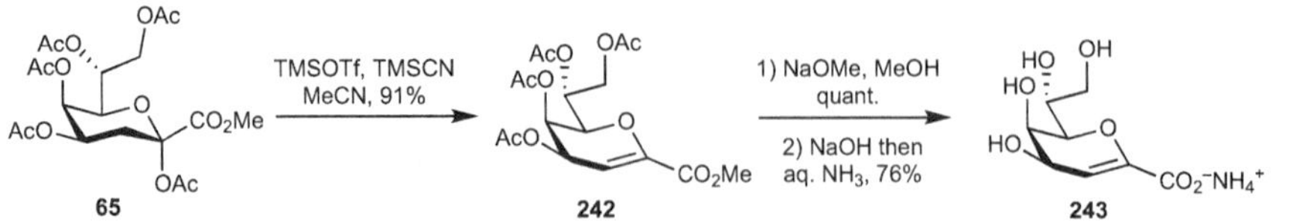

Scheme 4.38 ***Claesson synthesis of Kdo glycal* 243.**[114]

Following this work, Claesson and co-workers were interested in the synthesis of CMP-Kdo transition-state analogues as potent enzyme inhibitors.[115] Although no molecular mechanism details were available surrounding the formation of CMP-Kdo, it was proposed that the transition-state structure consisted of β-Kdo, cytidine as well as pyrophosphate moieties. They therefore synthesized an analogue of this structure by omitting the cytidine moiety, as it had already been shown that the lack of the pyrophosphate group, but inclusion of the cytidine moiety, rendered the compounds inactive.[103,116] Unfortunately, the synthesized intermediate did not exhibit any inhibitory activity.

As depicted in Scheme 4.39, Zamojski reported the preparation of 4-*O*-methyl Kdo pyranoside and furanoside **249** and **250**, respectively.[117] Photochemical reaction of *d*-gluconic acid derivative **244** with ethyl 2-trifluoroacetylacrylate **245** in the presence of *N*-hydroxy-2-thiopyridone (NHTP) and *N,N'*-dicyclohexylcarbodiimide (DCC), which occurred through the formation of an acylated *N*-hydroxy-2-thiopyridone Barton's ester, furnished α,β-unasturated ester **246**. Hydroxylation of the double bond was achieved through a two-step sequence consisting in an alkoxymercuration followed by a demetallation. This reaction occurred *via* a total control of regio – but not stereo – selectivity, as a mixture of diastereoisomers **247** and **248** was obtained. Additionally, during the alkoxymercuration, transesterification of the ethyl ester into a methyl ester occurred due to the presence of MeOH. Following a standard sequence of deprotection, cyclization, and peracetylation, isomers **247** and **248** were converted into pyranoside **249** and furanoside **250** as free acids.

With the aim of developing an efficient method for the α-selective glycosidation of Kdo, the Yamasaki group synthesized α-spiro compound **253** that could be employed for further coupling to other molecules (Scheme 4.40).[118] Kdo derivative **251**, bearing a free alcohol at the anomeric position, was employed as the starting material. When reacted with 2-chloroethyl isocyanate **252** in the presence of stoichiometric DCC in warmed toluene, α-spiro product **253** was predominantly formed, probably through

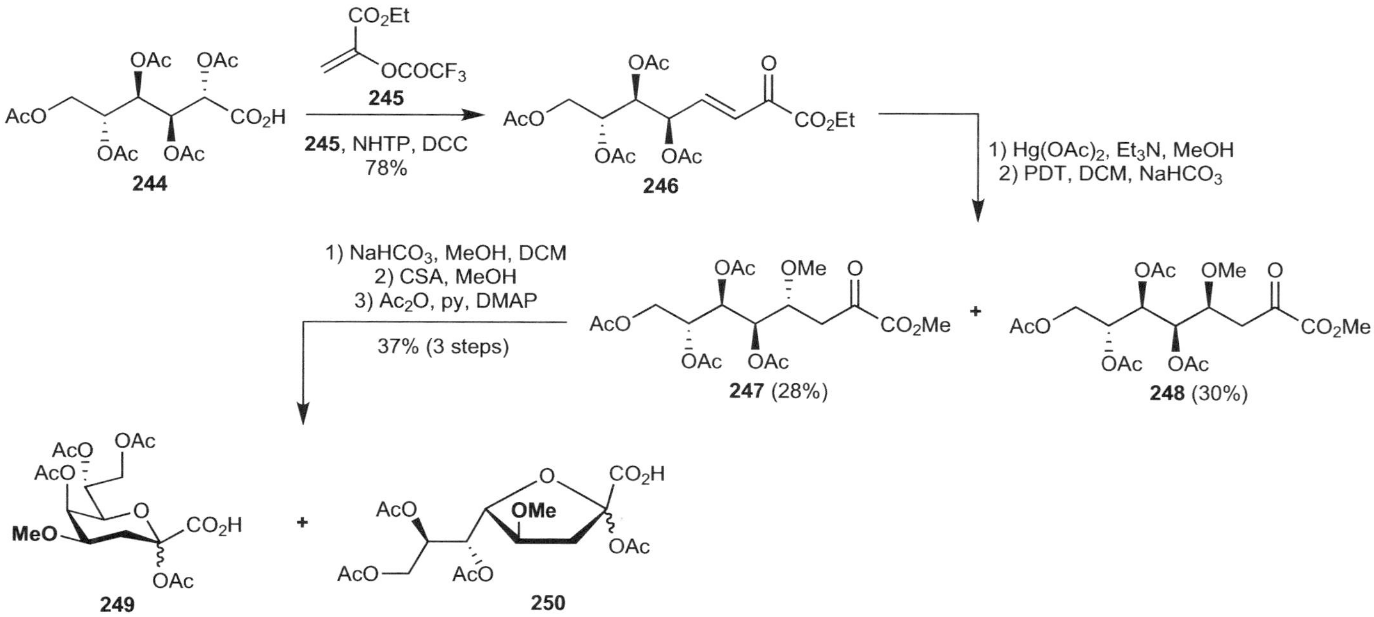

Scheme 4.39 ***Zamojski synthesis of 4-O-methyl-Kdo.***[117]

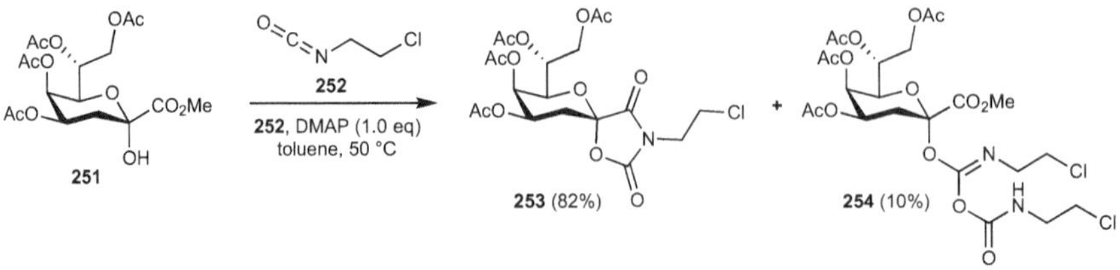

Scheme 4.40 ***Yamasaki synthesis of spiro Kdo derivatives.***[118]

the formation of a 2-*O*-carbamate intermediate followed by intramolecular cyclization. Interestingly, dimeric isocyanate derivative **254** was also isolated as a minor product (10% yield). The authors also tested 4-(chloromethyl) phenyl isocyanate as an electrophile and showed that, when conducted in THF, this reaction did not produce any dimeric isocyanate adduct. This observation was attributed to the high stability of the aryl spiro compound, which could enable faster cyclization reactions as compared to alkyl spiro compounds. Therefore, the authors demonstrated that anomeric *O*-acylation using an electrophilic isocyanate is α-specific and that the resulting spiro compound could be a useful tool for the development of higher glycosides.

With the aim of obtaining insights on the binding specificity of Kdo-processing enzymes, Kiefel and co-workers took advantage of the Cornforth procedure to efficiently synthesize Kdo derivatives.[119] More particularly, they prepared C8 modified Kdo analogs **258–260** since the additional Kdo moiety in LPS is usually attached to Kdo I at C4 or C8 (Scheme 4.41). Methoxy, azido, and thiol groups were chosen as the authors reasoned that they could participate in hydrogen bonding interactions with the enzyme. Compounds **258–260** were therefore prepared through standard base-catalyzed aldol condensation of C5-modified arabinose derivatives **255–257** with oxaloacetic acid followed by decarboxylation. However, rather than using a molar excess of arabinose, the authors instead used a molar excess

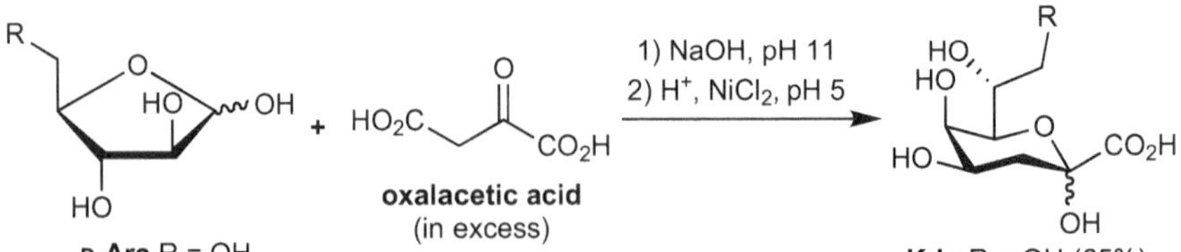

D-Ara R = OH
255 R = OMe
256 R = N_3
257 R = SH

Kdo R = OH (65%)
258 R = OMe (84%)
259 R = N_3 (87%)
260 R = SH (67%)

Scheme 4.41 ***Kiefel synthesis of Kdo derivatives modified at the C-8 position.***[119]

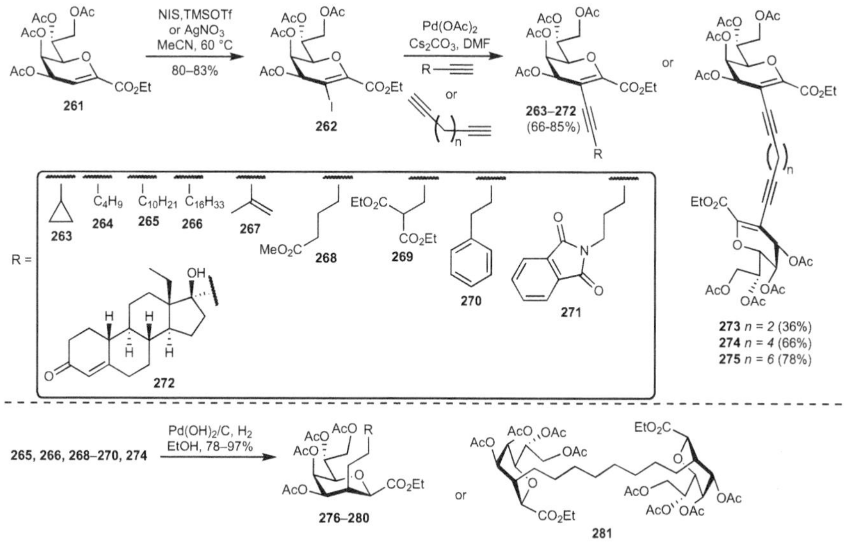

Scheme 4.42 ***Yang synthesis of C3 branched Kdo analogues through Sonogashira coupling.***[120]

of oxaloacetic acid, which improved the reaction yields. These conditions provided the target Kdo derivatives as mixtures of C-4 epimers, but favouring Kdo over 4-*epi*-Kdo.

Yang recently reported the first synthesis of 3-*C*-branched Kdo enyne derivatives based on Sonogashira coupling.[120] As shown in Scheme 4.42, iodination of glycal **261** with NIS in the presence of either TMSOTf or $AgNO_3$ furnished intermediate **262**. Installation of the new C—C bond at C3 was first attempted by the authors using Heck coupling but was not successful due to incomplete conversion of the starting material, which was attributed to the presence of the electron-withdrawing carboxylate group at C2. They therefore turned their attention to copper-free Sonogashira coupling. This approach enabled the attachment of terminal alkynes with various alkyl chain lengths and functionalities at C3 providing compounds **263–272** in good yields. Forcing the reaction conditions and using dialkynes additionally allowed the formation of 3-*C*-branched di-Kdo enynes **273–275.** Some of these 3-*C*-alkylated glycals were subjected to an asymmetric hydrogenation over Pearlman's catalyst, which not only reduced the alkyne bonds, but also triggered the *cis*-addition of hydrogen atoms from the bottom face. This reaction therefore provided 2-deoxy-β-Kdo analogs **276–281** with good to excellent yields.

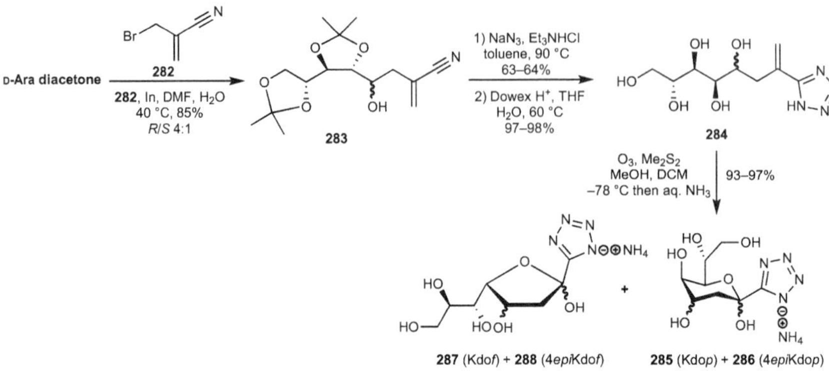

Scheme 4.43 ***Riedl synthesis of Kdo bioisosteres.***[121]

Another approach towards the discovery of inhibitors of Kdo-processing enzymes relies on the bioisosteric replacement of Kdo carboxylic acid moiety. As such, the Riedl group synthesized Kdo bioisosteres **285–288** in which the carbonyl group was substituted by a tetrazole group (Scheme 4.43).[121] As tetrazoles and carboxylic acids share some similarities regarding their pKa and planar structure, it was thought that they could exhibit interesting biological activities, especially considering that tetrazole moieties are lipophilic and can thus readily reach the cytoplasm. Their synthesis was accomplished from the indium-mediated allylation of *d*-arabinose diacetone with (bromomethyl) acrylonitrile **282**. A diastereoisomeric mixture of intermediates **283** was thus obtained, and the isomers were readily separated by flash chromatography so that the synthetic route could be performed in parallel on both compounds. Tetrazoles (4*R*)-**284** and (4*S*)-**284** were prepared through a [2 + 3] cycloaddition with sodium azide and subsequent deisopropylidenation. Ozonolysis of the C2 double bond introduced a ketone function. Treatment of the resulting intermediate with aqueous ammonia furnished Kdo derivatives **285** and **287**, obtained from (4*R*)-**284**, and 4-*epi*-Kdo derivatives **286** and **288**, obtained from (4*S*)-**284**. All of the products were obtained as furanose/pyranose mixtures along with the open chain form.

4 Kdo derivatives for LPS labeling of living organisms

Metabolic glycan labeling (MGL) is an imaging approach that has been recently developed for visualizing glycans found at the surface of living organisms such as zebrafish, mice, plants, and bacteria.[122] In MGL, a small chemical reporter (usually an azide tag) is covalently attached to a sugar of interest. Once incorporated and metabolized, the sugar bearing the chemical reporter is expressed at the surface of cells allowing its visualization by chemical ligation to a fluorescent probe *via* biorthogonal "click" reactions including metal-catalyzed and strain-promoted reactions.[123]

Kdo residues found in the LPS of GNB represent attractive targets for MGL of bacterial membranes. Indeed, Kdo is present in the LPS inner core of virtually all GNB with the exception of *Shewanella* genus, which has been shown to express 8-amino-8-deoxy-Kdo instead of Kdo.[2] Moreover, in the Kdo biosynthetic pathway (see Fig. 4.2), free Kdo is processed by CKS leading to CMP-Kdo prior to further LPS elaboration.[36] Recently, Vauzeilles and co-workers hypothesized that Kdo derivatives such as 8-azido-8-deoxy-Kdo ($KdoN_3$) could be processed by CKS by partially replacing endogenous Kdo in LPS while allowing bacterial membranes to be labelled using azide-alkyne click chemistry.[124] They also hypothesized that the presence of the azido group at the C8 position should prevent reverse metabolism by Kdo8P phosphatase thereby limiting its incorporation into other bacterial glycans.

They synthesized $KdoN_3$ using an approach similar to Kiefel and co-workers[119] through an optimized version of the Cornforth procedure in which 5-azido-5-deoxy-D-arabinofuranose was reacted with oxalacetic acid to furnish, following decarboxylation and anion-exchange purification, the ammonium salt of $KdoN_3$. Non-pathogenic *E. coli* strain K12 was cultured overnight in the presence of synthetic $KdoN_3$ allowing its incorporation into LPS by taking advantage of GNB biosynthetic machinery. Then, a copper-catalyzed azide-alkyne cycloaddition (CuAAC) reaction was performed with an Alexa Fluor 488 fluorophore containing a terminal alkyne group. Pleasingly, only after five min of incubation, the outer membranes of *E. coli* were label very brightly. This MGL approach involving $KdoN_3$ was also efficient with other GNB, including *S. typhimurium* and *Legionella pneumophila*. As expected, membranes of Gram-positive bacteria including *Bacillus subtilis* and *Staphylococcus aureus* as well as the Kdo-deficient *Shewanella oneidensis* were not labeled under such conditions. It is important to mention that this MGL strategy was recently applied by Chen and co-workers for the selective imaging of GNB microbiotas in the gut of mice.[125]

One of the main disadvantages of MGL enabled by CuAAC is the cellular toxicity of copper, which can kill GNB. In order to be able to visualize living GNB through MGL, Vauzeilles and co-workers reported the alternative use of copper-free strain-promoted alkyne-azide cycloaddition (SPAAC) (Scheme 4.44**A**).[126] To do so, $KdoN_3$ incorporation into *E. coli* was monitored with a cyclooctyne-biotin conjugate followed by staining with

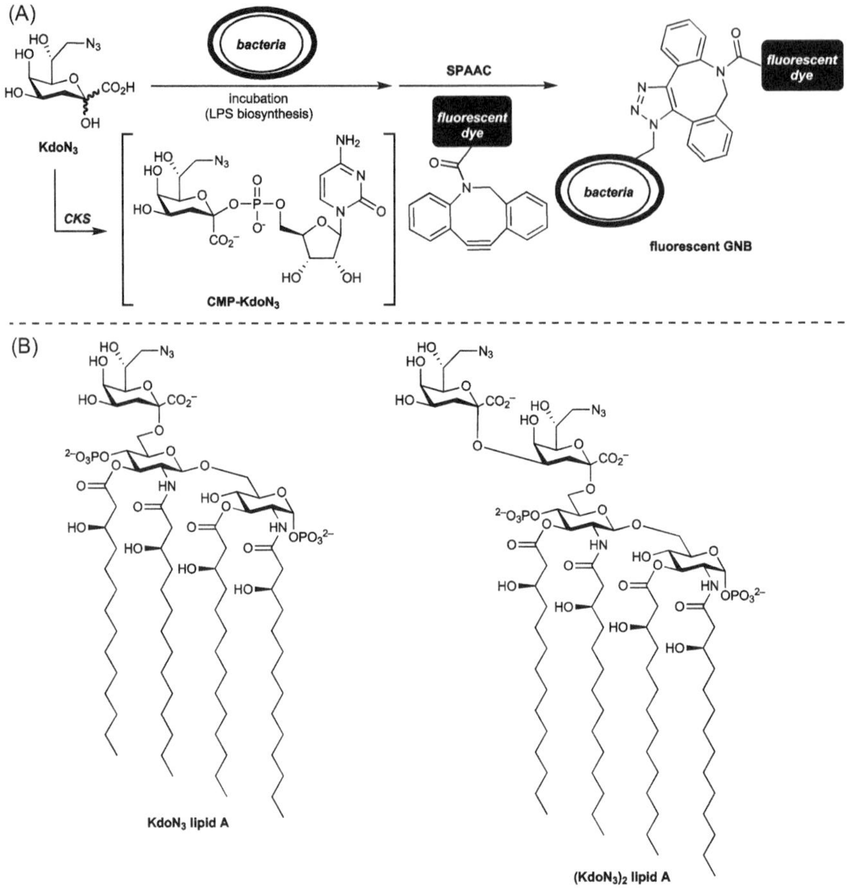

Scheme 4.44 (A) *In vivo* fluorescent labelling of GNB through a strain-promoted alkyne-azide cycloaddition (SPAAC) reaction of $KdoN_3$ embedded into LPS;[124] (B) Structure of $KdoN_3$ lipid A and $(KdoN_3)_2$ lipid A isolated from GNB incubated with $KdoN_3$.[127]

anti-biotin-A488 antibody. Microscopic analyses revealed that $KdoN_3$ could be efficiently detected at the outer membrane of *E. coli* without affecting bacterial growth. Interestingly, the authors were able to specifically labeled cultivable *E. coli* in the presence of dead *E. coli* with good specificity as only living GNB were able to metabolize $KdoN_3$ through the Kdo biosynthetic machinery. Impressively, they were also able to isolate *E. coli* with magnetic beads functionalized with streptavidin without significant interference of *B. subtilis*. This pioneering technology opens the possibility to enrich cultivable GNB directly from a sample mixture of microorganisms that do not express Kdo at their cell membranes while allowing their fast detection.

Capitalizing on this innovative way of labeling Kdo-containing glycans, Lerouge and co-workers recently reported *in planta* MGL of rhamnogalacturonan II using $KdoN_3$.[128] Rhamnogalacturonan II is an important structural component of plant pectins. It represents ~2–4% of the primary cell wall in plants and is composed of a homogalacturonan backbone that is substituted with four oligosaccharide side chains consisting of at least 12 different sugar residues including Kdo.[129] As the C8 position of Kdo is not involved in the linkage with rhamnose residues and Kdo residues are specific to rhamnogalacturonan II, performing MGL with $KdoN_3$ was found to be an exquisite approach for plant cell wall imaging. Actually, Arabidopsis seedlings treated with $KdoN_3$, then coupled with a fluorophore tag through CuAAC, exhibited strong cell wall labelling. In the presence of 2-deoxy-β-Kdo, a known inhibitor of CKS, a strong decrease in cell wall labelling was observed indicating that incorporation of $KdoN_3$ occurs through endogenous Kdo biosynthetic machinery.

In 2017, Nilsson et al. showed, using a combination of fluorescence microscopy and polyacrylamide gel electrophoresis (PAGE) analysis, that $KdoN_3$ is incorporated into rough and deep-rough LPS.[127] Using MALDI-TOF-MS analyses, they were the first to identify truncated LPS in the forms of $KdoN_3$-lipid A and $(KdoN_3)_2$-lipid A (Scheme 4.44B). The authors also revealed that $KdoN_3$ is incorporated less efficiently than Kdo itself. They demonstrated that this could be explained owing to the specific kinetic constant of the CKS with $KdoN_3$ compared with Kdo (six-fold difference in K_m constants). In a following paper, the same group showed that the sialic acid transporter NanT is the membrane receptor involved in the active diffusion of exogenous $KdoN_3$ into *E. coli*.[130] Their results suggest that the NanT receptor could enable GNB to scavenge exogenous Kdo in order to overcome inhibition of Kdo biosynthesis.

As an alternative to the CuAAC reaction, Pezacki and co-workers reported the *in vivo* fluorescent labelling of GNB using copper-catalyzed alkyne-nitrone cycloaddition followed by rearrangement (CuANCR, Scheme 4.45).[131] Also known as the Kinugasa reaction,[132] this bioorthogonal reaction involved the formation of a stable β-lactam ring from a terminal alkyne and a nitrone. As depicted in Scheme 4.45A, Kdo-HMMPO [2-(hydroxymethyl)-2-methyl-3,4-dihydro-2*H*-pyrrole 1-oxide] and Kdo-alkyne were synthesized in two steps from $KdoN_3$. Palladium oxide-catalyzed hydrogenation of $KdoN_3$ in the presence of acetic acid led to 8-amino-8-deoxy-Kdo ($KdoNH_2$), which was reacted either with HMMPO-*N*-hydroxysuccinimide (NHS) (**289**) or alkyne-NHS derivative **290** to furnish

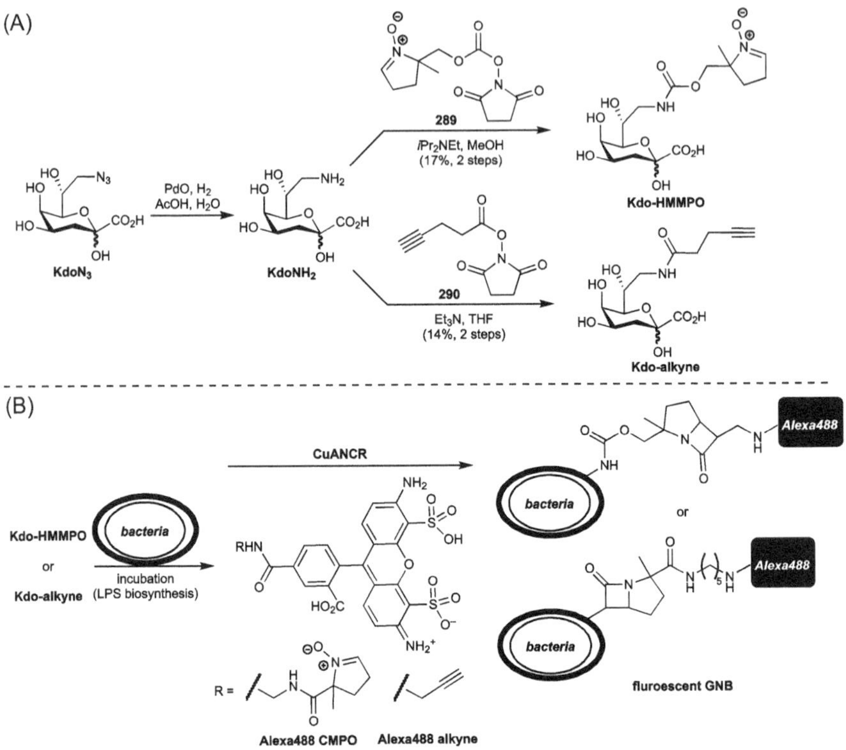

Scheme 4.45 Pezacki synthesis (A) of Kdo alkyne and nitrone derivatives and their use (B) for the *in vivo* fluorescent labelling of GNB through tandem copper-catalysed alkyne-nitrone cycloaddition rearrangement (CuANCR).[131]

Kdo-HMMPO or Kdo-alkyne, respectively, following purification by semi-preparative HPLC. *E. coli* was incubated with Kdo-alkyne, the cells were washed, and then treated with Alexa488-CMPO *via* the CuANCR reaction allowing fluorescent labelling of *E. coli* outer membrane (Scheme 4.45B). Alternatively, *E. coli* cells were treated with Kdo-HMMPO followed by treatment with Alexa488-alkyne using identical CuANCR conditions, which also enabled the fluorescent labelling of *E. coli*.

5 Conclusions and future perspectives

Considering that Kdo residues are ubiquitously found in the surface polysaccharides of GNB and that inhibition of their biosynthesis impedes cell viability, Kdo-processing enzymes stand as highly interesting targets for the development of therapeutics against emerging drug-resistant GNB. For these reasons and considering the synthetic challenges that such compounds represent, the synthesis of Kdo and its derivatives has attracted lots of attention. Thus, as illustrated in the present chapter, there has been tremendous progress reported on the stereoselective synthesis of α- and β-Kdo from chiral aldehydes and non-carbohydrate precursors since the publication of the Cornforth procedure. Following the synthesis of 2-deoxy-β-Kdo, a potent inhibitor of CKS, a wide range of Kdo derivatives were prepared, including 2-deoxy-8-amino-β-Kdo prodrugs, oligosaccharides, biosynthetic intermediate mimics as well as diverse epimers. Biological assays on these potential inhibitors of Kdo-processing enzymes highlighted important structural requirements for these compounds in order to exhibit potent inhibitory and/or antibacterial activities. These data will doubtlessly pave the way for the discovery of new prophylactic and therapeutic agents against pathogenic GNB based on the unique Kdo scaffold.

References

1. Levin, D.; Racker, E. *J. Biol. Chem.* **1959**, *234*, 2532–2539.
2. Knirel, Y. A.; Shevelev, S. D.; Perepelov, A.V. H. *Mendeleev Commun.* **2011**, *21*, 173–182.
3. Brandenburg, K.; Wiese, A. *Curr. Top. Med. Chem.* **2004**, *4*, 1127–1146.
4. Zähringer, U.; Lindner, B.; Rietschel, E. T. Chemical structure of lipid A: recent advances in structural analysis of biologically active molecules. *Endotoxin in Health and Disease*, 1, Brade, H., Opal, S. M., Vogel, S. N., Morrison, D. C., Eds.; pp. 93–114.
5. Holst, O. Chemical structure of the core region of lipopolysaccharides. *Endotoxin in Health and Disease*, 1, Brade, H., Opal, S. M., Vogel, S. N., Morrison, D. C., Eds.; pp. 115–154.
6. Tzeng, Y.-L.; Datta, A.; Strole, C.; Kolli, V. S. K.; Birck, M. R.; Taylor, W. P.; Carlson, R. W.; Woodard, R. W.; Stephens, D. S. *J. Biol. Chem.* **2002**, *277*, 24103–24113.
7. Dotson, G. D.; Dua, R. K.; Clemens, J. C.; Wooten, E. W.; Woodard, R. W. *J. Biol. Chem.* **1995**, *270*, 13698–13705.
8. Krosky, D. J.; Alm, R.; Berg, M.; Carmel, G.; Tummino, P. J.; Xu, B.; Yang, W. *BBA-Protein Struct. M.* **2002**, *1594*, 297–306.
9. Kumaran, G.; l Mootoo, D. R. *Tetrahedron Lett.* **2001**, *42*, 3783–3785.
10. Ray, P. *J. Bacteriol.* **1980**, *141*, 635–644.
11. Goldman, R. C.; Bolling, T.; Kohlbrenner, W.; Kim, Y.; Fox, J. L. *J. Biol. Chem.* **1986**, *261*, 15831–15835.
12. Goldman, R. C.; Kohlbrenner, W. E. *J. Bacteriol.* **1985**, *163*, 256–261.
13. Kohlbrenner, W.; Fesik, S. *J. Biol. Chem.* **1985**, *260*, 14695–14700.
14. Brozek, K. A.; Hosaka, K.; Robertson, A.; Raetz, C. *J. Biol. Chem.* **1989**, *264*, 6956–6966.
15. Ray, P. H.; Benedict, C. D. *J. Bacteriol.* **1980**, *142*, 60–68.
16. Lodowska, J.; Wolny, D.; Węglarz, L. T. *Can. J. Microbiol.* **2013**, *59*, 645–655.
17. Sirisena, D. M.; Brozek, K. A.; MacLachlan, P. R.; Sanderson, K. E.; Raetz, C. *J. Biol. Chem.* **1992**, *267*, 18874–18884.
18. Brabetz, W.; Muller-Loennies, S.; Holst, O.; Brade, H. *Eur. J. Biochem.* **1997**, *247*, 716–724.
19. Chen, L.; Coleman, W. *J. Bacteriol.* **1993**, *175*, 2534–2540.
20. Lin, C.-H.; Murray, B. W.; Ollmann, I. R.; Wong, C.-H. *Biochemistry* **1997**, *36*, 780–785.
21. Willis, L. M.; Whitfield, C. *Carbohydr. Res.* **2013**, *378*, 35–44.
22. Willis, L. M.; Stupak, J.; Richards, M. R.; Lowary, T. L.; Li, J.; Whitfield, C. *Proc. Natl. Acad. Sci.* **2013**, *110*, 7868–7873.

23. Willis, L. M.; Whitfield, C. *Proc. Natl. Acad. Sci.* **2013**, *110*, 20753–20758.
24. Clatworthy, A. E.; Pierson, E.; Hung, D.T. *Nat. Chem. Biol.* **2007**, *3*, 541–548.
25. Altman, E.; Brisson, J. R.; Gagné, S. M.; Perry, M. B. S. *Eur. J. Biochem.* **1992**, *204*, 225–230.
26. Altman, E.; Brisson, J. R.; Perry, M. B. *Eur. J. Biochem.* **1987**, *170*, 185–192.
27. Starr, K. F.; Porsch, E. A.; Heiss, C.; Black, I.; Azadi, P.; Geme, J.W. S., III. *PLoS One* **2013**, *8*, e75409.
28. Reistad, R.; Zähringer, U.; Bryn, K.; Alstad, J.; Bøvre, K.; Jantzen, E. *Carbohydr. Res.* **1993**, *245*, 129–136.
29. Duployez, C.; Loïez, C.; Ledoux, G.; Armand, S.; Jaillette, E.; Wallet, F. *IDCases.* **2017**, *8*, 12–13.
30. Nazarenko, E. L.; Komandrova, N. A.; Gorshkova, R. P.; Tomshich, S.V.; Zubkov, V. A.; Kilcoyne, M.; Savage, A.V. *Carbohydr. Res.* **2003**, *338*, 2449–2457.
31. Chance, T.; Chua, J.; Toothman, R. G.; Ladner, J.T.; Nuss, J. E.; Raymond, J. L.; Biot, F.V.; Demons, S.; Miller, L.; Halasohoris, S.; Mou, S.; Koroleva, G.; Lovett, S.; Palacios, G.; Vietri, N. J.; Worsham, P. L.; Cote, C. K.; Kijek, T. M.; Bozue, J. A. *PloS one* **2017**, *12*, e0174106.
32. Tytgat, H. L.; Lebeer, S. *Microbiol. Mol. Biol. Rev.* **2014**, *78*, 372–417.
33. Cipolla, L.; Gabrielli, L.; Bini, D.; Russo, L.; Shaikh, N. *Nat. Prod. Rep.* **2010**, *27*, 1618–1629.
34. Henderson, J. C.; Zimmerman, S. M.; Crofts, A. A.; Boll, J. M.; Kuhns, L. G.; Herrera, C. M.; Trent, M. S. *Annu. Rev. Microbiol.* **2016**, *70*, 255–278.
35. Gronow, S.; Xia, G.; Brade, H. *Eur. J. Cell Biol.* **2010**, *89*, 3–10.
36. Whitfield, C.; Trent, M. S. *Annu. Rev. Biochem.* **2014**, *83*, 99–128.
37. Yethon, J.; Whitfield, C. *Current Drug Targets-Infectious Disorders* **2001**, *1*, 91–106.
38. Wang, X.; Quinn, P. J.; Yan, A. *Biol. Rev.* **2015**, *90*, 408–427.
39. Wyckoff, T. J.; Raetz, C. R.; Jackman, J. E. *Trends Microbiol.* **1998**, *6*, 154–159.
40. Cipolla, L.; Polissi, A.; Airoldi, C.; Gabrielli, L.; Merlo, S.; Nicotra, F. *Curr. Med. Chem.* **2011**, *18*, 830–852.
41. Hansson, J.; Oscarson, S. *Curr. Org. Chem.* **2000**, *4*, 535–564.
42. Kosma, P.; Progress in Kdo-glycoside, chemistry. *Tetrahedron Lett.* **2016**, *57*, 2133–2142.
43. Pradhan, T. K.; Mong, K. K.T. *Isr. J. Chem.* **2015**, *55*, 285–296.
44. Li, L. S.; Wu, Y. L. *Cur. Org. Chem.* **2003**, 7, 447–475.
45. Schauer, R.; Kelm, S.; Reuter, G.; Roggentin, P.; Shaw, L. Biochemistry and role of sialic acids. In *Biology of the Sialic Acids*; Rosenbaerg, A., Ed.; Springer: Boston, 1995; pp. 7–67.
46. Cornforth, J. W.; Firth, M. E.; Gottschalk, A. *Biochem. J.* **1958**, *68*, 57–61.
47. Shirai, R.; Ogura, H. *Tetrahedron Lett.* **1989**, *30*, 2263–2264.
48. Mikula, H.; Blaukopf, M.; Sixta, G.; Stanetty, C.; Kosma, P. Synthesis of ammonium 3-deoxy-D-*manno*-oct-2-ulopyranosylate (Ammonium Kdo). *Carbohydrate Chemistry: Proven Synthetic Methods*, 2, van der Marel, G., Codée, J., Eds.; CRC Press, 2014; pp. 207–211.
49. Schmidt, R. R.; Betz, R. *Angew. Chem. Int. Ed. Engl.* **1984**, *23*, 430–431.
50. Dondoni, A.; Merino, P. *J. Org. Chem.* **1991**, *56*, 5294–5301.
51. Gao, J.; Härter, R.; Gordon, D. M.; Whitesides, G. M. *J. Org. Chem.* **1994**, *59*, 3714–3715.
52. Binder, W. H.; Prenner, R. H.; Schmid, W. *Tetrahedron* **1994**, *50*, 749–758.
53. Li, L.-S.; Wu, Y.-L. *Tetrahedron* **2002**, *58*, 9049–9054.
54. Yamamoto, T.; Kumazawa, H.; Inami, K.; Teshima, T.; Shiba, T. S. *Tetrahedron Lett.* **1992**, *33*, 5791–5794.
55. El-Sepelgy, O.; Schwarzer, D.; Oskwarek, P.; Mlynarski, J. *Eur. J. Org. Chem.* **2012**, *2012*, 2724–2727.
56. Molenda, M. A.; Bas, S.; Mlynarski, J. *Eur. J. Org. Chem.* **2016**, *2016*, 4394–4403.
57. Collins, P. M.; Overend, G.; Shing, T. *J. Chem. Soc. Chem. Commun.* **1981**, 1139–1140.
58. Imoto, M.; Kusumoto, S.; Shiba, T. *Tetrahedron Lett.* **1987**, *28*, 6235–6238.
59. Boons, G. J. P. H.; van der Klein, P. A. M.; van der Marel, G. A.; van Boom, J. H. *Recl. Trav. Chim. Pays-Bas* **1990**, *109*, 273–276.

60. van der Klein, P. A. M.; Boons, G. J. P. H.; Veeneman, G. H.; van der Marel, G. A.; van Boom, J. H. *Tetrahedron Lett.* **1989**, *30*, 5477–5480.
61. Reiner, M.; Schmidt, R. R. *Tetrahedron:Assymetr.* **2000**, *11*, 319–335.
62. Barton, D. H. R.; McCombie, S. W. *J. Chem. Soc., Perkin Trans.* **1975**, *1*, 1574–1585.
63. Wardrop, D. J.; Zhang, W.; Fritz, J. *Org. Lett.* **2002**, *4*, 489–492.
64. Wardrop, D. J.; Zhang, W. *Tetrahedron Lett.* **2002**, *43*, 5389–5391.
65. Kuboki, A.; Tajimi, T.; Tokuda, Y.; Kato, D.; Sugai, T.; Ohira, S. *Tetrahedron Lett.* **2004**, *45*, 4545–4548.
66. Hekking, K. F. W.; Moelands, M. A. H.; van Delft, F. L.; Rutjes, F. P. J. T. *J. Org. Chem.* **2006**, *71*, 6444–6450.
67. Hekking, K. F. W.; van Delft, F. L.; Rutjes, F. P. J. T. *Tetrahedron* **2003**, *59*, 6751–6758.
68. Ichiyanagi, T.; Sakamoto, N.; Ochi, K.; Yamasaki, R. *J. Carbohydr. Chem.* **2009**, *28*, 53–63.
69. Pradhan, T. K.; Lin, C. C.; Mong, K. K. T. *Synlett.* **2013**, *24*, 219–222.
70. Corey, E.; Fuchs, P. *Tetrahedron Lett.* **1972**, *36*, 3769–3772.
71. Feng, Y.; Dong, J.; Xu, F.; Liu, A.; Wang, L.; Zhang, Q.; Cha, Y. *Org. Lett.* **2015**, *17*, 2388–2391.
72. Tsukamoto, H.; Takahashi, T. *Tetrahedron Lett.* **1997**, *38*, 6415–6418.
73. Bräuer, N.; Kirschning, A.; Schaumann, E. *Eur. J. Org. Chem.* **1998**, *1998*, 2729–2732.
74. Burke, S. D.; Sametz, G. M. *Org. Lett.* **1999**, *1*, 71–74.
75. Burke, S. D.; Armistead, D. M.; Schoenen, F. J.; Fevig, J. M. *Tetrahedron* **1986**, *42*, 2787–2801.
76. Ireland, R. E.; Mueller, R. H. *J. Am. Chem. Soc.* **1972**, *94*, 5897–5898.
77. Danishefsky, S. J.; Pearson, W. H.; Segmuller, B. E. *J. Am. Chem. Soc.* **1985**, *107*, 1280–1285.
78. Lubineau, A.; Augé, J.; Lubin, N. *Tetrahedron* **1993**, *49*, 4639–4650.
79. Hu, Y.-J.; Huang, X.-D.; Yao, Z.-J.; Wu, Y.-L. *J. Org. Chem.* **1998**, *63*, 2456–2461.
80. Schlessinger, R. H.; Li, Y.-J.; Von Langen, D. J. *J. Org. Chem.* **1996**, *61*, 3226–3227.
81. Schlessinger, R. H.; Pettus, L. H. *J. Org. Chem.* **1998**, *63*, 9089–9094.
82. Williard, P. G.; Tata, J. R.; Schlessinger, R. H.; Adams, A. D.; Iwanowicz, E. J. *J. Am. Chem. Soc.* **1988**, *110*, 7901–7903.
83. Jiang, S.; Rycroft, A. D.; Singh, G.; Wang, X.-Z.; Wu, Y.-L. *Tetrahedron Lett.* **1998**, *39*, 3809–3812.
84. Kikelj, V.; Plantier-Royon, R.; Portella, C. *Synthesis* **2006**, *22*, 1200–1204.
85. Sugai, T.; Shen, G. J.; Ichikawa, Y.; Wong, C. H. *J. Am. Chem. Soc.* **1993**, *115*, 413–421.
86. Camci-Unal, G.; Mizanur, R. M.; Chai, Y.; Pohl, N. L. *Org. Biomol. Chem.* **2012**, *10*, 5856–5860.
87. Claesson, A.; Luthman, K.; Gustafsson, K.; Bondesson, G. A. *Biochem. Bioph. Res. Co.* **1987**, *143*, 1063–1068.
88. Claesson, A.; Jansson, A. M.; Pring, B. G.; Hammond, S. M.; Ekstroem, B. *J. Med. Chem.* **1987**, *30*, 2309–2313.
89. Hammond, S. M.; Claesson, A.; Jansson, A. M.; Larsson, L. G.; Pring, B.; Town, C. M.; Ekström, B. *Nature* **1987**, *327*, 730.
90. Goldman, R.; Kohlbrenner, W.; Lartey, P.; Pernet, A. *Nature* **1987**, *329*, 162.
91. Kondo, K.-i.; Doi, H.; Adachi, H.; Nishimura, Y. *Bioorg. Med. Chem. Lett.* **2004**, *14*, 467–470.
92. Ohrui, H.; Morita, M.; Meguro, H. *Carbohydr. Res.* **1992**, *224*, 319–325.
93. Lopez-Herrera, F. J.; Sarabia-Garcia, F. *Tetrahedron* **1997**, *53*, 3325–3346.
94. Sarabia-Garcia, F.; Lopez-Herrera, F. J.; Pino-Gonzalez, M. S. *Tetrahedron Lett.* **1994**, *35*, 6709–6712.
95. Molin, H.; Pring, B. *Tetrahedron Lett.* **1985**, *26*, 677–680.
96. Norbeck, D. W.; Kramer, J. B. *Tetrahedron Lett.* **1987**, *28*, 773–776.
97. Pring, B. G.; Jansson, A. M.; Persson, K.; Andersson, I.; Gagner-Milchert, I.; Gustafsson, K.; Claesson, A. *J. Med. Chem.* **1989**, *32*, 1069–1074.
98. Norbeck, D. W.; Rosenbrook, W.; Kramer, J. B.; Grampovnik, D. J.; Lartey, P. A. *J. Med. Chem.* **1989**, *32*, 625–629.

99. Fahey, R.; Brown, W.; Adams, W.; Worsham, M. *J. Bacteriol.* **1978**, *133*, 1126–1129.
100. Adachi, H.; Kondo, K. I.; Kojima, F.; Umezawa, Y.; Ishino, K.; Hotta, K.; Nishimura, Y. *Nat. Prod. Res.* **2006**, *20*, 361–370.
101. Adachi, H.; Kondo, K. I.; Kojima, F.; Umezawa, Y.; Ishino, K.; Hotta, K.; Nishimura, Y.; Akamatsu, Y. *Lett. Drug Des. Discov.* **2008**, *5*, 336–339.
102. Luthman, K.; Orbe, M.; Waglund, T.; Claesson, A. S. *J. Org. Chem.* **1987**, *52*, 3777–3784.
103. Norbeck, D. W.; Kramer, J. B.; Lartey, P. A. *J. Org. Chem.* **1987**, *52*, 2174–2179.
104. Waglund, T.; Claesson, A. *Acta Chem. Scand.* **1992**, *46*, 73–76.
105. Sarabia, F.; Lopez-Herrera, F. J. *Tetrahedron Lett.* **2001**, *42*, 8805–8809.
106. Mlynarski, J.; Banaszek, A. *Org. Lett.* **1999**, *1*, 1709–1711.
107. Mlynarski, J.; Banaszek, A. *Tetrahedron: Assymetr.* **2000**, *11*, 3737–3746.
108. Qian, Y.; Feng, J.; Parvez, M.; Ling, C.-C. *J. Org. Chem.* **2012**, *77*, 96–107.
109. Pradhan, T. K.; Lin, C. C.; Mong, K. K. *Org. Lett.* **2014**, *16*, 1474–1477.
110. Corey, E.; Winter, R. A. *J. Am. Chem. Soc.* **1963**, *85*, 2677–2678.
111. Mazur, M.; Barycza, B.; Andramboavonjy, H.; Lavoie, S.; Kenfack, M. T. Larous. *J. Org. Chem.* **2016**, *81*, 10585–10599.
112. Durka, M.; Buffet, K.; Iehl, J.; Holler, M.; Nierengarten, J. F.; Vincent, S. P. *Chem. Eur. J.* **2012**, *18*, 641–651.
113. Tikad, A.; Fu, H.; Sevrain, C. M.; Laurent, S.; Nierengarten, J. F.; Vincent, S. P. *Chem. Eur. J.* **2016**, *22*, 13147–13155.
114. Claesson, A.; Luthman, K. *Acta Chem. Scand.* **1982**, *36*, 719–731.
115. Wàglund, T.; Claesson, A. *Acta Chem. Scand.* **1990**, *44*, 1058–1061.
116. Orbe, M.; Claesson, A. *Eur. J. Med. Chem.* **1989**, *24*, 447–451.
117. Pakulski, Z.; Zamojski, A. *Tetrahedron* **1997**, *53*, 3723–3728.
118. Ichiyanagi, T.; Yamasaki, R. *Carbohydr. Res.* **2005**, *340*, 2682–2687.
119. Winzar, R.; Philips, J.; Kiefel, M. J. *Synlett.* **2010**, *4*, 583–586.
120. Fan, W.; Chen, Y.; Lou, Q.; Zhuang, L.; Yang, Y. *J. Org. Chem.* **2018**, *83*, 6171–6177.
121. Riedl, B.; Schmid, W. *Carbohydr. Res.* **2018**, *456*, 30–34.
122. Sletten, E. M.; Bertozzi, C. R. *y. Angewandte Chem Int Ed.* **2009**, *48*, 6974–6998.
123. Tra, V. N.; Dube, D. H. *Chem. Commun.* **2014**, *50*, 4659–4673.
124. Dumont, A.; Malleron, A.; Awwad, M.; Dukan, S.; Vauzeilles, B. *Angew. Chem. Int. Ed.* **2012**, *51*, 3143–3146.
125. Wang, W.; Zhu, Y.; Chen, X. *Biochemistry* **2017**, *56*, 3889–3893.
126. Fugier, E.; Dumont, A.; Malleron, A.; Poquet, E.; Pos, J. M.; Baron, A.; Vauzeilles, B.; Dukan, S. *PloS one* **2015**, *10*, e0127700.
127. Nilsson, I.; Grove, K.; Dovala, D.; Uehara, T.; Lapointe, G.; Six, D. A. *J. Biol. Chem.* **2017**, *292*, 19840–19848.
128. Dumont, M.; Lehner, A.; Vauzeilles, B.; Malassis, J.; Marchant, A.; Smyth, K.; Linclau, B.; Baron, A.; Pons, J. M.; Anderson, C. T.; Schapman, D.; Galas, L.; Mollet, J. C.; Lerouge, P. *Plant J.* **2016**, *85*, 437–447.
129. O'Neill, M. A.; Ishii, T.; Albersheim, P.; Darvill, A. G. *Annu. Rev. Plant Biol.* **2004**, *55*, 109–139.
130. Nilsson, I.; Prathapam, R.; Grove, K.; Lapointe, G.; Six, D. A. *Mol. Microbiol.* **2018**, *110*, 204–218.
131. Sherratt, A. R.; Chigrinova, M.; McKay, C. S.; Beaulieu, L. P. B.; Rouleau, Y.; Pezacki, J. P. *RSC Adv.* **2014**, *4*, 46966–46969.
132. Kinugasa, M.; Hashimoto, S. *J. Chem. Soc., Chem. Commun.* **1972**, 466–467.

CHAPTER FIVE

Sialic acid-containing molecules in drug discovery and development

Priyanka Bose*, Anand K. Agrahari*, Anoop S. Singh, Manoj K. Jaiswal, Vinod K. Tiwari
Department of Chemistry, Institute of Science, Banaras Hindu University, Varanasi, India

1 Introduction

Since long back, carbohydrate moiety is known to possesses many unique stereochemical and functional features that is incredibly essential for induction of selectivity and chiral discrimination in various chemical, metabolic, and recognition processes.[1–6] Hence, functionality and structural variations are the key features due to which sugar-based molecules prove to be the interesting scaffolds for various purposes particularly in drug discovery and development.[1,2] The five- and six-carbon sugars generally are one of the important constituents of the vertebrate's glycoconjugated glycan chains. However, the common salient feature in the composition of the glycan chains is the presence of sialic acid family (commonly referred as 'Sia' and structurally is 2-keto-3-deoxynonulosonic acids), which are negatively charged *R*-keto acids with a nine-carbon backbone.[7,8]

Comparison to five- or six- carbons monosaccharides, sialic acids having 9-carbon backbone are more structurally complex molecules, but possess very interesting chemical nature which is widely useful in medicinal chemistry. Besides, they have been predominantly found as the terminal carbohydrate units in glycoproteins and glycolipids of vertebrates, as well as in the components of capsular polysaccharides or lipo-oligosaccharides of pathogenic bacteria and represent the most important recognition elements as terminal sugars.

Because of the significant well ordered natural distribution on cell surfaces, sialic acids are known to mediate and modulate many important cellular interactions and play pivotal roles in many patho-physiological

*Both authors contributed equally.

Carbohydrates in Drug Discovery and Development. http://dx.doi.org/10.1016/B978-0-12-816675-8.00005-1

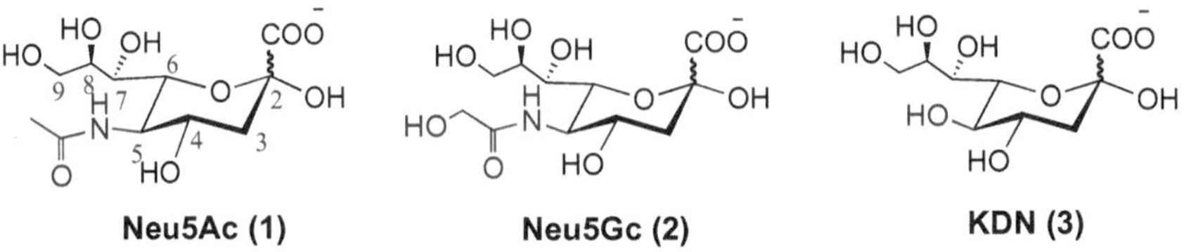

Figure 5.1 ***Three most common sialic acids N-acetylneuraminic acid (1, Neu5Ac), N-glycolylneuraminic acid (2, Neu5Gc) and deaminoneuraminic acid (3, KDN).***

processes such as cellular recognition, adhesion, migration, invasion, communication, infecting of bacteria and virus, and tumor metastasis, etc. These functions are majorly mediated by three most common form of sialic acids includes *N*-acetylneuraminic acid (**1**, Neu5Ac), *N*-glycolylneuraminic acid (**2**, Neu5Gc), and deaminoneuraminic acid (**3**, KDN) (Fig. 5.1) and their derivatives. Further, structural modification with different substitutions on these analogues lead to synthetically produced more than 50 different naturally occurring sialic acids that contributed in learning of the complexity of sialic acid-containing structures and their functionality.

Moreover, proteins that bind to sialic acids are implicated in a wide variety of biological processes and thus evaluation of carbohydrate-protein interaction and further their roles in vertebrates resulted to the enhanced understanding of normal physiology, disease and human evolution. In this chapter, we briefly highlight the significance of sialic acid with their impact, chemical diversity, recent advances and challenges at the interface of chemistry and biology focusing neurobiology in order to find out their promising scope in drug discovery and development.

2 Some representative characteristics of sialic acids

After the original discovery of sialic acid by Klenk and Blix, the molecule has been considered as an important scaffold to be investigated for a longer span, since last sixty years. Both 'Klenk and Blix' bear the trademark of their original discoveries: Gunnar Blix isolated sialic acid from submaxillary mucin (sialos: saliva in Greek) and Ernst Klenk isolated neuraminic acid derivative from brain glycolipids (neuro- + amine + acid).[9] In addition to the comprising of nine-carbon negatively charged backbone (commonly abbreviated as 'Sia' and 'Neu'), there are several other important features of sialic acids which are believed to be unusual, but are highly promising and interesting in the field of chemical biology that includes its diverse functionality as terminal sugar as well as its structural complexity.

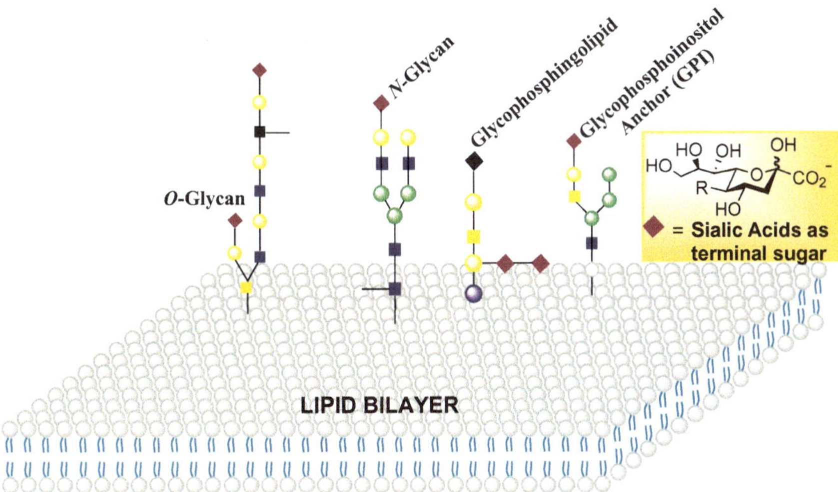

Figure 5.2 ***Schematic presentation of sialic acid and their derivatives as terminal sugars in lipid bilayer.***

2.1 Sialic acid as terminal sugars

The first notable feature of the sialic acid is the typical trait to occupy the distal end of glycan chains (Fig. 5.2). This way, sialic acids (**1-3**) and their derivatives are also termed as "***terminal sugars***". Thus, the 'outermost' location makes them suitable for interaction with the other cells.[8–11]

2.2 Sialic acid has great structural complexity

Second important feature of sialic acid is their diverse structural complexity.[11] The remarkable number of modifications with the distinct analogues of sialic acids majorly can be comprehended in three basic forms include *N*-acetylneuraminic acid (**1**, Neu5Ac, *N*-acetyl-5-amino-3,5-dideoxy-D-glycero-D-galacto-2-nonilosonic acid), *N*-glycolylneuraminic acid (**2**, Neu5Gc, having a hydroxyl group in place of the hydrogen at the *N*-acetyl group of Neu5Ac), deaminoneuraminc acid (**3**, KDN, 3-deoxy-D-glycero-D-galacto-2-nonilosonic acid, having a hydroxyl group in place of the *N*-acetamido group at the C-5 position of Neu5Ac) (Fig. 5.1). All the three forms have potential for the additional substitutions at the hydroxyl groups positioned at C-4, C-7, C-8 and C-9 including *O*-acetylation and the less frequent *O*-methylation, *O*-lactylation, *O*-sulfation, and also *O*-phosphorylation.[8,11]

Usually, the sia molecules are linked as acetal to the non-reducing end of carbohydrate chains of glycolipids and/or glycoproteins. Besides, some

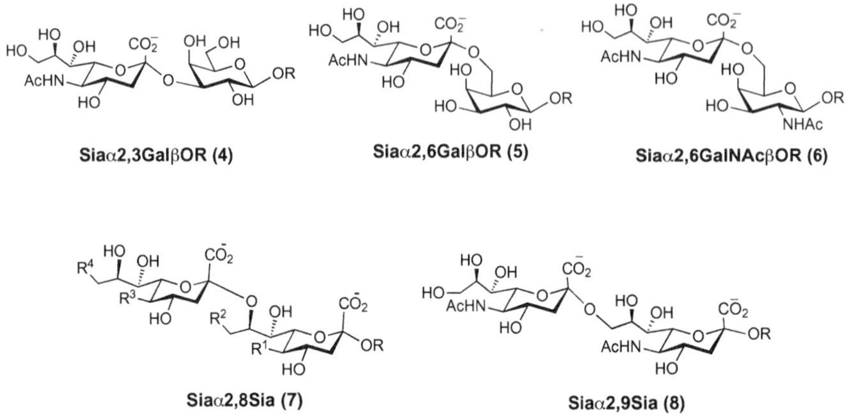

Figure 5.3 Some common glycosidic linkages of naturally occurring sialosides **(4-8).**

of the analogues contain variety of other functional groups as well as different substituents instead of the alcoholic —OH groups. They have been predominantly found as the terminal carbohydrate units in glycoproteins, glycolipids of vertebrates, components of capsular polysaccharides, or lipooligosaccharides of pathogenic bacteria.[8,11] In vertebrates, sialic acids are typically presented in Sia-α2,8Sia)n homopolymers or as the outermost units of glycan, for example, Sia-α-2,3-GalOR (**4**), Sia-α-2,6-GalOR (**5**), Sia-α-2,6-GalNAcOR (**6**), types of linked 'sia' molecules are most common structures to be found in nature. Their structural modifications with different substitutions (at C-4, C-5, C-7, C-8, and/or C-9 positions) lead to the development of versatile sets of sia (Fig. 5.3).

Out of these diverse structures, more than fifteen structures (of which mainly Neu5Ac and their derivatives, also few of KDN and *O*-acetylated KDN) have been found in human blood cell surfaces, saliva protein, and gastrointestinal mucins.[12–14] In bacteria, 'sia' are mostly found in the extracellular capsular polysaccharides and lipopolysaccharides (LPS) as internal structural unit having α-2,8- (**7**), α-2,9 (**8**), or altering α-2,8-/ α-2,9-linkages.[15] They also found as the terminal residues in the glycans of bacterial lipooligosaccharides of several Gram-negative bacteria.[16,17]

The major features which have been discussed here plays very important role in their chemical attribution and also in biological functioning. Obviously, the chemical features always vary with the substituent present on the parent molecular core system which gets directly influenced by the synthetic method or the enzymatic synthesis within the different organisms reported.

2.3 Occurrence of sialic acids

These nine carbon backbone sugars, dissimilar to the other vertebrate monosaccharides, are not ubiquitous in nature and seem to be predominantly found in two distinct branches of plant kingdom, on the other side in the deuterostome[8] lineage of animals as well as in different subclasses of animal kingdom (vertebrates and a few higher invertebrates, for example starfish, marine diatom *Nitzschia alba*[18] and in certain types of bacteria, like some strains of *E. coli*[19–21] *Neisseria meningitides*[22] (sia molecule as virulence factor), Salmonella O48 and Arizona O29 (contain sia as part of the O antigens).[23] Besides, some of viruses like, Sindbis virus,[24] Rous sarcoma,[25] vesicular stomatitis[26] posses 'sia' molecule as their membrane component.

2.4 Sialic acids have a great biological significance

Sialic acid is one of the major molecules found in the vertebrates having a versatile influence over the physiological functions. Moreover, human blood and serum are the major site containing different sia derivatives among them major portion up-to 80%, is Neu5Ac of 'sia' derivatives found.[27] Sia analogues play pivotal roles in variety of pathological and physiological process most notably includes, cell-cell interaction,[28,29] inflammation,[30] fertilization,[31] infection,[32] differentiation,[33] malignancy,[34,35] and cell signaling.[36] However, due to very complex structure and difficulties in their synthetic methods to achieve target, complete knowledge of their entire bio-chemical significance is yet to explore.[37] Thus, for clear understanding of their role in various important biological events experimental investigation should be encountered on the diverse glycoconjugates to achieve thorough chemical and biological expertise of the target molecule. Based on the above mentioned facts, development of homogenous libraries of diverse sialosides of chemotherapeutic values has been well investigated by various research groups. Although significance of 'sia' molecule is well explored in neurobiology has been comprehended in the Section 7.

2.5 Biosynthesis of diverse sialosides in eukarotic and prokaryotic system

Biosynthesis of sialic acids is quite diverse by nature among the varied organisms. The key step in the sialoconjugate biosynthetic pathways is the sialyltransferase (ST)-mediated transformation that transfers a sialic acid from CMP-sialic acid as activated donor to different acceptors, usually a galactoside or derivative (also may be another Sia-OR). The biosynthesis of

CMP-Neu5Ac (**12**) or their derivative is well documented in the several reports[38,39] and the synthesis procedure differs among the different genera and subclasses from bacteria to vertebrates.

De novo synthesis of Neu5Ac in eukaryotes occurs in the cytosol with three consecutive steps (**Path A,** Scheme 5.1).[40] However, in bacteria Neu5Ac is synthesized directly through condensation of ManNAc (**9**) with either phosphoenolpyruvate (PEP) or pyruvate catalyzed by Aldolase

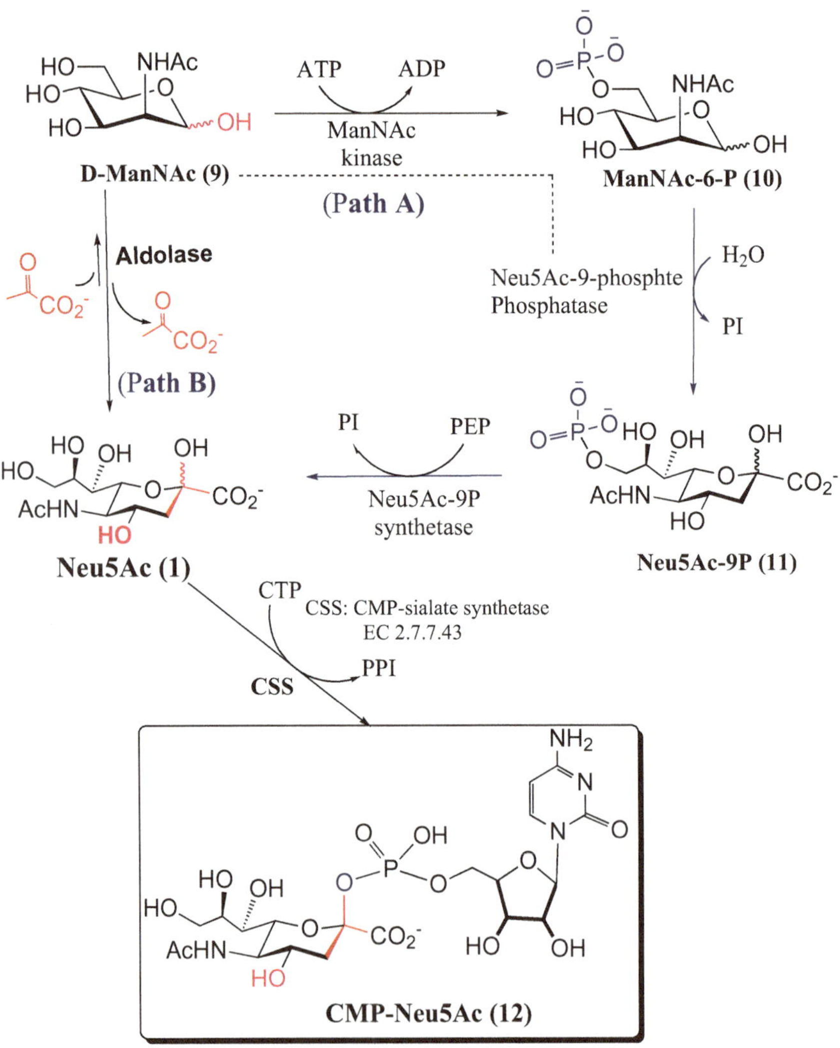

Scheme 5.1 ***Biosynthesis of sialosides in eukarotic and prokaryotic system for inspiration.***

(**Path B,** Scheme 5.1).[41,42] The biosynthetic pathway of CMP-KDN in bacteria is believed to follow a similar route.[8,39,43,44] Perhaps, CMP-Neu5Gc is not known in bacteria, it is believed to obtained from CMP-Neu5Ac (**12**) by CMP-Neu5Ac hydroxylase-mediated oxidative conversion of *N*-acetyl group to *N*-glycolyl group.[45] On the other side, in mammalian system, more advanced step of sialic acid modification, for example, *O*-acetylation, *O*-methylation, *O*-sulfation, *O*-lactylation etc, are believed to occur after the synthesis of sialoconjugates.[46–48]

However, reverse of the modification also takes place for the digestion of the molecule in cells, like sialic acid aldolases or *N*-acetylneuraminate lyases (NanAs) catalyze the reversible aldol cleavage of Neu5Ac (**1**) to form pyruvate and ManNAc (**9**). For investigation purpose, sialic acid and their derivatives were developed by exploiting the concept of naturally occurring aldolase-catalyzed asymmetric Aldol type of reactions for the six carbon monosaccharides (for example, *D*-mannose, ManNAc and their derivatives) with pyruvic acid as outlined in **Path B,** Scheme 5.1.

2.6 Enzymatic synthesis of sialylated glycans

Synthesis of sialic acids *via* condensation of a neutral six-carbon unit with the three-carbon pyruvate unit is the fourth important characteristic (*See, Biosynthesis of Sialic acid described in the next Section*).

Lastly, even as all the other vertebrate monosaccharides are activated in the form of uridine or guanine dinucleotides (e.g., GDP-Man and -Fuc, UDP-Glc, -Gal, -GlcNAc, -GalNAc, -GlcUA, and –Xyl, etc), ***sialic acids are activated as cytidine mononucleotides*** i.e., CMP-Sia molecules. Thus, sialic acids are unusual in the nature due to their high-energy nucleotide sugar donor form which is an important characteristic of these nine carbon sugars.[11] To explore the chemical nature of 'sia' different synthetic pathways have been developed to achieve the targets and in the next section the efforts of different group to achieve the aim has been discussed.

3 Common synthetic routes for sialic acid containing molecules

Sia plays a diverse function in the biological system as free sialic acids and derivatives, hetero-saccharides, glycoproteins, membrane glycol-saccharide chains, etc.[49] Some of these derivatives have been developed synthetically using different major pathways to achieve target moiety in the course of drug development. Among them glycosylation is one of the major synthetic route for achieving these molecular systems. Besides, chemo enzymatic synthesis tool and a click inspired synthetical route also have been employed for this purpose.

3.1 Chemical glycosylation

In order to develop the libraries of interesting sialic acid linked structures, a great endeavor has been made by number of chemist since last three decades by the direct or, the indirect chemical sialylation approaches.[30,31] Till today's date, the chemical synthesis of α-sialosides is still a challenging task in the carbohydrate field. For that purpose different synthons are being trialed over and over repeatedly to achieve α-anomer as the final major product with less thermodynamic product (β-anomer) as well as nullifying the side reactions like hydrolysis or, elimination (Scheme 5.2).[15,50–55]

In this process of sialylation, major reactions undergo *via* oxacarbenium ion intermediate which brings nucleophilic attack to the acceptor and to control that C-3 positioned group of sialic acid ring has been explored a lot. Besides, some other conditions take place like, positive charge of the anomeric carbon get destabilized by electron-withdrawing carboxylate group at C-1, acetamido group at C-5 (for *N*-acetylneuraminic acid, Neu5Ac, **1**) affects the reactivity of the donor and the stereoselectivity. In these following section versatile reactions which has explored α-anomeric derivatives of sialic acid has been focused majorly.

3.1.1 Modifications at C-1

Several groups have investigated versatile approaches for the nucleophilic attack toward the α-face of oxacarbenium ion to introduce the

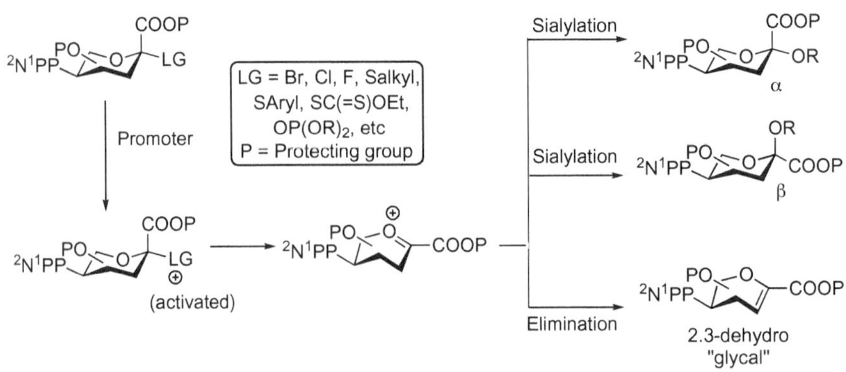

Scheme 5.2 ***Overview of a generic sialylation reaction.***

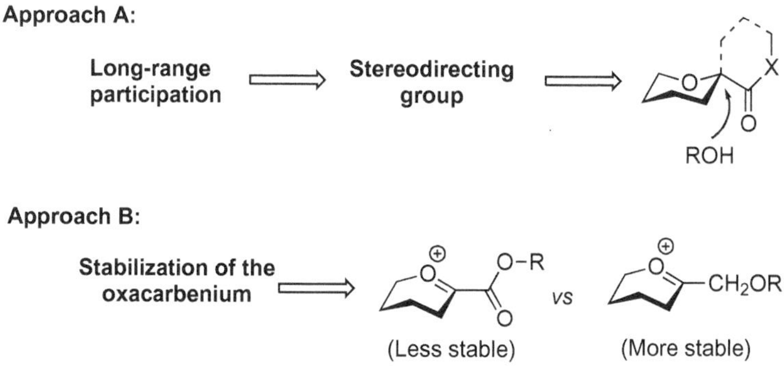

Figure 5.4 ***A common stereo-directing groups at C-1.***

stereodirecting groups at C-1 positioned. All these efforts can be jotted mainly under two categories majorly, (1) introducing a participating group at C-1 (**Approach A**, Fig. 5.4), and (2) stabilizing the oxacarbenium ion via reduction of carboxylate moiety (**Approach B**, Fig. 5.4). But the efforts doesn't concluded remarkable improved synthesis of α-sialosides rather these piles of report concluded better insight into the sialylation reactions.

Takahashi et al reported an improved stereoselective chemical glycosylation while converting carboxylic acid group at C-1 to 2-methylthioethyl ester **13**. The reaction includes glycosyl donor, 2-[2-(2-azidoethoxy) ethoxy]ethanol acceptor **14**, *N*-iodosuccinimide, trifluoromethanesulfonic acid (TfOH), acetonitrile as a solvent resulted in producing methyl ester-protected per-O-acetylated 2-methylthio Neu5Ac derivative **17** as final product with 62% of yield but, with poor stereoselectivity (α/β:1/1.5). However, usage of 2-methylthioethyl ester protected Neu5Ac **1** as donor reported to show better selectivity with lower yield (α/β:4/1, Scheme 5.3). The use of 1,2-dimethoxyethane (DME) as solvent although reported to afford the product in lower yield (45%), but achieve an excellent α-stereoselectivity (α/β:20/1). On the other hand, the utilization of methyl 2,4,6-tri-O-methoxybenzyl-β-D-galactopyranosyl **16** as acceptor for glycosylation in DME, reported to show better stereoselectivity but yields of the target product **19** lowered more (21% to 35%, Scheme 5.3).[56]

In case of Type B of reactions (Fig. 5.4) remarkable stability of the oxacarbenium ion has been reported by Wong while the stereodirecting reaction of sialyl donors to convert into various functionalized hydroxymethyl groups has been depicted. In despite of high yield β-product has been observed to predominate which is influenced by the solvent effect and the promoter system.[57]

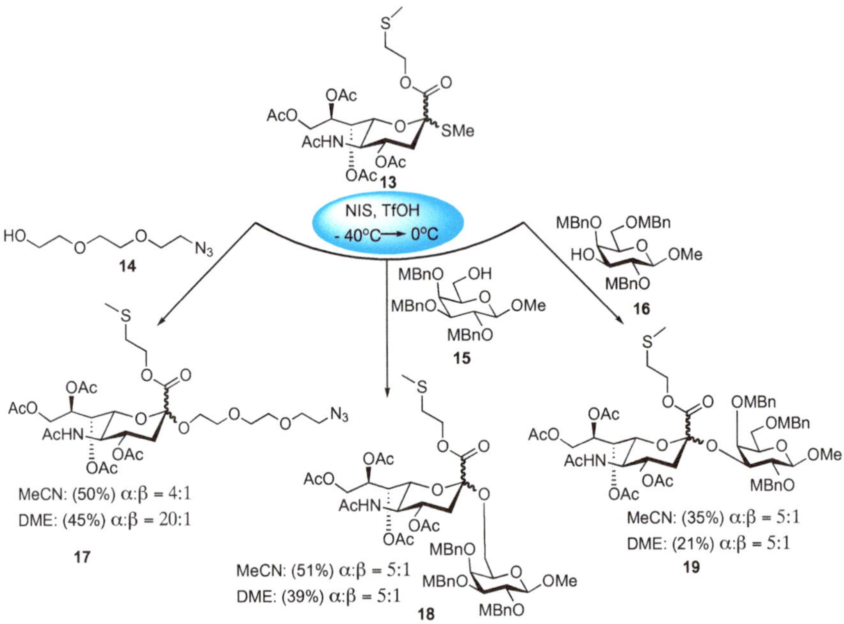

Scheme 5.3 ***Selected examples of glycosylation using 2-methylthioethyl ester.***

3.1.2 Modifications at C-2: leaving groups

C-2 is another positional variation site which has been greatly explored in anomeric sialation approach for sialic acid derivative synthesis. Majorly the route has been investigated for sialoside synthesis focused on the stereoselectivity. In these exploration halides, thiosialosides and trifluoroacetimidates have been majorly focused so the discussion also has been lined up in that approach only which is as follows.

In 1990s era, **sialyl chlorides** has been used as the donor majorly.[50,58–62] In fact as leaving groups these chemicals were the first to be utilized for the synthesis of *O*-sialosides.[63] In this course 2-Bromo[64] and 2-fluoro[65–67] derivatives have been synthesized but, found to be of very less effective for 2-chloro derivatives. Glycosylation reaction using sialyl halide undergo via the metal-induced activation, in case if, silver(I) salts were used method is known as Koenigs–Knorr approach with the advantage of faster reaction time and showed higher stereoselectivities, however mercury(II) salts also can be utilized which is known as Helferich method with the advantage of high yield.[52,68] Further the classic Koenigs–Knorr method reinvestigation using sialyl chloride donor in the presence of silver carbonate, galactopyranosyl diols and triols as acceptors to prodcue α(2→6)-glycosidic

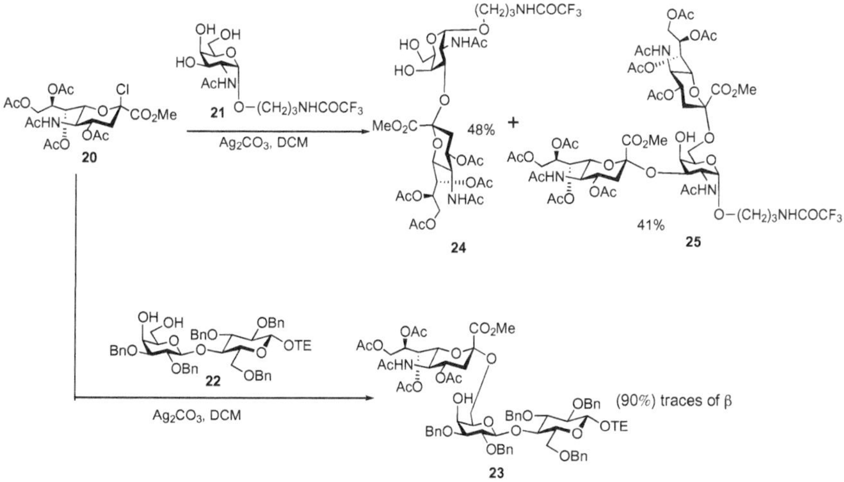

Scheme 5.4 ***Examples of sialylations using glycosyl chlorides.***

bonds, is reported to show moderate yields with the high stereoselectivities (Scheme 5.4).[68] More exploration of the reaction has been done using *N*-acetylgalactopyranosyl triol as glycosyl acceptor reported to show unusual regioselectivity resulting in a mixture of α(2→6)-linked disaccharide and α(2→3)-/α(2→6)-linked trisaccharide. In the last decade, expanded utilization of 2-chloro sialosides to convert other leaving groups and intermediates increased their demand which showed the advantage of high stereoselectivity.[69–74] In case of stereoselective apart from donor used choice of solvent also plays a crucial role like, high α-stereoselectivity has been reported by using dichloromethane or 1,2-dichloroethane (DCE) as solvent, on the other hand reaction in acetonitrile resulted to the lost of this selectivity.[75]

One of the mostly utilized and advantageous building blocks for the synthesis of sialosides, are the alkyl and aryl thiosialosides which include high stability, long shelf life, general reliability. Thus, **S-sialosides** has been explored as one of the most interesting donors for the generation of diverse sialosides of chemotherapeutic values. There are numerous publication explaining synthesis and performance of methyl-,[76–79] phenyl-[62,70] and ethyl-[70,80] thioglycosides as the donors[50–53] which clearly justifies their popularity (Scheme 5.5). Thioglycosides require promoter systems, like, dimethyl(methylthio)sulfonium trifluoro methanesulfonate (DMTST) or NIS/TfOH to get activated.[81,82]

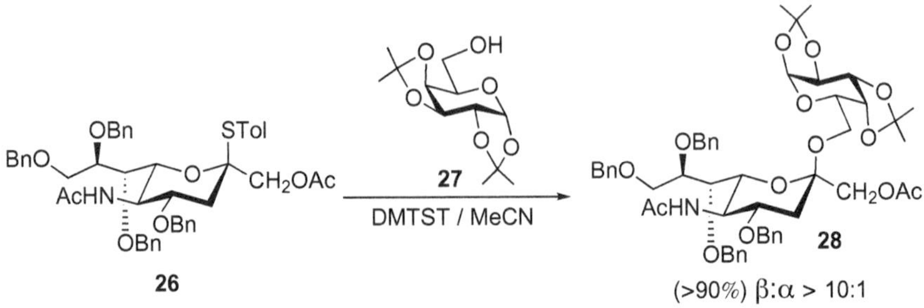

Scheme 5.5 *Examples of sialylations using thiosialosides.*

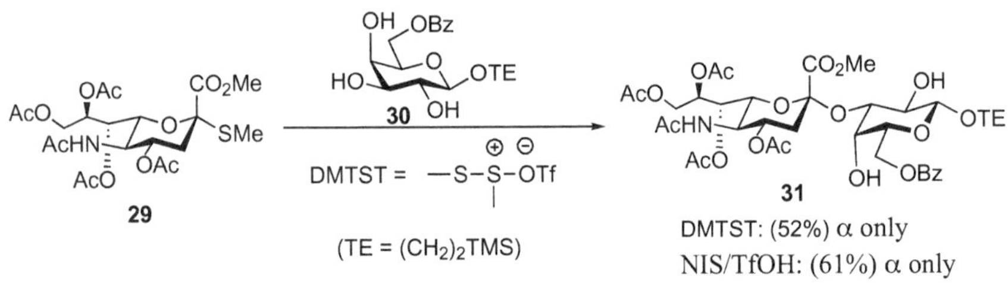

Scheme 5.6 *Representative applications of the methylthio sialyl donor.*

In 80s Hasegawa et al., reported usage of per-O-acetylated 2-methylthio Neu5Ac donor with DMTST and NIS/TfOH as promoter system in the presence of secondary galactopyranosyl and lactosyl acceptors. The same reaction is promoted using (2-trimethylsilyl) ethyl 6-O-benzoyl-β-D-galactopyranosyl acceptor **30**, DMTST in acetonitrile at -40°C, α(2→3)-linked disaccharide **31** was produced in 52% yield, however when the same donor–acceptor pair is promoted using NIS/TfOH reported to show increased yield (61%, Scheme 5.6).

Crich et al., reported the usage of adamantylthio group as thioglycoside donor for sialic acids.[83] The functional substituent can be added without the unpleasant odor which is the disadvantage associated with thio group involving reactions majorly, but the reaction showed moderate stereoselectivity. It is found that the reagent is more reactive than its phenylthio version. However no reaction proceed when it is used in the presence of an unreactive methyl 2,6-di-O-benzyl-β-D-galactopyranosyl **33** as acceptor, NIS/TfOH in acetonitrile at −40°C, an oxazolidinone protected β-phenylthio sialyl **32** as donor. But β-adamantylthio sialyl **34** usage as donor in the same sialylation process resulted in the desired (2→3)-linked disaccharide **35** with 87% of yield but with modestly high stereoselectivity (α/β = 10/1, Scheme 5.7).[83]

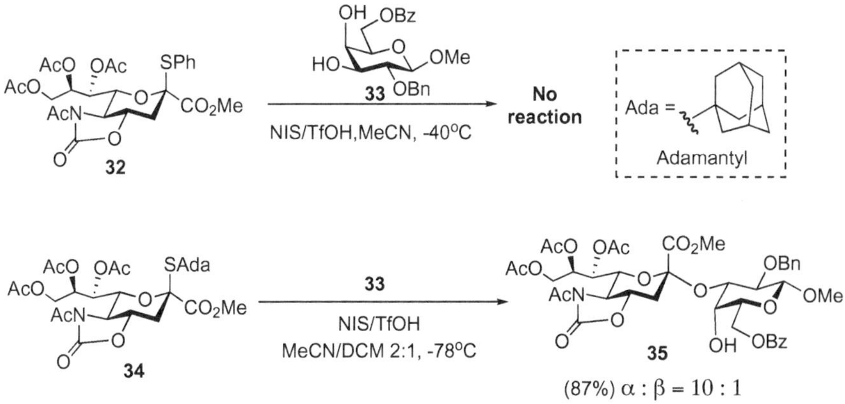

Scheme 5.7 *Some selected example of sialylation using S-adamantyl as donor.*

Among the other thio groups explored one must be discussed reported by Marra et al., i.e., anomeric xanthates which is obtained by the reaction of the sialyl chloride with potassium ethoxydithiocarbonate in ethanol.[70] The common promoters used for the thioglycosides can be utilized for the activation of xanthate as well as with selectively in the presence of thioglycosides.[84] Further, stereo-selectivity of the final product can be influenced with DMTST[85] or NIS/TfOH,[76,86,87] as well as with phenylsulfenyl trifluoromethanesulfonate (PhSOTf) and 2,6-di(tert butyl)pyridine (DTBP) at low temperature (−70°C).

In these context another functional group should be discussed, which is the thioimidates like, *S*-benzoxazolyl (SBox) group reported by De Meo et al. as leaving groups in chemical sialylations.[88] Thioimidates are associated with the advantage of desirable selective activation in the presence of thioglycosyl acceptors like, ethylthio group (Scheme 5.8). On reacting, using an ethylthio 2,3-di-O-benzoyl-β-D-galactopyranosyl acceptor **37** with AgOTf, the per-O-acetylated 2-S-benzoxazolyl Neu5Ac donor

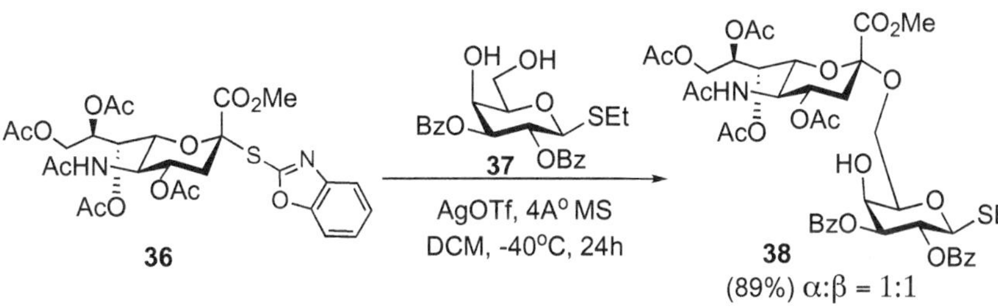

Scheme 5.8 *Selective sialylation using thioimidate as a sialyl donor.*

36 resulted in (2→6)-linked disaccharide **38** with 89% of yield but, as a racemic mixture. On the other side, using methyl 2,3,6-tri-O-benzyl-α-D-glucopyranosyl as acceptor and NIS/TfOH system favored to produce β(2→4)-linkage predominantly along with GM3 derivatives with 70% of yield.[88]

At the end, the third approach for the anomeric sialation reaction is the **O-sialosidation**. In early 2000 it was reported by Cai et al., in which an oxygen-based leaving group *N*-phenyltrifluoroacetimidates were utilized for achieving this purpose. On reacting per-O-acetylated 2-hydroxy-Neu5Ac derivative with 2,2,2-trifluoro-*N*-phenylacetimidoyl chloride and potassium carbonate resulted in the desired compound as the major product. But exceptional case is *N*-(*p*-methoxyphenyl) derivative which does not follow the same reaction order.[89] The author also reported that a strong solvent effect influences this leaving group activity like, while its activation by TMSOTf resulted different and complete inversion of the stereoselectivity occurred when dichloromethane (β-preferred) was switched with acetonitrile (α-preferred) as solvent. It is well explained while using 1,2,3,4-di-O-isopropylidene-α-D-galactopyranosyl as acceptor, the corresponding disaccharide was found in α/β ratio of 1/2.4 in DCM and 1.6/1 in MeCN respectively (Scheme 5.9). On the other side

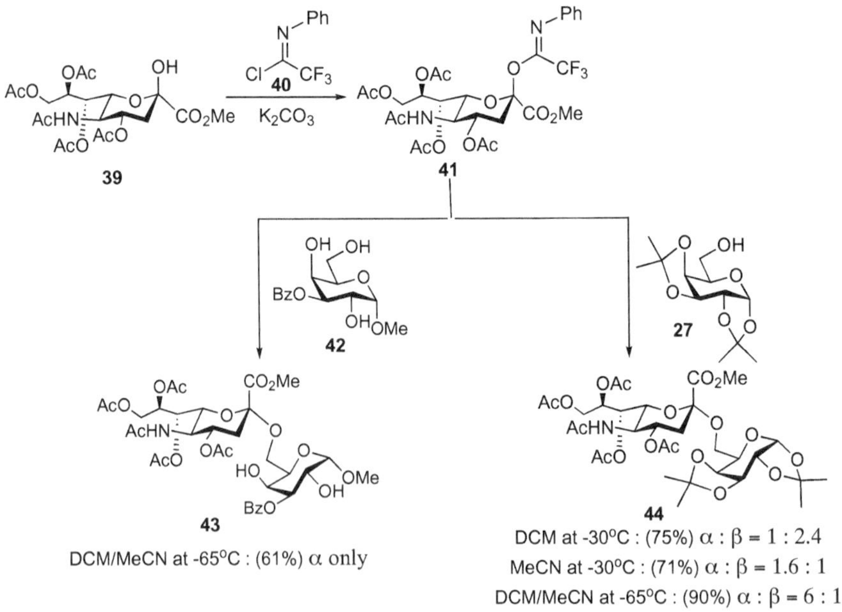

Scheme 5.9 *N-Phenyltrifluoroacetimidate synthesis and sialylations.*

when per-O-acetylated 2-*N*-phenyltrifluoroacetimidate sialyl is used as donor and methyl 3-O-benzoyl-α-D-galactopyranosyl as acceptor the reaction resulted in the α-anomer as final product only with 61% of yield (Scheme 5.9).

The structure of commonly used sialyl donors in chemical sialylation reactions is summarized in Scheme 5.10. Thioalkyl or thioaryl sialyl donors with various substitution pattern like *N*-trifluoroacetyl **45a**, *N*-TCA, *N*-Fmoc, *N*-2,2,2-trichloroethoxycarbonyl (*N*-Troc), *N*,*N*-diacetyl *N*-phthalimide group, or azido group at *C*-5 of the sialic acids were well utilized for the chemical sialylation method. The introduction of this group at the *C*-5 position in sialyl donors resulted in the enhancement of the donor reactivity in sialylation reaction with high stereoselectivity of the sialosides. Interestingly, some of these *N*-protecting groups may lead to Neu5Gc analogs simply by deprotection of the resulted glycosylated product followed by standard derivatization at *C*-5 amino group.[30]

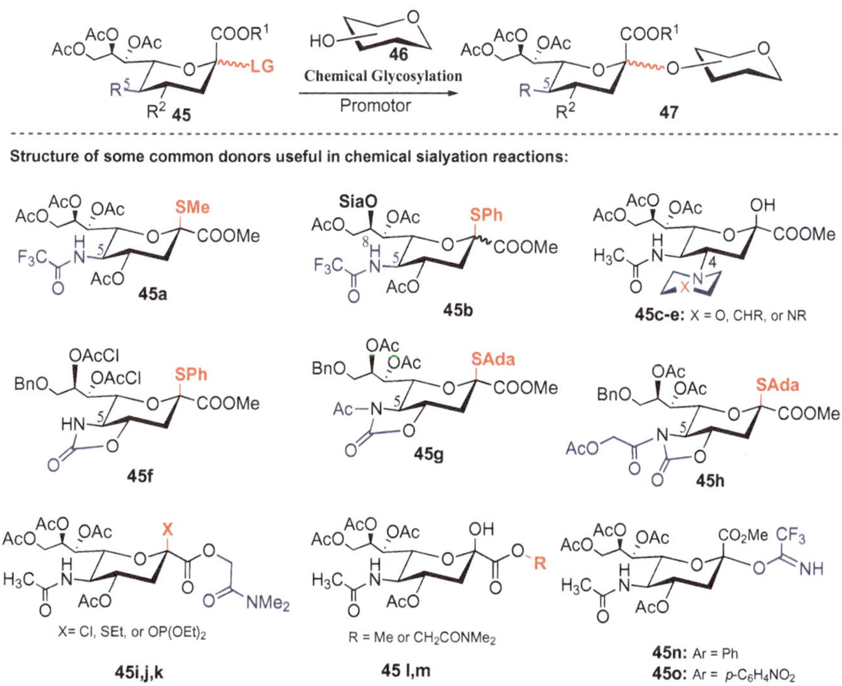

Scheme 5.10 ***Some selected common routes for chemical glycosylation in sialic acid donors.***[37]

3.2 Chemo-enzymatic synthesis for the designing sialoside libraries

Structurally modified carbohydrates is common in nature which is found to have versatile non-negligible biological significance.[1] For example, sugars available on the mammalian cell surface particularly sulfated sugar present in a number of proteoglycans, glycolipids, and glycoproteins, are believed to play important roles in specific molecular recognition processes. Furthermore, modification of the siaic acid monosaccharides (at specific position) such as acetylation, methylation, phosphorylation, sulfation, lactylation, etc leads to develop diverse structural analogues serving important bio-chemical funtions.[7,8,10] However, little is known about the SAR of these modified sialoconjugates, mainly due to the technical difficulties in obtaining homogeneous libraries of the structurally modified sialosides.

Due to the unique and complex structures of sialic acids, synthesis of the sialoconjugates or sialylated glycans is inherently difficult and makes the synthetic approach is far to be successful. The natural path for the synthesis of simple sialic acids and their derivatives encourages researcher to explore the chemoenzymatic route, simply by reacting *D*-mannose or their derivatives in presence of Aldolase (Scheme 5.11A). In 2008, Chen et al., developed a capillary electrophoresis assay to directly characterize the activities of NanAs in both Neu5Ac cleavage and Neu5Ac synthesis directions. The substrates specificity studies showed that 5-*O*-methyl-ManNAc (**48**) was

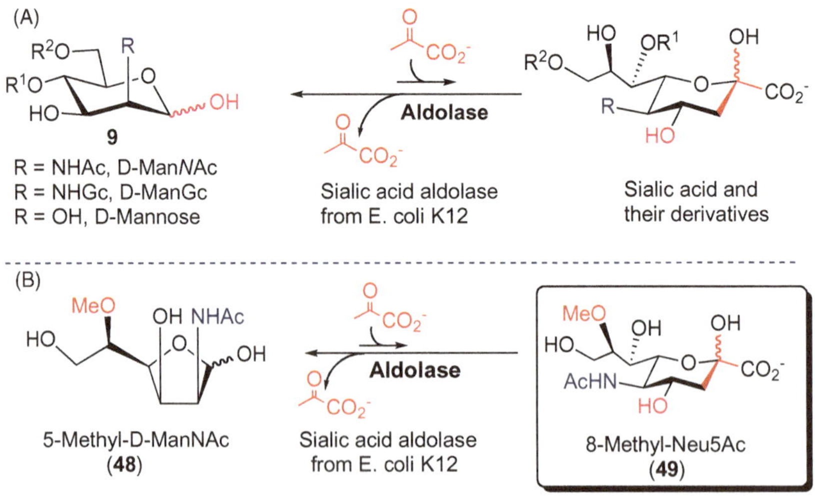

Scheme 5.11 (A) Aldolase mediated synthesis of sialic acids from respective mannose and their derivatives; (B) Aldolase catalyzed enzymatic synthesis of 8-methyl sialic acid.[90]

successfully be used efficiently as a substrate by PmNanA for the synthesis of 8-*O*-methyl Neu5Ac (**49**) (Scheme 5.11B).[90]

PmNanA mainly because of the higher expression level and a broader substrate tolerance identified as a better catalyst over EcNanA for the chemoenzymatic synthesis of 'sia' molecules. Chen and co-workers reported about the determination of two wild-type structures, in the absence of substrates, and also trapped in a Schiff base intermediate between Lys164 and pyruvate, respectively.[91] From the crystal structures of the wild-type *P. multocida* *N*-acetylneuraminate lyase and its K164A mutant, this was further concluded that both the sialic acids is known to bind to the active site of the open-chain ketone form of the monosaccharide. This further reveals that each hydroxyl group of the linear carbohydrates could formulate hydrogen bond interactions with the enzyme and the residues which finally determine the specificity.[91]

A large number of monosaccharides were further explored in order to access well substituted and well defined linked sialosides. In general, the substrate or ligand specificity of sia recognizing proteins may be estimated by three ways. The first important method can be accomplished by changing of the structure which is generally achieved by utilizing the various donors, preferably Mannose, ManNAc or their derivatives (*O*-methyl, deoxy, azido, acetate or other common groups) either at 2- or 4- or 5- or 6-position individually or as multiple substitution at the positions. The second way is to examine the substrate/ ligand specificity of the sialic acid recognizing proteins by observing the type of glycosidic linkage present, i.e., either α-2,3 or α-2,6 linkage of sialic acid with penultimate sugar. Next important feature can be examined is the structure of the penultimate sugar residue present (in general, Gal or GalNAc or Lactose or their derivatives). To investigate the biological significances of sialic acids, homogenous sialosides are needed. To achieve this target chemical method for the library synthesis of the well-defined sialosides is a really tedious work thus enzymatic method is well adopted around the globe. However, Prof. Xi Chen work is mainly focused on the chemoenzymatic synthesis of the complex sialic acids containing molecules and their application in chemical biology. Thus, there is no doubt that the most practical method to be considered for the synthesis of sialic acid-containing molecules is the '**Chen's Chemo-enzymatic Pne-Pot Multi-Enzyme (OPME) route**'.

Chemoenzymatic synthesis is a powerful approach to obtain the complex sialic acid-containing carbohydrates, however, the one-pot three enzymes approach is considered to be the best choice to obtain a library of sialoside with desired α-2,3 or α-2,6 linkages.[92–94] The method is greatly useful for the development of the sialic acid-containing complex architectures, including those with diverse naturally occurring and semi-synthetic or synthetic sialic acid forms, different sialyl linkages and different glycans that link to the sialic acid. In the first step, the reversible aldol cleavage of Neu5Ac or its derivatives to form pyruvate and ManNAc (**9**) is catalyzed by sialic acid aldolases/*N*-acetylneuraminate lyases (NanAs). Then, the subsequent activation by a CMP-sialic acid synthetase (lead to glycosyl donor) and finally transfer to a wide range of suitable acceptors, for example galactose, GalNAc, Lac, and their derivatives (**53–55**) by a respective sialyltransferase leads to the formation of sialosides containing natural and synthetic functionalities with the desired α-2,3- linked sialosides (**56**) or α-2,6-linked sialosides (**57**). The three-enzyme coupled synthesis of sialosides may be carried out in one-pot without the isolation of intermediates (Scheme 5.12). In a similar fashion, 5-azido modified hexoses could afford corresponding 8-azido sialic acids with the desired α-2,3- or α-2,6-linked sialyltrisaccharides, for example, Neu5Ac8N_3 and the other related azido-sialosides may be useful for the click diversification purpose.[5] Although, research dealing with sialylated-click is not well-explored.

In recent years, this one-pot three-enzyme tool of the synthetic chemistry has been well explored with different substitution patterns in both as diverse range of the donors as well as the galactose or lactose related acceptors (**53–55**) to achieve the diverse α-2,3-linked sialosides (**56**) and α-2,6-linked sialosides (**57**).[37,95,96] This is rather interesting that Chen's one-pot three-enzyme synthesis of α-2,3- and 2,6-linked sialosides equally work well for an easy synthesis of sialoside containing and 8-deoxy KDN and Neu5Ac8Me.[97] Naturally occurring 8-*O*-methylated sialic acids, including 8-*O*-Me-Neu-5Ac and 8-*O*-Me-Neu5Gc, along with Kdn8Me, 8-deoxy-Kdn and α-2,3-linked sialyltrisaccharides containing Neu5Ac8Me (**58**) and Kdn8-deoxy (**59**) in addition to α-2,6- linked sialyltrisaccharides containing Neu5Ac8Me (**60**) and Kdn8-deoxy (**61**) has been developed from corresponding 5-*O*-modified six-carbon monosaccharides using a one-pot multi-enzyme approach (Scheme 5.13). The strategy provides an efficient approach to produce glycans containing various C8-modified sialic acids for their biological evaluations.

Considering the biological significance of natural sialic acid variations on disialyl structures, Chen et al, reported a brilliant and practical two-step multi enzyme protocol for the synthesis of a series of GD3 ganglioside oligosaccharides and other disialyl glycans containing a terminal Siaα-2-8Sia

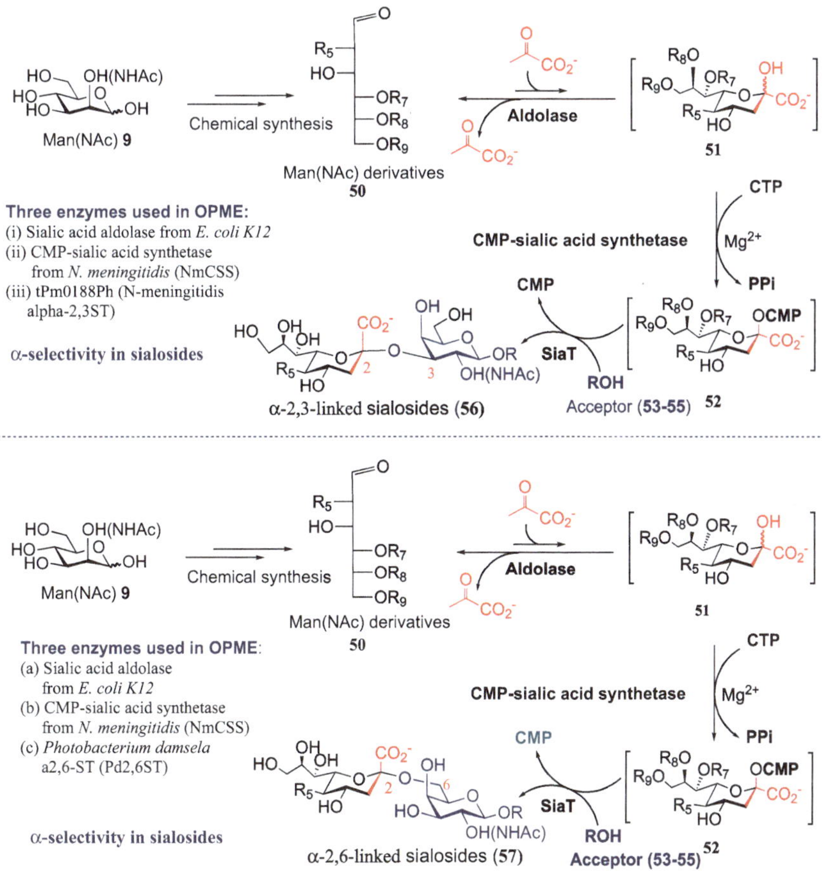

Scheme 5.12 ***Chen's one-pot synthesis of α-2,3-linked sialosides and α-2,6-linked sialosides.***[96]

component with various natural and un-natural 'sia' derivatives.[98] CstII mutant (CstIIΔ32I53S) which was cloned from a synthetic gene on investigation concluded that it has also the α-2,8-sialidase activity which simply known to catalyze the specific cleavage of the α-2,8-sialyl linkage of GD3-type oligosaccharides and α-2,8-trans-sialidase activity that resulted to the transfer of a sialic acid from a GD3 oligosaccharide to a different GM3 oligosaccharide.[94] In the first-step, α-2,3- or α-2,6-linked mono-sialylated oligosaccharides (**56** or **57**) were obtained in good to excellent yield by utilizing one-pot three-enzyme approach as depicted in Scheme 5.12. The resulted sialosides were then further used as acceptors for the α-2,8-sialyltransferase activity of a recombinant truncated multifunctional *Campylobacter jejuni* sialyltransferase CstII mutant to generate a number

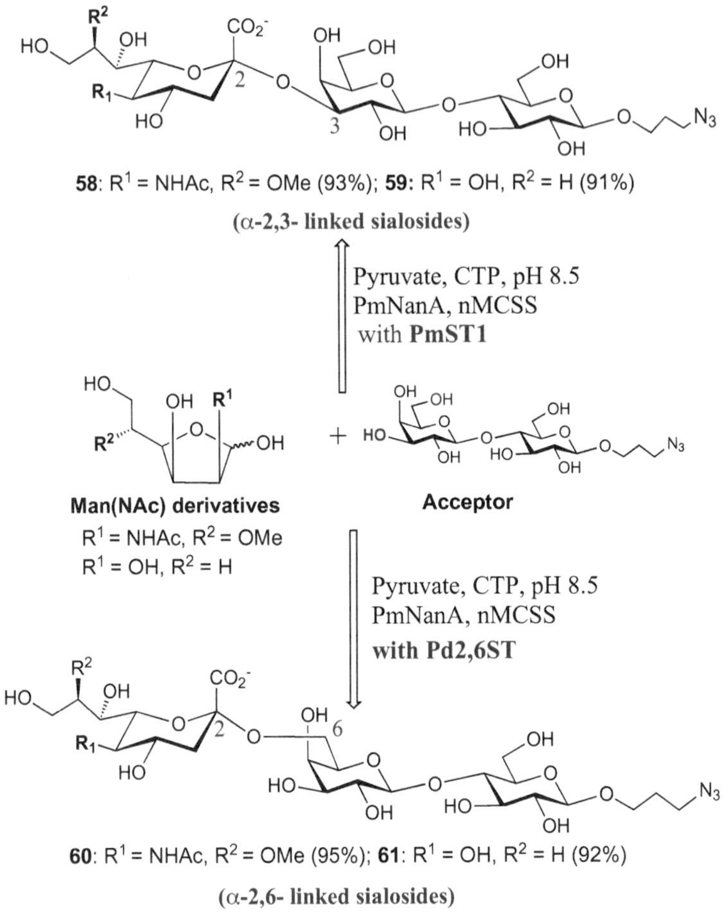

Scheme 5.13 ***One-pot three-enzyme synthesis of α-2,3- and 2,6-linked sialosides containing Neu5Ac8Me and 8-deoxy KDN.***[97]

of disialyl oligosaccharides (**62**).[98] Further the enzymatic preparation of GD3-type disialyl oligosaccharides, for example, Neu5Gc-α-2,8Neu5Gc-α-2,3LacProN_3, KDN-α-2,8-KDN-α-2,3-LacProN_3, and others related sialic acid-containing glycans of wide application in chemical biology were successfully achieved in good to excellent yields simply by utilizing the protocol depicted in Scheme 5.14.[98] Thus, one-pot three-enzyme (OPME) approach was first explored to get α-2,3- linked mono-sialylated oligosaccharides (**56**) or α-2,6-linked mono-sialylated oligosaccharides (**57**) (**Step 1**) and then the resulted sialosides were further used as the suitable acceptors for the α-2,8-sialyltransferase activity (**Step 2**) to develop a series

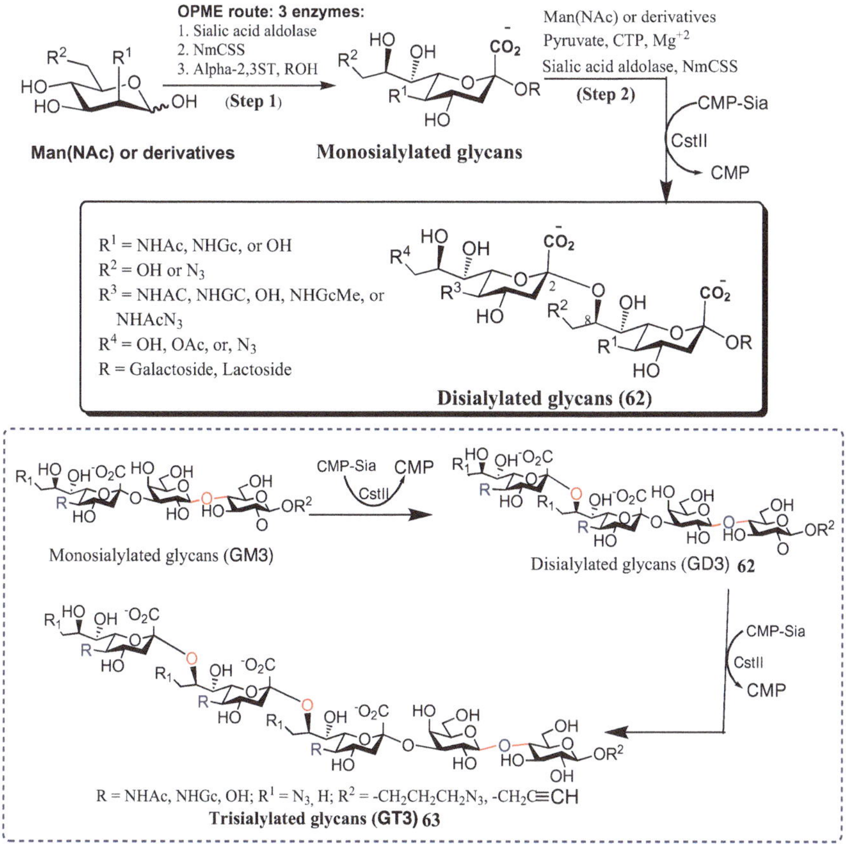

Scheme 5.14 ***Chen's chemoenzymatic protocol for oligosaccharides (both GD3 and GT3) of chemotherapeutic values, their diverse applications in chemical biology and study in carbohydrate-protein interaction.***[98]

of oligosaccharides having two sialyl units (**62**). In short, a brilliant and an efficient two-step multi-enzyme protocol has been reported by Chen Group for an easy access of a series of GD3 ganglioside oligosaccharides and other disialyl glycans **62** with an great ease (Scheme 5.14). The disialyl oligosaccharides (**62**) and trisialyl oligosaccharides (**63**) may be considered as a valuable probe to the evaluation of the biological importance of naturally occurring sialic acid modifications in disialyl structures and thus useful to help to the complete understanding the biological significance of variable sialic acid residues on disialyl structures present in nature.

Another efficient chemo-enzymatic approach has been discovered by same group to synthesize a series of biologically relevant 3-fluoro-sialosides.[99] The enzymatic synthesis of sialosides having fluorine atom

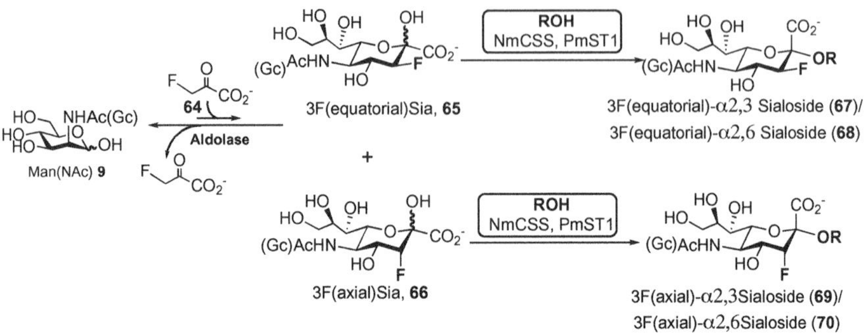

Scheme 5.15 ***Chemoenzymatic synthesis of fluorinated α-2,3- and 2,6-linked sialosides.***[99]

at the C-3 of sialic acid has been achieved using CMP-3FNeu5Ac and CMP-3FNeu5Gc as donor in one-pot two-enzyme system containing NmCSS and a multifunctional sialyltransferase from *Pasteurella multocida* (PmST1) (Scheme 5.15).[99] Fluorinated R-2,3-linked sialosides containing 3F(*equatorial*)Neu5Ac **(67)** and 3F(*equatorial*)Neu5Gc **(68)**, has been reported in good yields. Besides, fluorinated R-2,6-linked sialosides using NmCSS and an R-2,6-sialyltransferase from *Photobacterium damsela* (Pd2,6ST), were synthesized in low yield. Further, CMP-3F(*axial*)Neu5Ac and CMP-3F(*axial*)- Neu5Gc reported to be used as donors from PmST1 to produce fluorinated R-2,3-linked sialosides containing 3F(*axial*)Neu5Ac **(69)** and 3F(*axial*)Neu5Gc **(70)**, but in low yield (Scheme 5.15).[99] This presumably be the very first example of sialyltransferase-catalyzed synthesis of C3-fluorinated sialosides. The series is reported to be a potent inhibitors and mechanistic probes for the sialidases.

This is interesting to mention that in general the enzymatic synthesis is although highly efficient; stereo- and regio- specific, however finding of the right enzymes is the key point for the successful OPME synthetic approach. Thus, for an example, in order to get the promising enzyme, the key points may includes, (1) an easy expression of active and soluble enzyme in high yield (Obtainable), and also, (2) this can tolerate substrate modification (Flexible).

The gap between the synthesis of complex sialoconjugates and their application in chemical biology is narrowing through the emergence of synthetic methods. Furthermore, the substrate specificity of sialidase is considered to be a subject of considerable interest in terms of conjugation to Neu5Ac. Further utilization of the sialidase activity with varying sialic acid structure is under investigation. Besides, the biological importance of these synthesized analogues also should be kept in mind. Perhaps, 'sia' molecule is one of the major bio-molecules which are being investigated for more than three decades in search of novel moieties as it has a broad pharmacological significance in human physiology.

However, many glycan-binding proteins recognize sialic acid or its analogues, but their molecular recognition is not completely understood. A detailed report has investigated on the sialic acid recognition procedure using a novel sialylated glycan microarray containing modified sialic acids on different glycan backbones. Glycans contaning β-linked galactose terminal at the non-reducing end and alkylamine having fluorophore at the reducing end were sialylated to generate α-2,3- and α-2,6-linked sialyl glycans (**72**) resulted in the development of sialyl glycans (Fig. 5.5). The glycans terminating with Neu5Ac or Neu5Gc are found to interact with Influenza A virus and three human para-influenza viruses. The report concluded the

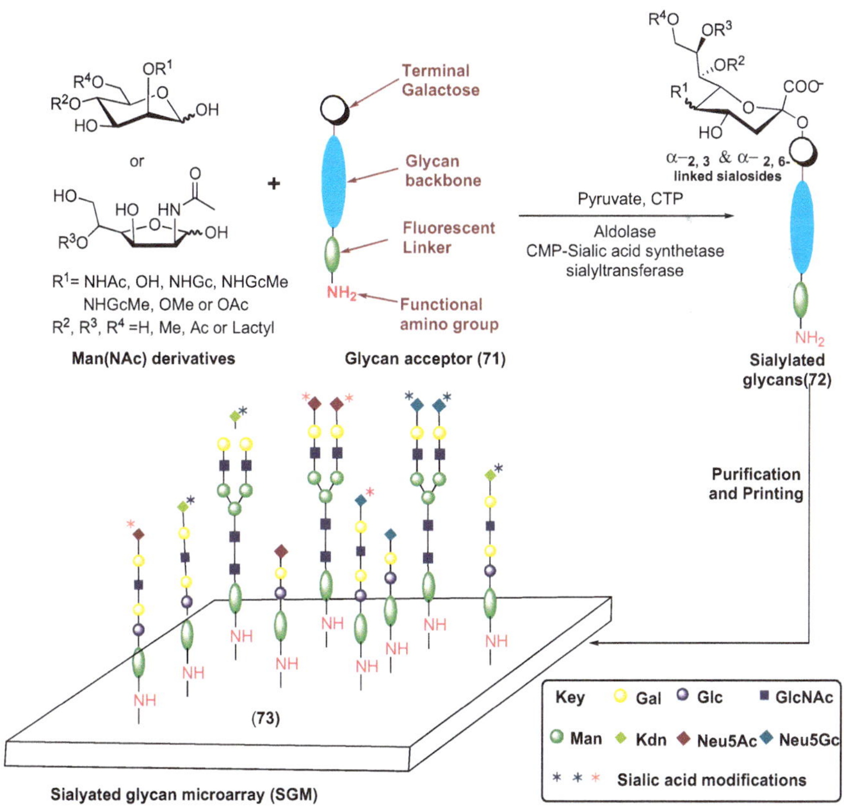

Figure 5.5 *The schematic design and preparation of a sialylated glycan microarray.*[100]

utility of this sialylated glycan microarray which are important to examine biological importance of the modified sialic acids in protein-glycan interactions.[100] The glycan microarray and their significance have been broadly discussed in the next chapter.

Besides chemical glycosylation and chemo-enzymatic tool approach in modern research other well known and easy methods are also being now employed for the target achievement, i.e., development of sialic acid analogues for the molecular library generation and finally drug like molecule and drug development.[101] This part dealing with Carbo-Click has been presented in **Kushwaha and Tiwari Chapter** of this book, thus omitted from this section.

4 Sialic acid-containing molecules in drug development

Investigators across the globe are focused on the role of sialic acid in biological system. The gap between the synthesis of complex sialoconjugates and their application in chemical biology is narrowing through the emergence of synthetic methods.

The influenza virus hemagglutinin recognizes and interacts with the Neu5Ac of the host cell and binding affinity is quite high. Considering the fact with the aim of drug development to cure influenza infection Neu5Ac scaffold based drug designing has been carried out. In the initiation period of this approach pendant α-sialosides were designed and evaluation was carried out against the viable virus which showed good binding interaction with the active site (Fig. 5.6).

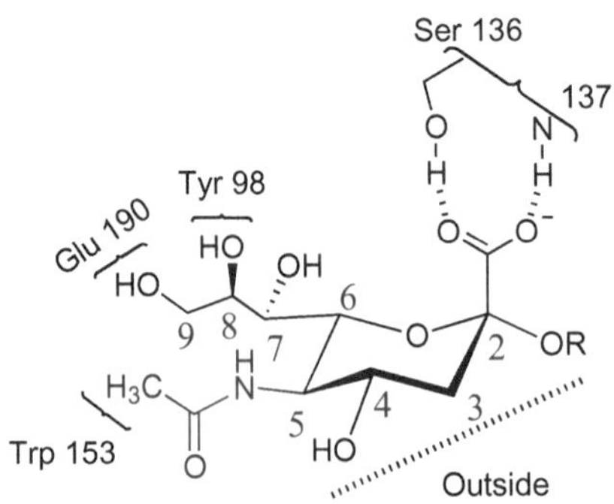

74

Figure 5.6 ***Schematic diagram showing the binding interaction of a α-sialoside in the binding site of influenza A virus hemagglutinin.***[102–104]

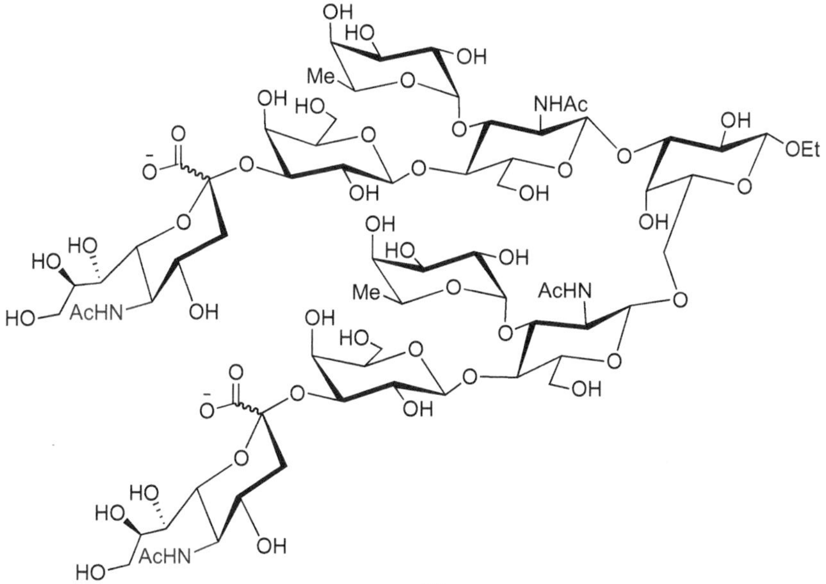

75

Figure 5.7 ***Schematic diagram of a nonasachharide scaffold derivative act as divalent influenza virus hemagglutinin inhibitor.***

It is a nonasachharide scaffold derivative **75** developed for the preparation of divalent influenza virus hemagglutinin inhibitors by the chemoenzymatic approach. The ligand contains ethyl β-D-galactoside core in its structure (Fig. 5.7). This inhibitors were developed on the basis of weak carbohydrate dependent rolling phenomena occur during the adhesion of leukocyte on activated endothelium in course of virus cell recognition and viral cell adhesion. [105,106]

In the development of divalent conjugated structures, slight procedure modification of the synthesis by using 1,4-butanediol or 1,5-pentanediol as tether leads to the formation of derivatives **76** and **77** from the gycosidal core (Fig. 5.8). However, the differences of orientation and distance between the monomeric domains directly influence their influenza virus hemagglutinin inhibition action. The pharmacological potential of these compounds are comparable with the compound **78** and the mode of action is also similar (Fig. 5.9). The chimeric analogue has been achieved through the amide bond coupling of the protected peptide with glycosylamine hemagglutinin inhibitor pharmacophore.

Another interesting approach for the drug design and development as anti-influenza was targeting the recognition site amino acid chain,

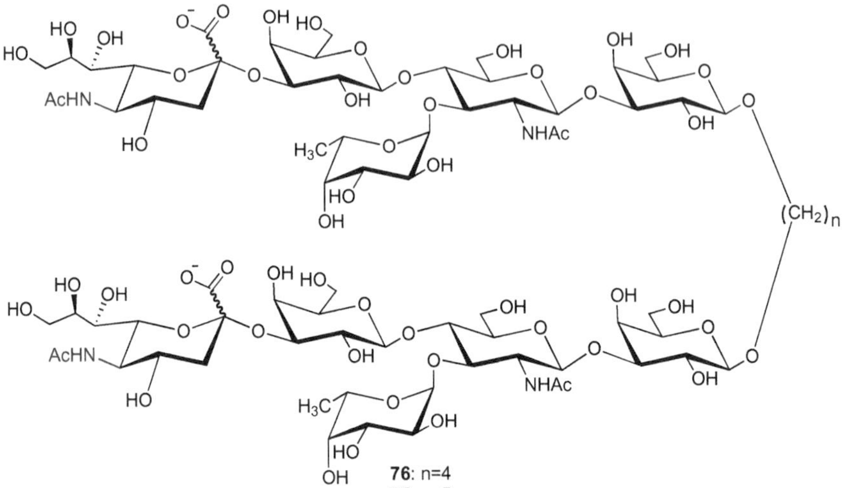

Figure 5.8 ***Schematic diagram of divalent conjugated structures act as influenza virus hemagglutinin inhibitor.***

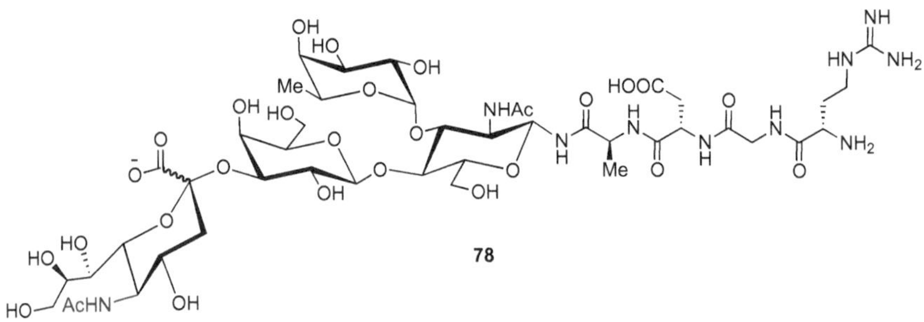

Figure 5.9 ***Schematic diagram for the chimeric analog of viral hemagglutinin inhibitor.***

commonly known as RGD. To achieve the better bioactivity, a of glycopeptides coupled with glycosylamine derivative was developed, for example compound **80** displayed an appreciable IC_{50} values (Scheme 5.16).[107]

For the target based drug designing study influenza virus sialadase enzyme has been under focus and many derivatives are developed for that purpose.

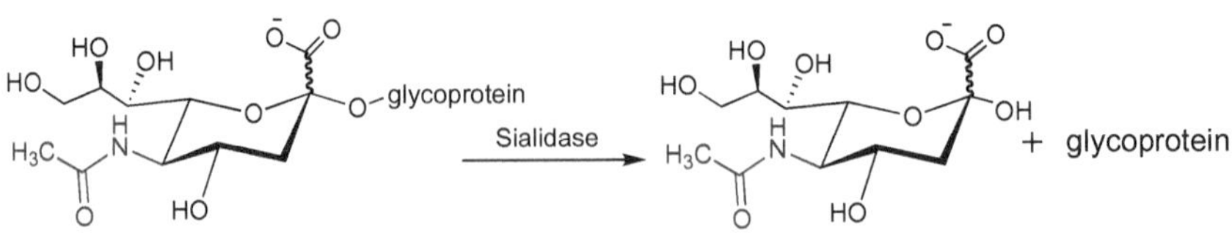

Scheme 5.16 ***Cleavage of terminal sialic acid residue via sialidase.***[107]

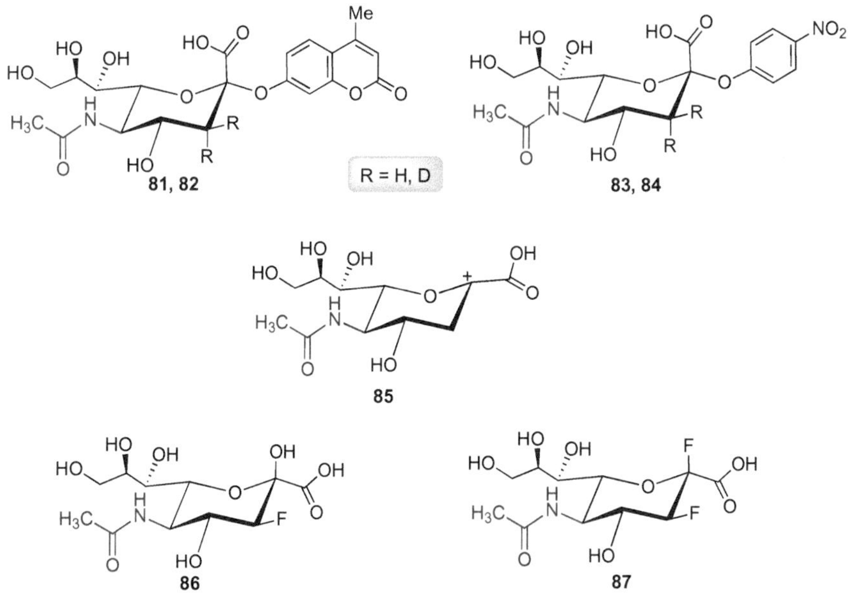

Figure 5.10 ***Schematic diagram of different synthetic sialosides.***

Among them 2,3-unsaturated derivatives of NeuAc5 inhibitors are mostly studied for their potency. In a kinetic isotope study of 4-methylumbelliferyl-*N*-actyl-α-D-neuraminic acid **81** and its C-3 perdeuterated derivative **82** was found to be potent motifs for the target (Fig. 5.10). At the site a cation intermediate **85** arises which is being stabilized by the negative charge at the environment of enzyme. But this proposed theory fails with the repetition of the approach with *p*-nitorphenyl-*N*-acetyl-α-D-neuraminic acid **83** and its perdeuterated derivative **84**. Further studies revealed the properties of the leaving group attached with the aglycon part of sialoside affects the degree of cation stabilization which varies with isotopes variants.

Among the derivatives two fluorinated derivatives of *N*-acetylneuraminic acid (**86** and **87**) are being reported and showed promising potency against the pre-discussed target.[108,109] So, a huge effort has been manifested to acquire drug like molecules among the different library genesis of 'sia' analogues. Versatile substituent groups have been tried out for achieving molecular set for drug design and development. In that procedure some anti-viral drugs have been developed and marketed for specific treatment and have acquired commercial success in treatment.

Besides, the medicinal application also has been explored in neurobiological disease and disorders which has been discussed in the following next section.

5 Synthesis of some commercially available drugs

To develop commercially available drug to treat disorders and diseases huge set of sialic acid derivatives have been screened and among them Zanamavir and Oseltamivir are the most successful drug indicated for anti-viral activity. Their commercially fruitful synthesis route has been reported by well-known pharma companies like Merck, Glaxo, etc.[110] In the following section the nature of drugs and their commercial synthesis have been discussed.

5.1 Zanamivir and oseltamivir as neuraminidase inhibitors

About fifty-five year back, 2,3-dehydroneuraminic acid (commonly known as DANA) **92** which is a well-known transition state analogue of 5-acetyl neuraminic acid was accepted as the successful inhibitor of neuraminidases.[111a] Latter on during 1990s, zanamivir [Relenza][111b] and oseltamivir [Tamiflu][111c] were although two more similar structure, but potency-wise even more potent and selective neuraminidase inhibitors (Fig. 5.11) were approved as anti-viral drug candidate.

Based on rational drug design, **Zanamivir 93** (commonly known as Relenza) was the first successful synthetic neuraminidase inhibitor. This drug was discovered in 1989 by Biota scientists in collaboration with CSIRO and Monash University, Australia.[115] Just next year in 1990, this drug was licensed to GlaxoSmithKline for the clinical development and after nine more years in 1999 this was finally approved by the FDA for commercialization in USA.[116]

The very first synthesis of Zanamivir drug was reported by the researchers of Monash University in Australia. The synthesis begins from Neu-4,5,7,8,9Ac52en1Me which chemically is methyl 5-acetamido-4,7,8,9-tetra-*O*-acetyl-2,6-anhydro-3,5-dideoxy-D-*glycero*-D*galacto*-non-2-enonate.

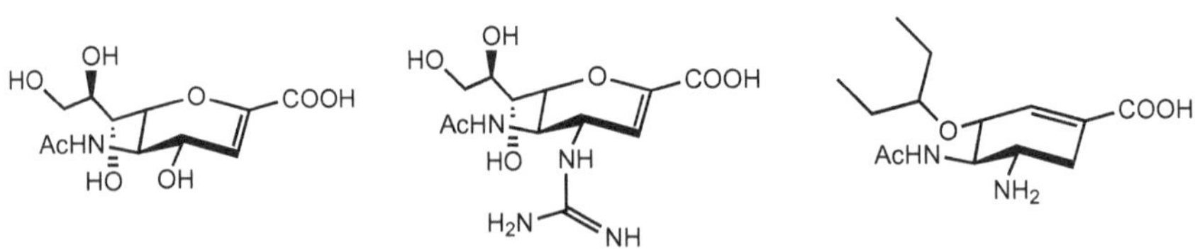

88
DANA

89
Zanamivir (Relenza)

90
Osteltamivir (Tamiflu)

Figure 5.11 ***Structure of Zanamivir and Oseltamivir, well known as neuraminidase inhibitors.***

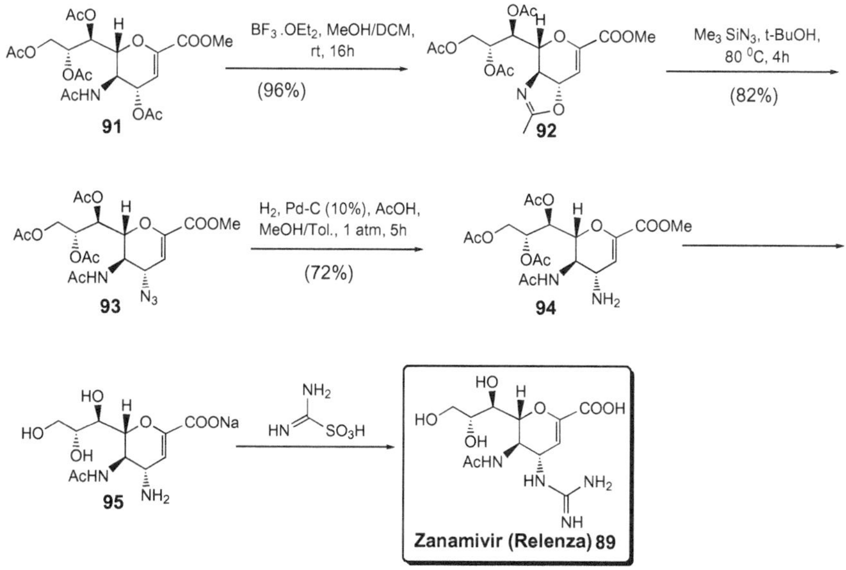

Scheme 5.17 ***Schematic representation of first synthesis path of Zanamivir (89) at Monash University, Australia.***[110,112]

The complete synthetic route is depicted in Scheme 5.17.[112] Neu-4,5,7,8,9Ac52en-1-Me **91**, was developed from NANA *via* several high-yielding steps[113] was first reacted with $BF_3 \cdot OEt_2$ to afford allylic oxazoline **92** to 96% yield. This oxazoline is prone to nucleophilic attack by azide group either as lithium azide or azidotrimethylsilane resulting in intermediate **93** of 82% yield with a very good stereoselective control. Hydrogenation of compound **93** under standard condition *i.e.* 10% Pd/C at atmospheric pressure resulted in high yield of corresponding amine **94** without any undesired side reactions such as double bond reduction or acetate migration. To obtain sodium salt of the compound **95**, ester hydrolysis of compound **94** was subjected with Amberlite-IRA 400 (OH^-) resin/aq.NaOH, and neutralized with Dowex-50W × 8 (H^+). Finally on coupling reaction of **95** with aminoiminomethanesulfonic acid afforded the target drug zanamivir **88**.

5.2 Synthesis of zanamivir from *N*-acetylneuraminic acid (NANA) by the Merck Frosst Centre, Canada

Research group of Monash University utilized Azidotrimethylsilane to introduce the azido group in the Zanamivir synthesis which has a limitation to generate explosive hydrazoic acid. Thus a group of researchers at the Merck Frosst Centre made some modification in the synthesis procedure

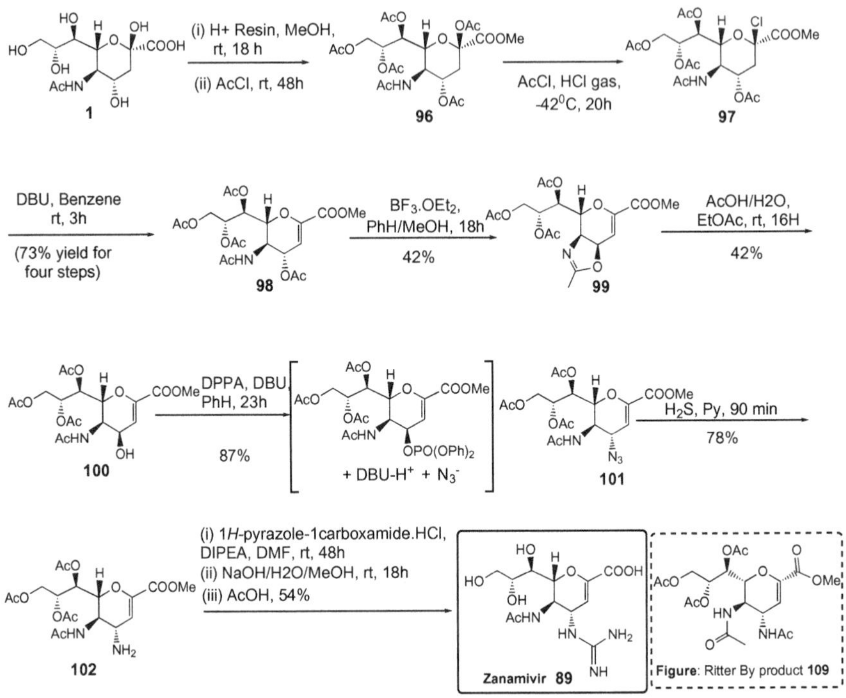

Scheme 5.18 ***Schematic synthesis pathway of Zanamivir at the Merck Frosst Centre, Canada.***[110,114]

like, they explore *N*-acetylneuraminic acid **1** (NANA) as starting material to establish a convenient synthetic route to zanamivir (Scheme 5.18).[114] *N*-acetylneuraminic acid **1** was first esterified using a cation exchange resin in dry methanol, further, subjected to acetylation of all the free hydroxy groups to produce compound **96**. The displacement of -OAc group at the anomeric position (2-acetoxy group) by chloride under pressurized condition resulted in respective pentaacetylated derivative **97**. Amidine base DBU was then successfully used for the elimination step to afford unsaturated ester **98**. It is important to mention that the acetic acid generated in the chlorination step must be removed completely in order to get the good yield of the elimination product **98**, otherwise compound **96** was also resulted in addition the desired one. The required epimerization of the 4*R*-center of the compound was achieved through oxazoline **99**. In the subsequent step, an alternative to the *Mitsunobu* condition, diphenylphosphoryl azide (DPPA) was successfully and presumably the first time applied in carbohydrate chemistry in order to get the desired compound **101**. Interestingly, the reaction is interestingly stereoselective. H_2S in pyridine was then used for the reduction of the azido group to afford respective amine **102**. At the final step of synthesis a guanidine group was introduced by reacting it with 1*H*-pyrazole-1-carboximidine hydrchloride with subsequent two more steps to give compound **89**.

Several intermediates involved in the synthesis of Zanamivir carried out at the Merck Frosst Centre are solids and their purification *via* recrystallization to rationalize the procedures provides its ease agreeable scale up.

5.3 Scalable route to zanamivir by Glaxo

To date the scalable route of zanamivir was commenced by Glaxo Group in UK (Scheme 5.19).[117] Thus, *N*-acetylneuraminic acid (NANA, **1**) was first

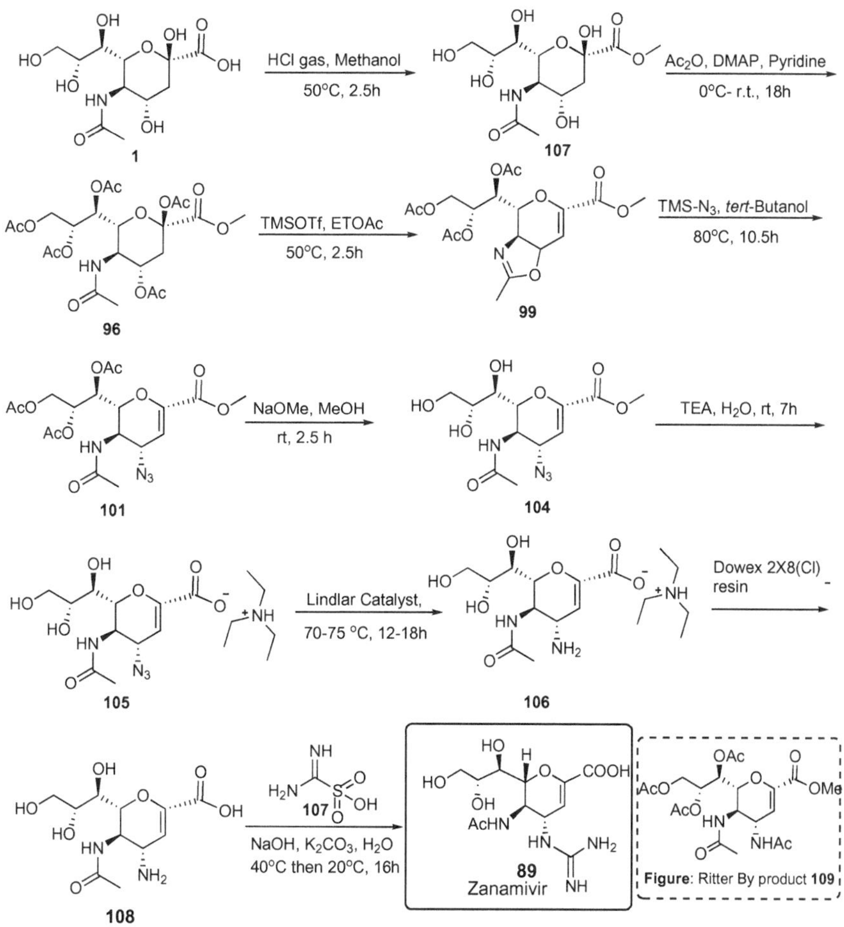

Scheme 5.19 ***Scalable route of Zanamivir 89 by Glaxo.***[110]

converted to respective methyl ester **107** by reacting it with methanolic HCl, which further on acylation under standard reaction condition (Ac_2O/DMAP/Pyridine) afforded penta-acetate derivative **96.** The compound **96** was reacted with TMSOTf in warm ethyl acetate resulted in cyclic oxazoline **99** as the sole product.[112] Notably, similar reaction in MeCN led to the Ritter product **109** of 10% yield only. Compound **99** on reaction with $TMSN_3$ in *t*-BuOH at 80°C afforded respective azido derivative **101** with the desired stereoselectivity. To overcome the problem of achieving the desired amine functionality from azido group without affecting the C—C double bond (as Pd/C condition reduced the double bond too), compound **101** was subjected to acetate removal using NaOMe in methanol (simply to improve the water solubility) to afford **104,** followed by TEA-mediated hydrolysis of ester to produce triethylammonium salt **105.** The resulted salt **105** on hydrogenation using Lindlar catalyst followed by treatment with Dowex 2 × 8 ion-exchange resin afforded desired free amine **108.**[118] At the end, free amine **108** was reacted with aminoiminomethanesulfonic acid **111** to introduce the guanidine functionality and gave zanamivir **109** as final product, which was successfully isolated by ion-exchange chromatography. The overall yield for this 9-step synthesis was although 8.3%, thus further optimized condition to improve the yield is still the prime objective.

5.4 First scalable synthesis of oseltamivir phosphate from (-)-quinic acid by Gilead Sciences Inc

After identification of the best drug candidate, next goal is to find out the scalable route to synthesize it in multikilogram quantities. Thus, a practical, new and scale up synthesis procedure of oseltamivir was developed at Gilead Sciences Inc starting from (-)-quinic acid **110** (Scheme 5.20).[110,111b]

Acetonide **111,** obtained from compound **110,** was first treated with sodium ethoxide (in catalytic amount) in ethanol and afforded mixture of two compounds **111** and **113** in 1:5 ratio. Separation of the desired compound from this mixture by fractional crystallization was difficult on large scale; therefore, the resulted mixture was subjected to mesylation using $MsCl/NEt_3$ in CH_2Cl_2 as solvent and gave respective mesylates **112** and **114** (in same 1:5 ratio). As the undesired mesylate **112** is crystalline in nature thus filtered off and the desired mesylate **114** was isolated from the filtrates. The compound **114** was reacted with sulfuryl chloride in CH_2Cl_2 at low temperature could not afforded the desired olefin, yet a mixture of alkenes (**115** and **116** in 4:1 ratio) along with respective chloride **117** (in 10%–15%) was obtained. Thus, the mixture was treated with pyrrolidine

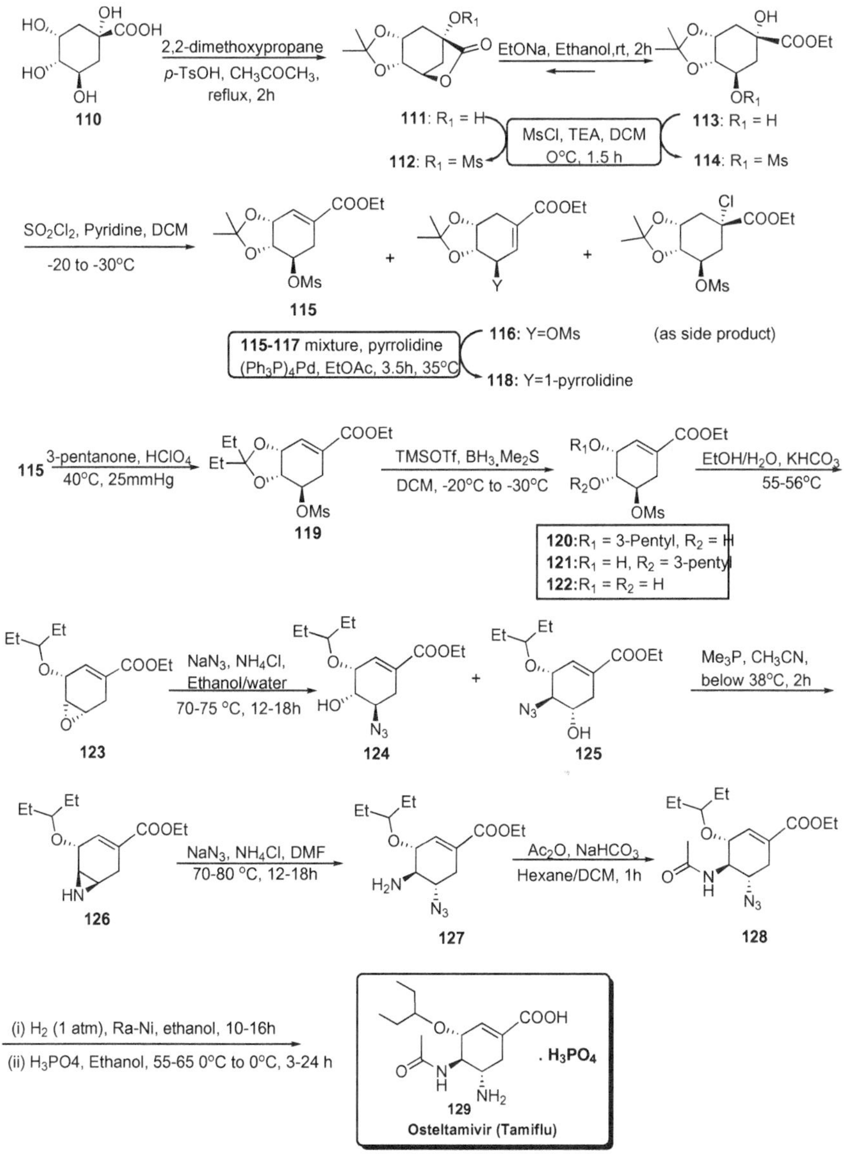

Scheme 5.20 ***First scalable synthesis of oseltamivir phosphate from (-)-quinic acid by Gilead Sciences Inc.***[110,112]

in presence of $(Ph_3P)_4Pd$, where mesylate **116** was converted to pyrrolidino anolog **118** which was easily free from the reaction mixture via extraction with aqueous H_2SO_4, and the required mesylate **115** was isolated simply by recrystallization in ethyl acetate/hexane. Perchloric acid-mediated trans-ketalization of acetonide **115** with 3-pentanone gave ketal **119**, which on ring opening while reacting with trimethylsilyl trifluoro- methanesulfonate and $BH_3{\cdot}Me_2S$ complex gave ethers **120** and **121** and diol **122** in a 10:1:1 ratio.

To achieve the desired regioselectivity, aqueous $NaHCO_3$ was added to the freshly prepared reaction mixture at -20°C. This was very difficult to separate **120–122** through distillation or fractional crystallization, thus the crude mixture was reacted with $KHCO_3$ in aqueous ethanol to give epoxide **123** that can be selectively extracted into hexanes without any problem. This epoxide **123** was reacted with NaN_3 in aqueous ethanol and afforded azido alcohols **124** and **125** (in 10:1 ratio). The crude azido alcohols was subjected to reduction followed by cyclization using Me_3P to produce aziridine **126** which on ring opening with sodium azide afforded azido amine **127.** At the end compound **127** was acylated using acetic anhydride (Schotten-Baumann) to give **128**, a highly crystalline intermediate. Thus, the oseltamivir phosphate salt **129** was obtained by reducing the azide functionality with H_2/Ni followed by the addition of phosphoric acid which precipitated the phosphate salt as feathery needles. This synthetic approach described in Scheme 5.20 consists of twelve steps successfully implemented in standard pilot plant utilized to develop kilogram quantities of oseltamivir phosphate with 4.4% overall yield and thus can be considered to be a truly scalable route.

Therefore, the well marketed anti viral drugs which are sia derivatives, are important molecules to treat viral flu diseases like Influenza A, Influenza B, bird flu etc. The library genesis of sia derivatives is still under progress to find more novel and potent molecules to treat neurological diseases as well because, sia molecules play different crucial roles in neuro system of human beings. Hence, it has great potentials to develop drug like molecules and drug compounds in future to treat different neurological disorders.

6 Sialic acid in neurobiology: opportunity and challenges

Ages of cell evolution served eukaryotic cell lines with much advance programming than the rest. For protection, functionality, diverse approach and extended modification have lead sugar molecules as the coating on the surface of cells. Depending on their structural motif they have been classified as glycans, glycoprotein, glycolipids.[119] Specifically, in case of mammalian cell surfaced glycoproteins and glycolipids are terminated with sia which is a nine-carbon backbone sugar with distinctive chemical properties that determine their functional activity. Besides, it plays a selective role in cell and molecular interactions of the glycans within the physiological system.[120] Natural origin of sia are quite interesting. It was isolated from submaxillary mucin as sialic acid and as neuramic acid from brain glycolipids during Second World War.[121] It is predominantly found in deuterostome lineage of animal kingdom according to Angata et al.,[8] which has further been mentioned in various reports as well, and in certain types of bacteria, including neuro-invasive bacteria contains homopolymers of α-2,8- and/or α-2,9-linked NeuAc known as polysialic acids,[122] pathogenic bacteria associated with the human brain diseases are able to express sialoglycans on lipooligosaccharides of their outer membranes,[123] *etc.* Though other resembled genera of Apes able to express significant amounts of *N*-Glycolylneuraminic acid (**2**) as well as *N*-Acetylneuramic acid (**1**) modern human only able to synthesize NeuAc physiologically[124] and it is because of exon deletion in the gene responsible for the conversion of *N*-acetyl to *N*-glycolyl form of sialic acid, i.e., CMP-*N*-acetylneuraminic acid hydroxylase (CMAH)[125] moreover, an anti-non-Gal antibodies of specific carbohydrate structures carrying terminally linked *N*-glycolylneuraminic acid (NeuGc, **2**), exists in human serum of more than 85% population. Further recent report claims the involvement of NeuGc in hyper acute rejection in case of xenotransplantation.[126] Besides, ProfVerki termed the evolutionary phenomena of NeuAc predominance and genetic selectivity as "sialoquake" which described the ultimatum effect of one gene mutation correspondence of sia biosynthesis in human physiology.[127] With timely investigation around 65 naturally occurring sias have been reported till date.[128] The structural unit or derivatives of sia present at the cell surface governs the functioning and serves the purpose of sia presence. The structural modification includes oligosaccharides, larger glycans, and multiglycan complexes. In 2010, a comprehensive report regarding a conceptual approach to discuss the cell surface "sialome", which is defined as complement of sialic acid types along with the linkages and the mode of presentation on a particular organelle, cell, tissue, organ or organism[129] on five hierarchical levels taking analogy of the forest canopy has been published by Cohen and Varki.[120] According to the phenomena, at the cell surface glycans are present as spatial organized forest canopy with dynamic micro domains responsible for mediating specific interaction and highly selective function. Next part is the trees of the forest i.e., the glycan class which get differentiated by terminating units attached to the chain, like, *N*-glycans, *O*-glycans, glycosphingolipids (gangliosides), or the side chain of glycosylphosphatidylinositol (GPI) anchors in turn affect sia presentation, localization and density. Trees generally get constituted of several branches, like that; the underlying sugar chain of the sialic acid structure which is constituted of multiple saccharides contributes to the binding specificity of the system. As it is a poly-saccharide chain with the specific linkage orientation, it is an essential part obviously serving major role in controlling the binding specificity. The branches of a tree are made of stems so here the α-linkage between the C-2 of the poly-saccharide system has been compared with and reported that α-2,3 or α-2,6 to various sugars, or α-2,8 to another sia are the most common linkages appear as random units. Besides some special cases like, α-2,9 linkage to another sia, and α-2,4 linkages etc can also be present depending on the system. The leaves of this canopy are the sialic acid residues present at the top of the arrangement. This entire structural unit in combination contributes to spatial arrangement and selective functionality of sialic acid attached to cell surface (Fig. 5.12).

The sialic acid present at different organ has different functionality and their molecular motif contributes to it, for example, in the blood circulation to prevent unwanted interactions of cells a negative charge is present on human erythrocytes and other cell types to provide charge repulsion. Next, podocalyxin, a sialomuscin proved to be correlated with the anti-adhesion

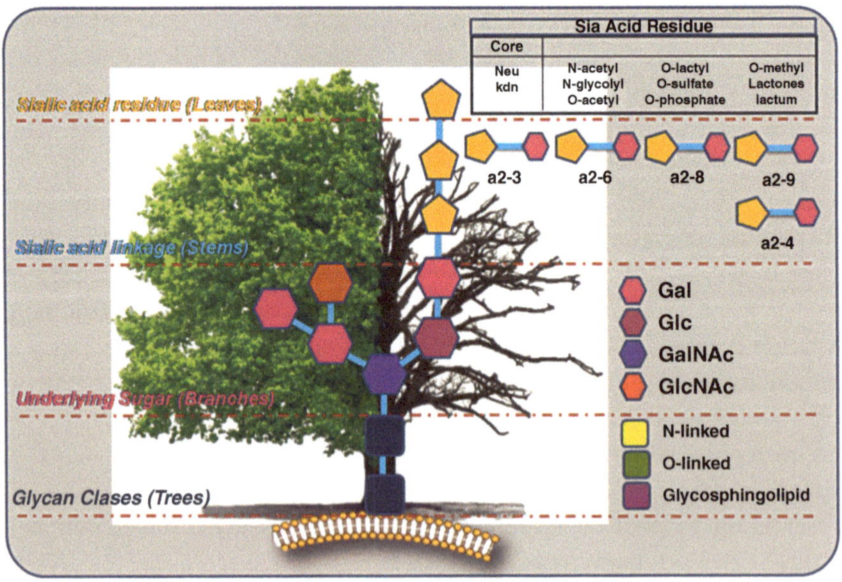

Figure 5.12 ***The compared structural arrangement of sia on the cell surface with the forest canopy concept.***

effect which in turn helpful in foot processes of glomerular system[130,131] and now being proved to be attached with Na^+/H^+-exchanger regulatory factor 2 (NHERF2) and ezrin, co-localize along with the apical plasma membrane of podocytes to form a co-immuno precipitable complex important in maintaining the unique organization of the kidney epithelium.[132] Besides the extended polysialic acid chain associated with the podocalyxin able to affect neuronal plasticity.[133–136] Furthermore, sia is able to regulate the half life of many circulatory protein molecules and effect over the circulatory phenomena directly.[137,138] However, NeuAc as 2-keto-3-deoxy-nononic acid presented on the surface of the cells and mucosa functions as a defense against oxidative damage additionally the mechanism controls the concentration of H_2O_2 not only for detoxification but also in the signal transduction which is found in recent literature.[139] Siglec (sialic acid-binding immunoglobulin-type lectins) are a type of immune regulatory receptors majorly found on the cells of the hematopoietic system and a V-set Ig-like domain mediates the recognition of different sialylated glycoconjugate which can lead to the activation or inhibition of the immune response, depending on the involved Siglecs.[140]

Among all the organ and physiological systems brain is the organ contains the highest level of sia, much of it in the form of sialylated glycolipids (gangliosides).[141] Simply, gangliosides are complex glycosphingolipids contain one or more Sia residues at terminal position. According to reports, it constitutes around 6%–10% of the total lipid mass of human brain, which in turn content 1/4th of the total conjugated saccharides with 70%–80% of the conjugated Sia residue.[142] Ganglioside chemically is amphipathic molecule that constituted of hydrophobic ceramide moiety, derivative of amino alcohol, sphingosine, and a single fatty acid chain and the fatty acyl chain remains attached to ceramide moiety via an amide bond.[143–145] Svennerholm et al., has discussed in a report about the ganglio series and their recognition, categorization depending on the attached group or lacking terminal galactose. Majorly four categories of gangliosides have been reported.[145] However, the sub classes of glycosphingolipids usually bear one to five sialic acid residues depending on the ganglio series tetrasaccharide core Galpb1–3GalpNAcb1–4Galpb1–4Glcpb which is reported to be highly conserved in vertebrates. Besides, the core also reported to be found in the other cell types as the mixtures of components having di-, tri-, and tetrasaccharide cores (Fig. 5.13).[146–148]

In cell sia is synthesized in cytoplasm and transported to the nucleus where it is activated by attaching CTP to form the activated nucleotide

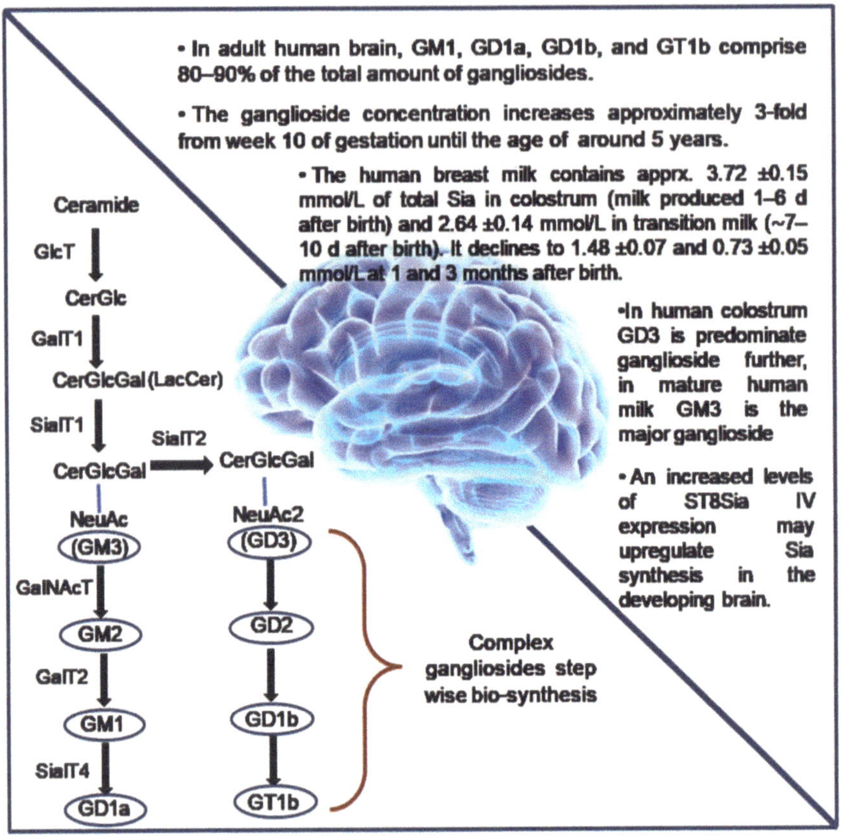

Figure 5.13 *Facts and synthesis of sialic acid in the physiological system.*[149]

donor, CMP-Sia which is then transported into the Golgi compartment. There it serves as the donor substrate for the sialyltransferases, including ST8Sia IV and ST8Sia II. CMP-Sia exerts negative feedback inhibition on GNE, the bifunctional cytoplasmic enzyme and helps in regulation of Sia biosynthesis.[150] Golgi body is the cellular apparatus where the biosynthesis of gangliosides[151] takes place which starts with the transfer of additional carbohydrate molecules to pre-existing lipid acceptor molecule i.e., sialic acid residues are being attached during the procedure to lactosylceramide in short, LacCer (Fig. 5.14) by specific sialyltransferases. For example, a simple ganglioside, GM3, is synthesized by addition of a sialic acid to LacCer by CMP-sialic acid: LacCer α-2,3 sialyltransferase (Silyl Transferase-I; ST-I or GM3 synthase), or, GD3 and GT3 are synthesized by sequential addition of sialic acids to GM3 and GD3 by CMP-sialic acid:

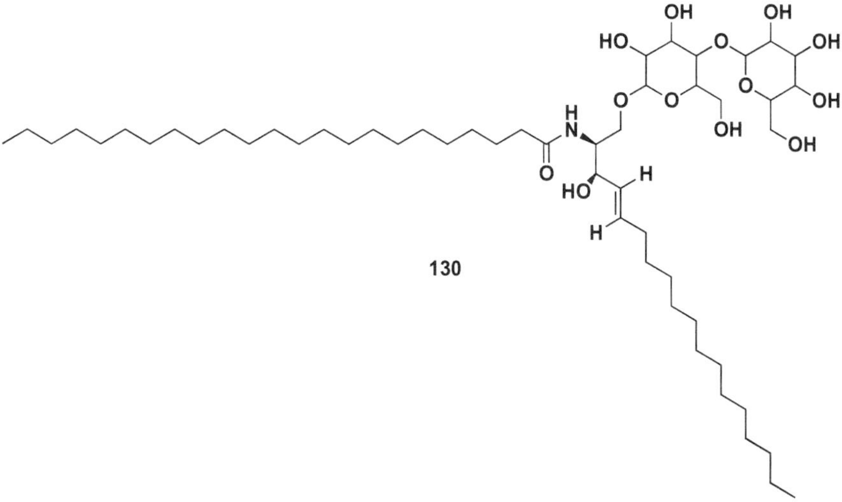

130

Figure 5.14 ***Molecular structure of lactosylceramide in short, LacCer.***

GM3 α-2,8 sialyltransferase (ST-II or GD3 synthase) and CMPsialic acid: GD3 α-2,8 sialyltransferase (ST-III or GT3 synthase), respectively. GM3 synthase, GD3 synthase, and GT3 synthase show high specificity towards the glycolipid substrates.[152]

The mentioned precursors, i.e., LacCer, GM3, GD3, and GT3 are specific for the 0-, a-, b-, and c-series of gangliosides respectively. In the biosynthesis pathway next step is the elongation of the simple gangliosides formed to the complex ganglioside via glycosyltransferases (GTs) in stepwise method. The type II membrane-anchored GTs takes part in biosynthesis of gangliosides which appear as gradient distribution in the Golgi apparatus. Furthermore, the expression of GT genes is tissue-specific and dependent on alternative promoters. The promoters control tissue-specific transcripts, differing in their 5'-untranslated region (5'UTR) but encode the same polypeptide. Their function is to regulate glycosylation and phosphorylation in the pathway.[153–155]

In a report further application has been described which states that the epigenetic up-regulation of ganglioside synthase genes during neural development is also a part GTs function.[156] Mainly, three galactosyltransferses, i.e., Gal T-I (UDP-galactose: glucosylceramide β-1,4-galactosyltransferase), Gal T-II (UDP-galactose:GD2 β-1,3-galactosyltransferase) and Gal T-III (ceramide UDP-galactosyltransferase, EC.2.4.1.45) have been mentioned in major reported till now involved in the biosynthetic pathway and conserved putative *N*-linked glycosylation site is found in all galactosyltransferase genes which predominates.[157] Finally the produced gangliosides after post transcriptional modification reaches its place of enrichment over the cell surface and serves the purpose of presence which is determined by its structural configuration.

In a well documented study of the relationship between the ganglioside synthesis and the development of axons and dendrites in polarized neurons report has been presented which states that the prominent modification arises in the levels and types of gangliosides during neuronal differentiation and development. In the study, neurons were labeled with [4,5-3H]dihydrosphingosine, which was rapidly incorporated into cells and metabolized to 3H labeled glycosphingolipids. During neuronal development of axonogenesis, a significant increase in the synthesis of complex gangliosides (i.e., GM1, GD1a, GD1b, and GT1b) with a corresponding reduction in the synthesis of glucosylceramide and ganglioside GD3 monitored by ^{3}H-labeled glycosphingolipid synthesis Further it was seen that rapid axon growth escalated with the ratio of a- to b-series gangliosides increment. However, increase in the ratio of a- to b-series gangliosides as well as increase in the synthesis of gangliosides GD1a and GT1b has been found while the process of dendritogenesis, dendrite growth, and synaptogenesis. So the development of the brain with time changes the preferable bio-synthetic predominance of the ganglioside variance in the body.[158]

GD3S transcripts have been found to be essential in the adult human brain tissues and the cDNA of it has been isolated by the expression cloning techniques.[159–163] The location of the gene is reported to be on chromosome 12, in p12.1–p11.2, along with a set of five coding exons spanning over 135 kbp of genomic DNA, are the major constituent of it. Study over the promoter region controlling the expression of the GD3S gene revealed that it lacks TATA or CCAAT boxes and instead contains several SP1 sites. A sequential repetition of GT/CG between −1200 and −1300 indicated it to be *Z*-type DNA, probably can regulate mRNA synthesis as well.[164] Moreover, further studies suggested that the 5'-untranslated region of the GD3S gene specifically, in the cell line of melanoma, glioblastoma, neuroblastoma, and breast cancer code an unique transcript and the location of the transcription starts with 450 to 690 bp upstream of the initiation codon at the first exon.[165]

In regular cell maintenance cycle, sia molecules are degraded in the cell system like other biological natural components' degradation with the purpose of cyclic synthesis and clearance process of the living cells. Sia-recognized pyruvate lyases break down it into N-acetyl-mannosamine and pyruvate molecules. Report suggests that there are two Sia-9-O-acetyltransferases in mammalian concerning major two cell organelles; one is cytoplasmic another, is lysosome dependent. The exported sia molecules from lysosome to cytoplasm induce the cytoplasmic pathway on the other hand, the sia molecules which are transported from ER-golgibody to lysysome for digestion is the other pathway. In case of other living organisms also this type of procedure has been observed but the whole mechanism is still out of experimental details.[166,167]

6.1 Disorders associated with sia

Sialic acid has a vast important functional approach in the physiological system specifically in neurobiology. So its disruption leads to several diseased conditions within the system depending on the depletion or shortage cell site or the organelle involved. In various possibilities of some are, free sialic acid accumulation or transportation disruption of the sialic acid within the neurological system leads to serious disorders need to treat with proper care. Dysfunctioning in the steps of biological synthesis as a result of genetic mutation also causes serious diseased conditions. In the following section the discussion has been elaborated.

6.1.1 Sialic acid storage disorder

The free sialic acid storage disease is one of the brain disorders occur in the early stage of life and suffer from developmental delay, weak muscle tone also known as, hypotonia and failure to gain weight and grow at the expected rate. Besides may have coarse, seizures, bone malformations, enlarged liver, spleen (hepato-splenomegaly) and enlarged heart in medical term known as cardiomegaly. The abdomen may be swollen due to the enlarged organs and an abnormal buildup of fluid in the abdominal cavity also known as ascites. Affected infants may suffer from excess fluid accumulation known as hydrops before birth. It is caused by the defective transport of free sialic acid outside the lysosome. A study was conducted over 12 French parents and found that fetal autopsy, or clinical examination showed prominent ascites, rarely progressing to complete hydrops, and highlighted the early severity of bone disease. Report concerns about the genetic modification factor as well and concluded seven novel mutations responsible for the syndrome i.e., three deletions as follows, del exon 7, del exons10 + 11 and c.1296delT, one splice site mutation, i.e., c.1350 + 1G→T, one nonsense mutation i.e., p.W339X, and two missense mutations i.e., p.R57C and p.G127E.[168,169]

6.1.2 Sialuria and salla disease

Among the rare genetic diseases, sialuria is one to be well-known, which is an abnormal condition characterized by uncontrollable overproduction of the free sialic acid within the physiological system along with massive excretion of free sia in the urine around 5–7 g/day.[170–172]

On the other side, in Salla disease, free sialic acid accumulation takes place within the lysosome caused by mutation in the chromosome 6 resulting in the cleavage of glycoproteins and glycolipids by lysosomal neuraminidase. Major problem in this condition is the system fails to transport out 'sia' from the lysosome where it accumulates. The physiological condition causes mental retardation which is autosomic recessive disorder become more severe with prolonged time and leads to early death as well. Still no scripted treatment is present, only supportive medication is available.[173–176]

6.1.3 Guillain–barré syndrome

Various researches on impairment of human brain and glycoconjugates in congenital athyroidism suggested that the protein-bound and ganglioside-bound sialic acids exist in the grey matter and a decrease of GD1b, GT1, and GQ1 in the brain ganglioside can hamper intelligence. Besides, decrement in glycoprotein sialic acid causes an impairment of cell maturation.[177]

Nowday's extensive studies revealed that among different auto-immune disorders Guillain–Barré syndrome is one in which antibodies against complexes consisting of two different gangliosides arises, rather than a single ganglioside which potentiate specific recognition of new conformational epitope formed by glycolipids. Moreover, some reports have suggested the presence of molecular mimicry between gangliosides and antecedent infectious agents for the development of the disease. As lipooligosaccharide is the major component of the outer membrane of *C. jejuni* and the bacterial isolation from the patients contains GM1-like or GD1a-like lipooligosaccharide. The pathophysilogy of the disease are muscle fatigue and pain, coordination problem in extrapyramidal parts due to the vesiculation of myelin in nerves further with time the conditions worsens towards the acute motor axonal neuropathy.[177,178] In 2018 report regarding a case study on a subject revealed that anti-ganglioside complex antibodies are present in the system who is suffering from GBS spectrum disorder and concluded that GQ1b-seronegative and GSC-sero positive profiles might be associated with recurrence of the disorder which can be further exploited for the establishment of detection study of the disease.[179]

6.1.4 Miller fisher syndrome

The disorder is a subset variation of Guillain–Barré syndrome first recognized by James Collier in 1932 and described it as trios of symptoms, i.e., ophthalmoplegia (weakness of the muscles within or surrounding the eye), ataxia (presence of abnormal, uncoordinated movements and poor balance with clumsy walking) and areflexia (loss of deep tendon reflexes on physical examination). Later in 1956, Miller Fisher rectified it as entity within the GBS spectrum. In molecular level, IgG anti-GQ1b antibodies cross reacts leads to the development of Miller Fisher syndrome (MFS). Furthermore, the anti-GQ1b antibodies, which act against GQ1b blocks acetylcholine release from the motor nerve terminals directly relates to the disease activity and may be used as a diagnostic marker on future perspective, although the marker is not unique to MFS, can be useful in serological confirmation to allow for more definite diagnostic apporach.[179–186]

6.1.5 Schrizophenia

In early 90s, various research studies on schizophrenia suggested that the lower level of sialic acid content in the glycoproteins of the cerebrospinal fluid and on successful treatment, the sialic acid content increased to the normal values. Same conclusion also drawn by an another report that is on chronic lithium treatment in animal models having schizophrenia which showed sialic acid content change of the rat synaptosomes. In late 90s more vast research reports on the topic at the molecular level suggested that the polySia or ST8Sia2 gene has a direct role in the disorder. In Ncam1- and ST8Sia2/4-KO genetically modified mice an experiment carried out and the result concluded that the number of polySia–NCAM immunostained cells derived from the HIP of the affected brains is less in number as compared with that normal condition. As well as, the lowered polySia–NCAM expression was in layers of IV and V of the dorsolateral PFC but no such difference was observed in AMG. The output clearly confirmed the location specificity of polySia impairment to be main feature for the disorder. Further study on genomic molecular level considering different mass population, explored that chromosome 15q26, the localized gene encoding region of ST8Sia2 is relatable as well and there is influential relationship exists between the single-nucleotide polymorphisms of the promoter region to the conditions arise.[187–194]

More recent studies using single-nucleotide polymorphisms SNP-7 (Glu141Lys) in the coding region of ST8Sia2 concluded that a sharp decrement occur in the enzymatic activity of ST8Sia2 *in vitro* and *in vivo* as well as affected the polySia in terms of quantity and quality. In addition, polySia functions were completely impaired in the patients sample code. Nextly, SNP-9 was focused and found to cause impairment of the quality and quantity of polySia, plausibly due to the change in the ST8Sia2 translation process. All the discussion supports the important role of polySia as a regulator in neural and brain functions which is governed by the help of regulating physiologically active molecules BDNF, FGF2 and dopamine.[195–199]

6.1.6 Autism spectrum disorder (ASD)

Neuronal heterogeneous condition characterized by early onset, like at the age of 2 years. The affected individual finds difficulties in social communication, has little interest in others, receptive movement, and difficulties in speech. Intense studies suggest that the insufficient development of Prefrontal cortex and Purkinje cells in the cerebellum is the reason for the disorder along with cortex, pons, and limbic area impairment. In case of genetic factor some are important like causative factor of ASD, accumulation of single nucleotide polymorphs due to DNA variations, mutations in the causative genes and impairment of epigenetic regulation alter the brain structural features. Besides recent studies says that a large number of *de novo* copy-number variations and single-base-pair mutations has been found in the patients which is increasing the estimated proportion of patients with identifiable genetic mutations of 20%–40%.[200–204]

6.1.7 Alzheimer's disease (AD)

Among the major neurodegenerative diseases, Alzheimer's is a quite common name. The majorly implicated factor in the pathogenesis is amyloid-beta peptide (Aβ) which is reported to be the main constituent of senile plaques. In the brain of suffering individuals polySia percentage has been reported to increase at the sub ventricular zone of the lateral ventricle, sub granular zone predominantly. Furthermore, studies on the rats revealed that the injection of Aβ into the rat's hippocampus leads to the impairment of memory and learning, and increased staining of polySia at CA1 and DG in hippocampus. But the study failed to recognize underlying mechanism of action of polySia towards the neurodegenerative disorder as the staining procedure and increased load caused disorganization of the tissues. Unless its functional role is clarified as a cleaned picture, the detailing is missing so the effort is for the future work dependent only. Moreover, to treat the disordered condition the target molecules which either sequester Aβ or interfere with Aβ interaction/binding to the cells are under investigation as a means to reduce the pathological effects. Some sialic acid derivatives and its analogs are able to show binding interaction with Aβ. It will help for more specific and detailed improvements in the therapeutic polysaccharide structures that can be developed and modified to overcome other shortcomings of AD therapeutic development, particularly of penetrating the blood–brain barrier.[205–207]

6.2 Uses in neurobiology

Sia analogs has a diverse function in physiological system and influences many important pharmacological factors and govern major biological functions. Considering those major points it can be explored as biomarkers and as drug molecule to treat different diseases.

6.2.1 As biomarker

In human serum, total sialic acid means the overall sum of protein-bound sialic acid, lipid-bound sialic acid, and free sialic acid and its normal concentration in serum or plasma is 1.58-2.22 mmol/L (52-73 mg/dL). A well known fact is chronic alcoholism leads to inhibition of glycosylation of many proteins, such as transferrin, fibrinogen, and complement proteins, so directly affects the total sialic acid concentration which recommended to be increased in individuals who are regular alcohol consumers. However, the factor is an indirect alcohol biomarker and trial study report comprehended sensitivity and specificity of sialic acid as an alcohol biomarker to be 95.5% in women and 81.3% in men respectively.

Total sialic acid in plasma may be increased in a variety of diseases, which are like to be noticed incase of renal disease, diabetes, bacterial infection, inflammatory disorder, various types of cancers etc. expected to show higher values but non-biased model study with plasma concentration remark is yet to be discovered.[208,209]

6.2.2 'Sia' as drug for disorders

As the sia variation between the animal kingdom and the human physiology is quite vast. The enzymes concerning the system are potential targets for drug development includes as follow, sialic acid synthases, CMP-sialic acid synthetases, sialyltransferases, sialidases, and sialic acid modification enzymes.

Various human pathogenic bacteria, *e.g. Vibrio cholerae, Anthrobacter ureafaciens, Clostridium perfringens, Salmonella typhimurium, and Streptococcus pneumoniae*, etc., viruses *e.g.* new castle disease virus and influenza virus etc not able to biosynthesize sia or sialylglycoconjugates, instead they produce sialidases or neuraminidases which are known as virulence factors for the human physiology. So this sensitivity can be targeted for the drug development. Taking that concept in account using protein crystal structure of human influenza virus enzymes some drugs have been established like Zanamivir (Relenza) and Oseltamivir (Tamiflu). These drugs have been commercialized and used as effective anti-influenza virus drugs and more studies are ongoing for more potential candiadte.[210,211]

A 2011 report discussed about the sialidase substrate specificity-based inhibitor design. It concluded the approach for the identification of selective inhibitors against certain sialidases. Further in a Caecal Ligation and Puncture (CLP) model using mice a study has been carried to evaluate two certain bacterial sialidase inhibitors efficacy for pathogenic bacterial sialidases which disrupted the repressive immune-regulation of sialic acid-based interaction causing server damage to the host tissues during bacterial sepsis. So the enzyme is a booster target for the development of drugs and drug like molecule to achieve the treatment of the disease condition.[212,213]

In a more recent report regarding influenza virus sialidase, has found the viral enzyme to be essential in the virus' life cycle. The report comprehended about two distinct groups of influenza A virus sialidases which differ in the flexibility of the 150-loop and concern about a more open active site in the *apo* form of the group-1 compared to group-2 enzymes. So the novel sialic acid-based derivatives can be useful to exploit the structural difference and selective inhibition of activity of the group-1 sialidases. Moreover, group-1 sialidases from drug-resistant mutant influenza viruses are sensitive to some designed compounds reported in literature. Besides, the protein *X*-ray crystallography of the enzyme molecule demonstrated that the inhibitors lock opens the group-1 sialidase flexible 150-loop supported by molecular modeling approach. So the approach is useful towards the development of selective target based drug development like, pandemic A/H1N1, avian A/H5N1 and other group-1 sialidase-containing viruses, further based on an open 150-loop conformation of the enzyme.[214]

Drug resistance is a major cause of relapse in acute lymphoblastic leukemia in which has strong induction of 9-*O*-acetylated Neu5Ac including 9-*O*-acetyl GD3 has been reported in recent literature and the phenomena caused resistance against vincristine or nilotinib. Moreover, the authors claimed that the removal of the intracellular and cell surface–resident of 9-*O*-acetyl Neu5Ac by lentiviral transduction of esterase has been lethal to the cells in the presence of stromal protection also, on the other hand for the normal cells no such phenomena occurred. Further, experiments on the transplanted mice were conducted to induce expression of the *O*-acetylesterase in the affected cells and reported a reduction of leukemia to minimal cell numbers along with increased survival. The result clarified that Neu5Ac 9-*O*-acetylation is essential for survival of the cells and can be targeted for the drug designing and development research.[215]

Another report regarding a new sia transporter, i.e., *Sa*SiaT stated the transporter to show higher binding affinity for Neu5Gc than Neu5Ac due to altered substrate specificity which is defined by three residues unique to the *Sa*SiaT substrate-binding site (Tyr79, Asn83, and Asn244) in turn helps to achieve higher affinity. The phenomena have the ability to draw advantage to *S. aureus* in specific niches as functional sia transporter is essential for the uptake and utilization of sialic acid. These types of specific transporters are the new field of focus for drug design when drug resistance is becoming a daily trouble to have a cure.[216,217]

But still a long way to go to have the crystal clear picture and details of all various naturally existing sia oriented targets for the drug development and designing strategy due to the analytical challenges in recognizing and determining the sialic acid-dependent interactions and synthetic difficulties in obtaining homogenous sialic acid-containing oligosaccharides and glycol-conjugates, specifically those natural huge diversity. It is crystal clear in future metabolic engineering approaches are required to recognize the numerous sia targeted structure along with the research effort to establish simple efficient method to synthesize them and crystallize them so the drug design and development can have a more focused and effective approach.

7 Conclusions and future perspective

The biological relevance of sialic acid-containing molecules/glycans to various interactions at the cell surface has lead to their increased demand for the synthesis of structurally defined sialoconjugates. This chapter briefly highlighted the way to design sialic acid containing natural and synthetic compounds that are water-soluble, easy to take as a formulation dose and bio-chemical potent as anti-cancer, anti-viral and other means of medicinal glycoconjugates. This is obvious to conclude that sialic acid-containing molecules are considered to be promising and useful for the study of carbohydrate-protein interaction which in fact is the basis of drug discovery and development. Interest is also directed towards the potential impact of the exclusively human sialic acid profile on the brain and thus a series of sialoglycoconjugates developed has been briefed to achieve some potent ones to be useful in neurobiology. The sialoglycoconjugates particularly 2,8-linked polysialic acids, produced by post translational modification of the neural cell adhesion is important for investigating the role of glycoconjugates in brain (i.e. neuroglycobiochemistry), tumor development as well as general ganglioside functions in the nervous system. These specific research directions have also lead to an extensive attention in explaining the origins and functioning of the human neurological system.

Acknowledgments

VKT gratefully acknowledge Council of Scientific & Industrial Research, New Delhi (Grant No.: P-25/370) for the funding. Authors sincerely thank to Prof. Xi Chen, Professor at UC-Davis and Dr Hai Yu for their useful discussion during preparation of chapter.

References

1. Varki, A. *Glycobiology* **1993**, *3*, 97–130.
2. Bertozzi, C. R.; Kiessling, L. L. *Science* **2001**, *291*, 2357–2364.
3. Zhu, X.; Schmidt, R. R. *Angew. Chem. Int. Ed.* **2009**, *48*, 1900–1934.
4. Tiwari, V. K.; Mishra, R. C.; Sharma, A.; Tripathi, R. P. *Mini-Rev. Med. Chem.* **2012**, *12*, 1497–1519.
5. Tiwari, V. K.; Mishra, B. B.; Mishra, K. B.; Mishra, N.; Singh, A. S.; Chen, X. *Chem. Rev.* **2016**, *116*, 3086–3240.
6. Mishra, A.; Mishra, N.; Tiwari, V. K. *Curr. Org. Synth.* **2016**, *13*, 176–219.
7. Dwek, R. A. *Chem. Rev.* **1996**, *96*, 683–720.
8. Angata, T.; Varki, A. *Chem. Rev.* **2002**, *102*, 439–469.
9. Blix, G.; Gottschalk, A.; Klenk, E. *Nature* **1957**, *179*, 1088.
10. Honke, K.; Taniguchi, N. *Med. Res. Rev.* **2002**, *22*, 637–654.
11. Schauer, R. *Glycoconjigate J.* **2000**, *17*, 485–499.
12. Schauer, R. *Curr. Opin. Struct. Biol.* **2009**, *19*, 507–514.
13. Robbe, C.; Capon, C.; Maes, E.; Rousset, M.; Zweibaum, A.; Zanetta, J.-P.; Michalski, J. -C. *J. Biol. Chem.* **2003**, *278*, 46337–46348.
14. Zanetta, J. -P.; Pons, A.; Iwersen, M.; Mariller, C.; Leroy, Y.; Timmerman, P.; Schaue, R. *Glycobiology* **2001**, *11*, 663–676.
15. Kiefel, M. J.; Itzstein, M.V. *Chem. Rev.* **2002**, *102*, 471–490.
16. Hood, D. W.; Cox, A. D.; Gilbert, M.; Makepeace, K.; Walsh, S.; Deadman, M. E.; Cody, A.; Martin, A.; Mansson, M.; Schweda, E. K.; Brisson, J. R.; Richards, J. C.; Moxon, E. R.; Wakarchuk, W. W. *Mol. Microbial.* **2001**, *39*, 341–350.
17. Wakarchuk, W. W.; Watson, D.; Michael, F. S.; Li, J.; Wu, Y.; Brisson, J. -R.; Young, N. M.; Gilbert, M. *J. Biol. Chem.* **2001**, *276*, 12785–12790.
18. Sullivan, C. W.; Volcani, B. E. *Arch. Biochem. Biophys.* **1974**, *163*, 29–45.
19. De Witt, C. W. *Bact. Proc.* **1958**, *75* abstract M61.
20. DeWitt, C. W.; Rowe, J. A. *Nature* **1959**, *184*, 381–382.
21. DeWitt, C. W.; Rowe, J. A. *J. Bacteriol.* **1961**, *82*, 838–848.
22. Watson, R. G.; Marinetti, G.V.; Scherp, H. W. *Immunol.* **1958**, *81*, 337–344.
23. Barry, G. T.; Abbot, V.; Tsai, T. *J. Gen. Microbiol.* **1962**, *29*, 335–352.
24. Burge, B. W.; Strass, J. H. *J. Mol. Biol.* **1970**, *47*, 449–477.
25. Krantz, M. J.; Lee, Y. C.; Hung, P. P. *Nature* **1974**, *248*, 684–686.
26. Schloemer, R. H.; Wagner, R. R. *J. Virol.* **1974**, *14*, 270–281.
27. Corfield, A. P.; Wember, M.; Schauer, R.; Rott, R. *Eur. J. Biochem.* **1982**, *124*, 521–525.

28. Crocker, P. R. *Curr. Opin. Struct. Boil.* **2002**, *12*, 609–615.
29. Brusés, J. L.; Rutishauser, U. *Biochemie.* **2001**, *83*, 635–643.
30. Simanek, E. E.; McGarvey, G. J.; Jablonowski, J. A.; Wong, C. -H. *Chem. Rev.* **1998**, *98*, 833–862.
31. Hughes, R. C. *Biochim. Biophys. Acta* **1999**, *1473*, 172–185.
32. Kim, C. U.; Lew, W.; Williams, M. A.; Liu, H.; Zhang, L.; Swaminathan, S.; Bischofberger, N.; Chen, M. S.; Mendel, D. B.; Tai, C.Y.; Laver, W. G.; Stevens, R. C. *J. Am. Chem. Soc.* **1997**, *119*, 681–690.
33. Eckhardt, M.; Bukalo, O.; Chazal, G.; Wang, L.; Goridis, C.; Schachner, M.; Gerardy-Schahn, R.; Cremer, H.; Dityatev, A. *J. Neurosci.* **2000**, *20*, 5234–5244.
34. Takano, R.; Muchmore, E.; James, W. *Glycobiology* **1994**, *4*, 665–674.
35. Malykh, Y. N.; Schauer, R.; Shaw, L. G. *Biochimie* **2001**, *83*, 623–634.
36. Munday, J.; Floyd, H.; Crocker, P. R. *J. Leukoc. Boil* **1999**, *66*, 705–711.
37. Chen, X.; Varki, A. *ACS Chem. Biol.* **2010**, *5*, 163–176.
38. Hinderlich, S.; Stäsche, R.; Zeitler, R.; Reutter, W. *J. Biol. Chem.* **1997**, *272*, 24313–24318.
39. Angata, T.; Nakata, D.; Matsuda, T.; Kitajima, K.; Troy, F. A. *J. Biol. Chem.* **1999**, *274*, 22949–22956.
40. Kean, E. L. *Glycobiology* **1991**, *1*, 441–447.
41. Comb, D. G.; Roseman, J. *J. Biol. Chem.* **1960**, *235*, 2529–2537.
42. Vann, W. F.; Tavarez, J. J.; Crowley, J.; Vimr, E.; Silver, R. P. *Glycobiology* **1997**, 7, 697–701.
43. Terada, T.; Kitazume, S.; Kitajima, K.; Inoue, S.; Ito, F.; Troy, F. A.; Inoue, T. *J. Biol. Chem.* **1993**, *268*, 2640–2648.
44. Asahina, S.; Sato, C.; Kitajima, K. *Glycoconjugate J.* **1994**, *11*, 493–499.
45. Kawano, T.; Koyama, S.; Takematsu, H.; Kozutsumi, Y.; Kawasaki, H.; Kawasimal, S.; Kawasaki, T.; Suzuki, A. *J. Biol. Chem.* **1995**, *270*, 16458–16463.
46. Butor, C.; Diaz, S.; Varki, A. *J. Biol. Chem.* **1993**, *268*, 10197–10206.
47. Shi, W. X.; Chammas, R.; Varki, A. *J. Biol. Chem.* **1996**, *271*, 15130–15138.
48. Lewis, A. L.; Nizet, V.; Varki, A. *Proc. Nat. Acad. Sc.* **2004**, *101*, 11123–11128.
49. Ng, S.-S.; Dain, J.A. The Natural Occurrence of Sialic Acids: *Biological Roles of Sialic Acid, Eds. Springer,* **1976**, 59–102.
50. Boons, G. J.; Demchenko, A.V. *Chem. Rev.* **2000**, *100*, 4539–4565.
51. De Meo, C.; Boons, G. J.; Demchenko, A.V. Synthesis of glycosides of sialic acid: chemical aspects (scope and limitations). In *Comprehensive Glycoscience—From Chemistry to Systems Biology*; Kamerling, J. P., Ed.; Elsevier: Amsterdam: The Netherlands, 2007.
52. Ress, D. K.; Linhardt, R. J. *Curr. Org. Synth.* **2004**, *1*, 31–46.
53. Navaluri, C.; Crich, D. Stereocontrolled Synthesis of Sialosides. In *Glycochemical Synthesis Strategies and Applications*; Hung, S. -C., Zulueta, M. M. L., Eds.; John Wiley & Son, Inc: Hoboken: New Jersey, 2016.
54. Magnusson, G.; Nilsson, U. J.; Fraser-Reid, B.; Kuniaki, T.; Thiem, J.; Eds. *Springer: Berlin,* **2001**, 2, 1543-1587.
55. Ando, T.; Ando, H.; Kiso, M. *Trends Glycosci. Glycotechnol.* **2001**, *13*, 573–586.
56. Takahashi, T.; Tsukamoto, H.; Yamada, H. *Tetrahedron Lett.* **1997**, *38*, 8223–8226.
57. Ye, X. -S.; Huang, X.; Wong, C. H. *Chem. Commun. (Cambridge)* **2001**, 974–975.
58. Okamoto, K.; Goto, T. *Tetrahedron* **1990**, *46*, 5835–5857.
59. Deninno, M. P. *Synthesis* **1991**, 583–593.
60. Paulsen, H.; Tietz, H. *Angew. Chem., Int. Ed. Engl.* **1982**, *21*, 927–928.
61. Marra, A.; Sinaÿ, P. *Carbohydr. Res.* **1989**, *190*, 317–322.
62. Kirchner, E.; Thiem, F.; Dernick, R.; Heukeshoven, J.; Thiem, J. *Carbohydr. Chem.* **1988**, 7, 453–486.
63. Kuhn, R.; Lutz, P.; MacDonald, D. L. *Chem. Ber.* **1966**, *99*, 611–617.
64. Byramova, N. E.; Tuzikov, A. B.; Bovin, N.V. *Carbohydr. Res.* **1992**, *237*, 161–175.

65. Sharma, M. N.; Eby, R. *Carbohydr. Res.* **1984**, *127*, 201–210.
66. Kunz, H.; Waldmann, H. *J. Chem. Soc., Chem. Commun.* **1985**, (10), 638–640.
67. Mugunthan, G.; Kartha, K. P. R. *Tetrahedron Lett.* **2012**, *53*, 5631–5634.
68. Pazynina, G.; Tuzikov, A.; Chinarev, A.; Obukhova, P.; Bovin, N. *Tetrahedron Lett.* **2002**, *43*, 8011–8013.
69. Tropper, F. D.; Andersson, F. O.; Cao, S.; Roy, R. *J. Carbohydr. Chem.* **1992**, *11*, 741–750.
70. Marra, A.; Sinaÿ, P. *Carbohydr. Res.* **1989**, *187*, 35–42.
71. Martichonok, V.; Whitesides, G. M. *J. Org. Chem.* **1996**, *61*, 1702–1706.
72. Gregar, T. Q.; Gervay-Hague, J. *J. Org. Chem.* **2004**, *69*, 1001–1009.
73. Sun, X. -L.; Kanie, Y.; Guo, C. -T.; Kanie, O.; Suzuki, Y.; Wong, C. -H. *Eur. J. Org. Chem.* **2000**, (14), 2643–2653.
74. Shpirt, A. M.; Kononov, L. O.; Torgov, V. I.; Shibaev, V. N. *Russ. Chem. Bull.* **2004**, *53*, 717–719.
75. Orlova, A. V.; Shpirt, A. M.; Kulikova, N. Y.; Kononov, L. O. *Carbohydr. Res.* **2010**, *345*, 721–730.
76. Nifantiev, N. E.; Tsvetkov, Y. E.; Shashkov, A. S.; Kononov, L. O.; Menshov, V. M.; Tuzikov, A. B.; Bovin, N. V. *J. Carbohydr. Chem.* **1996**, *15*, 939–953.
77. Tsvetkov, Y. E.; Schmidt, R. R. *Carbohydr. Lett.* **1996**, *2*, 149–156.
78. Simeoni, L. A.; Byramova, N. E.; Bovin, N. V. *Bioorg. Khim.* **1999**, *25*, 62–69.
79. Cheshev, P. E.; Khatuntsvesa, E. A.; Gerbst, A. G.; Tsvetkov, Y. E.; Shashkov, A. S.; Nifant'ev, N. E. *J. Bioorg. Chem.* **2003**, *29*, 372–381.
80. Cao, S.; Meunier, S. J.; Andersson, F. O.; Letellier, M.; Roy, R. *Tetrahedron: Asymmetry* **1994**, *5*, 2303–2312.
81. Oscarson, S. Thioglycosides. *Carbohydrates in Chemistry and Biology*, 1, Ernst, B., Hart, G. W., Sinay, P., Eds.; Wiley-VCH: Weinheim: New York, 2000; pp. 93–116.
82. Oscarson, S. General synthetic methods: reactions at the anomeric center: *S*-glycosylation. *Glycoscience: Chemistry and Chemical Biology*, 1, Fraser-Reid, B., Tatsuta, K., Thiem, J., Eds.; Springer: New York, 2001; pp. 643–671.
83. Crich, D.; Li, W. *J. Org. Chem.* **2007**, *72*, 7794–7797.
84. Dziadek, S.; Brocke, C.; Kunz, H. *Chem. Eur. J.* **2004**, *10*, 4150–4162.
85. Marra, A.; Sinay, P. *Carbohydr. Res.* **1990**, *195*, 303–308.
86. Schmidt, R.R.; Kovac, P. (Ed.). *ACS Symposium Series, American Chemical Society: Washington, DC*, 1994; 560, 276–296.
87. Greilich, U.; Brescello, R.; Jung, K. H.; Schmidt, R. R. *Liebigs Ann.* **1996**, (10), 663–672.
88. De Meo, C.; Parker, O. *Tetrahedron: Asymmetry* **2005**, *16*, 303–307.
89. Cai, S.; Yu, B. *Org. Lett.* **2003**, *5*, 3827–3830.
90. Li, Y.; Yu, H.; Cao, H.; Lau, K.; Muthana, S.; Tiwari, V. K.; Son, B.; Chen, X. *Appl. Microb. Biotech.* **2008**, *79*, 963–970.
91. Huynh, N.; Aye, A.; Li, Y.; Yu, H.; Cao, H.; Tiwari, V. K.; Shin, D. -W.; Chen, X.; Fisher, A. *J. Biochem.* **2013**, *52*, 8570–8579.
92. Yu, H.; Huang, S.; Chokhawala, H.; Sun, M.; Zheng, H.; Chen, X. *Angew. Chem. Int. Ed.* **2006**, *45*, 3938–3944.
93. Huang, S.; Yu, H.; Chen, X. *Angew. Chem. Int. Ed.* **2007**, *46*, 2249–2253.
94. Cheng, J.; Yu, H.; Lau, K.; Huang, S.; Chokhawala, H. A.; Li, Y.; Tiwari, V. K.; Chen, X. *Glycobiology* **2008**, *18*, 686–697.
95. Xiao, A.; Li, Y.; Li, X.; Santra, A.; Yu, H.; Li, W.; Chen, X. *ACS Catalysis* **2019**, *8*, 43–44.
96. Yu, H.; Chen, X. *Org. Bio. Mol. Chem.* **2016**, *14*, 2809–2818.
97. Yu, H.; Cao, H.; Tiwari, V. K.; Li, Y.; Chen, X. *Bio. Org. Med. Chem. Lett.* **2011**, *21*, 5037–5040.
98. Yu, H.; Cheng, J.; Ding, L.; Khedri, Z.; Chen, Y.; Lau, K.; Tiwari, V. K.; Chen, X. *J. Am. Chem. Soc.* **2009**, *131*, 18467–18477.

99. Chokhawala, H.A.; Cao, H.;Yu, H.; Chen, X. *J. Am. Chem. Soc.* **2007**, *129*, 10630–10631.
100. Song, X.; Yu, H.; Chen, X.; Lasanajak, Y.; Tappert, M. M.; Air, G. M.; Tiwari, V. K.; Cao, H.; Chokhawala, H. A.; Zheng, H.; Cummings, R. D.; Smith, D. F. *J. Biol. Chem.* **2011**, *286*, 31610–31622.
101. Slack, T. J.; Li, W.; Shi, D.; Mcarthur, J. B.; Zhao, G.; Li, Y. *Bioorg. Med. Chem.* **2018**, *26*, 5751–5757.
102. Sauter, N. K.; Hanson, J. E.; Glick, G. D.; Brown, J. H.; Crowther, R. L.; Park, S. J.; Skehel, J. J.; Wiley, D. C. *Biochemistry* **1992**, *31*, 9609–9621.
103. Gamian, A.; Chomik, M.; Laferrsre, C. A.; Roy, R. *Can. J. Microbiol.* **1991**, *37*, 233–237.
104. Sabesan, S.; Duus, J. O.; Fukunaga, T.; Bock, K.; Ludvigsen, S. *J. Am. Chem. Soc.* **1991**, *113*, 3236–3246.
105. Swarte, V.V.; Joziasse, D. H.; Mebius, R. E.; Van den Eijnden, D. H.; Kraal, G. *Cell. Adhes. Commun.* **1998**, *6*, 311–322.
106. Shawn, A.; Frees, De.; Kosch, W.; Way, W.; Paulson, J. C.; Sabesan, S.; Halcomb, R. L.; Huang, D. -H.; Ichikawa, Y.; Wong, C. -H. *J. A. Chem. Soc.* **1995**, *117*, 66–79.
107. Chem, D.; Sprengard, U.; Kretzschmar, G.; Bartnik, E.; Hüls, C.; Kunz, H. *Angew. Chem. Int. Ed.* **1995**, *34* (9), 990–993.
108. (a) Chong, A. K.; Pegg, M. S.; Taylor, N. R.; Von Itzstein, *M. Eur. J. Biochem.*, **1992**, *207*, 335–343; (b) Guo, X., Laver, W. G.; Vimr, E.; Sinnott, M. L. *J. A. Chem. Soc.*, **1994**, *116*, 5572–5578.
109. (a) Tiralongo, J.; Pegg, M. S.; von, *Itzstein. M. FEBS Lett.* **1995**, *372*,148–150; (b) Nakajima, Tatsuo.; Hori, H.; Ohrui, H.; Meguro, H.; Ido, *T. Agric. Biol. Chem.*, **1988**, *52*,1209–1215; (c)Teruhiko, H.; Kijima-Suda, I.; Ido, T.; Ohrui, H.; Tomita, *K. Carbohydr Res.*, **1994**, *263*, 167–72.
110. Magano, J. *Chem. Rev.* **2009**, *109*, 4398–4438.
111. (a) Meindl, P.; Tuppy, H.; *Über Hoppe-Seyler's Z. Physiol. Chem.* **1969**, *350*, 1088−109; (b) Michiels, B.; Van Puyenbroeck, K.; Verhoeven, V.; Vermeire, E.; Coenen, S. *PLoS One*, **2013**, *8*, e60348; (c) Burch, J.; Corbett, M.; Stock, C.; Nicholson, K.; Elliot, A.J.; Duffy, S.; Westwood, M.; Palmer, S.; Stewart, *L. Lancet Infect. Dis.* **2009**, *9*, 537−545.
112. Schreiner, E.; Zbiral, E.; Kleineidam, R. G.; Schauer, R. Liebigs. *Ann. Chem.* **1991**, 129–134.
113. von Itzstein, M.; Wu, W. -Y.; Jin, B. *Carbohydr. Res.* **1994**, *259*, 301–305.
114. Chandler, M.; Bamford, M. J.; Conroy, R.; Lamont, B.; Patel, B.; Patel, V.K.; Steeples, I. P.; Storer, R.; Weir, N. G.; Wright, M.; Williamson, C. J. Chem. Soc., *Perkin Trans.* 1, **1995**, 1173–1179.
115. Von Itzstein, M.; Wu, W. -Y.; Kok, G. B.; Pegg, M. S.; Dyason, J. C.; Jin, B.; Phan, T. V.; Smythe, M. L.; White, H. F.; Woods, J. M.; Bethell, R. C.; Hotham, V. J.; Cameron, J. M.; Penn, C. R. *Nature* **1993**, *363*, 418.
116. McNicholl, R.; McNicholl, J. J. *Ann Pharmacother.* **2001**, *35*, 57.
117. Koketsu, M.; Nitoda, T.; Sugino, H.; Juneja, L. R.; Kim, M.; Yamamoto, T.; Abe, N.; Kajimoto, T.; Wong, C. -H. *J. Med. Chem.* **1997**, *40*, 3332–3335.
118. Tamai, H.; Ando, H.; Ishida, H.; Kiso, M. *Org. Lett.* **2012**, *14*, 6342–6345.
119. Varki, A. *C.S.H. Perspect. Biol.* **2011**, *3*, a005462.
120. Cohen, M.; Varki, A. *OMICS* **2010**, *14*, 455–464.
121. (a) Blix, G.; Gottschalk, A.; Klenk, E., *Nature*, **1957**, *179*, 1088; (b) Gottschalk, A. *University Press: Cambridge,* **1960**; (c) Faillard, H. *TIBS*, **1989**, *14*, 237.
122. Ziak, M.; Qu, B.; Zuo, X.; Zuber, C.; Kanamori, A.; Kitajima, K.; Inoue, S.; Inoue, Y.; Roth, J. *Proc Natl. Acad. Sci. U. S. A.* **1996**, *93*, 2759–2763.
123. Yuki, N. *Muscle Nerve* **2007**, *35*, 691–711.
124. (a) Suzuki, A. Proc. Jpn. Acad. Ser. B. *Phys. Biol. Sci.*, **2006**, *82*, 93–103; (b) Chou, H. H.; Hayakawa, T.; Diaz, S.; Krings, M.; Indriati, E.; Leakey, M.; Paabo, S.; Satta, Y.; Takahata, N.; Varki, A. *Proc. Natl. Acad. Sci. U. S. A.*, **2002**, *99*, 11736–11741.

125. Irie, A.; Koyama, S.; Kozutsumi, Y.; Kawasaki, T.; Suzuki, A. *J. Biol Chem.* **1998**, *273*, 15866–15871.
126. Zhu, A.; Hurst, R. *Xenotransplantation* **2002**, *9*, 376–381.
127. Varki, A. *Glycoconj. J.* **2009**, *26*, 231–245.
128. Kamerling, J. P.; Gerwig, G. J. *Methods Mol. Biol.* **2006**, *347*, 69–91.
129. Varki, A.; Angata, T. *Glycobiology* **2006**, *16*, 1R–27R.
130. Dekan, G.; Gabel, C.; Farquhar, M. G. *Proc. Natl. Acad. Sci. U. S. A.* **1991**, *88*, 5398–5402.
131. Gelberg, H.; Healy, L.; Whiteley, H.; Miller, L. A.; Vimr, E. *Lab. Invest.* **1996**, *74*, 907–920.
132. Takeda, T. *Clin. Exp. Nephrol.* **2003**, 7, 260–269.
133. Weinhold, B.; Seidenfaden, R.; Röckle, I.; Mühlenhoff, M.; Schertzinger, F.; Conzelmann, S.; Marth, J. D.; Gerardy-Schahn, R.; Hildebrandt, H. *J. Biol. Chem.* **2005**, *280*, 42971–42977.
134. Johnson, C. P.; Fujimoto, I.; Rutishauser, U.; Leckband, D. E. *J. Biol. Chem.* **2005**, *280*, 137–145.
135. El Maarouf, A.; Petridis, A. K.; Rutishauser, U. *Proc. Natl. Acad. Sci. U. S. A.* **2006**, *103*, 16989–16994.
136. Rutishauser, U. *Nat. Rev. Neurosci.* **2008**, *9*, 26–35.
137. Ashwell, G.; Harford, J. *Annu. Rev. Biochem.* **1982**, *51*, 531–554.
138. Weigel, P. H.; Yik, J. H. *Biochim. Biophys. Acta* **2002**, *1572*, 341–363.
139. Iijima, R.; Takahashi, H.; Namme, R.; Ikegami, S.; Yamazaki, M. *FEBS Lett.* **2004**, *561*, 163–166.
140. Bornhöfft, K. F.; Goldammer, T.; Rebl, A.; Galuska, S. P. *Dev. Comp. Immunol.* **2018**, *86*, 219–231.
141. Varki, A. *Trends Mol. Med.* **2008**, *14*, 351–360.
142. Tettamanti, G.; Bonali, F.; Marchesini, S.; Zambotti, V. *Biochim. Biophys. Acta* **1973**, *296*, 160–170.
143. Ledeen, R. W.V 8. *J. Supramol Struct.* **1978**, 1–17.
144. Wang, B. *Annu. Rev. Nutr.* **2009**, *29*, 177–222.
145. Svennerholm, L. *Prog. Brain Res.* **1994**, *101*, XI–XIV.
146. Rosenberg, A.; Noble, E. P. *Alcohol* **1994**, *11*, 565–569.
147. Sjoberg, E. R.; Manzi, A. E.; Khoo, K. H.; Dell, A.; Varki, A. *J. Biol. Chem.* **1992**, *267*, 16200–16211.
148. Dasgupta, S.; van Halbeek, H.; Spicer, S.; Hogan, E. L. *Glycobiology* **2000**, *10*, 1–9.
149. Wang, B.V 3. *Adv. Nutr.* **2012**, 465S–472S.
150. Bork, K.; Reutter, W.; Gerardy-Schahn, R.; Horstkorte, R. *FEBS Lett.* **2005**, *579*, 5079–5083.
151. Maccioni, H. J. F.V 103. *J. Neurochem.* **2007**, 81–90.
152. Groux-Degroote, S.; Guérardel, Y.; Delannoy, P. *Chem. Bio. Chem.* **2017**, *18*, 1146–1154.
153. Zeng, G.; Yu, R. K. *Curr. Drug Targets* **2008**, *9*, 317–324.
154. Gu, X.; Preuss, U.; Gu, T.; Yu, R. K. *J. Neurochem.* **1995**, *64*, 2295–2302.
155. Bieberich, E.; Freischetz, B.; Liour, S. S.; Yu, R. K. *J. Neurochem.* **1998**, *71*, 972–979.
156. Tsai, Y. T.; Yu, R. K. *J. Neurochem.* **2014**, *128*, 101–110.
157. Van Echten-Deckart, G.; Sandhoff, K. *Pergamon Press; New York*, 1999, pp. 87–106.
158. Hirschberg, K.; Zisling, R.; Van Echten-Deckert, G.; Futerman, A. H. *J. Biol. Chem.* **1996**, *271*, 14876–14882.
159. Nakayama, J.; Fukuda, M. N.; Hirabayashi, Y.; Kanamori, A.; Sasaki, K.; Nishi, T.; Fukuda, M. *J. Biol. Chem.* **1996**, *271*, 3684–3691.
160. Nara, K.; Watanabe, Y.; Maruyama, K.; Kasahara, K.; Nagai, Y.; Sanai, Y. *Proc. Natl. Acad. Sci. USA* **1994**, *91*, 7952–7956.
161. Sasaki, K.; Kurata, K.; Kojima, N.; Kurosawa, N.; Ohta, S.; Hanai, N.; Tsuji, S.; Nishi, T. *J. Biol. Chem.* **1994**, *269*, 15950–15956.

162. Haraguchi, M.; Yamashiro, S.; Yamamoto, A.; Furukawa, K.; Takamiya, K.; Lloyd, O. K.; Shiku, H.; Furukawa, K. *Proc. Natl. Acad. Sci. USA* **1994**, *91*, 10455–10459.
163. Zhu, J.; Chen, G.; Zhu, S.; Li, S.; Wen, Z.; Li, B.; Zheng, Y.; Shi, L. *Sci Rep.* **2016**, *6*, 28400.
164. Furukawa, K.; Horie, M.; Okutomi, K.; Sugano, S.; Furukawa, K. Biochim. *Biophys. Acta. Gene Struct. Expression* **2003**, *1627*, 71–78.
165. Liu, J.; Zheng, X.; Pang, X.; Li, L.; Wang, J.; Yang, C.; Du, G. *Acta Pharm. Sin. B.* **2018**, *8*, 713–720.
166. Varki, A.; Schauer, R.; Cumings, R. D.; Esko, J. D.; Freeze, H. H.; Stanley, P.; Bertozzi, C. R.; Hart, G. W.; Etzler, M. E.. 2nd ed.; *Essentials of Glycobiology*, Cold Spring Harbor Laboratory Press: Cold Spring Harbor (NY), 2009; pp. 171–182 Chapter 14.
167. Traving, C.; Schauer, R. *CMLS Cell. Mol. Life Sci.* **1998**, *54*, 1330–1349.
168. Adams, D.; Gahl, W. A.; Free Sialic Acid Storage Disorders. In: Pagon RA, Adam M. P.; Ardinger, H. H.; Wallace, S. E.; Amemiya, A.; Bean, L. J. H.; Bird, T. D.; Ledbetter, N.; Mefford, H. C.; Smith, R. J. H., Stephens, K. *GeneReviews®: University of Washington, Seattle*, **2003**, 1993–2017.
169. Froissart, R.; Cheillan, D.; Bouvier, R.; Tourret, S.; Bonnet, V.; Piraud, M.; Maire, I. *J. Med. Genet.* **2005**, *42*, 829–836.
170. Fontaine, G.; Biserte, G.; Montreuil, J.; Dupont, A.; Farriaux, J. P. *Helv. Paediatr. Acta. Suppl* **1968**, *17*, 1–32.
171. Seppala, R.; Tietze, F.; Krasnewich, D.; Weiss, P.; Ashwell, G.; Barsh, G.; Thomas, G. H.; Packman, S.; Gahl, W. A. *J. Biol. Chem.* **1991**, *266*, 7456–7461.
172. Prenatal Diagnosis, 2006, 26, 655–658; (b) Aula, P.; Autio, S.; Raivio, K.O.; Rapola, J.; Thodén, C. J.; Koskela, S. *l.*; Yamashina, I. Archiv. *Neurology*, **1979**, 36, 88–94.
173. Renlund, M.; Tietze, F.; Gahl, W. A. *Science* **1986**, *232*, 759–762.
174. Mancini, G. M.; Hu, P.; Verheijen, F. W.; Van Diggelen, O. P.; Janse, H. C.; Kleijer, W. J.; Beemer, F. A.; Jennekens, F. G. I. *Eur. J. Pediatr.* **1992**, *151*, 590–595.
175. Sonninen, P.; Autti, T.; Varho, T.; Hämäläinen, M.; Raininko, R. *AJNR Am. J. Neuroradiol.* **1999**, *20*, 433–443.
176. Alajoki, L.; Varho, T.; Posti, K.; Aula, P.; Korhonen, T. *Dev. Med. Child Neurol.* **2004**, *46*, 832–837.
177. Annunziata, P.; Federico, A.; D'Amore, I.; Corona, R. M.; Guazzi, G. C. *Early Hum. Dev.* **1983**, *8*, 269–278.
178. Yuki, N.; Hartung, H. -P. *N. Engl. J. Med.* **2012**, *366*, 2294–2304.
179. Ito, H.; Hatanaka, Y.; Fukami, Y.; Harada, Y.; Kobayashi, R.; Okada, H.; Uchibori, A.; Chiba, A.; Okuda, S. *BMC Neurol.*, **2018**, 23; 18(1), 72.
180. Chiba, A.; Kusunoki, S.; Shimizu, T.; Kanazawa, I. *Ann. Neurol.* **1992**, *31*, 677–679.
181. Bukhari, S.; Taboada, J. *Cureus.* **2017**, *9*, e1048.
182. Chae, C. S.; Kwon, K. M.; Lee, J. S.; Kim, Y. H. *Neurologist.* **2018**, *23*, 128–130.
183. Vojdani, A.; Vojdani, E. *World J. Gastroenterol.* **2019**, *25*, 1050–1066.
184. De Bruyn, A.; Poesen, K.; Bossuyt, X.; Heremans, I. P.; Claeys, T.; Depuydt, C. E.; Van Damme, P.; Claeys, K. G. *Acta Neurol. Belg.* **2019**, *119*, 29–36.
185. Yoon, L.; Kim, B. R.; Kim, H. Y.; Kwak, M. J.; Park, K. H.; Bae, M. H.; Lee, Y.; Nam, S. O.; Choi, H. Y.; Kim, Y. M. *J. Neuroimmunol.* **2019**, *330*, 170–173.
186. Cabrero, F.R.; Morrison, E.H. *StatPearls Publishing*, 2019.
187. Campbell, R. J.; Bogoch, S.; Scolaro, M. J.; Belval, P. C. *Am. J. Psychiatry* **1967**, *123*, 952–962.
188. Edelfors, S. *Acta Pharmacol. Toxicol.* **1981**, *48*, 61–64.
189. Barbeau, D.; Liang, J.; Robitalille, Y.; Quirion, R.; Srivastava, L. *Proc. Natl. Acad. Sci. U. S. A.* **1995**, *92*, 2785–2789.
190. Gilabert-Juan, J.; Varea, E.; Guirado, R.; Blasco-Ibáñez, J. M.; Crespo, C.; Nacher, J. *Neurosci. Lett.* **2012**, *530*, 97–102.

191. Varea, E.; Guirado, R.; Gilabert-Juan, J.; Martı, U.; Castillo-Gomez, E.; BlascoIbánez, J. M.; Crespo, C.; Nacher, J. *J. Psychiatr. Res.* **2012**, *46*, 189–197.
192. Maziade, M.; Roy, M.; Chagnon, Y.; Cliche, D.; Fournier, J.; Montgrain, N.; Dion, C.; Lavallee, J.; Garneau, Y.; Gingras, N.; Nicole, L.; Pire's, A.; Ponton, A.; Potvin, A.; Wallot, H.; Merette, C. *Mol. Psychiatry* **2005**, *10*, 486–499.
193. Arai, M.; Yamada, K.; Toyota, T.; Obata, N.; Haga, S.; Yoshida, Y.; Nakamura, K.; Minabe, Y.; Ujike, H.; Sora, I.; Ikeda, K.; Mori, N.; Yoshikawa, T.; Itokawa, M. *Biol. Psychiatry* **2006**, *59*, 652–659.
194. Tao, R.; Li, C.; Zheng, Y.; Qin, W.; Zhang, J.; Li, X.; Xu, Y.; Shi, Y. Y.; Feng, G.; He, L. *Schizophr. Res.* **2007**, *90*, 108–114.
195. Hane, M.; Sumida, M.; Kitajima, K.; Sato, C. *Pure Appl. Chem.* **2012**, *84*, 1895–1906.
196. Hane, M.; Matsuoka, S.; Ono, S.; Miyata, S.; Kitajima, K.; Sato, C. *Glycobiology* **2015**, *25*, 1112–1124.
197. Sato, C.; Hane, M.; Kitajima, K. *Biochim. Biophys. Acta* **2016**, *1860*, 1739–1752.
198. Mori, A.; Hane, M.; Niimi, Y.; Kitajima, K.; Sato, C. *Glycobiology* **2017**, *27*, 834–846.
199. Sato, C.; Kitajima, K. *Adv. Carbohydr. Chem. Biochem.* **2019**, *76*, 1–64.
200. Constantino, J. N.; Charman, T. *Lancet Neurol.* **2015**, *15*, 279–291.
201. Itahashi, T.; Yamada, T.; Nakamura, M.; Watanabe, H.; Yamagata, B.; Jimbo, D.; Shioda, S.; Kuroda, M.; Toriizuka, K.; Kato, N.; Hashimoto, R. *Neuroimage Clin.* **2015**, 7, 155–169.
202. Lai, M. C.; Lombardo, M. V.; Ecker, C.; Chakrabarti, B.; Suckling, J.; Bullmore, E. T.; Happe, F.; Murphy, D. G.; Baron-Cohen, S.; Consortium, M. A. *Cereb. Cortex* **2015**, *25*, 3613–3628.
203. Ronemus, M.; Iossifov, I.; Levy, D.; Wigler, M. *Nat. Rev. Genet.* **2014**, *15*, 133–141.
204. Yu, X.; Qiu, Z.; Zhang, D. *Neurosci Bull.* **2017**, *33*, 125–129.
205. Murray, H. C.; Low, V. F.; Swanson, M. E.; Dieriks, B. V.; Turner, C.; Faull, R. L.; Curtis, M. A. *Neuroscience* **2016**, *330*, 359–375.
206. Dhavale, D.; Henry, J. E. *Biochim. Biophys. Acta.* **2012**, *1820*, 1475–1480.
207. Leshchynska, I.; Sytnyk, V. *Neural Plast.* **2016**, 6427537. doi: 10.1155/2016/6427537.
208. Sillanaukee, P.; Ponnio, M.; Jaaskelainen, I. P. *Eur. J. Clinic. Invest.* **1999**, *29*, 413–425.
209. Dasgupta, A. *Alcohol Biomark.* **2015**, 91–120. doi: 10.1016/B978-0-12-800339-8.00004-3.
210. Von Itzstein, M. *Nat. Rev. Drug Discov.* **2007**, *6*, 967–974.
211. Paulson, J. C.; Weinstein, J.; Dorland, L.; van Halbeek, H.; Vliegenthart, J. F. *J. Biol. Chem.* **1982**, *257*, 12734–12738.
212. Li, Y.; Cao, H.; Yu, H.; Chen, Y.; Lau, K.; Qu, J.; Thon, V.; Sugiarto, G.; Chen, X. *Mol. Biosyst.* **2011**, 7, 1060–1072.
213. Chen, G. Y.; Chen, X.; King, S.; Cavassani, K. A.; Cheng, J.; Zheng, X.; Cao, H.; Yu, H.; Qu, J.; Fang, D.; Wu, W.; Bai, X. F.; Liu, J. Q.; Woodiga, S. A.; Chen, C.; Sun, L.; Hogaboam, C. M.; Kunkel, S. L.; Zheng, P.; Liu, Y. *Nat Biotechnol.* **2011**, *29*, 428–435.
214. Rudrawar, S.; Dyason, J. C.; Rameix-Welti, M. A.; Rose, F. J.; Kerry, P. S.; Russell, R. J.; van der Werf, S.; Thomson, R. J.; Naffakh, N.; von Itzstein, M. *Nat Commun.* **2010**, *16* (1), 113.
215. Parameswaran, R.; Lim, M.; Arutyunyan, A.; Abdel-Azim, H.; Hurtz, C.; Lau, K.; Müschen, M.; Yu, R. K.; von Itzstein, M.; Heisterkamp, N.; Groffen, J. *J. Exp. Med.* **2013**, *210*, 805–819.
216. North, R. A.; Wahlgren, W. Y.; Remus, D. M.; Scalise, M.; Kessans, S. A.; Dunevall, E.; Claesson, E.; Soares de Costa, T. P.; Perugini, M. A.; Ramaswamy, S.; Allison, J. R.; Indiveri, C.; Friemann, R.; Dobson, R. C. *J. Front. Chem.* **2018**, *6*, 233.
217. Wahlgren, W. Y.; Dunevall, E.; North, R. A.; Paz, A.; Scalise, M.; Bisignano, P.; Bengtsson-Palme, J.; Goyal, P.; Claesson, E.; Caing-Carlsson, R.; Andersson, R.; Beis, K.; Nilsson, U. J.; Farewell, A.; Pochini, L.; Indiveri, C.; Grabe, M.; Dobson, R. C. J.; Abramson, J.; Ramaswamy, S.; Friemann, R. *Nat. Commun.* **2018**, *9*, 1753.

CHAPTER SIX

Glycan microarray: Toward drug discovery and development

Saddam M. Muthana
Department of Chemistry, College of Sciences & General Studies, Alfaisal University, Riyadh, Kingdom of Saudi Arabia

1 Introduction

Carbohydrates and their glycoconjugates are known to play important roles in a variety of biological processes and a wide range of diseases. They are involved in numerous biological recognition events, inflammation, cancer development and metastasis, and bacterial and viral infections.[1–4] Studying the biological significances of carbohydrates, in particularly their roles in disease progression, are proven very challenging due in part to their structural diversity, the limited access to complex carbohydrate-containing molecules, and the lack of proper tools that enable the high-throughput analysis of carbohydrates interactions with other biomolecules such as glycan-binding proteins (GBP). Conventional methods such as isothermal titration calorimetry (ITC), surface plasmon resonance (SPR), and enzyme-linked lectin assay (ELLA) can be time consuming and require significant amounts of material.[5–7] Moreover, multivalent interactions between carbohydrate ligands and GBPs are generally required to achieve good binding.

The emerging of glycan microarrays as high-throughput technology for studying carbohydrate interactions have overcome some of these challenges, and have greatly contributed to our understanding of the biological roles of carbohydrates or glycoconjugates and their interactions with a variety of macromolecules.[8–13] Glycan microarrays allow for the rapid screening of thousands of binding interactions in one experiment using minimal amounts of precious material. Therefore, glycan microarray has become a powerful tool for studying carbohydrate protein interaction. Herein, we describe the use of glycan microarrays in biomedical applications and their potential role in drug discovery and development.

Carbohydrates in Drug Discovery and Development. http://dx.doi.org/10.1016/B978-0-12-816675-8.00006-3

2 Fabrication of glycan microarrays

The key step in the fabrication of glycan microarrays involves the immobilization of the glycans into solid support. Various solid supports such is glass microscope slides, microtiter plates, gel beads, and nitrocellulose membranes have been utilized in the construction of glycan microarrays. The two main methods used for glycan array fabrications are non-covalent immobilization and covalent immobilization (Fig. 6.1). The most commonly used solid support is microscope glass slides due to their compatibility with optical detection systems, being relatively inexpensive, and ease of surface functionalization.

Non-covalent immobilization techniques are mainly through hydrophobic and electrostatic interactions. Free or modified glycans are absorbed into underivatized or derivatized solid surfaces. For example, polysaccharides and neoglycolipids have been passively adsorbed on nitrocellulose-coated glass slides.[14–16] Negatively charged heparin polysaccharides have been immobilized into lysine-coated slides via electrostatic interactions.[17] Alternative non-covalent immobilization method involves flourophilic interaction between fluorous-tagged glycans and slides coated with fluoroalkylsilane.[18,19] Moreover, other non-covalent immobilization such as using biotinylated sugars to streptavidin-coated surfaces and DNA-based glycan arrays have been reported.[20–22]

Covalent immobilization of glycan arrays usually involves having reactive groups at the end of spacer moieties that reacted with functionalized surfaces to form a covalent bond. Covalent immobilization methods often involved amine and thiol chemistry. For example, amine-terminated glycans are immobilized into functionalized glass slide surfaces by reacting

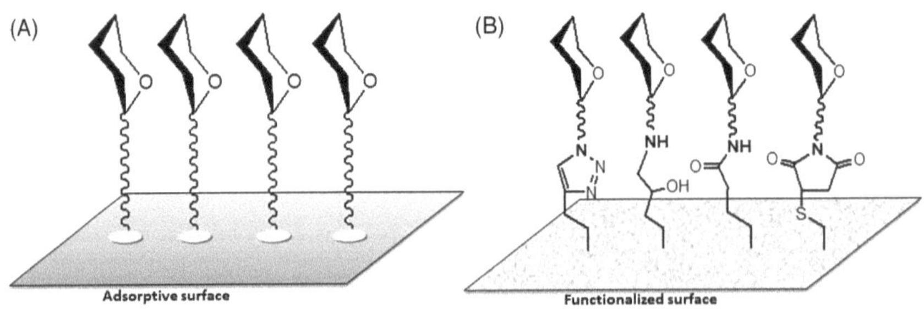

Figure 6.1 ***Glycan microarrays immobilization.*** (A) Noncovalent immobilization techniques based on hydrophobic absorption or by other interactions (electrostatic, fluorophilic, etc); (B) Covalent immobilization by linking a reactive group to the functionalized surface via (cycloaddition, epoxide opening, amide, thioether, etc).

with *N*-hydroxysuccinamide (NHS) activated ester, epoxide, or aldehyde via covalent bonds.[23–26] Other glycan arrays have been constructed using thiol chemistry by reacting maleimide functional group via disulfide bond formation or reacting with gold surfaces.[27–31] Other methods used to construct glycan arrays include cycloaddition reaction of azide with an alkyne, amine with epoxide functionalized surfaces, and free glycans with amino-oxy or hydrazide surfaces.[32–35]

3 Detection of glycan microarrays

Various methods have been developed for glycan microarray detection including fluorescent-based methods, surface plasmon resonance, and mass spectrometry (Matrix-Assisted Laser Desorption/Ionization-Time of Flight, MALDI-TOF).[36] Because of their high sensitivity and availability, fluorescent-based methods are the most commonly used methods for the detection of glycan microarray. Fluorophores such as fluorescein isothiocyanate, Cy3, and Cy5 are often coupled to glycan binding proteins (GBP) or secondary antibodies for detecting interactions with glycan microarrays. In general, detection of glycan microarray involves incubating the glycan array with GBPs, washing unbound proteins, subsequent incubation with detection reagents if necessary, followed by multiple washing steps after incubation, detection and quantification of the fluorescent signals (Fig. 6.2).

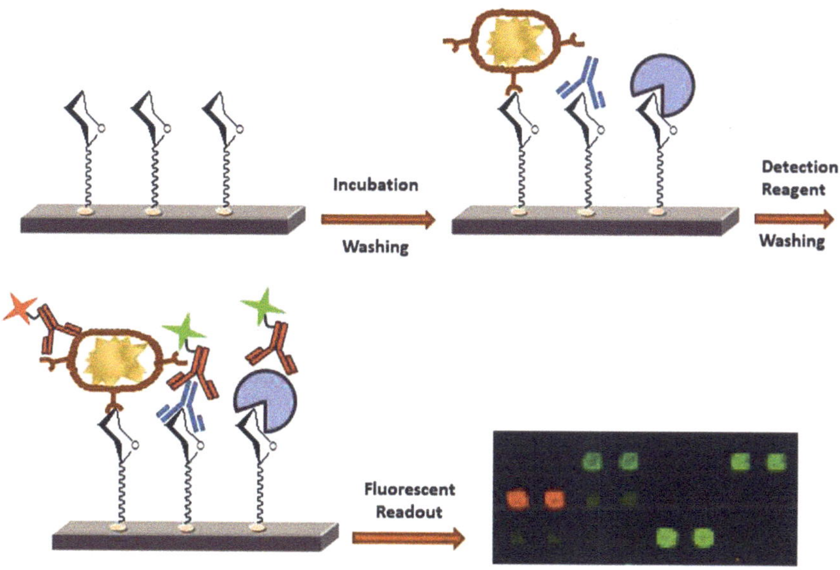

Figure 6.2 ***Fluorescent-based method for glycan microarray detection.***

For glycan-binding molecules labeled with a fluorescent tag, readout of the fluorescence intensities from the microarray can be obtained directly. For unlabeled GBP, a second incubation step with labeled detection reagents such as antibodies is commonly used. Alternatively, label-free detection methods, such as SPR and mass spectrometry have been employed to evaluate the interactions of GBPs. The use of gold surfaces as array support for immobilizing thiol-linked glycans allows the use of glycan arrays for SPR studies. Using multi-channel SPR instruments hundreds of glycans microspots can be analyzed simultaneously.[37,38]

4 Biomedical applications of glycan microarrays

Glycan microarrays have been used for studying the binding interactions of a variety of glycan- binding molecules such as proteins, antibodies, cells, bacteria, and viruses. In the past two decades, glycan microarrays have become the ideal format for studying carbohydrate protein interactions. Hundreds of glycans can be screened to study the binding interactions, activities, and specificities of various glycan-binding proteins. Therefore, glycan microarray has emerged as a powerful high-throughput tool for studying the biological role of glycans and as a potential tool for disease detection and vaccine development (Fig. 6.3). It has been utilized for the

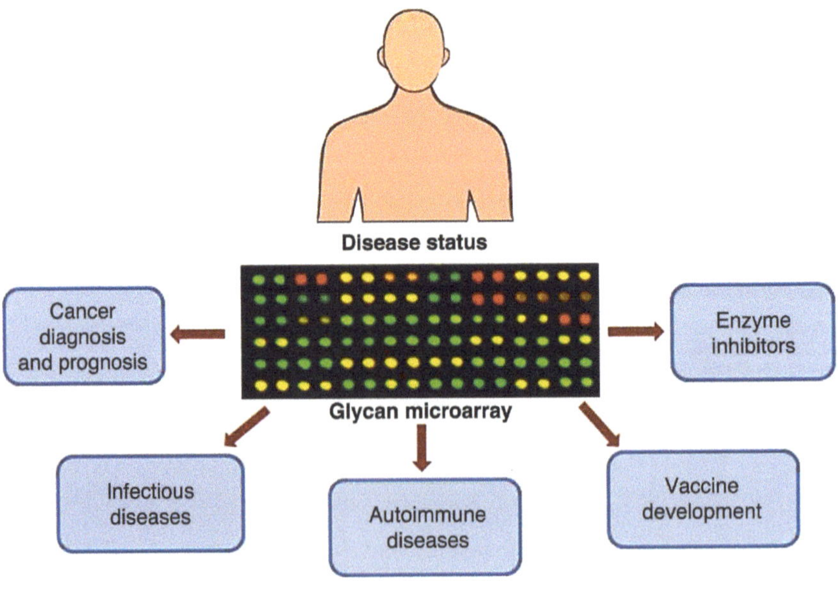

Figure 6.3 *Biomedical applications of glycan microarray.*

screening for biomarkers for a wide variety of diseases including cancer, infectious diseases, autoimmune diseases, and for monitoring immune response to vaccines.[39] Additionally, glycan microarray can be employed in the high-throughput screening of potential new inhibitors for enzymes that are known to be complicated in the biosynthesis of disease-related glycans.

4.1 Cancer

Glycan microarrays have been used in evaluating cancer antigens as potential cancer biomarkers for cancer detection, prognosis, and response to treatment. Expression of altered glycans on cell surfaces is one of the hallmarks of cancer, and the alternation in glycosylation patterns can be explored for the diagnosis of cancer, drug targeting treatment, and differentiating cancerous cells from normal cells. For example, higher-ordered branching of *N*-linked glycans with an increased level of fucosylation and sialylation has been reported to be associated with cancer development.[40–44] On the contrary, the structure of the *O*-linked glycans are often much more truncated.[45] The use of these glycans as potential biomarkers has been extensively studied, but they are not ideal for the early detection of cancer as they are difficult to isolate from the tumor site. Alternatively, in the promise that aberrant glycosylation in tumor cells may give rise to changes in antibody levels aberrant glycans, many research groups have focused on profiling antibodies to tumor-associated carbohydrate antigens (TACAs) as potential biomarkers for the early detection, diagnosis, and prognosis of cancer.

Prior to the development of glycan microarray technology studies have focused on profiling antibodies to a very small number of available carbohydrate antigens, including Tn antigen, TF antigen, and some gangliosides. Glycan microarrays allow for the high throughput screening of hundreds of glycans using a miniscule amount of material. A number of groups have demonstrated the use of glycan arrays to identify anti-glycan antibodies in serum as potential biomarkers for the early detection, diagnosis, and prognosis of cancer. For example, glycan microarrays have been utilized to profile antibody levels to Globo H and related structures in patients with breast cancer.[46,47] They reported significantly higher levels of antibodies against Globo H in patients *versus* the healthy control group ($P < 0.0001$).

From profiling antibodies of patients with breast cancer and healthy controls, Wandall et al. reported the presence of higher levels of antibodies to MUC1 in patients with breast cancer compared to that of healthy controls.[48] The antibodies against MUC1 were specific to the glycan moiety and peptide sequence, and no cross-reactivity was observed with other glycopeptides containing the same glycan epitopes. In addition, the same group profiled antibodies in serum of patients with colorectal cancer and identified cancer-associated IgG and IgA antibodies to a set of altered MUC1 and MUC4 glycopeptides.[49] The tumor-associated Tn and sialylated Tn glycans were the main glycan antigens associated with these epitopes. Moreover, the tumor- associated Tn and TF antigens have been greatly examined as potential diagnostic and prognostic biomarkers as well as therapeutic targets for cancer.[50,51] Tn/TF antigen-based vaccines for prostate cancer and breast cancer were developed and have progressed into clinical trials.[52–54] Blixt and co-workers also reported a significantly higher levels of antibodies to tumor-associated MUC1 in patients with early-stage breast cancer, but not in late-stage breast cancer patients.[55] They reported that the levels of IgG antibodies to MUC1 carrying core 3 glycans and sialylated Tn antigens are associated with reduced incidence and delay in metastases, which suggest that these antibodies may play a role in the progression of cancer and can be used as potential prognostic biomarkers.

A study using glycan microarray for profiling antibodies of non-mucinous ovarian cancers reported that antibodies against a set of glycans can differentiate between non-mucinous borderline or ovarian cancer from healthy controls and that P_1 (Galα1-4Galβ1-4GlcNAcβ) was the best candidate ($P < 0.001$) for detecting ovarian cancer.[56] In another study, Vuskovic et al. used glycan microarray to profile antibodies in patients with mesothelioma and high-risk subjects that were exposed to asbestos.[57] They reported that glycan arrays can be used for the diagnosis and prognosis of mesothelioma. The antigen Neu5Acα2-3Galβ1-4Glcβ has the best correlation for diagnosis (P 0.00005), while Glcα1-4Glcβ showed the best correlation for prognosis ($P > 0.005$). A Comparison study of serum antibody levels in classical Hodgkin's lymphoma and healthy controls demonstrated that antibody levels to GalNAcα-Ser/Thr (Tn) were significantly higher in patients with classical Hodgkin's lymphoma.[58]

Several studies reported that dietary non-human glycans, such as *N*-glycolylneuraminic acid (Neu5Gc) incorporation into human tissue cell surfaces is associated with the development of cancer.[59,60] Comparing the antibody profiles of patients with carcinomas and other diseases showed that antibodies to Neu5Gcα2-6Tn were prominent in patients with carcinomas. These antibodies can mediate the killing of Neu5Gcα2-6Tn-expressing tumors indicating that antibodies against Neu5Gc might serve as diagnostic and prognostic biomarkers or as immunotherapeutic agents in human carcinomas.[61] Additionally, antibodies to *N*-glycans such as high mannose and multi-antennary type II chains have been detected in human sera of patients with prostate cancer.[62]

4.2 Infectious diseases

Many pathogens contain specific glycans on their cell surfaces that can elicit immune responses in infected individuals producing specific antibodies to the invading pathogens. Therefore, various glycan microarrays have been developed and used to evaluate immune responses to infectious diseases to detect pathogen-specific diagnostic biomarkers. For example, glycan microarray was used to profile the sera of individuals infected with the parasite, *T. spiralis* and reported that GalNAcβ1-4(Fucα1-3)GlcNAc (LDNF) antigen presented on BSA (bovine serum albumin) as a potential diagnostic biomarker for trichinellosis with very good sensitivity (96%) but modest specificity (67%).[63] Seeberger and co-workers used glycan microarray to profile antibodies in malaria patients and healthy controls, and reported that exposure to malaria can affect the antibody levels to glycosylphosphatidylinositol (GPI) and their reactivity pattern control.[64] They reported that the minimal epitope required for binding with anti-GPI antibodies is the pentasaccharide (Man_3-GPI, Manα1-2Manα1-6Manα1-4GlcNH_2α1-6-*myo*-inositol-1-PO_4). Another study indicated that specific antibodies that recognize glycopeptides bearing the Tn antigen is potentially useful for the diagnosis of *Cryptosporidium parvum* infection.[65]

Glycan microarrays have also been utilized to find biomarkers for viral infections. The initial step in viral cell invasion is the attachment of the virus to cell surface antigens that are often times glycans. The adhesion of the virus to the cell surface is mediated by viral surface proteins such as hemagglutinin (HA) in the case of influenza viruses. Glycan microarrays containing sialylated glycans have been used to profile the binding preferences of several human and avian hemagglutinins.[66–68] These studies indicated that human H3N2 viruses preferentially bind to alpha2-6-linked *N*-acetylneuraminic acid (Neu5Acα2-6-linked) glycans, while the avian H5N1 viruses prefer binding to Neu5Acα2-3-linked glycans. In the recent outbreak of the H1N1 influenza virus in 2009, Tumpey's and colleagues demonstrated that the virus exhibited a dose-dependent binding to only Neu5Acα2-6-diLacNAc (LacNAc = Galβ1-4GlcNAc) [69]. In another study, a sialylated glycan array was used to study the interactions of modified sialic acids using proteins and viruses.[70] The results demonstrated that influenza viruses H1N1 and H3N2 have preferences for binding to only the alpha2-6-linked glycans containing Neu5Ac or Neu5Ac9Lt (9*O*-Lactoyl-*N*-acetylneuraminic acid). Moreover, Blixt and colleagues constructed a glycopeptide array to study the immune responses to the herpes simplex viruses (HSV-1 and HSV-2).[71] They reported the presence of IgG antibodies to the glycopeptide P-P-A-(GalNAc)T-A-P-G in HSV-2 infected individuals but not in healthy controls, or those infected with HSV-1. Furthermore, glycan microarray has been used to detect anti α-fetoprotein fraction L3 (AFP-L3) for early prediction of Hepatitis B hepatocelluar carcinoma.[72] They demonstrated that AFP-L3 antibody levels is better for differentiating between hepatocelluar carcinoma (HCC) and chronic hepatitis B (CHB) and could potentially be used as a biomarker for the early diagnosis of HCC.

Glycan microarrays have also been used to identify biomarkers in detecting and differentiating bacterial infections. Microarray platform containing polysaccharide has been constructed to identify potential markers for microbial infections by profiling antibodies in serum of infected human or animal subjects.[14,73] Parthasarathy et. al. probed the use of polysaccharide array to detect antibodies against *Francisellatularensis* (causative agent of tularemia), *Burkholderiapseudomallei* (causative agent of melioidosis), and *Bacillus anthracis* (causative agent of anthrax).[74] The results from this study demonstrated that glycan arrays can specifically detect and differentiate subjects infected with tularemia, melioidosis, and anthrax. Another study evaluated the use of lipopolysaccharide (LPS) for detecting antibodies in canine serum collected from tularemia positive and control subjects.[75] The results indicated that LPS microarrays can detect anti-LPS antibodies at low concentrations and better sensitivity than the conventional immunofluorescence assays. Moreover, Blixt and co-workers used glycan microarray to profile antibodies in sera of patients diagnosed with salmonellosis and detected *Salmonella* specific antibodies in infected subjects.[76] In addition to detecting anti-glycan antibodies in sera, glycan microarrays have also been used to study direct interactions of bacteria and array glycans.[77] The results demonstrated that glycan microarrays can be used to discriminate different bacteria strains and that bacterial detection can be accomplished using complex mixtures. Using shotgun glycomics, Song et al. profiled sera from individuals infected with Lyme disease.[78] The study showed that higher antibody levels to disialylated ganglioside (GD1b-lactone) were present in subjects with Lyme disease and can potentially be used as diagnostic markers for Lyme disease.

4.3 Autoimmune diseases

Autoimmune diseases arise from complex interactions of various factors such as genetic and environmental factors and affect millions of individuals worldwide.[79,80] Autoimmune diseases occur when the immune system mistakenly attacks its own cells or healthy tissues. Diagnosis and prognosis of autoimmune diseases can be very challenging due to the lack of specific symptoms to a particular autoimmune disease. Human cells and a variety of macromolecules in nature are covered with dense complex glycans, also known as glycocalyx. These glycans play an important role in facilitating cell communications, pathogen recognition, and modulating both innate and adaptive immunity. Higher anti-glycan antibody levels in patients with autoimmune diseases are often correlated to disease progression. Therefore, anti-glycan antibodies can potentially be used in the detection and prognostic testing of autoimmune diseases.

The development of glycan microarray technology has allowed for the systematic screening of serum samples of patients with autoimmune diseases. This has led to the discovery of anti-glycan antibodies as potential diagnostic and prognostic markers for some autoimmune diseases such as Crohn's disease (CD) and multiple sclerosis. Profiling anti-glycan antibodies in patients with chronic inflammatory bowel disease (IBD) has led to the identification of anti-glycan antibodies as biomarkers for CD. Dotan et al. reported the use of glycan arrays to identify novel anti-glycan antibodies against laminaribioside ($P < 0.001$) and chitobiosides ($P < 0.05$) that are specifically associated with CD.[81] Antibodies against laminaribioside, chitobiosides, or mannan can be sued to differentiated CD patients from ulcerative colitis (UC) patients with over 99% specificity when two of these markers are combined. The results of this study were further validated using an enzyme-linked immunosorbent assay.

Multiple sclerosis (MS) is another autoimmune disease that affects the central nervous system as a result of damage to the myelin sheath.[82,83] A number of research groups have attempted to identify potential biomarkers for the diagnosis and prognosis of MS. Miller and colleagues used glycan microarray to profile serum antibodies of patients with relapsing-remitting multiple sclerosis (RRMS). They reported significantly higher IgM anti-glycan antibodies to Glcα1-4Glcα ($P < 0.0001$) in patients with RRMS in comparison to patients suffering from other neurological diseases.[84] The results from this study were further validated in a separate study using enzyme-linked immunoassay suggesting that anti-Glcα1-4Glcα IgM can be used as a potential biomarker for the diagnosis or prognosis of MS.[85] Furthermore, Grader-Beck et al. reported the presence of distinct antibodies to 4-sulfated LacNAc [4S-LacNAc, 4-(OSO_3)Galβ1-4GlcNAc] in patients with systemic sclerosis ($P = 0.02$).[86] The presence of antibodies to 4S-LacNAc was also associated with a high prevalence of pulmonary hypertension. This suggests that post-glycosylation modifications of glycans can possibly be immunogenic and may play a key role in disease progression.

4.4 Vaccine development

Glycan microarrays are very useful tools for profiling immune responses induced by therapeutics glycan conjugate or glycan-based vaccines. Blixt and co-workers used glycan microarrays to monitor immune response to MUC1 conjugated vaccine in cancer patients and detect the presence of vaccine-induced antibodies to MUC1 glycopeptides.[87] Reactive antibodies to Tn-MUC1 epitope were only present in vaccinated patients. Globo-H based vaccines have been developed and made it to clinical trials. A glycan microarray study indicated that antibody levels against Globo-H were higher in patients with breast cancer.[23] Comparing immune response to Globo-H vaccine candidates in mice showed higher antibody levels for the glycoconjugates with enhanced specificity.[88] Zhang et al. utilized glycan microarrays to evaluate serum antibodies in patients with advanced prostate cancer before vaccination and 2–3 months after vaccination with a poxvirus-based vaccine (PROSTVAC-VF).[89] The results indicated changes in antibody levels to a number of antigens including forssman antigen and blood group A antigens. The PROSTVAC-VF vaccine has reached advanced stages in clinical trials.[90–92] A follow-up study reported that the pre-vaccination IgM levels to blood group A trisaccharide may be used to predict survival for PROSTVAC-VF vaccine.[93]

The development of vaccines against pathogens is an ongoing effort. For example, anthrax-producing bacterium (*Bacillus anthracis*) contains a tetrasaccharide composed of three rhamnose and terminal anthrose on the spore surface of BC1A glycoprotein. Wang et al. used glycan microarrays to show that vaccine-induced rabbit anti-anthrax antibodies bind strongly to the trisaccharide (Antβ1-3-L-Rhapα1-3-L-Rhap) and tetrasaccharides (Antβ1-3-L-Rhapα1-3-L-Rhapα1-2-L-Rhap).[94] The results illustrated that anthrose-containing oligosaccharides are immunogenic and are potential candidates for developing anthrax vaccines. Glycan microarrays were also utilized in monitoring immune responses to inactivated SARS-coronavirus (SARS-CoV) vaccine.[95] They reported significant levels of IgG antibody levels to the human serum glycoprotein asialo-orosomucoid, indicating that the vaccine can elicit a glycan-dependent autoimmune response. Moreover, glycan microarrays have been used to monitor immune responses in mice to a synthetic fragment of the capsular polysaccharide (PS-II) of Gram-positive bacteria *Clostridium difficile* that was conjugated to an immunogenic carrier protein.[96] The vaccine-induced antibodies in mice that specifically interacted with the synthetic fragment of PS-II. The same array was utilized to detect IgA antibodies in the stool of infected hospital patients suggesting that the hexasaccharide fragment of PS-II is a key target of the immune response produced in infected subjects.

Viral envelope glycoproteins are known to mask the virus from the host immune system and are believed to play a key role in viral infections.[97] Therefore, developing vaccines to fight viral infections is a continuing effort. For example, broadly neutralizing antibodies including 2G12 have shown anti-HIV activity and can protect against HIV infection in macaques.[98,99] Glycan microarray containing high-mannose oligosaccharides revealed that 2G12 specifically binds high mannose *N*-linked glycans containing Manα1-2Man determinants including Man8 and Man9.[100] Glycan microarrays were used to monitor immune responses to see if 2G12-like antibodies are produced in vaccinated animals.[101] Rabbits immunized with a mutant strain of *Saccharomyces cerevisiea* induced antibodies that recognized a broad range of HIV-1 and SIV envelope glycoproteins. Additionally, rabbits immunized with *Saccharomyces cerevisiea* deficient with α1-3mannosyltransferase produced sera that contained antibodies that with similar specificity to that of 2G12.[102] Another study evaluating anti-glycan antibody responses to an SIV vaccine and SIV infection demonstrate the potential use of glycan microarrays in vaccine development.[103]

4.5 Enzyme inhibitors

Although it's not well explored, another important application of glycan microarrays is for the screening of potential inhibitors for glycan-processing enzymes including those that are known to be complicated in the biosynthesis of disease-related glycans. For example, a microtiter array was used to screen for fucosyl transferase inhibitors.[104] The reported four potential fucosyltransferase inhibitors with inhibition constants in the nanomolar range. Fucosyltransferases transfer fucose to sialylLacNAc to form sialyl Lewis X, which is an antigen that is involved in inflammation. Blixt et al. used glycan microarray to screen various recombinant sialyltransferase to determine acceptor specificities and made a series of unnatural sialosides.[105]

5 Conclusions

Glycan microarray is a powerful and transformative high throughput tool for studying protein-carbohydrate interactions. It has been extensively used for study the binding interactions, activities, and specificities of various glycan-binding proteins. Glycan microarrays have been used in a variety of biomedical applications, however, the potential of this technology has not been optimized. This is in part due to the limited number of glycans available compare to the glycan diversity found in nature. For example, only small fractions of the human glycan repertoire are incorporated into the arrays. With developments of new chemical and enzymatic methods for the synthesis of complexed carbohydrates and more advances of current microarray technology, more applications will be explored.

References

1. Seeberger, P. H. *Nature* **2005**, *437*, 1239.
2. Koeller, K. M.; Wong, C. H. *Nature Biotech.* **2000**, *18*, 835–841.
3. Dube, D. H.; Bertozzi, C. R. *Nat. Rev, Drug Disc.* **2005**, *4*, 477–488.
4. Varki, A.; Etzler, M. E.; Cummings, R. D.; Esko, J. D. Discovery and classification of glycan-binding proteins. In *Essentials of Glycobiology*; Varki, A., Cummings, R. D., Esko, J. D., Freeze, H. H., Stanley, P., Bertozzi, C. R., Hart, G. W., Etzler, M. E., Eds.; Cold Spring Harbor (NY), 2009.
5. Shinohara, Y.; Kim, F.; Shimizu, M.; Goto, M.; Tosu, M.; Hasegawa, Y. *Eur. J. Biochem./FEBS* **1994**, *223*, 189–194.
6. Dam, T. K.; Brewer, C. F. *Methods Enzymol.* **2004**, *379*, 107–128.
7. Hirabayashi, J.; Hashidate, T.; Arata, Y.; Nishi, N.; Nakamura, T.; Hirashima, M.; Urashima, T.; Oka, T.; Futai, M.; Muller, W. E.; Yagi, F.; Kasai, K. *Biochim. Biophy. Acta* **2002**, *1572*, 232–254.
8. Oyelaran, O.; Gildersleeve, J. C. *Curr. Opin. Chem. Biol.* **2009**, *13*, 406–413.
9. de Paz, J. L.; Seeberger, P. H. *Methods Molecular Biol.* **2012**, *808*, 1–12.
10. Rillahan, C. D.; Paulson, J. C. *Annual Rev. Biochem.* **2011**, *80*, 797–823.
11. Lonardi, E.; Balog, C. I.; Deelder, A. M.; Wuhrer, M. *Exp. Rev. Proteomics* **2010**, 7, 761–774.
12. Park, S.; Gildersleeve, J. C.; Blixt, O.; Shin, I. *Chem. Soc. Rev.* **2013**, *42*, 4310–4326.
13. Kline, K. A.; Falker, S.; Dahlberg, S.; Normark, S.; Henriques-Normark, B. *Cell Host Microbe* **2009**, *5*, 580–592.
14. Wang, D.; Liu, S.; Trummer, B. J.; Deng, C.; Wang, A. *Nat. Biotech.* **2002**, *20*, 275–281.
15. Fukui, S.; Feizi, T.; Galustian, C.; Lawson, A. M.; Chai, W. *Nat. Biotech.* **2002**, *20*, 1011–1017.
16. Feizi, T.; Chai, W. *Nat. Rev. Mol. Cell. Biol.* **2004**, *5*, 582–588.
17. Shipp, E. L.; Hsieh-Wilson, L. C. *Chem. Biol.* **2007**, *14*, 195–208.
18. Ko, K. S.; Jaipuri, F. A.; Pohl, N. L. *J. Am. Chem. Soc.* **2005**, *127*, 13162–13163.
19. Chen, G. S.; Pohl, N. L. *Org. Lett.* **2008**, *10*, 785–788.
20. Guo, Y.; Feinberg, H.; Conroy, E.; Mitchell, D. A.; Alvarez, R.; Blixt, O.; Taylor, M. E.; Weis, W. I.; Drickamer, K. *Nat. Struct. Mol. Biol.* **2004**, *11*, 591–598.
21. Bochner, B. S.; Alvarez, R. A.; Mehta, P.; Bovin, N.V.; Blixt, O.; White, J. R.; Schnaar, R. L. *J. Biol. Chem.* **2005**, *280*, 4307–4312.

22. Chevolot, Y.; Bouillon, C.; Vidal, S.; Morvan, F.; Meyer, A.; Cloarec, J. P.; Jochum, A.; Praly, J. P.; Vasseur, J. J.; Souteyrand, E. *Angew. Chem. Int. Ed. Engl.* **2007**, *46*, 2398–2402.
23. Wang, C. C.; Huang, Y. L.; Ren, C. T.; Lin, C. W.; Hung, J. T.; Yu, J. C.; Yu, A. L.; Wu, C. Y.; Wong, C. H. *Proc. Natl. Acad. Sci. U. S. A.* **2008**, *105*, 11661–11666.
24. Song, X.; Xia, B.; Stowell, S. R.; Lasanajak, Y.; Smith, D. F.; Cummings, R. D. *Chem. Biol.* **2009**, *16*, 36–47.
25. Xia, B.; Kawar, Z. S.; Ju, T.; Alvarez, R. A.; Sachdev, G. P.; Cummings, R. D. *Nat. Methods* **2005**, *2*, 845–850.
26. Gama, C. I.; Tully, S. E.; Sotogaku, N.; Clark, P. M.; Rawat, M.; Vaidehi, N.; Goddard, W. A., 3rd.; Nishi, A.; Hsieh-Wilson, L. C. *Nat. Chem. Biol.* **2006**, *2*, 467–473.
27. Park, S.; Shin, I. *Angew. Chem. Int. Ed. Engl.* **2002**, *41*, 3180–3182.
28. Park, S.; Lee, M. R.; Pyo, S. J.; Shin, I. *J. Am. Chem. Soc.* **2004**, *126*, 4812–4819.
29. Smith, E. A.; Thomas, W. D.; Kiessling, L. L.; Corn, R. M. *J. Am. Chem. Soc.* **2003**, *125*, 6140–6148.
30. Harris, L. G.; Schofield, W. C.; Doores, K. J.; Davis, B. G.; Badyal, J. P. J. *Am. Chem. Soc.* **2009**, *131*, 7755–7761.
31. Love, J. C.; Estroff, L. A.; Kriebel, J. K.; Nuzzo, R. G.; Whitesides, G. M. *Chem. Rev.* **2005**, *105*, 1103–1169.
32. Calarese, D. A.; Lee, H. K.; Huang, C. Y.; Best, M. D.; Astronomo, R. D.; Stanfield, R. L.; Katinger, H.; Burton, D. R.; Wong, C. H.; Wilson, I. A. *Proc. Natl. Acad. Sci. U. S. A.* **2005**, *102*, 13372–13377.
33. Manimala, J. C.; Li, Z.; Jain, A.; VedBrat, S.; Gildersleeve, J. C. *Chem. Biochem.* **2005**, *6*, 2229–2241.
34. Lee, M. R.; Shin, I. *Org. Lett.* **2005**, 7, 4269–4272.
35. Zhi, Z. L.; Powell, A. K.; Turnbull, J. E. *Anal. Chem.* **2006**, *78*, 4786–4793.
36. Song, E. H.; Pohl, N. L. *Curr. Opin. Chem. Biol.* **2009**, *13*, 626–632.
37. Liang, P. H.; Wang, S. K.; Wong, C. H. *J. Am. Chem. Soc.* **2007**, *129*, 11177–11184.
38. de Paz, J. L.; Noti, C.; Bohm, F.; Werner, S.; Seeberger, P. H. *Chem. Biol.* **2007**, *14*, 879–887.
39. Muthana, S. M.; Gildersleeve, J. C. *Cancer Biomark* **2014**, *14*, 29–41.
40. Miyoshi, E.; Moriwaki, K.; Nakagawa, T. *J. Biochem.* **2008**, *143*, 725–729.
41. Burchell, J.; Poulsom, R.; Hanby, A.; Whitehouse, C.; Cooper, L.; Clausen, H.; Miles, D.; Taylor-Papadimitriou, J. *J. Glycobiol.* **1999**, *9*, 1307–1311.
42. Sewell, R.; Backstrom, M.; Dalziel, M.; Gschmeissner, S.; Karlsson, H.; Noll, T.; Gatgens, J.; Clausen, H.; Hansson, G. C.; Burchell, J.; Taylor-Papadimitriou, J. *J. Biol. Chem.* **2006**, *281*, 3586–3594.
43. Chiricolo, M.; Malagolini, N.; Bonfiglioli, S.; Dall'Olio, F. *J. Glycobiol.* **2006**, *16*, 146–154.
44. Granovsky, M.; Fata, J.; Pawling, J.; Muller, W. J.; Khokha, R.; Dennis, J. W. *Nat. Med.* **2000**, *6*, 306–312.
45. Brockhausen, I. *Biochim. Bio. Acta* **1999**, *1473*, 67–95.
46. Huang, C. Y.; Thayer, D. A.; Chang, A. Y.; Best, M. D.; Hoffmann, J.; Head, S.; Wong, C. H. *Proc. Natl. Acad. Sci. U. S. A* **2006**, *103*, 15–20.
47. Wang, C. C.; Huang, Y. L.; Ren, C. T.; Lin, C. W.; Hung, J. T.; Yu, J. C.; Yu, A. L.; Wu, C. Y.; Wong, C. H. *Proc. Natl. Acad. Sci. U. S. A.* **2008**, *105*, 11661–11666.
48. Wandall, H. H.; Blixt, O.; Tarp, M. A.; Pedersen, J. W.; Bennett, E. P.; Mandel, U.; Ragupathi, G.; Livingston, P. O.; Hollingsworth, M. A.; Taylor-Papadimitriou, J.; Burchell, J.; Clausen, H. *Cancer Res.* **2010**, *70*, 1306–1313.
49. Pedersen, J. W.; Blixt, O.; Bennett, E. P.; Tarp, M. A.; Dar, I.; Mandel, U.; Poulsen, S. S.; Pedersen, A. E.; Rasmussen, S.; Jess, P.; Clausen, H.; Wandall, H. H. *Int. J. Cancer* **2011**, *128*, 1860–1871.
50. Springer, G. F. *J. Mol. Med.* **1997**, *75*, 594–602.

51. Desai, P. R. *Trans. Med. Rev.* **2000**, *14*, 312–325.
52. Springer, G. F.; Desai, P. R.; Spencer, B. D.; Tegtmeyer, H.; Carlstedt, S. C.; Scanlon, E. F. *Cancer Detect. Prev.* **1995**, *19*, 374–380.
53. Slovin, S. F.; Ragupathi, G.; Musselli, C.; Olkiewicz, K.; Verbel, D.; Kuduk, S. D.; Schwarz, J. B.; Sames, D.; Danishefsky, S.; Livingston, P. O.; Scher, H. I. *J. Clinic. Oncol.* **2003**, *21*, 4292–4298.
54. Slovin, S. F.; Ragupathi, G.; Fernandez, C.; Diani, M.; Jefferson, M. P.; Wilton, A.; Kelly, W. K.; Morris, M.; Solit, D.; Clausen, H.; Livingston, P.; Scher, H. I. *Cancer Immunol., Immunotherapy:CII* **2007**, *56*, 1921–1930.
55. Blixt, O.; Bueti, D.; Burford, B.; Allen, D.; Julien, S.; Hollingsworth, M.; Gammerman, A.; Fentiman, I.; Taylor-Papadimitriou, J.; Burchell, J. M. *Breast Cancer Res.:BCR* **2011**, *13*, R25.
56. Jacob, F.; Goldstein, D. R.; Bovin, N.V.; Pochechueva, T.; Spengler, M.; Caduff, R.; Fink, D.; Vuskovic, M. I.; Huflejt, M. E.; Heinzelmann-Schwarz, V. *J. Int. Cancer* **2012**, *130*, 138–146.
57. Vuskovic, M. I.; Xu, H.; Bovin, N.V.; Pass, H. I.; Huflejt, M. E. *Int. J Bioinformatics Res. Appl.* **2011**, 7, 402–426.
58. Lawrie, C. H.; Marafioti, T.; Hatton, C. S.; Dirnhofer, S.; Roncador, G.; Went, P.; Tzankov, A.; Pileri, S. A.; Pulford, K.; Banham, A. H. *J. Int. Cancer* **2006**, *118*, 3161–3166.
59. Malykh, Y. N.; Schauer, R.; Shaw, L. *Biochimie* **2001**, *83*, 623–634.
60. Tangvoranuntakul, P.; Gagneux, P.; Diaz, S.; Bardor, M.; Varki, N.; Varki, A.; Muchmore, E. *Proc. Natl. Acad. Sci. U. S. A.* **2003**, *100*, 12045–12050.
61. Padler-Karavani, V.; Hurtado-Ziola, N.; Pu, M.; Yu, H.; Huang, S.; Muthana, S.; Chokhawala, H. A.; Cao, H.; Secrest, P.; Friedmann-Morvinski, D.; Singer, O.; Ghaderi, D.; Verma, I. M.; Liu, Y. T.; Messer, K.; Chen, X.; Varki, A.; Schwab, R. *Cancer Res.* **2011**, *71*, 3352–3363.
62. Wang, D. *J. Proteomics Bioinform.* **2012**, *5*, 90–95.
63. Aranzamendi, C.; Tefsen, B.; Jansen, M.; Chiumiento, L.; Bruschi, F.; Kortbeek, T.; Smith, D. F.; Cummings, R. D.; Pinelli, E.; Van Die, I. *Exp. Parasitol.* **2011**, *129*, 221–226.
64. Kamena, F.; Tamborrini, M.; Liu, X.; Kwon, Y. U.; Thompson, F.; Pluschke, G.; Seeberger, P. H. *Nat. Chem. Biol.* **2008**, *4*, 238–240.
65. Heimburg-Molinaro, J.; Priest, J. W.; Live, D.; Boons, G. J.; Song, X.; Cummings, R. D.; Mead, J. R. *Int. J. Parasitol.* **2013**, *43*, 901–907.
66. Stevens, J.; Blixt, O.; Paulson, J. C.; Wilson, I. A. *Nat. Rev. Microbiol.* **2006**, *4*, 857–864.
67. Stevens, J.; Blixt, O.; Glaser, L.; Taubenberger, J. K.; Palese, P.; Paulson, J. C.; Wilson, I. A. *J. Mol. Biol.* **2006**, *355*, 1143–1155.
68. Stevens, J.; Blixt, O.; Chen, L. M.; Donis, R. O.; Paulson, J. C.; Wilson, I. A. *J. Mol. Biol.* **2008**, *381*, 1382–1394.
69. Maines, T. R.; Jayaraman, A.; Belser, J. A.; Wadford, D. A.; Pappas, C.; Zeng, H.; Gustin, K. M.; Pearce, M. B.; Viswanathan, K.; Shriver, Z. H.; Raman, R.; Cox, N. J.; Sasisekharan, R.; Katz, J. M.; Tumpey, T. M. *Science* **2009**, *325*, 484–487.
70. Song, X.; Yu, H.; Chen, X.; Lasanajak, Y.; Tappert, M. M.; Air, G. M.; Tiwari, V. K.; Cao, H.; Chokhawala, H. A.; Zheng, H.; Cummings, R. D.; Smith, D. F. *J. Biol. Chem.* **2011**, *286*, 31610–31622.
71. Clo, E.; Kracun, S. K.; Nudelman, A. S.; Jensen, K. J.; Liljeqvist, J. A.; Olofsson, S.; Bergstrom, T.; Blixt, O. *J. Virol.* **2012**, *86*, 6268–6278.
72. Wu, C. S.; Lee, T. Y.; Chou, R. H.; Yen, C. J.; Huang, W. C.; Wu, C. Y.; Yu, Y. L. *PLoS One* **2014**, *9*, e99959.
73. Parthasarathy, N.; DeShazer, D.; England, M.; Waag, D. M. *Diagn. Micr. Infec. Dis.* **2006**, *56*, 329–332.
74. Parthasarathy, N.; Saksena, R.; Kovac, P.; Deshazer, D.; Peacock, S. J.; Wuthiekanun, V.; Heine, H. S.; Friedlander, A. M.; Cote, C. K.; Welkos, S. L.; Adamovicz, J. J.; Bavari, S.; Waag, D. M. *Carbohydr. Res.* **2008**, *343*, 2783–2788.

75. Thirumalapura, N. R.; Morton, R. J.; Ramachandran, A.; Malayer, J. R. *J. Immunol. Methods* **2005**, *298*, 73–81.
76. Blit, O.; Hoffmann, J.; Svenson, S.; Norberg, T. *J. Glycoconj.* **2008**, *25*, 27–36.
77. Disney, M. D.; Seeberger, P. H. *Chem. Biol.* **2004**, *11*, 1701–1707.
78. Song, X.; Lasanajak, Y.; Xia, B.; Heimburg-Molinaro, J.; Rhea, J. M.; Ju, H.; Zhao, C.; Molinaro, R. J.; Cummings, R. D.; Smith, D. F. *Nat. Methods* **2011**, *8*, 85–90.
79. Firestein, G. S. *Nature* **2003**, *423*, 356–361.
80. Hueber, W.; Robinson, W. H. *Proteomics* **2006**, *6*, 4100–4105.
81. Dotan, I.; Fishman, S.; Dgani, Y.; Schwartz, M.; Karban, A.; Lerner, A.; Weishauss, O.; Spector, L.; Shtevi, A.; Altstock, R. T.; Dotan, N.; Halpern, Z. *Gastroenterology* **2006**, *131*, 366–378.
82. Trapp, B. D.; Peterson, J.; Ransohoff, R. M.; Rudick, R.; Mork, S.; Bo, L. *N. Engl. J. Med.* **1998**, *338*, 278–285.
83. Lassmann, H.; Bruck, W.; Lucchinetti, C. *Trends Mol. Med.* **2001**, *7*, 115–121.
84. Schwarz, M.; Spector, L.; Gortler, M.; Weisshaus, O.; Glass-Marmor, L.; Karni, A.; Dotan, N.; Miller, A. *J. Neurol. Sci.* **2006**, *244*, 59–68.
85. Brettschneider, J.; Jaskowski, T. D.; Tumani, H.; Abdul, S.; Husebye, D.; Seraj, H.; Hill, H. R.; Fire, E.; Spector, L.; Yarden, J.; Dotan, N.; Rose, J. W. *J. Neuroimmunol.* **2009**, *217*, 95–101.
86. Grader-Beck, T.; Boin, F.; von Gunten, S.; Smith, D.; Rosen, A.; Bochner, B. S. *Ann. Rheum. Dis.* **2011**, *70*, 2218–2224.
87. Blixt, O.; Clo, E.; Nudelman, A. S.; Sorensen, K. K.; Clausen, T.; Wandall, H. H.; Livingston, P. O.; Clausen, H.; Jensen, K. J. *J. Proteome Res.* **2010**, *9*, 5250–5261.
88. Huang, Y. L.; Hung, J. T.; Cheung, S. K.; Lee, H. Y.; Chu, K. C.; Li, S. T.; Lin, Y. C.; Ren, C. T.; Cheng, T. J.; Hsu, T. L.; Yu, A. L.; Wu, C. Y.; Wong, C. H. *Proc. Natl. Acad. Sci. U.S.A.* **2013**, *110*, 2517–2522.
89. Zhang, Y.; Campbell, C.; Li, Q.; Gildersleeve, J. C. *Mol. Biosyst.* **2010**, *6*, 1583–1591.
90. Gulley, J. L.; Arlen, P. M.; Madan, R. A.; Tsang, K. Y.; Pazdur, M. P.; Skarupa, L.; Jones, J. L.; Poole, D. J.; Higgins, J. P.; Hodge, J. W.; Cereda, V.; Vergati, M.; Steinberg, S. M.; Halabi, S.; Jones, E.; Chen, C.; Parnes, H.; Wright, J. J.; Dahut, W. L.; Schlom, J. *Cancer Immunol., Immunotherapy: CII* **2010**, *59*, 663–674.
91. Lechleider, R. J.; Arlen, P. M.; Tsang, K. Y.; Steinberg, S. M.; Yokokawa, J.; Cereda, V.; Camphausen, K.; Schlom, J.; Dahut, W. L.; Gulley, J. L. *Clin. Cancer Res.* **2008**, *14*, 5284–5291.
92. Madan, R. A.; Gulley, J. L.; Schlom, J.; Steinberg, S. M.; Liewehr, D. J.; Dahut, W. L.; Arlen, P. M. *Clin. Cancer Res.* **2008**, *14*, 4526–4531.
93. Campbell, C. T.; Gulley, J. L.; Oyelaran, O.; Hodge, J. W.; Schlom, J.; Gildersleeve, J. C. *Clin. Cancer Res.* **2013**, *19*, 1290–1299.
94. Wang, D.; Carroll, G. T.; Turro, N. J.; Koberstein, J. T.; Kovac, P.; Saksena, R.; Adamo, R.; Herzenberg, L. A.; Steinman, L. *Proteomics* **2007**, *7*, 180–184.
95. Wang, D.; Lu, J. *Physiol. Genomics* **2004**, *18*, 245–248.
96. Oberli, M. A.; Hecht, M. L.; Bindschadler, P.; Adibekian, A.; Adam, T.; Seeberger, P. H. *Chem. Biol.* **2011**, *18*, 580–588.
97. Olofsson, S.; Hansen, J. E. *J. Infect. Dis.* **1998**, *30*, 435–440.
98. Trkola, A.; Purtscher, M.; Muster, T.; Ballaun, C.; Buchacher, A.; Sullivan, N.; Srinivasan, K.; Sodroski, J.; Moore, J. P.; Katinger, H. *J. Virol.* **1996**, *70*, 1100–1108.
99. Mascola, J. R.; Stiegler, G.; VanCott, T. C.; Katinger, H.; Carpenter, C. B.; Hanson, C. E.; Beary, H.; Hayes, D.; Frankel, S. S.; Birx, D. L.; Lewis, M. G. *Nat. Med.* **2000**, *6*, 207–210.
100. Adams, E. W.; Ratner, D. M.; Bokesch, H. R.; McMahon, J. B.; O'Keefe, B. R.; Seeberger, P. H. *Chem. Biol.* **2004**, *11*, 875–881.
101. Luallen, R. J.; Lin, J.; Fu, H.; Cai, K. K.; Agrawal, C.; Mboudjeka, I.; Lee, F. H.; Montefiori, D.; Smith, D. F.; Doms, R. W.; Geng, Y. *J. Virol.* **2008**, *82*, 6447–6457.

102. Dunlop, D. C.; Bonomelli, C.; Mansab, F.; Vasiljevic, S.; Doores, K. J.; Wormald, M. R.; Palma, A. S.; Feizi, T.; Harvey, D. J.; Dwek, R. A.; Crispin, M.; Scanlan, C. N. *J. Glycobio.* **2010**, *20*, 812–823.
103. Campbell, C. T.; Llewellyn, S. R.; Damberg, T.; Morgan, I. L.; Robert-Guroff, M.; Gildersleeve, J. C. *PloS one* **2013**, *8*, e75302.
104. Bryan, M. C.; Lee, L.V.; Wong, C. H. *Bioorg. Med. Chem. Lett.* **2004**, *14*, 3185–3188.
105. Blixt, O.; Han, S.; Liao, L.; Zeng, Y.; Hoffmann, J.; Futakawa, S.; Paulson, J. C. *J. Am. Chem. Soc.* **2008**, *130*, 6680–6681.

CHAPTER SEVEN

Recent developments in the synthesis of biologically relevant inositol derivatives

Mysore S. Shashidhar[a,b], Nivedita T. Patil[a]
[a]Division of Organic Chemistry, National Chemical Laboratory, Pune, India
[b]The Academy of Scientific and Innovative Research, Ghaziabad, India

1 Introduction

Inositol was an inconspicuous chemical entity in scientific literature at the beginning of the 20th century but became conspicuous in the 1960s and very conspicuous by the end of the century. Fig. 7.1 show the variation in the number of papers published (related to inositols) in journals between 1900 and 2018. Currently the publishing activity centered around inositol seems to have leveled off. The chemistry and biology of phosphoinositols began appearing in reviews and text books, perhaps 3–4 decades ago.[1–5]

The reason for this increase in the pace of research related to inositols and their derivatives was mainly due to the realization of the involvement of phosphoinositols in cellular processes, primarily, signal transduction mechanisms, in eukaryotic cells. Cellular signal transduction processes are crucial for the survival and maintenance of balance in living organisms and to make sure that different groups of cells in a multicellular organism work in coordination and don't hurt each other. However, if communication between cells does get affected, corrective measures are required to restore the optimum level of communication that is necessary to maintain the balance. These corrective measures invariably interfere with the communication systems constructively and restore the balance. When the corrective measure used is external (to a living cell) chemical entity, the process is referred to as "pharmacological intervention" of a biological process. Research related to the biology and chemistry of inositols led more or less to an explosive growth in research, in the last few decades, due to hope and optimism that certain inositol derivatives or their analogs could have the ability to interfere beneficially in cellular signal transduction mechanisms and hence result in the discovery of drugs of the future, for the treatment of a variety

Carbohydrates in Drug Discovery and Development. http://dx.doi.org/10.1016/B978-0-12-816675-8.00007-5

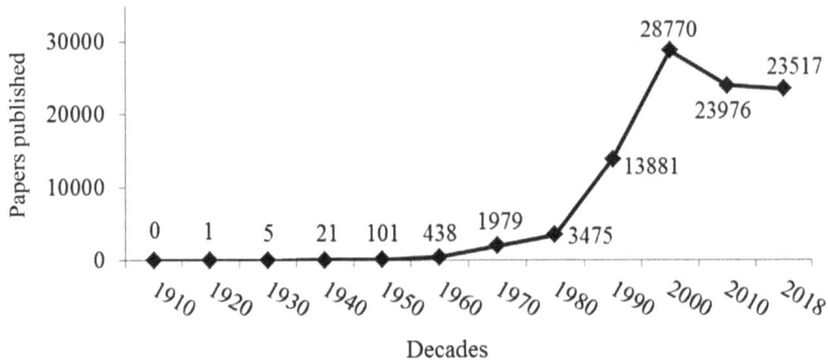

Figure 7.1 ***Inositol related publications between 1900 and present showing vigorous activity after 1980.*** Each year on the *X*-axis represents the end of the decade in which the number of papers shown on each square were published. The number of publications were estimated using 'inositol' as the search word in Scifinder.

of ailments such as diabetes, cancer, *etc.* as well as certain genetic disorders. Due to the current high level of interest in the chemistry and biology of inositols, especially *myo*-inositol and its derivatives, this chapter sketches the recent developments in the chemistry and synthesis of inositol derivatives and analogs. A short account of the molecular structure and nomenclature of *myo*-inositol and its derivatives is included for the uninitiated. Original journal citations are included as extensively as possible, but not necessarily exhaustively.

2 Structure, nomenclature of *myo*-inositol and associated implications for synthesis

myo-Inositol (**1**), a 1,2,3,4,5,6-hexahydroxy cyclohexane (molecular formula – $C_6H_{12}O_6$) is an isomer of glucose. *myo*-Inositol is a carbocyclic compound while glucose (and other analogous sugars) is a heterocycle - pyran derivative. By virtue of its molecular formula $C_6(H_2O)_6$ *myo*-inositol can be considered as a carbohydrate. Eight other isomeric inositols are known to exist; the difference between these molecules being changes in the relative disposition of the six hydroxyl groups on the cyclohexane ring (Scheme 7.1).

Inositols are also referred to as cyclitols, since the former constitute a subgroup of the latter. Cyclitols are derivatives of cyclohexane containing at least three hydroxyl groups. Hence conversely, cyclitols containing five or lesser number of hydroxyl groups can be considered as deoxygenated inositols. For example, mono deoxygenated inositols (pentahydroxy

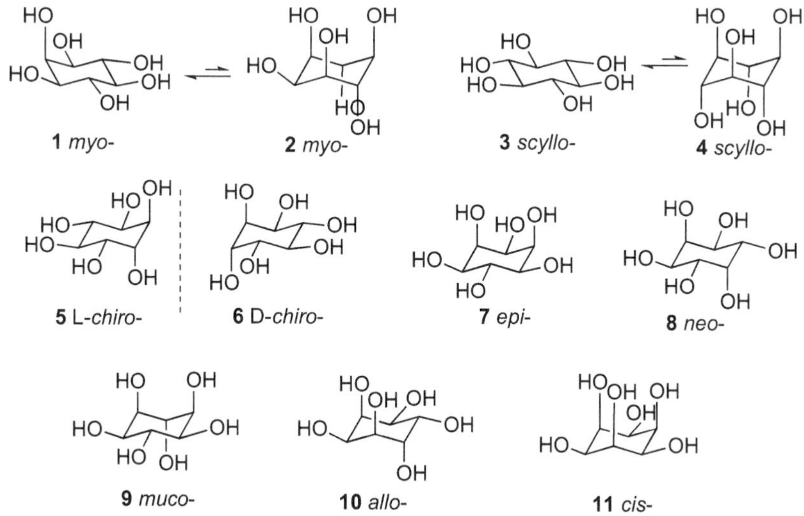

Scheme 7.1 ***Structure of isomeric inositols and ring inverted forms of myo- and scyllo-isomers revealing 'equatorial rich' (1, 3) and 'axial rich' (2, 4) conformations.***

cyclohexanes) are known as quercitols. *myo*-Inositol (**1**) has one axial and five equatorial hydroxyl groups, while other inositols have zero, two or three axially oriented hydroxyl groups. In all the nine known inositols the carbocyclic ring has the chair conformation, in which there are a minimum number of axial hydroxyl groups (and hence they are 'equatorial rich'). The corresponding inverted chair which has minimum number of equatorial hydroxyl groups (and hence they are 'axial rich') is usually higher in energy and hence not usually stable. *cis*-Inositol (**11**) is a special case, in which there are three axial and equatorial hydroxyl groups each; hence inversion of the carbocyclic ring results in a degenerate structure. However, certain derivatives of *myo*- and *scyllo*-inositols exist in the 'axial rich' conformation (Scheme 7.2).[6,7] Phytic acid (**15** *myo*-inositol hexakisphosphate) exists as an equilibrium mixture of two conformations, relative proportion of which is pH dependent.[8] Based on ENDOR spectroscopic studies, binding of *myo*-inositol in an axial-rich conformation to *myo*-inositol oxygenase has been suggested.[9] The change in conformation (from 'equatorial rich' to 'axial rich') of the carbocyclic ring leads to change in relative reactivity of the hydroxyl groups, which has been exploited for the synthesis of inositol derivatives.[10]

As expected, inositol isomers (Scheme 7.1) are chemically interconvertible, with the aid of suitable protection – deprotection strategy of the hydroxyl groups, but most such reports pertain to the conversion of the

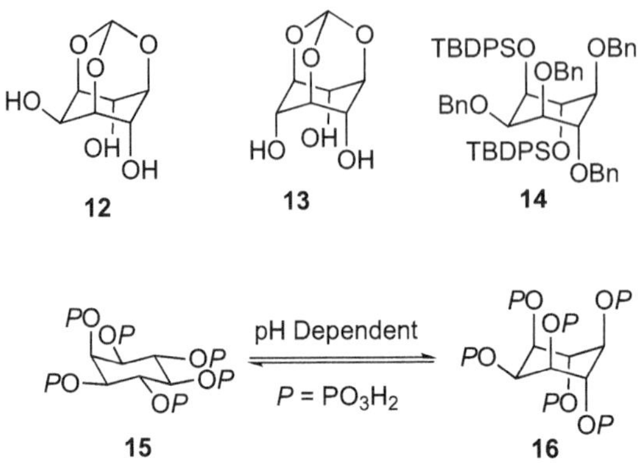

Scheme 7.2 ***Selected inositol derivatives known to exist in the 'axial rich' conformation.***

abundantly available *myo*-isomer to other isomers.[11–15] However, direct epimerization of inositols and their derivatives have also been reported[16–19] although, they might not be of practical synthetic utility. Interestingly, inositols as a family of molecules provide a platform to examine how subtle variations in molecular structure could markedly influence the (physical, chemical, biological and material) properties.[20,21] Inositols can be represented by three different structural formulae: (1) Haworth formula **(17)**; (2) planar carbocyclic ring **(18)**; (3) the chair form **(1)**. Different molecular representations of *myo*-inositol **(1)** are shown as an example, since it is the most frequently encountered inositol in the literature (Fig. 7.2).

The Haworth formula **(17)** shows the relative disposition of the hydroxyl groups (above or below the plane of the ring) and hence their relative stereochemistry; this representation is currently the least encountered in the literature. The planar cyclohexane representation **(18)** is encountered relatively more frequently, but the (axial / equatorial) orientation of the hydroxyl groups with respect to the ring is not obvious. The chair form of the cyclohexane **(1)** clearly reveals the stereochemical disposition of the hydroxyl groups and hence this representation will be used in this chapter. For an interesting and educative discussion on the molecular structure, conformation of *myo*-inositol **(1)**, and the associated conventions for the nomenclature of *myo*-inositol and its derivatives, the reader is referred to the work of Parthasarathy and Eisenberg.[22,23] All the inositols except the *chiro*-isomers **5** and **6**, have the *meso*-configuration (*i.e.* presence of a mirror plane within the molecule, as indicated for **1** in Fig. 7.2) and hence are achiral. Symmetric substitution **21–24** (or derivatization) of the hydroxyl groups of *myo*-inositol (*i.e.* any substitution or derivatization of one or more

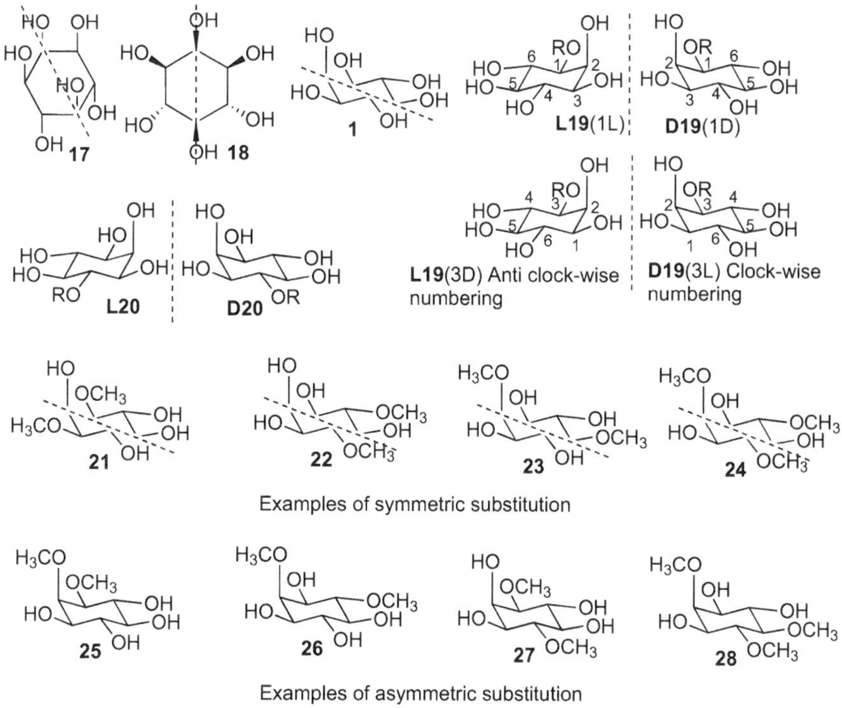

Figure 7.2 ***Different representations of myo-inositol (Haworth* 17*, planar* 18 *and chair* 1*) and generation of enantiomers by asymmetric O-substitution of its different hydroxyl groups* (25-28).** Conventional numbering of the ring carbon atoms for the *D*-configuration shown in **D19**(1D), **L19**(3D) will be adhered to (where necessary) for chiral inositol derivatives. The discontinuous line represents the mirror plane.

hydroxyl groups that leave the mirror plane intact) does not disturb the *meso*-configuration. Asymmetric substitution **25–28** (*i.e.* any substitution or derivatization of one or more hydroxyl groups that eliminate the mirror plane) results in the formation of a chiral molecule due to break down of the molecular symmetry. The products of such reactions are racemic, unless otherwise mediated by chiral agents. In other words, unsymmetrically substituted *myo*-inositol derivatives could exist as enantiomers and as expected this has implications for both the chemistry (laboratory synthesis) and biology (biological activity) of *myo*-inositol derivatives. The convention for designating chiral derivatives of *myo*-inositol as D- or L- is based on the anticlock-wise **(D19)** or clock-wise **(L19)** numbering of the inositol ring carbon atoms.[22,23] A clock-wise numbering [as in **L19**(3D)] leads to designation of the molecular configuration of a chiral inositol derivative as L- while an anticlock-wise numbering [as in **D19**(1D)] leads to designation of the molecular configuration as D-. Consequently, D-1-*O*-substituted *myo*-inositol derivative can be designated as L-3-*O*-substituted *myo*-inositol derivative and *vice-versa*.This could lead to some confusion in the naming of *myo*-inositol derivatives. Hence in order to avoid such ambiguities, IUBAC recommended designation of all optically active *myo*-inositol derivatives with the D- configuration [*i.e.* anticlock-wise numbering as in **D19**(1D)]. Hence, D-1-*O*-substituted *myo*-inositol derivative and D-3-*O*-substituted *myo*-inositol derivatives constitute a pair of enantiomers (**D19** and **L19**, Fig. 7.2). Similarly, D-4-*O*-substituted *myo*-inositol derivative and D-6-*O*-substituted *myo*-inositol derivatives constitute a pair of enantiomers (**D20** and **L20**, Fig. 7.2).

3 Inositols and their derivatives: the biological and medicinal context

Several inositol derivatives have been isolated from plant as well as animal sources and new derivatives continue to appear in the literature,[24–26] but inositols are not known to occur in nature in the free-state. Inositol derivatives are known to occur widely in plants and to a relatively lesser extent in animals. Galactosyl derivatives of inositols are widely distributed in plant seeds.[27] There are indications that their function is protection of membranes and other cellular structures during seed desiccation or storage in the dry state.[28] Certain living systems are capable of accepting isomeric inositol derivatives as substrates and convert them to other products (*e.g.* degradation of *chiro*- and *scyllo*-inositol phosphates by soil bacteria). Illustrative examples of inositol derivatives isolated from natural sources are shown in Fig. 7.3.[26] Mycothiol **(39)** has generated considerable interest due to its role in the cellular physiology of mycobacteria, such as *Mycobacterium tuberculosis*, and is thought to play role in antimycobacterial drug resistance.[29–31]

Inositol containing glyco lipids are known to exhibit anti-inflammatory activity.[24] Interestingly *chiro*- (**5**, **6**) and *scyllo*-inositols (**3**) are known to have beneficial medicinal properties. *scyllo*-Inositol (**3**) is known to decrease the symptoms of the disease and improve cognitive functions in Alzheimer's patients. *scyllo*-inositol (**3**) also inhibits the Aβ42 peptide[32,33] and is classified as an α-syn aggregation inhibitor (ASI).[34] D-*chiro*-Inositol (**6**) is known to act as insulin mimetic and aid those suffering from polycystic ovary syndrome.[35] D-*chiro*-Inositol (**6**) and *myo*-inositol (**1**) are being advocated and sold as dietary supplements.[36–38] *myo*-Inositol (**1**) and D-*chiro*-inositol (**6**) also seem to help maintenance of good health when used as food supplements.

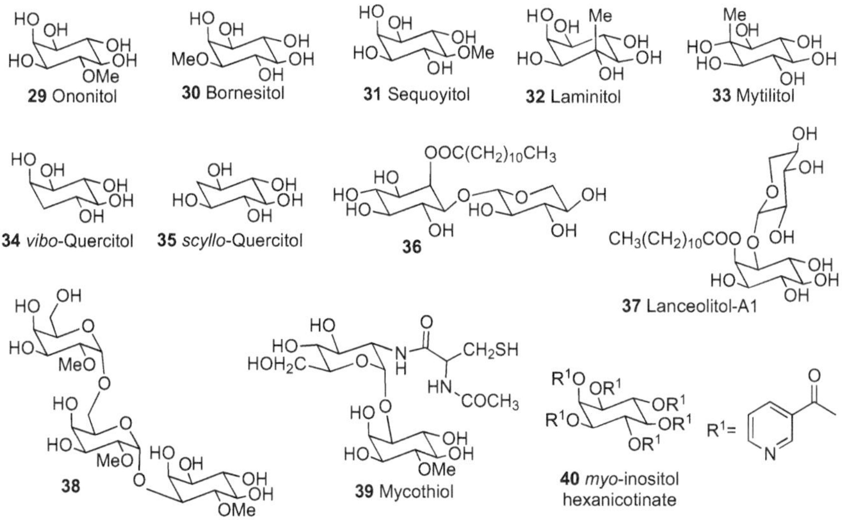

Figure 7.3 ***Naturally occurring* myo-*inositol derivatives showing structural diversity.***

myo-Inositol (**1**) is an ingredient in cosmetics such as moisturizers, sunscreens, hair growth promoters and anti-aging agents. D-*chiro*-Inositol (**6**) helps control gestational diabetes mellitus when used as a food supplement. *myo*-Inositol hexanicotinate (**40**) is an interesting example of an inositol derivative that is being used as a food supplement for the delivery of niacin (vitamin B3). Illustrative examples of cyclitols derivable from inositols other than the *myo*-isomer are shown in Fig. 7.4. *D*-Pinitol (**41**) is an inhibitor of cancer growth.[39] Quinic acid (**47**) is a starting material for the synthesis of pharmaceuticals (*e.g.* Tamiflu)[40] and shikimic acid (**46**) is a biosynthetic intermediate, hygromycincs (**44**) and kanamycin (**48**) are antibiotics.[41,42]

Aminocyclitols, a class of compounds exhibiting a wide range of biological properties (such as antibiotic and antiglycosidase activities) and having the potential to function as sugar mimics are frequently pursued as synthetic targets.[43–48] There is evidence to suggest that problems associated with carbohydrate metabolizing enzymes are linked to the progression of several diseases. Selective inhibitors for carbohydrate metabolizing enzymes promise to be potential therapeutic agents.[49–51] Deoxystreptamine is present in several antibiotics (*e.g.* kanamycin, **48**) and the *neo*-inosamine derivative is present in hygromycin (**44**).[52–54] Investigations on the structure activity relationships of hygromycin (**44**) revealed that the aminocyclitol moiety was crucial for the antibacterial activity.[55–58] Methoxyhygromycin which contains the methoxyaminocyclitol moiety exhibits herbicidal activity.[59]

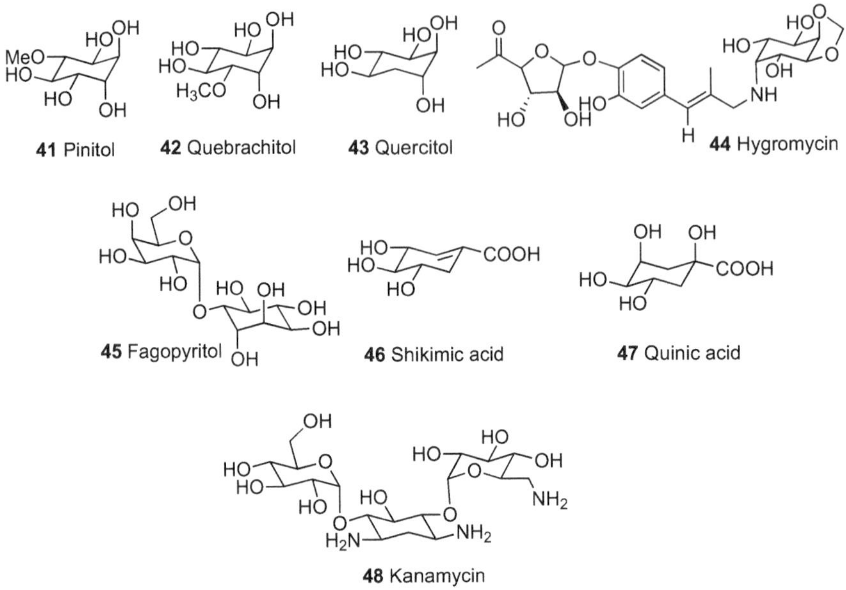

Figure 7.4 ***Naturally occuring cyclitols derived from inositols other than the*** **myo-*****isomer.***

Certain aminocyclitols have shown promise as candidates for the development of therapeutic agents, enzyme inhibitors suitable for biomedical applications, and tools for the delineation of the *myo*-inositol cycle and related cellular processes.[60–65] The aminocyclitol moiety has served as a scaffold for drug design and development; for instance, aminocyclitol containing analogs of nucleosides have been tested as metabolically stable analogs for antiviral drug development.[66] *myo*-Inosamines are known to be potent inhibitors of glucocerebrosidase. The decreased activity of this β-glucosidase enzyme causes the Gaucher disease and hence its inhibitors could be of use in pharmacological chaperone therapy.[67,68] Hence efforts to isolate aminocyclitols from natural sources as well as to synthesize aminocyclitols and their derivatives continue, with the hope of obtaining enhanced activity and selective biological profile suitable for the targeting and intervention of cellular processes.[66,69]

Recent research suggests that *myo*-inositol derivatives, especially inositol phosphates and their lipid derivatives occur in almost all cells, although in smaller amounts. Phosphorylated *myo*-inositol derivatives are ubiquitous in eukaryotic cells and are crucial for their existence and survival. Lipid derivatives of *myo*-inositol are present in the cell membranes while

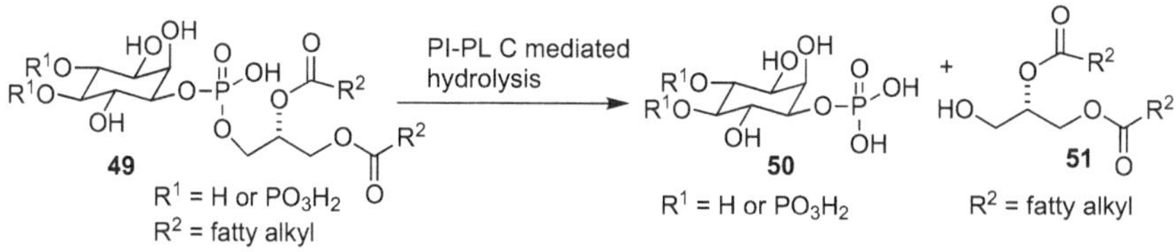

Scheme 7.3 ***Hydrolysis of phosphoinositides* 49 *mediated by PhosphatidylInositol – specific PhosphoLipase C (PI-PLC) generates the corresponding hydrophilic myo-inositol phosphate* 50 *and the hydrophobic diacyl glycerol* 51.**

inositol phosphates are intracellular. Although, phosphatidylinositols (seven are known to occur in nature) form about 15% of total phospholipids in eukaryotic cells, they are metabolically very active. More than thirty inositol phosphates have been identified in cells.[70] Phosphoinositol metabolism is controlled by several kinases, phosphatases and phospholipases and some of the products (*e.g.* D-*myo*-inositol-1,4,5-trisphosphate, **50**, R^1 = PO_3H_2) of the reactions mediated by enzymes are known to function as second messengers in cell signaling pathways. Action of phospholipases on phosphoinositides generates an inositol phosphate and a diacylglycerol (Scheme 7.3) which are poles apart in their nature – the former is a hydrophilic carbohydrate derivative while the latter is a hydrophobic lipid derivative. Consequently, the inositol phosphate is released into the cytoplasm whereas the diacylglycerol (**51**) generated remains in the cell membrane.

Hence this reaction in which a phosphoinositide is hydrolyzed, plays key role in trans-membrane cellular signal transduction pathways. The *X*-ray crystal structure of an engineered phospholipase C enzyme from *Bacillus thuringiensis* has been solved in order to better understand the role of calcium in its catalytic mechanism.[71] Surprisingly, PI-PLC from *Streptomyces antibioticus* cleaves phosphatidylinositol to *myo*-inositol-1,6-cyclic phosphate and diacylglycerol.[72] Phosphoinositols **(52-58)** and *myo*-inositol phosphates (Fig. 7.5) are involved in diverse cellular aspects such as regulation of ion channels, pumps, and transporters as well as in endocytic and exocytic processes. Binding of D-*myo*-inositol -1,4,5-trisphosphate (**50**, R^1 = PO_3H_2) to its receptor is known to mobilize calcium from intracellular sources,[73–76] and the structure of the receptor has been reported;[77] but the exact mechanism of this process is yet to be completely unraveled. D- and L-*myo*-inositol-3,4,5,6-tetrakisphosphates, are bioactive molecules and play important roles in cellular signaling.[78]

Phosphatidylinositol is also involved in the anchoring of certain proteins to cell membranes (*e.g.* variant surface glycoprotein of trypanosomes).[80,81]

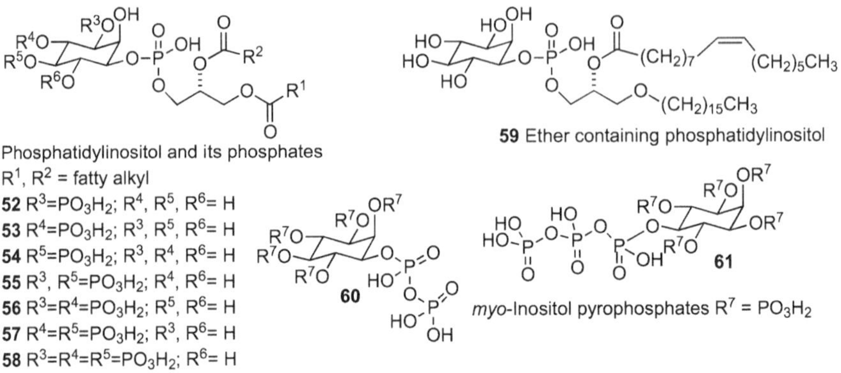

Figure 7.5 ***Illustrative examples of phosphorylated myo-inositol derivatives known to occur in cells.*** An Unusual ether containing phosphatidylinositol **59** was isolated from the slime mold *Dictyostelium discoideum.* [25,79]

These anchors contain several carbohydrate residues (glycosylinositols) between the anchored protein and the lipid embedded in the cell membrane. Lipophosphoglycans are known to play role in the virulence factor of certain parasites, *e.g.* the ceramide phosphoinositol glycan core (CPI-GC) of the lipophosphoglycan of *Trichomonas vaginalis.* These cell surface glycoconjugates are suspected to play key roles in cell - cell interactions, host cell invasion and evasion of host responses.[82–84] Phosphatidylinositol mannosides (PIMs) which are present in mycobacteria are known to have the ability to (a) affect human T-cell proliferation; (b) recruit natural killer T cells; (c) function as immune adjuvants and (d) suppress allergic airway disease.[85] *myo*-Inositol derived pyrophosphates (**61**) have been found to act as allosteric effectors of human hemoglobin, inducing enhanced capacity for oxygen release. Since hypoxia is involved in many types of diseases, the exploration of *myo*-inositol phosphate derivatives opens up possibilities for the discovery of drugs. It is interesting to note that phytic acid (**15**) is a very potent allosteric effector of hemoglobin.[86] *myo*-Inositol-1,3,4,5,6-pentakisphosphate is reported to inhibit PI 3-K/Akt pathway and exhibit pro-apoptotic and antiangiogenic properties. Hence it is projected to be a potential anticancer agent.[87] D-3-Deoxyphosphatidylinositol derivatives exhibit cytotoxic activity against several human cancer cell lines.[88] Diphosphoinositols (inositol-derived pyrophosphates) have been identified in cells, and now are known to be ubiquitous in eukaryotic cells. Diphosphoinositols are biosynthesized by the phosphorylation of phytic acid (**15**). Research in the last two decades suggests that these high energy molecules are involved in the regulation of production of ATP (and hence regulate cellular energy metabolism)[89–91] and metastatic growth[92] although all the biological implications related to *myo*-inositol pyrophosphates are not clearly understood. Inositol derived pyrophosphates as well as their non-hydrolysable analogs were synthesized subsequently. The latter are useful tools (as inhibitors of kinases) for binding studies with enzymes and to understand the process of phosphotransfer from the former pyrophosphates.[93,94]

Investigation of the structure activity relationship for the binding of human diphosphoinositol pentakisphosphate kinase with several *myo*-inositol-1,3,4,6-tetrakisphosphate analogs resulted in the identification of an initial binding site crucial for the capture of the substrate. Such possibilities offer fertile ground for new drug development.[92,93] 1,2-Anhydro-*myo*-inositol (conduritol-B epoxide) is an irreversible inhibitor of various β-glucosidases and can be exclusively obtained from *l*-quebrachitol.[95] Conduramines are known to have glycosidase inhibitory activity.[96] Larner *et al.* have recently isolated a putative insulin mediator from beef liver. A structural analysis has revealed that it has a pseudo-disaccharide Mn^{2+} chelate complex containing pinitol (**41**) and galactosamine moieties.[97] All in all, phosphoinositides control several aspects of cells from their birth to death. This diversity of involvement in cellular process makes the phosphoinositide regulation prone for attack by infectious agents; the latter (e.g. *Listeria, Mycobacterium, Shigella, Salmonella, Legionella* and *Yersinia*) are known to amend these processes for their use, such as intracellular movements, replication, and assembly. On the brighter side, the study of pathogen interactions with host signaling cascades, (in the present case, phosphoinositide signaling pathways) allows us to understand the molecular basis of infection. In other words, bacterial intracellular pathogens can be visualized as tools to discover novel cellular processes.

Hence pharmacological intervention of the phosphoinositide related pathways are thought to have high potential for drug discovery, development, and therapy for many common ailments such as cancer, obesity, and diabetes as well as genetic disorders, such as myopathies and neurological disorders. Furthermore, since cellular signal transduction requires clustering and declustering of phosphoinositides, understanding their dynamics (change in concentrations on the cell membrane) becomes important. Phosphoinositide dynamics can in principle afford a handle to understand and manipulate cell-signaling processes. Almost all of the aspects mentioned above have been extensively reviewed.[91,98,99–107]

From the biochemical perspective, isomeric inositol phosphates and their derivatives are useful in gaining insight into the substrate/agonist/

antagonist - structure – activity relationship of enzymes and receptors, which is crucial for identifying new drug candidates for diseases that are a consequence of impairment of the *myo*-inositol cycle. For example, four 5-position-modified *myo*-inositol trisphosphate derivatives have been synthesized and tested for agonist / antagonist activity.[108] Most phosphorylated *myo*-inositol derivatives are rapidly metabolized in cells which make *in vivo* studies (including pharmacological studies) difficult. This created a need for synthetic *myo*-inositol derivatives which are not metabolized by enzymes. Phosphorothioate, phosphonate and other analogs have been synthesized as enzyme-resistant analogs of *myo*-inositol trisphosphates and phosphatidylinositols and have also been used to derive mechanistic information of enzymes (such as phospholipases, phosphatases) action.[109–114] Such mechanistic studies also indicated that catalytic efficiency of PI-PLC for the cleavage of phosphodiester could be a thousand fold higher than that of RNase A.[115,116] On the other hand there have been attempts to develop peptide based sensors for the detection and monitoring of inositol phosphates.[117]

As expected, due to the scanty availability of phosphoinositols, inositol phosphates and their analogs necessary to delineate the cellular process and to understand the *myo*-inositol cycle, development of laboratory synthesis of inositol derivatives are essential. Synthetic analogs of inositols and their derivatives are also useful for biological screening / testing leading to discovery of new lead molecules for the development of drugs. Synthetic analogs of *myo*-inositol and their derivatives are also essential for screening as potential candidates for the developments of enzyme inhibitors and drug candidates. For instance, synthetic ether analogue of dipalmitoyl PIM2 was significantly more active (at enhancing IL-12 production by immature bovine dendritic cells) than dipalmitoyl PIM2.[118]

In essence, any phsophorylated *myo*-inositol derivative is a potential starting point for the discovery of drugs since inositol phosphates are naturally occurring small molecules that interfere in a large number of cellular processes. Initial synthetic efforts involving inositols mostly targeted the development of selective reactions of *myo*-inositol, phosphorylated derivatives of which form the head group of phosphoinositols. Modification of the diacylglycerol part of phosphoinositols is necessary for the study of phosphoinositol binding proteins as well as their functional elucidation. Phosphatidylinositols carrying 2,3-diacylthreitol backbone, instead of the 1,2-diacylglycerol backbone, were synthesized for the facile incorporation of labels such as biotin, fluorophores, spin label.[119] Availability of synthetic analogs of inositols and their derivatives are crucial for the identification and isolation of phosphoinositol binding proteins and delineating their role in cell signaling processes.[120,121]

Another area of activity where inositols figure is the investigations related to glycosyl inositol derived lipids. Since major portion of these molecules are oligosaccharides, most of the synthetic effort is directed towards building the oligosaccharide units and the glycosidic bond connecting the inositol ring. However, suitably protected enantiomeric inositol derivatives are required to connect the oligosaccharide to the inositol moiety stereoselectively (to achieve the α- or the β-glycosidic link), depending on the purpose for which the glycosylated inositol (or its analog) is being synthesized. Since GPIs consist of carbohydrate, lipid, phosphate, and inositol moieties, the covalent linking of these units in a well defined manner is a formidable task. Most of the glycosyl units are connected to the C6-oxygen of *myo*-inositol and the C1-oxygen carries the lipidic phosphatidic acid unit. Some of the key *myo*-inositol derivatives used for the synthesis of various GPI anchors are shown in Fig.7.6[35,97,122–129,130–134]

GPIs function as anchors for many proteins and locate them at the cell surface (*e.g.* variant surface glycoprotein of *trypanosomes*), for various functions either while they are anchored to the cell surface or after their cleavage

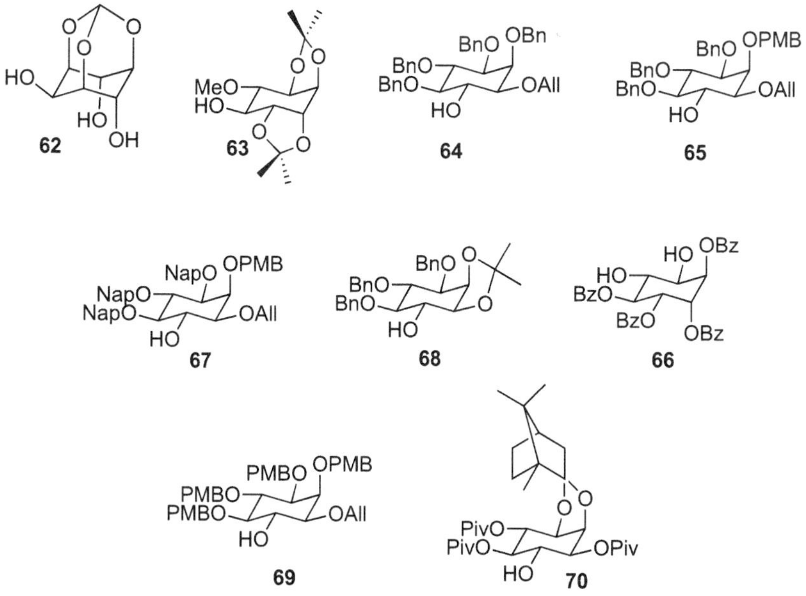

Figure 7.6 ***Selected inositol derivatives used for the synthesis of glycosylphosphatidylinositol anchors and their analogs.*** Piv = Pivalate; Nap = 2-naphthylmethyl.

from the cell surface and release into the surroundings. Biosynthesis of these anchors involves many enzymes which are thought to be potential targets for drug development (for diseases such as malaria, leishmeniasis, sleeping sickness). Since biosynthetic and metabolic pathways and the associated proteins and enzymes could vary widely in parasites and their hosts, GPI anchors are targets for pharmacological intervention of diseases. In mammals, GPI anchoring plays vital roles in embryogenesis, fertilization, neurogenesis and immune system. GPI biosynthesis has implications for virulence of infectious agents such as fungi and hence GPIs are potentially targets for the treatment of infection and consequently, inhibitors of GPI biosynthesis are being identified and tested.[135] Attempts are being made to develop antibiotic treatment for tuberculosis, by targeting the enzymes that are essential for the biosynthesis of mycobacterial cell wall.[136] Antigenic glycolipids found on the cell wall of *M. tuberculosis* and synthetic arabinomannan oligosaccharides are thought to aid the development of carbohydrate antigens for vaccinations against tuberculosis.[137,138] Similarly, GPI glycan analogues from plasmodium might be useful as tools to deleneate biochemical pathways involved in malaria pathophysiology and also aid vaccine development against malaria.[139,140] All in all, an understanding of the structure and function of GPIs and the associated proteins is expected to be a fertile ground for the discovery of drugs for many diseases.

Discovery of the presence of an unexpectedly large number of phosphoinositols in living cells, the associated receptors and enzymes have raised the hope of discovering small organic molecules of medicinal interest and utility for the intervention of diseases implicated with the impairment of the *myo*-inositol cycle.[141] Search for compounds with potential utility include *myo*-inositol derivatives, other cyclitol derivatives as well as compounds having widely different structures (as compared to inositols). The intended application or use of these novel synthetic small molecules could be any one or more of the following: (a) agonists or antagonists for known receptors of phosphoinositols; (b) ligands for the identification and / or isolation of inositol binding proteins; (c) substrates for enzymes involved in the *myo*-inositol cycle; (d) inhibitors for enzymes involved in the *myo*-inositol cycle; (*e*) charge masked phosphoinositols capable of crossing cell membranes (*i.e.* membrane permeant analogs); (f) compounds having biological properties necessary to function as drugs or lead molecules for discovery of drugs; (g) molecules for the delineation of the structure – activity relationships (with intention to develop antagonisits, inhibitors, *etc.*) and (*h*) molecular tools and probes to understand the *myo*-inositol cycle. Synthetic molecules of

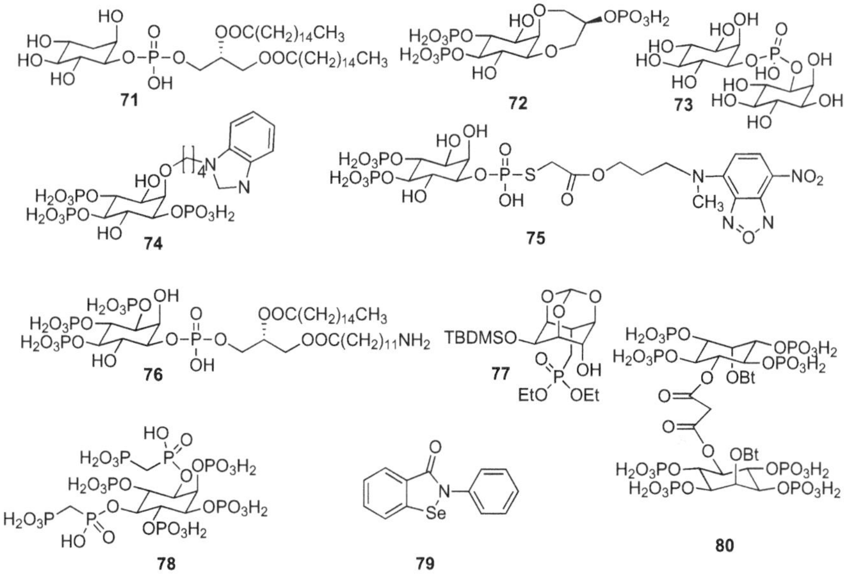

Figure 7.7 ***Selected synthetic small molecules having application potential in biology and medicine related to the myo-inositol cycle.*** **71**: An inhibitor of cancer cell colony formation,[142] **72**: releases Ca^{2+} from intracellular stores,[143] **73**: a stablizer of enzyme activity for hyperthermophillic organisms,[144] **74**: agonist at the D-*myo*-inositol-1,4,5-tris(phosphate) receptor,[145] **75**: a fluoroscent tag,[146] **76**: ligand for affinity chromatography,[147] **77**: exhibits antitumor activity,[148] **78**: competitive antagonist for Ins(1,4,5)P3 receptor,[149] **79**: *myo*-inositol pyrophosphate mimic,[150] **80**: inhibitor of inositol monophosphatase, helps in the treatment of neurodegenerative disorders.[151,152]

these kinds were immensely useful in probing the cellular second messenger pathways and advance our knowledge of the *myo*-inositol cycle and the role played by phosphoinositols in plant and animal cells. Hopefully, the breadth and depth of knowledge accumulated through these small molecular probes lead to the discovery of therapeutic agents for dreadful diseases. Illustrative examples of compounds that were synthesized and used for the purposes mentioned above are shown in Fig. 7.7.

4 Strategies for the synthesis of inositol derivatives

4.1 Necessity *versus* current state of art

Periodically reviews on the synthesis of inositols and their derivatives have been published. These emphasize the following aspects: (1) development of synthetic strategies for inositol derivatives aimed at use as tools for the study of cellular processes[153,154] as well as scaffold for the development of other

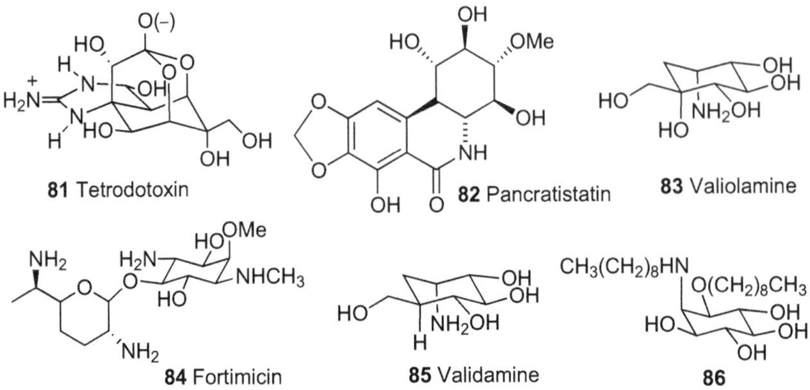

Figure 7.8 ***Illustrative examples of natural products containing the inositol/cyclitol moieties.***

interesting compounds with potential applications in biology, medicine or material science;[155–167] (2) building of inositol (in general cyclitol) scaffolds for use as key intermediates for the synthesis of natural and unnatural products containing the inositol moiety;[168–174] (3) synthesis of inositol derivatives starting from commercially available *myo*-inositol.[19,154,175–184]

Synthetic methods that built the inositol scaffold also potentially allow the synthesis of natural products containing the cyclitol moiety (Fig. 7.8) as well as unusual molecules such as *O*-linked diinositols and oligo-inositols which could open up unforeseen possibilities for drug discovery.[185,186] However, these ideologically different approaches complement and supplement each other and augment the progress in synthetic aspects of cyclitol chemistry, which in turn accelerates research aimed at unraveling the consequences of the *myo*-inositol cycle. This is illustrated by the synthesis of 5-deoxy-5-fluoro-*myo*-inositol (**88**, Scheme 7.4).

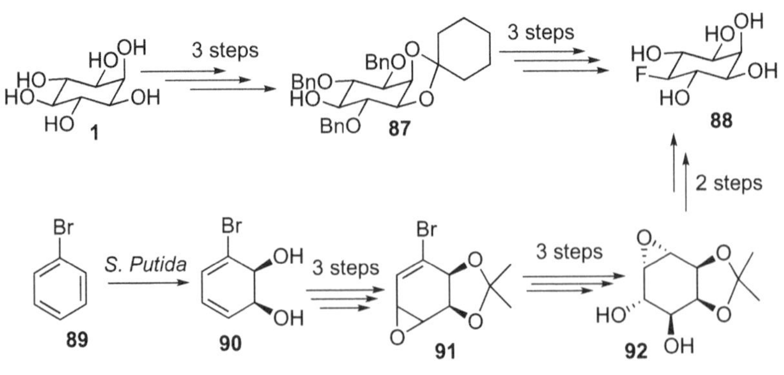

Scheme 7.4 ***Different approaches to the synthesis of deoxyfluoro inositol* 88.**

The fluoro inositol **88** was first prepared from commercially available *myo*-inositol using the classical protection – deprotection protocol.[187] Subsequently **88** was also synthesized starting from bromobenzene (**89**) *via* the chemoenzymatic route (Scheme 7.4). This route has the flexibility to be used for the synthesis of several other ring substituted *myo*-inositols as well as their diastereomeric analogs (diversity oriented synthesis).[188] The synthesis from *myo*-inositol (**1**) on the other hand does not require any specialized reagents, enzymes and conditions. Many flouroinositols were used to probe the phosphatidylinositol signaling pathway and tested as drug candidates for cancer, but unfortunately none turned out to be good at inhibiting cell proliferation.[189]

Synthesis of water soluble inositol phosphates and other derivatives invariably involve many protection – deprotection steps and hence is labor and cost intensive. Furthermore, these synthetic manipulations often change the hydrophobic / hydrophilic nature of synthetic intermediates, making the process of purification and isolation of desired (water soluble) product tedious. Typically, the last step in the synthesis of a phosphorylated inositol derivative involves deprotection of all the protecting groups to generate an inositol phosphate, which is hydrophilic, while its precursor is usually hydrophobic. This variation in physical properties often calls for specialized or unaccustomed methods, reagents and isolation procedures. Bru *et al.*, offered a solution to this situation, by carrying out the synthesis of water soluble analogs of phosphatidylinositol-3,4,5-trisphosphate (**58**) using a polymer (solid phase) support.[190] This strategy was expected to circumvent the synthetic issues mentioned above. Synthesis of other water soluble inositol derivatives is yet to be explored using this strategy.

In the context of biological and pharmaceutical research, the hydrophilicity of inositol phosphates makes them impermeable to cell membranes, and hence their delivery into cells becomes inefficient. One way to circumvent this issue is to make hydrophobic membrane-permeant analogs (of hydrophilic molecules) which when metabolized by intracellular enzymes (often hydrolases), are converted to the parent hydrophilic molecules. Several hydrophobic analogs of inositol phosphates have been synthesized and tested (some examples are shown in Scheme 7.5). However, preparation of these derivatives required their total synthesis, with suitably *O*-protected inositol derivatives as key intermediates. This total synthetic approach also aids in the tuning of hydrophobicity (lipophilicity) of the desired phosphoinositol and could pave the way for the discovery of inositol based pro-drugs.[191] Membrane-permeant derivatives of phosphatidylinositol-3-phosphate (**52**), from which the parent phosphoinositide could be generated (within cells)

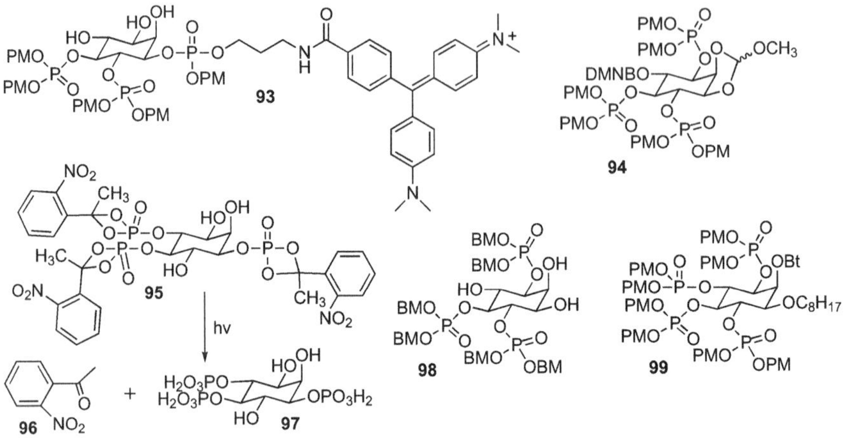

Scheme 7.5 ***Examples of charge-masked inositol phosphates.*** BM- butyryloxymethoxy; PM - Propionyloxymethyl; DMNB - Dimethoxy nitrobenzyl; Bt - butyrate.

was used to show that **52** induced rapid endosomal fusion in living cells, thus indicating its role as a sufficient signal for driving endosomal fusion.[192] Another approach to masking- de-masking of hydrophilic molecules for delivery into cells is by the use of photolabile groups. For example, the 2-nitrobenzyl group has been used as a photolabile protecting group (as in **95**, Scheme 7.5) in *myo*-inositol 1,4,5-trisphosphate.[191,193–195]

4.2 Synthesis of inositol derivatives from *myo*-inositol–relative reactivity of the hydroxyl groups

Since inositols have six secondary hydroxyl groups and all the known isomers (except the *chiro*-isomer) are symmetric, selective functionalization or derivatization of the inositol hydroxyl groups is an arduous task. Most reactions on inositols generate isomeric products, separation of which is labor, time and cost intensive. An overwhelming majority of the synthetic methods aimed at the synthesis of naturally occurring inositol derivatives and their analogs used *myo*-inositol (**1**) as the starting material. This was mainly due to the fact that *myo*-inositol (**1**) is relatively inexpensive and large amounts can be commercially procured, while other isomeric inositols are prohibitively expensive to serve as starting materials for routine synthesis. Although initial synthetic methods starting from *myo*-inositol (**1**) often yielded mixture of products, they did yield significant amount of the final desired products (often phosphoryalted inositol derivatives). However, due to the mixture of products formed early in the synthetic sequences, overall

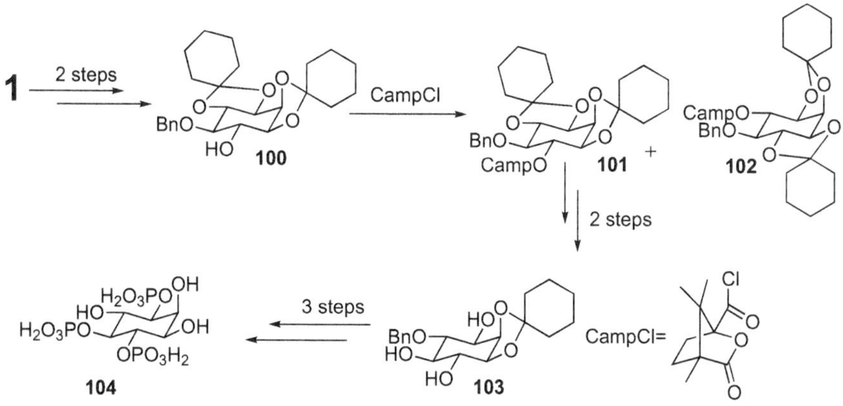

Scheme 7.6 ***Synthesis of L-myo-inositol-1,4,5-trisphosphate (104) using acetal protecting groups; the trisphosphate 104 was obtained in about 2% overall yield from myo-inositol (1).***

yield of the final product suffered severely. This is illustrated by a synthesis of L-*myo*-inositol-1,4,5-trisphosphate (**104**, Scheme 7.6) with the use of ketals for the initial protection of the hydroxyl groups.[196] One of the three ketals formed, which was useful for the synthesis, was separated by column chromatography and converted to **100** and resolved as camphanate esters (**101**, **102**). The trisphosphate **104** was prepared from the diketal **101**.

Reactions, such as ketalization, usually do not help to chemically distinguish between the hydroxyl groups as all the six hydroxyl groups of *myo*-inositol have subtle differences in reactivity. An additional issue in ketal forming reactions with polyols is the possibility of formation of oligomeric ketals, which could hinder the process of separation (of isomeric ketals) as well as lower the yield of the expected product. Conditions for the preferential formation of only the *cis*-ketal[197,198] and the hydrolysis of the *trans*-ketal[199] have been worked out, but this leaves four other hydroxyl groups to be distinguished later. Difference in solubility between benzoate derivatives of the three isomeric *iso*-propylidene ketals of *myo*-inositol (**1**) has been exploited for the isolation of one of the ketals **105** (without the aid of chromatography) in gram quantities (Scheme 7.7).[200,201]

Use of heterogeneous catalytic conditions for the formation of ketals appeared to give better yield and selectivity.[201] Reagents and conditions which exclusively led to the formation of *trans*-ketals (**108**) have been used to restrict the formation of isomeric ketals.[15,202,203] This approach led to an efficient synthesis of the isomeric *neo*-inositol (**8**).[180] Disiloxane derivatives of inositols have also been used to protect vicinal *trans*-diols in *myo*-inositol

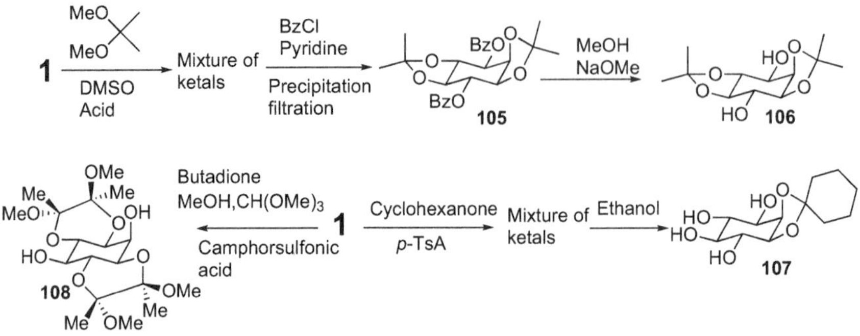

Scheme 7.7 ***Methods for the selective ketalization of myo-inositol (1).***

(**1**). Disiloxane and ketal protecting groups constitute a pair of orthogonal protecting groups since silyl ethers can be cleaved using tetrabutylammonium fluoride.[204] Other reactions, such as etherification, esterification, sulfonylation, *etc.* on *myo*-inositol give rise to a mixture of products and hence are of little use for synthesis. Intramolecular and intermolecular migration of acyl groups leading to the formation of mixture of products discourages chemists from using esters for the protection of inositol hydroxyl groups. However, there are instances of the exploitation of intramolecular migration of acyl groups for the divergent synthesis of inositol phosphates.[205] Interestingly, the first instance of intermolecular acyl migration in the crystalline state was observed in crystals of an inositol orthoformate derivative **108** (Scheme 7.8).[206] Further investigations on this reaction in the solid state allowed the identification of other molecular crystals capable of supporting intermolecular acyl group transfer reactions in them.[164]

The selective protection and deprotection of *myo*-inositol hydroxyl groups have been reviewed.[154,175] This compilation of the data suggested that the sequence of reactivity of *myo*-inositol hydroxyl groups are in the order C1,C3-OH > C4,C6-OH > C5-OH > C2-OH. This order of reactivity however does not imply exclusivity in product formation, but could be useful while designing syntheses starting from partially protected *myo*-inositol

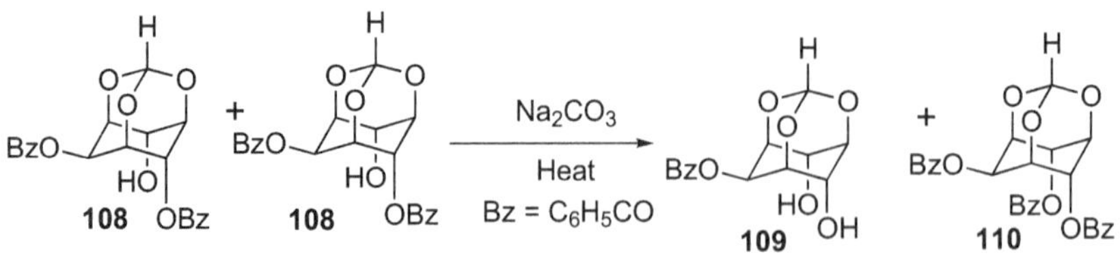

Scheme 7.8 ***Intermolecular benzoyl group migration in crystals of the racemic dibenzoate* 108.**

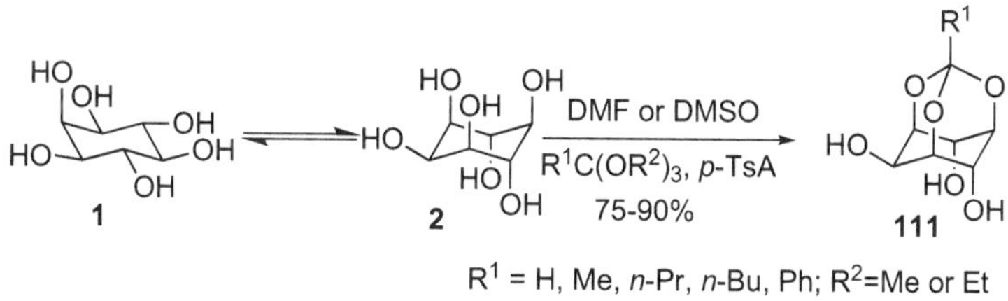

Scheme 7.9 ***myo-Inositol (1) orthoester formation is accompanied by inversion of the carbocyclic ring of 1.***

derivatives, to predict the major product.[207] There are very few reactions of *myo*-inositol (or its isomers) which lead to the formation of a single product. Tosylation and acylation of *myo*-inositol in *N,N-di*methylacetamide - lithium chloride has been reported to give the corresponding 1,3-di-*O*-substituted derivative as the major product.[208,209] Glycosylation of D-*chiro*-, *muco*-, and *allo*-inositol (**6, 9, 10**) with cyclodextrin glucosyl transferase from *Thermoanaerobacter sp.* and subsequent hydrolysis of the by products with *Aspergillus niger* glucoamylase, gave the corresponding monoglucosylated inositol.[210,211] Reaction of *myo*-inositol (**1**) with trialkyl orthoesters gives the corresponding *myo*-inositol-1,3,5-orthoester **111** as the only product in very good yields (Scheme 7.9).

Formation of *myo*-inositol orthoester is an efficient reaction, the yield ranges between 60 and 90%, with the formation of a single product.[6,212–216] This reaction has been exploited for the selective esterification of the C2-hydroxyl group (see below).[217,218] The *myo*-inositol orthoester formation leads to the simultaneous protection of C1, C3, and C5-hydroxyl groups. This is an unusual reaction in the sense that in the product the inositol ring has the inverted (axial-rich) conformation. As a result, the C4-and C6-hydroxyl groups are axial, while the C2-hydroxyl group is equatorially oriented with respect to the inositol ring. Furthermore, since these orthoesters are analogs of adamantane (they are in fact trioxa-adamantanes), the molecules are inherently rigid. These conformational aspects of *myo*-inositol orthoester molecules, impart certain unusual chemical properties, which have been exploited for the synthesis of inositol derivatives. Interestingly, aryl sulfonate groups could be used as hydroxyl protecting groups in these orthoesters, providing convenient synthesis of several inositol derivatives.[13,219–221] 1,3-Diaxial orientation of the C4- and C6-hydroxyl groups in *myo*-inositol orthoesters (**111**, Scheme 7.10) result in a strong hydrogen bond between them (as evidenced by crystal structures).[222,223] This results in an increase in the acidity of one of the hydroxyl groups

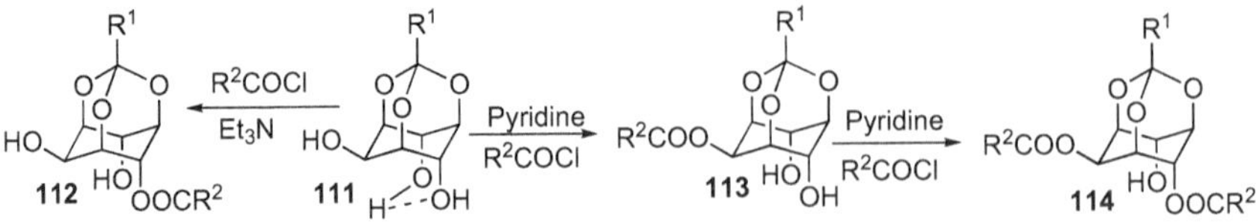

Scheme 7.10 ***Regio-selective esterification of myo-inositol orthoesters.***

(consequently reduced nucleophilicity) and hence it can be deprotonated with greater facility as compared to the other two hydroxyl groups.[224] This has been exploited for the selective derivatization of the hydroxyl groups of *myo*-inositol orthoesters **111** (Scheme 7.10).[225]

Proximity of the C4- and C6-oxygen atoms in **111** also imparts them the ability to chelate metal ions, which stabilizes the oxy-anion at the C4- or the C6-position (**115**, **118**, Scheme 7.11), relative to the oxy-anion at the C2-position. This allowed selective derivatization of the axial hydroxyl group in preference to the equatorial hydroxyl group in the axial monoether **117** (Scheme 7.11). Alternately, both the C4- and C6-axial hydroxyl groups could be derivatized as ethers (**119**) without disturbing the C2-hydroxyl group, in one step (using lithium hydride as the base). These possibilities allow differential and selective protection of the three hydroxyl groups in *myo*-inositol orthoesters **111**.[226,227]

Incidentally, chelation effects also seem to induce a very efficient and unusual intramolecular migration of the acyl-group from C4-*O*-position to the C2-*O*-position (*i.e.* isomerization of **112** to **113**, Scheme 7.10), providing an indirect route for the exclusive acylation of the C2-hydroxyl group.[228] The ability of inositol orthoester derivatives to chelate with metals also resulted

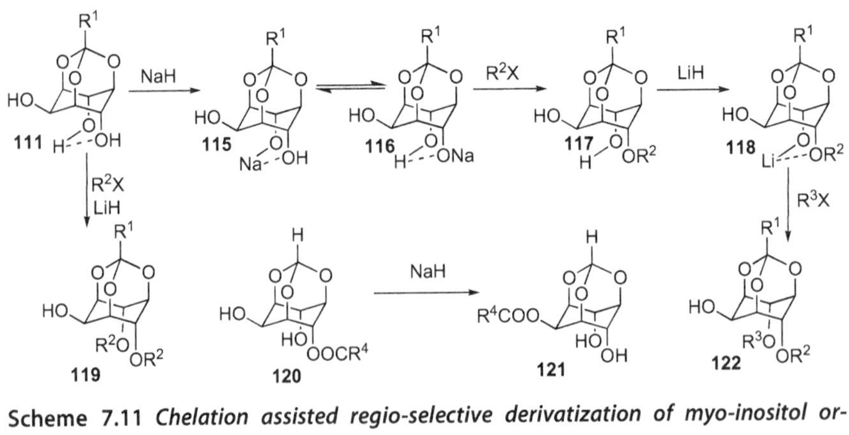

Scheme 7.11 ***Chelation assisted regio-selective derivatization of myo-inositol orthoesters.***

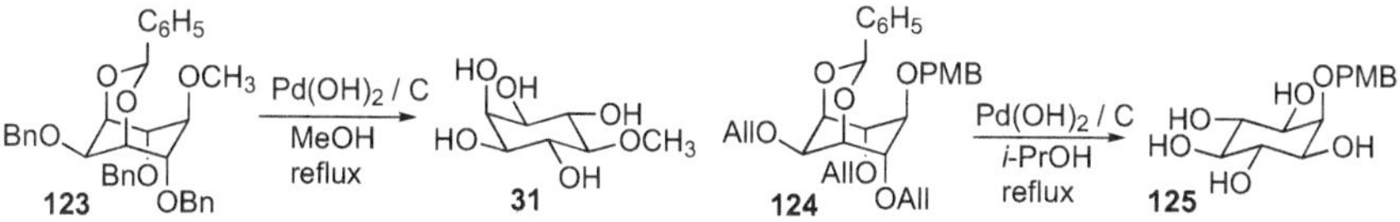

Scheme 7.12 ***Palladium hydroxide mediated cleavage of 1,3-acetal, allyl and benzyl ethers in* 123, 124.** Note that no hydrogen gas was required for the cleavage of benzyl ethers in **123.**

in unprecedented reactivity patterns and these have been exploited in certain synthetic protocols.[229] For instance, a method for the non-hydrogenolytic cleavage of benzyl ethers and selective one-step cleavage of allyl ethers could be developed.[230] These possibilities allowed efficient synthesis of sequoyitol (**31**) as well as a precursor (**125**) for a pyrophosphate derivative of *myo*-inositol (Scheme 7.12). This synthetic Scheme provided sequoyitol in 81% yield,[229] while the yield in previously reported method did not exceed 8%.[231]

The orthoester group in inositol orthoesters **111** can be cleaved to release the C1-, C3-, C5-hydroxyl groups as necessary during a synthetic Scheme either by hydrolysis or by reductive cleavage. The number of hydroxyl groups released can be two or three, depending on the orthoester being hydrolyzed. Hydrolysis of an inositol orthoformate (**111**, R^1 = H) derivative results in the formation the corresponding 1,3,5-triol, while hydrolysis of an inositol orthoacetate (**111**, R^1 = CH_3) or inositol orthobenzoate (**111**, R^1 = C_6H_5) derivative results in the formation of the corresponding diol. This is because the acetate and the benzoate formed on hydrolysis of the corresponding orthoester are far more stable to the conditions of hydrolysis as compared to the formate ester formed from the orthoformate.[232] Consequently, *myo*-inositol orhoformate could be used as a key intermediate for the synthesis of α-D-glucopyranosyl-(1→4)-(DL)-*myo*-inositol and α-D-galactopyranosyl-(1→4)-(DL)-*myo*-inositol, as the conditions for hydrolysis of the orthoformate moiety was mild enough to leave the glycosidic linkages undisturbed (Scheme 7.13). These two conjugates were found to be substrates for inositol dehydrogenase.[127]

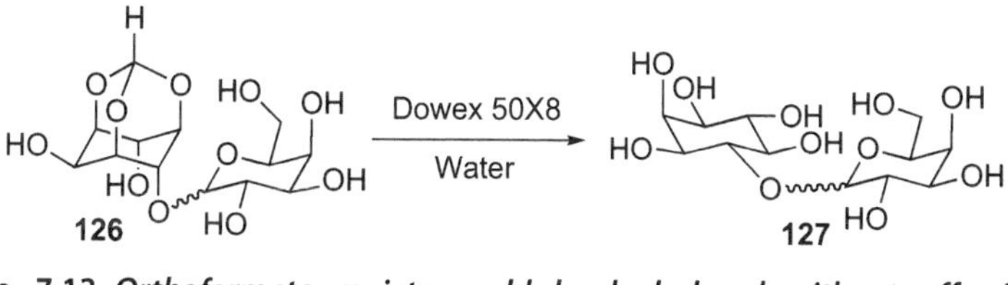

Scheme 7.13 ***Orthoformate moiety could be hydrolyzed without affecting the glycosidic linkage.***

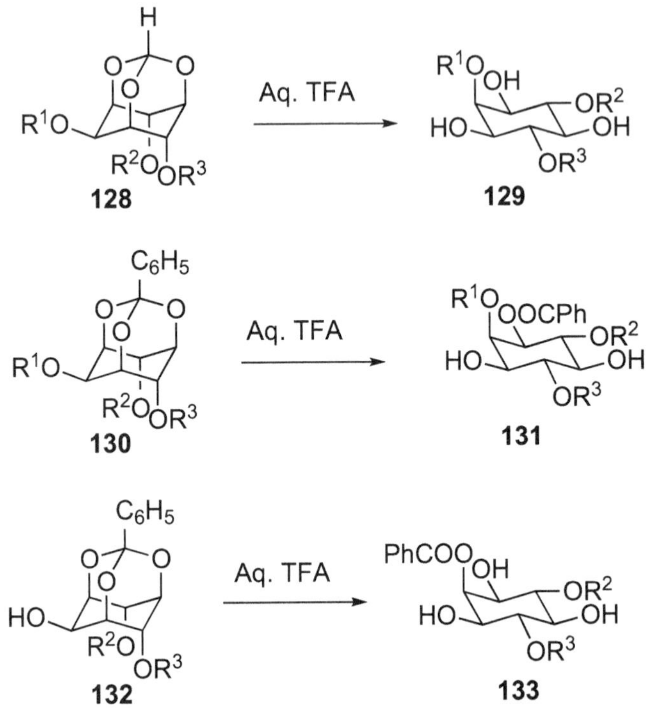

Scheme 7.14 ***Hydrolysis reactions of myo-inositol orthoester derivatives; R[1], R[2], R[3] are protecting groups stable to acid hydrolysis; TFA = trifluoroacetic acid.***

Hydrolysis of *myo*-inositol orthobenzoate (**132** $R^2 = R^3 = H$, Scheme 7.14) yields exclusively the corresponding C2-benzoate which was used as a precursor for the preparation of *myo*-inositol-1,3,4,5,6-pentakis phosphate, in gram quantities.[233] Selective formation of the C2-acetate has also been observed on hydrolysis of the 1,2-orthoacetate derivative of *myo*-inositol.[234]

Rigidity of the *myo*-inositol orthoesters and their ability to coordinate with metal ions also enable the partial and selective reductive cleavage of the orthoester moiety, to release the C1, C3 or C5-hydroxyl groups (Scheme 7.15). For instance, DIBAL-H and trimethylaluminum are known to cleave the C5-O and the C1—O (or the C3—O since they are equivalent) bonds respectively to release the C5- and the C1(or C3)-hydroxyl groups.[235] The regioselectivity for the reductive alkylation of *myo*-inositol orthoesters with Grignard reagents is similar to that observed for the cleavage with trimethylaluminum. It is interesting to note that this approach could be exploited for the selective protection of four of the six hydroxyl groups of *myo*-inositol.[10,11] Selective DIBAL-H mediated cleavage of a

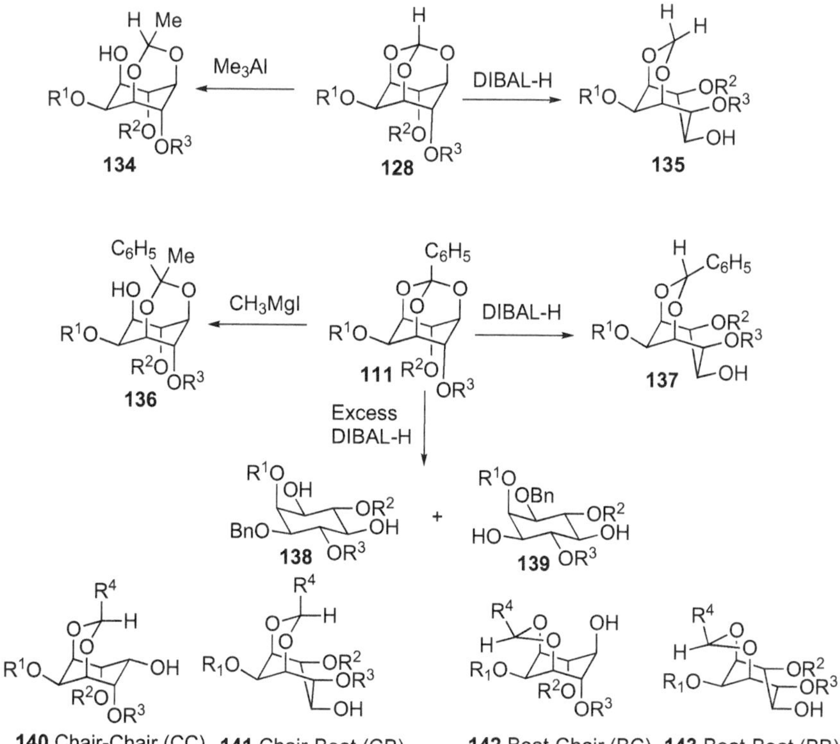

Scheme 7.15 ***Reductive cleavage of myo-inositol orthoesters and the possible conformations of the two rings in the generated 1,3-acetals. R^1, R^2, R^3 are protecting groups stable to reductive cleavage; $R^4 = H$ or C_6H_5.***

myo-inositol orthobenzoate derivative was exploited for the synthesis of a *C*-methyl *myo*-inositol-trisphosphate and *myo*-inositol tetrakisphosphate, as well as isomeric inositols and their derivatives.[236,237]

The partially cleaved *myo*-inositol orthoesters contain the unusual 1,3-acetal moiety, which cannot be obtained by the direct acetalization of *myo*-inositol. Molecular frame of these 1,3-acetals of *myo*-inositol is not as rigid as molecules of the *myo*-inositol orthoesters. Hence this gives rise to interesting conformational aspects. The two rings in these bicylic systems can have different conformations, which raises four possibilities as shown in Scheme 7.15 (**140–143**). DFT calculations have shown that the difference in stability between three (**140–142**) of the four conformations is not large, while the fourth conformation in which both the rings have the boat conformation (**143**) is unstable.[10] Hence the actual conformation of

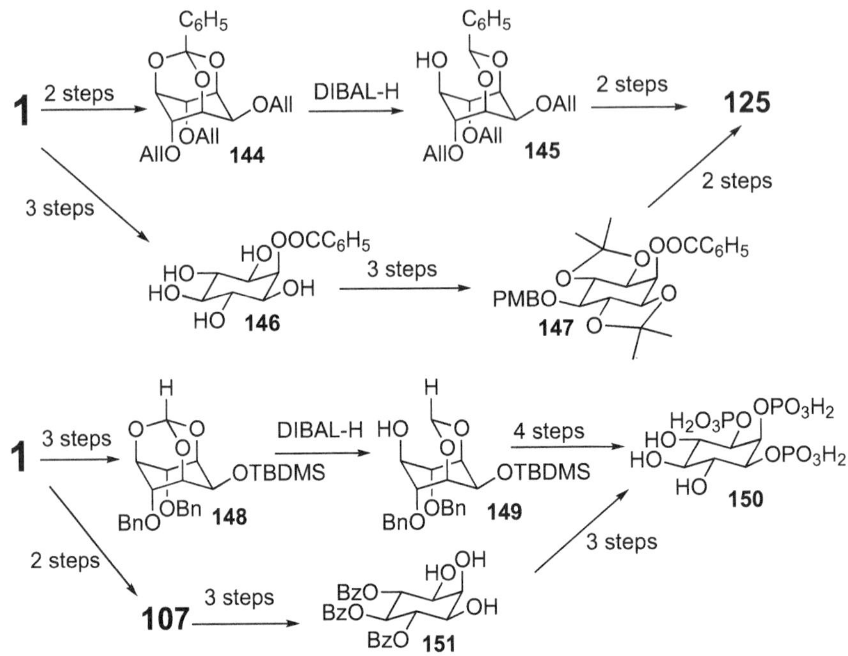

Scheme 7.16 ***Selected examples to illustrate the advantage of using 1,3-acetals for the synthesis of phosphorylated myo-inositol derivatives.*** The PMB ether **125** was obtained in 45% yield when synthesized using the 1,3 acetal intermediate **145**, while the yield was 12% when prepared using the 1,2 acetal intermediate **147**. The trisphosphate **150** was obtained in 15% yield when synthesized through the 1,3 acetal intermediate **149** while the yield was 5% when prepared using the 1,2 acetal intermediate **107**.[160,230,242,243]

these 1,3-acetal molecules could depend on the substituents on the two rings or even on the phase in which they are present. In fact the conformation of some of these 1,3-acetals were different in the fluid and solid states. These aspects opened up possibilities for selective reactions on these molecular systems and opened up new avenues for the synthesis of inositol derivatives.[10] Few examples of the synthesis of inositol derivatives shown in Scheme 7.16 and Scheme 7.17 illustrate the advantages of using *myo*-inositol 1,3-acetals as synthetic intermediates. Other products that could be accessed from *myo*-inositol orthoesters (often with improved yields over other intermediates) include C7-cyclitols, phosphatidylinositols, aminocyclitols, *etc.* [238–241]

As expected, most of the inositol derivatives involved in cellular processes and cyclitol based natural products are chiral and usually only one of the enantiomers is biologically active. However, optical isomers of the

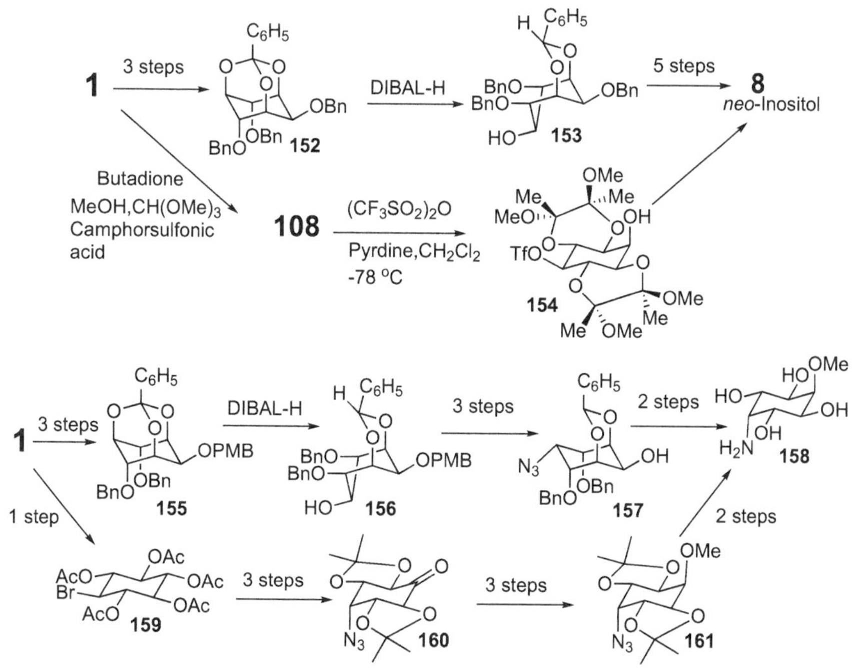

Scheme 7.17 ***Selected examples to illustrate the advantage of using 1,3-acetals for the synthesis of cyclitol derivatives.*** *neo*-Inositol (**8**) could be obtained in 60–65% yield when synthesized through the 1,3 acetal intermediate **153**, but the yield was 5–17% when synthesized through the 1,2 acetal intermediate **154**. The yield of the aminocyclitol **158** was 61% when 1,3 acetal intermediate **156** was used while the yield was about 1% when 1,2 acetal intermediate **160** was used. [180,184,244–247]

biologically active naturally occurring inositol derivatives could be of significance as tools for unraveling the intricacies of the biological processes as well as drug candidates or leads for drug discovery. Hence synthetic routes for the preparation of naturally occurring inositol derivatives as well as their enantiomers are essential.

4.3 Synthesis of enantiomeric *myo*-inositol derivatives from chiral precursors

Chiral inositol derivatives have been synthesized from chiral starting materials as well as from achiral starting materials such as *myo*-inositol, benzene and its derivatives. More or less all the known methods and approaches for the preparation of chiral organic compounds have been used for the synthesis of inositol derivatives. Various chiral starting materials that have

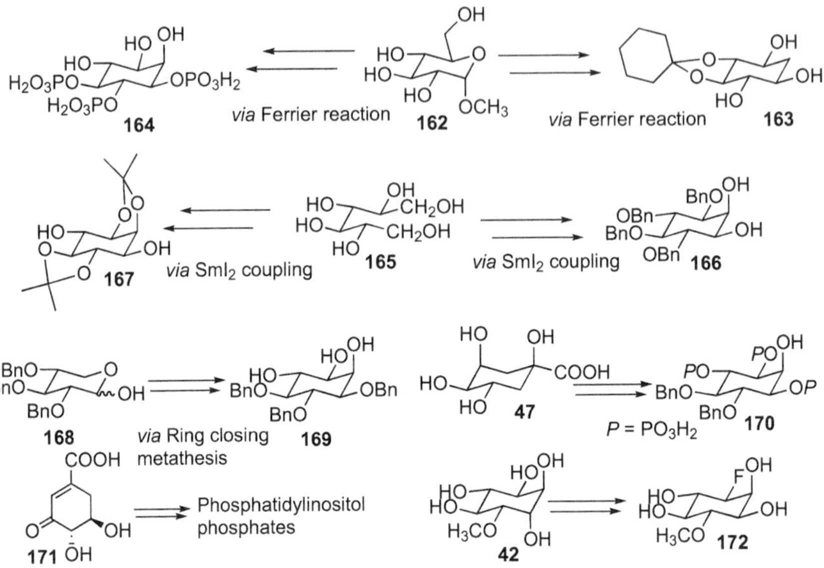

Scheme 7.18 ***Illustrative examples of chiral pool molecules that have been used for the synthesis of chiral inositol derivatives.***

been used for the synthesis of enantiomeric inositol derivatives are shown in Scheme 7.18.

The basic requirement in this approach to synthesize inositol derivatives is the conversion of the non-cyclitol starting materials of varied structures to the required inositol derivatives, often to the *myo*-isomer. In case of pyranose sugars (such as glucose, galactose) their conversion to the inositol framework was achieved by the Ferrier reaction (II).[248–254] Olefin metathesis, aldol reaction, and samarium iodide mediated cyclization reactions have been used to generate the carbocycle from acyclic starting materials (*e.g.* **165**).[255–265] Computer aided retrosynthetic analysis approach for the synthesis of chiral inositol derivatives to identify carbohydrate starting materials has also been attempted.[266] When the chiral starting material is a cyclitol (other than inositol, *e.g* **47**, **171**) or its derivative, introduction of more hydroxyl groups with required stereochemistry is essential to obtain an inositol derivative.[267–271] Use of naturally occurring chiral inositol derivative (*e.g.* **42**) circumvents the introduction of additional hydroxyl groups, but could require inversion of at least one hydroxyl group to obtain the *myo*-configuration. However, naturally occurring chiral inositol derivatives are expensive since their natural abundance is low, and hence this approach

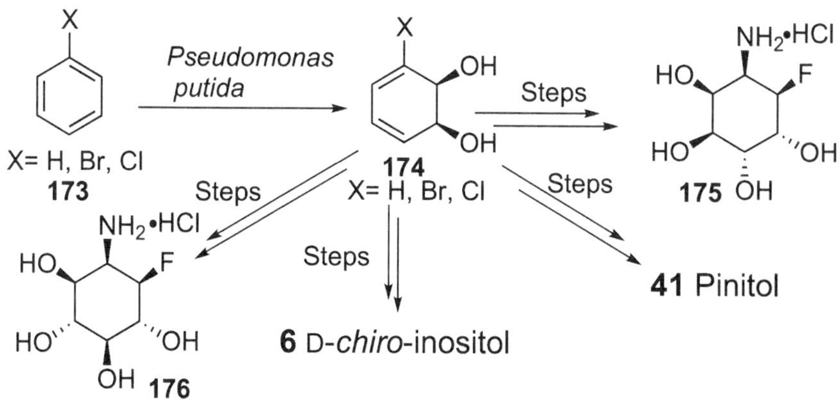

Scheme 7.19 ***Chiral inositol derivatives could be synthesized from benzene and its derivatives.***

does not appear to be practical at present.[272–275] Conversion of benzene and its halogenated derivatives (**173**, Scheme 7.19) to inositol derivatives rely on microbial oxidation (dihydroxylation of one of the three double bonds) of the aromatic ring. [168,276–279,185,188] Other hydroxyl groups are subsequently introduced by chemical dihydroxylation of the two double bonds of the vicinal dihydroxy cyclohexadiene **174** generated. However, often such chemical transformations give rise to diastereomeirc product mixtures.

Each of these elegant approaches to the synthesis of inositol derivatives, particularly the *myo*-isomer, has its own purpose, advantages and disadvantages. Hence it is the preference of the end user, to weigh out these aspects and decide on a synthetic course to obtain the chiral products required.

4.4 Enantiomeric inositol derivatives from *myo*-inositol

myo-Inositol (**1**) is a frequently used starting material for the synthesis of its phosphoryalted derivatives as well as other analogs, since it is available in large quantities and is inexpensive compared to other commercially available inositols and their derivatives. *myo*-Inositol has also been used as a starting material for the synthesis of natural products containing the cyclitol moiety (Scheme 7.20). However, many of the end products obtained were not enantiomeric, possibly due to the non-availability of suitable methods for the resolution of racemic intermediates or enantioselective methods of synthesis.[247,280,281]

Since *myo*-inositol (**1**) is achiral, its use as a starting material for the preparation of enantiomeric products requires the intervention of other chiral molecular entities. These chiral entities could be resolving agents

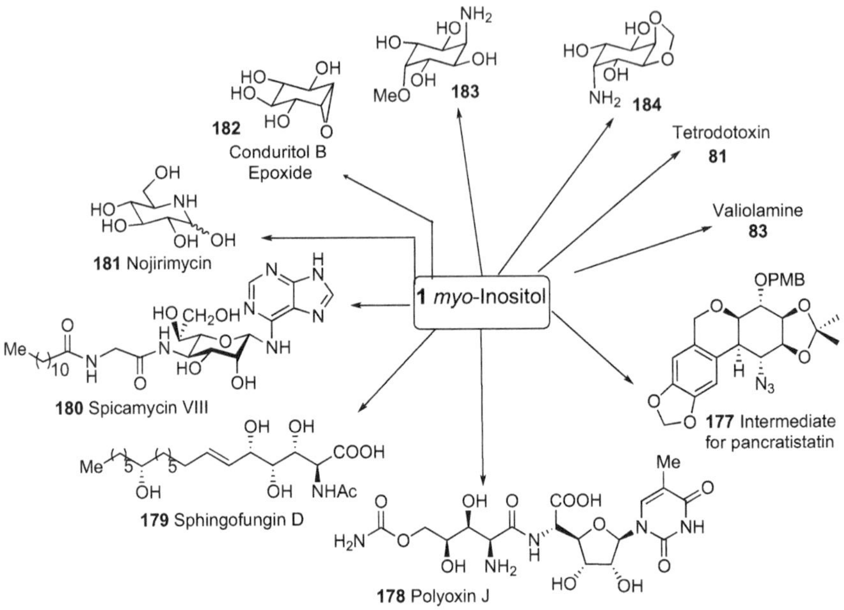

Scheme 7.20 ***Natural products (other than phosphoinositols) or their moieties synthesized using myo-inositol as the starting material.***

(in the case of classical resolution of racemic *myo*-inositol derivatives and desymmetrization of symmetric inositol derivatives) or chiral catalysts (for asymmetric synthesis), or enzymes (for enantioselective functionalization). All these approaches have been utilized for the synthesis of enantiomeric *myo*-inositol derivatives.

4.5 Resolution of racemic *myo*-inositol derivatives by conversion to separable diastereomers

A host of resolving agents has been used to obtain separable *myo*-inositol derived diastereomers. Often the separations were carried out by chromatography, but there are instances of separation of diastereomers by crystallization. Scheme 7.21 shows the resolving agents that have been frequently used to obtain chiral *myo*-inositol derivatives.

By far, resolution of a racemate by conversion to diastereomers appears to be the most frequently used method to obtain chiral *myo*-inositol derivatives, due to its simplicity and wide applicability. Often, in this approach the 'unwanted' enantiomer was discarded. However, due to the *meso*-configuration of *myo*-inositol, it is possible to convert both the resolved enantiomers to the desired chiral end product, by judicious use of the protection – deprotection strategy.[204,285] Whether this technique is economical or not

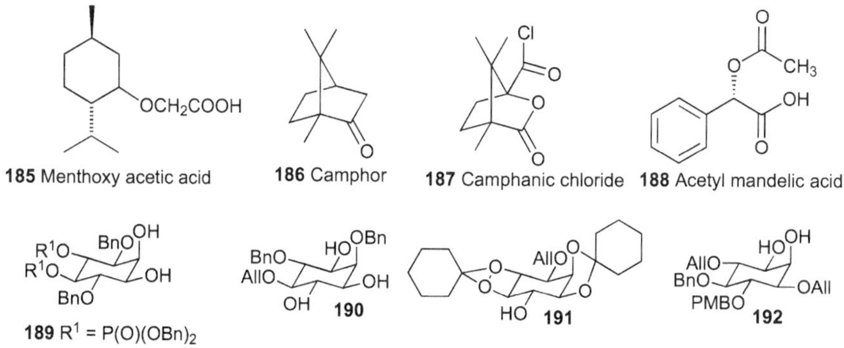

Scheme 7.21 ***Frequently used resolving agents (185-188) for the resolution of racemic myo-inositol derivatives and selected examples of the derivatives resolved (189-192).***[183,282–284]

(in terms of labor, cost, time, *etc.*) could depend on the end product, purpose of the overall synthetic scheme and aim of the investigation. This classical method of resolution generally makes the synthetic scheme lengthy and compromises the yield, unless the resolution is done early on during a synthesis and the resolved intermediate is synthetically versatile (see below). The effort can be minimized if the two diastereomers obtained from a racemate during resolution are separable by fractional crystallization. For instance, several isomeric inositol derivatives were obtained using enantiomers of 1,2:4,5-di-*O*-isopropylidene-*myo*-inositol (**106**). The latter was resolved as diastereomeric diacetyl mandalates, which were separable by crystallization.[286] Another example is the resolution of racemic 4-allyl ethers **193** of *myo*-inositol orthoesters (Scheme 7.22).[287]

These racemic diols **193** were acylated with camphanic acid chloride and the diastereomers **194** and **dia194** obtained were separated by crystallization. Development of the procedures for preferential crystallization of diastereomers was aided by a comparison of the crystal structure of the

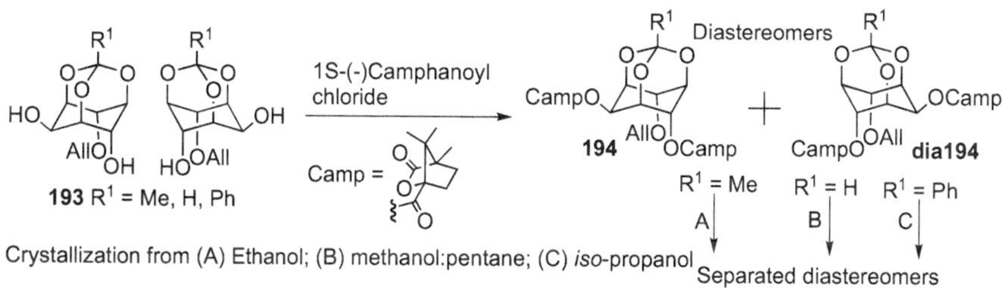

Scheme 7.22 ***Resolution of myo-inositol orthoester derived racemic ally ethers*** **193** ***via separation of diastereomeric derivatives*** **194** ***and*** **dia194** ***by fractional crystallization.***

individual diastereomers and their solvates. This method provided diastereomeric allyl ethers in several gram quantities and in >99% purity. Since these racemic allyl ethers can be obtained in three steps from *myo*-inositol, their scope for the synthesis of various inositol derivatives in high.

4.6 Enzyme mediated resolution of racemic *myo*-inositol derivatives

Several *myo*-inositol derivatives have been resolved using enzymes, illustrative examples of these have been shown in Scheme 7.23. As is evident from the variety of inositol derivatives that have been resolved, early intermediates (such as **198**) as well as late intermediates (such as **204**) *i.e.* *O*-protected inositol derivatives containing varied number of free hydroxyl groups in a synthetic Scheme, are amenable to resolution with the assistance of enzymes. This allows great flexibility during synthesis of target molecules and provides opportunity to maximize the yield and purity of the enantiomeric derivatives at different stages of the synthetic scheme. Lipases are frequently used for the resolution of polyols since they catalyze esterification as well as

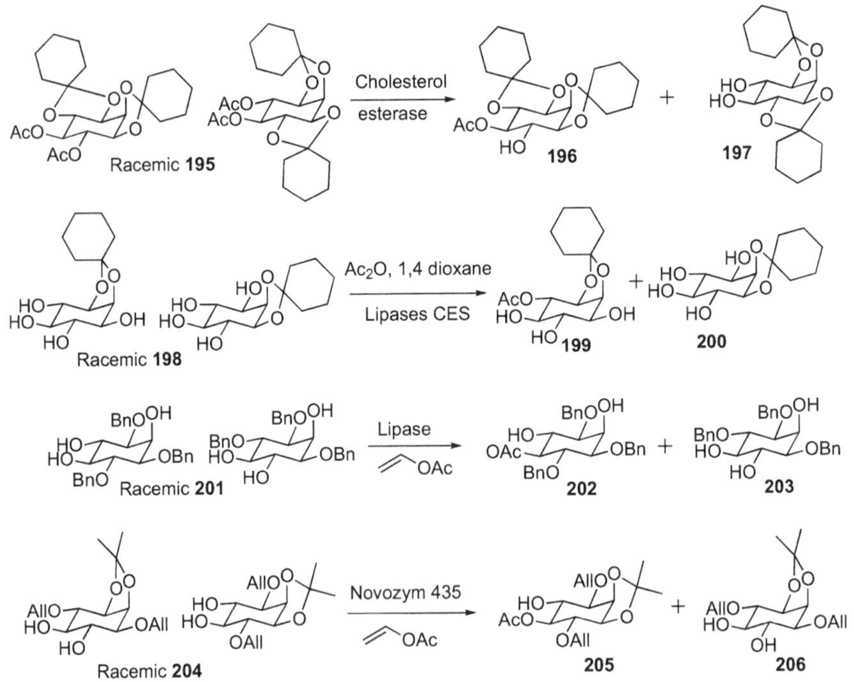

Scheme 7.23 ***Enzyme assisted resolution of partially protected racemic myo-inositol derivatives.***

ester hydrolysis and many lipases are commercially available. In addition, (1) lipases do not require co-factors for their action; (2) are even amenable for use in non-aqueous reaction systems;[288] (3) reactions can be carried out in conventional glassware. These advantages are attractive to synthetic organic chemists.[275,289–296] More relevant to the current discussion, enzyme assisted resolution of inositol derivatives led to the synthesis of several phosphatidylinositol mannosides and their derivatives.[297]

4.7 Desymmetrization of symmetric *myo*-inositol derivatives

Desymmetrization of a symmetric *myo*-inositol derivative could generate either diastereomers or enantiomers depending on the reagent (chiral or achiral) and the method used for the desymmetrization reaction (Scheme 7.24). For instance, a chiral acid can be used to convert a symmetric *myo*-inositol derivative to diastereomeric esters (such as **207–210**, R^1 is chiral) or chiral catalyst can be used to carry out enantioselective esterification to generate one enantiomer preferentially (such as **207–210**, R^1 is achiral). Both of these approaches have been realized to generate chiral *myo*-inositol derivatives. Scheme 7.24 illustrates the possible products from *myo*-inositol (for reaction at C1/C3 hydroxyl groups) and for its orthoformate (**12**) (for reaction at C4/C6 hydroxyl groups). It is pertinent to mention that (a) reaction at C4/C6 hydroxyl groups of *myo*-inositol (**1**) with similar

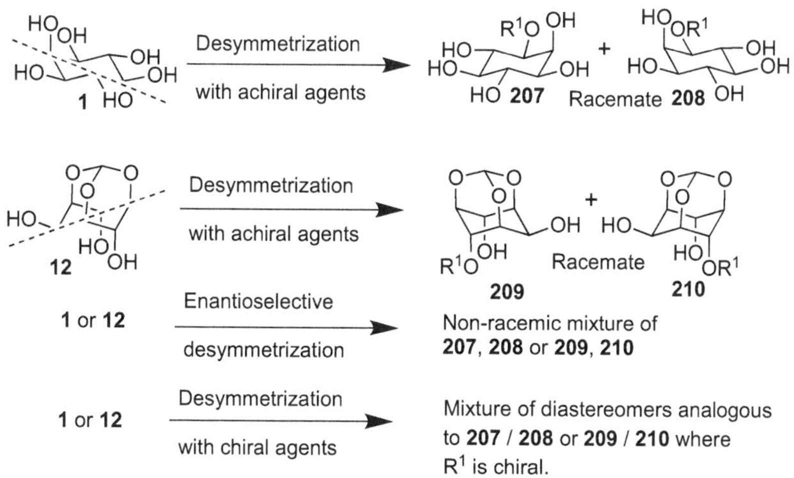

Scheme 7.24 ***Desymmetrization of myo-inositol (1) and its orthoformate* 12.** Expected outcomes depending on the reagents are shown for the *O*-substitution of the C1- and C3-hydroxyl groups in **1** and C4- and C6-hydroxyl groups in **12**. When R^1 is chiral, diastereomers are generated.

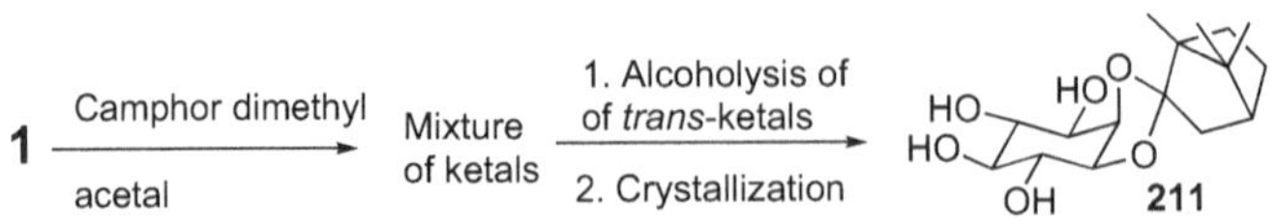

Scheme 7.25 ***Desymmetrization of myo-inositol (1) by formation of camphor acetals; the diastereomer 211 could be obtained in 31% yield.***

possibilities also exist and (b) *O*-substitution at the C2- and C5-hydroxyl groups of *myo*-inositol (**1** and its symmetric derivatives) generate (achiral) products with *meso*-configuration, as mentioned earlier (Section 2).

Desymmetrization of symmetric *myo*-inositol derivatives using chiral resolving agents provides diastereomers (which can be converted to enantiomeric *myo*-inositol derivatives). However this method also requires the separation of diastereomeric intermediates just as in classical resolution of racemates. *myo*-Inositol has been desymmetrized by ketal formation with camphor or its dimethyl acetal (Scheme 7.25). This facilitated a short synthesis of D-*chiro*-inositol, several naturally occurring phosphorylated *myo*-inositol derivatives and their analogs, from *myo*-inositol (**1**).[298–302]

myo-Inositol orthoformate (**12**) could be desymmetrized with several chiral acids or their derivatives as well as by glycosylation (Scheme 7.26). The latter glycosylation reaction provided one of the diastereomers in excess (64%). This was utilized for the synthesis of mycobacterial triacylated phosphatidylinositol dimannoside. Acylation of *myo*-inositol orthoformate (**12**) with a proline based anhydride in the presence of yitturbium triflate yielded predominantly one diastereomer **214**. Similarly, other orthoesters of *myo*-inositol have been converted to diastereomeric derivatives.[260–261,303,304] These approaches were used to synthesize several inositol phosphates.[305,306]

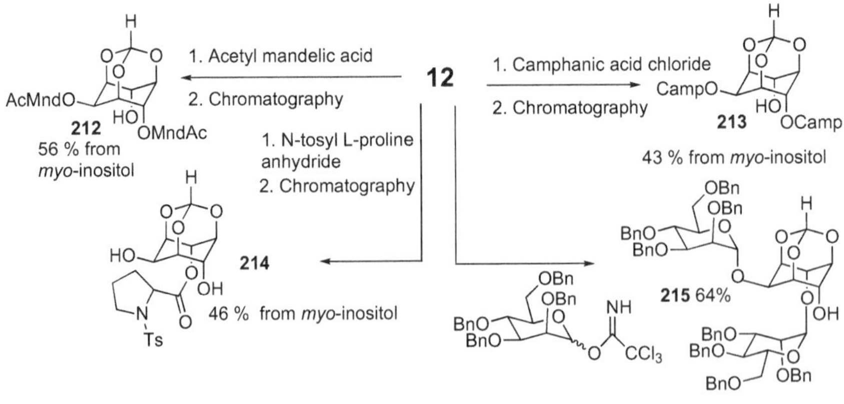

Scheme 7.26 ***Desymmetrization of myo-inositol orthoformate (12) with chiral reagents.***

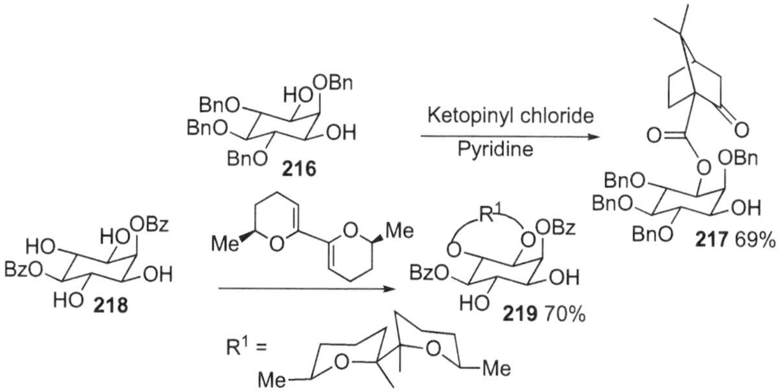

Scheme 7.27 ***Desymmetrization of symmetric myo-inositol derivatives.***

Acylation of the symmetric diol **216** with ketopinyl chloride gave the 3-*O*-acyl derivative **217** predominantly (Scheme 7.27), which was used for the synthesis of mycothiol.[307] Chiral dispiroketals (such as **219**) have been prepared by desymmetrization of *myo*-inositol derived tetrols.[308]

Asymmetric catalytic *O*-substitution of symmetric *myo*-inositol derivatives (**220**, **222**, **224;** Scheme 7.28) led to the synthesis of several phosphoinositols. For example, 4,6-di-*O*-substituted *myo*-inositol derivatives

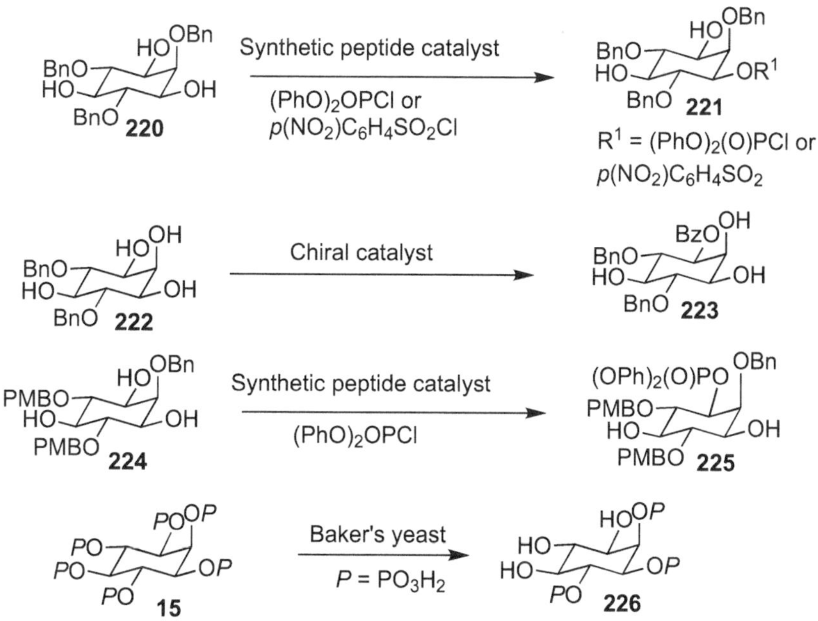

Scheme 7.28 ***Enantioselective desymmetrization of symmetric myo-inositol derivatives.***

have been enantioselectively functionalized at the C1 (or C3) position with 90-99% enantiomeric excess, using chiral peptide catalysts. This approach allowed the synthesis of eight inositol phosphates from a common intermediate. Attempts have been made at the synthesis of inositol derivatives from naturally occurring phytic acid (**15**).[309–317]

4.8 Phosphorylation of inositol derivatives

Phosphorylation of inositol derivatives is normally the penultimate step during the synthesis of inositol phosphates and the associated lipids (Scheme 7.29).

It is generally observed that phosphorus(III) reagents give better yields than phosphorus(V) reagents, although the former requires an additional oxidative step for P(III) to P(V) conversion. In addition, the P(III) approach is flexible to allow the synthesis of phosphodiesters and thiophosphates, if necessary.[72,110] Recently, a novel chiral phosphorylating agent has been used to synthesize enantiomeric *myo*-inositol derivatives.[318] Irrespective of the phosphorylation method used, the alkyl groups used to protect the phosphate needs to be cleaved subsequently to obtain the corresponding inositol phosphate or its lipid derivative. These conditions should be mild enough so as not to disturb the phosphate groups on the inositol ring. Often, benzyl groups are preferred since they can be cleaved by hydrogenolysis. However, hydrogenolysis might not be suitable for the synthesis of natural phosphatidylinositols as they could carry an unsaturated fatty acid on the diacylglycerol moiety. Hence phosphorylating agents which provide a protected phosphate that can be deprotected without the use of hydrogen are preferred for the synthesis of natural phosphatidylinositols.[319] In the biochemical perspective, phosphate group provides a marker to track phosphorylated inositol derivatives by ^{31}P NMR spectroscopy. This approach provided valuable information on the mechanism of action of phosphatidylinositol specific phospholipase C on its substrate. ^{31}P NMR studies showed that bacterial phospholipase C has the ability to catalyze the

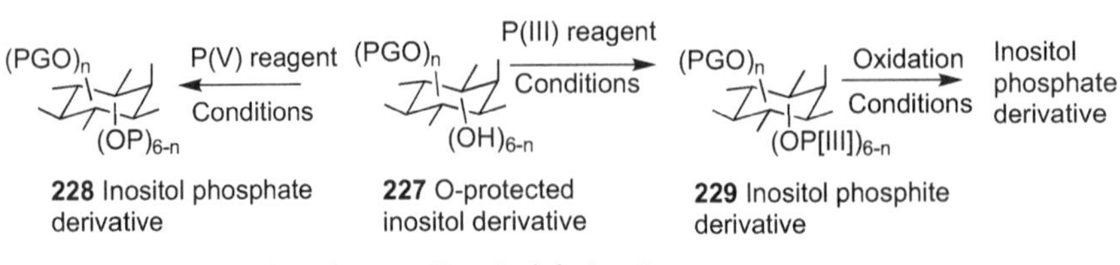

Scheme 7.29 ***Phosphorylation of inositol derivatives.***

cleavage of phosphatidylinositol rapidly to *myo*-inositol-1,2-cyclic phosphate and diacylglycerol as well as hydrolyze the cyclic phosphate exclusively to *myo*-inositol-1-phosphate.[320,321] Hence, phospholipase C could also be used to catalyze the formation of inositol derived phosphodiesters by the reaction of *myo*-inositol-1,2-cyclicphosphate with alcohols. [322]

5 Conclusions and future outlook

Revival of interest in the chemistry of inositols was due to the realization of the role of phosphoinositols in various cellular processes, primarily signal transduction pathways. This revival was augmented by the organic chemists who attempted to use *myo*-inositol as a starting material for the synthesis of natural products and devised good methods for the synthesis of phosphoinositols, glycosylphosphoinositols and their analogs. The latter aided the efforts in unraveling certain aspects of the *myo*-inositol cycle as well as biological roles played by glycosylated inositol derivatives. Apart from being biologically significant, inositols as a class of molecules are interesting as they can be used as scaffolds or starting materials for the synthesis of natural products, catalysts,[323] metal-complexing agents, gelators, and supramolecular assemblies.[167,324–326] The foregoing reveals that a variety of synthetic approaches have been developed starting from commercially available inositols, sugars, aromatic compounds, chiral acids, tetrahydrobenzoquinone, cyclohexene derivatives, and norbornene. But it is the opinion of the authors that the most frequently encountered methods of synthesis of phosphoinositols bank on using *myo*-inositol as the starting material due to its easy availability and well developed protection-deprotection strategies. Initial methods of preparation of protected *myo*-inositol derivatives were heavily dependent on the use of acetal formation which was not very regioselective. The use of inositol orthoesters was a significant improvement due to non-formation of isomeric products and provided a window for newer methods of regioselective synthesis of inositol derivatives. The methods which used starting materials other than *myo*-inositol, contributed to the synthesis of carbocyclic ring-modified inositol derivatives such as deoxyinositols (quercitols), aminoinositols, carbasugrs, pseudosugars, fluorocarbohydrates *etc*, which widened the scope for drug discovery. The recent focus appears to be on the development of enantioselectvie reactions of inositol derivatives which is crucial for access to large amounts of chiral and synthetically versatile derivatives with high enantiomeric purity. Versatile chiral inositol derivatives have the potential to be modified into biologically interesting and pharmacologically promising compounds required to unravel the biological processes mediated by phosphoinositols and discover newer tools for the pharmacological intervention of dreadful diseases.

References

1. Shvets, V. I. *Russ. Chem. Rev.* **1974**, *43*, 488–502.
2. Dennis, J.; Cosgrove, G. C. J. *Inositol Phosphates: Their Chemistry, Biochemistry, and Physiology*. Elsevier Scientific Pub. Co, 1980.
3. Voet, D.; Voet, J. G. *Biochemistry*. Wiley, 1990.
4. White, P. J. *J. Agric. Sci.* **2007**, *145*, 651–652.
5. Majumder, A. L.; Biswas, B. B. *Biology of Inositols and Phosphoinositides*. Springer, 2011.
6. Lee, H. W.; Kishi, Y. *J. Org. Chem.* **1985**, *50*, 4402–4404.
7. Yamada, H.; Okajima, K.; Imagawa, H.; Mukae, T.; Kawamura, Y.; Nishizawa, M. *Tetrahedron Lett.* **2004**, *45*, 3157–3160.
8. Bauman, A. T.; Chateauneuf, G. M.; Boyd, B. R.; Brown, R. E.; Murthy, P. P. N. *Tetrahedron Lett.* **1999**, *40*, 4489–4492.
9. Kim, S. H.; Xing, G.; Bollinger, J. M.; Krebs, C.; Hoffman, B. M. *J. Am. Chem. Soc.* **2006**, *128*, 10374–10375.
10. Gurale, B. P.; Sardessai, R. S.; Shashidhar, M. S. *Carbohydr. Res.* **2014**, *399*, 8–14.
11. Jagdhane, R. C.; Shashidhar, M. S. *Eur; J. Org. Chem.* **2010**, *2010*, 2945–2953.
12. Patil, M. T.; Krishnaswamy, S.; Sarmah, M. P.; Shashidhar, M. S. *Tetrahedron Lett.* **2011**, *52*, 3756–3758.
13. Sarmah, M. P.; Shashidhar, M. S. *Carbohydr. Res.* **2003**, *338*, 999–1001.
14. Hense, A.; Ley, S. V.; Osborn, H. M. I.; Owen, D. R.; Poisson, J.-F.; Warriner, S. L.; Wesson, K. E. *J. Chem. Soc. Perkin Trans.* **1997**, *1*, 2023–2032.
15. Montchamp, J.-L.; Tian, F.; Hart, M. E.; Frost, J. W. *J. Org. Chem.* **1996**, *61*, 3897–3899.
16. Miethchen, R.; Sowa, C.; Frank, M.; Michalik, M.; Reinke, H. *Carbohydr. Res.* **2002**, *337*, 1–9.
17. Miethchen, R. *J. Carbohydr. Chem.* **2003**, *22*, 801–825.
18. Miethchen, R.; Schmidt, A.; Neitzel, K.; Michalik, M.; Pundt, T.; Ruth, W. *Carbohydr. Res.* **2005**, *340*, 741–748.
19. Christian Husson, P. J. A. V.; Odier, L. *Carbohydr. Res.* **1998**, *307*, 163–165.
20. Simperler, A.; Watt, S. W.; Bonnet, P. A.; Jones, W.; Motherwell, W. D. S. *CrystEngComm* **2006**, *8*, 589.
21. Yang, P.; Murthy, P. P. N.; Brown, R. E. *J. Am. Chem. Soc.* **2005**, *127*doi: 10.1021/JA053371U.
22. Parthasarathy, R.; Eisenberg, F. *Biochem. J.* **1986**, *235*, 313–322.
23. J. Biochem. J. 1989, 258, 1–2
24. Herrera-Salgado, Y.; Garduño-Ramírez, M. L.; Rios, L. V. A. M. Y.; Alvarez, L. *J. Nat. Prod.* **2005**, *68*, 1031–1036.
25. Clark, J.; Kay, R. R.; Kielkowska, A.; Niewczas, I.; Fets, L.; Oxley, D.; Stephens, L. R.; Hawkins, P. T. *EMBO J.* **2014**, *33*, 2188–2200.
26. Peterbauer, T.; Brereton, I.; Richter, A. *Carbohydr. Res.* **2003**, *338*, 2017–2019.
27. Horbowicz, M.; Obendorf, R. L. *Seed Sci. Res.* **1994**, *4*, 385–405.
28. Obendorf, R. L. *Seed Sci. Res.* **1997**, *7*, 63–74.
29. Hand, C. E.; Auzanneau, F.-I.; Honek, J. F. *Carbohydr. Res.* **2006**, *341*, 1164–1173.
30. Ajayi, K.; Thakur, V. V.; Lapo, R. C.; Knapp, S. *Org. Lett.* **2010**, *12*, 2630–2633.
31. Lo, H. J.; Chang, Y. K.; Ananthan, B.; Lih, Y. H.; Liu, K. S.; Yan, T. H. *J. Org. Chem.* **2019**, *84*, 10065–10075.

32. Nitz, M.; Fenili, D.; Darabie, A. A.; Wu, L.; Cousins, J. E.; McLaurin, J. *FEBS J.* **2008**, *275*, 1663–1674.
33. Sun, Y.; Zhang, G.; Hawkes, C. A.; Shaw, J. E.; McLaurin, J.; Nitz, M. *Bioorg. Med. Chem.* **2008**, *16*, 7177–7184.
34. Savelieff, M. G.; Nam, G.; Kang, J.; Lee, H. J.; Lee, M.; Lim, M. H. *Chem. Rev.* **2019**, *119*, 1221–1322.
35. Cid, M. B.; Alfonso, F.; Martín-Lomas, M. *Carbohydr. Res.* **2004**, *339*, 2303–2307.
36. Nestler, J. E.; Jakubowicz, D. J. *J. Pediatr. Endocrinol Metab.* **2000**, *13*, 1295–1298.
37. Laganà, D. R.; Vitagliano, A. S.; Noventa, A.; Ambrosini, M. G. *Arch. Gynecol. Obstet.* **2018**, *298* 675-668.
38. Diamanti-Kandarakisa, E.; Argyrakopouloua, G.; Economoua, F.; Kandarakia, E.; Koutsilieris, M. *J. Steroid Biochem. Mol. Biol.* **2008**, *109*, 242–246.
39. Rengarajan, T.; Nandakumar, N.; Rajendran, P.; Ganesh, M. K.; Balasubramanian, M. P.; Nishigaki, I. *J. Physiol. Biochem.* **2015**, *71*, 191–204.
40. Federspiel, M.; Fischer, R.; Hennig, M.; Mair, H. J.; Oberhauser, T.; Rimmler, G.; Albiez, T.; Bruhin, J.; Estermann, H.; Gandert, C., et al. *Org. Process Res. Dev.* **1999**, *3*, 266–274.
41. Mahmud, T.; Flatt, P. M.; Wu, X. *J. Nat. Prod.* **2007**, *70*, 1384–1391.
42. Trapero, A.; Egido-Gabás, M.; Bujons, J.; Llebaria, A. *J. Org. Chem.* **2015**, *80*, 3512–3529.
43. Da Silva Pinto, S.; Davies, S. G.; Fletcher, A. M.; Roberts, P. M.; Thomson, J. E. *J. Org. Chem.* **2018**, *83*, 9939–9957.
44. Asano, N.; Nash, R. J.; Molyneux, R. J.; Fleet, G. W. J. *Tetrahedron Asymm.* **2000**, *11*, 1645–1680.
45. Ille, F.; Sommer, L. *Cell. Mol. Life Sci.* **2005**, *62*, 1100–1108.
46. DeAngelis, P. L. *J. Am. Chem. Soc.* **1999**, *121*, 6972.
47. Mahmud, T. *Nat. Prod. Rep.* **2003**, *20*, 137–166.
48. Girard, E.; Desvergnes, V.; Tarnus, C.; Landais, Y. *Org. Biomol. Chem.* **2010**, *8*, 5628–5634.
49. Asamizu, S. *Biosci Biotechnol Biochem.* **2017**, *81*, 871–881.
50. Mahmud, T. *Curr. Opin. Chem. Biol.* **2009**, *13*, 161–170.
51. Diaz, A. D. L. *Curr. Med. Chem.* **2010**, *17*, 2393–2418.
52. Koizumi, K.; Nishimoto, Y. 1979, XXXIII.
53. Uyeda, M.; Mizukami, M.; Yokomizo, K.; Suzuki, K. *Biosci. Biotechnol. Biochem.* **2001**, *65*, 1252–1254.
54. Lo, H. J.; Chang, Y. K.; Yan, T. H. *Org. Lett.* **2012**, *14*, 5896–5899.
55. Palaniappan, N.; Dhote, V.; Ayers, S.; Starosta, A. L.; Wilson, D. N.; Reynolds, K. A. *Chem. Biol.* **2009**, *16*, 1180–1189.
56. Cooper, M. L.; Blair, C. B.; Jones, K. T.; Minich, C. S. *Bioorg.Med. Chem. Lett.* **1997**, 7, 1747–1752.
57. Jaynes, B. H.; Elliott, N. C.; Jefson, M. R.; Koss, D. A.; Schicho, D. L. *J. Org. Chem.* **1994**, *59*, 1224–1225.
58. Visser, M. S.; Freeman-Cook, K. D.; Brickner, S. J.; Brighty, K. E.; Le, P. T.; Wade, S. K.; Monahan, R.; Martinelli, G. J.; Blair, K. T.; Moore, D. E. *Bioorganic Med. Chem. Lett.* **2010**, *20*, 6730–6734.
59. Lee, H. B.; Kim, C. J.; Kim, J. S.; Hong, K. S.; Cho, K. Y. *Lett. Appl. Microbiol.* **2003**, *36*, 387–391.
60. Zheng, Y. G.; Shentu, X. P.; Shen, Y. C. *J. Enzyme Inhib. Med. Chem.* **2005**, *20*, 49–53.
61. Chang, Y. T.; Choi, G.; Bae, Y. S.; Burdett, M.; Moon, H. S.; Lee, J. W.; Gray, N. S.; Schultz, P. G.; Meijer, L.; Chung, S. K., et al. *ChemBioChem* **2002**, *3*, 897–901.
62. Mills, S. J.; Potter, B. V. L. *Bioorganic Med. Chem.* **2003**, *11*, 4245–4253.
63. Fauq, A. H.; Kozikowski, A. P.; Powis, G.; Aksoy, I. A.; Melder, D. C.; Aksoy, S.; Eichinger, H. *Cancer Chemother. Pharmacol.* **1991**, *29*, 95–104.
64. Panayotou, G. J. H.; Waterfield, G.; Clarke, M. D. *Trends Cell Biol.* **1992**, *2*, 358–360.
65. Clarke, J. H. *Curr. Biol.* **2003**, *13*, R815−R817.

66. Díaz, L.; Bujons, J.; Casas, J.; Llebaria, A.; Delgado, A. *J. Med. Chem.* **2010**, *53*, 5248–5255.
67. Trapero, A.; González-Bulnes, P.; Butters, T. D.; Llebaria, A. *J. Med. Chem.* **2012**, *55*, 4479–4488.
68. Trapero, A.; Llebaria, A. *ACS Med. Chem. Lett.* **2011**, *2*, 614–619.
69. Yñigez-Gutierrez, A. E.; Bachmann, B. O. *J. Med. Chem.* **2019**, *62*, 8412–8428.
70. Irvine, R. F.; Schell, M. J. *Nat. Rev. Mol. Cell Biol.* **2001**, *2*, 327–338.
71. Apiyo, D.; Zhao, L.; Tsai, M.-D.; Selby, T. L. *Biochemistry* **2005**, *44*, 9980–9989.
72. Bai, C.; Zhao, L.; Tsai, M.-D.; Bruzik, K. S. *J. Am. Chem. Soc.* **2010**, *132*, 1210–1211.
73. Berridge, M. J.; Irvine, R. F. *Nature* **1989**, *341*, 197–205.
74. Kozikowski, A. P.; Fauq, A. H.; Aksoy, I. A.; Seewald, M. J.; Powis, G. *J. Am. Chem. Soc.* **1990**, *112*, 7403–7404.
75. Safrany, S. T.; Wilcox, R. A.; Liu, C.; Potter, B. V. L.; Nahorski, S. R. *Eur. J. Pharmacol. Mol. Pharmacol.* **1992**, *226*, 265–272.
76. Stricker, R.; Chang, Y.-T.; Chung, S.-K.; Reiser, G. *Biochem. Biophys. Res. Commun.* **1996**, *228*, 596–604.
77. Fan, G.; Baker, M. L.; Wang, Z.; Baker, M. R.; Sinyagovskiy, P. A.; Chiu, W.; Ludtke, S. J.; Serysheva, I. I. *Nature* **2015**, *527*, 336–341.
78. Zonia, L.; Cordeiro, S.; Tupý, J.; Feijó, J. A. *Plant Cell* **2002**, *14*, 2233–2249.
79. Kenta Goto, K.; Tamai, H.; Takeda, Y.; Tanaka, H.-N.; Mizuno, T.; Akihiro, M.; Hideharu, I.; Makoto, I.; Kiso, I.; Ando, I. *Org. Lett.* **2019**, *21*, 4054–4057.
80. Ferguson, M. A. J.; Williams, A. F. *Annu. Rev. Biochem.* **1988**, *57*, 285–320.
81. Martín-Lomas, M.; Flores-Mosquera, M.; Khiar, N. *Eur. J. Org. Chem.* **2000**, 1539–1545.
82. Ferguson, M. A.; Kinoshita, T.; Hart, G. W. *Glycosyl phosphatidyl inositol Anchors*. Cold Spring Harbor Laboratory Press, 2009.
83. Heiss, C.; Wang, Z.; Black, I.; Azadi, P.; Fichorova, R. N.; Singh, B. N. *Carbohydr. Res.* **2016**, *419*, 51–59.
84. Li., C.; Wanh, L. X. *Chem Rev.* **2018**, *118*, 8359–8413.
85. Ainge, G. D.; Martin, W. J.; Compton, B. J.; Hayman, C. M.; Larsen, D. S.; Yoon, S.; Wilson, I. A.; Harper, J. L.; Painter, G. F. *J. Med. Chem.* **2011**, *54*, 7268–7279.
86. Koumbis, A. E.; Duarte, C. D.; Nicolau, C.; Lehn, J.-M. *ChemMedChem* **2011**, *6*, 169–180.
87. Maffucci, T.; Piccolo, E.; Cumashi, A.; Iezzi, M.; Riley, A. M.; Saiardi, A.; Godage, H. Y.; Rossi, C.; Broggini, M.; Iacobelli, S., et al. *Cancer Res.* **2005**, *65*, 8339–8349.
88. Wang, Y. K.; Chen, W.; Blair, D.; Pu, M.; Xu, Y.; Miller, S. J.; Redfield, A. G.; Chiles, T. C.; Roberts, M. F. *J. Am. Chem. Soc.* **2008**, *130*, 7746–7755.
89. Szijgyarto, Z.; Garedew, A.; Azevedo, C.; Saiardi, A. *Science* **2011**, *334*, 802–805.
90. Shears, S. B. *Mol. Pharmacol.* **2009**, *76*, 236–252.
91. Wilson, M. S. C.; Livermore, T. M.; Saiardi, A. *Biochem. J.* **2013**, *452*, 369–379.
92. Wang, H.; Godage, H. Y.; Riley, A. M.; Weaver, J. D.; Shears, S. B.; Potter, B. V. L. *Chem. Biol.* **2014**, *21*, 689–699.
93. Riley, A. M.; Wang, H.; Weaver, J. D.; Shears, S. B.; Potter, B. V. L. *Chem. Commun.* **2012**, *48*, 11292.
94. Wu, M.; Dul, B. E.; Trevisan, A. J.; Fiedler, D. *Chem. Sci.* **2013**, *4*, 405–410.
95. Falshaw, A.; Hart, J. B.; Tyler, P. C. *Carbohydr. Res.* **2000**, *329*, 301–308.
96. Łysek, R.; Vogel, P. *Tetrahedron* **2006**, *62*, 2733–2768.
97. Larner, J.; Price, J. D.; Heimark, D.; Smith, L.; Rule, G.; Piccariello, T.; Fonteles, M. C.; Pontes, C.; Vale, D.; Huang, L. *J. Med. Chem.* **2003**, *46*, 3283–3291.
98. Michell, R. H. *Biochim. Biophys. Acta–Rev. Biomembr.* **1975**, *415*, 81–147.
99. Balla, T. *Physiol. Rev.* **2013**, *93*, 1019–1137.
100. Schink, K. O.; Tan, K.-W.; Stenmark, H. *Annu. Rev. Cell Dev. Biol.* **2016**, *32*, 143–171.
101. Wallroth, A.; Haucke, V. *J. Biol. Chem.* **2018**, *293*, 1526–1535.
102. Mondal, S.; Rakshit, A.; Pal, S.; Datta, A. *ACS Chem. Biol.* **2016**, *11*, 1834–1843.

103. Pizarro-Cerdá, J.; Kühbacher, A.; Cossart, P. *Biochim. Biophys. Acta–Mol. Cell Biol. Lipids* **2015**, *1851*, 911–918.
104. Dieck, C. B.; Boss, W. F.; Perera, I.Y. *Front. Plant Sci.* **2012**, *3*doi: 10.3389/fpls.2012.00050.
105. Hancock, J. T.; Desican, R.; Neil, S. J. *Biochem. Society Trans.* **2001**, *29*, 345–350.
106. Kinoshita, T.; Fujiita, M. *J. Lipid Res.* **2016**, *57*, 6–24.
107. Takeda, T.; Miyata, J.; Kawagoe, T.; Takahashi, K.; Kitani, M.; Kinoshita, T. *Trends Biochem. Sci.* **1995**, *20*, 367–371.
108. Keddie, N. S.; Ye, Y.; Aslam, T.; Luyten, T.; Bello, D.; Garnham, C.; Bultynck, G.; Galione, A.; Conway, S. J. *Chem. Commun.* **2011**, *47*, 242–244.
109. Fauq, A. H.; Zaidi, J. H.; Wilcox, R. A.; Varvel, G.; Nahorski, S. R.; Kozikowski, A. P.; Erneux, C. *Tetrahedron Lett.* **1996**, *37*, 1917–1920.
110. Huang, G. D.; Zhang, W.; Davrazou, H.; Kutateladze, E.; Shi, T. G.; Gozani, X.; Prestwich, O. *J. Am. Chem. Soc.* **2007**, *129*, 6498–6506.
111. Zhang, H.; Xu, Y.; Zhang, Z.; Liman, E. R.; Prestwich, G. D. *J. Am. Chem. Soc.* **2006**, *128*, 5642–5643.
112. Xu, Y.; Lee, S. A.; Kutateladze, T. G.; Sbrissa, D.; Shisheva, A.; Prestwich, G. D. *J. Am. Chem. Soc.* **2006**, *128*, 885–897.
113. Zhao, L.; Liu, Y.; Bruzik, K. S.; Tsai, M.-D. *J. Am. Chem. Soc.* **2003**, *125*, 22–23.
114. Gajewiak, J.; Xu, Y.; Lee, S. A.; Kutateladze, T. G.; Prestwich, G. D. *Org. Lett.* **2006**, *8*, 2811–2813.
115. Kubiaak, R. J.; Bruzik, K. S. *J. Am. Chem. Soc.* **2001**, *123*, 1760–1761.
116. Roberts, M. F.; Khan, H. M.; Goldstein, R.; Reuter, N.; Gershenson, A. *Chem. Rev.* **2018**, *118*, 8435–8473.
117. Morii, T.; Sugimoto, K.; Makino, K.; Otsuka, M.; Imoto, K.; Mori, Y. *J. Am. Chem. Soc* **2002**, 1138–1139.
118. Ainge, G. D.; Parlane, N. A.; Denis, M.; Dyer, B. S.; Härer, A.; Hayman, C. M.; Larsen, D. S.; Painter, G. F. *J. Org. Chem.* **2007**, *72*, 5291–5296.
119. Rzepecki, P. W.; Prestwich, G. D. *J. Org. Chem.* **2002**, *67*, 5454–5460.
120. Prestwich, G. D. *Chem. Biol.* **2004**, *11*, 619–637.
121. Rowland, M. M.; Bostic, H. E.; Gong, D.; Speers, A. E.; Lucas, N.; Cho, W.; Cravatt, B. F.; Best, M. D. *Biochemistry* **2011**, *50*, 11143–11161.
122. Tsai, Y.-H.; Götze, S.; Azzouz, N.; Hahm, H. S.; Seeberger, P. H.; Varon Silva, D. *Angew. Chemie Int. Ed.* **2011**, *50*, 9961–9964.
123. Ding, N.; Li, X.; Chinoy, Z. S.; Boons, G.-J. *Org. Lett.* **2017**, *19*, 3827–3830.
124. Hederos, M.; Konradsson, P. *J. Org. Chem.* **2005**, *70*, 7196–7207.
125. Mensah, E. A.; Yu, F.; Nguyen, H. M. *J. Am. Chem. Soc.* **2010**, *132*, 14288–14302.
126. Swarts, B. M.; Guo, Z. *J. Am. Chem. Soc.* **2010**, *132*, 6648–6650.
127. Daniellou, R.; Palmer, D. R. J. *Carbohydr. Res.* **2006**, *341*, 2145–2150.
128. Swarts, B. M.; Guo, Z. *Adv. Carbohydr. Chem. Biochem.* **2012**, 137–219.
129. Stallforth, P.; Lepenies, B.; Adibekian, A.; Seeberger, P. H. *J. Med. Chem.* **2009**, *52*, 5561–5577.
130. McConnell, M. S.; Mensah, E. A.; Nguyen, H. M. *Carbohydr. Res.* **2013**, *381*, 146–152.
131. Ainge, G. D.; Compton, B. J.; Hayman, C. M.; Martin, W. J.; Toms, S. M.; Larsen, D. S.; Harper, J. L.; Painter, G. F. *J. Org. Chem.* **2011**, *76*, 4941–4951.
132. Xue, Z.; Guo, J. *J. Am. Chem. Soc.* **2003**, *125*, 16334–16339.
133. Dyer, B. S.; Jones, J. D.; Ainge, G. D.; Denis, M.; Larsen, D. S.; Painter, G. F. *J. Org. Chem.* **2007**, *72*, 3282–3288.
134. Pekari, K.; Schmidt, R. R. *J. Org. Chem.* **2003**, *68*, 1295–1308.
135. Phalipon, P.; Mulard, A.; Sansonetti, L. A. *J. Microbes Infect.* **2008**, *10*, 1057–1062.
136. Cramer, J. P.; Lepenies, B.; Kamena, F.; Hölscher, C.; Freudenberg, M. A.; Burchard, G. D.; Wagner, H.; Kirschning, C. J.; Liu, X.; Seeberger, P. H. *Microbes Infect.* **2008**, *10*, 1259–1265.

137. Liu, X.; Stocker, B. L.; Seeberger, P. H. *J. Am. Chem. Soc.* **2006**, *128*, 3638–3648.
138. Hölemann, A.; Stocker, B. L.; Seeberger, P. H. *J. Org. Chem.* **2006**, *71*, 8071–8088.
139. Kwon, Y.-U.; Liu, X.; Seeberger, P. H. *Chem. Commun.* **2005**, 2280–2282.
140. Liu, X.; Kwon, Y.-U.; Seeberger, P. H. *J. Am. Chem. Soc.* **2005**, *127*, 5004–5005.
141. Kim, H.; McGrath, B. M.; Silverstone, P. H. *Hum. Psychopharmacol.* **2005**, *20*, 309–326.
142. Kozikowski, A. P.; Kiddle, J. J.; Frew, T.; Berggren, M.; Powis, G. *J. Med. Chem.* **1995**, *38*, 1053–1056.
143. Riley, A. M.; Correa, V.; Mahon, M. F.; Taylor, C. W.; Potter, B. V. L. *J. Med. Chem.* **2001**, *44*, 2108–2117.
144. Longo, C. M.; Wei, Y.; Roberts, M. F.; Miller, S. J. *Angew. Chemie–Int. Ed.* **2009**, *48*, 4158–4161.
145. Moris, M. A.; Caron, A. Z.; Guillemette, G.; Rognan, D.; Schmitt, M.; Schlewer, G. *J. Med. Chem.* **2005**, *48*, 1251–1255.
146. Conway, S. J.; Miller, G. J. *Nat. Prod. Rep.* **2007**, *24*, 687.
147. Conway, S. J.; Gardiner, J.; Grove, S. J. A.; Johns, M. K.; Lim, Z.-Y.; Painter, G. F.; Robinson, D. E. J. E.; Schieber, C.; Thuring, J. W.; Wong, L. S.-M., et al. *Org. Biomol. Chem.* **2010**, *8*, 66–76.
148. Song, F.; Zhang, J.; Cui, Q.; Wang, T.; Chen, W.; Li, L.; Xi, Z. *Tetrahedron Lett.* **2012**, *53*, 1102–1104.
149. Santhosh, K. S. K.; Nair, C. P. R.; Radhakrishnan, T. S.; Ninan, K. N. *Eur. Polym. J.* **2007**, *43*, 2504–2514.
150. Riley, A. M.; Wang, H.; Shears, S. B.; Potter, B. V. L. *Chem. Commun.* **2015**, *51*, 12605–12608.
151. https://doi.org/10.1021/acs.jmedchem.9b01152.
152. Dudev, T.; Mazmanian, K.; Weng, W.-H.; Grauffel, C.; Lim, C. *Acc. Chem. Res.* **2019**, *52*, 2960–2970.
153. Thomas, M. P.; Mills, S. J.; Potter, B. V. L. *Angew. Chemie–Int. Ed.* **2016**, *55*, 1614–1650.
154. Kılbaş, B.; Balci, M. *Tetrahedron* **2011**, *67*, 2355–2389.
155. Trapero, A.; Egido-Gabás, M.; Llebaria, A. *Medchemcomm* **2013**, *4*, 1584.
156. Peng, Z.; Maxwell, D.; Sun, D.; Ying, Y.; Schuber, P. T.; Bhanu Prasad, B. A.; Gelovani, J.; Yung, W.-K. A.; Bornmann, W. G. *Tetrahedron* **2014**, *70*, 984–990.
157. Kadirvel, M.; Abudalal, S. A.; Rajendran, R.; Gbaj, A.; Demonacos, C.; Freeman, S. *Carbohydr. Res.* **2012**, *355*, 13–18.
158. Catelani, G.; Andrea, A.; Guazzelli, L.; Griselli, A.; Testi, N.; Chiacchio, M. A.; Legnani, L.; Toma, L. *Carbohydr. Res.* **2017**, 443–444 29-36.
159. Okauchi, T.; Nakamura, S.; Tsubaki, K.; Asakawa, M.; Kitamura, M. *Carbohydr. Res.* **2017**, *448*, 24–27.
160. Šala, M.; Kolar, J.; Strlič, M.; Kočevar, M. *Carbohydr. Res.* **2006**, *341*, 897–902.
161. Šala, M.; Makuc, D.; Kolar, J.; Plavec, J.; Pihlar, B. *Carbohydr. Res.* **2011**, *346*, 488–494.
162. Day, G. M.; van de Streek, J.; Bonnet, A.; Burley, J. C.; Jones, W.; Motherwell, W. D. S. *Cryst. Growth Des.* **2006**, *6*, 2301–2307.
163. Krishnaswamy, S.; Shashidhar, M. S. *J. Org. Chem.* **2018**, *83*, 3952–3959.
164. Shashidhar, M. S.; Krishnaswamy, S. *Acc. Chem. Res.* **2019**, *52*, 437–446.
165. Catimel, B.; Yin, M.-X.; Schieber, C.; Condron, M.; Patsiouras, H.; Catimel, J.; Robinson, D. E. J. E.; Wong, L. S.-M.; Nice, E. C.; Holmes, A. B., et al. *J. Proteome Res.* **2009**, *8*, 3712–3726.
166. Hosoda, A.; Nomura, E.; Mizuno, K.; Taniguchi, H. *J. Org. Chem.* **2001**, *66*, 7199–7201.
167. Hosoda, A.; Miyake, Y.; Nomura, E.; Taniguchi, H. *Chem. Lett.* **2003**, *32*, 1042–1043.
168. Duchek, J.; Adams, D. R.; Hudlicky, T. *Chem. Rev.* **2011**, *111*, 4223–4258.
169. Vitelio, C.; Bellomo, A.; Brovetto, M.; Seoane, G.; Gonzalez, D. *Carbohydr. Res.* **2004**, *339*, 1773–1778.
170. Pilgrim, S.; Kociok-Köhn, G.; Lloyd, M. D.; Lewis, S. E. *Chem. Commun.* **2011**, *47*, 4799.

171. Angyal, S. J.; Odier, L.; Tate, M. E. *Carbohydr. Res.* **1995**, *266*, 143–146.
172. Lightfoot, M. P.; Mair, F. S.; Pritchard, R. G.; Warren, J. E. *Chem. Commun.* **1999**, 1945–1946.
173. Mehta, G.; Ramesh, S. S. *Tetrahedron Lett.* **2001**, *42*, 1987–1990.
174. Mehta, G.; Pujar, S. R.; Ramesh, S. S.; Islam, K. *Tetrahedron Lett.* **2005**, *46*, 3373–3376.
175. Sureshan, K. M.; Shashidhar, M. S.; Praveen, T.; Das, T. *Chem. Rev.* **2003**, *103*, 4477–4504.
176. Estevez, V. A.; Prestwich, G. D. *J. Am. Chem. Soc.* **1991**, *113*, 9885–9887.
177. Chiara, J. L.; Martín-Lomas, M. *Tetrahedron Lett.* **1994**, *35*, 2969–2972.
178. Kiddle, J. J. *Chem. Rev.* **1995**, *95*, 2189–2202.
179. Gigg, J.; Gigg, R. *Carbohydr. Res.* **1997**, *299*, 77–83.
180. Riley, A. M.; Jenkins, D. J.; Potter, B. V. L. *Carbohydr. Res.* **1998**, *314*, 277–281.
181. Liu, C.; Riley, A. M.; Yang, X.; Shears, S. B.; Potter, B. V. L. *J. Med. Chem.* **2001**, *44*, 2984–2989.
182. Kubiak, R. J.; Bruzik, K. S. *J. Org. Chem.* **2003**, *68*, 960–968.
183. Bello, D.; Aslam, T.; Bultynck, G.; Slawin, A. M. Z.; Roderick, H. L.; Bootman, M. D.; Conway, S. J. *J. Org. Chem.* **2007**, *72*, 5647–5659.
184. Murali, C.; Gurale, B. P.; Shashidhar, M. S. *Eur. J. Org. Chem.* **2010**, 755–764.
185. Paul, B. J.; Willis, J.; Martinot, T. A.; Ghiviriga, I.; Abboud, K. A.; Hudlicky, T. *J. Am. Chem. Soc.* **2002**, *124*, 10416–10426.
186. Paul, B. J.; Martinot, T. A.; Willis, J.; Hudlicky, T. *Synthesis* **2001**, 952–956.
187. Jiang, C.; Schedler, D. J. A.; Morris, P. E.; Zayed, A.-H.; Baker, D. C. *Carbohydr. Res.* **1990**, *207*, 277–285.
188. Nguyen, B. V.; York, C.; Hudlicky, T. *Tetrahedron* **1997**, *53*, 8807–8814.
189. Schedler, D. J. A.; Baker, D. C. *Carbohydr. Res.* **2004**, *339*, 1585–1595.
190. Bru, M.; Kotkar, S. P.; Kar, N.; Köhn, M. *Chem. Sci.* **2012**, *3*, 1893–1902.
191. Dinkel, C.; Schultz, C. *Tetrahedron Lett.* **2003**, *44*, 1157–1159.
192. Subramanian, D.; Laketa, V.; Müller, R.; Tischer, C.; Zarbakhsh, S.; Pepperkok, R.; Schultz, C. *Nat. Chem. Biol.* **2010**, *6*, 324–326.
193. Walker, D. R.; Feeney, J. M.; Trentham, J. *Biochemistry* **1989**, *28*, 1157–1159.
194. Walker, J. W.; Somlyo, A. V.; Goldman, Y. E.; Somlyo, A. P.; Trentham, D. R. *Nature* **1987**, *327*, 249–252.
195. Li, W.; Llopis, J.; Whitney, M.; Zlokarnik, G.; Tsien, R. Y. *Nature* **1998**, *392*, 936–941.
196. Vacca, J. P.; Desolms, S. J.; Huff, J. R. *J. Am. Chem. Soc.* **1987**, *109*, 3478–3479.
197. Desai, N.; Gigg, T.; Gigg, J.; Martin-Zamora, R.; Schnetz, E. *Carbohydr. Res.* **1994**, *258*, 135–144.
198. Bakera, G.R.; Billingtonb, D.C.; Gani Tetrahedron, D. 1991, 41, 3895-3908.
199. Vidyasagar, A.; Pathigoolla, A.; Sureshan, K. M. *Org. Biomol. Chem.* **2013**, *11*, 5443–5453.
200. Gigg, J.; Gigg, R.; Payne, S.; Conant, R. *Carbohydr. Res.* **1985**, *142*, 132–134.
201. Vibhute, A. M.; Sureshan, K. M. *RSC Adv.* **2013**, *3*, 7321.
202. Lence, E.; Castedo, L.; González-Bello, C. *Chem. Soc. Rev.* **2008**, *37*, 1689–1708.
203. Mills, S. J.; Riley, A. M.; Liu, C.; Mahon, M. F.; Potter, B. V. L. *Chem. Eur. J.* **2003**, *9*, 6207–6214.
204. Han, F.; Hayashi, M.; Watanabe, Y. *Chem. Lett.* **2003**, *32*, 724–725.
205. Chung, S. K.; Kwon, Y. U.; Shin, J. H.; Chang, Y. T.; Lee, C.; Shin, B. G.; Kim, K. C.; Kim, M. J. *J. Org. Chem.* **2002**, *67*, 5626–5637.
206. Praveen, T.; Samanta, U. K.; Das, T.; Shashidhar, M. S.; Chakrabarti, P. *J. Am. Chem. Soc.* **1998**, *120*, 3842–3845.
207. Dimakos, V.; Taylor, M. S. *Chem. Rev.* **2018**, *118*, 1457–11517.
208. Watanabe, Y.; Kawamoto, M.; Shintaku, H.; Tanabe, K.; Ohta, H.; Hayashi, M. *Asian J. Org. Chem.* **2013**, *2*, 927–930.
209. Watanabe, Y.; Uemura, T.; Yamauchi, S.; Tomita, K.; Saeki, T.; Ishida, R.; Hayashi, M. *Tetrahedron* **2013**, *69*, 4657–4664.

210. Miranda-Molina, A.; Marquina-Bahena, S.; López-Munguía, A.; Álvarez, L.; Castillo, E. *Carbohydr. Res.* **2012**, *360*, 93–101.
211. Hart, J. B.; Kröger, L.; Falshaw, A.; Falshaw, R.; Farkas, E.; Thiem, J.; Win, A. L. *Carbohydr. Res.* **2004**, *339*, 1857–1871.
212. Garrett, S. W.; Liu, C.; Riley, A. M.; Potter, B.V. L. *J. Chem. Soc. Perkin Trans.* **1998**, *1* (57), 1367–1368.
213. Baudin, G.; Glänzer, B. I.; Swaminathan, K. S.; Vasella, A. *Helv. Chim. Acta* **1988**, *71*, 1367–1378.
214. Riley, A.M.; Potter, B.V. L. **1998**, *39*, 6769–6772.
215. Praveen, T.; Shashidhar, M. S. *Carbohydr Res.* **2001**, *5*, 409–411.
216. Bhosekar, G.; Murali, C.; Gonnade, R. G.; Shashidhar, M. S.; Bhadbhade, M. M. *Cryst. Growth Des.* **2005**, *5*, 1977–1982.
217. Vibhute, A. M.; Vidyasagar, A.; Sarala, S.; Sureshan, K. M. *Chem. Commun.* **2012**, *48*, 2448.
218. Schlueter, U.; Lu, J.; Fraser-Reid, B. *Org. Lett.* **2003**, *5*, 255–257.
219. Sarmah, M. P.; Shashidhar, M. S.; Sureshan, K. M.; Gonnade, R. G.; Bhadbhade, M. M. *Tetrahedron* **2005**, *61*, 4437–4446.
220. Sureshan, K. M.; Das, T.; Shashidhar, M. S.; Gonnade, R. G.; Bhadbhade, M. M. *Eur. J. Org. Chem.* **2003**, *4*, 1035–1041.
221. Sureshan, K. M.; Shashidhar, M. S. *Tetrahedron Lett.* **2001**, *42*, 3037–3039.
222. Samanta, U.; Puranik, V. G.; Chakrabarti, P.; Thoniyot, P.; Shashidhar, M. S. *Acta Crystallogr. Sec-C* **1998**, *54*, 1289–1291.
223. Uhlmann, A.; Vasella, A. *Helv. Chim. Acta* **1992**, *75*, 1979–1994.
224. Chen, X.; Brauman, J. I. *J. Am. Chem. Soc.* **2008**, *130*, 15038–15046.
225. Sureshan, K. M.; Subramanian, D.; Shashidhar, M. S. *Tetrahedron* **2009**, *65*, 2703–2710.
226. Devaraj, S.; Jagdhane, R. C.; Shashidhar, M. S. *Carbohydr. Res.* **2009**, *344*, 1159–1166.
227. Devaraj, S.; Shashidhar, M. S.; Dixit, S. S. *Tetrahedron* **2005**, *61*, 529–536.
228. Sureshan, K. M.; Shashidhar, M. S. *Tetrahedron Lett.* **2000**, *41*, 4185–4188.
229. Murali, C.; Shashidhar, M. S.; Gopinath, C. S. *Tetrahedron* **2007**, *63*, 4149–4155.
230. Mart, A.; Shashidhar, M. S. *Tetrahedron* **2012**, *68*, 9769–9776.
231. Anderson, L.; DeLuca, E. S.; Bieder, A.; Post, G. G. *J. Am. Chem. Soc.* **1957**, *79*, 1171–1174.
232. Sardessai, R.; Krishnaswamy, S.; Shashidhar, M. S. *CrystEngComm* **2012**, *14*, 8010–8016.
233. Godage, H.Y.; Riley, A. M.; Woodman, T. J.; Potter, B.V. L. *Chem. Commun.* **2006**, 2989–2991.
234. Nkambule, C. M.; Kwezi, N. W.; Kinfe, H. H.; Nokwequ, M. G.; Gammon, D. W.; Oscarson, S.; Karlsson, E. *Tetrahedron* **2011**, *67*, 618–623.
235. Gilbert, R. C.; Holmes, I. H.; Pestchanker, A. B.; Young, M. J. *Carbohydr. Res.* **1992**, *234*, 117–130.
236. Swarbrick, J. M.; Cooper, S.; Bultynck, G.; Gaffney, P. R. J. *Org. Biomol. Chem.* **2009**, *7*, 1709.
237. Swarbrick, J. M.; Gaffney, P. R. J. *J. Org. Chem.* **2010**, *75*, 4376–4386.
238. Mondal, S.; Prathap, A.; Sureshan, K. M. *J. Org. Chem.* **2013**, *78*, 7690–7700.
239. Conway, S. J.; Gardiner, J.; Grove, S. J. A.; Johns, M. K.; Lim, Z.Y.; Painter, G. F., et al. *Org. Biomol. Chem.* **2010**, *8*, 66–76.
240. Jagdhane, R. C.; Shashidhar, M. S. *Tetrahedron* **2011**, *67*, 7963–7970.
241. Ogawa, S.; Ohishi, Y.; Asada, M.; Tomoda, A.; Takahashi, A.; Ooki, Y.; Mori, M.; Itoh, M.; Korenaga, T. *Org. Biomol. Chem.* **2004**, *2*, 884–889.
242. Zhang, H.; Thompson, J.; Prestwich, G. D. *Org. Lett.* **2009**, *11*, 1551–1554.
243. Spiers, I. D.; Freeman, S.; Poyner, D. R.; Schwalbe, C. H. *Tetrahedron Lett.* **1995**, *36*, 2125–2128.
244. Wessig, P.; Möllnitz, K.; Hübner, S. *Synlett* **2010**, 1497–1500.

245. Chida, N.; Nakazawa, K.; Ohtsuka, M.; Suzuki, M.; Ogawa, S. *Chem. Lett.* **1990**, *19*, 423–426.
246. McCasland, G. E.; Horswill, E. C. *J. Am. Chem. Soc.* **1953**, *75*, 4020–4026.
247. Gurale, B. P.; Shashidhar, M. S.; Gonnade, R. G. *J. Org. Chem.* **2012**, *77*, 5801–5807.
248. Takahashi, H.; Kittaka, H.; Ikegami, S. *J. Org. Chem.* **2001**, *66*, 2705–2716.
249. Keddie, N. S.; Bultynck, G.; Luyten, T.; Slawin, A. M. Z.; Conway, S. J. *Tetrahedron Asymmetry* **2009**, *20*, 857–866.
250. Guidot, J. P.; Gall, T. L.; Mioskowski, C. *Tetrahedron Lett.* **1994**, *35*, 6671–6672.
251. Jenkins, D. J.; Potter, B.V. L. *J. Chem. Soc. Perkin Trans.* **1998**, *1*, 41–50.
252. Chen, J.; Feng, L.; Prestwich, G. D. *J. Org. Chem.* **1998**, *63*, 6511–6522.
253. Clive, D. L. J.; He, X.; Postema, M. H. D.; Mashimbye, M. J. *J. Org. Chem.* **1999**, *64*, 4397–4410.
254. Suzuki, T.; Suzuki, S. T.; Yamada, I.; Koashi, Y.; Yamada, K.; Chida, N. *J. Org. Chem.* **2002**, *67*, 2874–2880.
255. Catelani, G.; Corsaro, A.; Andrea, F. D.; Mariani, M.; Pistarà, V. *Bioorg. Med. Chem. Lett.* **2002**, *12*, 3313–3315.
256. Marnera, G.; D'Alarcao, M. *Carbohydr. Res.* **2006**, *341*, 1105–1116.
257. Heo, J.-N.; Holson, E. B.; Roush, W. R. *Org. Lett.* **2003**, *5*, 1697–1700.
258. Conrad, R. M.; Grogan, M. J.; Bertozzi, C. R. *Org. Lett.* **2002**, *4*, 1359–1361.
259. Kornienko, A.; Alarcao, M. D. *Tetrahedron: Asymmetry* **1999**, *10*, 827–829.
260. Nishikawa, A.; Saito, S.; Hashimoto, Y.; Koga, K.; Shirai, R. *Tetrahedron Lett.* **2001**, *42*, 9195–9198.
261. Rameh, L. E.; Cantley, L. C. *J. Biol. Chem.* **1999**, *274*, 8347–8350.
262. Saito, S.; Shimazawa, R.; Shirai, R. *Chem. Pharm. Bull. (Tokyo).* **2004**, *52*, 727–732.
263. Sawada, T.; Shirai, R.; Iwasaki, S. *Tetrahedron Lett.* **1996**, *37*, 885–886.
264. Reddy, K. K.; Ye, J.; Falck, J. R.; Capdevila, J. H. *Bioorg. Med. Chem. Lett.* **1997**, *7*, 2115–2116.
265. Colobert, F.; Tito, A.; Khiar; N.; Denni, D.; Medina, M. A.; Martin-Lomas, M.; Ruano, J. L. G.; Solladié, G. *J. Org. Chem.* **1998**, *63*, 8918–8921.
266. Nuzillard, J.-M.; Banchet, A.; Haudrechy, A. *J. Chem. Inf. Model.* **2007**, *47*, 1979–1985.
267. Lee, Y. J.; Lee, K.; Jung, S. I.; Jeon, H. B.; Kim, K. S. *Tetrahedron* **2005**, *61*, 1987–2001.
268. Falck, J. R.; Yadagiri, P. *J. Org. Chem.* **1989**, *54*, 5851–5852.
269. Reddy, K. K.; Saady, M.; Falck, J. R.; Whited, G. *J. Org. Chem.* **1995**, *60*, 3385–3390.
270. Falck, J. R.; Krishna, U. M.; Capdevila, J. H. *Tetrahedron Lett.* **1999**, *40*, 8771–8774.
271. Arjona, O.; de Dios, A.; de la Pradilla, R. F.; Plumeta, J. *Tetrahedron Lett.* **1991**, *32*, 7309–7312.
272. Akiyama, T.; Takechi, N.; Ozaki, S. *Tetrahedron Lett.* **1990**, *31*, 1433–1434.
273. Mercier, D.; Barnett, J. E. G.; Géro, S. D. *Tetrahedron* **1969**, *25*, 5681–5687.
274. Kozikowski, A. P.; Fauq, A. H.; Rusnak, J. M. *Tetrahedron Lett.* **1989**, *30*, 3365–3368.
275. Johnson, S. C.; Tagliaferri, F.; Baker, D. C. *Carbohydr. Res.* **1993**, *250*, 315–321.
276. Hudlicky, T. *Chem. Rev.* **1996**, *96*, 3–30.
277. Desjardins, M.; Brammer, L. E.; Hudlicky, T. *Carbohydr. Res.* **1997**, *304*, 39–42.
278. Jung, P. M. J.; Motherwell, W. B.; Williams, A. S. *Chem. Commun.* **1997**, 1283–1284.
279. Freeman, S.; Hudlicky, T. *Bioorg. Med. Chem. Lett.* **2004**, *14*, 1209–1212.
280. Sato, K.; Akai, S.; Sugita, N.; Ohsawa, T.; Kogure, T.; Shoji, H.; Yoshimura, J. *J. Org. Chem.* **2005**, *70*, 7496–7504.
281. Chida, N.; Yoshinaga, M.; Tobe, T.; Ogawa, S. *Chem. Commun.* **1997**, 1043–1044.
282. Watanabe, Y.; Ogasawara, T.; Nakahira, H.; Matsuki, T.; Ozaki, S. *Tetrahedron Lett.* **1988**, *29*, 5259–5262.
283. Aneja, R.; Aneja, S. G.; Parra, A. *Tetrahedron Lett.* **1996**, *37*, 5081–5082.
284. Mills, S. J.; Potter, B.V. L. *J. Org. Chem.* **1996**, *61*, 8980–8987.
285. Xue, J.; Shao, N.; Guo, Z. *J. Org. Chem.* **2003**, *68*, 4020–4029.

286. Sureshan, K. M.; Ikeda, K.; Asano, N.; Watanabe, Y. *Tetrahedron* **2008**, *64*, 4072–4080.
287. Patil, N. T.; Shashidhar, M. S.; Tamboli, M. I.; Gonnade, R. G. *Cryst. Growth Des.* **2017**, *17*, 5432–5440.
288. Klibanov, A. M. *Nature* **2001**, *409*, 241–246.
289. Ghanem, A.; Aboul-Enein, H. Y. *Tetrahedron Asymmetry* **2004**, *15*, 3331–3351.
290. Bezborodov, A. M.; Zagustina, N. A. *Appl. Biochem. Microbiol.* **2014**, *50*, 313–333.
291. Liu, Y.-C.; Chen, C.-S. *Tetrahedron Lett.* **1989**, *30*, 1617–1620.
292. Gou, D.-M.; Chen, C.-S. *Tetrahedron Lett.* **1992**, *33*, 721–724.
293. Ling, L.; Watanabe, Y.; Akiyama, T.; Ozaki, S. *Tetrahedron Lett.* **1992**, *33*, 1911–1914.
294. Ling, L.; Ozaki, S. *Carbohydr. Res.* **1994**, *256*, 49–58.
295. Manoel, E. A.; Pais, K. C.; Cunha, A. G.; Coelho, M. A. Z.; Freire, D. M. G.; Simas, A. B. C. *Tetrahedron: Asymmetry* **2012**, *23*, 47–52.
296. Bian, J.; Schneider, S. R.; Maguire, R. J. *Tetrahedron Lett.* **2011**, *52*, 5417–5420.
297. Lee, A. M. M.; Painter, G. F.; Compton, B. J.; Larsen, D. S. *J. Org. Chem.* **2014**, *79*, 10916–10931.
298. Takahashi, Y.; Nakayama, H.; Katagiri, K.; Ichikawa, K.; Ito, N.; Takita, T.; Takeuchi, T.; Miyake, T. *Tetrahedron Lett.* **2001**, *42*, 1053–1056.
299. Bruzik, K. S.; Salamończyk, G. M. *Carbohydr. Res.* **1989**, *195*, 67–73.
300. Pietrusiewicz, K. M.; Salamończyk, G. M.; Bruzik, K. S.; Wieczorek, W. *Tetrahedron* **1992**, *48*, 5523–5542.
301. Bruzik, K. S.; Tsai, M. D. *J. Am. Chem. Soc.* **1992**, *114*, 6361–6374.
302. Anderson, R. J.; Osborne, S. L.; Meunier, F. A.; Painter, G. F. *J. Org. Chem.* **2010**, *75*, 3541–3551.
303. Padiyar, L. T.; Wen, Y.-S.; Hung, S.-C. *Chem. Commun.* **2010**, *46*, 5524–5526.
304. Riley, A. M.; Godage, H. Y.; Mahon, M. F.; Potter, B. V. L. *Tetrahedron: Asymmetry* **2006**, *17*, 171–174.
305. Padiyar, L. T.; Zulueta, M. M. L.; Sabbavarapu, N. M.; Hung, S.-C. *J. Org. Chem.* **2017**, *82*, 11418–11430.
306. Patil, P. S.; Hung, S.-C. *Org. Lett.* **2010**, *12*, 2618–2621.
307. Chung, C. C.; Manuel, M. L.; Laxmansingh, Z.; Padiyar, T.; Hung, S.-C. *Org. Lett.* **2011**, *13*, 5496–5499.
308. Edwards, P. J.; Entwistle, D. A.; Genicot, C.; Kim, K. S.; Ley, S. V. *Tetrahedron Lett.* **1994**, *35*, 7443–7446.
309. Gonzalez-Bulnes, P.; Casas, J.; Delgado, A.; Llebaria, A. *Carbohydr. Res.* **2007**, *342*, 1947–1952.
310. Sculimbrene, B. R.; Miller, S. J. *J. Am. Chem. Soc.* **2001**, *123*, 10125–10126.
311. Morgan, A. J.; Komiya, S.; Xu, Y.; Miller, S. J. *J. Org. Chem.* **2006**, *71*, 6923–6931.
312. Matsumura, Y.; Maki, T.; Tsurumaki, K.; Onomura, O. *Tetrahedron Lett.* **2004**, *45*, 9131–9134.
313. Fiori, K. W.; Puchlopek, A. L. A.; Miller, S. J. *Nat. Chem.* **2009**, *1*, 630–634.
314. Lauber, M. B.; Daniliuc, C.-G.; Paradies, J. *Chem. Commun.* **2013**, *49*, 7409–7411.
315. Morgan, A. J.; Wang, Y. K.; Roberts, M. F.; Miller, S. J. *J. Am. Chem. Soc.* **2004**, *126*, 15370–15371.
316. Blum, C.; Rehnberg, N.; Spiess, B.; Schlewer, G. *Carbohydr. Res.* **1997**, *302*, 163–168.
317. Metrano, A. J.; Miller, S. J. *Acc. Chem. Res.* **2019**, *52*, 199–215.
318. Durantie, E.; Huwiler, S.; Leroux, J.-C.; Castagner, B. *Org. Lett.* **2016**, *18*, 3162–3165.
319. Watanabe, Y. *J. Synth. Org. Chem. Japan* **2000**, *58*, 1057–1065.
320. Volwerk, J. J.; Shashidhar, M. S.; Kuppe, A.; Griffith, O. H. *Biochemistry* **1990**, *29*, 8056–8062.
321. Hondal, R. J.; Zhao, Z.; Kravchuk, A. V.; Liao, H.; Riddle, S. R.; Yue, X.; Bruzik, K. S.; Tsai, M.-D. *Biochemistry* **1998**, *37*, 4568–4580.
322. Bruzik, K. S.; Guan, Z.; Riddle, S.; Tsai, M.-D. *J. Am. Chem. Soc.* **1996**, *118*, 7679–7688.

323. Zhou, Q.; Du, F.; Chen, Y.; Fu, Y.; Sun, W.; Wu, Y.; Chen, G. *J. Org. Chem.* **2019**, *84*, 8160–8167.
324. Akiyama, T.; Hara, M.; Fuchibe, K.; Sakamoto, S.; Yamaguchi, K. *Chem. Commun.* **2003**, 1734.
325. Sureshan, K. M.; Shashidhar, M. S.; Varma, A. J. *J. Org. Chem.* **2002**, *67*, 6884–6888.
326. Sureshan, K. M.; Shashidhar, M. S.; Gonnade, R. G.; Puranik, V. G.; Bhadbhade, M. M. *Chem. Commun.* **2001**, 881–882.

CHAPTER EIGHT

Iminosugars

Namakkal G. Ramesh
Department of Chemistry, Indian Institute of Technology Delhi, New Delhi, India

1 General introduction

Carbohydrates are one of the most abundant classes of natural products and are present in the form of monomers, oligomers and polymers. They are found as constituents of biopolymers and other naturally occurring substances, and also bound covalently to other non-carbohydrate biomolecules to form glycoconjugates, the size of which may vary from small molecules to large biopolymers. Smaller glycoconjugates, in the form of anthracyclines, macrolides, alkaloids etc. possess antibiotic activities, while those made of oligosaccharides play important roles in biological storage and transport properties.[1] On the other hand, glycoconjugates obtained through attachment of complex carbohydrates to proteins and lipids, termed as glycoproteins and glycolipids respectively, are generally expressed on the cell surfaces and hence play crucial roles in various cell-mediated processes such as cell-cell recognition, signal transduction, cell-adhesion etc.[2] While the importance of nucleic acids and proteins in providing biochemical targets have been extensively investigated, resulting in the discovery of several marketed drugs, they alone could not account for the biology and function of pathway, cells, tissues etc. The role of carbohydrates in major diseases such as diabetes, cancer, infection etc. have been well-understood and these studies have paved the way for further research on their functions in neurodegenerative and rare diseases as well.[3] Most of the rare diseases have a genetic component and certain genomes are believed to encode enzymes associated with glycan formation and degradation. It is therefore of no wonder to comprehend that disorders of glycosylation and glycan degradation is one of the causes of rare diseases. Unlike natural carbohydrates, their mimics (due to their metabolic stability, affinity and selectivity) are ideally suited to address diseases caused by carbohydrate-related disorders.

Carbohydrates in Drug Discovery and Development. http://dx.doi.org/10.1016/B978-0-12-816675-8.00008-7

1.1 Glycoconjugate processing enzymes

Glycoconjugate processing enzymes are enzymes that are involved in a wide range of important biological processes such as degradation of polysaccharides, intestinal digestion, post-translational modifications, quality-control systems in the endoplasmic reticulum (ER), the lysosomal catabolism of glycoconjugates, etc.[4] Interest in the chemistry and biology of glycoconjugates processing enzymes has grown exponentially during the last few decades due to important roles played by them in controlling the structures of carbohydrates present on the cell surface as well as their functions.[5] These enzymes catalyze the coupling of carbohydrate moieties covalently to various biomolecules such as lipids, proteins or peptides on cell surfaces and thus facilitate the assembly of complex glycans.[5] For instance, they catalyze the biosynthesis of *N*-linked oligosaccharide chains through two distinct series of reactions.[6] First series involves synthesis of a precursor oligosaccharide which is then transferred co-translationally to the protein chain, while the protein is being synthesized on membrane-bound polysomes. In the second series, modifications of the oligosaccharide chain by removal or addition of sugars to give a large number of a variety of oligosaccharidic structures. The oligosaccharides then begin to undergo a number of processing and trimming reactions. The initial reactions catalyzed by glycosidases result in the removal of glucose and mannose residues, while the processing reactions later catalyzed by glycosyltransferases involve an external addition of a number of other sugars, such as *N*-acetylglucosamine, galactose, neuraminic acid, *L*-fucose, *N*-acetylgalactosamine, etc.

Glycosidases (glycoside hydrolases) catalyze the hydrolysis of oligo- and/or polysaccharides through the facilitation of attack of water molecule at the anomeric carbon of sugars.[7] Depending on the overall stereochemical outcome of the anomeric configuration, they can be classified into two categories: (a) retaining glycosidases in which the anomeric stereochemistry of the oligosaccharide is retained after the hydrolysis, and (b) inverting glycosidases wherein there is an inversion of stereochemistry at the anomeric carbon.

Glycosyltransferases are enzymes that are involved in the biosynthesis of oligosaccharides and glycoconjugates.[8] They have also been classified into two types based on the mechanistic pathways they follow: those of the Leloir pathway or of the non-Leloir pathway.[9] While Leloir pathway glycosyltransferases utilize nucleotide mono- or di-phosphate sugar as monosaccharide donors,[10] non-Leloir pathway glycosyltransferases use glycosyl phosphate sucrose or glycolipid phosphate for the synthesis of oligosaccharides and glycoconjugates.[11] Leloir glycosyltransferases catalyze the transfer of sugar moiety from an activated nucleotide sugar to the hydroxyl group of an acceptor, which may be an oligosaccharide, a lipid or a protein.[12] Depending on the glycosyltransferase involved, the reaction can proceed either with an inversion or with retention of configuration at the anomeric centre like glycosidases.

1.2 Inhibitors of glycosidases

Since hydrolysis of glycosidic bonds is an ubiquitous biological process, inhibitors of glycosidases have many potential applications including their use as agrochemicals and therapeutic agents.[5] For example, α-glucosidase plays an important role in controlling blood glucose level in human and in the transport of glucose in insects and fungi.[13b] Hence, inhibitors of α-glucosidase have found their use in the treatment of type II diabetes[13a,e,f] and as insecticides.[13c,d] As glycosidases participate in trimming of complex glycans present on the cell surfaces and viruses, biosynthesis of complex carbohydrates can be disrupted by them and this, in turn, can interfere in various cell-mediated processes.[14] This principle is the basis for discovery of inhibitor based drugs towards the treatment of various diseases such as diabetes,[15] tumor metastasis,[16] viral infections[17] (including HIV, hepatitis B and C virus) and lysosomal storage disorders.[18] Carbohydrate-based inhibitors (glycomimetics) of glycoconjugate processing enzymes are expected to be potent molecules against such diseases due to their structural resemblance to naturally occurring carbohydrates. Their ability to mimic the oxo-carbenium ion transition state offers unparalleled advantages over other classes of inhibitors.[19] Among them, several reversible inhibitors that are able to block glycoconjugate formations/functions have emerged as versatile tools for biochemists and cell biologists, especially for those in quest for the development of novel and new therapeutic agents.[5]

Marketed drugs of this class include, Zanamivir® **1**,[20f] Tamiflu® **2**,[20d] Topamax® **3** (epilepsy),[20g] Zavesca® **4** (Gaucher's disease),[20b] Glyset® **5** (against type-II diabetes),[20c] Orthovisc® **6** (osteoarthritis),[20a] Fragmin® **7**[20e] (as anti-infective agents) and Fraxiparin® **8** (for the treatment of thrombosis)[20e] (Fig. 8.1).

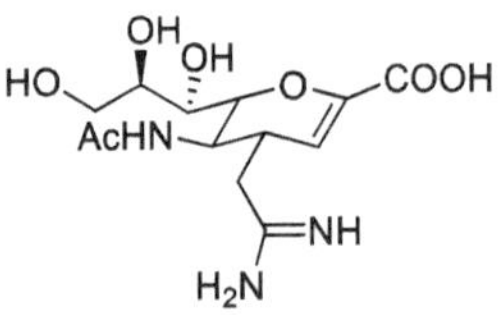

1
Zanamivir (Relenza®)

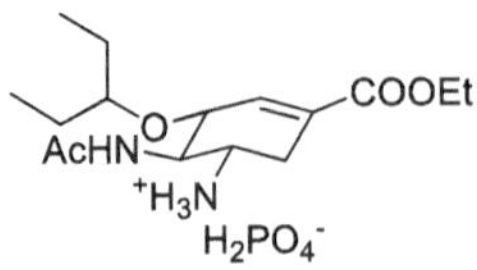

2
Oseltamivir phosphate (Tamiflu®)

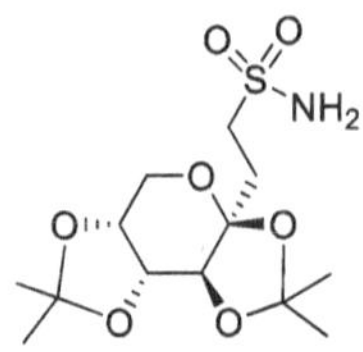

3
Topiramate (Topamax®)

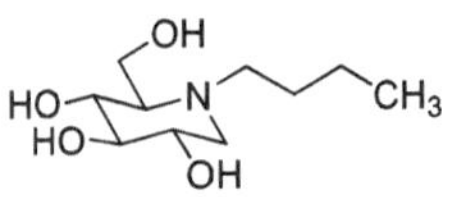

4
Miglustat (Zavesca®)

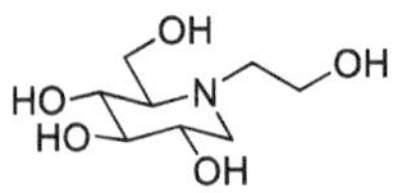

5
Miglitol (Glyset®)

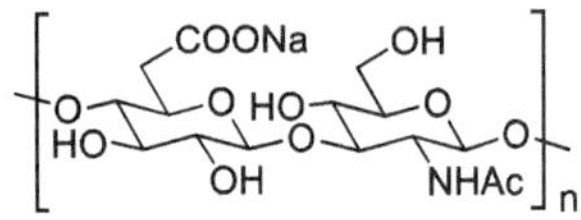

6
Sodium hyaluronate (Orthovisc®)

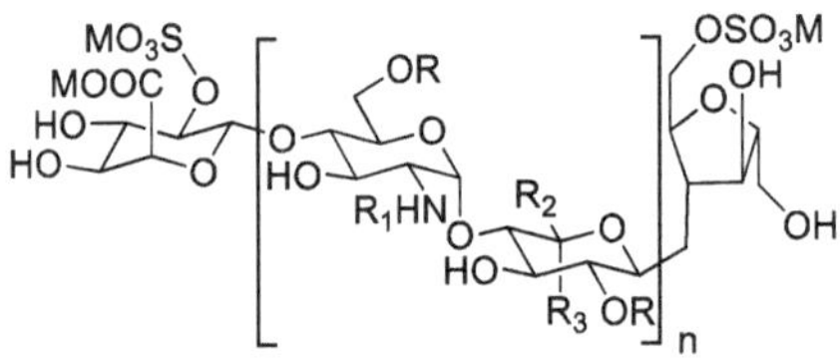

$n = 3\text{-}20$; R = H or SO_3M; $R^1 = SO_3M$ or Ac
$R^2 = H$; $R^3 = CO_3M$ or $R^2 = CO_3M$, $R^3 = H$

7 Dalteparin sodium (Fragmin®)(M = Na)
8 Nadroparin calcium (Fraxiparin®)(M = 1/2Ca)

Figure 8.1 ***Selected examples of glycomimetic based drugs (trade names are given parenthesis).***

1.3 Classifications of carbohydrate-derived inhibitors

Carbohydrate-derived inhibitors against glycoconjugates processing enzymes generally belong to three major classes: iminosugars **II**, thiosugars **III** and carbaglycosylamines (aminocyclitols) **IV** where the ring oxygen atom of naturally occurring carbohydrates has been replaced by nitrogen, sulphur and carbon atoms respectively (Fig. 8.2).[5,19,20]

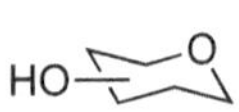

sugars
I

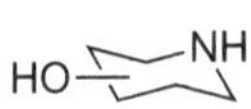

iminosugars
II

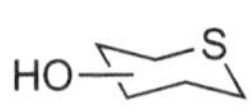

thiosugars
III

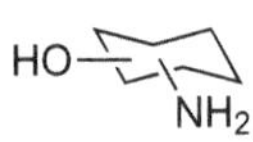

aminocyclitols
IV

Figure 8.2 ***General classes of inhibitors of glycoconjugates processing enzymes based on their structures.***

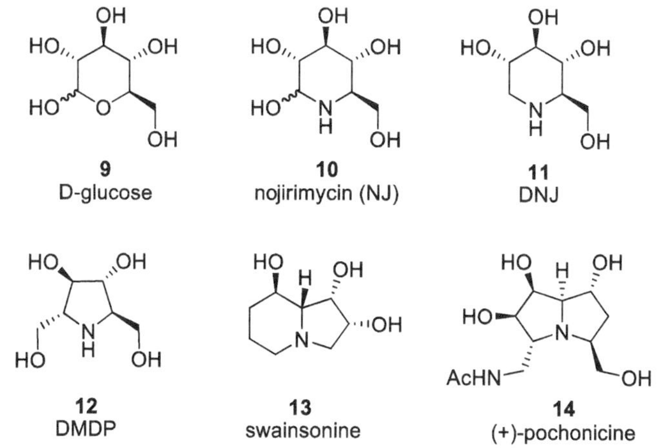

Figure 8.3 ***Structure of D-glucose and representative examples of naturally occurring iminosugars.***

1.4 Iminosugars

Among the various classes of inhibitors mentioned above, iminosugars single out to be the most exploited category. Iminosugars (also called as iminocyclitols or erroneously as azasugars) are polyhydroxylated alkaloids wherein the endocyclic oxygen atom of natural carbohydrates has been replaced by a basic nitrogen atom.[21] A few examples of naturally occurring iminosugars (**10–14**) are given in Fig. 8.3. Iminosugars have been found to be of great medicinal interest largely due to their inhibition of glycoconjugate processing enzymes.[22]

The first therapeutic use of iminosugars dates back to 17th century when medicinal Haarlem oil, which was extracted from leaves of *Morus alba* (the white mulberry) with an extremely rich source of iminosugars, was recommended for the purpose of weight loss, whitening of the skin and for the treatment of diabetes.[23]

Since the initial reports of the synthesis of sugar derivatives containing a nitrogen atom in the ring by the groups of Paulson, Jones, and Hanessian in the early 1960s, the chemistry and biology of iminosugars have witnessed exponential growth.[24]

2 Iminosugars as inhibitors glycosidases

Iminosugars mimic the structure, conformation and chirality of their carbohydrate counterparts but at the same time are stable towards further processing and possess the ability of blocking the activities of glycosidases.

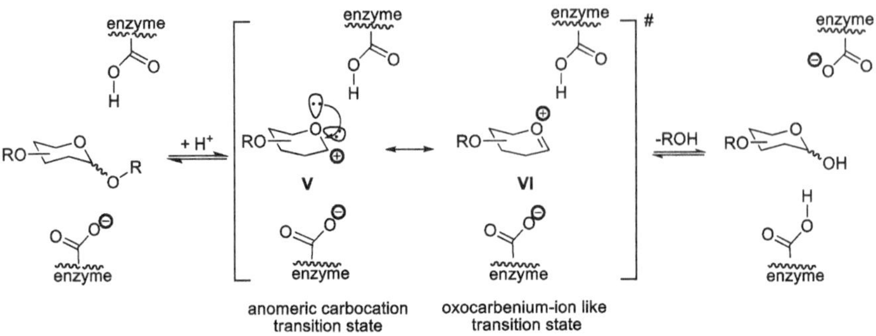

Scheme 8.1 ***Possible transition states involved during the hydrolysis of glycosides by glycosidases.***

Because of this unique combination, such compounds and their derivatives have found enormous therapeutic potential against many diseases.[21,25] Iminosugars behave as inhibitors of glycosidases as they mimic the charge and shape of the anomeric carbocation **V** or oxocarbenium-ion like transition state **VI** involved during the hydrolysis of glycosides (Scheme 8.1). The ring nitrogen of iminosugars being basic in nature would get protonated under physiological pH to form ammonium ions such as **VII** and **VIII** (Fig. 8.4), that mimic the positive charge build up associated with carbocation and oxocarbenium-ion transition states **V** and **VI,** respectively, involved during the process of enzymatic glycoside bond cleavage and thereby form strong electrostatic interactions with the catalytic carboxylate residue of the enzymes. Choice of the iminosugars as selective inhibitors of a particular enzyme arises mainly due to the similarity in the absolute

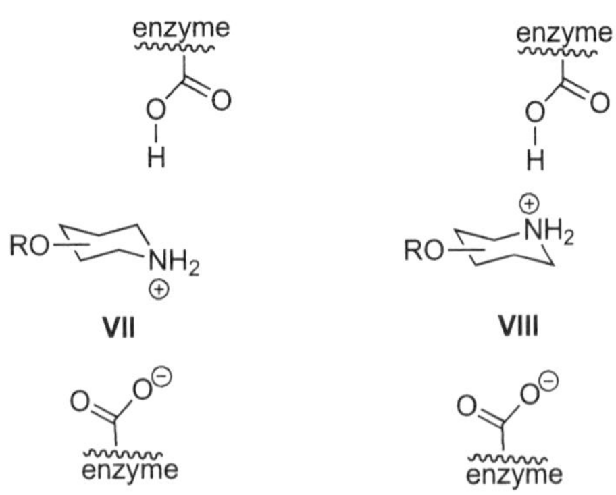

Figure 8.4 ***Mimicry of transition states involved in the glycosidic hydrolysis by protonated iminosugars VII and VIII.***

configurations at various stereocentres to that of the natural substrates as well as charge distribution.

2.1 Mode of action

Six-membered as well as five-membered iminosugars have been found to be selective inhibitors of glycosidases. While the six membered iminosugar (for instance 1-deoxynojirimycin DNJ **11**) closely resembles the ground state of natural substrates in an unexpected chair conformation, the five-membered ring of pyrrolidine iminosugar (for instance 2,5-dihydroxymethyl-3,4-dihydroxypyrrolidine DMDP **12**) is assumed to mimic the half-chair conformation involved in the transition state (Fig. 8.5).[26,27] This has prompted the search for new analogues that mimic the transition state more closely both electronically and structurally, with the aim of developing more potent and selective inhibitors. Interestingly, to mimic the distorted structure involved in the transition state, introduction of a double bond or an epoxide into the ring, or a bicyclic system have been considered.[28]

Mechanistic studies of the hydrolysis of glycosides by enzymes evidenced a substantial bond cleavage and development of positive charge at the anomeric carbon (Fig. 8.5A).[29] It is generally accepted that protonation occurs reversibly at the anomeric oxygen atom followed cleavage of *C*1 oxygen bond (Scheme 8.1). A detailed study of kinetic isotopic effects has shown that the resulting positive charge at *C*1 on glycan is not completely

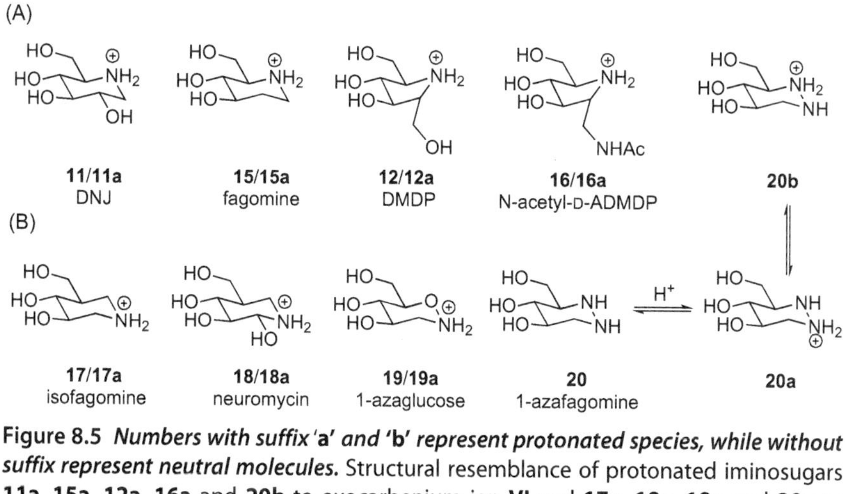

Figure 8.5 ***Numbers with suffix 'a' and 'b' represent protonated species, while without suffix represent neutral molecules.*** Structural resemblance of protonated iminosugars **11a**, **15a**, **12a**, **16a** and **20b** to oxocarbenium ion **VI** and **17a**, **18a**, **19a** and **20a** to anomeric carbocation **V**.

delocalized by one of lone pairs of electrons of ring oxygen atom.[30,31] As a result, iminosugars with a nitrogen atom at *C*1 position (as in iminosugars **17-20**) are also expected to function as glycosidase inhibitors. Such iminosugars are termed as isoiminosugars (Fig. 8.5B).[32]

Isofagomine **17**, a synthetic isomer of fagomine **15**, in which the location of the nitrogen atom corresponds to the anomeric carbon and the ring oxygen being replaced by a carbon atom, is a much more potent glycosidase inhibitor than fagomine **15** (Fig. 8.5).[32] Isofagomine **17** and its analogues, when protonated on the nitrogen atom, do not resemble the oxocarbenium ion (transition state **VI**, Scheme 8.1), but rather the resonance form, the anomeric carbocation (transition state **V**, Scheme 8.1). As a resonance form of oxocarbenium ion, the anomeric carbocation is also a reasonable intermediate in the cleavage of glycosidic bonds by enzymes.[32]

The presence of *C*2-OH group similar to the substrate, though expected to compromise on the binding of the ring nitrogen with enzyme, due to the electronegativity, it is not the case in some cases. For example, neuromycin **18** is 4000 times more potent inhibitor of ∝-glucosidase from yeast than isofagomine **17** (Fig. 8.5).[33] However, mutarotation at *C*2 allows the compound to bind to proteins in both *gluco*- and *manno*-configurations, showing that the –OH group is certainly not superfluous. Moreover, the hemiaminal at *C*2 of **18** results in the compound decomposition at neutral pH, which is similar to what has been found with NJ **10**.[34] 1-Azaglucose analogue **19** in which the ring oxygen has been retained but a nitrogen atom replaced the anomeric carbon, inhibited almond β-glucosidase with a K_i value of 60 μM (Fig. 8.**5**).[35] Thus it is a 500 fold weaker inhibitor of β-glucosidase than isofagomine **17** and a 100 fold weaker inhibitor than 1-azafagomine **20**. The explanation for this notable difference is due to the weak basicity of **19**.

1-Azafagomine **20**, a compound with a hydrazine incorporated into the ring, upon protonation, is capable of electronically resembling both the resonance forms viz., oxocarbenium ion (transition state **VI**, Scheme 8.1) or anomeric carbocation (transition state **V**, Scheme 8.1). Hence, it was found to inhibit both α– and β–glucosidases in micromolar range (Fig. 8.5).[35]

2.2 Structural basis for glycosidase inhibition by iminosugars

Caines and co-workers, in their efforts towards providing structural basis of glycosidase inhibition, obtained the crystal structure of cellobiose-like isofagomine **22** - endoglycoceramidase (II) (EGC) complex (**26**) and observed that the isofagomine moiety of **22** was located in the –1 subsite of the catalytic active site residues of EGC (Fig. 8.**7**).[36] In line with earlier observations by

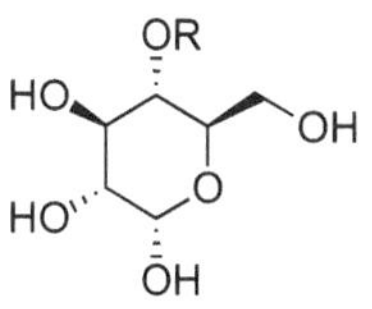

21
R = β-1,4-Gal
(lactose)

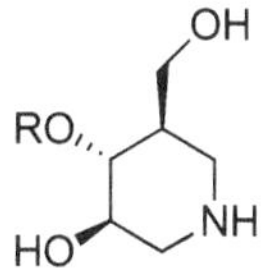

17 R = H
22 R = β-1,4-Glu

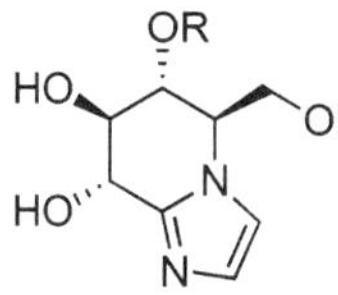

23 R= H
24 R = β-1,4-Glu

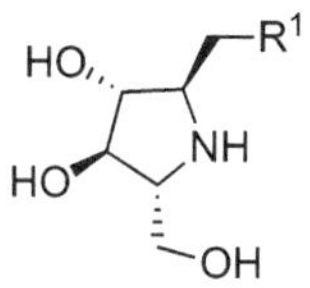

12 R^1 = OH

25 R^1 =

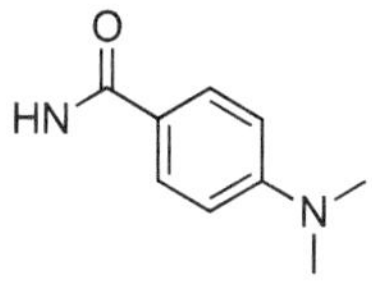

Figure 8.6 ***Inhibitors of endoglycocermidase (II) (ECG) utilized by Caines et al for structural basis of glycosidase inhibition.***

them and other groups, the isofagomine moiety of **22** (Fig. 8.6) assumes an undistorted $^4C_1$ conformation.[37] A comparison of the lactosyl-EGC complex **27** with that of **26** revealed that the endocyclic nitrogen atom of isofagomine moiety of **26** superimposes with the anomeric carbon of complex **27** there by mimicking the anomeric cation and not the oxocarbenium-ion-like half-chair transition state (Fig. 8.7). On the other hand, the imidazole fused iminosugar **24** was found to be ten times more potent than **22**. It was reasoned that the relatively higher activity of **24** could be due to the imidazole induced "oxocarbenium-ion-like" half chair conformation that is not possible in **22**. Further the increased interaction of 2-hydroxy group of **24** with the active site residues of EGC in complex **28** provides additional stability to the transition states, a factor that was absent in compound **26**.

The authors have also provided, for the first time, the structural basis of interaction of a five-membered iminosugar **25** with EGC based on the crystal structure of inhibitor-enzyme complex. Though an inhibitor of EGC, it was found to be less potent than the isofagomine derivative **22** and imidazole fused iminosugar **24**. In this case, the iminosugar moiety adapts an envelope conformation and binds in the −1 subsite (**29**). The ring nitrogen superimposes with the anomeric carbon of **27** and the ring nitrogen of **26**. This positioning of nitrogen mimics charge development at the transition state of retaining β-glycosidase while the –OH of hydroxymethylgroup mimics the 2-hydroxy position of **27** and **28**, thereby **29** capitalizes the important transition state stabilizing interactions (Fig. 8.7). However due to

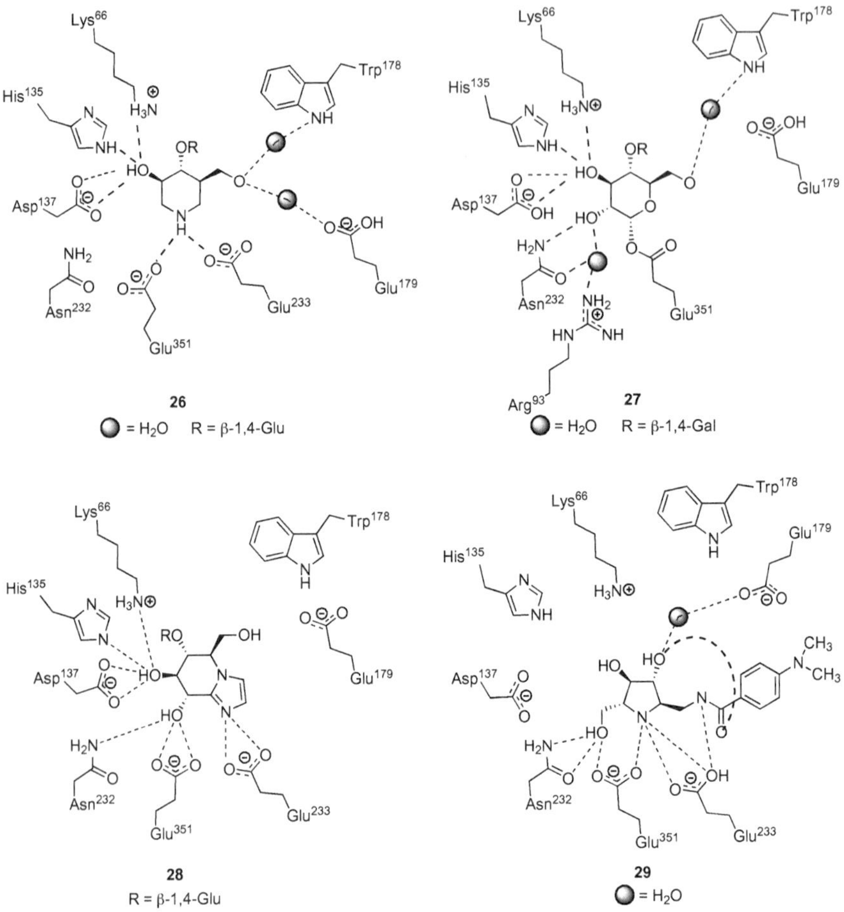

Figure 8.7 ***A schematic representation of interactions of lactose, cellobiose-like isofagomine, imidazole fused iminosugar and five-membered iminosugar with endoglycocermidase (II).***

lack of interaction between **25** and other residues such as Lys[66], His[135] and Asp[137] as well as the minimal contact of 3- and 4-hydroxy groups of **25** with the active sites residues of EGC presumably make this molecule less potent than the others.[36]

3 Classification of naturally occuring iminosugars

Iminosugars have been found as secondary metabolites primarily in plants, but also found in microorganisms, fungi, and more recently, they have been isolated from insects and sea sponges as well.[38] Naturally occurring iminosugars can generally be classified based on their ring size and nature such as piperidines, pyrrolidines, pyrrolizidines, indolizidines, nortropanes etc.[39]

3.1 Natural polyhydroxypiperidines and their source of isolation

Six membered iminosugars possess hydroxyl groups whose absolute configurations are similar to that of natural sugars. In 1966, Paulsen published the first synthesis of 1-deoxynojirimycin (DNJ **11**, Fig. 8.5).[34] Sometime later in the same year, Inouye et al. isolated nojirimycin NJ **10** from bacteria (*Streptomyces*) and studied its antibiotic properties.[38e] NJ **10** being a hemi-aminal is fairly unstable under basic and neutral conditions, undergoing a facile elimination reaction to give a biologically inactive compound.[34d] This instability acts as a deterrent for its use as a drug molecule. 1-Deoxynojirimycin (DNJ) **11** which is void of the anomeric hydroxyl group was also later isolated from the roots of mulberry trees (*Moracae*) in 1976. DNJ **11** was found to be more stable than NJ **10**.[40] Concurrent with the synthesis of DNJ **11**, several publications appeared on the isolation, synthesis and biological studies of its other stereo-analogues and *N*-substituted piperidine derivatives.[21,40,41] A few *N*-methylated derivatives of polyhydroxylated piperidines have been isolated from plant species. While *N*-Methyl DNJ **30**, the first naturally occurring *N*-alkylated derivative of DNJ **11** was isolated from the leaves of *Morus bomycis* in 1994,[42] its *manno*-isomer, *N*-methyl-D-DMJ **33** was later isolated in 2001 from *Angylocalyx pynaertii* (Fig. 8.8).[43] 1-Deoxy-D-*manno*jirimycin (DMJ) **32** was isolated from *Lonchocarpus sericeus* in 1979.[44] Its unstable oxygenated parent compound D-*manno*jirimycin **31**was isolated from *Streptomyces lavandulae* later in 1984.[45]D-*Galacto*nojirimycin **34**[46] was isolated in 1988 from *Streptomyces lydicus*. D-*Altro*-DNJ **36** and L-*gulo*-DNJ **37**[43] [from *A. pynaertii* (*Leguminosae*), 2001] and D-*allo*-DNJ **35**[47] [from *Connarus ferruginens* (Combretaceae), 2005] were isolated by Asano and co-workers. In 2000, the same authors also isolated 1-deoxyadenophorine **38** from *Adenophorea radix*.[48]

Fagomine **15** (*Fagopyrum esculentum*, 1974), which is a 1,2-dideoxy analogue of nojirimycin **10**, as well as its epimers, 3-*epi*-fagomine **41** and 3,4-di-*epi*-fagomine **42** (*Xanthocercis zambesiaca*, 1997) have also been found in natural sources (Fig. 8.9). [38,49,50,51]

Examples of homo analogues of DNJ, containing an additional methylene group at C-2 position of the piperidine ring, along with their source of isolation include α-homo-NJ **43** (α-HNJ, *Omphalea diandra*, 1988),[52] β-homo-NJ **44** (β-HNJ, *Aglaonema treubii*, 1997) (Fig. 8.10).[53]

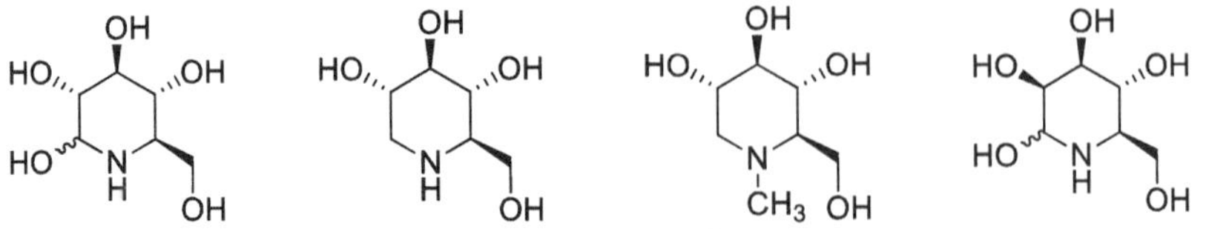

10 nojirimycin | **11** 1-deoxynojirimycin | **30** N-methyl-DNJ | **31** mannojirimycin (MJ)

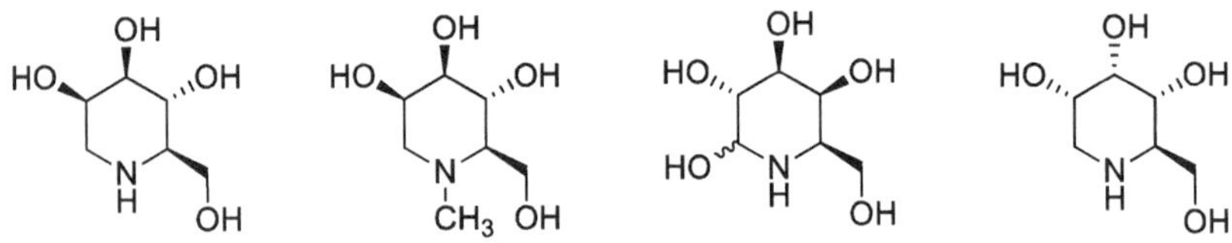

32 *manno*-DNJ (DMJ) | **33** N-methyl-D-DMJ | **34** D-galactostatin | **35** D-*allo*-DNJ

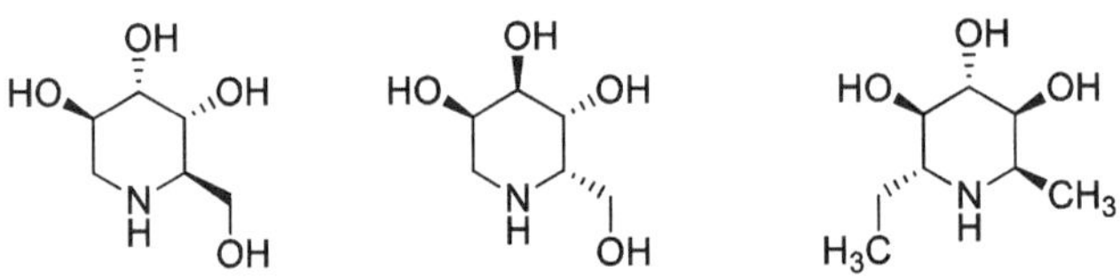

36 D-*altro*-DNJ | **37** L-*gulo*-DNJ | **38** 1-deoxyadenophorine

Figure 8.8 ***Examples of naturally occurring piperidine iminosugars.***

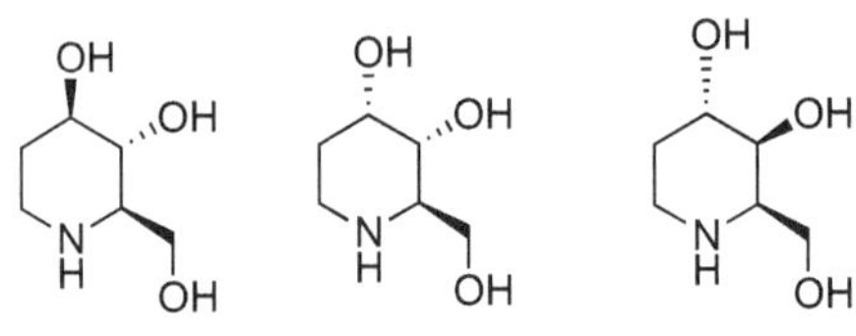

15 D-fagomine | **41** 3-*epi*-fagomine | **42** 3,4-di-*epi*-fagomine

Figure 8.9 ***Structures of D-fagomine and its stereoisomers.***

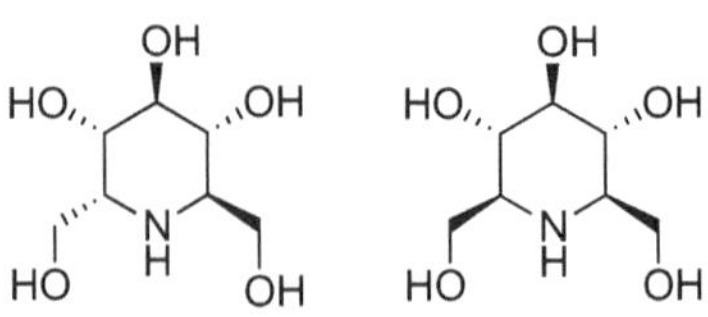

43 α-D-HNJ | **44** β-D-HNJ

Figure 8.10 ***Structures of α-D-HNJ and β-D-HNJ.***

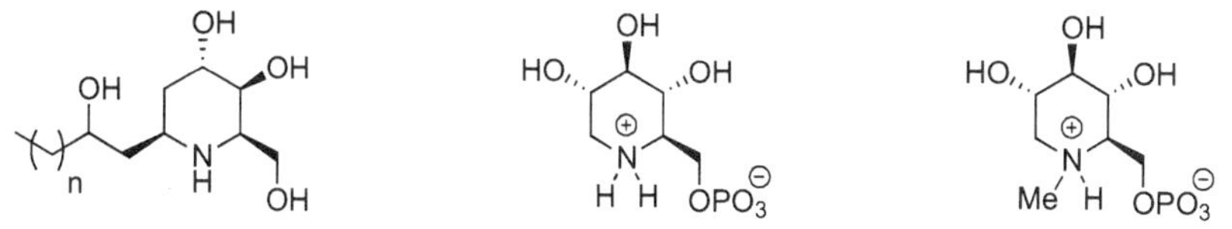

45 n = 9, (+)-Batzellaside **A**
46 n = 8, (+)-Batzellaside **B**
47 n = 10, (+)-Batzellaside **C**

48
1-deoxynojirimycin-6-phosphate

49
N-methyl-1-deoxynojirimycin-6-phosphate

Figure 8.11 ***Structures of batzellasides* A-C *and phosphate iminosugars.***

Batzellasides A-C **45–47** were the first examples of naturally occurring piperidine iminosugars isolated from marine sponge *Batzella* sp in 2005.[54] Subsequently 1-deoxynojirimycin-6-phosphate **48** and its *N*-methyl derivative **49** (Fig. 8.11) were isolated from the marine sponge *Lendenfeldia chondrodes*.[55]

3.2 Natural polyhydroxypyrrolidines and their source of isolation

Welter et al, in 1976, reported the isolation of first five-membered iminosugar, 2,5-dihydroxymethyl-3,4-dihydroxypyrrolidine (DMDP) **12** possessing all *R*-configurations and identified it as a glycosidase inhibitor.[56] DMDP **12** and other five membered iminosugars have been found to mimic flattened half-chair conformation of the oxocarbenium ion transition state involved during the glycosidic hydrolysis. 2-Hydroxymethyl-3-hydroxypyrrolidine (CYB-3) **50** was isolated in 1985 from *Castanospermum australe*.[57] 1,4-Dideoxy-1,4-imino-D-arabinitol (DAB-1) **51**, five-membered iminosugar that is void of a hydroxymethyl group as compared to DMDP **12**, was isolated in 1985 from fruits of *Angylocalyx boutiqueanus* (Leguminosae).[58] In 2000, 2-hydroxymethyl-3,4-dihydroxy-5-methylpyrrolidine **52** (6-deoxy-DMDP) was isolated from seeds of the *Angylocalyx pynaerii*.[42,43] In 1991, 2,5-dideoxy-2,5-imino-D-glucitol (DGDP) **53**, a *C2* epimer of DMDP **12**, was first synthesized.[59] Subsequently, in 2005, DGDP **53**, was also isolated from the Thai traditional medicine "Non tai yak" (*Stemona tuberosa*).[60] 3,4-Dihydroxy-5-hydroxymethyl-1-pyrroline **54** (nectrisine) was isolated from *Nectria lucida* in 1988.[61] 2,5-Dideoxy-2,5-imino-D-glycero-D-manno-heptitol, **55** a homologue of DMDP **12**, was isolated from *Hyacinthoides nonscripta* in 1997 (Fig. 8.12).[62]

Broussonetines and broussonetinines belong to a family, with more than 30 members, of naturally occurring polyhydroxypyrrolidines (Fig. 8.13–8.14) that were isolated from the branches of deciduous tree *Broussonetia Kazinoki*

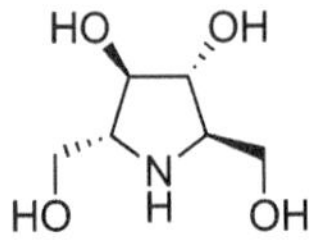

12 DMDP

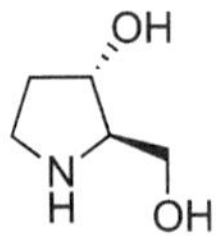

50 CYB-3

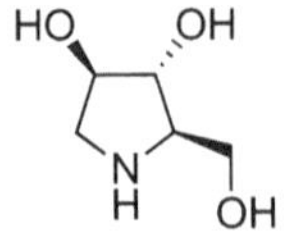

51 DAB-1

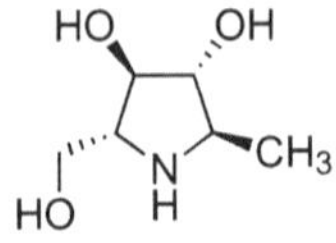

52 6-deoxy-DMDP

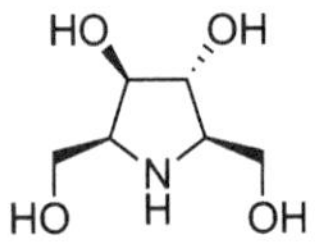

53 DGDP

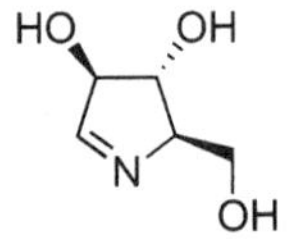

54 nectrisine

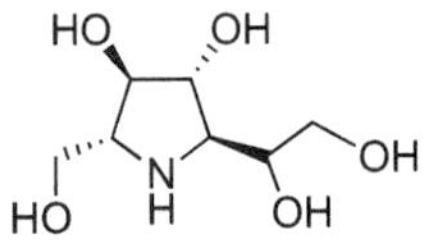

55 homo-DMDP

Figure 8.12 ***Examples of naturally occurring pyrrolidine iminocyclitols.***

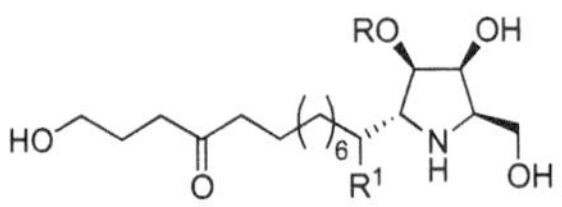

56 R = R^1 = H; broussonetinine A
57 R = β-D-glucopyranosyl, R^1 = H; broussonetine A

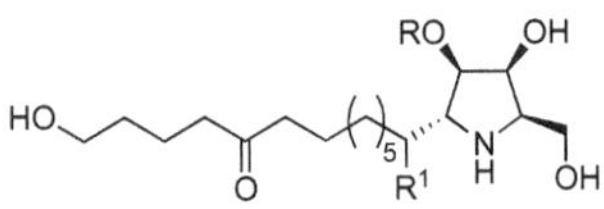

58 n = 5, R = R^1 = H; broussonetinine B
59 n = 5, R = β-D-glucopyranosyl, R^1 = H; broussonetine B

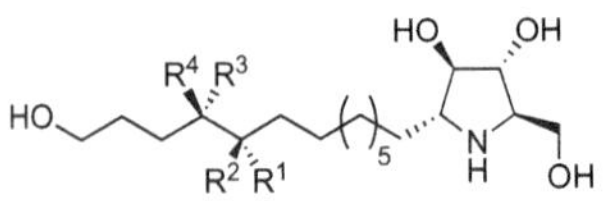

60 R^1 = R^2 = H, R^3,R^4 =O; broussonetine C
61 R^1,R^2 =O; R^3 = R^4 = H; broussonetine D

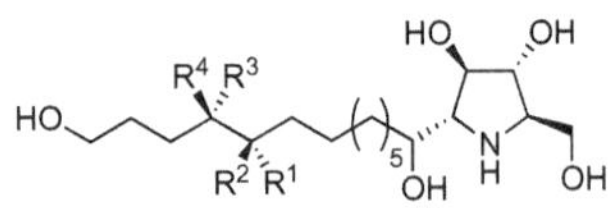

62 R^1 = R^2 = H, R^3,R^4 =O; broussonetine E
63 R^1,R^2 =O; R^3 = R^4 = H; broussonetine F

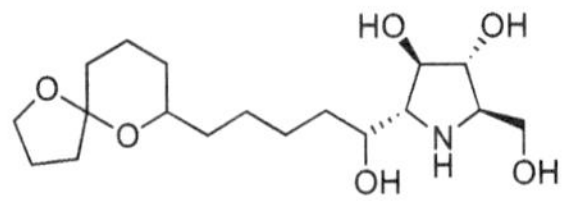

64 broussonetine G

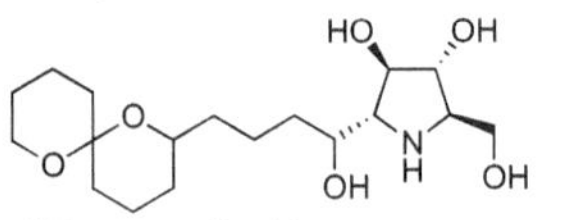

65 broussonetine H

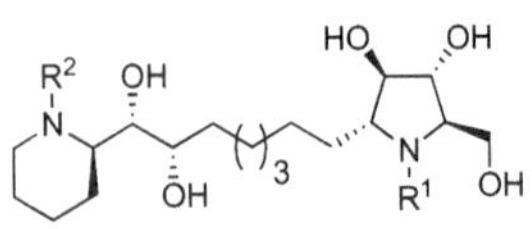

66 R^1 = H, R^2 = Ac; broussonetine I
67 R^1 = R^2 = Ac; broussonetine J
68 R^1 = Ac, R^2 = H; broussonetine J_1
69 R^1 = R^2 = H; broussonetine J_2

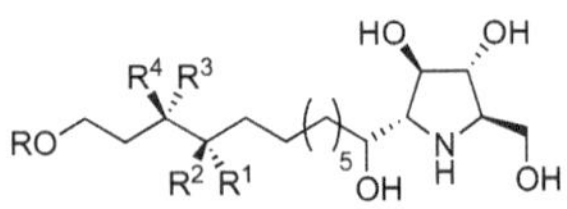

70 R^1 = R^2 = H, R^3,R^4 =O, R = β-D-glucopyranosyl; broussonetine K
71 R^1,R^2 =O, R^3 = R^4 = H, R = β-D-glucopyranosyl; broussonetine L

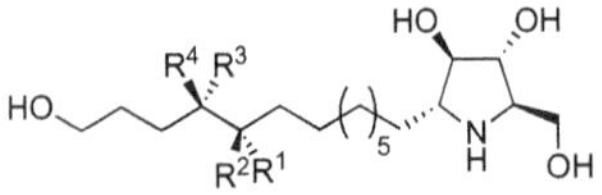

72 R^1 = R^2 = R^3 = H, R^4 = OH; broussonetine M
73 R^1 = R^3 = R^4 = H, R^2 = OH; broussonetine M_1

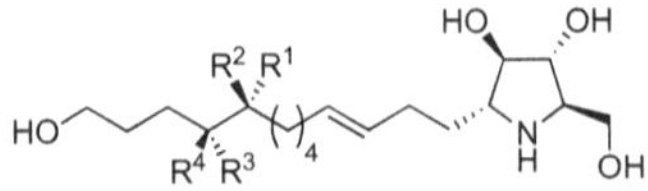

74 R^1 = R^2 = H, R^3 = R^4 = O; broussonetine O
75 R^1 = R^2 = O, R^3 = R^4 = H, broussonetine P

Figure 8.13 ***Structures of naturally occurring broussonetines A-P.***

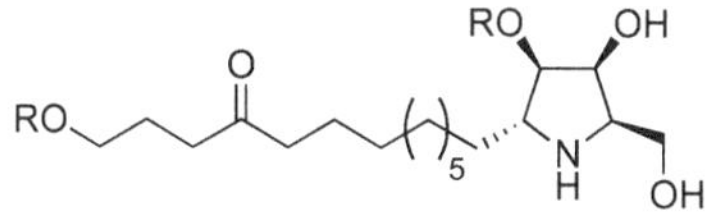

76 R = β-D-glucopyranosyl; broussonetine Q

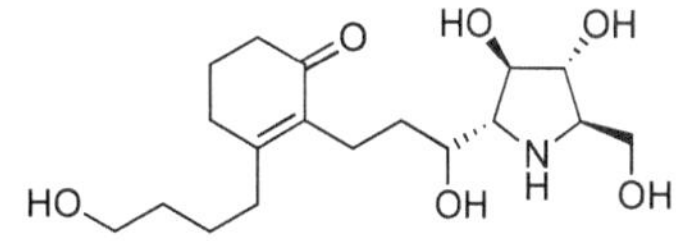

77 broussonetine R

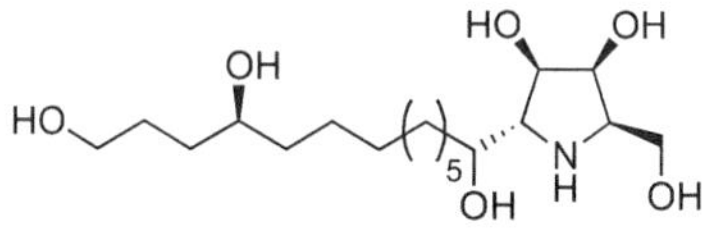

78 broussonetine S

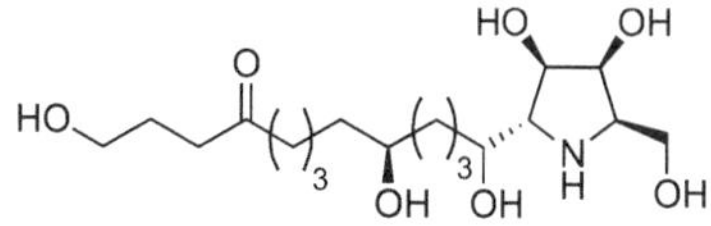

79 broussonetine T

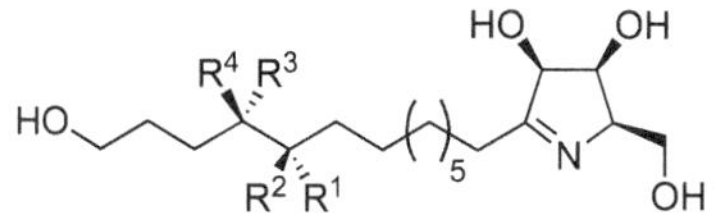

80 R^1,R^2 =O, R^3 = R^4 = H; broussonetine U
81 R^1 = R^2 = H, R^3,R^4 =O; broussonetine U_1

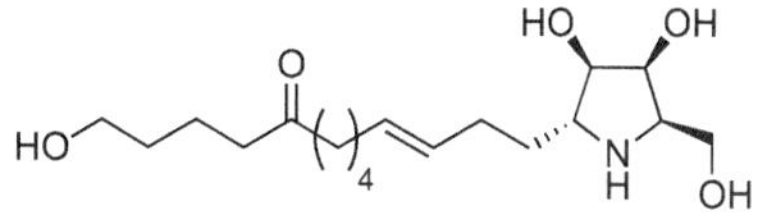

82 broussonetine V

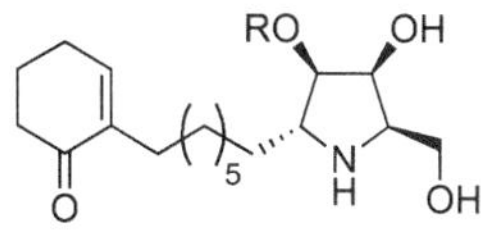

83 R = H; broussonetine W
84 R = β-D-glucopyranosyl; broussonetine X

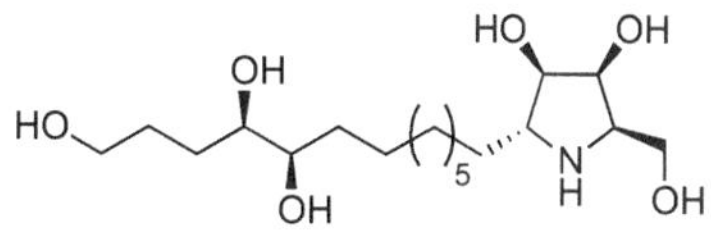

85 broussonetine Y

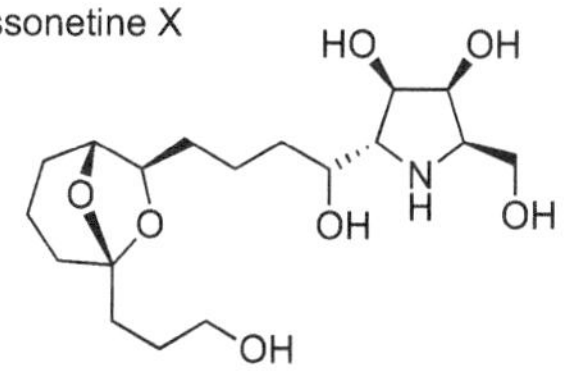

86 broussonetine Z

Figure 8.14 ***Structures of naturally occurring broussonetines Q-Z.***

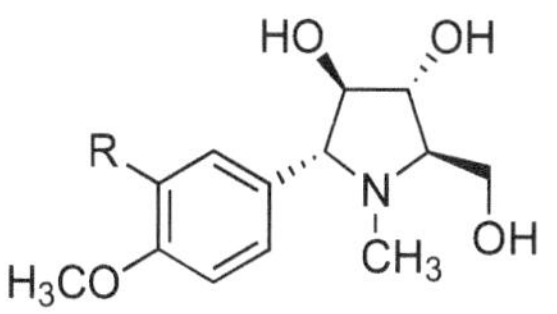

87 R = H; (–)-Codonopsinine
88 R = OCH_3; (–)-Codonopsine

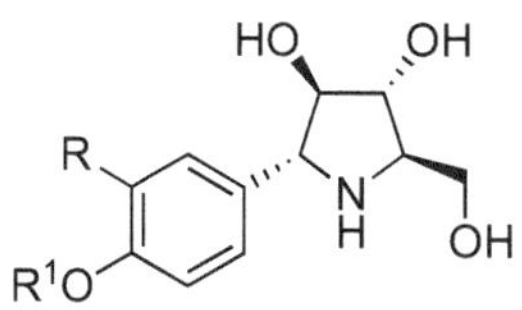

89 R = OH; R^1 = CH_3; radicamine A
90 R = R^1 = H; radicamine B

Figure 8.15 ***Examples of naturally occurring aryl substituted pyrrolidine iminosugars.***

SIEB.[63] Most of the Broussonetines share a common (2*R*,3*R*,4*R*,5*R*)-3,4-dihydroxy-2-hydroxymethylpyrrolidine with a long alkyl chain possessing varied functional groups at C-5 position of the pyrrolidine ring, though compounds with varied stereoisomers are also known. In general, though these compounds can be structurally regarded as C-5 alkylated analogues of DAB-1 **51**, their glycosidase inhibition properties closely resemble that of DMDP **12**.

Polyhydroxypyrrolidines possessing an aromatic substituent on the pyrrolidine ring represents a unique class of iminosugars found in nature. (-)-Codonopsinine **87** and (-)-codonopsine **88** are the first two examples of this category whose isolation was reported in 1971 (Fig. 8.15).[64] In 2001, Kusano and co-workers isolated two new polyhydroxypyrrolidines, radicamines A (**89**) and B (**90**), having an aromatic ring from *Lobelia chinensis* LOUR (Campanulaceae), a plant widely grown in China, Korea and Japan (Fig. 8.15).[65] However the absolute configuration of the nature products proposed by Kusano was later revised by Yu and co-workers based on their synthesis.[66]

3.3 Naturally occurring polyhydroxyindolizidines and their source of isolation

Indolizidines are bicyclic ring systems possessing piperidine and pyrrolidine rings fused together with a nitrogen atom at the ring junction (Fig. 8.16). They include swainsonine **13** which was isolated from plants of genus *Swainsona canescens* in 1979 and from the plants of species *Astralagus spp.* and *Oxytropis spp.* in 1982.[67] Its glycosidase inhibition activities are attributed to the similarity of its hydroxyl group stereochemistry to that of mannose in the furanose form. Swainsonine **13** has been shown to be a more potent inhibitor of mannosidase than DMJ **32**, and this may due to the bicyclic ring which imposes a rigidity of structure resembling a common

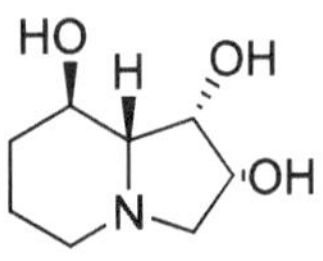

13
swainsonine

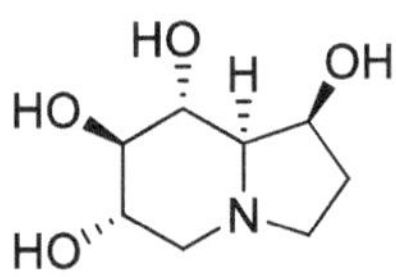

91
castanospermine

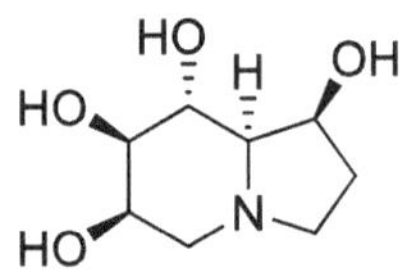

92
6-epi-castanospermine

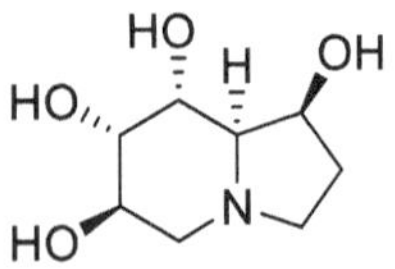

93
6,7-di-*epi*-castanospermine

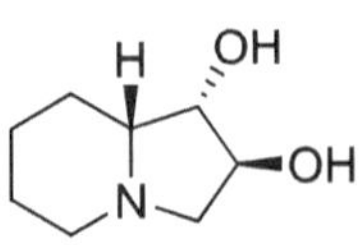

94
lentiginosine

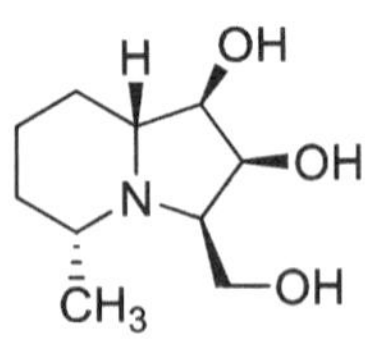

95
(−)-steviamine

Figure 8.16 *Examples of naturally occurring polyhydroxylated indolizidine alkaloids.*

intermediate in the glycolysis pathway.[59,68] Castanospermine **91** is a polyhydroxylated indolizidine alkaloid in which the hydroxyl groups of its piperidine moiety resembles that of glucose in the pyranose form. It was isolated from *Castanospermum australe* in 1981.[69] Later, 6-*epi*-castanospermine **92** and 6,7-di-*epi*-castanospermine **93** were also isolated from the same source.[70] Lentiginosine **94**, the least hydroxylated naturally occurring indolizidine derivative was isolated from *Astragalus lentiginosus* in 1990.[71] Recently, in 2010, (−)-steviamine **95** was isolated from *Stevia rebaudiana* (Asteraceae) by Fleet and coworkers (Fig. 8.**16**)[72] and it is the only naturally occurring indolizidine alkaloid that possesses a methyl group in one of the rings.

3.4 Naturally occurring polyhydroxypyrrolizidines and their source of isolation

Pyrrolizidine derivatives are fused pyrrolidine bicyclic ring systems and they have been shown to exhibit several inhibitory properties (Fig. 8.17). Casuarine **96**, the most oxygenated bicyclic iminosugar reported so far from natural sources, was isolated from *Casuarina equisetifolia* in 1994.[73] Alexine **97** was isolated from *Alexa spp.* in 1989.[74] Australine **98** which is regarded as a ring-contracted form of castanospermine **91**, resembles DMDP **12**, was

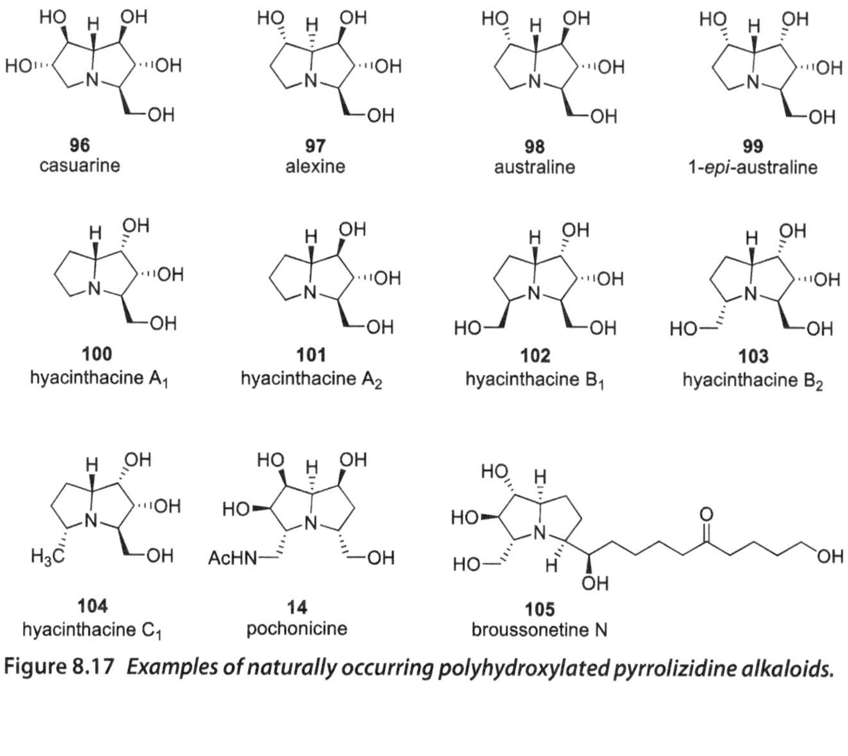

Figure 8.17 *Examples of naturally occurring polyhydroxylated pyrrolizidine alkaloids.*

isolated from *Castanospermum australe*, in 1988.[75] Australine was shown to be a potential glucosidase inhibitor. 1-*Epi*-australine **99** was isolated from *Castanospermum australe* in 1990.[74] Hyacinthacine A_1**100** and hyacinthacine A_2 **101** were isolated from *Muscari armeniacum* in 2000.[76] Hyacinthacine B_1**102** and hyacinthacine B_2 **103** were isolated from *Scilla campanulata* in 1999.[76,77] In the same year, hyacinthacine C_1**104** was isolated from *Hyacinthoids nonscipta* (Fig. 8.**17**).[76,77] Pochonicine **14** is a new polyhydroxylated pyrrolizidine alkaloid, isolated from a solid fermentation culture of the fungal strain *Pochonia suchlasporia var. suchlasporia TAMA 87*.[78] It is the first natural iminosugar found to possess a side-chain aminomethyl functionality. The inhibition studies showed pochonicine **14** as a potent inhibitor against β-*N*-acetylglucosaminidases (jack bean) with a K_i value of 0.162 nM. Broussonetine N **104** was isolated in 1999 by Kusano and co-workers from the branches of *Broussonetia kazinoki* SIEB. (Moraceae) and was found to be an inhibitor of β-glucosidase, β-galactosidase and β-mannosidase.[79]

3.5 Naturally occurring nortropanes and their source of isolation

Nortropanes belong to a class of polyhydroxylated bridged bicyclic systems that were isolated from the plants of the convolvulaceae, solanaceae and moraceae families since 1988 (Fig. 8.18).[80] They have also been found in a variety of vegetables and fruits such as table potato cultivars, tomato etc. These compounds are related to tropane alkaloids which include medicinally important derivatives such as cocaine, scopolamine and atropine. They are divided into three categories depending on the number of hydroxyl groups present, namely, calystegine A (three hydroxyl groups), calystegine B (four hydroxyl groups) and calystegine C (five hydroxyl groups). A common feature among them is that they all contain a tertiary hydroxyl group as part of a hemiaminal functionality. In case of calystegine N_1 **114**, the bridge-head hydroxyl group has been replaced with an amino group. A few examples of glycoside derivatives of calystegines (such as **115**) were also found in nature. Calystegines **106–115** have been shown to be competitive glycosidase inhibitors (Fig. 8.18).[42,81-85]

4 Iminosugars as inhibitors of glycosidases

4.1 Iminosugars as inhibitors of α-glucosidases

α-Glucosidase is a carbohydrate-hydrolase that hydrolyses the terminal non-reducing (1→4) linked α-glucose residues resulting in the release of

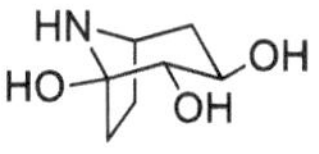

106
calystegine A_3

107
calystegine A_5

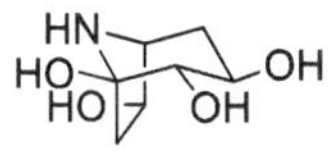

108
calystegine B_1

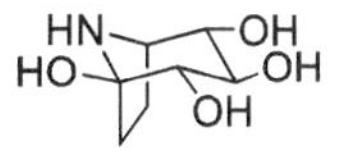

109
calystegine B_2

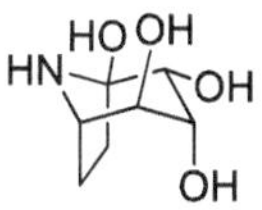

110
calystegine B_3

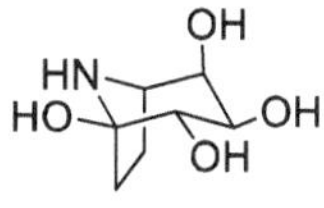

111
calystegine B_4

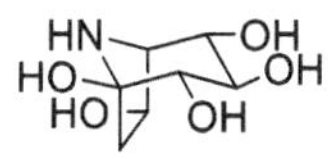

112
calystegine C_1

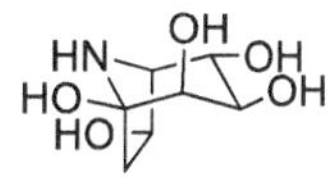

113
calystegine C_2

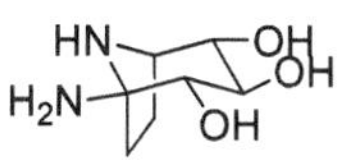

114
calystegine N_1

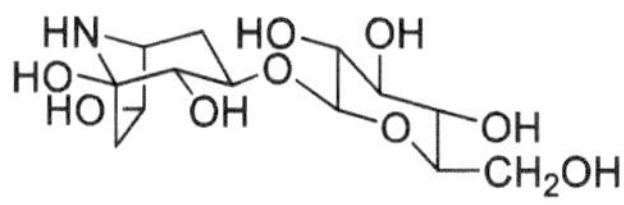

115
calystegine B_1-3-*O*-β-D-glucopyranoside

Figure 8.18 ***Examples of naturally occurring bridged bicyclic iminosugars.***

α-glucose. These enzymes (such as sucrase, maltase, isomaltase, lactase, trehalase etc.) are responsible for the digestion of dietary carbohydrate and convert them into monosaccharides which are then absorbed through the intestinal wall. Inhibition of these enzymes regulates the absorption of carbohydrates and release of monosaccharides. For instance, α-glucosidase plays a vital role in regulating blood glucose level in human and in the transport of glucose in insects and fungi, and hence inhibitors of α-glucosidase could find application as therapeutics for non-insulin dependent diabetes mellitus, especially for patients with postprandial hyperglycemia.[13b] DNJ **11** was found to be a very strong inhibitor of α-glucosidase from rice, human lysosomal and calf liver.[38,86] Subsequently, a number of iminosugars have been identified as inhibitors of α-glucosidases. *N*-methyl DNJ **30** was found to be an even stronger inhibitor of α-glucosidase type I from calf liver with a K_i value of 0.07 μM as compared to DNJ **11** (K_i = 1 μM) itself.[86,87] Such an observation has elicited great interest among research groups to develop *N*-alkylated derivatives of DNJ with the aim of further improving the inhibitory activity against α-glucosidase. The sucrase (from rat intestine) inhibitory activity of α-D-HNJ **43** with an IC_{50} value of 0.17 μM is comparable to that of DNJ **11**.[49,88] Though equally or less potent than DNJ **11** towards α-glucosidase, both α-D-HNJ and β-D-HNJ[49,53,88] surpass DNJ in terms of specificity as these two enzymes inhibit only α-glucosidase and not any other glycosidases. In the pyrrolidine series, DMDP **12** was though found to be a very strong inhibitor of α-glucosidase from *Crustacean zooplankton* with an IC_{50} value of 0.019 μM, it also inhibits β-glucosidase from the same source (IC_{50}= 0.049 μM).[62, 86f, 88c, 89,90] Nectrisine **54** is a very strong inhibitor of α-glucosidase from yeast (IC_{50} = 0.048 μM).[61,91] However, in general, most of the naturally occurring iminosugars, are found to be less potent towards α-glucosidase than DNJ, thus warranting the development of synthetic iminosugars that are more potent than natural ones towards α-glucosidase. Continued research in this area has thus resulted in the development of miglitol (*N*-hydroxyethyl DNJ **5**) as a FDA approved drug for type-II diabetes. Several publications have appeared on the development of novel molecules for efficient inhibition of α-glucosidases.

In general, natural D-iminosugars have been found to be strong inhibitors than their unnatural L-enantiomers.[92] However, striking examples of unnatural enantiomers (L-iminosugars) to be more potent than their natural counterparts are not uncommon. These L-iminosugars behave as non-competitive inhibitors in contrast to the competitive nature of natural iminosugars. For instance, 1,4-dideoxy-1,4-imino-L-arabinitol **116** (LAB-1) (Fig. 8.19) is a powerful and more potent inhibitor of α-glucosidase from rice, rat intestinal maltase, rat intestinal sucrase and rat intestinal isomaltase (IC_{50} = 1.7, 1.3, 1.7, 0.08 μM respectively) than the natural DAB-1 **51** having the corresponding IC_{50} values as 250, 55, 230, 5.8 μm.[90a] Similarly, L-DMDP **117** (Fig. 8.19) is also more potent than DMDP **12**. The IC_{50} values of inhibition of α-glucosidase from rice, rat intestinal maltase, rat intestinal sucrase and rat intestinal isomaltase by L-DMDP **117** 1.5, 1.4, 0.1, 0.05 μM respectively, are 250–1500 times more potent than DMDP **12**.[90a] Moreover L-DMDP was found to be a more specific inhibitor of only α-glucosidases, where as DMDP is a broad spectrum inhibitor of various glycosidases.[89] These observations, subsequently, have spurred research towards the synthesis of unnatural L-iminosugars and investigate their glycosidase inhibition activities.[93]

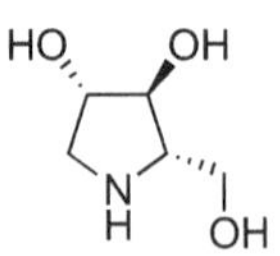

116 LAB-1

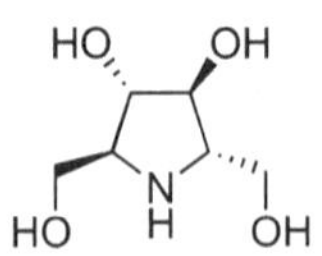

117 L-DMDP

Figure 8.19 ***Structures of LAB-1 and L-DMDP as inhibitors of α-glucosidases.***

4.2 Iminosugars as inhibitors of β-glucosidases

β-Glucosidases are enzymes that are responsible for the hydrolysis of β-glucosidic linkages in aryl-, amino- and alkyl-β-D-glucosides, disaccharides and oligosaccharides. In the enzymatic hydrolysis of cellulose, degradation of cellulose to cellobiose was carried out by endoglucanase and cellobiohydrolase, while β-glucosidase, in a rate-determining process, further hydrolyzes cellobiose to free glucose molecules. Depending upon the substrate specificity, β-glycosidases can further be classified as aryl-β-glucosidases, cellobioses and broad-specificity-β-glucosidases. Three types of β-glycosidases, namely lysosomal glucocerebrosidase (GCase), intestinal lactase-phlorizin hydrolase (LPH) and cytosolic β-glucosidase have been characterized in mammals. Gcase is responsible for the hydrolysis of glycosylceramide derived from endogenous membrane glycolipids to glucose and ceramide, while LPH plays a crucial role in the digestion of dietary lactose and β-glycosides. A deficiency in Gcase would result in the accumulation of non-degraded substrate leading to clinical manifestations that include anemia, bone lesions, respiratory failure etc. Inhibitors of Gcases are expected to restore its activity and hence can be potent lead against such diseases. While nojirimycin **10** itself is a moderate inhibitor of human GCase (IC_{50} = 19 μM),[38] DNJ **11**[38,86] and DMDP **12**[62,86f,88c,90] are even weaker towards Gcase. However introduction of an alkyl chain on the ring nitrogen of DNJ enhanced its activity against Gcase. Research in this direction led to the development of *N*-butyl DNJ **4** (miglustat or zavesca®), which has an IC_{50} value of 270 μM, as a FDA approved drug for Gaucher's disease. Further increase in the length of the *N*-alkyl chain to nine carbon, *N*-nonyl DNJ **118**, resulted in a 400 fold increase in its activity against β-glucocerebrosidase.[94] Introduction of an adamantyl group (**119**) resulted in a two-fold decrease in its activity as compared to **118**, but still very active against Gcase. Most of the nortropanes (calystegines) are found to be inhibitors of Gcases with IC_{50} values ranging from 1.0 to 76 μM, notable among them being calystegine A_3 **106** (IC_{50} = 3.1 μM), calystegine B_1 **108** (IC_{50} = 2.5 μM), calystegine B_2 **109** (IC_{50} = 1.0 μM) and calystegine C_1 **112** (IC_{50} = 2.5 μM).[81] 1,5-Dideoxy-1,5-iminoxylitol (DIX) **120** (Fig. 8.20) has been found to be a very specific inhibitor of Gcase with an IC_{50} value of 1.9 μM. Taking cue from *N*-nonyl DNJ **118** that led to highly enhanced inhibitory activity against Gcases, similar alkylation, with a nine carbon chain, at C-1 position of DIX **120** was carried out and its Gcase inhibitory activity tested. α-1-*C*-Nonyl-DIX **121** was found to be not only a very powerful inhibitor of Gcase with an IC_{50} value of 6.8 nM and a K_i value

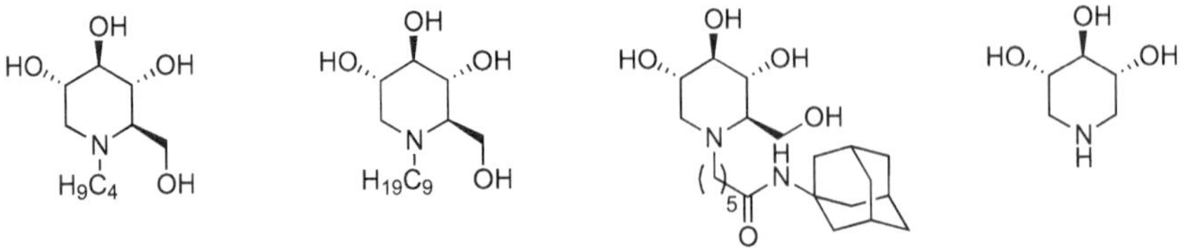

4 *N*-butyl DNJ IC_{50} = 270 μM

118 *N*-nonyl DNJ IC_{50} = 0.66 μM

119 AD-DNJ IC_{50} = 1.2 μM

120 DIX IC_{50} = 1.9 μM

Figure 8.20 ***Derivatives of nojirimycin and structure of DIX that are strong inhibitors of β-glucosidases.***

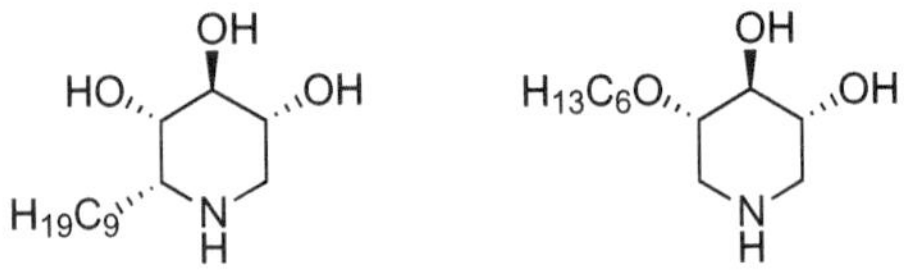

121 1-*C*-nonyl-DIX IC_{50} = 6.8 nM

122 2-*O*-hexyl-DIX IC_{50} = 9.0 nM

Figure 8.21 ***Structures of 1-C-nonyl-DIX and 2-O-hexyl-DIX, and their IC_{50} values against β-glucosidase.***

of 2.2 nM but also was very specific as it did not inhibit α-glucosidases. 2-*O*-Hexyl-DIX **122** also exhibited inhibition of Gcase almost to the same extent (IC_{50} = 9 nM) as that of **121**(Fig. 8.21). As discussed earlier, isofagomine **17**, which is a *C*-3 hydroxymethyl analogue of DIX **120** is also a powerful inhibitor of Gcase with an IC_{50} value of 40 nM.[94]

Recently, a set of four synthetic sp^2-iminosugars **123–126** (Fig. 8.22) were reported to be powerful and competitive inhibitors of human Gcase with K_i values ranging from 0.013 μM to 54 μM.[95]

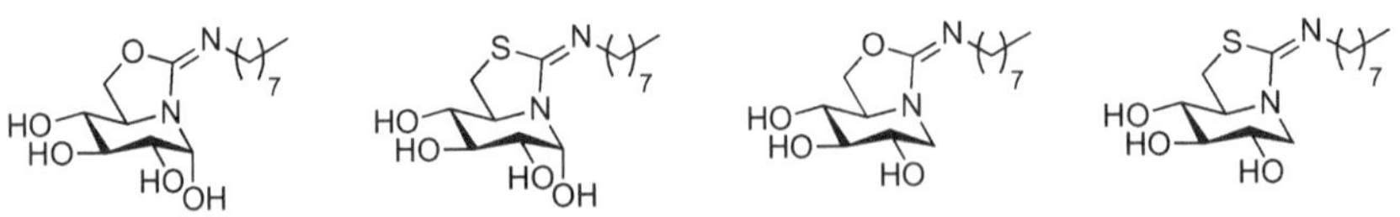

123 pH = 7; K_i = 15.1 μM pH = 5; K_i = 54.4 μM

124 pH = 7; K_i = 0.26 μM pH = 5; K_i = 1.05 μM

125 pH = 7; K_i = 1.7 μM pH = 5; K_i = 6.3 μM

126 pH = 7; K_i = 0.013 μM pH = 5; K_i = 0.059μM

Figure 8.22 ***sp^2-Iminosugars along with their K_i values against β-glucosidases.***

4.3 Iminosugars as inhibitors of α-galactosidases

α-Galactosidase, encoded by *GLA* gene, is a homodimeric protein that hydrolyses the terminal α-galactosyl moieties from glycolipids and glycoproteins, predominantly involved in the hydrolysis of trihexoside. Hydrolysis of melibiose to galactose and glucose is also catalyzed by α-galactosidase. In human, α-galactosidase is present in two recombinant forms called agalsidase α and agalsidase β. Fabry disease is a rare lysosomal storage disorder, caused due to the deficiency of α-galactosidase, resulting in the accumulation of α-D-galactosyl glycolipid moieties. In addition to several characteristic symptoms such as pain, skin spots, decreased vision, hearing loss, tinnitus etc. caused by Fabry disease, it may lead to life-threatening situations in rare cases. Inhibitors of α-galactosidases are expected to find applications as chaperones for the treatment of Fabry disease. D-Galactostatin **34** (D-galactonojirimycin, GJ) is a powerful inhibitor of α-D-galactosidase especially from coffee beans (K_i = 0.0007 μM).[96] 1-Deoxygalactonojrimycin **127** (DGJ), synthetic analogue, obtained through $NaBH_4$ reduction of DJ **34** is also a powerful inhibitor of α-galactosidase from coffee beans with a K_i value of 0.0016 μM. Fan et al. identified DGJ **127** as a potential chaperone for Fabry's disease that has been approved by FDA as a drug in 2018, in the name of migalastat (GALAFOLD™).[18i] The homo analogues of DNJ, α-homogalactonojirimycin **129** (α-HGJ) is an inhibitor of α-galactosidase (IC_{50} = 0.21 μM from human; IC_{50} = 0.06 μM from coffee beans) and β-homogalactonojirimycin **130** (β-HGJ) is not a promising candidate.[97] β-1-*C*-Butyl-deoxygalactonojirimycin (β-1-*C*-butyl-DGJ) **128**, synthesized prior to its isolation from *Adenophora sp.* (Campanulaceae) is a moderate inhibitor of α-galactosiadse from coffee beans with a K_i value of 0.71 μM.[48] A few other naturally occurring iminosugars such as **11**, **35**, **106**, and **112** are very weak inhibitors (IC_{50} values in the range of 15–900 μM) whereas **109** is a better inhibitor of α-galactosidase from coffee beans (IC_{50} = 1.9 μM) and *Aspergillus niger* (IC_{50} = 3.9 μM).[81a] *N*-Methyl derivatives of calystegines A_3, B_3, B_4 and C_1 were shown to be moderate inhibitors of α-galactosidase from coffee beans while *N*-methylcalystegine B_2 is a strong inhibitor with a K_i of 0.47 μM.[82a] Conformationally locked sp^2 iminosugars **131–133** (Fig. 8.23), in the form of thiourea derivatives of DGJ **127**, were found to be very strong inhibitors of α-galactosiadse from coffee beans, notable among them being **131** with a K_i value of 1.9 nM.[98] On the other hand, conformationally locked bicycliciminosugars **134–136** (Fig. 8.23) were found to be less potent than DGJ **127**.Though none of the naturally occurring pyrrolidine iminosugars display potent inhibition against

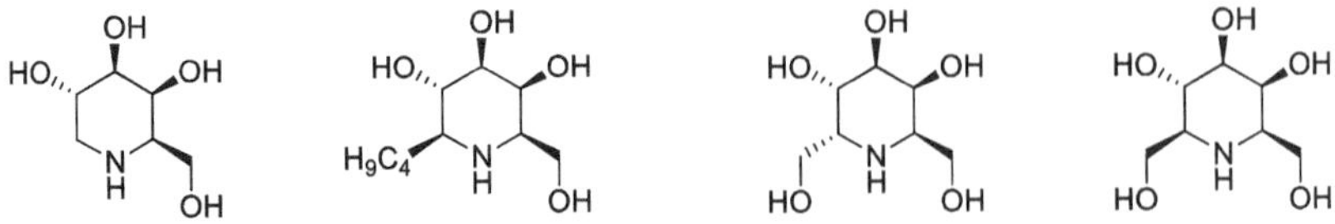

127 DGJ $K_i = 0.016\ \mu M$

128 β-1-C-Butyl-DGJ $K_i = 0.71\ \mu M$

129 α-HGJ $IC_{50} = 0.21\ \mu M$, human; $IC_{50} = 0.06\ \mu M$, coffee beans

130 β-HGJ

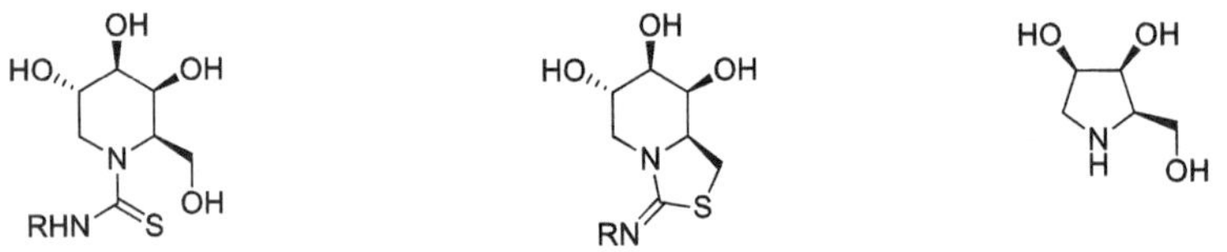

131 R = Ph, $K_i = 0.0019\ \mu M$
132 R = *n*-Bu, $K_i = 0.044\ \mu M$
133 R = *n*-octyl, $K_i = 0.029\ \mu M$

134 R = Ph, $K_i = 18\ \mu M$
135 R = *n*-Bu, $K_i = 134\ \mu M$
136 R = *n*-octyl, $K_i = 81\ \mu M$

137 DIL $IC_{50} = 0.2\ \mu M$

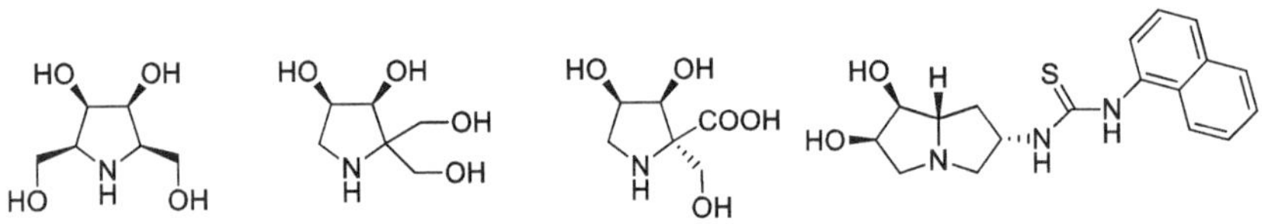

138 $IC_{50} = 0.05\ \mu M$

139 $IC_{50} = 0.016\ \mu M$

140 $IC_{50} = 0.079\ \mu M$

141 $IC_{50} = 0.37\ \mu M$

Figure 8.23 ***Structures of iminosugars that are strong inhibitors of a-galactosidases.***

α-galactosidase, synthetic derivatives **137** (1,4-dideoxy-1,4-imino-D-lyxitol) and **138** (2,5-dideoxy-2,5-imino-D-galactitol) were found to be sub micromolar inhibitors of α-galactosidse from coffee beans.[99] Introduction of an additional methyl group at C-3 or C-4 position of **137** reduced the inhibitory activity.[100] Geminal disubstituted pyrrolidine iminosugars **139** ($IC_{50} = 0.016\ \mu M$) and **140** ($IC_{50} = 0.079\ \mu M$) were identified as strong and specific inhibitors of α-galactosiadse from coffee beans.[101] Recently, pyrrolizidine alkaloids possessing aryl thio urea moieties were found to be strong inhibitors of α-galactosidase from coffee beans, notable among them is compound **141** that has an IC_{50} value of 0.37 μM.[102]

4.4 Iminosugars as inhibitors of β-galactosidases

β-Galactosidase (also called lactase, β-gal) is an exoglycosidase that catalyzes the hydrolysis of glycoside bonds in β-galactosides. Substrates of β-galactosides are lactose, lactosyl ceramides, ganglioside GM1 and glycoproteins. β-Galactosidase has three enzyme active sites- (i) it can make allolactose through a process termed as transgalactosylation reaction, (ii) it can hydrolyse lactose in to galactose and glucose, and (iii) it can cleave allolactose to the monosaccharides. Deficiency in the lysosomal enzyme β-galactocerebrosidase is responsible for a neurodegenerative lysosomal storage disorder known as Krabbe disease.[103] Inhibitors of β-galactosidases are thus expected to act as pharmacological chaperones for such a disease. DMDP **12** inhibits β-galactosidase from bovine liver (IC_{50} = 3.3 μM) and rat intestinal lactase (IC_{50} = 7.9 μM).[89] D-Galactostatin **34** is a strong inhibitor of bacterial and fungal β-galactosidases.[21c, 46,96,104] Broussonetinine A **56**, broussonetinine B **58** and broussonetines **60, 61, 62, 63, 64, 65, 70, 71,74, 75, 76** and **83** were also found to be strong and in most cases specific inhibitors of β-galactosidase from bovine liver with IC_{50} values ranging from 0.36 μM to 0.002 μM.[63] Several synthetic iminosugars have been developed as β-galactosidase inhibitors. A few among them that display very strong inhibition are discussed here. 4-*epi*-Isofagomine **142**, an iminosugar with a nitrogen at the anomeric position and having absolute configurations similar to galactose, was found to be a nanomolar inhibitor of β-galactosidase from *Aspergillus orizae* with an IC_{50} value of 12 nM.[105]On the other hand, the β-galactosidase inhibitory activity of **142** from human peripheral blood mononuclear cells was found to be 240 nM.[106] Introduction of an additional hydroxyl group at C-5 position of **142**, *i.e.* iminosugar **143**, led to a drastic decrease in its inhibitory activity of β-galactosiadse from *Aspergillus orizae* (IC_{50} = 17.5 μM for **143**).[107] Interestingly, with an added hydrophobic *n*-pentyl group at C-5 position of **142**, the β-galactosidase inhibitory activity of **144,** which is 5-*C*-pentyl-4-*epi*-isofagomine, (IC_{50} = 8 nM) was found to be even higher than **142** (Fig. 8.24). Compound **144** was identified as a promising pharmaceutical chaperone based drug candidate for the treatment of GM1-gangliosiodosis and Morquio disease type B.[106] 4-*epi*-Fagomine **145** displayed only moderate inhibition against β-galactosidase.[108] The sp^2 iminosugar **136** has been found to be a sub micromolar inhibitor of β-galctosidase from *E.coli* (K_b = 0.65 μM) and bovine (K_b = 0.2 μM).[98] Fluorinated pyrrolidine **146** was also reported to be a sub micromolar inhibitor of β-galactosidase from *E.coli.* (IC_{50} = 0.065 μM).[101]

4.5 Iminosugars as inhibitors of mannosidases

Mannosidase is an enzyme that hydrolyses mannosides. α-1,2-Mannosidases are essential for the formation of complex *N*-glycans on mammalian glycoproteins. In the endoplasmic reticulum (ER) where the synthesis of *N*-glycans begins, there are two classes of processing α-mannosidases.

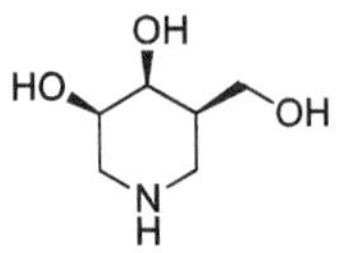

142
4-*epi*-isofagomine
IC_{50} = 12 nM, *Aspergillus orizae*
IC_{50} = 240 nM, human glycosidases

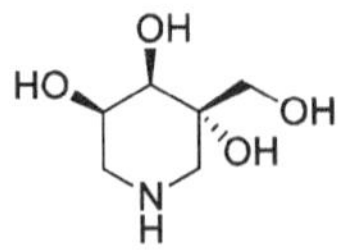

143
IC_{50} = 17.5 μM

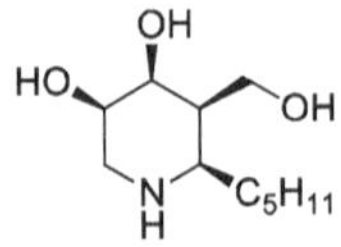

144
IC_{50} = 8 nM, human glycosidases

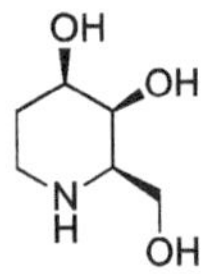

145
4-*epi*-fagomine
6.8 mg/mL

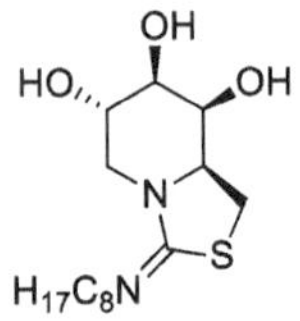

136
K_b = 0.65 μM, *E.coli*
K_b = 0.2 μM, bovine

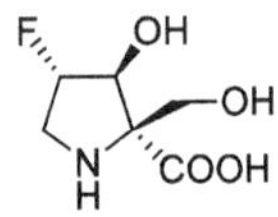

146
IC_{50} = 0.065 μM, *E.coli*

Figure 8.24 ***Structures of compounds that are strong inhibitors of β-galactosidases.***

Class I α-1,2-mannosidases (Man I) belong to glycosylhydrolase family 47 and possess subgroups that have different enzymatic properties or specialities. One sub group cleaves a single mannose from $Man_9GlcNAc_2$ to form $Man_8GlcNAc_2$ isomer B while the other subgroup trims $Man_9GlcNAc_2$ to $Man_8GlcNAc_2$ isomer A and/or C intermediates enroute to the formation of $Man_5GlcNAc_2$.[109] DMJ **32**, isolated from *Lonchocarpus sericeus* in 1979 and later from other plants and species, was found to be a weak inhibitor of α-mannosidases from almond (IC_{50} = 840 μM), human lysosomal (IC_{50} = 560 μM) and jack beans (IC_{50} = 150 μM).[68] Kifunensine **147** (Fig. 8.25), isolated from the culture broth of actinomycete, *Kitasatosporia kifunense* 9482 is also a weak inhibitor of α-mannosidase from Jack beans ((IC_{50} = 100 μM).[21c] Man I promotes proteosomal degradation of misfolded glycoproteins and hence inhibitors of Man I are expected to find applications as potential pharmacological chaperones for human genetic disorders. MJ **31**, DNJ **11** and L-*gulo*-DNJ **37** are weak inhibitors of α-mannosidase with 90.3% inhibition at 1 mM, 54.3% inhibition at 1 mM and 52% inhibition at 770 μM, respectively.[38,68,86] In the pyrrolidine series, while DAB1 **51** is a weak inhibitor of α-mannosidase from jack beans (IC_{50} = 100 μM),[68a] nectrisine **54** is a strong inhibitor (IC_{50} = 2 μM).[61,91] Broussonetinines A **56** and B **58** both are equally and the most potent naturally occurring iminosugars in inhibiting α-mannosidase from jack beans (IC_{50} = 0.3 μM).[65] Swainsonine **13** has an IC_{50} value of 1.75 μM against α-mannosidase from

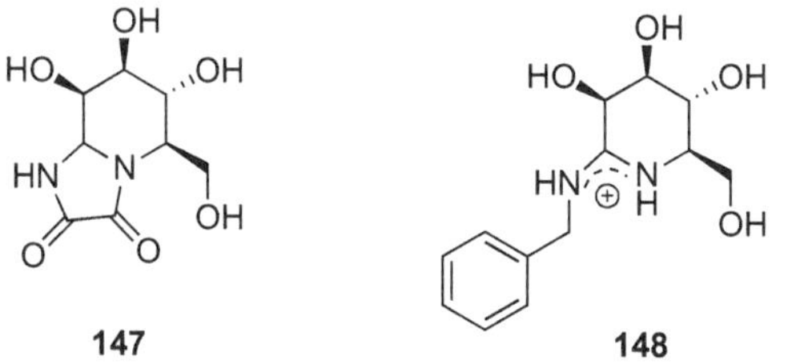

147
IC_{50} = 100 μM
α-mannosidase, jack beans

148
IC_{50} = 550 nM
α-mannosidase, jack beans

Figure 8.25 ***Structures of compounds that are strong inhibitors of α-mannosidases.***

jack beans.[68a, 86a,110] Amidine tethered DMJ derivative **148** (Fig. 8.**25**) was found to be a nanomolar inhibitor of α-mannosidase from jack beans (K_i = 550 nM).[111] A few iminosugar clusters,[112] iminosugar based glycopeptides,[113] and multivalent iminosugar based glycosidase inhibitors[114] have been developed that were found to be strong inhibitors of α-mannosidase.

Golgi α-mannosidases II belong to glycosylhydrolase family 38 are less specific. They cleave α1,2-, α1,3- and α1,6-linked mannose residues. They utilize aryl α-mannopyranosides as substrates.[109] Swainsonine **13** is a very strong inhibitor of Golgi α-mannosidase II with a K_i of 40 nM against *Drosophila melanogaster* enzyme.[109] It is a poisonous alkaloid and has been identified as a causative agent in locoism. However swainsonine **13** has attracted the attention of pharmaceuticals for investigation as an effective anticancer agent. By inhibiting Golgi α-mannosidase II very effectively, it stops the trimming of the mannose after the initiation of β-1,6-GlcNAc-linked chain. Thus it inhibits tumour growth, tumour metastatis and pro-apoptotic activity in a number of tumour types. However its clinical development is restricted due to its inhibition of lysosomal α-mannosidase.[21c]

4.6 Iminosugars as inhibitors of α-L-fucosidases

α-L-Fucosidases are exo-glycosidases that catalyse, during the last stages of glycoprotein biosynthesis, the hydrolysis of terminal α-L-fucose residues present in the cell surface oligosaccharides. They are also able to catalyse glycosylation reaction and hence are useful in the synthesis of fucosylated glycans. Deficiency of α-L-fucosidases leads to the accumulation of fucose in the tissues resulting in an autosomal recessive lysosomal disorder storage disease known as fucosidosis. The phenotypes of this disease include neurologic deterioration, growth retardation, seizures etc. Several symptoms such as inflammation, cystic fibrosis, tumour cell growth reveal

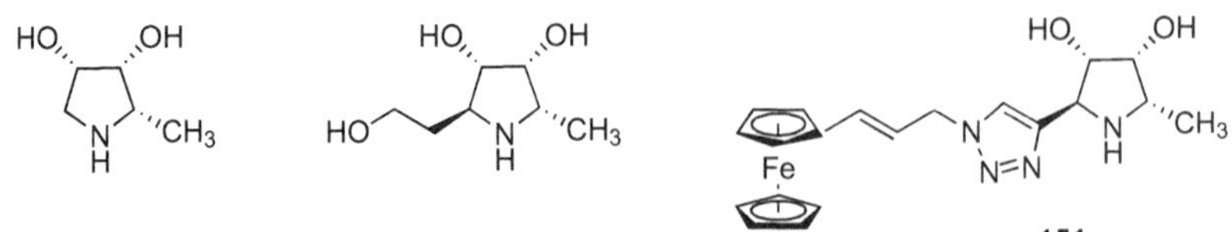

149 K_i = 50 nM, bovine kidney

150 K_i = 8 nM, bovine kidney

151 K_i = 23 nM, bovine kidney

Figure 8.26 ***Structures of compounds that are inhibitors of α-L-fucosidases.***

an abnormal distribution of α-L-fucosidases in them. Inhibitors of α-L-fucosidases may find applications as probes for correlating them with their biological functions and also may lead to the development of therapeutic drugs against fucosidosis. However, due to the presence of α-L-fucosidases in the membranes of human sperm cells and their role in facilitating sperm transport and sperm-egg interactions, their inhibitors may be counter active as contraceptives.[115]

While DNJ **11** is a very weak inhibitor of α-fucosidase (IC_{50} = 500 μM) from bovine kidney,[38,86] DMJ **32** is a much stronger inhibitor of the same (IC_{50} = 2.2 μM).[68] *N*-Methyl DMJ **33**[43] is weaker than DMJ **32**, while L-*gulo*-DNJ **37** has an IC_{50} value of 22 μM against α-fucosidase from bovine kidney.[43] Broussonetine I **66** inhibits α-fucosidase with an IC_{50} value of 52 μM. Iminosugars that resemble fucose, especially those possessing a methyl group at C-6 position of a piperidine ring or at C-5 position of a pyrrolidine ring with *fuco* configuration (such as **149**) were found to be very strong inhibitors of α-L-fucosidase. A detailed list of such inhibitors of α-L-fucosidase along with their inhibitory activities has been well documented in the review article by Robina and coworkers in 2011.[115] Subsequently, though several modified iminosugars have been identified as α-L-fucosidase inhibitors, those with high potency only are presented here. The 1-*C*-hydroxyethylpyrrolidine **150** has been shown to be a nanomolar inhibitor of α-L-fucosidase from bovine kidney with a K_i value of 8 nM.[116] A few divalent and trivalent iminosugars containing the pyrrolidine core of **149** have been found to be micromolar/sub micromolar inhibitors of α-L-fucosidase.[117] A triazole fused analogue of **149** tethered with a ferrocenyl moiety **151** (Fig. 8.26) has been found to be a nanomolar inhibitor of α-L-fucosidase from bovine kidney with a K_i of 23 nM.[118]

4.7 Iminosugars as inhibitors of α-L-rhamnosidoses

α-L-Rhamnosidases are enzymes that hydrolyse the terminal non-reducing α-L-rhamnose residues from a large number of natural products.[119]

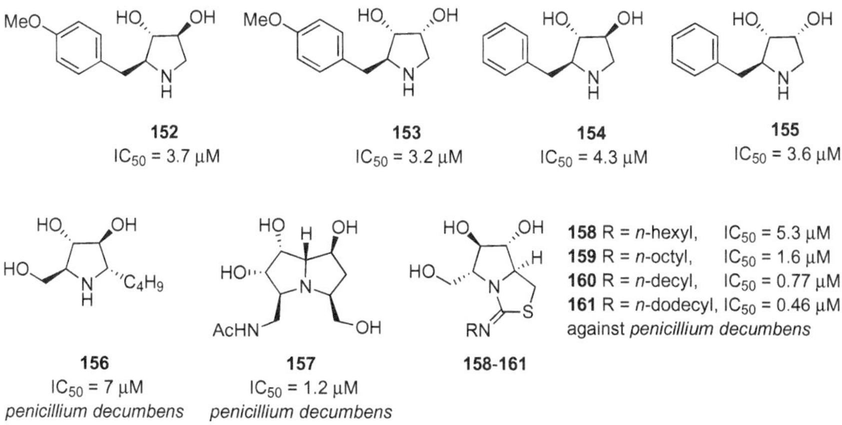

Figure 8.27 ***Structures of compounds that are inhibitors of α-L-rhamnosidases.***

α-Rhamnosidase is involved in the cleavage of *O*-antigen tetrasaccharides which is the first phase of bacterial invasion of host cell. It also hydrolyzes rhamnogalacturonans which act as immunomodulators thereby the cytotoxicity of human NK cells is enhanced. Inhibitors of α-L-rhamnosidases are expected to behave as potential therapeutic agents for basillary dysentery and cancer. Various stereoisomers of anisomycin and their deacetylated analogues have been investigated for their role as inhibitors of α-L-rhamnosidase. Pyrrolidines **152-153** as well as demethoxy analogues **154–155** (Fig. 8.27) are stronger inhibitors with K_i in the range of 2–5 μM. The *p*-methoxy group in **152** and **153** seems to have hardly any effect on their glycosidase inhibition against α-L-rhamnosidase.[120] C-5 Butyl substituted pyrrolidine **156** was found to be a strong inhibitor of α-L-rhamnosidase from *Penicillium decumbens* with an IC_{50} value of 7 μM.[121] 3-*epi*-(-)-Pochonicine **157**, the C-3 epimer of enantiomer of natural (+)-pochonicine **14**, inhibits α-L-rhamnosidase from *Penicillium decumbens* with an IC_{50} value of 1.2 μM.[122] A couple of *N*-alkylated polyhydroxyazepanes were shown to be moderate inhibitors of α-L-rhamnosidase.[123] L-DMDP cyclic isothioureas **158–161**, bicyclic compounds with L-pyrrolidine moieties have so far been the most potent inhibitors of α-L-rhamnosidase from *Penicillium decumbens*.[124]

4.8 Iminosugars as inhibitors of β-*N*-acetylhexosaminidases

β-*N*-Acetylhexosaminidase is a lysosomal enzyme found in most body tissues especially in kidneys. It is a complex glycoprotein lysosomal isoenzyme that is responsible for the release of *N*-acetylhexosamine from the non-reducing end of oligosaccharide moieties of glycoconjugates.

β-*N*-Acetylhexosaminidase breaks down GM2 ganglioside and hence deficiency of this enzyme would result in the accumulation of GM2 ganglioside in nerve cells of the brain, spinal cord leading to a rare neurodegenerative disease called Tay-Sachs disease. Sandhoff disease is another neurodegenerative rare lipid storage disorder caused by the deficiency of β-*N*-acetylhexosaminidase. Hence inhibitors of β-*N*-acetylhexosaminidase are expected to act as pharmacological chaperones for restoring the enzyme activity and thus find applications against these diseases. Among the natural iminosugars, pochonicine **14** is the only compound that has been found to inhibit β-*N*-acetylglucosaminidase in nM range (K_i = 0.192 nM)[78] from various sources including insects, fungi, mammals and plants. Though not as active as **14** some of its stereoisomers are sub micromolar inhibitors of β-*N*-acetylglucosaminidases.[122] Several synthetic iminosugars possessing an additional acetamido group have been found to be selective and strong inhibitors of this enzyme. Wong et al. observed that a synthetic iminosugar **162** (Fig. 8.28) obtained by replacement of one of the side chain hydroxyl groups of naturally occurring DMDP **12** by an acetamido group had a pronounced inhibitory activity against *N*-acetylglucosaminidase with a K_i of 9.8 μM from bovine kidney and 1.9 μM from Jack beans.[125,126] A high throughput screening of a library of such 1-aminodeoxy-DMDP analogues resulted in the identification of novel structures for antivirals and arthritis.[126] In general, though presence of an additional acetamido group on the side chain of iminosugars make them potential inhibitors of β-*N*-acetylhexosaminidase,[39] the same is not true for azetidine iminosugars containing an acetammido group as they do not inhibit these enzymes.[127] A detailed review on the synthesis of amino-modified iminosugars and their inhibition against β-*N*-acetylhexosaminidases has appeared from our group recently.[39]

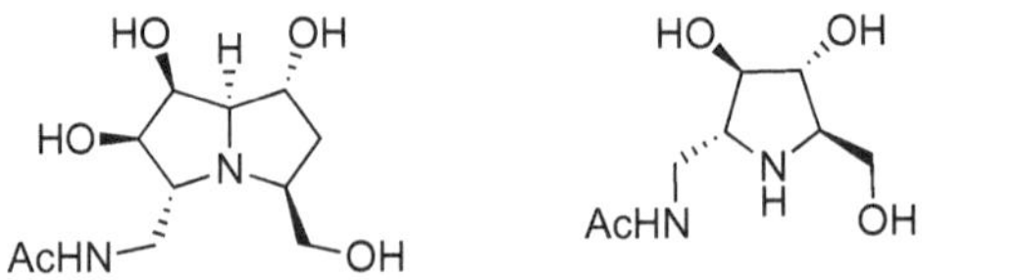

14 K_i = 0.192 nM

162
K_i = 9.8 μM, bovine kidney
K_i = 1.9 μM, Jack beans

Figure 8.28 ***Structures of compounds that are strong inhibitors of β-N-acetylhexosaminidases.***

4.9 Iminosugars as inhibitors of glycogen phosphorylase

Glycogen phosphorylase is an enzyme that catalyzes the phosphorylsis of glycogen resulting in the release of glycogen-1-phosphate. In patients suffering with type 2 diabetes, the glucose production in the liver increases. Inhibition of glycogen phosphorylase, thereby controlling the release of glucose from the liver glycogen's supplies, has been considered as a possible way to treat type 2 diabetes.[15] Naturally occurring iminosugar DAB1 **51** was reported to be a powerful inhibitor of mammalian glycogen phosphorylase with a K_i of 0.35 μM.[128] Synthetic isofagomine **17** has been reported to be a very powerful inhibitor of glycogen phosphorylase from pig liver (IC_{50} = 0.7 μM), rat liver (0.697 μM) and rabbit muscle (0.7 μM). Compound **17** inhibited, in hepatocytes, both basal and glucagon-induced glucose production dose-dependently (IC_{50} = 2-3 μM).[50a] However the inhibitory activity could not be further improved by introducing substituents on the ring nitrogen as such compounds decreased the activity. On the other hand, fagomine **15** and 2,3-di-*epi*-fagomine **42** are only very weak inhibitors of glycogen phosphorylase.[38,88e,50a,51] Structural insights in to the molecular basis for the inhibition of glycogen phosphorylase by iminosugars have been presented by Oikonomakos et al.[129] and in the review by Vidal and coworkers.[130]

5 Iminosugars as antivirals

Iminosugars that inhibit ER α-glucosidase I (α-Glu I) and II (α-Glu II) are potential candidates for host-directed antivirals as they have the capability to target a host process required for viral replication.[21c] ER α-Glu I and α-Glu II are the host enzyme targets for antiviral activity as they control entry to the calnexin cycle that is required for the folding of many glycoproteins. *N*-Linked glycoproteins play vital role in the viral life cycle and hence inhibition of biosynthesis of *N*-linked oligosaccharides, involving both glycosylation and glycoprotein processing, has been construed as a possible way for the development of antivirals.[17e,131], The success of iminosugar antivirals rely on their ability to display toxicity towards the virus selectively, though they avoid the problem of resistance mutations arising in the viral target. Recently, Zitzmann and co-workers, through crystal structures of trypsinolytic fragment of maurine α-Glu II alone and in complex with key catalytic cycle ligands and broad spectrum antiviral iminosugars, investigated the structural insights in to the antiviral target ER α-Glu II.[132] Based on the crystal structures, it was evidenced that the activity and substrate specificity are due to the portions of the enzyme outside its catalytic activity. Iminosugars that inhibit α-Glu

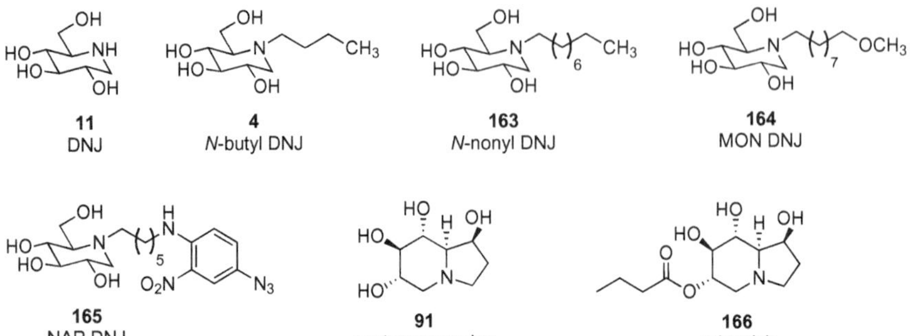

11 DNJ; 4 *N*-butyl DNJ; 163 *N*-nonyl DNJ; 164 MON DNJ; 165 NAP DNJ; 91 castanospermine; 166 celgosivir

Figure 8.29 ***Structures of iminosugars that are potential antiviral agents.***

I and α-Glu II have been found to have antiviral activity against a range of viruses including influenza, dengue virus (DENV), hepatitis C virus (HCV) and human immunodeficiency virus (HIV).[132] DNJ **11** was the first iminosugar, a strong inhibitor of ER glu I and Glu II, to be identified to supress the formation of complex glycan and thus was realized as a potential antiviral agent. Subsequently, several modified derivatives of DNJ such as *N*-butyl DNJ (miglustat) **4**, *N*-nonyl DNJ **163**, *N*-(9-methoxynonyl)-1-deoxynojirimycin (MON-DNJ) **164** and *N*-(6′-[4″-azido-2″-nitrophenylamino]hexyl)-1-DNJ (NAP-DNJ) **165** (Fig. 8.29) have been developed to increase the antiviral activity and all of them displayed potential inhibition against both the percentage of cells infected with dengue virus and release of infectious virus from primary human monocyte derived macrophages.[133] MON DNJ **164** and celgosivir **166** (the prodrug of natural indolizidine alkaloid castanospermine **91**) are also currently in phase I and phase II clinical trials, respectively, against dengue.[132]

Compounds **4**, **11**, **91** and **166** also inhibit, *in vitro*, HIV replication and HIV-mediated syncytial counts. All of them are inhibitors of processing α-glucosidase but not processing α-mannosidase. The *in vivo* studies do not find them to be promising candidates against HIV. Thus the α-glucosidase I inhibitory activities of iminosugars correlate well with their antiviral activities but their efficacy as anti HIV agents was not clinically promising.[21c]

6 Iminosugars as pharmacological chaperones for lysosomal storage diseases

Lysosomals storage disorders (LSD) are disorders caused by the accumulation of complex non-metabolized molecules in the lysosomal region. Such accumulation leads to cellular dysfunction and clinical syndromes.

Mutations in enzymes lead to destabilization of their native forms and impair their trafficking. The catalytic activities of the enzyme is reduced and this results in premature ER associated degradation deficiencies of specific hydrolytic functions. Consequence of this is the storage of metabolites in the lysosomes giving rise to abnormalities referred to as lysosomal storage disorders. One way to address this issue of tackling LSDs is to stabilize the native conformation of the mutant enzyme in the ER and restore its catalytic activities. Pharmacological chaperones (PCs) are small molecules that stabilize the native conformation of the mutant enzyme in the ER. The PC-enzyme complex is then transported to the Golgi apparatus for maturation and reach the lysosome. In the lysosome, the stored substrates stabilize the mutant enzyme by displacing the PC and the catalytic activities of the enzyme is restored/increased. LSDs generally encompassed enzyme deficiencies of only lysosome hydrolases which are soluble acidic glycosidases produced in the ER and the accumulated substrate is an oligosaccharide, a polysaccharide or a glycoconjugate. Iminosugars being competitive inhibitors of glycosidases are expected to behave as promising PCs for various LSDs.[134]

Gaucher's disease is caused by the mutations in the glucocerebrosidase gene leading to deficiency in the activity of β-glucocerebrosidase, an enzyme that is responsible for the hydrolysis of β-glycopyranosyl linkage in glucosylceramide. This results in the accumulation of glucocerebroside (also known as glucosylcermide, a sphingolipid) in cells, macrophages and certain organs such as liver, kidney, spleen, lungs, brain etc. Symptoms of Gaucher's disease include enlarged spleen and liver, liver malfunction, skeletal disorders, bone lesions and severe neurolical manisfestations as well. Gaucher's disease is the most common among all LSDs. Even though isofagomine **17**, which is a powerful inhibitor of Gcase though reached a more advanced stage as a drug for Gaucher's disease developed by Amicus Therapeutics, it failed Phase-II clinical trials. *N*-Butyl DNJ **11** (known as Miglustat) developed for type- I Gaucher's disease by Oxford Glycosciences and marketed by Actelion, was approved as a drug by US FDA in 2003. Several synthetic iminosugars are currently being developed that are more potent inhibitor of β-glucocerebrosidase than **11**.[3,134]

Fabry disease is a rare genetic disorder caused by mutations in the *GLA* gene that encodes lysosomal α-galactosidase A. This results in the deficiency of α-galactosidase A, an enzyme that is essential for the daily breakdown of globotriaosylceramide (GL-3 or GB-3), due to which GL-3 gets accumulated in the majority of cells throughout the body and leads to progressive cell damage. Clinical manifestations of Fabry disease include kidney failure, heart attack, stroke

Table 8.1 Iminosugar based marketed drugs[3].

Structure	Therapeutic application	Trade Name	Company Name
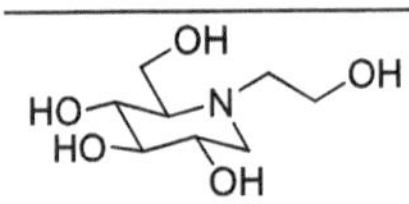 **5** Miglitol	diabetes	Glyset®	Bayer Healthcare Pharmaceuticals Inc.
		Mignar	Glenmark Pharmeceuticals Ltd.
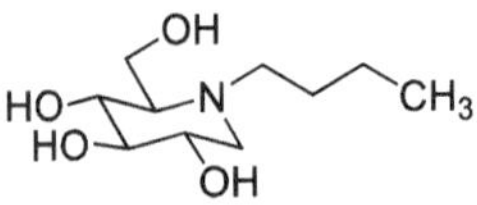 **4** Miglustat	Gaucher's disease	Zavesca®	Actelion
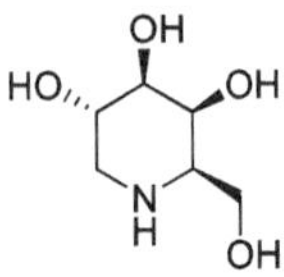 **127** DGJ	Fabry's disease	Galafold®	Amicus Therapeutics

Table 8.2 Selected examples of iminosugar therapeutics in clinical trials[3,131]

Structure	Potential therapeutic application	Clinical status
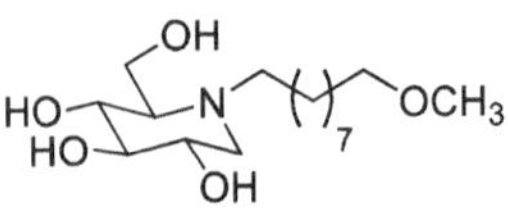 **164** MON DNJ	Dengue virus	Phase 1
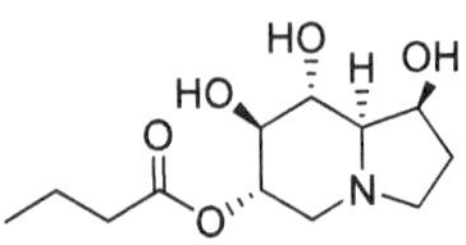 **166** celgosivir	Dengue virus	Phase 2

and neurological symptoms. 1-Deoxygalactonojirimycin (DGJ **127**, migalastat) with trade name Galafold, which is a powerful inhibitor of α-galactosidase, developed by Amicus Therapeutics was approved as a drug in 2016.[3,134]

The glycosidase inhibitory activities of some of the naturally occurring iminosugars reported by various groups are given in the Table 8.3.

Table 8.3 Glycosidase inhibitory activities of natural iminosugars

Imino-sugar	Enzyme	Source	IC_{50} (μM) / % inhibition	Ref.
10	α-mannosidase	jack beans	19% (at 1 mM)	38
	β-glucuronidase	bovine liver	37.2% (at 1 mM)	
11	α−glucosidases	rice	0.03	38, 86
		human lysosomal	0.65	
	α−glucosidase type I	calf liver	K_i = 1.0 μM	
	isomaltase	rat intestine	0.3	
	α-mannosidase		54.3% (at 1 mM)	
	β-glucocerebrosidase		240	
	hepatitis B virus HepG2 2.2.15 cells		100-500	
	trehalase	porcine kidney	4.3	
	β-glucosidases	almond	80	
		human lysosomal	240	
	α-fucosidase	bovine kidney	500	
	α-galactosidases	coffee bean	880	
30	isomaltase	rat intestine	4.4	86, 87
	α-glucosidase type-I	calf liver	K_i = 0.07 μM	
	trehalase	rat intestine	28	
43	maltase	rat intestine	0.34	49, 88
	sucrase	rat intestine	0.17	
	trehalase	porcine kidney	34.0	
44	maltase	rat intestine	15	49, 53, 88
	sucrase	rat intestine	7.2	
31	α-mannosidase	jack beans	90.3% (at 1 mM)	38, 45
32	α-fucosidases	bovine epididymis	39	68
		bovine kidney	2.2	
	isomaltase	rat intestine	110	
	α-mannosidases	almond	840	
		human lysosomal	560	
		jack beans	150	
	golgi α-mannosidase II		400	
33	α-L-fucosidase	bovine epididymis	190	43
		bovine palcenta	53	
34	β-D-glucosidase	*Asp. Wentii*	K_i = 400 μM	46, 96, 104
		almonds	K_i = 22.4 μM	
	α-D-galactosidase	coffee beans	K_i = 0.0007 μM	
		E.coli	K_i = 0.17 μM	
	β-galactosidase	penicillium multicolor	K_i = 0.0182 μM	
		E.coli	K_i = 0.045 μM	
		Asp. wentii	K_i = 0.011 μM	

(*Continued*)

Table 8.3 Glycosidase inhibitory activities of natural iminosugars. (*Cont.*)

Imino-sugar	Enzyme	Source	IC_{50} (μM) / % inhibition	Ref.
35	α-galactosidase	coffee bean	260	38, 47
	α-fucosidase	bovine epididymis	194	
	maltase	rat intestine	560	
	isomaltase	rat intestine	34	
	lactase	rat intestine	27	
36	α-glucosidase	rice	NI	38, 47
	maltase	rat intestine	NI	
	β-glucosidase	almond	NI	
	α-mannosidase	jack bean	NI	
		rat epididymis	NI	
	β-mannosidase	rat epididymis	NI	
	α-galactosidase	coffee bean	NI	
		Aspergillus niger	NI	
	β-galactosidase	bovine liver	NI	
		rat epididymis	NI	
	α-L-fucosidase	bovine epididymis	NI	
37	α-mannosidase	jack bean	52% (at 770 μM)	
	α-fucosidase	bovine epididymis	156	
		bovine kidney	22	
		human lysosomal	150	
38	maltase	rat intestine	110	48
	isomaltase	rat intestine	280	
	lactase	rat intestine	110	
15	isomaltase	rat intestine	460	38, 50,
	glycogen phosphorylase	pig liver	200	88e
41	maltase	rat intestine	500	38, 50a, 51
	isomaltase	rat intestine	6.4	
	lactase	rat intestine	4.0	
42	glycogen phosphorylase	pig liver	>200	38, 50a, 51
12	isomaltase	rat intestine	91	62, 86f, 88c,
	trehalase	porcine kidney	200	89, 90
	β-glucocerebrosidase	human placenta	340	
	α-glucosidase	Crustacean zooplankton	0.019	
	sucrase	rat intestine	40	
	β-glucosidase	Crustacean zooplankton	0.049	
		almond	10	
		bovine liver	9.7	
	cellobiose	rat intestine	24	
	β-galactosidase	bovine liver	3.3	
	lactase	rat intestine	7.9	

Table 8.3 Glycosidase inhibitory activities of natural iminosugars. (*Cont.*)

Imino-sugar	Enzyme	Source	IC_{50} (μM) / % inhibition	Ref.
50	disaccharidase	mammalian digestive	NI	38, 57, 88e
51	isomaltase	rat intestine	5.8	57, 68a,
	trehalase	porcine kidney	2.5	86c,
	α-mannosidase	jack bean	100	88e,
	β-glucocerebrosidase	human placenta	160	90a
	glycogen phosphorylase	mammalian	$K_i = 0.35$	
52	α-L-fucosidase	bovine epididymis	150	43, 75
		bovine kidney	370	
	β-mannosidase		380	
53	α-glucosidase	almond	300	60
	sucrose and isomaltase	rat intestine	1000	
54	α-glucosidase	yeast	0.048	61, 91
	α-mannosidase	jack bean	2.0	
	β-mannosidase	snail	12	
	β-glucosidase	almond	41	
	β-*N*-acetylhexosaminidase	bovine kidney	500	
55	galactopyranose mutase	*E.Coli* K 12 UDP	64% inhibition of UDP-Gal*p* to UDP-Gal*f* at 200 μg/mL	62, 77
56	α-glucosidase		0.445 mg/mL	65b
	α-glucosidase	from yeast	NI	
	β-glucosidase	sweet almond	NI	
	β-galactosiadse	bovine liver	0.016	
	α-mannosidase	jack bean	0.3	
	β-mannosidase	snail	NI	
57	α-glucosidase		0.53 mg/mL	65b
	α-glucosidase	yeast	NI	
	β-glucosidase	sweet almond	NI	
	β-galactosiadse	bovine liver	NI	
	α-mannosidase	jack bean	NI	
	β-mannosidase	snail	NI	
58	α-glucosidase	yeast	NI	65b
	β-glucosidase	sweet almond	NI	
	β-galactosidase	bovine liver	0.010	
	α-mannosidase	jack bean	0.29	
	β-mannosidase	snail	NI	

(*Continued*)

Table 8.3 Glycosidase inhibitory activities of natural iminosugars. (*Cont.*)

Imino-sugar	Enzyme	Source	IC_{50} (μM) / % inhibition	Ref.
59	α-glucosidase		0.46 mg/mL	65b
	α-glucosidase	yeast	NI	
	β-glucosidase	sweet almond	NI	
	β-galactosidase	bovine liver	NI	
	α-mannosidase	jack bean	NI	
	β-mannosidase	snail	NI	
60	α-glucosidase	yeast	NI	65b
	β-glucosidase	sweet almond	NI	
	β-galactosidase	bovine liver	0.36	
	α-mannosidase	jack bean	NI	
	β-mannosidase	snail	0.1	
61	α-glucosidase	yeast	NI	65b
	β-glucosidase	sweet almond	NI	
	β-galactosidase	bovine liver	0.29	
	α-mannosidase	jack bean	NI	
	β-mannosidase	snail	0.1	
62	α-glucosidase	yeast	3.3	65b
	β-glucosidase	sweet almond	0.055	
	β-galactosidase	bovine liver	0.002	
	α-mannosidase	jack bean	N	
	β-mannosidase	snail	0.02	
63	α-glucosidase	yeast	1.5	65b
	β-glucosidase	sweet almond	0.01	
	β-galactosidase	bovine liver	0.0041	
	α-mannosidase	jack bean	NI	
	β-mannosidase	snail	0.028	
64	α-glucosidase	yeast	NI	65b
	β-glucosidase	sweet almond	0.024	
	β-galactosidase	bovine liver	0.002	
	α-mannosidase	jack bean	NI	
	β-mannosidase	snail	0.76	
65	α-glucosidase	yeast	NI	65b
	β-glucosidase	sweet almond	0.036	
	β-galactosidase	bovine liver	0.0032	
	α-mannosidase	jack bean	NI	
	β-mannosidase	snail	0.32	
66	α-glucosidase	yeast	NI	65b
		rice	NI	
	maltase	rat intestine	NI	
	β-glucosidase	almond	652	
		bovine liver	2.9	
	α-galactosidase	coffee bean	NI	
	β-galactosidase	bovine liver	408	
	α-mannosidase	jack bean	NI	

Table 8.3 Glycosidase inhibitory activities of natural iminosugars. (*Cont.*)

Imino-sugar	Enzyme	Source	IC_{50} (μM) / % inhibition	Ref.
	β-mannosidase	*Helix pomatia*	NI	
	α-L-fucosidase	bovine kidney	52	
	amyloglucosidase	*Aspergillus niger*	NI	
	α-L-rhamnosidase	*Penicillium decumbens*		
69	α-glucosidase	yeast	NI	65b
		rice	NI	
	maltase	rat intestine	NI	
	β-glucosidase	almond	1000	
		bovine liver	10	
	α-galactosidase	coffee bean	NI	
	β-galactosidase	bovine liver	10	
	α-mannosidase	jack bean	NI	
	β-mannosidase	*Helix pomatia*	NI	
	α-L-fucosidase	bovine kidney	NI	
	amyloglucosidase	*Aspergillus niger*	54	
	α-L-rhamnosidase	*Penicillium decumbens*	NI	
70	α-glucosidase	yeast	NI	65b
	β-glucosidase	sweet almond	0.026	
	β-galactosidase	bovine liver	0.005	
	α-mannosidase	jack bean	NI	
	β-mannosidase	snail	0.3	
71	α-glucosidase	yeast	NI	65b
	β-glucosidase	sweet almond	0.017	
	β-galactosidase	bovine liver	0.004	
	α-mannosidase	jack bean	NI	
	β-mannosidase	snail	0.2	
72	α-glucosidase	Yeast	NI	65b
		rice	NI	
	maltase	rat intestine	NI	
	β-glucosidase	almond	NI	
		bovine liver	6.3	
	α-galactosidase	coffee beans	NI	
	β-galactosidase	bovine liver	2.3/8.1	
	α-mannosidase	jack bean	NI	
	β-mannosidase	snail	NI	
	α-L-fucosidase	bovine kidney	NI	
	trehalalse	porcine kidney	NI	
	amyloglucosidase	*Aspergillus niger*	NI	
	α-L-rhamnosidase	*Penicillium decumbens*	NI	
	β-glucuronidase	*E.coli*	86	
		Bovine liver	NI	

(*Continued*)

Table 8.3 Glycosidase inhibitory activities of natural iminosugars. (*Cont.*)

Imino-sugar	Enzyme	Source	IC_{50} (μM) / % inhibition	Ref.
74	α-glucosidase	yeast	NI	65b
	β-glucosidase	sweet almond	1.4	
	β-galactosidase	bovine liver	0.17	
	α-mannosidase	jack bean	NI	
	β-mannosidase	snail	8.2	
75	α-glucosidase	yeast	NI	65b
	β-glucosidase	sweet almond	2.4	
	β-galactosidase	bovine liver	0.2	
	α-mannosidase	jack bean	NI	
	β-mannosidase	snail	7.6	
76	α-glucosidase	yeast	NI	65b
	β-glucosidase	sweet almond	1.4	
	β-galactosidase	bovine liver	0.6	
	α-mannosidase	jack bean	NI	
	β-mannosidase	snail	20	
83	α-glucosidase	yeast	22	65b
		rice	119	
	maltase	rat intestine	67	
	isomaltase	intestine	NI	
	sucrose	rat intestine	216	
	ER α-glucosidase II		NI	
	β-glucosidase	almond	9.8	
		bovine liver	0.12	
	α-galactosidase	coffee beans	NI	
	β-galactosidase	bovine liver	0.03	
	α-mannosidase	jack bean	NI	
	β-mannosidase	snail	282	
	α-L-fucosidase	bovine kidney	NI	
	trehalase	porcine kidney	NI	
	amyloglucosidase	*Aspergillus niger*	30	
	α-L-rhamnosidase	*Penicillium decumbens*	NI	
	β-glucuronidase	*E.coli*	3.3	
		bovine liver	NI	
89	α-glucosidase	yeast	6.7	65b
90	α-glucosidase	yeast	9.3	65b
13	α-mannosidase	jack beans	1.75	86a,
	golgi α-mannosidase-II		20	88e,
		Drosophila melanogaster	$K_i = 0.04$	88c, 109, 110

Imino-sugar	Enzyme	Source	IC_{50} (μM) / % inhibition	Ref.
91	α-glucosidase	human liver lysosomal	100% (at 1 mM)	70, 88b,
	β-glucosidase	human liver lysosomal	96% (at 1 mM)	88c
	sucrase		100% (at 2 μM)	
	trehalase	pig lever	2.5	
92	α-glucosidase	human liver lysosomal	28% (at 1 mM)	70, 71,
	β-glucosidase	human liver lysosomal	14% (at 1 mM)	110a
	α-fucosidase	human liver	15% (at 1 mM)	
	amyloglucosidase	*Aspergillus niger*	2 μg/mL	
93	amyloglucosidase	*Aspergillus niger*	84	70b
94	amyloglucosidase	*Aspergillus niger*	5 μg/mL	71
95	β-glucosidase	almond	454	72
	β-galactosidase	*C. saccharolyticum*	739	
		rat intestine	35	
96	isomaltase	rat intestine	3.9	51a
	trehalase	porcine kidney	12.0	
97	isomaltase	rat intestine	540	74, 86a
	trehalase	porcine kidney	55	
98	isomaltase	rat intestine	97	51a, 75a,
	sucrase	rat intestine	4.6	75c, 88a
	trehalase	porcine kidney	160	
99	sucrase	rat intestine	470	51a, 74
100	lactase	rat intestine	4.4	76
101	lactase	rat intestine	260	76
	trehalase	porcine kidney		
102	lactase	rat intestine	270	76,77
103	lactase	rat intestine	3.6	76,77
104	lactase	rat intestine	18	76,77
14	β-*N*-acetylglucosa minidase	jack beans	0.288	78
105	β-glucosidase		6.7	65, 79
	β-galactosidase		2.9	
	β-mannosidase		3.3	
106	lactase	rat intestine	110	81
	trehalase	pig kidney	13.0	
		rat small intestine	12.0	
	β-glucosidase	almond	26	
		Caldocellum Saccharolyticum	37	
	α-galactosidase	green coffee bean	160	
		Aspergillus niger	180	
	β-galactosidase	bovine liver	270	
		rat intestine	110	
	glucocerebrosidase	wild-type human	3.1	

(*Continued*)

Table 8.3 Glycosidase inhibitory activities of natural iminosugars. (*Cont.*)

Imino-sugar	Enzyme	Source	IC_{50} (μM) / % inhibition	Ref.
107	rat digestive glycosidases		NI	81
	trehalase		NI	
	β-glucosidase	almond	NI	
		Caldocellum Saccharolyticum	NI	
	α-galactosidase	green coffee bean	NI	
		Aspergillus niger	NI	
	β-galactosidase	bovine liver	NI	
		rat intestine	NI	
	glucocerebrosidase		31	
108	lactase	rat intestine	2.6	81
	trehalase	rat small intestine	260	
	β-glucosidase	almond	4	
		Caldocellum Saccharolyticum	1	
	β-galactosidase	bovine liver	9.8	
		rat intestine	2.6	
	glucocerebrosidase	wild-type human	2.5	
109	maltase	rat intestine	640	81, 82a
	sucrase	rat intestine	500	
	trehalase	porcine kidney	10.0	
	α-glucosidase	rice	75	
	β-glucosidase	almond	2.6	
		Caldocellum Saccharolyticum	2.4	
	α-galactosidase	green coffee bean	1.9	
		Aspergillus niger	3.9	
		rat liver	21	
	β-galactosidase	bovine liver	240	
		rat intestine	7.8	
	glucocerebrosidase	wild-type human	1.0	
110	α-glucosidase	rice	NI	81
	β-glucosidase	almond	720	
		Caldocellum Saccharolyticum	390	
	α-galactosidase	green coffee bean	NI	
		Aspergillus niger	220	
	β-galactosidase	bovine liver	NI	
	Trehalase	rat intestine	NI	
		porcine kidney	200	
	glucocerebrosidase	wild-type human	76	
111	glucocerebrosidase	wild-type human	82	

Table 8.3 Glycosidase inhibitory activities of natural iminosugars. (*Cont.*)

Imino-sugar	Enzyme	Source	IC_{50} (μM) / % inhibition	Ref.
112	sucrase	rat intestine	160	42, 80, 81, 82
	trehalase	porcine kidney	270	
	α-glucosidase	rice	420	
	β-glucosidase	almond	0.82	
		Caldocellum	0.86	
		Saccharolyticum	360	
	α-galactosidase	green coffee bean	440	
		Aspergillus niger	16	
	β-galactosidase	bovine liver	0.38	
		rat intestine	2.5	
	glucocerebrosidase	wild-type human		
113	Trehalase	rat intestine	600	74
		porcine kidney	460	
114	sucrase	rat intestine	1000	76
	lactase	rat intestine	360	
	trehalase	rat intestine	210	
		porcine kidney	100	

7 Conclusions and future scope

Chemistry and biology of iminosugars have been dominating the area of carbohydrate chemistry for the last four decades. Their potential applications as drug molecules for various diseases are being realized with the development of iminosugar based drugs that have already hit the market. Having several advantages such as high water solubility, low molecular weight, stability and ability to access selective targets, these classes of compounds are the most sought after by pharma companies today in their quest to develop new drug molecules. Their unique combination of structural resemblance to natural sugars, ability to mimic the transition states and charge involved during the hydrolysis of glycosidases but remain inert for further processing offer enormous advantages in the area of drug development. They have high water-solubility, are least toxic and are excreted in urine without any change. Though iminosugars, in all aspects, possess privileged drug-like properties, success in this area has been often hampered due to various reasons that include inadequate potency, substrate specificity as well as side effects. From the lessons learnt out of failure of some of these molecules in their advanced stages of clinical trials, cutting-edge research has been continuing to overcome these problems. The chemistry and biology of iminosugars would thus remain a fertile area for scientists to explore and exploit the full potential of this unique class of compounds. In this direction, Seglin™ (Second Generation Leads from Iminosugars) platform provides a great opportunity to exploit this area for drug discovery in the near and medium term.[135]

References

1. Lindhorst, T. K. *Essentials of Carbohydrate Chemistry and Biochemistry*, 2nd ed.; WILEY-VCH: Germany, 2003.
2. Wong, C.-H, Ed. *Carbohydrate Based Drug Discovery*, 1 & 2; WILEY-VCH: Germany, 2003.
3. Horne, G. *Top. Med. Chem.* **2014**, *12*, 23–52 and references cited therein.
4. (a) de Melo, E. B.; da Silveira Gomes, A.; Carvalho, I. *Tetrahedron* **2006**, *62*, 10277–10302. (b) Wolfenden, R.; Lu, X.; Young, G. *J. Am. Chem. Soc.*, **1998**, *120*, 6814-6815.
5. (a) Varki, A.; Cummings, R. D.; Esko, J. D.; Freeze, H. H.; Stanley, P.; Bertozzi, C. R.; Hart, G. W.; Etzler, M. E. Eds., *Essentials of Glycobiology*, 2nd ed. Cold Spring Harbor, NY, 2009. (b) Fraser-Reid, B. O.; Tatsuta, K.; Thiem, J. Eds., *Glycoscience*, Springer: Berlin, 2001.
6. (a) Kornfeld, R.; Kornfeld, S. *Annu. Rev. Biochem.* **1985**, *54*, 631–664; (b) Elbein, A. D.; Molyneux, R. J. In *Iminosugars as Glycosidase Inhibitors*, Stütz, A. E., Ed.; Wiley-VCH: Weinheim, 1999; pp. 216-251.
7. (a) Zechel, D. L.; Withers, S. G. *Acc. Chem. Res.* **2000**, *33*, 11–18. (b) Rye, C. S.; Withers, S. G. *Curr. Opin. Chem. Biol.* **2000**, *4*, 573–580. (c) Sinnott, M. L. *Chem. Rev.* **1990**, *90*, 1171–1202. (d) Bojarova, P.; Křen, V. *Trends Biotechnol.* **2009**, *27*, 199-209.
8. Lairson, L. L.; Henrissat, G. J.; Davis, G. J.; Withers, S. G. *Annu. Rev. Biochem.* **2008**, *77*, 521–555.
9. (a) Leloir, L. F. *Science* **1971**, *172*, 1299–1303. (b) Leloir, L. F. *Methods Enzymol.* **1972**, *83*, 458-514.
10. Kornfeld, R.; Kornfeld, S. *Annu. Rev. Biochem.* **1985**, *54*, 631–664.
11. Wong, C. H.; Halcomb, R. L.; Ichikawa, Y.; Kajimoto, T. *Angew. Chem. Int. Ed.* **1995**, *34*, 412–432 and 521-546.
12. Paulson, K.; Colley, K. J. *J. Biol.chem* **1989**, *264*, 17615–17618.
13. (a) Krentz, A. J.; Bailey, C. J. *Drugs* **2005**, *65*, 385–411. (b) Elbein, A. D.; Pan, Y. T.; Pastuszak, I.; Carroll, D. *Glycobiology* **2003**, *13*, 17R–27R. (c) Ando, O.; Kifune, M.; Nakajima, M. *Biosci. Biotechnol. Biochem.* **1995**, *59*, 711–712. (d) Asano, N. Takeuchi, M.; Kameda, Y.; Matsui, K.; Kono, Y. *J. Antibiot.* **1990**, *43*, 722–726. (e) Joubert, P. H.; Venter, H. L.; Foukaridis, G. N. *Br. J. Clin. Pharmacol.* **1990**, *30*, 391–396. (f) Joubert, P. H.; Foukaridis, G. N.; Bopape, M. L. *Eur. J. Clin. Pharmacol.* **1987**, *31*, 723-724.
14. (a) Bucior, I.; Burger, M. M. *Curr. Opin. Struct. Biol.* **2004**, *14*, 631–637. (b) Kim, C. U.; Lew, W.; Williams, M. A.; Liu, H. T.; Zhang, L. J.; Swaminathan, S.; Bischofberger, N.; Chen, M. S.; Mendel, D. B.; Tai, C.Y.; Laver, W. G.; Stevens, R. C. *J. Am. Chem. Soc.* **1997**, *119*, 681-690.
15. Somsak, L.; Nagy, V.; Hadady, Z.; Docsa, T.; Gergely, P. *Curr. Pharm. Design* **2003**, *9*, 1177–1189.
16. (a) Humphries, M. J.; Matsumoto, K.; White, S. L.; Olden, K. *Cancer Res.* **1986**, *46*, 5215–5222. (b) Goss, P. E. Baker, M. A.; Carver, J. P.; Dennis, J. W., *Clin, Cancer Res.*

1995, *1*, 935–944. (c) Nishimura, Y.; Satoh, T.; Kudo, T.; Kondo, S.; Takeuchi, T. *Bioorg. Med. Chem.* **1996**, *4*, 91–96. (d) Wrodnigg, T. M.; Steiner, A. J.; Ueberbacher, B. J. *Anti-Cancer Agents Med. Chem.* **2008**, *8*, 77-85.

17. (a) Steinmann, E.; Whitfield, T.; Kallis, S.; Dwek, R. A.; Zitzmann, N.; Pietschmann, T.; Bartenschlager, R. *Hepatology* **2007**, *46*, 330–338. (b) Augustin, L. A.; Fantini, J.; Mootoo, D. R. *Bioorg. Med. Chem.* **2006**, *14*, 1182–1188. (c) Inmaculada, R.; Moreno-Vargas, A. J.; Carmona, A. T.; Vogel, P. *Curr. Drug. Metabol.* **2004**, *5*, 329-61.(d) Metha, A. S.; Gu, B.; Conyers B.; Ouzounov, S.; Wang, L; Moriarty, R. M.; Dwek, R. A.; Block, T. M. *Antimicrob. Agents Chemother.* **2004**, *48*, 2085–2090. (e) Greimel, P.; Spreitz, J.; Stutz, A. E.; Wrodnigg, T. M. *Curr. Med. Chem.* **2003**, *3*, 513–523. (f) Pavlovic, D.; Neville, D. C. A.; Argaud, O.; Blumberg, B.; Dwek, R. A.; Fischer, W. B.; Zitzmann, N. *Proc. Natl. Acad. Sci.*, U.S.A. **2003**, *100*, 6104–6108; (g) Mehta, A.; Conyers, B.; Tyrrell, D. L. J.; Walters, K.-A.; Tipples, G. A.; Dwek, R. A.; Block, T. M. *Antimicrob. Agents Chemother.* **2002**, *46*, 4004–4008, (*h*) Durantel, D.; Branza-Nichita, N.; Carrouee-Durantel, S.; Butters, T. D.; Dwek, R. A.; Zitzmann, N. *J. Virol.* **2001**, *75*, 8987–8998. (i) Block, M.; Jordan, R. *Antiviral Chem, Chemotherapy* **2001**, *12*, 317–325. (j) Weber, K. T. Hammache, D. Fantini, J. Ganem, B. Bioorg. *Med. Chem. Lett.* **2000**, *10*, 1011–1014. (*k*) Mehta, A. *FEBS Lett.* **1998**, *430*, 17–22. (*l*) Fischer, P. B.; Karlsson, G. B.; Butters, T. D.; Dwek, R. A.; Platt, F. M. *J. Virol.* **1996**, *70*, 7143–7152. (m) Fischer, P. B. Karlsson, G. B.; Butters, T. D.; Dwek, R. A.; Platt, F. M. *J. Virol.* **1996**, *70*, 7153–7160. (n) Asano, N. *J. Med. Chem*, **1995**, *38*, 2849–2856. (o) Fischer, P. B. *J. Virol.* **1995**, *69*, 5791–5797. (p) Ahmed, S. P.; Nash, R. J.; Bridges, C. G.; Taylor, D. L.; Kang, M. S.; Porter, E. A.; Tyms, A. S. *Biochem. Biophysics. Res. Commun.* **1995**, *208*, 267–273. (q) Brodges, C. G.; Ahmed, S. P.; Kang, M. S.; Nash, R. J.; Porter, E. J.; Tymes, A. S. *Glycobiology*, **1995**, *5*, 249–253. (r) Block, T. M.; Lu, X.; Platt, F. M.; Foster, G. R.; Gerlich, W. H.; Blumberg, B. S.; Dwek, R. A. *Proc. Natl. Acad. Sci.* U.S.A. **1994**, *91*, 2235-2239.

18. (a) Wang, G.-N.; Reinkensmeier, G.; Zhang, S.-W.; Zhou, J.; Zhang, L.-R.; Zhang, L.-H.; Butters, T. D.; Ye, X.-S. *J. Med. Chem.* **2009**, *52*, 3146–3149. (b) Benjamin, E. R.; Flanagan, J. J.; Schilling, A.; Chang, H. H.; Agarwal, L.; Katz, E.; Wu, X.; Pine, C.; Wustman, B.; Desnick, R. J.; Lockhart, D. J.; Valenzano, K. J. J. *Inherited Metab. Dis.* **2009**, *32*, 424–440. (c) Yu, Z.; Sawkar, A. R.; Whalen, L. J.; Wong, C.-H.; Kelly, J. W. *J. Med. Chem.* **2007**, *50*, 94–100. (d) Butters, T. D, Dwek, R. A. Platt, F. M. *Glycobiology* **2005**, *15*, 43R–52R. (e) Butters, T. D.; Mellor, H. R.; Narita, K. Dwek, R. A.; Platt, F. M *Curr. Med. Chem.* **2003**, *358*, 92–945. (f) Butters, T. D. Dwek, R. A.; Platt. F. M. *Curr. Med. Chem.* **2003**, *3*, 56–574. (g) Sawker, A. R,; Cheng, W.-C, Beutler, E.; Wong. C.–H.; Balch, W. E.; Kelly, J. W. *Proc. Natl. Acad. Sci.*, **2002**, *99*, 1542–15433. (*h*) Meikle, P. J.; Hopewood, J. J.; Clague, A. E.; Carey, W. F. *J. Am. Med. Ass.*, **1999**, *281*, 24–254. (i) Fan, J.-Q.; Ishii, S.; Asano, N.; Suzuki, Y. *Nat. Med.* **1999**, *5*, 112–115.

19. (a) Wu, C.-Y.; Wong. C.-H. *Chem. Commun.* **2011**, 6201–6207. (b) Seeberger, P, H. *Chem. Commun.* **2003**, 1115–1121. (c) Bertozzi, C. R.; Kiessling, L. L. Science 2001, 291, 2357–2364. (d) Maeder, T. *Sci. Am.* **2002**, *287*, 40–47. (e) Shriver, Z.; Raguram, S.; Sasisekharan, R. *Nat. Rev.* **2004**, *3*, 863–873. (f) Dwek, R. A. *Chem. Rev.* **1996**, *96*, 683. (g) Hirabayashi, J.; Kasai, K.-I. *Trends Glycosci. Glycotechnol.* **2000**, *12*, 1–5. (f) Feizi, T.; Mulloy. B. *Curr. Opin. Struct. Biol.* **2003**, *13*, 602–604. (*h*) Asano, N. *Glycobiology* **2003**, *13*, 93R–104R. (i) El Ashry, E. S. H.; Rashed, N. Shobier, A. H. S. *Pharmazie* **2005**, *55*, 403–415. (j) Ghazarian, H.; Idoni, B.; Oppenheimer S. B. *Acta Histochemica* **2011**, *113*, 236–247. (*k*) Sears, P. Wong, C.-H. Angew. *Chem. Int. Ed.* **1999**, *38*, 2300-2324.

20. (a) Ernst, B.; Magnani, J. L. *Nat. Rev.* **2009**, *8*, 661–677. (b) Koeller, K. M.; Wong, C.-H. *Nat. Biotech.* **2000**, *18*, 835–841. (c) Bencomo, V.V. *Science* **2004**, *305*, 522-525.

21. (a) McNaught, A. D. *Pure Appl. Chem.* **1996**, *68*, 1919–2008; (b) A. E. Stütz, *Iminosugars as Glycosidase Inhibitors: Nojirimycin and Beyond;* Wiley-VCH: Weinheim, **1999**. (c) Compain, P.; Martin, O. R. Iminosugars-From Synthesis to Therapeutic Applications;

John Wiley & Sons Ltd: *West Sussex England*, **2007**; (d) A. E. Stütz; T. M. Wrodnigg, Adv. Carbohydr. *Chem. Biochem.* **2011**, *66*, 187-298.

22. (a) Gloster, T. M.; Davies, G. J. *Org. Biomol. Chem.* **2010**, *8*, 305–320. (b) Dwek, R. A. *Chem. Rev.* **1996**, *96*, 668–720. (c) Lillelund, V. H.; Jensen, H. H; Liang, X.; Bols, M. *Chem. Rev.* **2002**, *102*, 51–553. (d) Durantel, D.; Branza-Nichita, N.; Carrouée-Durantel, S.; Butters, T. D.; Dwek, R. A.; Zitzmann, N. J. *Virol.* **2001**, *75*, 8987-8998.
23. (a) Aerts, J. M. F. G. 2006, The remarkable story of Hermann Boerhaave, the mulberry and iminosugar drugs. Proc. European Working Group for the Study of Gaucher Disease. http://www.ewggd.co.uk/EWGGD%20Abstracts.pdf. (b) Scott, L. J. and Spenser, C. M. *Drugs* **2000**, *59*, 521-549.
24. (a) Hanessian, S. Chem. Commun., 1966, 796–798. (b) Hanessian, S.; Haskell, T. H. J. *Org. Chem.*, **1963**, *28*, 2604–2610. (c) Jones, J. K. N.; Szarek, W. A. *Can. J. Chem.*, **1963**, *41*, 636–640. (d) Paulsen, H. Angew. *Chem. Int. Ed. Engl.*, **1962**, *1*, 597. (d) Jones, J. K. N.; Turner, J. C. J. *Chem. Soc.* **1962**, 4699–4703. (e) Paulsen, H. Angew. *Chem. Int. Ed. Engl.* **1962**, *1*, 454.
25. (a) Cox, T. M.; Platt, F. M.; Aerts, J. M. F. G. Medicinal use of iminosugars. In Iminosugars: From Synthesis to Therapeutic Applications, Compain, P.; Martin, O. R., Eds. pp. 295–326, John Wiley and Sons Ltd, West Sussex England, **2007**; (b) Winchester, B. G. *Tetrahedron Asymmetry* **2009**, *20*, 645–651; (c) D'Alonzo, D.; Guaragna, A.; Palumbo, G. *Curr. Med. Chem.* **2009**, *16*, 473-505.
26. Lillehund, V. H.; Jensen, H. H.; Liang, X.; Bols, M. *Chem. Rev.* **2002**, *102*, 515–553.
27. Sears, P.; Wong, C.-H. *Angew. Chem. Int. Ed.* **1999**, *38*, 2301–2324.
28. (a) Marco-Contelles, J. Eur. J. *Org. Chem.*, **2001**, 1607–1618. (b) Knapp S., Vocadlo D., Gao, Z., Kirk, B., Lou, J., Withers, S. G. J. *Am. Chem. Soc.* **1996**, *118*, 6804-6805.
29. Tull, D.; Withers, S. G. *Biochemistry* **1994**, *33*, 6363–6370.
30. Heightman, T. D.; Vasella, A. T. *Angew. Chem. Int. Ed.* **1999**, *38*, 750–770.
31. Bennett, A. J.; Sinnott, M. L. *J. Am. Chem. Soc.* **1986**, *108*, 7287–7294.
32. (a) Varrot, A.; Tarling, C. A.; Macdonald, J. M.; Stick, R. V.; Zechel, D. L.; Withers, S. G.; Davies, G. J. J. *Am. Chem. Soc.*, **2003**, *125*, 7496–7497. (b) Ichikawa, Y.; Igarashi, Y.; Ichikawa, M.; Suhara, Y. J. *Am. Chem. Soc.*, **1998**, *120*, 3007–3018. (c) Jespersen, T. M.; Dong, W. L.; Sierks, M. R.; Skrydstrup, T.; Lundt, I.; Bols, M. Angew. *Chem. Int. Ed.* **1994**, *33*, 1778–1779. (d) Jespersen, T. M.; Bols, M.; Sierks, M. R.; Skrydstrup, T. *Tetrahedron* **1994**, *50*, 13449-13460.
33. Liu, H.; Liang, X.; Søhoel, H.; Bülow, A.; Bols, M. *J. Am. Chem. Soc.* **2001**, *123*, 5116–5117.
34. (a) Paulsen, H.; Todt, K. Chem. Ber., 1967, 100, 3385–3396. (c) Paulsen, H.; Todt, K. *Chem. Ber.* **1967**, *100*, 3397–3407. (c) Paulsen, H. Angew. *Chem. Int. Ed. Engl.*, **1966**, *5*, 495–511. (d) Paulsen, H. *Liebigs Ann. Chim.*, **1965**, *683*, 187-198.
35. (a) Gerber-Lrmaire, Sandrine, Juilleret-Jeanneret, Lucienne. *Mini rev. Med. Chem.* **2006**, *6*, 1043-1052. (b) Liu, H. Z.; Lillelund, V. H.; Andersch, J.; Liang, X. F.; Bols, M. J. *Carbohydr. Chem.* **2004**, *23*, 223–238. (c) Dwek, R. A., Butters, T. D.; Platt, F. M.; Zitzmann, N. *Nature Rev. Drug discovery* **2002**, *1*, 65–75. (d) Kondo, K.; Adachi, H.; Shitara, E.; Kojima, F.; Nishimura, Y. *Biooorg, Med. Chem.* **2001**, *9*, 1091–1095.; (e) Bach, P.; Bols, M. *Tetrahedron Lett.* **1999**, *40*, 3461–3464. (f) Dennis, J. W.; Granovsky, M.; Warren, C. *Biochim. Biophys. Acta* **1999**, 1473, 21-34.
36. Caines, M. E. C.; Hancock, S. M.; Tarling, C. A.; Wrodnigg, T. M.; Stick, R. V.; Stutz, A. E.; Vasella, A.; Withers, S. G.; Strynadka, N. C. J. *Angew. Chem. Int. Ed.* **2007**, *46*, 4474–4476.
37. Caines, M. E. C.; Vaughan, M. D.; A., T. C.; Hancock, S. M.; Warren, R. A. J.; Withers, S. G.; Strynadka, N. C. *J. J. Biol. Chem.* **2007**, *282*, 14300-14308.
38. (a) Tsuruoka, T.; Fukuyasu, H.; Ishii, M.; Usui, T.; Shibahara, S.; Inouye, S. J. Antibiot., **1996**, *49*, 155–161. (b) Legler, G. *Adv. Carbohydr. Chem. Biochem.* **1990**, *48*, 319–384.

(c) Niwa, T.; Inouye, S.; Tsuruoka, T.; Koaze, Y.; Niida, T. Agric. *Biol. Chem.*, **1970**, *4*, 966–968. (d) Inouye, S.; Tsuruoka, T.; Koaze, Y.; Niida, T. *Tetrahedron* **1968**, *24*, 2125–2144. (e) Inouye, S.; Tsuruoka, T.; Niida, T. J. *Antibiot. Ser. A*, **1966**, *19*, 288-292.

39. Harit, V. K.; Ramesh, N. G. *RSC Advances* **2016**, *6*, 109528–109607 and references cited therein.
40. Yagi, M.; Kouno, T.; Aoyagi, Y.; Murai, H. *Nippon Nogeik. Kaishi* **1976**, *50*, 571–572.
41. For recent reviews see: (a) Davies, S. G.; Fletcher, Ai. M.; Roberts, P. M.; Thomson, J. E. *Tetrahedron* **2019**, *75*, 130727. (b) Prichard, K.; Campkin, D.; O'Brien, N.; Kato, A.; Fleet, G. W. J.; Simone, M. I. *Chemical Biology & Drug Design* **2018**, *92*, 1171–1197. (c) Sousa, C. E.; Mendes, R. R.; Costa, F. T.; Duarte, V. C. M.; Fortes, A. G.; Alves, M. J. *Curr. Org. Synth.* **2014**, *11*, 182–203 and references cited in these reviews.
42. Asano, N.; Tomioka, E.; Kizu, H.; Matsui, K. *Carbohydr. Res.* **1994**, *253*, 235–245.
43. Asano, N.; Yasuda, K.; Kizu, H.; Kato, A.; Fan, J.-Q.; Nash, R. J.; Fleet, G. W. J.; Molyneux, R. J. *Eur. J. Biochem* **2001**, *268*, 35–41.
44. (a) Fellows, L. E.; Bell, A.; Lynn, D. G.; Pilkiewicz, F.; Miura, I.; Nakanishi, K. J. Chem. *Soc. Chem. Commun.* **1979**, 977–978. (b) Watson, A. A.; Fleet, G. W. J.; Asano, N.; Molyneux, R. J.; Nash, R. J. *Phytochemistry* **2001**, *56*, 265-295.
45. Niwa, T.; Tsuruoka, T.; Goi, H.; Kodama, Y.; Itoh, J.; Inouye, S.; Yamada, Y.; Niida, T.; Nobe, M.; Ogawa, Y. *J. Antibiot* **1984**, *37*, 1579–1586.
46. Miyake, Y.; Ebata, M. *Agric. Biol. Chem.* **1988**, *52*, 661–666.
47. Kato, A.; Kato, N.; Kano, E.; Adachi, I.; Ikeda, K.; Yu, L.; Okamoto, T. Y.; Banba, Y.; Ouchi, H.; Takahata, H.; Asano, N. *J. Med. Chem.* **2005**, *48*, 2036–2044.
48. Ikeda, K.; Takahashi, M.; Nishida, M.; Miyauchi, M.; Kizu, H.; Kameda, Y.; Arisawa, M.; Watson, A. A.; Nash, R. J.; Fleet, G. W. J.; Asano, N. *Carbohydr. Res.* **2000**, *323*, 73–80.
49. (a) Asano, N.; Nishida, M.; Kato, A.; Kizu, H.; Matsui, K.; Shimada, Y.; Itoh, T.; Baba, M.; Watson, A. A.; Nash, R. J.; Lilley, P.M. de Q.; Watkin, D. J.; Fleet, G. W. J. J. *Med. Chem.*, **1998**, *41*, 2565-2571.
50. (a) Jakobsen, P.; Lundbeck, J. M,; Kristiansen, M.; Breinholt, J.; Demuth, H.; Pawlas, J.; Candela, M. P. T; Andersen, B.; Westergaard, N.; Lundgren K.; Asano, N. Bioorg. *Med. Chem.*, **2001**, *9*, 733–744. (b) Asano, N.; Nishida, M.; Miyauchi, M.; Ikeda, K.; Yamamoto, M.; Kizu, H.; Kameda, Y.; Watson, A. A.; Nash R. J.; Fleet, G. W. J. *Phytochemistry* **2000**, *53*, 379-382.
51. (a) Kato, A.; Kano, E.; Adachi, I.; Molyneux, R. J.; Watson, A. A.; Nash, R. J.; Fleet, G. W. J.; Wormald, M. R.; Kizu, H.; Ikeda, K.; Asano, N. *Tetrahedron: Asymmetry*, **2003**, *14*, 325–331. (b) Kato, A.; Asano, N.; Kizu, H.; Matsui, K.; Watson, A. A.; Nash, R. J. J. *Nat. Prod*, **1997**, *60*, 312-314.
52. Kite, G. C.; Fellows, L. E.; Fleet, G. W. J.; Liu, P. S.; Scofield, A. M.; Smith, N. G. *Tetrahedron Lett.* **1988**, *29*, 6483–6486.
53. Asano, N.; Nishida, M.; Kizu, H.; Matsui, K.; Watson, A. A.; Nash, R. J. *J. Nat. Prod* **1997**, *60*, 98–101.
54. Segraves, N. L.; Crews, P. *J. Nat. Prod.* **2005**, *68*, 118–121.
55. Sengoku, T.; Wierzejska, J.; Takahashi, M.; Yoda, H. *Nat. Prod. Commun.* **2013**, *8*, 1011–1019.
56. Welter, A.; Jadot, J.; Dardenne, G.; Marlier, M.; Casimir, J. *Phytochemistry* **1976**, *15*, 747–749.
57. Nash, R. J.; Bell, E. A.; Fleet, G. W. J.; Jones, R. H.; Williams, J. M. *J. Chem. Soc., Chem. Commun* **1985**, 738–740.
58. Nash, R. J.; Bell, E. A.; Williems, J. M. *Phytochemistry* **1985**, *24*, 1620–1622.
59. Liu, K. K.-C.; Kajimoto, T.; Chen, L.; Zhong, Z.; Ichikawa, Y.; Wong, C.-H. *J. Org. Chem.* **1991**, *56*, 6280–6289.
60. Asano, N.; Yamauchi, T.; Kagamifuchi, K.; Shimizu, N.; Takahashi, S.; Takatsuka, H.; Ikeda, K.; Kizu, H.; Chuakul, W.; Kettawan, A.; Okamoto, T. *J. Nat. Prod.* **2005**, *68*, 1238–1242.

61. Shibata, T.; Nakayama, O.; Tsurumi, Y.; Okuhara, M.; Terano, H.; Kohsaka, M. *J. Antibiot.* **1988**, *41*, 296–301.
62. Watson, A. A.; Nash, R. J.; Wormald, M. R.; Harvey, D. J.; Dealler, S.; Lees, E.; Asano, N.; Kizu, H.; Kato, A.; Griffiths, R. C.; Cairns, A. J.; Fleet, G. W. J. *Phytochemistry* **1997**, *46*, 255–259.
63. (a) Shibano, M.; Kitagawa, S.; Nakamura, S.; Akazawa, N.; Kusano, G. *Chem. Pharm. Bull.* **1997**, *45*, 700–705. (b) Shibano, M.; Kitagawa, S.; Kusamno, G. *Chem. Pharm. Bull.* **1997**, *45*, 505–508. (c) Shibano, M.; Nakamura, S.; Akazawa, N.; Kusano, G. *Chem. Pharm. Bull.* **1998**, *46*, 1048–1050. (d) Shibano, M.; Nakamura, S.; Motoya, N.; Kusano, G. *Chem. Pharm. Bull.* **1999**, *47*, 472–476. (e) Shibano, M.; Tsukamoto, D.; Fujimoto, R.; Masui, Y.; Sugimoto, H.; Kusano, G. *Chem. Pharm. Bull.* **2000**, *48*, 1281–1285. (f) Tsukamoto, D.; Shibano, M.; Okamoto, R.; Kusano, G. *Chem. Pharm. Bull.* **2001**, *49*, 492–496. (g) Song, Y.-Y.; Kinami, K.; Kato, A.; Jia, Y.-M.; Li, Y.-X.; Fleet, G. W. J.; Yu, C.-Y. *Org. Biomol. Chem.* **2016**, *15*, 5157-5174.
64. Khanov, M. T.; Sultanov, M. B.; Egorova, T. A. Farmakol. Alkaloidov. Serdech. Glikoyidov. 1971, 210–212; *Chem. Abs.* **1972**, 77, 135091r,.
65. (a) Shibano, M.; Tsukamoto, D.; Masuda, A.; Tanaka, Y.; Kusano, G. *Chem. Pharm. Bull.* **2001**, *49*, 1362–1365. (b) Shibano, M.; Tsukamoto, D.; Kusano, G. Heterocycles **2002**, *57*, 1539-1553.
66. Yu, C.-Y.; Huang, M.-H. *Org. Lett.* **2006**, *8*, 3021–3024.
67. (a) Colegate, S. M.; Darling, P. R.; Huxtable, C. R. *Aust. J. Chem.* **1979**, *32*, 2257–2264. (b) Molyneux, R. J.; James, L. F. *Science*, **1982**, *216*, 190–191. (c) Dorling, P. R.; Huxtable, C. R.; Colegate, S. M. *Neuropathol. Appl. Neurobio* **1978**, *4*, 285–295. (d) Davis, D.; Schwarz, P.; Hernandez, T.; Mitchell, M.; Warnock, B.; Elbein, A. D. *Plant Physiol* **1984**, *76*, 972-975.
68. (a) Winkler, D. A.; Holan, G. J. Med. Chem., 1989, 32, 2084–2089. (b) Tulsiani, D. R. P.; Harris, T. M.; Touster, O. J. *Biol. Chem.*, **1982**, *257*, 7936-7939.
69. Hohenschutz, L. D.; Bell, E. A.; Jewess, P. J.; Lowerthy, D. P.; Pryce, R. J.; Arnold, E.; Clardy, J. *Phytochemistry* **1981**, *20*, 811–814.
70. (a) Molyneux, R. J.; Roitman, J. N.; Dunnheim, G.; Szumilo, T.; Elbein, A. D. *Arch. Biochem. Biophys.* **1986**, *257*, 450–457. (b) Aoyagi, T.; Suda, H.; Uotani, K.; Kojima, F.; Aoyama, T.; Horiguchi, K.; Hamada, M.; Takeuchi, T. *J. Antibiot.* **1992**, *45*, 1404–1408. (c) Molyneux, R. J.; Pan, Y. T.; Tropea, J. E.; Benson, M.; Kaushal, G. P.; Elbein, A. D. *Biochemistry* **1991**, *30*, 9981-9987.
71. Pastuszak, I.; Molyneux, R. J.; James, L. F.; Elbein, A. D. *Biochemistry* **1990**, *29*, 1886–1891.
72. Michalik, A.; Hollinshead, J.; Jones, L.; Fleet, G. W. J.; Yu, C.-Y.; Hu, X. G.; van Well, R.; Horne, G.; Wilson, F. X.; Kato, A.; Jenkinson, S. F.; Nash, R. J. *Phytochem. Lett.* **2010**, *3*, 136–138.
73. Nash, R. J.; Thomas, P. I.; Waigh, R. D.; Fleet, G. W. J.; Wormald, M. R.; Lilley, P. M.; de, Q.; Watkin, D. J. *Tetrahedron Lett.* **1994**, *35*, 7849–7852.
74. Nash, R. J.; Fellows, L. E.; Bring, J.V.; Fleet, G. W. J.; Girdhar, A.; Ramsden, N. G.; Peach, J. M.; Hegarty, M. P.; Scofield, A. M. *Phytochemistry* **1990**, *29*, 111–114.
75. (a) Molyneux, R. J.; Benson, M.; Wong, R. Y.; Tropea, J. E.; Elbein, A. D. *J. Nat. Prod.*, **1988**, *57*, 1198–1206. (b) Tropea, J. E.; Molyneux, R. J.; Kaushal, G. P.; Pan, Y. T.; Mitchell, M.; Elbein, A. D. *Biochemistry* **1989**, *28*, 2027–2034. (c). Molyneux, R. J.; Pan, Y. T.; Tropea, J. E.; Elbein, A. D.; Lawyer, C. H.; Hughes, D. J.; Fleet, G. W. *J. J. Nat. Prod.* **1993**, *56*, 1356-1364.
76. Asano, N.; Kuroi, H.; Ikeda, K.; Kizu, H.; Kameda, Y.; Kato, A.; Adachi, I.; Watson, A. A.; Nash, R. J.; Fleet, G. W. J. *Tetrahedron: Asymmetry* **2000**, *11*, 1–8.
77. Kato, A.; Adachi, I.; Miyauchi, M.; Ikeda, K.; Komae, T.; Kizu, H.; Kameda, Y.; Watson, A. A.; Nash, R. J.; Wormald, M. R.; Fleet, G. W. J.; Asano, N. *Carbohydr. Res.* **1999**, *316*, 95–103.

78. Usuki, H.; Toyooka, M.; Kanzaki, H.; Okuda, T.; Nitoda, T. *Bioorg. Med. Chem.* **2009**, *17*, 7248–7253.
79. Shibano, M.; Tsukamoto, D.; Kusano, G. *Chem. Pharm. Bull.* **1999**, *47*, 907–908.
80. Dräger, B. *Nat. Prod. Rep.* **2004**, *21*, 211–223.
81. (a) Asano, N.; Kato, A.; Oseki, K.; Kizu, H.; Matsui, K. *Eur. J. Biochem.*, **1995**, *229*, 369–376. (b) Molyneux, R. J.; Pan, Y. T.; Goldmann, A.; Tepfer, D. A.; Elbein, A. D. *Arch. Biochem. Biophys.* **1993**, *304*, 81-88.
82. (a) Asano, N.; Kato, A.; Miyauchi, M.; Kizu, H.; Tomimori, T.; Matsui, K.; Nash, R. J.; Molyneux, R. J. *Eur. J. Biochem.*, **1997**, *248*, 296–303. (b) Asano, N.; Kato, A.; Kizu, H.; Matsui, K.; Watson, A. A.; Nash, R. J. *Carbohydr. Res.*, **1996**, *293*, 195-204.
83. (a) Kato, A.; Asano, N.; Kizu, H.; Matsui, K.; Suzuki, S.; Arisawa, M. *Phytochemistry* **1997**, *45*, 425–429. (b) Asano, N.; Oseki, K.; Tomioka, E.; Kizu, H.; Matsui, K. Carbohydr. Res., 1994, 259, 243-255.
84. (a) Asano, N.; Kato, A.; Kizu, H.; Matsui, K.; Griffiths, R. C.; Jones, M. G.; Watson, A. A.; Nash, R. J. *Carbohydr. Res.* **1997**, *304*, 173–178.(b) Griffiths, R. C.; Watson, A. A.; Kizu, H.; Asano, N.; Sharp, H. J.; Jones, M. G.; Wormald, M. R.; Fleet, G. W. J.; Nash, R. J. *Tetrahedron Lett.* **1996**, *37*, 3207-3208.
85 Asano, N.; Kato, A.; Yokoyama, Y.; Miyauchi, M.; Yamamoto, M.; Kizu, H.; Matsui, K. *Carbohydr. Res.* **1996**, *284*, 169–178.
86. (a) Compain, P.; Martin, O. R.; Boucheron, C.; Godin, G.; Yu, L.; Ikeda, K.; Asano, N. *ChemBioChem* **2006**, 7, 1356–1359. (b) Metha, A.; Ouzounov, S.; Jordan, R.; Simsek, E.; Lu, X.; Moriarty, R. M.; Jacob, G.; Dwek, R. A.; Block, T. M. *Antimicrob. Agents Chemother.* **2002**, *13*, 315–320. (c) Asano, N.; Kizu, H.; Oseki, K.; Tomioka, E.; Matsui, K.; Okamoto, M.; Baba, M. *J. Med. Chem.* **1995**, *38*, 2349–2356. (d) Kyosseva, S.V.; Kyossev, Z. N.; Elbein, A. D. *Arch. Biochem. Biophys.*, **1995**, *316*, 821–826. (e) Asano, N.; Oseki, K.; Kizu, H.; Matsui, K. J. *Med. Chem.*, **1994**, *37*, 3701–3706. (f) Kajimoto, T.; Liu, K.; Pederson, R. L.; Zhong, Z.; Ichikawa, Y.; Porco, J. A.; Wong, C.-H. *J. Am. Chem. Soc.* **1991**, *113*, 6187–6196. (g) Evans, S.V.; Fellows, L. E.; Shing, T. K. M.; Fleet, G.W. J. *Phytochemistry*, **1985**, *24*, 1953–1955. (*h*) Murao, S.; Miyata, S. *Agric. Biol. Chem.*, **1980**, *44*, 219-221.
87. (a) Lesur, B.; Ducep, J.-B.; Lalloz, M.-N.; Ehrhard, A.; Danzin, C. *Bioorg. Med. Chem. Lett.*, **1997**, 7, 355–360. (b) Yoshikuni, Y. *Agric. Biol. Chem.*, **1988**, *52*, 121–128. (c) Hettkamp, H.; Legler, G.; Bause, E. *Eur. J. Biochem.*, **1984**, *142*, 85-90.
88. (a) Zeng, Y.; Pan, Y. T.; Asano, N.; Nash, R. J.; Elbein, A. D. *Glycobiology* **1997**, 7, 297–304. (b) Scofield, A. M.; Witham, R.; Nash, R. J.; Kite, G. C.; Fellows, L. E. *Comp. Biochem. Physiol.*, **1995**, 112A, 187–196. (c) Scofield, A. M.; Witham, P.; Nash, R. J.; Kite, G. C.; Fellows, L. E. *Comp. Biochem. Physiol.*, **1995**, 112A, 197–205. (d) Kite, G. C.; Horn, J. M.; Romeo, J. T.; Fellows, L. E.; Lees, D. C.; Scofield, A. M.; Smith, N. G. *Phytochemistry* **1990**, *29*, 103–105. (e) Scofield, A. M.; Fellows, L. E.; Nash, R. J.; Fleet, G. W. J. *Life Sci.*, **1986**, *39*, 645-650.
89. Best, D.; Wang, C.; Weymouth-Wilson, A. C.; Clarkson, R. A.; Wilson, F. X.; Nash, R. J.; Miyauchi, S.; Kato, A.; Fleet, G. W. J. *Tetrahedron Asymmetry* **2010**, *21*, 311–319.
90. (a) Asano, N.; Ikeda, K.; Yu, L.; Kato, A. Takebayashi, K.; Adachi, I.; Kato, I.; Ouchi, H.; Takahata, H.; Fleet, G. W. J. *Tetrahedron: Asymmetry* **2005**, *16*, 223–229. (b) Asano, N.; Kato, A.; Miyauchi, M.; Kizu, H.; Kameda, Y.; Watson, A. A.; Nash, R. J.; Fleet, G. W. J. *J. Nat. Prod.*, **1998**, *61*, 625–628. (c) Watanabe, S.; Kato, H.; Nagayama, K.; Abe, H. *Biosci. Biotech. Biochem.*, **1995**, *59*, 936–937. (d) Winchester, B. *Biochem. Soc. Trans.*, **1992**, *142*, 699-705.
91. Kayakiri, H.; Nakamura, K.; Takase, S.; Setoi, H.; Uchida, I.; Terano, H.; Hashimoto, M.; Tada, T.; Koda, S. *Chem. Pharm. Bull.* **1991**, *39*, 2807–2812.
92. (a) Takahata, H.; Banba, Y.; Sasatani, M.; Nemoto, H.; Kato, A.; Adachi, I. *Tetrahedron*, **2004**, *60*, 8199–8205. (b) Takahata, H.; Banba, Y.; Ouchi, H.; Nemoto, H. *Org. Lett.* **2003**, *5*, 2527-2529.

93. Natori, Y.; Sakuma, T.; Watanabe, H.; Wakamatsu, H.; Kato, A.; Adachi, I.; Takhata, H.; Yoshimura, Y. *Tetrahedron* **2019**, *75*, 2866–2876 and references cited therein.
94. Trapero, A.; Liebaria, A. *Future Med. Chem.* **2013**, *5*, 573–590 and references cited therein.
95. Mena-Barragán, T.; García-Moreno, M. I.; Sevšek, A.; Okazaki, T.; Nanba, E.; Higaki, K.; Martin, N. I.; Pieters, R. J.; Fernández, J. M. G.; Mellet, C. O. *Molecules* **2018**, *23*, 927–944.
96. Legler, G.; Pohl, S. *Carbohydr. Res.* **1986**, *155*, 119–129.
97. (a) Asano, N.; Ishii, S.; Kizu, H.; Ikeda, K.; Yasuda, K.; Kato, A.; Martin, O. R.; Fan, J.-Q. *Eur. J. Biochem.* **2000**, *267*, 4179–4186. (b) Martion, O. R.; Xie, F.; Liu, L. Tetrahedron Lett. **1995**, *36*, 4027-4030.
98. Aguilar-Moncayo, M.; Takai, T.; Higaki, K.; Mena-Barragán, T.; Hirano, Y.; Yura, K.; Li, L.; Yu, Y.; Ninomiya, H.; Garcia-Moreno, M. I.; Ishi, S.; Sakakibara, Y.; Ohno, K.; Nanba, E.; Mellet, C. O.; Fernández, J. M. G.; Suzuki, Y. *Chem. Commun.* **2012**, *48*, 6514–6516.
99. (a) Fleet, G. W. J.; Nicholas, S. J.; Smith, P. W.; Evans, S. V.; Fellows, L. E.; Nash, R. J. *Tetrahedron Lett.* **1985**, *26*, 3127–3130. (b) Wang, Y.-F.; Takaoka, Y.; Wong, C.-H. *Angew. Chem. Int. Ed. Engl.* **1994**, *33*, 1242-1244.
100. Blanco, M. J.; Sardina, F. J. *J. Org. Chem.* **1998**, *63*, 3411–3416.
101. Gavale, K. S.; Chavan, S. R.; Kumbhar, N.; Kawade, S.; Doshi, P.; Khan, A.; Dhavale, D. D. *Bioorg. Med. Chem.* **2017**, *25*, 5148–5149.
102. Elías-Rodríguez, P.; Pingitore, V.; Carmona, A. T.; Moreno-Vargas, A. J.; Ide, D.; Miyawaki, S.; Kato, A.; Álvarez, E.; Robina, I. *J. Org. Chem.* **2018**, *83*, 8863–8873.
103. Biela-Banaś, A.; Oulaïdi, F.; Front, S.; Gallienne, E.; Ikeda-Obatake, K.; Asano, N.; Wenger, D. A.; Martin, O. R. *ChemMedChem* **2014**, *9*, 2647–2652.
104. (a) Miyake, Y.; Ebata, M. *Agric. Biol. Chem.*, **1988**, *52*, 153–158. (b) Miyake, Y.; Ebata, M. *Agric. Biol. Chem.*, **1988**, *52*, 1649-1654.
105. Ichikawa, Y.; Igarashi, Y. *Tetrahedron Lett.* **1995**, *36*, 4585–4586.
106. Front, S.; Biela-Banaś, A.; Burda, P.; Balhausen, D.; Higaki, K.; Caciotti, A.; Morrone, A.; Charollais-Thoenig, J.; Gallienne, E.; Demotz, S.; Martin, O. R. *Eur. J. Med. Chem.* **2017**, *126*, 160–170.
107. Ichikawa, M.; Ichikawa, Y. *Bioorg. Med. Chem.* **1995**, *3*, 161–165.
108. Kumar, K. S. A.; Rathee, J. S.; Subramaniam, M.; Chattopadhyay, S. *J. Org. Chem.* **2013**, *78*, 7406–7413.
109. (a) Herscovics, A. *Biochimie* **2001**, *83*, 757–762. (b) Shah, N.; Kuntz, D. A.; Rose, D. R. *PNAS*, **2008**, *105*, 9570-9575.
110. (a) Winchester, B. G.; Cenci di Bello, I.; Richardson, A. C.; Nash, R. J.; Fellows, L. E.; Ramsden, N. G. Fleet, G. *Biochem. J.* **1990**, *269*, 227–231. (b) Cenci di Bello, I.; Fleet, G.; Son, J. C.; Tadano, K.-I.; Winchester, B.; 1989. James, L. F.; Elbein, A. D.; Molyneux, R. J.; Warren, C. D. (Eds.), In Swainsonine and Related Glycosidase Inhibitors. Iowa State University Press, Ames, IA, pp.367-381. (c) Cenci di Bello, I.; Fleet, G.; Namgoong, K.; Tadano, K.-I.; Winchester, B. *Biochem. J.* **1989**, *259*, 855–861. (d) Trugnan, G.; Rousset, M.; Zweibaum, A. *FEBS Lett.*, **1986**, *195*, 28–32. (e) Dorling, P. R.; Huxtable, C. R.; Colegate, S. M. *Biochem. J.* **1980**, *797*, 649-651.
111. Blériot, Y.; Genre-Grandpierre, Tellier, C. *Tetrahedron Lett.* **1994**, *35*, 1867–1870.
112. (a) Pichon, M. M.; Stauffert, F.; Bodlenner, A.; Compain, P. *Org. Biomol. Chem.* **2019**, *17* 5801–5817. (b) Lepage, M. U.; Meli, A.; Bodlenner, A.; Tarnus, C.; Riccardis, F. D.; Izzo, I.; Compain, P. Beils. *J. Org. Chem.* **2014** *10*, 1406–1412. (c) Marra, A.; Zelli, R.; D'Orazio, G.; Ferla, B. L.; Dondoni, A. *Tetrahedron*, **2014**, *70*, 9387-9393.
113. Bonduelle, C.; Huang, J.; Mena-Barragán, T.; Mellet, C. O.; Decroocq, C.; Etame, E.; Heise, A.; Compain, P.; Lecommandoux, S. *ChemComm* **2014**, *50*, 3350–3352.
114. (a) Nieregarten, J.-F.; Schneider, J. P.; Trinh, T. M. N.; Joosten, A.; Holler, M.; Lepage, M. L.; Bodlenner, A.; García-Moreno, M. I.; Mellet, C. O.; Compain, P. *Chem. Eur.*

J. **2018**, *24*, 2483–2492. (b) Devroocq, C.; Joosten, A.; Sergent, R.; Barragán, T. M.; Mellet, C. O.; Compain, P. *ChemBioChem* **2013**, *14*, 2038-2049.
115. (a) Moreno-Clavijo, E.; Carmona, A. T.; Moreno-Vargas, A. J.; Molina, L.; Robina, I. *Curr. Org. Synth.* **2011**, *8*, 102–133. (b) Moreno-Clavijo, E.; Carmona, A. T.; Moreno-Vargas, A. J.; Molina, L.; Robina, I. *Chimia* **2011**, 40-44.
116. Chevrier, C.; Nouën, D. L.; Defoin, A.; Tarnus, C. *Carbohydr. Res.* **2011**, *346*, 1202–1211.
117. Moreno-Clavijo, E.; Carmona, A. T.; Moreno-Vargas, A. J.; Molina, L.; Wrught, D. W.; Davies, G. J.; Robina, I. *Eur. J. Org. Chem.* **2013**, 7328–7336.
118. Hottin, A.; Scandolera, A.; Duca, L.; Wright, D. W.; Divies, G. J.; Behr, J.-B. *Bioorg. Med. Chem. Lett.* **2016**, *26*, 1546–1549.
119. Yadav, V.; Yadav, P. K.; Yadav, S.; Yadav, K. D. S. *Process Biochemistry* **2010**, *45*, 1226–1235.
120. Kim, J. H.; Curtis-Long, M. J.; Seo, W. D.; Lee, J. H.; Lee, B. W.; Yoon, Y. J.; Kang, K. Y.; Park, K. H. *Bioorg. Med. Chem. Lett.* **2005**, *15*, 4282–4285.
121. Calveras, J.; Egido-Gabás, M.; Gómez, L.; Casas, J.; Parella, T.; Joglar, J.; Bujons, J.; Clapés, P. *Chem. Eur. J.* **2009**, *15*, 7310–7328.
122. Zhu, J. –S.; Nakagawa, S.; Chen, W.; Adachi, I.; Jia, Y. –M.; Hu, X. –G.; Fleet, G. W. J.; Wilson, F. X.; Nitoda, T.; Horne, G.; van Well, R.; Kato, A.; Yu, C. –Y. *J. Org. Chem.* **2013**, *78*, 10298–10309.
123. Désiré, J.; Mondon, M.; Fontelle, N.; Nakagawa, S.; Hirokami, Y.; Adachi, I.; Iwaki, R.; Fleet, G. W. J.; Alonzi, D. S.; Twigg, G.; Butters, T. D.; Bertrand, J.; Cendret, V.; Becq, F.; Norez, C.; Marrot, J.; Kato, A.; Blériot, Y. *Org. Biomol. Chem.* **2014**, *12*, 8977–8996.
124. Miyawaki, S.; Hirokami, Y.; Kinami, K.; Hoshino, M.; Minehira, D.; Miyamoto, D.; Nash, R. J.; Fleet, G. W. J.; Adachi, I.; Toyooka, N.; Kato, A. *Bioorg. Med. Chem.* **2017**, *25*, 107–115.
125. Takaoka, Y.; Kajimoto, T.; Wong, C. –H. *J. Org. Chem.* **1993**, *58*, 4809–4812.
126. Liang, P.-H.; Cheng, W.-C.; Lee, Y.-L.; Yu, H.-P.; Wu, Y.-T.; Linand, Y.-L.; Wong, C.-H. *ChemBioChem* **2006**, *7*, 165–173.
127. Liu, Z.; Jenkinson, S. F.; Kato, A.; Nakagawa, S.; Wormald, M. R.; Yu, C. –Y.; Fleet, G. W. J. *Tetrahedron Asymmetry* **2016**, *27*, 872–881.
128. Fosgerau, K.; Westergaard, N.; Quistorff, B.; Grunnet, N.; Kristiansen, M.; Lundgren, K. *Arch. Biochem. Biophys.* **2000**, *380*, 274–284.
129. Oikonomakos, N. G.; Tiraidis, C.; Leonidas, D. D.; Zographos, S. E.; Kristiansen, M.; Jessen, C. U.; Nørskov-Lauritsen, L.; Agius, L. *J. Med. Chem.* **2006**, *49*, 5687–5701.
130. Praly, J.-P.; Vidal, S. *Mini-Reviews in Medicinal Chemistry* **2010**, *10*, 1102–1126.
131. Alonzi, D. S.; Scott, K. A.; Dwek, R. A.; Zitzmann, N. *Biochem. Soc. Trans.* **2017**, *45*, 571–582 and references cited therein.
132. Caputo, A. T.; Alonzi, D. S.; Kiappes, J. L.; Struwe, W. B.; Cross, A.; Basu, S.; Darlot, B.; Roversi, P.; Zitzmann, N.. *In Dengue and Zika: Control and Antiviral Treatment Strategies. Advances in Experimental Medicine and Biology*Springer: Singapore, 2018; pp. 1062.
133. Miller, J. L.; Lachica, R.; Sayce, A. C.; Williams, J. P.; Bapat, M.; Dwek, R.; Beatty, P. R.; Harris, E.; Zitzmann, N. *Antimicrobial Agents and Chemotherapy* **2012**, *56*, 6379–6386.
134. Sanchez-Fernández, E. M.; Fernández, J. M. G.; Mellet, C. O. *ChemComm* **2016**, *52*, 5497–5515.
135. Horne, G.; Wilson, F. X.; Tinsley, J.; Williams, D. H.; Storer, R. *Drug Discovery Today* **2011**, *16*, 107–118.

CHAPTER NINE

Carbohydrate-protein interactions: Enhancing multivalency effects through statistical rebinding

Jordi van Heteren, Roland J. Pieters
Department of Chemical Biology & Drug Discovery, Utrecht Institute for Pharmaceutical Sciences, Utrecht University, Utrecht, The Netherlands

1 Introduction

Inhibitors of protein-carbohydrate recognition are an emerging group of therapeutics applicable to a wide range of diseases.[1] However, these inhibitors often show undruggable affinities at a first glance. This well-known problem can be addressed with multivalency approaches in many cases. Multivalent structures can have enhanced affinities or inhibitory potencies of the presented ligands beyond the sum of that of the individual ligands, which is the multivalency effect.[2] Thus, the relative potency of the construct over its monovalent analogue per binding unit (rp/n) is higher than one.[3]

The currently known origins of multivalency effects in ligand binding are chelation, statistical rebinding, and also subsite binding, a special case of chelation (Fig. 9.1).[4] Chelation occurs when a multivalent structure binds two binding sites on one protein at the same time. This leads to a considerable multivalency effect because the translational and rotational entropic penalty of the second binding event has already been paid for at the first binding event.[5] Subsite binding is very similar with the difference being that the second site is not identical to the primary binding site. This can be recognised by a smaller enthalpic contribution of the second binding compared to the first binding and, just as is the case with chelation, leads to a smaller entropic penalty of the second binding event. Statistical rebinding is an increase in affinity due to an increase in local concentration generated from the proximity of other (sub)ligands during the binding event. Aggregation is often the consequence of multivalent binding and may contribute in cases to the potency. Of these modes of multivalency, that can also

Carbohydrates in Drug Discovery and Development. http://dx.doi.org/10.1016/B978-0-12-816675-8.00009-9

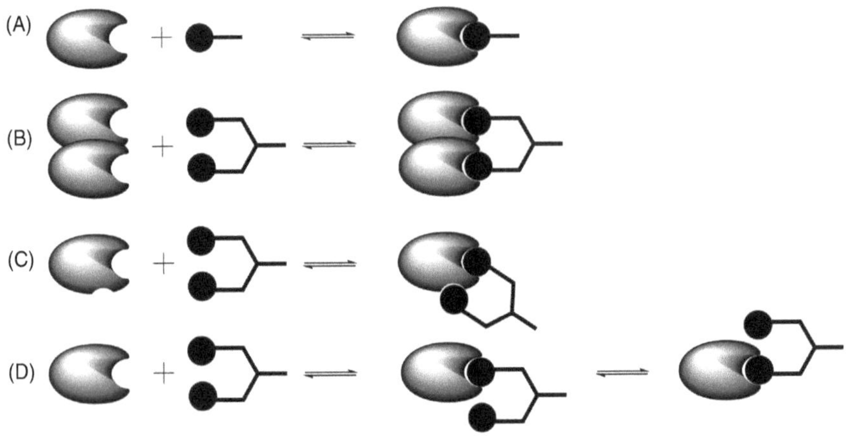

Figure 9.1 (A) Monovalent binding and multivalent effect mechanisms, (B) chelation, (C) subsite binding, and (D) statistical rebinding.

be operative at the same time, chelation has generally the largest effect on the affinity between ligand and protein.[2,6] Be that as it may, it is becoming increasingly clear that statistical rebinding contributes considerably to an increase in affinity in many cases.

Statistical rebinding is a process that arises from a high local concentration artificially generated by a multivalent structure. The close proximity of a second ligand often results in reassociation of the multivalent structure as a whole after the inevitable dissociation of a (sub) ligand. The result is a prolonged dissociation process. Statistical rebinding theoretically lowers the dissociation rate (k_d) and thus the dissociation constant (K_D). This should for instance be observable by surface plasmon resonance (SPR). One such binding event involves many associations of different epitopes of the multivalent structure. Therefore, the rotational and translational entropic penalties among multivalent structures are lower compared to their monovalent analogue. While this type of binding is characterised by a lower $T\Delta S$ penalty compared to its monovalent analogue, the same ΔH contribution might be gained.

To indicate the relevance of statistical rebinding an overview of relevant literature is presented here. Multivalency effects are compared on a per sugar basis as an *x*-fold increase, *i.e.* the relative potency per ligand (rp/n) in either K_D, K_i, or IC_{50}. Note that, the presence of one multivalency mechanism does not exclude the simultaneous occurrence of the others. Therefore, in order to isolate statistical rebinding effects, circumstances need to be found were the other mechanisms are ruled out. The occurrence of statistical rebinding will be discussed on a per protein basis. The first protein discussed in the present review is the Jack bean lectin ConA. ConA is well studied with regards to multivalency effects. Next, the occurrence of statistical rebinding will be brought to a wider perspective, which are the proteins: dendritic cell-specific intercellular adhesion molecule-grabbing nonintegrin (DC-SIGN), peanut agglutinin (PNA), and the enzyme Jack bean α-mannosidase (JbαMan).

1.1 ConA

ConA has affinity for both mannose and glucose and is a highly studied lectin with respect to multivalency effects.[2] It is a tetramer with binding sites *ca* 72Å apart. This large distance precludes chelation for most synthetic multivalent structures as they rarely reach that far. Note that, without any cluster depriving conditions, ConA will form clusters, which can also lead to additional beneficial binding effects.[7] Cluster formation which leads to aggregation in combination with statistical rebinding can be a very powerful affinity enhancing scenario. For instance, Sleiman et al. showed a very large enhancement of inhibitory potency of 3750rp/n between hexavalent α-mannoside**1** and ConA[8] in a hemagglutinin inhibition assay (HIA). The fact that aggregates were formed of *ca.* 850 nm between ConA and **1** (Fig. 9.2) was confirmed by dynamic light scattering (DLS).

It seems that statistical rebinding and aggregation are linked in a disjoint manner with each other, meaning that each mechanism partially takes away the contribution of the other. Dam et al. investigated the thermodynamics between α-mannose-conjugates **2,3** and **4** and ConA (Fig. 9.3).[9,10] The results of the study seem to indicate this non-cooperative relationship. With isothermal calorimetry (ITC) multivalency effects and the binding ratios of lectin *versus* α-mannose-conjugates were measured. As valence increased so did the multivalency effect and so did the ratio lectin: α-mannose-conjugates. Strikingly, the ΔH contribution was a multiplication of the ΔH contribution of the representative monomeric ligand and the valency of the multimeric presented ligands. This indicates that every epitope of each α-mannose-conjugate was bound to identical binding sites. Investigation of the binding between the divalent α-mannose-conjugate **2** and ConA with reverse-ITC showed that the observed macroscopic Gibbs free energy (ΔG) was the average of the two microscopic ΔGs of each epitope in a negative cooperative manner, meaning that the first binding event is more favourable than the second. Again the ΔHs of both microscopic binding events

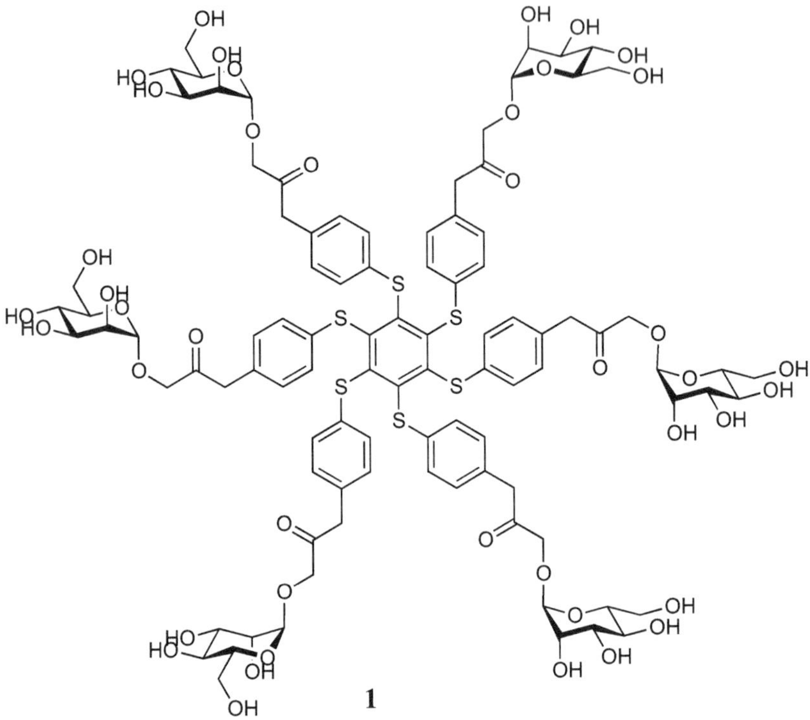

Figure 9.2 ***Hexavalent α-mannoside 1 showed amultivalency effect of 3750 rp/n with ConA based on a HIA assay.*** Aggregation was involved as detected with DLS by Sleiman et al.[8]

were very similar to that of the monovalent ligand. The profit in ΔG was from a decrease in the TΔS penalty from which binding of the first epitope benefited the most. Thus, strongly indicating the presence of statistical rebinding being the main driver of the multivalency effect of the first binding. The fact that the first binding event seems to benefit more than the second shows that the effect is due to proximity of the second ligand since the second binding can no longer benefit from the same proximity of another ligand.

If we can create aggregation-depriving conditions in ConA binding, then a multivalency effect can only be driven by either or both statistical rebinding and/or subsite binding, as long interbinding site distances of ConA rule out chelation. Considering the similar ΔH contributions of each epitope of **2**, **3**, and **4**,[11,9] subsite binding seems very unlikely for ConA. This makes statistical rebinding the main cause of multivalency under those conditions. Kinetic information of binding between **5**, **6**, and **7** (Fig. 9.4) and

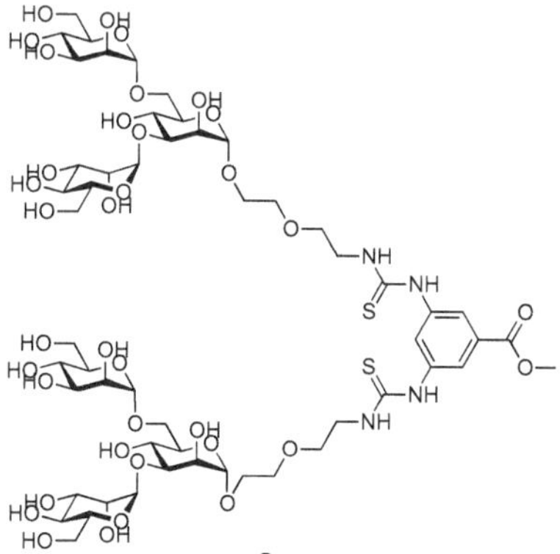

2

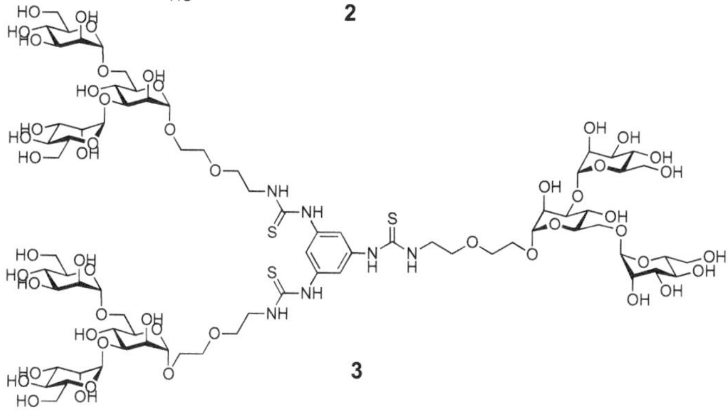

3

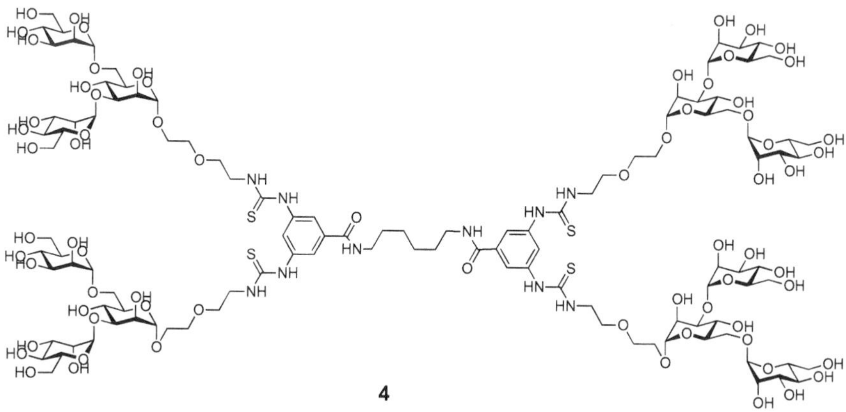

4

Figure 9.3 ***Dam et al. multivalency effect determined with ITC di-, tri-, and tetra-valent α-mannose-conjugates* 2*,* 3*, and* 4 *showed a multivalency effect of 3.2rp/n, 3.8rp/n, 8.7rp/n with Con Arespectively,*[9,10] *as a consequence of both statistical rebinding and aggregation.*** Binding of the first epitope was mainly driven by statistical rebinding.

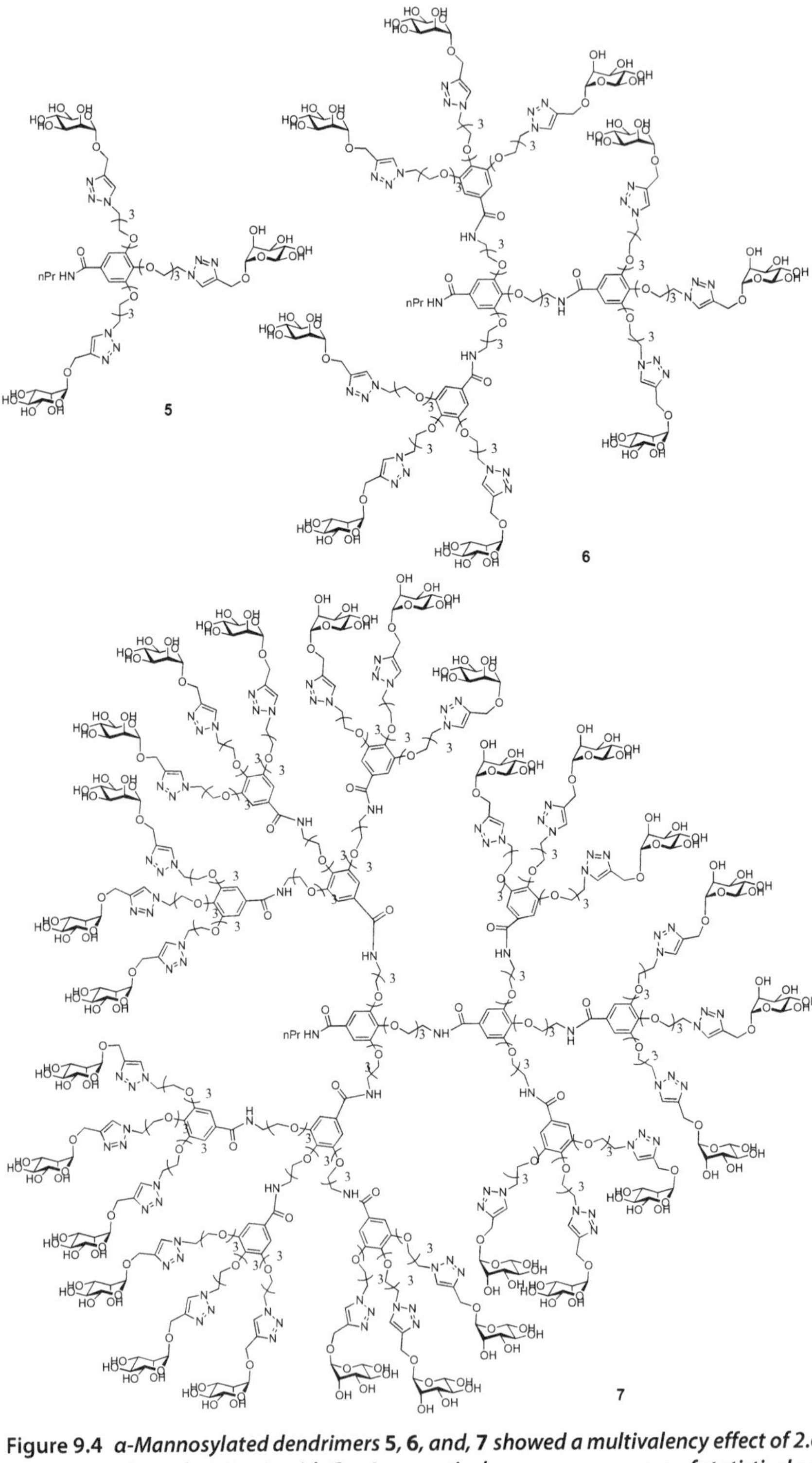

Figure 9.4 ***α-Mannosylated dendrimers 5, 6, and, 7 showed a multivalency effect of 2.6 rp/n, 12.5 rp/n, and 13.8 rp/n with ConA respectively, as a consequence of statistical rebinding.*** Affinities were measured by Munoz et al. with SPR.[12]

ConA supports this hypothesis. The three dendrimers of increasing size and valency were tested for their kinetics of ConA binding by surface plasmon resonance (SPR).[12] ConA was immobilized on a sensor chip at both a high and low density. When comparing the results of the dendrimers on the low and high density chips it became apparent that only the largest dendrimer was able to bind to two ConA tetramers simultaneously and only in the high ConA density setting. This ruled out any chelate effect in the other cases. The remaining dissociation constants (K_Ds) of the low density settings on a per sugar basis decreased 2.6 fold, 12.5 fold, and 13.8 fold respectively for **5**, **6**, and **7**. Strikingly, the decreases in K_D sresulted from decreases in dissociation rates (k_ds), which supports the idea of it being a characteristic of statistical rebinding. The highest multivalency effect under aggregation depriving conditions with ConA occurred with dodecavalent fullerene **8**, coated with mannose units, *i.e.* 42 rp/n (Fig. 9.5).[7] An enzyme-linked lectin assay (ELLA) was performed assessing the inhibitory activity of the fullerene-ball on horseradish peroxidase (HRP) labeled ConA against mannan coated on the plates. The HRP label was reported to prevent ConA from the option to form clusters.[13,14] However, this was not the case for larger multivalent ligands.[15] Fullerene-ball **8** is in this context relatively compact, and the authors argue against a role for aggregation as the cause of the multivalency effect. Note, the steric hindrance of the HRP label could possibly also hinder the statistical rebinding mechanism.

Similar to and often overlapping with the multivalency effect is the heterocluster effect. The heterocluster effect is a general term for an affinity increase due to glycoheterogeneity.[16] Gómez-García et al. revealed that a heterocluster effect at ConA is a process composed of a "high local concentration" effect anda "bind and slide" mechanism,[17] thus driven by statistical rebinding. An ELLA was performed testing multi-antennary homo- and heteroglycoclusters, based on a β-cyclodextrin (β-CD) core, on HRP labelled ConA against mannan coated on the wells. Every β-CD core had 7 arms attached which had three possible sites for a ligand. Thus, every glycocluster had a maximum capacity of 21 ligands. The best result was achieved with a glycocluster with 21 α-mannose ligands **14**, *i.e.* 7.5 rp/n, and the worst binding occurred with a glycocluster with 21 β-glucose ligands **11**, *i.e.* no binding detected (Fig. 9.6). Strikingly, when a glycocluster with 7 α-mannose ligands **12** was compared with a glycocluster with 7 α-mannose ligands and 14 β-glucose ligands **13** the multivalent effect increased, *i.e.* from 1.8 rp/n to 6.9 rp/n (n being the number of α-mannoses), but not beyond that of **14**. The number of β-glucose ligands was not taken into

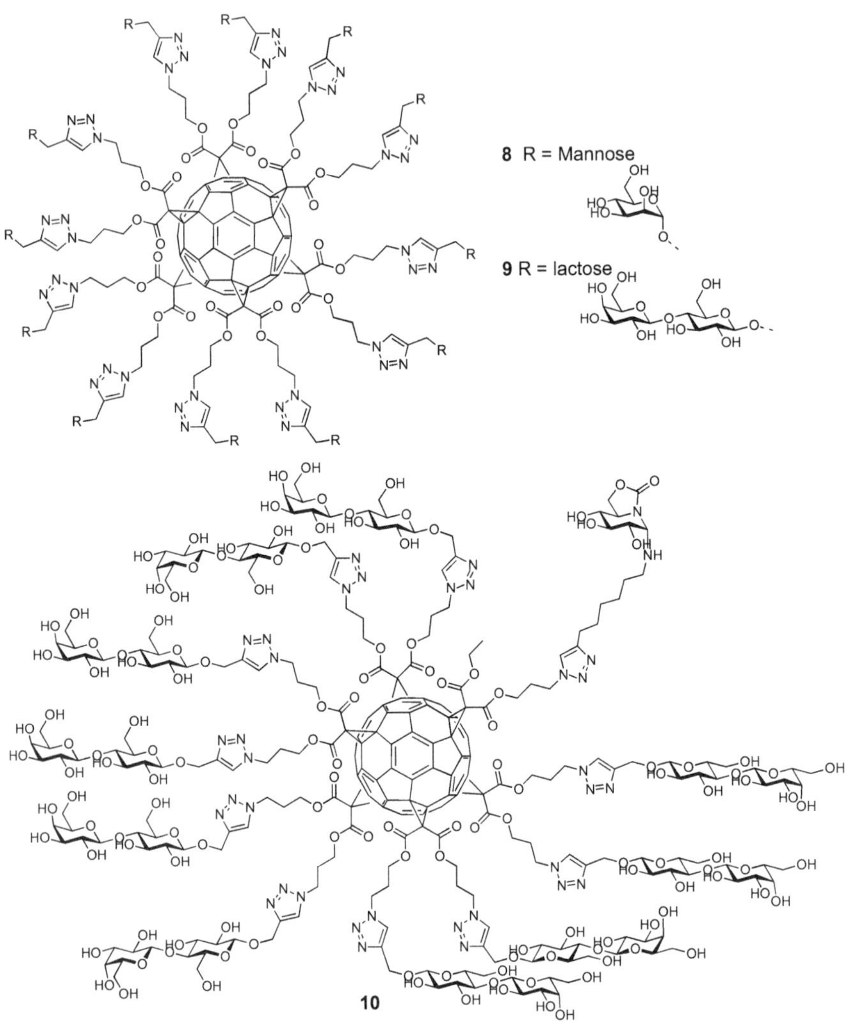

Figure 9.5 ***Homo-dodecane-valent α-mannosefullerene* 8 *showed amultivalency effect of 42 rp/n at ConA as a consequence of statistical rebinding.*** Homo-dodecane-valent α-lactosefullerene **9** showed amultivalency effect of 16.7 rp/n with PNA as a consequence of statistical rebinding. Heterovalent 10:1 lactose/1N-ONJ fullerene **10** showed a multivalency effect of 68.2 rp/n at PNA relative to lactose as a consequence of statistical rebinding. Affinity was measured with an ELLA and aggregation was excluded by the HRP-label on both ConA and PNA as described by Abellán Flos et al.[7]

consideration in the calculation of the multivalency effect since the glycocluster with only β-glucose ligands **11** showed no binding. At this point their contribution could be explained by either statistical rebinding or subsite binding. Energetic contributions from aggregation were excluded since the

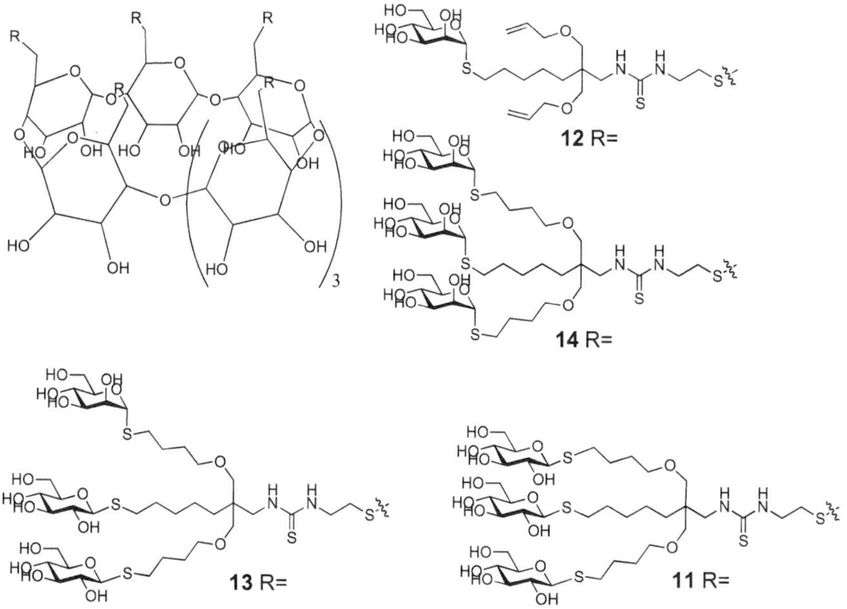

Figure 9.6 ***Homo- and hetro-glycoclusters* 11, 12, 13, *and* 14 *showed a multivalency effect of respectively 0 rp/n, 1.8 rp/n, 6.9 rp/n, and 7.5 rp/n at ConA as a consequence of statistical rebinding, which for heterovalent clusters is called the "heterocluster effect".*** Affinity was measured with ITC and aggregation was excluded by the presence of HRP on ConA, see Gómez-García et al.[17]

HRP label deprived ConA from the option to form clusters. Additionally, ITC experiments were conducted revealing the thermodynamic contributions to the binding events. Again HRP labelled ConA was used, thus, excluding any effects from aggregation. A trend was revealed. When combined with α-mannoses the β-glucoses had a negative TΔS contribution to the free binding energy, thus enhancing the binding. This indicates that β-glucose facilitated a bind and slide mechanism. **14** is hypothesized to have benefited more from statistical rebinding due to α-mannoses present in a high local concentration instead of β-glucoses, *i.e.* ΔG = -33.3 kJ/mol, ΔH = −152.8 kJ/mol, TΔS = −119.1 kJ/mol. β-glucose is hypothesised to temporarily take α-mannose's place providing more freedom of movement to the molecule and thus an entropic contribution to the binding event. For **13** this is even so extreme that its binding is mainly entropically driven, *i.e.* ΔG = −32.4 kJ/mol, ΔH = −8.3 kJ/mol, TΔS = 24.1 kJ/mol.

1.2 DC-SIGN

DC-SIGN is a pathogen recognition receptor expressed by dendritic cells, which might unfortunately also be involved in the dissemination and immune suppression of numerous pathogens.[18] DC-SIGN is a tetrameric C-type lectin whose binding sites are *ca.* 40 Å apart.[19] These inter-binding site distances do allow for chelation to take place at the protein. Be that as it may, when there are more epitopes than the number of binding sites that the multivalent structure can reach, then statistical rebinding and chelation could both simultaneously contribute to a multivalency effect. Ordanini et al. showed the relevance of exploiting statistical rebinding of multivalent structures binding to DC-SIGN even under conditions that allow for chelation.[19] Two identical monovalent sugar like ligands were linked through two rod like rigid spacers of different lengths, resulting in compounds **15** and **16** respectively (Fig. 9.7). Additionally, two trivalent ligands were linked through the longest spacer of the two, resulting in compound **17**. With the use of SPR it was shown that a perfect fit of the spacers length resulted in the best affinity, for **15** (IC_{50} 8 μM, rp/n = 17). It was hypothesized that the longer spacer resulted in a higher

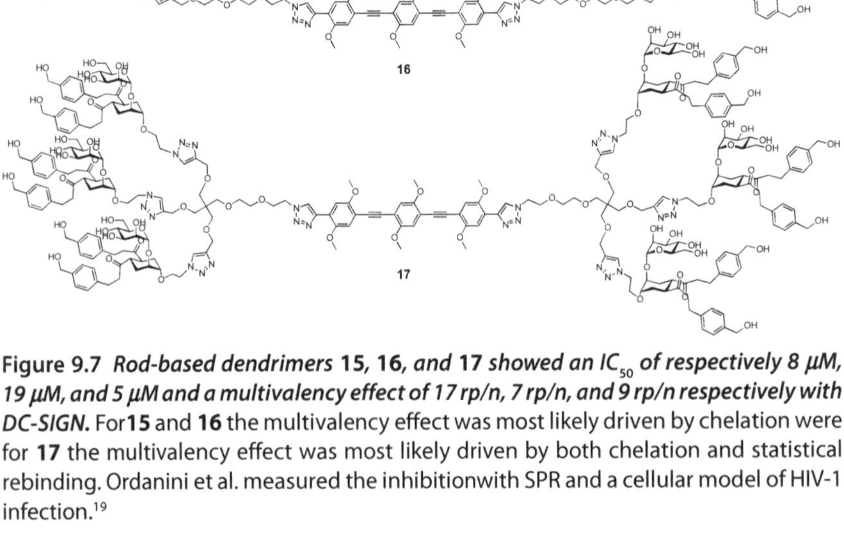

Figure 9.7 ***Rod-based dendrimers* 15, 16, *and* 17 *showed an*** ***IC_{50} of respectively 8 μM, 19 μM, and 5 μM and a multivalency effect of 17 rp/n, 7 rp/n, and 9 rp/n respectively with DC-SIGN.*** For **15** and **16** the multivalency effect was most likely driven by chelation were for **17** the multivalency effect was most likely driven by both chelation and statistical rebinding. Ordanini et al. measured the inhibitionwith SPR and a cellular model of HIV-1 infection.[19]

entropic loss upon binding and thus a lower affinity, see *e.g.* **16** (IC_{50} 19 μM, rp/n = 7). Strikingly, this loss in affinity could be compensated by the use of a trivalent ligand instead of a monovalent ligand at either end of the spacer, resulting in the binding of **17** (IC_{50} 5 μM, rp/n =9). Statistical rebinding is the likely driver behind the affinity compensation. In the SPR experiment, aggregation could not be excluded since mannoside ligands, were attached to the SPR chip instead of DC-SIGN. Additionally, a cellular model of HIV-1 infection was used to assess the efficacy of the compounds. In the cellular model of HIV-1 infection, aggregation is less likely since DC-SIGN is immobilised on the cellular membrane. The resulting inhibition concentration ranges from the cellular model of HIV-1 infection showed a similar trend as in the SPR experiments, *i.e.* 0.01–10 μM, 1–50 μM, and 0.01–10 μM respectively for **15**, **16**, and **17**. Thus, the most likely driver behind the multivalency effect increase of **17** over **16** is statistical rebinding as an addition to the already occurring chelation effect.

1.3 PNA

PNA is a homotetrameric galactose specific lectin with inter binding site distances between 54 Å and 74 Å (PDB ID: 2PEL).[6] These distances basically exclude chelation as a driver of multivalency effects for PNA when using small molecules as the sites would be hard to reach. It should be noted that, without aggregation depriving conditions, PNA will form clusters when presented with a multivalent inhibitor.[7] Two studies with interesting multivalency effects deprived PNA from the option to form clusters by labelling PNA with HRP in order to perform an ELLA to assess the inhibitory potential of fullerenes **18**, **19**, **9**,and **10** (Fig. 9.8) .[7,20] HRP deprives molecules of this size from forming aggregates with PNA,[11] which renders statistical rebinding as the only remaining multivalency mechanism. The 1N-ONJ coated fullerene balls with a short (**18**) and long (**19**) spacer had a multivalency effect of 13 rp/n and 31 rp/n for PNA, respectively.[16] Dodecavalent lactose coated fullerene **9** showed a multivalency effect of 16.7 rp/n.[7] Strikingly, when presented in a heterovalent manner, *i.e.* fullerene **10** coated with 10 lactoses and 1 1N-ONJ, the effect went up to a total of 68.2 rp/n relative to 10 lactoses. This is argued to be due to a synergistic enhancement of lactose and 1N-ONJ binding, *i.e.* a heterocluster effect. Note, this rp/n is an over-estimation since the affinity of 1N-ONJ for PNA is not taken into consideration. Be that as it may, the increase in the multivalency effect is still remarkable and a positive omen for heterocluster related research.

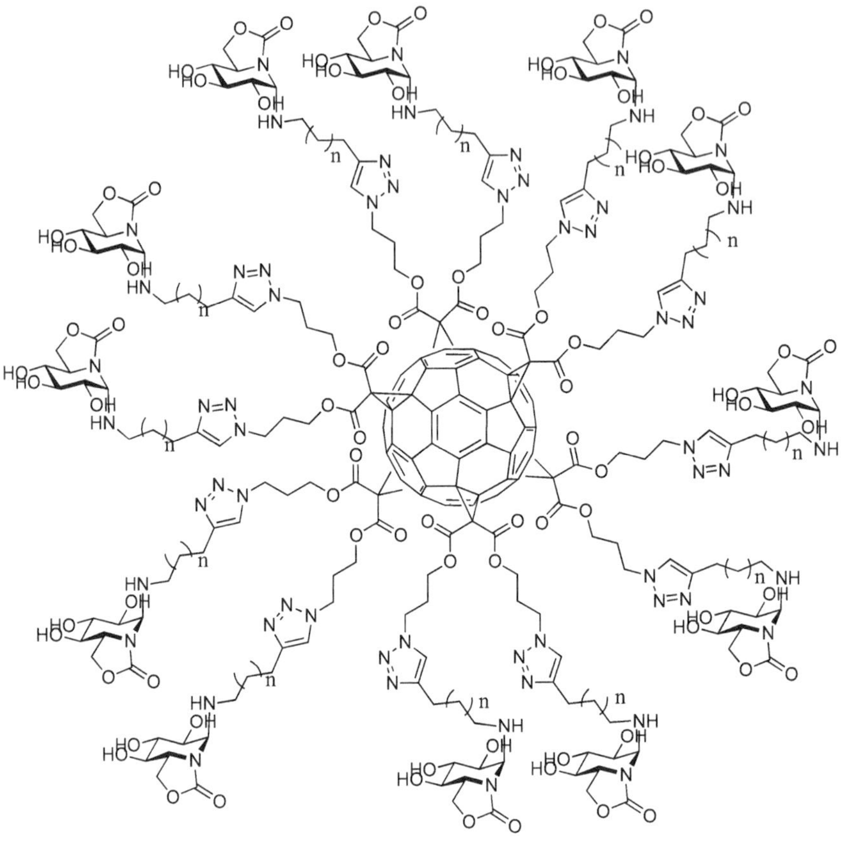

18: n=6
19: n=9

Figure 9.8 ***Fullerenes* 18 *and* 19 *showed a multivalency effect of 13rp/n and 31 rp/n respectively as a consequence of statisticalrebinding.*** Affinity was measured with an ELLA and aggregation was excluded by a HRP-label on PNA by Rísquez-CuadroS et al.[20]

1.4 Jack bean α-mannosidase

Jack bean α-mannosidase (JbαMan) is part of the glycoside hydrolase (GH) family. Its function is the hydrolysis of terminal non-reducing mannose residues during *N*-linked glycosylation. JbαMan is also considered of interest as inhibitors of its close relatives could serve as anti-cancer agents.[21] In general GHs have deeply buried binding sites which makes the occurrence of statistical rebinding unlikely.[7] However α-mannosidases seem to be an exception to this general phenomena since they have shallow and wide binding sites, which correlates with their susceptibility to multivalency effects.[20] Several studies support this theory in the study of several GHs by presenting ligands in a multivalent fashion but only in the case of JbαMan a multivalency effect is typically observed.[20,22] Furthermore, it should be noted that these proteins have sugar sub-binding sites near their catalytic site.[23] This makes subsite bindinga likely driver behind multivalency effects with these proteins.The most studied alpha-mannosidase is the one from the jack bean, *i.e.* JbαMan. JbαMan is a homodimer of heterodimers, *i.e.* a tetramer,[17] which means that JbαMan could theoretically form aggregates with multivalent compounds, and benefit from chelation effects if its binding sites are close enough to each other, which seems to be the case with respect to the size of several multivalent compounds described below.

JbαMan cluster formation with multivalent structures leading to aggregation has shown to be greatly dependent on the topology of the multimer.[24] Inhibition studies of multivalent glycoclusters **20**, **21**, **22**, **23**, **24**, **25**, and **26** showed a variable multivalency effect of 5 rp/n, 67 rp/n, 1.3 rp/n, 2.4 rp/n, 200 rp/n, 2.4 rp/n, and 2.2 rp/n respectively (Fig. 9.9). Atomic force microscopy (AFM) showed a correlation between the magnitude of the observed multivalency effects and the size and degree of aggregate formation. When the size of the multimer increased it was shown that both cluster formation and chelation can be present forJbαMan.[25] The interactions between the 36-valent inhibitor **27** and JbαMan were studied using an inhibition assay, EM, AUC-SV, and native ESI-MS (Fig. 9.10). The resulting multivalency effect of the inhibition study was a very large 4747 rp/n. Native ESI-MS, EM, and AUC-SV revealed a 2:1 ratio of the JbαMan tetramer in complex with inhibitor **27** as the major species. The binding event seems to be driven by a combination of chelation and cluster formation. Likely, the chelation mode geometrically restricts JbαMan to form larger clusters, as was seen for the compounds in the previous study, especially compound **24**, which were less likely to engage in chelation due to their smaller size, also resulting in much smaller enhancements.

Since the crystal structure is unknown it is hard to know what the interbinding site distances of JbαMan are. Nevertheless, it is certain that its smallest inter binding site distance is smaller than the largest interligand distances in **27**. It seems that this distance is somewhere between the interbinding site distances of 1-deoxynojirimycin based cyclodextrins (DNJ-βCD) **28** and **29** (Fig. 9.11). The difference between the two is a difference of three carbon atoms in the spacer. With these 14-valent dendrimers a combination of chelation, clustering, and statistical rebinding effects at JbαMan through ITC was shown.[26] The smaller DNJ-βCD **28**

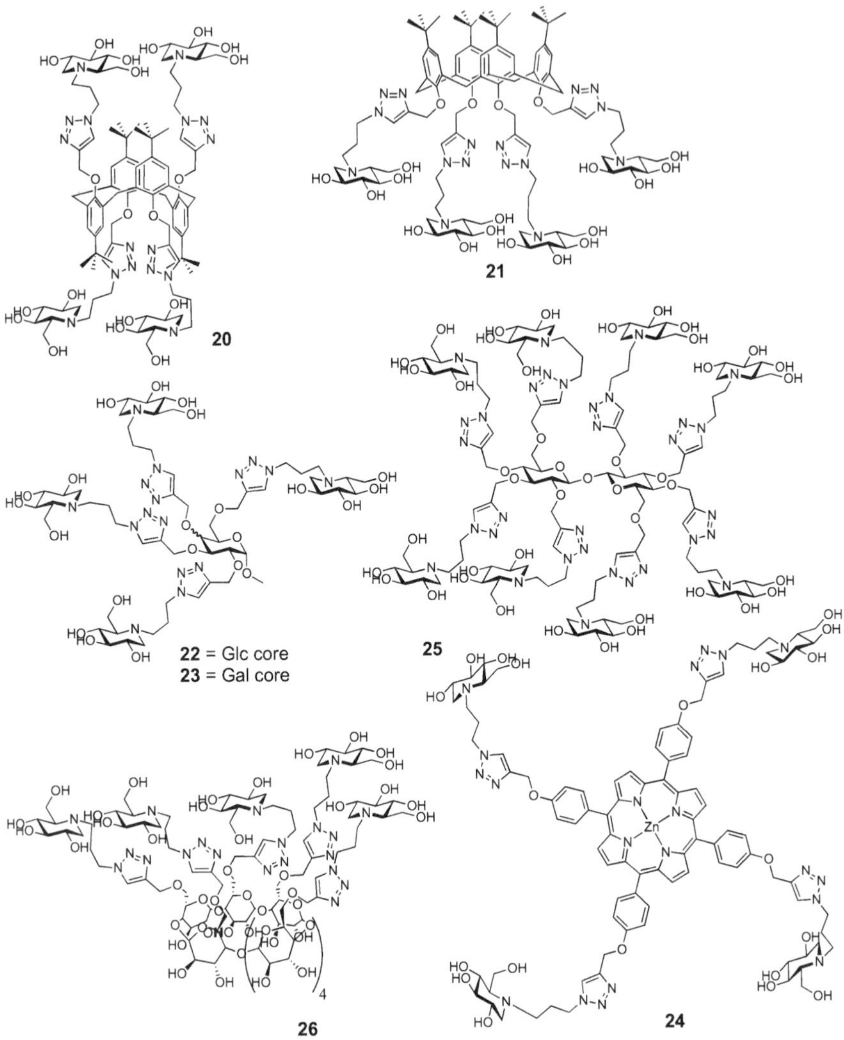

Figure 9.9 ***Brissonnet et al. multivalency effect assured inhibitory activity of tetravalent glycoclusters 20, 21, 22, 23, and 24 and octovalentglycoclusters 25, and 26 which showed a multivalency effect of 5 rp/n, 67 rp/n, 1. 2 rp/n, 2.5 rp/n, 200 rp/n, 2.4 rp/n, and 2.1 rp/n respectively.***[24] A correlation was shown between the multivalency effects and the size and degree of aggregate formation.

showed a ΔG of -31.2 kJ/Mol (34 rp/n) which mainly comprised of a ΔH contribution and a small beneficial TΔS contribution. Strikingly, the larger dendrimer had a larger ΔG of -39.9 kJ/Mol (197 rp/n). This number was comprised of more than four times the amount of ΔH contribution and a large TΔS subtraction. Noteworthy, the binding stoichiometries as deduced

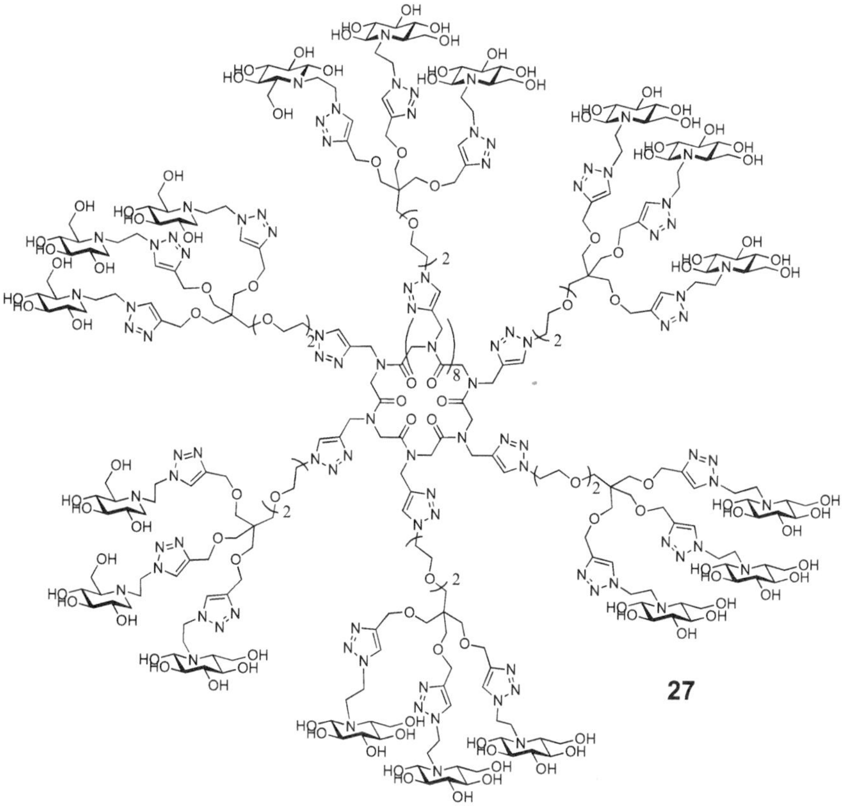

27

Figure 9.10 ***Multivalent inhibitor 27 showed a multivalency effect of 4747 rp/n forJbαMan,[25] as a consequence of aggregation.***

by ITC for the small and large dendrimers were 1.0 and 2.3, respectively. Considering these numbers, the binding of the larger dendrimer might be driven by different forces than the smaller dendrimer. The aggregation observed for the larger dendrimer is entropically disadvantageous and that is what is observed, with the large TΔS subtraction for the larger dendrimer. The binding of the smaller dendrimer is entropically favoured and thus likely driven by statistical rebinding where movement and dynamics are part of the mechanism. Additionally, since JbαMan has sugar sub-binding sites near their catalytic site,[20] subsite binding is also likely to contribute to the observed multivalency effect.

Several structures smaller than **28**, thus most likely incapable of inducing chelation with JbαMan, showed noteworthy multivalency effects with JbαMan but not nearly as high as that of **27**. That is, trivalent compound **30**(Fig. 9.12) had 2.1 rp/n,[22] fullerene **19** exhibited 46 rp/n,[20] for tetravalent

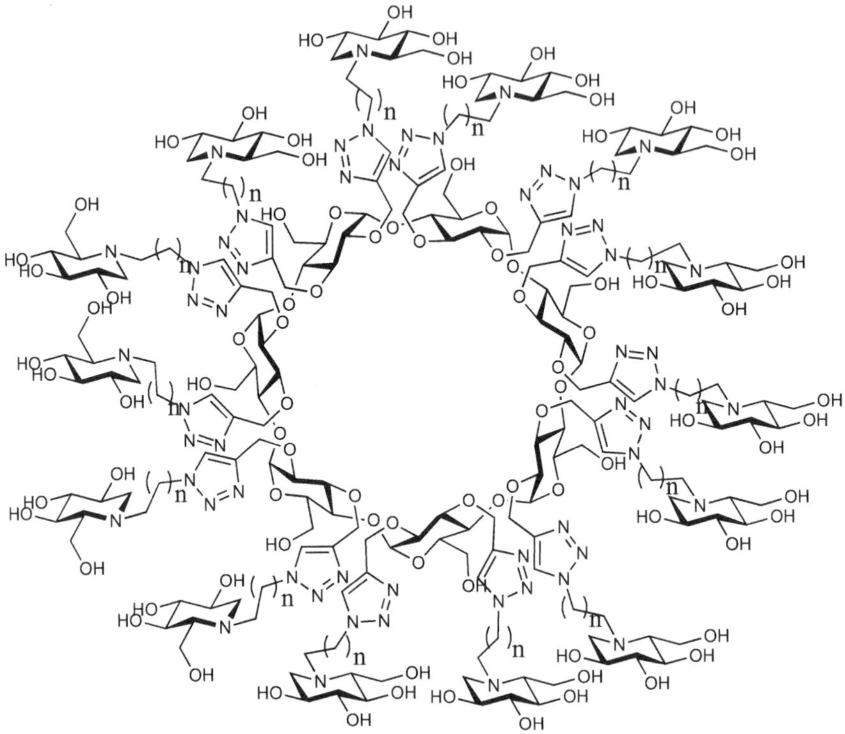

28: n=5
29: n=8

Figure 9.11 ***Tetradecavalent DNJ-βCD* 28 *and* 29 *showed multivalency effects of respectively 34 rp/n and 194 rp/n for JbαMan.*[26]** For **28** the main driver is likely statistical rebinding. For **29** the main driver is more likely chelation.

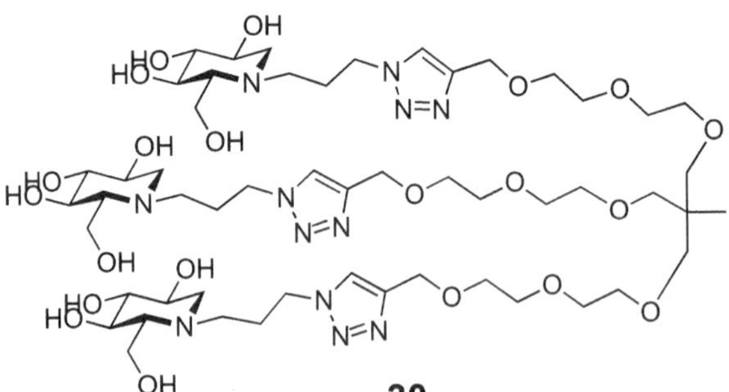

30

Figure 9.12 ***Multivalent iminosugar* 30 *showed a multivalency effect of 2.1 rp/n for JbαMan,*[22] *as the likely consequence of a combination of statistical rebinding, subsite binding and aggregation.***

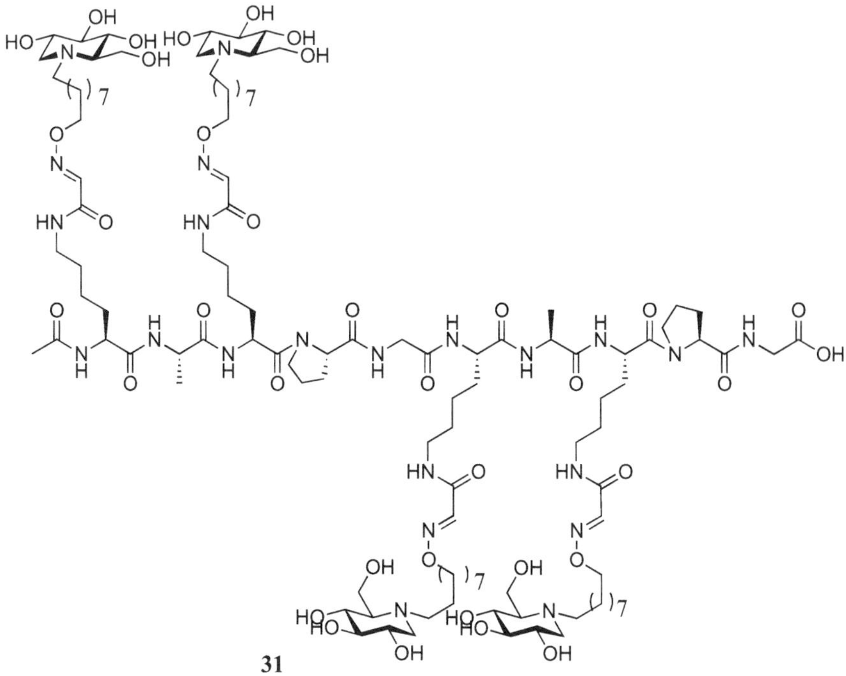

31

Figure 9.13 ***Iminosugar cluster 31 showed a multivalency effect of 77 rp/n for JbαMan,[27] as a consequence of a likely combination of statistical rebinding, subsite binding and aggregation.***

iminosugar cluster **31** (Fig. 9.13) the number was 77 rp/n,[27] and for fullerene ball **32** 179 rp/n (Fig. 9.14).[28] These observed effects are likely due to a combination of aggregation, statistical rebinding and subsite binding considering that no aggregation depriving conditions were present, the size of these molecules, and the fact that JbαMan has an open superficial catalytic site with sub-binding sites near it.

2 Conclusion and future perspectives

In conclusion, the current described literature provides a case for statistical rebinding. For ConA it seems to be the main driver when aggregation is prevented. Although, it does not provide as high of a multivalency effect compared to combinations of effects including aggregation and chelation, it can still provide a noteworthy enhancement. For PNA statistical rebinding was also shown to be present when aggregation is prevented. For DC-SIGN chelation and statistical rebinding were shown to contribute

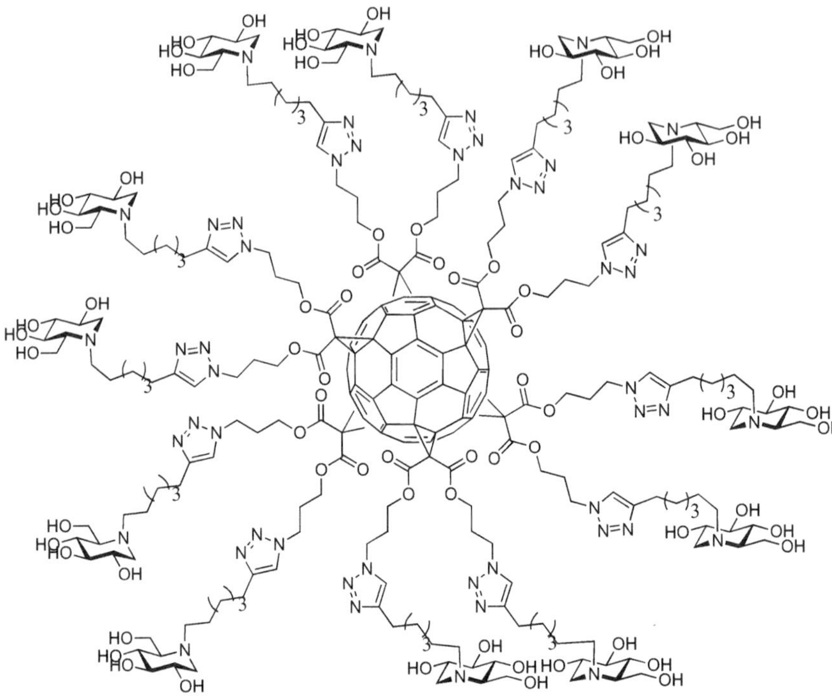

32

Figure 9.14 ***Fullerene* 32 *showed a multivalency effect of 179 rp/n for JbαMan,*[28] *as a consequence of probably a combination of statistical rebinding, subsite binding and aggregation.***

simultaneously to the multivalency effect. JbαMan seems to have inter binding site distances at the border of a possible chelation effect. When multivalent structures smaller than this distance are presented both statistical rebinding and subsite binding seem to be drivers of the observed multivalency effects. Additionally, aggregation can be and is often also a mechanism contributing to affinity enhancements. Literature of multivalency effects at GHs seem to indicate that JbαMan is the only GH that can benefit from statistical rebinding since, in contrast to the others, it has a wide, superficial, and open binding site where others have deeply buried binding sites.[20,22] This is a similarity with the other proteins discussed in the present report since they also have wide and open binding sites. The binding sites of ConA have a binding area bigger than just for one mannose ligand.[29] Possibly, this extra binding potential could facilitate the bind and slide mechanism of statistical rebinding. When considering the different types of multimeric structures described in the present report it seems that with respect to statistical rebinding functionalized fullerenes provide the highest multivalency effects. Likely, there spherical shape facilitates the statistical rebinding mechanism.

In rational drug design multivalency can be very helpful to bring certain drugs to the right inhibitory concentration. Likely, a possible requisite is an open, superficial binding site present on the target protein. Note that, the multivalency related research extends beyond just sugar (like) ligands.[30,31] Aggregation as an affinity enhancer in assays should be viewed with care since in physiological conditions it may or may not be relevant,[1] and it might even lead to undesired side effects. The current report shows that aggregate formation is not rare among such assays. In many ELLAs aggregation is reduced by the use of the HRP label. However, care should be taken since the label will not prevent aggregation in all cases.[15] Detection of aggregates is possible for instance with SPR, DLS, AFM, or analytical ultracentrifugation. Chelation is the strongest multivalency driver.[2,6] Therefore, when chelation can be exploited this should be the first mechanism to strive for when rationally designing drugs or biological tools. However, when circumstances do not allow for chelation, *e.g.* due to too long inter binding site distances or because the target protein only possesses one binding site, then statistical rebinding is a good mechanism to strive for. A good first choice to start with is to present the ligand of interest in a multimeric fashion on a fullerene ball or another round or spherical entities since these have shown to be the most efficient with respect to multivalency effects through statistical rebinding.

References

1. Dimick, S.M.; Powell, S.C.; McMahon, S.A.; Moothoo, D.N.; Naismith, J.H.; Toone, E.J. *J. Am. Chem. Soc.* **1999**, *121*, 10286–10296.
2. Pieters, R.J. *Org. Biomol. Chem.* **2009**, 7, 2013–2025.
3. Matassini, C.; Parmeggiani, C.; Cardona, F.; Goti, A. *Tetrahedron Lett.* **2016**, *57*, 5407–5415.
4. Kiessling, L.L.; Gestwicki, J.E.; Strong, L.E. *Angew. Chem. Int. Ed. Engl.* **2006**, *45*, 2348–2368.
5. Wittmann, V.; Pieters, R.J. *Chem. Soc. Rev.* **2013**, *42*, 4492–4503.
6. Pera, N.P.; Branderhorst, H.M.; Kooij, R.; Maierhofer, C.; Kaaden, M. van der; Liskamp, R.M. J.; Wittmann, V.; Ruijtenbeek, R.; Pieters, R.J. *ChemBioChem* **2010**, *11*, 1896–1904.
7. Abellán Flos, M.; García Moreno, M.I.; Ortiz Mellet, C.; García Fernández, J.M.; Nierengarten, J.F.; Vincent, S.P. *Chem. Eur. J.* **2016**, *22*, 11450–11460.
8. Sleiman, M.; Varrot, A.; Raimundo, J.-M.; Gingras, M.; Goekjian, P.G. *Chem. Commun.* **2008**, 6507–6509.
9. Dam, T.K.; Roy, R.; Das, S.K.; Oscarson, S.; Brewer, C.F. *J. Biol. Chem.* **2000**, *275*, 14223–14230.
10. Dam, T.K.; Roy, R.; Pagé, D.; Brewer, C.F. *Biochemistry* **2002**, *41*, 1351–1358.
11. Dam, T.K.; Talaga, M.L.; Fan, N.; Brewer, C.F. *Methods Enzymol.* **2016**, *567*, 71–95.

12. Munoz, E.M.; Correa, J.; Fernandez-Megia, E.; Riguera, R. *J. Am. Chem. Soc.* **2009**, *131*, 17765–17767.
13. Gómez-García, M.; Benito, J.M.; Rodríguez-Lucena, D.; Yu, J.X.; Chmurski, K.; Ortiz Mellet, C.; Gutiérrez Gallego, R.; Maestre, A.; Defaye, J.; García Fernández, J.M. *J. Am. Chem. Soc.* **2005**, *127*, 7970–7971.
14. Lundquist, J.J.; Toone, E.J. *Chem. Rev.* **2002**, *102*, 555–578.
15. Almant, M.; Mastouri, A.; Gallego-Yerga, L.; Fernandez, J.M. G.; Mellet, C. O.; Kovensky, J.; Morandat, S.; Kirat, K. El.; Gouin, S.G. *Chem. Eur. J.* **2013**, *19*, 728–737.
16. Jiménez Blanco, J.L.; Ortiz Mellet, C.; García Fernández, J.M. *Chem. Soc. Rev.* **2013**, *42*, 4518–4531.
17. Gómez-García, M.; Benito, J.M.; Gutiérrez-Gallego, R.; Maestre, A.; Mellet, C. O.; Fernández, J.M.G.; Blanco, J.L.J. *Org. Biomol. Chem.* **2010**, *8*, 1849–1860.
18. Van Kooyk, Y.; Geijtenbeek, T.B.H. *Nat. Rev. Immunol.* **2003**, *3*, 697–709.
19. Ordanini, S.; Varga, N.; Porkolab, V.; Thépaut, M.; Belvisi, L.; Bertaglia, A.; Palmioli, A.; Berzi, A.; Trabattoni, D.; Clerici, M., et al. *Chem. Commun.* **2015**, *51*, 3816–3819.
20. Rísquez-Cuadro, R.; García Fernández, J.M.; Nierengarten, J.F.; Ortiz Mellet, C. *Chem. Eur. J.* **2013**, *19*, 16791–16803.
21. Kumar, A.; Gaikwad, S.M. *Int. J. Biol. Macromol.* **2011**, *49*, 1066–1071.
22. Diot, J.; García-Moreno, M.I.; Gouin, S.G.; Mellet, C.O.; Haupt, K.; Kovensky, J. *Org. Biomol. Chem.* **2009**, *7*, 357–363.
23. Shah, N.; Kuntz, D.A.; Rose, D.R. *Proc. Natl. Acad. Sci.* **2008**, *105*, 9570–9575.
24. Brissonnet, Y.; Ortiz Mellet, C.; Morandat, S.; Garcia Moreno, M.I.; Deniaud, D.; Matthews, S.E.; Vidal, S.; Šesták, S.; El Kirat, K.; Gouin, S.G. *J. Am. Chem. Soc.* **2013**, *135*, 18427–18435.
25. Lepage, M.L.; Schneider, J.P.; Bodlenner, A.; Meli, A.; De Riccardis, F.; Schmitt, M.; Tarnus, C.; Nguyen-Huynh, N.T.; Francois, Y.N.; Leize-Wagner, E., et al. *Chem. Eur. J.* **2016**, *22*, 5151–5155.
26. Decroocq, C.; Joosten, A.; Sergent, R.; Mena Barragán, T.; Ortiz Mellet, C.; Compain, P. *ChemBioChem* **2013**, *14*, 2038–2049.
27. Zelli, R.; Bartolami, E.; Longevial, J.F.; Bessin, Y.; Dumy, P.; Marra, A.; Ulrich, S. *RSC Adv.* **2016**, *6*, 2210–2216.
28. Compain, P.; Decroocq, C.; Iehl, J.; Holler, M.; Hazelard, D.; Barragan, T.M.; Mellet, C. O.; Nierengarten, J.-F. *Angew. Chem. Int. Ed.* **2010**, *49*, 5753–5756.
29. Ramström, O.; Lehn, J.-M. *ChemBioChem* **2002**, *1*, 41–48.
30. Kanfar, N.; Tanc, M.; Dumy, P.; Supuran, C.T.; Ulrich, S.; Winum, J.Y. *Chem. Eur. J.* **2017**, *23*, 6788–6794.
31. Carta, F.; Osman, S.M.; Vullo, D.; AlOthman, Z.; Supuran, C.T. *Org. Biomol. Chem.* **2015**, *13*, 6453–6457.

CHAPTER TEN

Carbo-click in drug discovery and development: Opportunities and challenges*

Divya Kushwaha[a,b], **Sumit K. Singh**[b], **Vinod K. Tiwari**[b]
[a]Department of Chemistry (MMV), Banaras Hindu University, Varanasi, India
[b]Department of Chemistry, Institute of Science, Banaras Hindu University, Varanasi, India

1 Introduction

Cu(I)-catalyzed azide-alkyne cycloaddition (CuAAC) is an extremely efficient chemical reaction that conveniently establishes a CN covalent connection between two diversely functionalized building blocks (Scheme 10.1).[1,2] This simple and high yielding reaction is also referred as "Click Reaction", because it requires very mild reaction conditions and tolerates a wide range of functional groups, pH and solvents including water. Furthermore, the starting materials of reaction azide and an alkyne can be easily inserted into organic molecules and remain considerably inactive to general reagents and functional groups, therefore, this attributes more efficiency to the approach.[3] Click reaction has gained immense popularity since its discovery and has been extensively used to construct a large number of organic molecules that are applicable in various areas of science; for instance, material sciences,[4] supramolecular chemistry,[5] polymer chemistry,[6] medicinal chemistry,[7] and bioconjugation studies,[8] etc.

Carbohydrate constitutes a major class of biomolecules lying inside or on the surface of cells in the form of glycoconjugates, and have been identified to play critical roles in various pathologically and physiologically vital biological processes such as cellular recognition, adhesion, migration, invasion, communication, bacterial/viral infection, tumor metastasis, post-translational modifications of protein etc.[9,10] Therefore, in order to explore inhibitors or promoters for these biological events, there is an increased demand to design library of glycosyl mimics and glycoconjugates.[11,12]

*This chapter is dedicated to Prof. (Dr) R. P. Tripathi, Former Senior Scientist at CSIR-Central Drug Research Institute, Lucknow for his excellent contribution on impact of glycohybrid molecules on drug development against tuberculosis.

Carbohydrates in Drug Discovery and Development. http://dx.doi.org/10.1016/B978-0-12-816675-8.00010-5

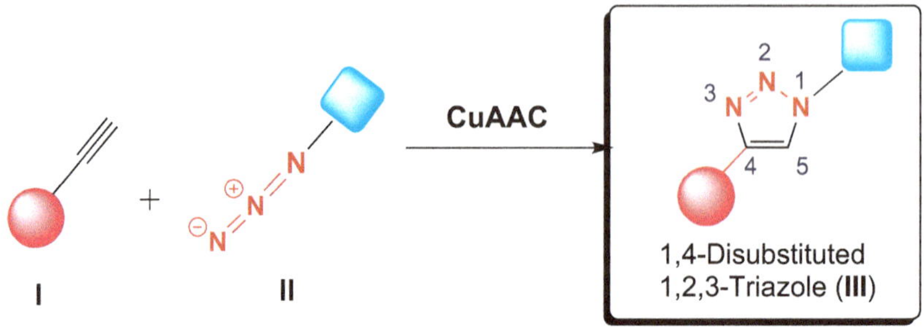

Scheme 10.1 ***Cu(I)-catalyzed azide-alkyne cycloaddition (CuAAC) mediated regioselective synthesis of 1,4-disubstituted 1,2,3-triazoles.***

Cu(I)-catalyzed click reaction is such a powerful conjugation strategy that rapidly generates a plethora of new compounds with apparent ease, thereby facilitates the process of lead discovery and optimization.[13]

The significance of click strategy also lies in product of the reaction 1,4-disubstituted 1,2,3-triazole, which itself is a potent pharmacophore and shows chemical and electronic resemblance with amide bond.[14] The therapeutic potential of the triazoles is attributed to their high dipole moment that enables strong hydrogen bonding interaction with various receptors. Additionally, triazole moiety is non-toxic and extremely stable towards hydrolysis, oxidation and reduction conditions, which makes the moiety biologically more relevant and eventually registers CuAAC a more valuable approach. A large number of biologically active glycosylated 1,2,3-triazole moieties with good pharmacological properties have been described in literature.[13–16]

2 Carbo-click in drug discovery and development

As carbohydrate based interactions govern many essential metabolic processes such as cell-cell interaction, cell migration, pathogen defence etc., this makes sugars a potential candidate for the drug development.[17,18] However, certain factors regarding carbohydrates such as their complex synthetic modification particularly in relation to anomeric stereochemistry, limited pharmacological properties, and moderate affinity for various receptors confines their utilization as drug leads.[19,20] In this respect, the compatibility of Cu(I) catalyzed click coupling of carbohydrate based azides and alkynes to pre-appended protecting groups, as well as relative inertness of triazole ring towards string of protection/deprotection and glycosylation steps establish this approach as a potent tool in accessing diverse carbohydrate mimetics with improved affinity.[13–16]

Identification of new and more effective lead compounds by the structure modification of already established drugs is a fast growing drug discovery process in recent years. Covalent linking of sugars with biologically privileged heterocycles and natural products to create bioconjugates with blended features of both skeletons is one approach for the creation of carbohydrate-based therapeutics. Presence of sugars in conjugates increases hydrophilicity and bioavailability of the functional molecule, while reduces the toxicity and side effects.[21] Additionally, since carbohydrates are involved in various interactions, their incorporation to various simpler molecules enables a fine tuning with targeting mechanism of action and/or pharmacology.[22]

In quest to develop potential therapeutics, click reaction has been immensely employed to construct a large number of glycohybrids that finds their applications in various aspects of drug discovery ranging from lead discovery to target-templated *in situ* chemistry,[23] to tagging of biological systems, such as proteins,[24] nucleotides,[25] and whole organisms such as bacteria and viruses using bioconjugation reactions.[8,26] This chapter highlights the applications of carbo-click derived triazolyl glycoconjugates in drug discovery and development.

2.1 Traizolyl glycoconjugates as enzyme inhibitors

Enzymes are the proteins that speed up specific chemical reactions taking place in the biological systems. As they regulate multitudes of crucial physiological processes, consequently their inhibitors constitute an excellent target for the pharmacological intervention. An enzyme inhibitor selectively disrupts the metabolic pathway by blocking the function of that particular enzyme or a group of enzymes, thereby acts as a therapeutic agent for the treatment of ailments. Currently, several anti-bacterial,[27] anti-fungal,[28] anti-cancer,[29]and anti-parasitic[30] drugs available in the market are enzyme inhibitors.

The process of drug discovery and development initially involves screening of a large number of compounds to find leads and then the lead compounds had to undergo several stages of clinical trial before coming to the market as a drug. In this regard, click reaction presents an efficient method that enables ready assembly of variety of compounds having diverse functionalizations. Also, the compatibility of high yielding CuAAC with aqueous conditions, in most cases, allows direct screening of compounds without any purification.[14–15]

2.1.1 Carbonic anhydrase inhibitor

Carbonic anhydrases (CA, EC 4.2.1.1) are wide spread Zn (II) metallo-enzymes that regulates the pH and CO_2: HCO_3^- homeostasis owing to their role in catalysis of reversible hydration of carbon dioxide (CO_2) to give bicarbonate anion (HCO_3^-) and a proton (H^+). Humans encode many active isozymes of CA (also known as hCA), which play critical physiological functions therefore; they have been targeted for the intervention of several diseases.[31–32] For example, the membrane-bound isozymes CA IX and CA XII support a pH-regulating system that enables hypoxic tumor cell survival and proliferation; hence these isoforms have been recognized as a potential target for the anticancer therapy. CA IX and CA XII are transmembrane proteins with extracellular enzyme active site and Zn^{2+}cation of the metallo-enzyme has been found to play a significant catalytic role in CA's action.[33] Thus, the CA inhibitors mostly comprise of a zinc binding group for example; an arylsulfonamide ($ArSO_2NH_2$) motif, that has been recognized to inhibit CA enzyme by its Zn^{2+} binding ability. In literature, a number of CA inhibitors having improved potency and desirable selectivity profiles have been described that possess sulfonamide ($ArSO_2NH_2$) moiety appended with various "tails". The development of CA-based cancer therapy requires specific inhibitors that targets tumor-associated CA isozymes hCA IX and hCA XII selectively.[32] However, it is challenging to develop isoenzyme selective inhibitors as the active site topology remains conserved within this enzyme class.

Supuran et al. employed "the tail approach" to obtain CA-isozyme selective inhibitors and demonstrated that tethering sugar tails onto sulphonamide pharmacophore increases the potency and selectivity of inhibitors for different CA isoforms due to their stereochemical diversity.[34] Their group used the facile click reaction to append the sugar tails onto the $ArSO_2NH_2$ group and thus obtained CA inhibitors showed lower cLog p, better aqueous solubility and limited passive membrane permeability. The library of glycoconjugated benzenesulfonamides containing diverse carbohydrates was obtained by the coupling of terminal alkyne/ azide equipped glycosides with the respective click partner acetylenic/ azido functionalized Ar-SO_2NH_2.[35] The *in vitro* study of the developed compounds for their ability to inhibit various hCA isozymes using the CO_2 hydration assay verified many of them as potent inhibitors with K_is < 10 nM for hCA IX & XII and also some of them demonstrated exceptional selectivity too (Fig. 10.1; results summarized in Table 10.1). Triazole bearing amide linked glucuronic acid derivative **1** (K_i = 23 nM) showed strong hCA IX

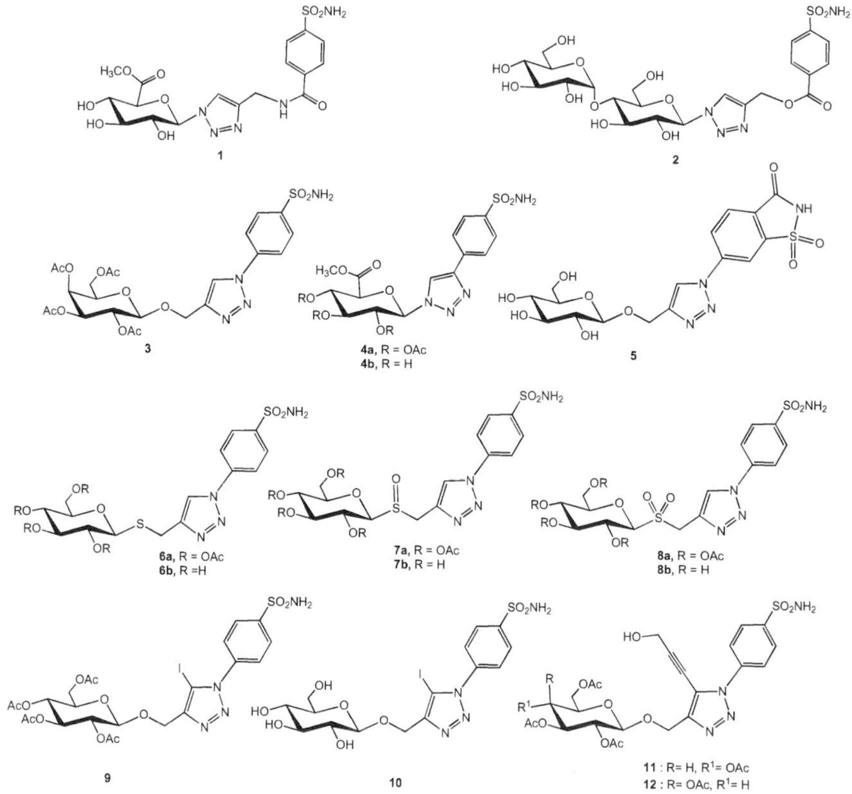

Figure 10.1 ***Carbohydrate based carbonic anhydrase inhibitors prepared by the "the click tailing approach" using CuAAC.***

inhibition, comparable to indisulam (K_i = 24 nM; anticancer agent in phase II trial) having 100-, 16.4-, 16.8- and 4.6-fold selectivity for hCA IX over hCA I, II, XII, and XIV, respectively. Triazole equipped ester linked derivatives exhibited moderate to weak activity against hCA IX, however many of them were found excellent inhibitor of hCA I and XII (**2**) (Fig. 10.1, Table 10.1).[36] Galactose derivative **3** and methyl D-glucuronates **4a** and **4b** (**3**, K_i = 9.7 nM; **4a**, K_i = 9.9 nM; **4b**, K_i = 8.4 nM) displayed more potency than the standard sulphonamides (which are used as reference for isozyme studies) and were among one of the strongest glycoconjugated sulphonamides developed for hCA IX (Fig. 10.1).[37,38]

Another series of Glycosyl-triazole conjugated benzene sulphonamides was prepared by the replacement of *N*- and *O*-glycosides with *S*-linked glycosides containing sulphur in three different oxidation states (-*S*-, -*SO*-, -SO_2-). Enzyme inhibition data obtained by assaying the CA catalyzed

Table 10.1 Enzyme inhibition data of CA isozymes I, II, IX and XII with triazolyl glycoconjugates **1-12**.[37–41]

	K_i(nM)			
Compound	hCA I	hCA II	hCA IX	hCA XII
1	2400	378	23	388
2	3500	7.3	39	1.0
3	6.9	6.8	9.7	n.d.
4a	48	13	9.9	n.d.
4b	560	13	8.4	n.d
5	>50,000	>50,000	50	600
6a	91	5.3	8.6	9.5
6b	114	5.6	257	9.5
7a	103	5.4	9.9	8.4
7b	102	7.8	9.8	10.2
8a	97	6.9	9.3	9.1
8b	101	7.6	9.5	10.3
9	112	5.3	6.2	10.0
11	105	11.9	72	7.9
12	6860	219	91	9.0

hydration of CO_2, showed *S*-glycosides as potent inhibitor against hCA II (K_is 2.9–9.9 nM), IX (K_is 6.1–9.9 nM) and XII (K_is 8.4–11.9 nM) (**6a–8a**, **6b–8b**) (Fig. 10.1, Table 10.1). Notably, no significant effect on the CA inhibition constants or selectivity for hCAs was observed with the change in the oxidation state of the sulfur.[39]

In further development, Poulsen's group studied the CA inhibitory activity (for the cancer associated isozymes CA IX and CA XII) in a new click inspired library of glycoconjugated sulphonamides in CA relevant cell and animal based models. The developed molecules inhibited CA IX and XII within a narrow range of low nanomolar K_i values (5.3-11.2 nM). Particularly compounds **9** and **10** (with K_i = 6.2 and 6.5 nM respectively, for CA IX) were recognized as blockers for CA IX-induced survival and had shown potential for development *in vivo* cancer cell selective inhibitors. The trisubstituted compounds, **11** and **12**, each bearing a short hydroxyl group at the 5-position on the triazole, demonstrated some remarkable SAR regarding isozyme selectivity for CA IX and XII. Though, galactose derivatve **12** exhibited much better selectivity (over cystolic CAs) and lower inhibition for cancer linked CAs, especially for CA XII (K_i is 9.0 nM) as compared to all other compounds of series (Fig. 10.1, Table 10.1).[40]

Recently, the potential of saccharin (SAC; a benzoic sulfimide) variant (**5**) has been explored to achieve CA isoform specificity where; SAC, a zinc binding group (ZBG) was combined to carbohydrate tail through click product a triazole linker. SAC was found to display both nanomolar affinity and preferential binding, for CA IX compared to CA II (>50-fold for SAC and >1000-fold when SAC is conjugated to a carbohydrate moiety). Therefore, the study established that SAC derivatives as ZBG in combination with various sugar tails may provide an avenue to overcome CA isoform specificity (Fig. 10.1, Table 10.1).[41]

In order to understand the topology of active site of CA II, Poulsen and Supuran's group have recently synthesized the ZBG acetazolamide (AZA) substituted molecules using click reaction that carried dual-tail combinations (i) two hydrophobic moieties, (ii) two hydrophilic moieties, and (iii) one hydrophobic and one hydrophilic moiety. The synthesized compounds were found to be weaker inhibitors than the parent AZA, concluding that $-NH_2$ protons in AZA were significant and their substitution with alkyne groups (**13**) hampered its activity. Thus, the synthesized compounds having dual tails whether having hydrophobic or hydrophilic groups had showed weaker inhibition. The loss of activity in **13** (caused due to loss of $-NH_2$) was compensated in case of **14** (K_i = 83 nM) as the presence of two phenyl moieties supposedly increases the hydrophobic interactions with CA II pocket. Compound **15** (K_i = 114 nM) having one hydrophobic and one hydrophilic sugar residue showed improved activity than **13** while compound **16** armed with two sugars had K_i = 201 nM. The K_i values for single-tail compounds **17** and **18** were determined to be 9.1 and 2.7 nM, respectively. The study concluded that presence hydrophobic residues enhanced inhibitory activity and the hydrophilic sugars did not impair the inhibitory activity however, the hydrophilic interactions can be further optimized with different sugar tails (Fig. 10.2).[42]

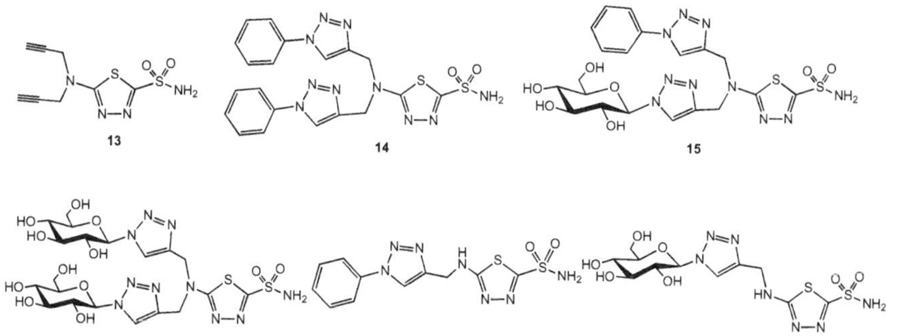
13 14 15 16 17 18

Figure 10.2 ***Acetazolamide derived single and dual tail inhibitors synthesized*** via ***click reaction.***

2.1.2 Glycosyl transferase inhibitors

Glycosyl transferase enzymes (GTs; EC 2.4) catalyzes the glycosylation of protein and lipid, which is a crucial post-translational modification and governs several molecular recognition events including bacterial and viral infections, cell adhesion, inflammation, embyogenesis, fertilization, immune response, cellular differentiation, cancer metastasis etc.[43,44] Consequently, the development of specific and powerful inhibitors of GTs can serve as potential tools for the study of regulatory mechanism of glycan biosynthesis and can lead to the discovery of new class of therapeutics.

Fucosyltransferase enzymes promotes the final glycosylation step in the biosynthesis of important saccharides sialyl Lewis x (sLe^x) and sialyl Lewis a (sLe^a) that resides on glycolipids and glycoproteins of the cell wall and are crucial molecules for the cell-cell recognition processes such as fertilization, embryogenesis, lymphocyte trafficking, immune responses, and cancer metastasis. The enzyme catalyzes the last step of glycosylation of sLe^x and sLe^a by facilitating transfer of an L-fucose moiety from guanosine diphosphate *β*-L-fucose (GDP-fucose) to a specific hydroxyl group of sialyl *N*-acetyl-lactosamine. With the aim to identify inhibitor against this enzyme, Wong *et al.* designed a library of 85 GDP-triazoles by a click reaction between alkyne functionalzed GDP with a variety of azide equipped hydrophobic group and linker chain. The screening of the library for inhibitory activity directly in microtiter plates, using the pyruvate kinase/lactate dehydrogenase coupled-enzyme assay demonstrated **19** to be a strong inhibitor of human α-1,3-fucosyltransferase VI (Fuc-T VI) enzyme with IC_{50} 0.15 μM (Fig. 10.3). Evaluation of inhibition property against other glycosyl transferases and nucleotide binding enzyme established **19** as a potent and highly selective inhibitor of Fuc-TVI (K_i = 62 nM).[45]

Nishimura et al. developed potent and selective inhibitors of glycosyl transferases by high throughput quantitative MALDI-TOFMS-based

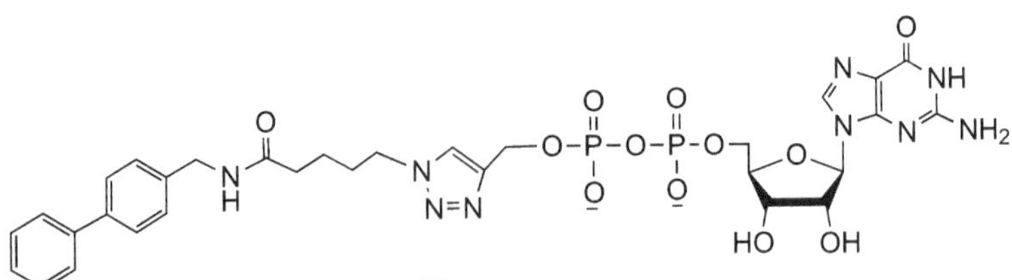

19

α-1,3-FucT VI inhibitor (IC_{50} = 0.15 μM)

Figure 10.3 ***Fucosyltransferase VI inhibitor prepared through CuAAC.***

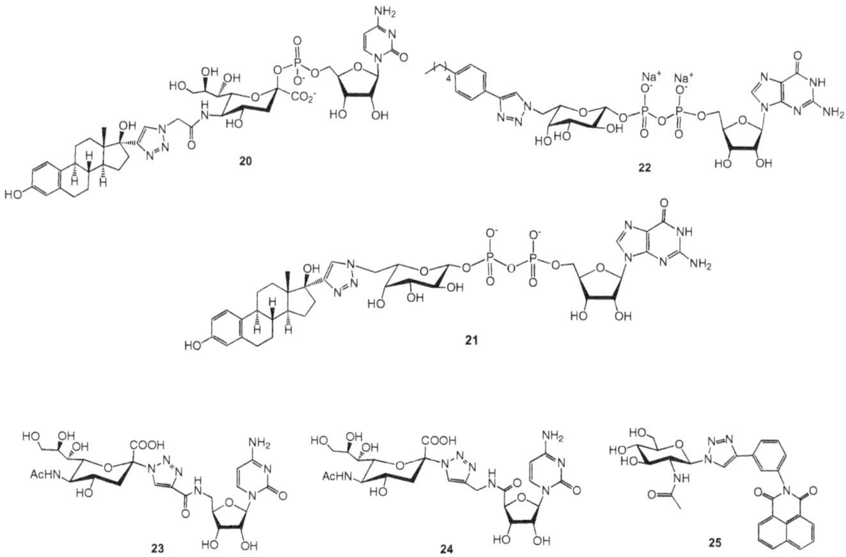

Figure 10.4 ***Triazlole glycoconjugates as glycosyltransferase inhibitors.***

screening of focused compound libraries constructed *via* CuAAC between azide appended sugar nucleotides and various alkynes. A synthetic sugar nucleotide carrying a steroidal skeleton **20** was found to be the first highly specific inhibitor for mammalian α2,3-(*N*)-sialyltransferase (α2,3ST, IC_{50} = 8.2 μM) and interestingly, **20** also served a good donor substrate for α2,6-(*N*)-sialyltransferase (α2,6ST) with K_m = 125 μM (Fig. 10.4). They established the versatility of their strategy and identified two highly potent and selective inhibitors **21** and **22** for two important classes of human FucTs; α1,3-fucosyltransferase V (α1,3-FucT, K_i = 293 nM) and α1,6-fucosyltransferase VIII (α1,6-FucT, K_i = 13.8 μM) respectively (Fig. 10.4).[46]

In order to optimize the binding of sialyltransferase inhibitors at the active site of enzyme, Zou et al. prepared and tested analogs of transition state inhibitors. The first type of molecules were accessed through click reaction and had an uncharged 1,2,3-triazole moiety in CMP-Neu5NAc (Cytidine 5'-monophospho-β-D-*N*-acetylneuraminic acid: a substrate for sialyltransferase) in place of a phosphate linkage. The other class of variants prepared had a 2-deoxy-2,3-dehydro-acetylneuraminic moiety connected to cytidine through its carboxylic acid and amide linkers. In the third class of molecules the sialyl phosphate was replaced by an aryl sulfonamide which was further connected to cytidine. Inhibition study of the conjugates against *Campylobacter jejuni* sialyltransferase Cst 06 depicted that the combination of two active site binder's cytidine and a neuraminyl group without charge interaction (phosphate group) is not enough to exhibit an effective binding at the active site. Only the first type of molecules containing triazole units was found competitive inhibitors, whereas the other two could only inhibit the enzyme non-competitively. Compound **23** and **24** containing a triazole linkage as mimic of phosphodiester bond with an α-anomeric configuration of sialic acid showed more than 50% inhibition at 500 μM. The K_i for **24** is 160 μM, which is twice lower than the K_m of CMP-Neu5Ac (400 μM) (Fig. 10.4).[47]

O-GlcNAcylation, the covalent addition of *N*-acetylglucosamine unit onto serine or threonine residues of proteins *via* a β-*O*-glycosidic linkage is one of the crucial post-translational modifications related to cellular regulation and signal transduction. Enzyme *O*-GlcNAcase (OGA) regulates this metabolism by promoting the cleavage of *O*-linked β-*N*-acetylglucosamine (*O*-GlcNAc) from protein residues. As a consequence, potent OGA inhibitors can serve as a useful tool for understanding the cellular processes of *O*-GlcNAc, and may be developed as drugs for the treatment neurodegenerative diseases. Wang and his group recently utilized the Cu(I)-catalyzed click reaction between glycosyl azides and alkynes to generate a library of triazole linked glycoconjugates that were identified as potential *O*-GlcNAcase inhibitors. Enzymatic kinetic screening of the series confirmed the compound **25** as most powerful competitive inhibitor of human *O*-GlcNAcase (K_i = 185.6 μM) (Fig. 10.4).[48]

2.1.3 Trypanosoma cruzi trans-sialidase (TcTS) inhibitor

A protozoan parasite *Trypanosoma cruzi*, causative agent of one of the most widespread neglected tropical disease, Chagas' disease affects millions population in latin America. The host cell invasion and infection process of the parasite is regulated by a cell surface *Trypanosoma cruzi trans-sialidase* (TcTS), which is a glycosyl transferase enzyme that transfers sialic acid from host cell to mucin-like glycoproteins to modify its cell surface glycocalyx and thus helps parasite to escape human immune response.[49,50] The development of TcTS inhibitors offers potential for therapeutic intervention of Chagas disease.[51] The wide scope of Cu(I)-catalyzed click reaction has been utilized to obtain *Trypanosoma cruzi trans sialidase* (TcTS) enzyme inhibitor.[52]

Campo et al. synthesized a library of 46 galactose 1,2,3-triazole derivatives by the click cycloaddition of either *C*-6 or *C*-1 azide functionalized galactose moiety with a panel of 23 different terminal alkynes under microwave conditions. *In vitro* evaluation of the sugar-triazoles for their inhibition potential against TcTS suggested them to be moderate to weak inhibitors (<40% inhibition at 1 mM concentration), though most compounds evaluated were able to act as acceptor substrates for TcTS-catalyzed trans-sialylation. Despite this modest inhibitory activity, *in vitro* trypanocidal activity assays against the trypomastigote form of T. cruzi Y strain revealed several compounds to be active in the low 100s of μM range. Assessment of the most active TcTS inhibitors for cytotoxicity against mammalian cells established selective anti-parasitic activity for some compounds, whereas a few were found toxic too. *N*-Methyl benzylamine derivative **26** (50% parasite kill at ~200 μM; <<10% mouse cell kill at 1 mM) was found most promising derivative for further studies (Fig. 10.5).[53]

Likewise, their group explored TcTS inhibitory activity in 1,2,3-triazole sialylmimetic neoglycoconjugates that they obtained by the click reaction of different azido-derived sugars (Gal, Glc, Gul)/amino acid (Thr) with

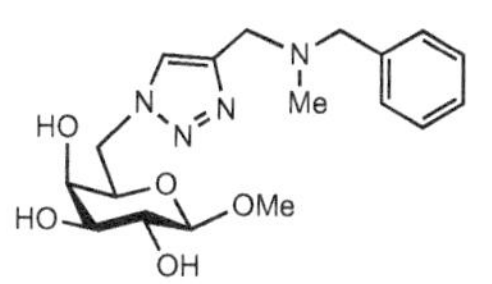

26, TcTS inhibitor (17% inhibition at 1 mM)

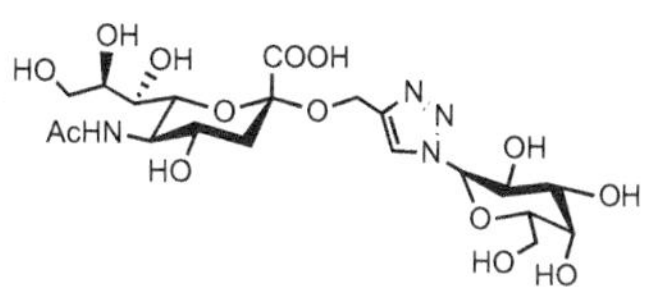

27, Trypanocidal activity IC_{50} = 475 μM)

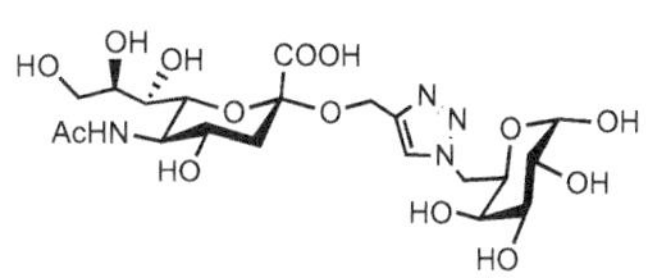

28, Trypanocidal activity IC_{50} = 260 μM)

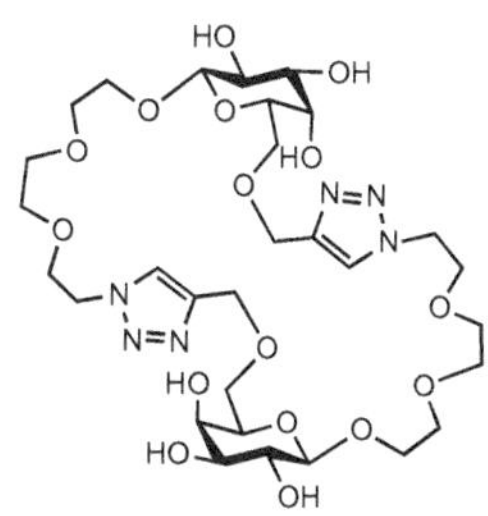

29

TcTS inhibitor

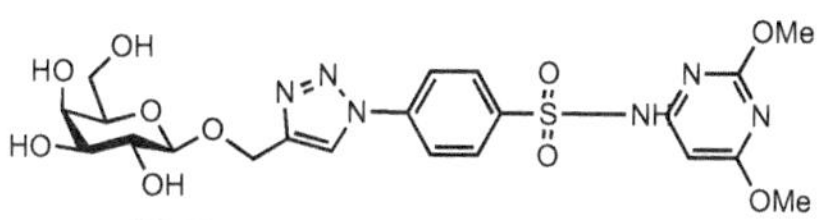

30, Trypanocidal activity IC_{50} = 44 μM)

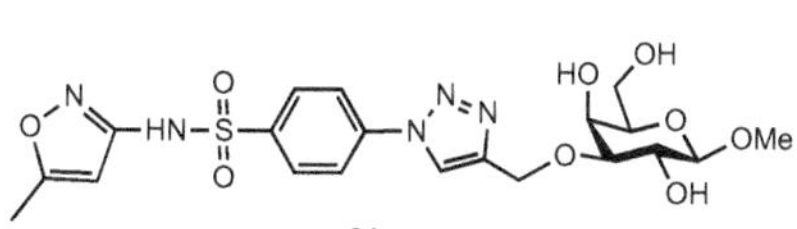

31

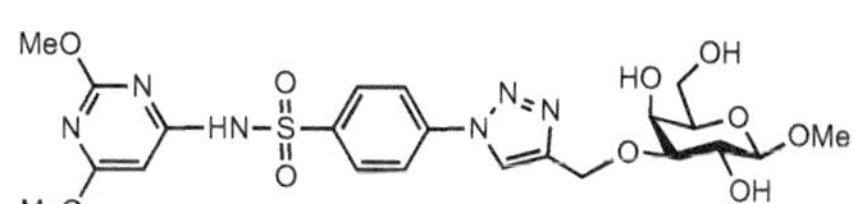

32

TcTS inhibitors; binding affinity for galectin-3 (EC_{50} ~17–18 μM)

Figure 10.5 ***Click reaction derived*** **Trypanosoma cruzi trans-*****sialidase (TcTS) inhibitors.***

alkyne functionalized sialic acid/galactose as precursors. In biological testing, 1,2,3-triazole linked sialic acid-6-*O*-galactose **27** and the sialic acid-galactopyranoside **28** displayed high TcTS inhibitory activity at 1.0 mM (approx. 90%), but assessment of trypanocidal anti-parasite activity verified only compound **27** as powerful inhibitor (IC_{50} 260 μM). Therefore, results evidenced the necessity of sialic acid and galactosyl unit in development of more effective inhibitors for chagas' disease (Fig. 10.5).[54] Click reaction was further explored by the same group to derive triazole-linked linear oligomers and macrocycles by the cyclic oligomerization of galactose based azido-alkyne-bifunctionalised precursors **29** (Fig. 10.5). Thus obtained pseudo-galacto oligomers were established as acceptor substrates for the cell surface trans-sialidase of the human parasite *T. cruzi* and were also found to be able to block macrophage invasion of the parasite.[55,56]

With the aim of obtaining potential inhibitors of *T. cruzi* cell invasion, Campo and co-workers further synthesized the triazole-linked galactosyl arylsulfonamides by employing click reaction between different azide arylsulfonamides and the alkyne-based β galactose precursor. Anti-trypanosomal assay performed on the library displayed compound **30** to be most promising with IC_{50} of 44 μM. The inhibition assays of *T. cruzi* cell invasion revealed compounds **31** and **32** bearing the 5-methylisoxazole and 2,4-dimethoxypyrimidine groups respectively, to be the most effective in blocking parasite invasion, with reduced values of infection index (~20). Additionally, compounds **31** (EC_{50} 17.2 μM) and **32** (EC_{50} 18.3 μM) displayed the highest binding affinity for galectin-3 experimentally as well as theoretically, suggesting that **31** and **32** block *T. cruzi* cell invasion by means of galectin-3 binding-related mechanism (Fig. 10.5). Thereby, the work highlighted galectin-3 as an important host target for design of novel anti-trypanosomal agents.[57,58]

2.1.4 Glycosidase inhibition activity

A wide variety of physiological processes such as intestinal digestion, post-translational modification of glycoproteins and the lysosomal catabolism of glycoconjugates etc., which involve the breaking or construction of glycosidic linkages, are being regulated by several glycosidase enzymes.[59] Therefore, the development of inhibitors against these glycosidase enzymes serves a promising strategy to obtain therapeutics for the treatment of diseases such as diabetes,[60] viral infections[61] or HIV[62] and lysosomal storage diseases.[63,64] In recent years, Cu(I)-catalyzed click reaction has been successfully utilized for accessing glycosidase inhibitors.

In search of potent α-glucosidase inhibitors, the mimetics of anti-diabetic drug acarbose has been prepared by Perion et al. A library of saccharidyl-triazole conjugates was synthesized using Huisgen 1,3-dipolar cycloaddition of azide and alkyne that exhibited significant inhibition for α-glucosidase enzyme. 6-deoxyglucosyl triazole (linked via *C*4 of the sugar unit) derivative showed a 2-fold higher inhibitory activity against yeast α-glucosidase (K_i = 73 μM) than acarbose.[65]

Basu and Rossi reported click inspired synthesis of 1-β-D-glucosyl-4-phenyl triazole and 1-β-D-galactosyl-4-phenyl triazole having triazole unit connected to anomeric position of sugar.The assessment of inhibitory activity against three β-glycosidase enzymes sweet almond glucosidase (SAG), *Escherichia coli* galactosidase (ECG), and bovine liver galactosidase (BLG) using a percentage assay revealed them a moderate glycosidase inhibitor. In comparison to glycosidase inhibitors deoxygalactonojirimycin (DNJGal) and deoxynojirimycin (DNJ, gluco configuration), galactose derivative **33** showed moderate inhibition for ECG, whereas glucosyl triazole **34** was observed to be better inhibitor of BLG than the **33,** although none of them showed good inhibiton for SAG (Fig. 10.6).[66]

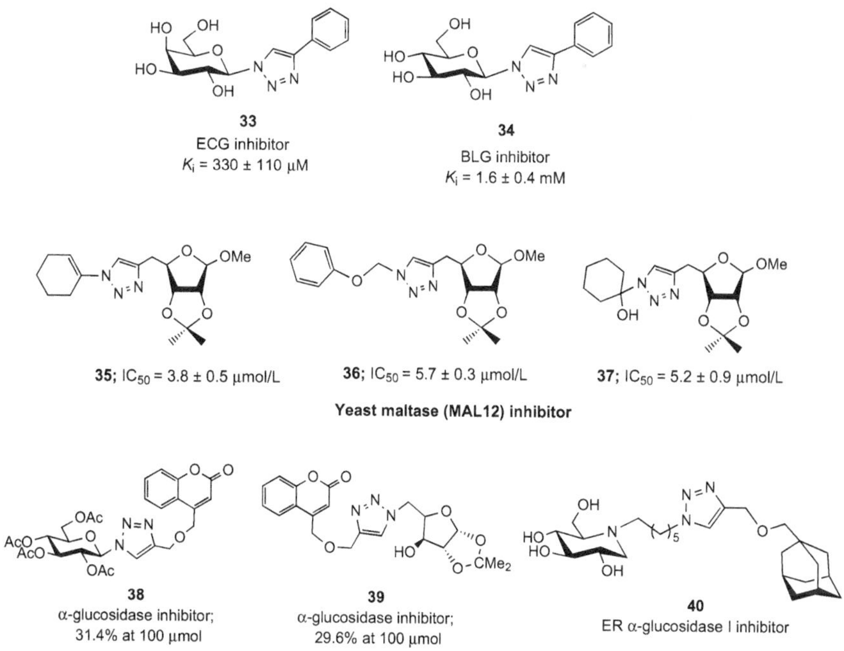

Figure 10.6 ***1,2,3-triazolyl glycohybrids, inhibitors of various glycosidase enzymes.***

Ferreira et al. reported the synthesis of a series of 1,4-disubstituted 1,2,3-triazoles conjugated with several sugars, including D-xylose, D-galactose, D-allose, and D-ribose. These glycoconjugates were obtained by the Cu(I) catalyzed click coupling of isopropylidene protected azido sugars with terminal alkynes. The evaluation α-glucosidase inhibition activity of compounds at 500 μM using yeast maltase (MAL12) as a model enzyme, showed all of them active; however the triazoles linked with D-ribose were in general found more efficient than the other triazoles bearing D-xylose, D-galactose, and D-allose sugars. The determination of the IC_{50} for the β-D-ribosyl derivatives demonstrated that compounds **35**, **36** and **37** carrying respectively, the 1-cyclohexenyl, phenoxymethyl, and 1-cyclohexanol moieties at *C*4 of the triazole ring, have around 20-fold higher potency in yeast maltase inhibition than the standard drug acarbose (Fig. 10.6).[67]

Field and co-workers studied the glycosidase inhibition potential of a library of triazole linked α-D- and β-D-glucopyranosides prepared by the CuAAC reaction of α-D- and β-D-glucopyranosyl azides with various commercially available terminal alkynes having lipophilic, hydrophilic, acidic or basic functional groups. The developed α- and β-linked triazolyl glucosides displayed inhibition for both the assayed enzymes sweet almond β-glucosidase (GH1) and yeast α-glucosidase (GH13) in 100 μM range. Furthermore, the yeast α-glucosidase showed a strong preference for α-linked glucosyl triazoles over their β-anomers, irrespective of the substitution present over triazole ring, which evidenced the role of anomeric configuration on inhibiton.[68] Tripathi's group developed a series of 1,2,3-triazolyl glycohybrids of two or more than two sugar units or a chromenone moiety *via* CuAAC reaction of glycosyl azides with 2,3-unsaturated alkynyl glycosides or propargyloxy coumarins. Synthesized glycosyl triazoles were assayed for their α-glucosidase, glycogen phosphorylase, and glucose-6-phosphatase inhibitory activities. Some representative triazoles glycohybrids **38** and **39** showed a significant 31.4% and 29.6% inhibition respectively, against α-glucosidase at 100 μmol concentrations with respect to control acarbose (60% inhibition at 100 μmol concentrations) (Fig. 10.6).[69]

Imino sugars, a sugar mimic where oxygen of the hemiacetal ring is replaced by a nitrogen atom, constitute one of the most promising class of therapeutic agents.[70] They have shown their potential in the treatment of carbohydrate-mediated diseases such as viral infections,[71] cancer,[72] diabetes,[73] and lysosomal storage disorder (gauchers diasease).[74a] Among imino sugars, deoxynojirimycin (DNJ) a mimic of glucose is of particular interest as its *N*-hydroxyethyl analogue, miglitol (marketed as Glyset) is an approved drug for the treatment of type II diabetes and a *N*-butyl derivative (NBDNJ) miglustat (marketed as Zavesca) is a drug for the treatment of Gaucher disease. Moreover, NBDNJ is in phase II trials for the cure of cystic fibrosis.[73]

DNJ variants are well known to exhibit antiviral activity by the inhibition of endoplasmic reticulum (ER) α-glucosidases I and II, consequently in order to obtain novel antiviral agents, Butters and coworkers synthesized deoxynojirymicin neoglycoconjugates *via* CuAAC of azide-armed deoxygalactonojirymicin with alkyne functionalized admantane skeleton. The compounds showed α-glucosidase inhibition (compound **40** was most potent) as they were found to disrupt the viral glycoprotein processing, with increased glucosylation on bovine viral diarrhoea virus BVDV E2 species (Fig. 10.6). It has been demonstrated that with increase in the length of alkyl chain between DNJ and triazole groups in DNJ-admantyl hybrids, the α-glucosidase inhibition increases and the production of viral progeny RNA and the maturation of the envelope polypeptide reduces.[74b]

In order to study the effect of multivalency on selectivity and affinity of glycosidases, Diot *et al.* synthesized multivalent imino sugars derived from the glycosidase inhibitor 1-DNJ by exploiting click reaction. Azide carrying DNJ derivatives upon click coupling with alkynyl-armed oligo(ethyleneglycol) (EG) resulted in mono-, di- and tri-valent imino sugars. These analogs when assayed against different glycosidases showed good inhibition though they were found less active than parent nucleus DNJ. The inhibition constants (K_i) of the compounds **41–45** showed an increase of inhibitory potency with increasing valency in the case of α-mannosidase and amyloglucosidase enzymes, whereas reverse order of K_i was found for β-glucosidase. In case of other glycosidases screened no regular variation of inhibition activity K_i with valences was obtained (Fig. 10.7, Table 10.2).[75,14]

A library of novel piperidine and azepane based pseudo-disaccharide iminosugars was assessed for the glycosidase inhibitor activity against α-glucosidase, β-glucosidase and barley β-glucosidase. Compound **46** (IC_{50} = 53 μM) showed a marked yeast α-glucosidase inhibitory activity with IC_{50} ~3 fold lower than that of DNJ (IC_{50} 180 μM) (Fig. 10.7). In general piperidine series compounds were found to be more active than azepane variants and iminosugar pseudo-disaccharides showed no inhibitory activity against almond β-glucosidase.[76]

Gauchers disease, the most widespread lysosomal storage disorder is characterized by accumulation of a sphingolipid glucosylceramide in macrophages.[77] This disease is caused by the deficient activity of a

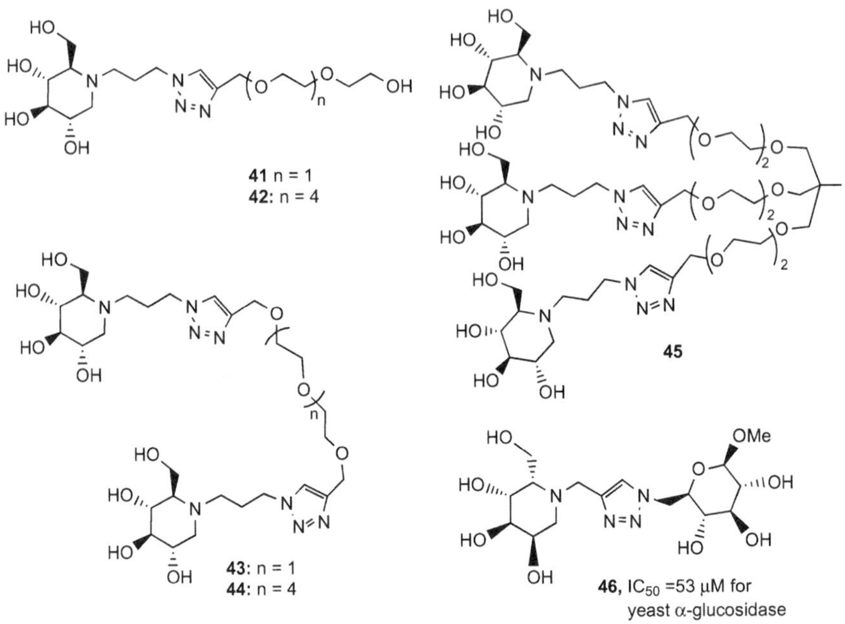

Figure 10.7 ***Synthetic multivalent DNJ derivatives prepared via click reaction.***

Table 10.2 Glycosidase inhibition activities (inhibition constant K_i, μM) for 1-DNJ and its derivatives **41-45**[75].

Enzyme	1-DNJ	41	42	43	44	45
β- Galactosidase	42	950 ±20	n.i.	n.i.	760 ±10	790 ±10
α- Galactosidase	-	120 ±5	95 ±5	215 ±5	105 ±5	225 ±5
β- Glucosidase	47	47 ±2	100 ±5	70 ±2	70 ±2	400 ±10
α- Glucosidase	25	n.i.	n.i.	n.i.	n.i.	n.i.
β- Mannosidase	-	n.i.	n.i.	n.i.	772 ±10	n.i.
α- Mannosidase	270	230 ±10	265 ±10	120 ±5	90 ±3	35 ±2
Amyloglucosidase	2.1	28 ±1	30 ±2	17 ±1	20 ±2	11 ±1

n.i. = no inhibition detected for 2 mM.

β-glucocerebrosidase (β-GCase) enzyme that hydrolyzes glucosylceramide into glucose and ceramide and prevents its deposition in lysozyme.[78] For the treatment of gauchers disease, the two important therapies available are Substrate reduction thepary (SRT)[79] and Chaperone-mediated therapy (CMT).[80] The former one prevents glycolipid storage by inhibiting the glucosylceramide synthase (GCS) enzyme, which is essential for the biosynthesis of glucoylceramide and this way balances defective β-GCase activity.[79] CMT involves the use of pharmacological chaperones (small molecules) that bind to the active site of β-GCase to stabilize it and assist in the correct folding of enzyme leading to revival of β-GCase activity.[80]

The iminosugar *N*-butyl-1-deoxynojirimycin (NB-DNJ) is a well established inhibitor for glucosylceramide synthase (GCS) enzyme, whereas the hydrophobic iminocyclitols have been found to behave as a pharmacological chaperones.[79,80] Inspired with these observations, several groups have prepared many deoxynojirimycin derivatives using the very robust click reaction. A series of DNJ-admantane conjugates having triazole moiety flanked through alkyl chains of various length have been reported as a potent glycosidase inhibitors. All the DNJ derivatives of series were found more potent β-glucocerebrosidase inhibitor than NB-DNJ, **47** being the most active with IC_{50} 0.30 ± 0.04. Regarding Gaucher disease and glucosylceramide synthase inactivation, the DNJ-adamantane hybrids **48** and **49** bearing the longest arms were found approximately 20-fold more potent than model *N*-Butyl-DNJ and even more potent than reference compound AMP-DNJ (*N*-[5-(adamantan-1-yl-methoxy)-pentyl]-1-deoxynojirimycin (Fig. 10.8).[81a]

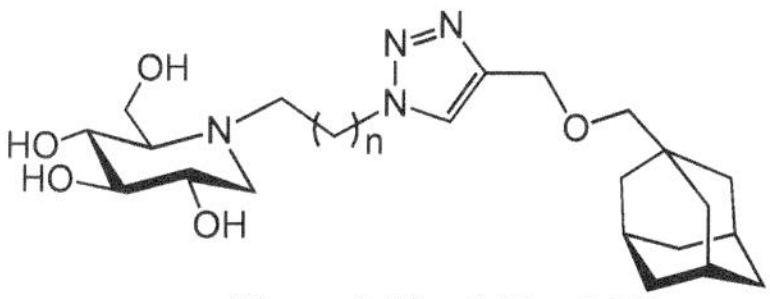

47: n = 3 (IC_{50} 0.30 ± 0.04)
48: n = 7
49: n = 9

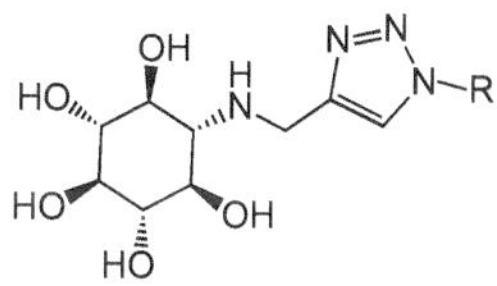

51: R = decyl [IC_{50} = 0.09 (pH 7.4)
52: R = dodecyl [IC_{50} = 0.06 (pH 7.4)

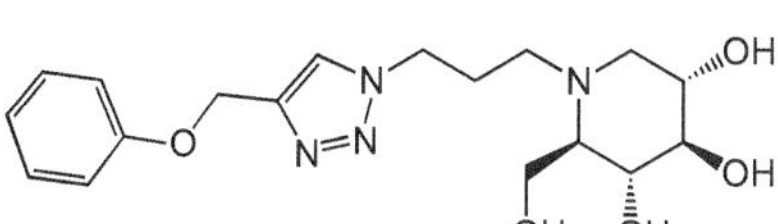

50

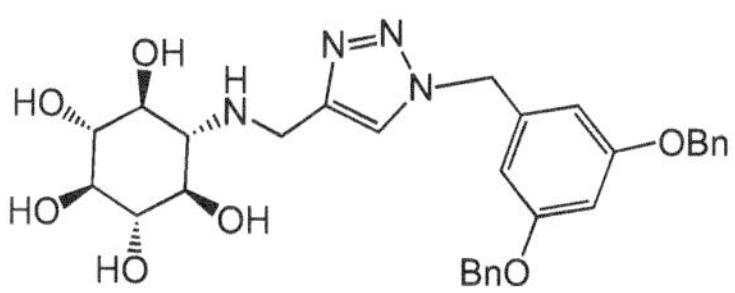

53: GCase inhibitor
IC_{50} = 0.06 mM (pH 7.4)
IC_{50} = 0.05 mM (pH 5.2)]

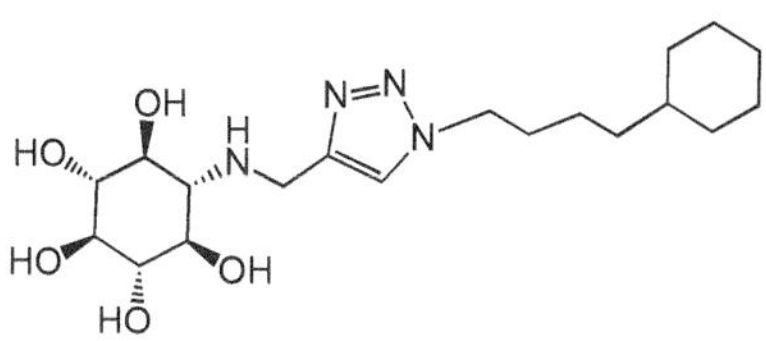

54: GCase inhibitor
IC_{50} = 0.10 mM (pH 7.4)
IC_{50} = 0.12 mM (pH 5.2)]

Figure 10.8 ***Click reaction derived imino sugars active for Gauchers disease.***

Following similar strategy, Gouin et al. explored click reaction for accessing a library of 14 iminosugar derivatives by the coupling of *N*-azidopropyl-1-deoxynojirimycin with diverse alkynyl-armed hydrophobic scaffolds. Most of the compounds were found micro molar inhibitors of commercial β-glucosidase and amyloglucosidase and showed a considerable selectivity for these enzymes toward α-glucosidase. *In vitro* assay for N370S β-GCase inhibition established the DNJ derivatives having aromatic aglycons as strong inhibitors. In contrast, the hydrophilic aglycon containing derivatives showed poor inhibition. Further, the study of the chaperone activity of potent β-GCase inhibitors for mutant gaucher cells suggested compound **50** to be the most active with 2-fold increased activity at 20 μM in comparison to the 1.33-fold recorded for the reference compound *N*-nonyl-1-deoxynojirimycin (Fig. 10.8).[81b]

Apart from DNJ, *N*-alkylaminocyclitol deviatives carrying 1,2,3-triazole system at different positions of the alkyl chain were developed as potential glucocerebrosidase pharmacological chaperones using CuAAC. Compounds **51** and **52**, having the shorter spacer (n = 1) between the alkyltriazolyl system and the aminocyclitol core, were the most active ones as GCase inhibitors, suggesting significance of the position of the triazole ring on the activity (Fig. 10.8). Also, all the compounds of the series exhibited high selectivity for GCase showing insignificant inhibition towards glucosyl ceramide synthase (GCS) and other commercially available glycosidases.[82] In extension to this work, recently their group reported the synthesis of new aminocyclitol analogues using click coupling of *N*-propargyl aminocyclitol with diverse set of azides under aqueous conditions. CuAAC performed in aqueous medium allowed to skip additional protection/ deprotection steps on aminocyclitol scaffold and facilitated the *in situ* screening of compounds against GCase inhibitors just by simple dilution of the crude reaction mixtures in the ample amount of buffer. The developed approach was conveniently performed into a microtiter plate, therefore, found suitable for combinatorial protocols. A library of 25 diverse reactants was chosen to be linked to aminocyclitol skeleton, out of collection potential 343 reactants by the application of a proper selection algorithm. Library members when evaluated for GCase inhibition and also for their ability to induce enzyme thermal stabilization, the compound **53** and **54** were found very promising candidates. In fact **53** turned out to be the most potent of the library, not only as GCase inhibitor (K_i = 0.05 μM), but also as *in vitro* pharmacological chaperone due to its ability to stabilize GCase after thermal denaturation (Fig. 10.8).[83]

2.1.5 Neuraminidase inhibitors

Neuraminidases (NAs; EC 3.2.1.18) belong to a crucial family of glycoside hydrolases that cleave the terminal sialic acid (Neu5Ac; *N*-acetylneuraminic acid) residue from the glycoproteins and glycolipids. They play essential role in the regulation of diverse processes including cellular communication, signalling, adhesion, inflammation, infection as well as the apoptosis and metastasis of malignant cells.[84] Several pathogens for instance; viruses, bacteria, and protozoa etc. utilize their own NAs for the various stages of host infection such as invasion, nutrition, and escape. As a result, pathogenic NA inhibitors have become an attractive target for drug discovery, most notably the influenza viral neuraminidase.[84]

Influenza virus comprises of two unique surface glycoproteins a hemagglutinin and a neuraminidase. The hemagglutinin recognizes sialic acid moiety of host and thereby, facilitates adhesion of virus to the target cell. Whereas, neuraminidase cleaves these virus attached sialic acid residues and allows virus entry into the cells.[85] A transition state analogue of sialic acid, 2,3-dehydro-*N*-acetylneuraminic acid (DANA, **55**)[86] has been recognized as inhibitor of neuraminidases in 1969. During 1990s, sialic acid modified highly potent and selective neuraminidase inhibitors zanamivir [Relenza, **56**] and oseltamivir [Tamiflu, **57**] (Fig. 10.9) were approved as drugs.[87,88] Though, rapid emergence of resistant influenza viruses due to surface glycoprotein modification demands development of new vaccines. Over the past few years, several groups have modified this sialic acid mimetic in order to formulate better inhibitors, which can effectively bind to enzyme's active site to cure influenza directly by blocking the synthesis of neuraminidases.

2.1.5.1 Zanamivir based neuraminidase inhibitors

Jiang *et al.* employed click reaction to design analogues of drug zanamivir, which is a neuraminidase inhibitor and effective drug for the treatment of avian influenza (H5N1). A library of sixteen *C*-4 triazole substituted

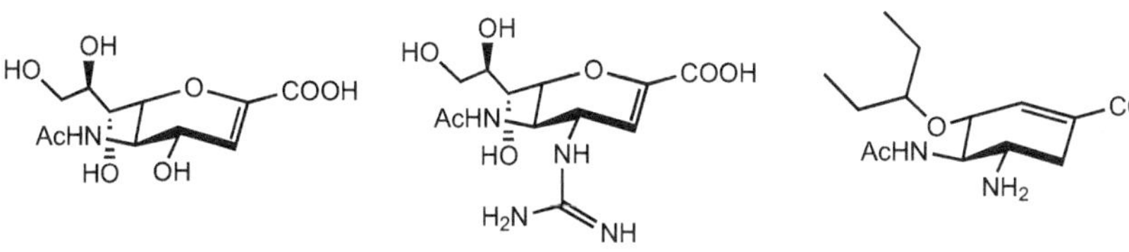

55
DANA

56
Zanamivir (Relenza)

57
Osteltamivir (Tamiflu)

Figure 10.9 ***Structure of approved neuraminidase inhibitors.***

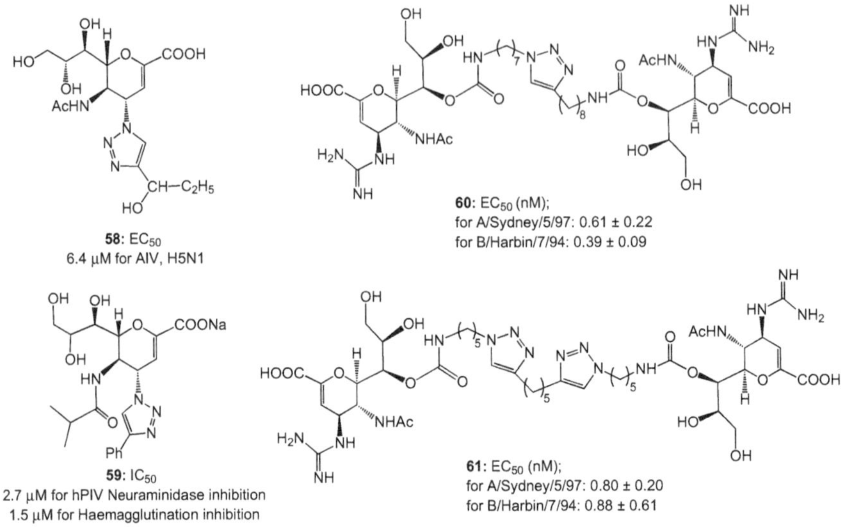

Figure 10.10 ***Click inspired zanamivir analogs as potent Neuraminidase inhibitors (NIs).***

zanamivir analogues was reported by the click combination of a common precursor azide functionalized zanamivir with variously substituted terminal alkynes. Developed zanamivir glycoconjugates were subsequently screened against avian influenza virus (AIV, H5N1), where compound **58** substantially showed maximum anti-AIV activity (EC_{50} = 6.4 µM) comparable to zanamivir (EC_{50} = 2.8 µM) (Fig. 10.10).[89]

Guillon and his group designed *C*4-functionalized triazole unsaturated *N*-acetylneuraminic acid derivatives through a structure guided approach and identified them as effective inhibitor for human parainfluenza viruses hPIV type 3 haemagglutinin–neuraminidase (hPIV-3HN). Inhibitors were found effective blockers of hPIV cell entry and virion progeny release both, example **59** (Fig. 10.10).[90]

Dimers of zanamivir carrying variable linker length and triazole units were synthesized by Fraser et al. and were tested for their anti-viral activity against influenza A/Sydney/5/97 and B/Harbin/7/94 in a cytopathic effect (CPE) assay. All the dimers were found significantly more potent than standard oseltamivir carboxylate and zanamivir against influenza B/Harbin/7/94. Interestingly, the dimers with the longer linker size (no. of linker atoms 18 to 25) were around 100 times more potent than oseltamivir carboxylate and 1000–3000 fold more potent than zanamivir against influenza A/Sydney/5/97, in comparison to the dimers with shorter linker (12-16 atoms) that were less potent, example **60–61** (Fig. 10.10).[91]

2.1.5.2 DANA based neuraminidase inhibitors

DANA, another transition state mimic of sialidase enzyme has been modified by several groups to obtain potent inhibitors for neuraminidases.[86] Zou et al. designed a library of DANA analogues by modifying it at *C*9 and *N*5Ac positions using click reaction. Since the molecular docking studies of DANA in human neuraminidase 3 (NEU3) active site evidenced the presence of large hydrophobic pocket adjacent to the *C*9 position of the sialic acid; the authors attached a variety of substituents on *C*9 and *N*5Ac positions of DANA to explore binding site topology of NEU3 and to identify new inhibitors. *N*5-azidoacetyl- and 9-azido-9-deoxy DANA based precursors were prepared from *N*5-acetylneuraminic acid (Neu5Ac) in multi steps, and then click reaction was performed to append different alkyne functionalized substituents. Evaluation of enzymatic inhibition showed distinctly improved activity in case of *C*9 triazolyl derivatives with phenyl **62** (20 ± 10 μM), hexyl **63** (23 ± 4 μM), and phenoxymethyl **64** (45 ± 3 μM) whereas, none of the derivatives made at the N5Ac side-chain were active (Fig. 10.11). Moreover, the enhanced inhibition potency of the DANA derivatives was shown to be contributed by insertion of triazole moiety.[92]

A novel two-step strategy was adapted by Nishimura and coworkers to obtain potent inhibitor for *Vibrio cholerae* neuraminidase (VCNA), that included a target amino acid validation using mechanism-based labeling information, and a potent inhibitor search by using a focused library. The

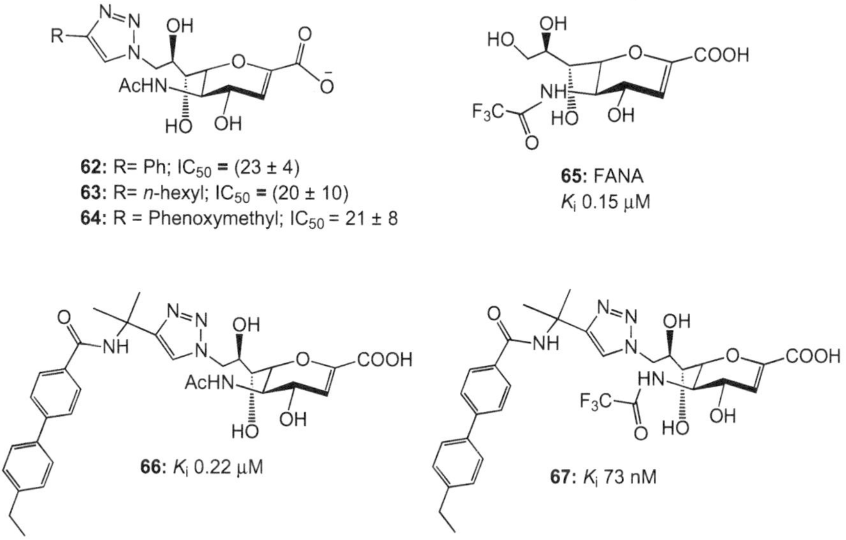

Figure 10.11 ***Triazole linked DANA based Neuraminidase inhibitors.***

labeling study informed about the loop structure ofVCNA that might be potential target of the novel inhibitor. A focused library composed of 187 compounds was prepared from a 9-azide derivative of DANA to interrupt the function of the loop of the labeled residues. Inhibitor **66** (K_i = 0.22 μM) showed potent inhibition comparable to the strongest inhibitor FANA **65** (FANA = a *N*-trifluoroacetyl derivative of DANA; K_i: 0.15 μM). Furthermore, **66** and FANA were combined to afford a novel compound **67** (K_i = 73 nm, mixed inhibition), which was found most potent inhibitor ofVCNA with high selectivity among the tested viral, bacterial, and mammal neuraminidases (Fig. 10.11).[93]

Chen and co-workers synthesized peptide conjugated C9-modified Neu5Ac2en analogs using CuAAC and studied their inhibitory potential against sialidases. Monoamino acid-conjugate (TriazoleNeu5Ac2en)-A (**68**) and tetrapeptide-conjugate E-(TriazoleNeu5Ac2en)-AKE (**70**) (Fig. 10.12) were found to be selective inhibitors for *V. cholerae* sialidase with IC_{50} value of 13.5 ± 0.5 μM and 28.9 ± 3.2 μM, respectively, while a dipeptide-conjugate (TriazoleNeu5Ac2en)-AdE (**69**) selectively inhibited *V. cholerae* and *A. ureafaciens* sialidases.[94]

1,2,3-Triazole linked pentacyclic triterpene-Neu5Ac2en derivatives were synthesized and evaluated for their *in vitro* anti-influenza virus activity in MDCK cells by Cytopathic effect reduction assay. An amide linked triterpene-Neu5Ac2en displayed highest inhibitory activity against influenza A/WSN/33 (H1N1) with IC_{50} of 8.3 μM comparable to the curcumin, a known anti-influenza entry inhibitor. However, insertion of (1,2,3-triazol-4-yl) methyl linker between amide and Neu5Ac2en lead to a significant decrease in activity indicating that the linker between pentacyclic triterpene and Neu5Ac2en played an important role in their antiviral activity.

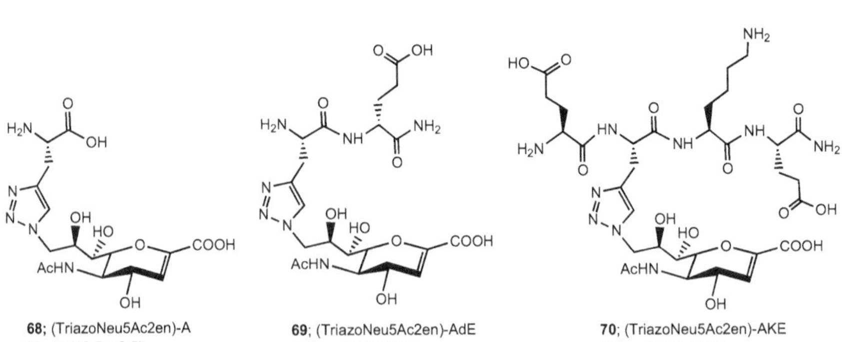

68; (TriazoNeu5Ac2en)-A
IC_{50} = (13.5 ± 0.5)

69; (TriazoNeu5Ac2en)-AdE
IC_{50} = (10.9 ± 0.8)

70; (TriazoNeu5Ac2en)-AKE
IC_{50} = (28.9 ± 3.2)

Figure 10.12 *Triazole-linked transition state analogs as V. cholerae sialidase inhibitors.*

Interestingly, a detailed study on the most potent compound did not show neuraminidase inhibition, although showed inhibition of the hemagglutination of red blood cells, suggesting the antiviral activity in compound to be mediated by interaction with hemagglutinin at the initial stage of viral infection.[95]

2.1.5.3 Sialic Acid based Neuraminidase inhibitors

Linhardt *et al.* reported a straightforward route for the synthesis of triazolyl sialic acid derivatives by using CuAAC between azido containing sialic acid [methyl-(5-acetamido-3,5-dideoxy-2-azido-D-glycero-α-D-galactonon-2-ulopyranoside)onate] and different alkynes. Subsequent screening of library members for the neuraminidase inhibitory activity using 96-well plate fluorescence assay, displayed compound **71**; which was accessed by the click reaction of sialyl azide with dec-1-yne, as a potential lead having neuraminidase inhibitory activity (IC_{50} = 28 μM) comparable to the known sialidase inhibitor *N*-Acetyl-2,3-dehydro-2-deoxyneuraminic acid (Neu5Ac2en; IC_{50} = 67 μM). Further, some interesting multivalent derivatives, including a 1,2,3-triazole-linked disaccharide mimic of sialic acid **72** (IC_{50} = 17 μM), and a dendrimeric structure containing four sialic acid residues **73** (IC_{50} = 20 μM) were also developed using click approach that possessed better neuraminidase inhibitory profile than the Neu5Ac2en (Fig. 10.13).[96] Herczegh et al. explored the click reaction to obtain fullerene

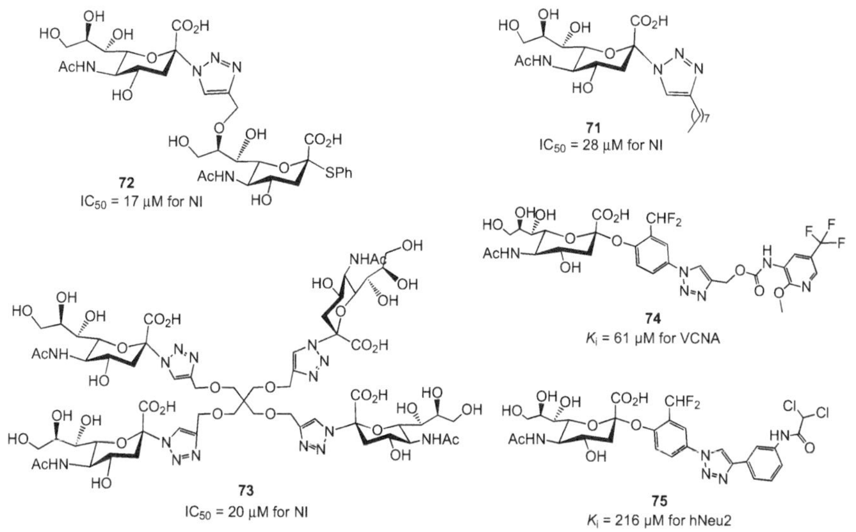

Figure 10.13 ***Developed sialosides as potent neuraminidase inhibitors (NI) inhibitors.***

conjugate containing a thiosialosyl-α(2,6)-galactose, which showed moderate inhibitory activity towards the influenza virus neuraminidase however, failed to inhibit the influenza virus hemagglutinin.[94]

Nishimura and coworkers designed an aglycone-focused library by the click coupling of 4-azide-2-difluoromethylphenyl sialoside and an alkyne-containing compound library. Study of the inhibition patterns revealed two leads **74** and **75** as the potent and selective inhibitors for *vibrio cholerae* neuraminidase (VCNA) and human neuraminidase 2 (hNeu2), respectively (Fig. 10.13).[98]

2.1.6 Glycogen phosphorylase inhibitor

Glycogen phosphorylase (GP, EC: 2.4.1.1) enzyme catalyzes complete breakdown of glycogen into glucose and related metabolites to supply energy.[99] Due to the involvement of GP in the modulation of glycogen metabolism, inhibition of GP has been considered as a valuable remedy for the treatment of diseases caused by irregularities in glycogen metabolism, such as type 2 diabetes,[100] myocardial ischemia,[101] and tumors[102] In literature, numerous glucose derivatives have been documented as GP inhibitors that bind effectively at the catalytic site of the enzyme.[103,104]

Xie and co-workers reported oleanolic acid (OA) and related pentacyclic triterpenes as a new class of GP inhibitors.[105] Several oleanoic acid glycoconjugates were synthesized with the aid of click chemistry and later evaluated for glycogen phosphorylase inhibitory activity, where compound **76** showed excellent inhibition with an IC_{50} value of 1.14 μM (Fig. 10.14).[106]

Employing the same chemistry pentacyclic triterpene heteroglycoconjugates were prepared through the coupling of propargyl esters of *C*-28 carboxylic acids of triterpenes (oleanolic, ursolic, and maslinic acids) with 2,3,4,6-tetra-*O*-acetyl-β-D-glucopyranosyl azide or *N*-(ω-azido-[*C*-2, *C*-6, and *C*-11]alkanoyl)-β-D-glucopyranosylamines. When screened against rabbit muscle glycogen phosphorylase (RMGP) compound **77** was found the best inhibitor (IC_{50} = 26 μM), however some other compounds exhibited inhibition in low micro molar range (IC_{50} = 40–70 μM) (Fig. 10.14).[107]

Furthermore, Xie et al. synthesized novel sugar nucleoside conjugates by adjoining uridine and *N*-acetylglucosamine or oleanolic acid derivatives using click reaction.These novel molecules displayed micro molar inhibition towards glycogen phosphorylase and compounds **78** and **79** were the most potent inhibitors of the series (Fig. 10.14).[108] Loganathan and co-workers identified simple glycoconjugates, 4-phenyl-*N*-(β-D-glucopyranosyl)-1*H*-1,2,3-triazole-1-acetamide **80** (K_i = 180 μM) and hydroxymethyl substituted β-D-glucopyranosyl triazole **81** (K_i = 14 μM) as potent leads to

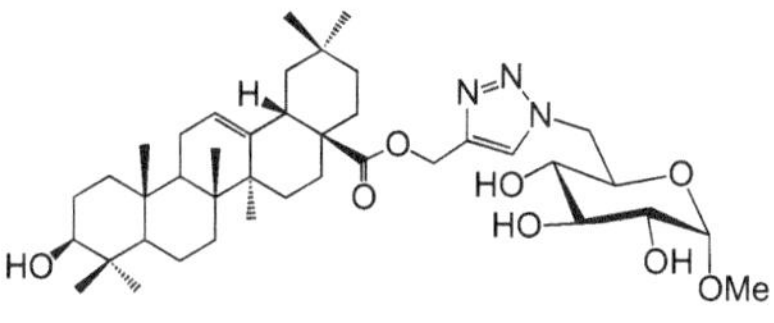

76: GPa inhibitor (IC_{50} = 1.14 μM)

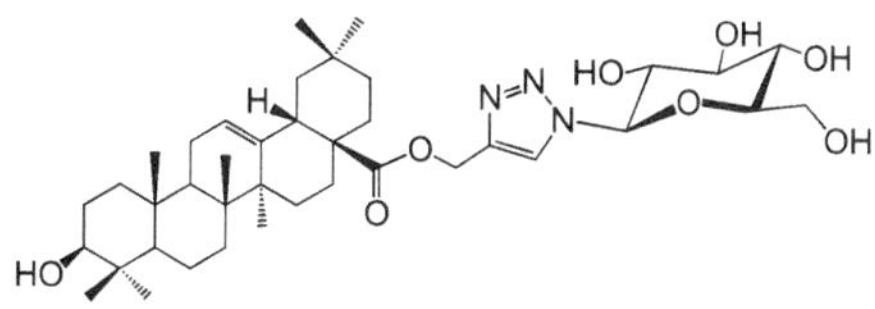

77: GPa inhibitor (IC_{50} = 26 μM)

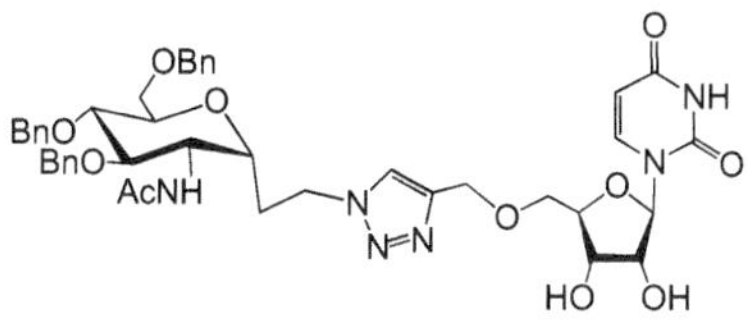

78: GPa inhibitor (IC_{50} = 13.6 μM)

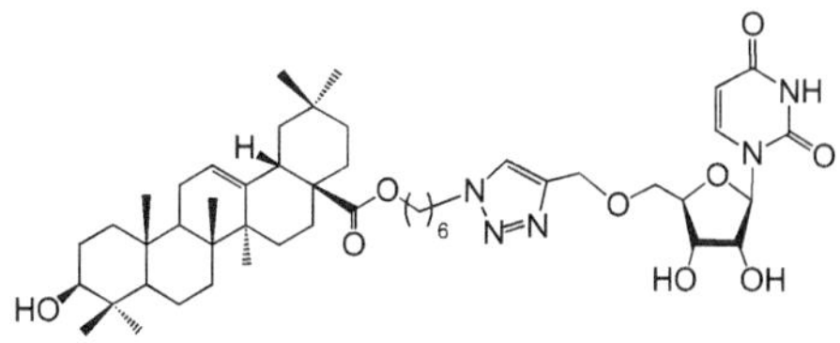

79: GPa inhibitor (IC_{50} = 15.5 μM)

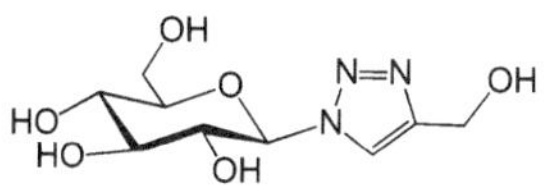

80: GPb lead (K_i = 14 μM)

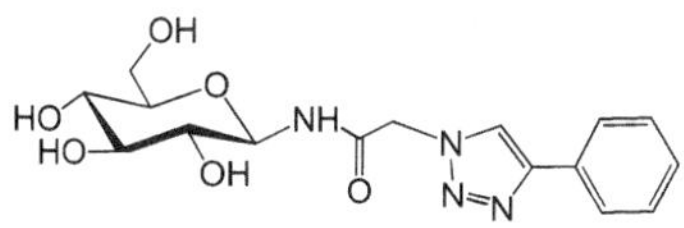

81: GPb lead (K_i = 180 μM)

Figure 10.14 ***Triazolyl glycoconjugate as glycogen phosphorylase inhibitors.***

design inhibitor against glycogen phosphorylase b by carrying out kinetic and crystallographic experiments (Fig. 10.14).[109]

2.1.7 Protein tyrosine phosphatases (PTPs) inhibitors

Protein tyrosine phosphatase (PTPs) belongs to a large family of enzymes that promotes dephosphorylation of phosphorylated tyrosine residues on proteins. These enzymes govern several physiological processes by playing a crucial role in controlling cell signalling pathways. PTP1B enzyme, a member of the family has been established as a major therapeutic target for many human diseases including cancer, diabetes,[110] obesity,[111] and neurodegenerative diseases.[112]

Click reaction was used by Xie *et al.* to develop dimeric acetylated and benzoylated β-*C*-D-glucosyl and β-*C*-D-galactosyl 1,4-dimethoxybenzenes or naphthalenes. These compounds were further oxidized into the corresponding β-*C*-D-glycosyl-1,4-benzoquinone and β-*C*-D-glycosyl-naphthaquinone derivatives by CAN oxidation. The *in vitro* inhibition study of the compounds revealed the dimeric benzoylated glycosyl 1,4-dimethoxybenzenes **82–83** and 1,4-benzoquinones **84–85** as potent inhibitors of PTP1B (IC_{50} = 0.62-0.88 μM), having almost similar activity for both gluco and galacto derivatives (Fig. 10.15).[113]

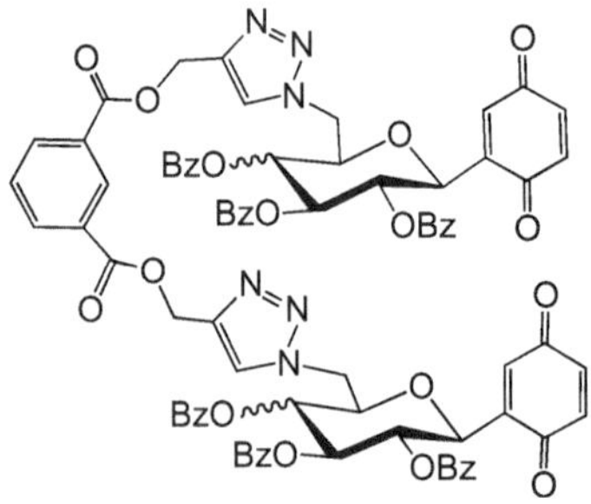

82: IC_{50} (μM) Gluco: 0.62
83: Galacto: 0.72

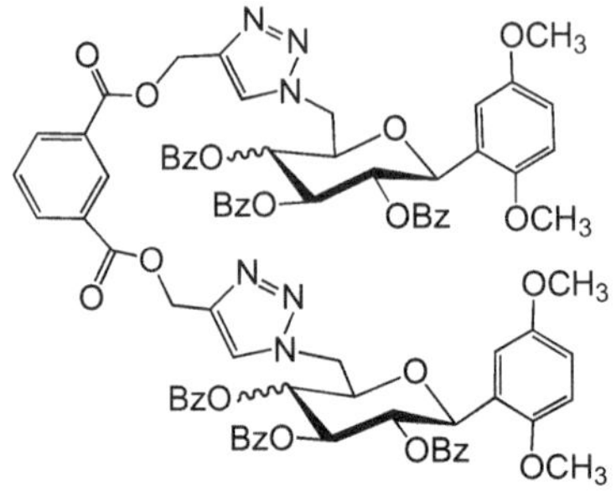

84: IC_{50} (μM) Gluco: 0.87
85: Galacto: 0.88

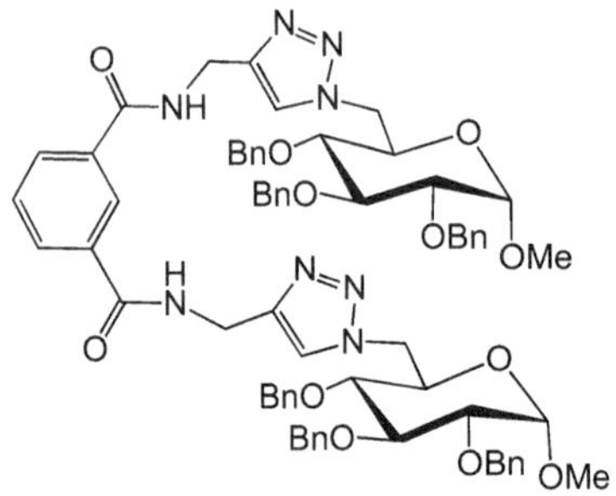

86: IC_{50} = 1.5 μM for PTP1B
= 15.5 μM for TCPTP

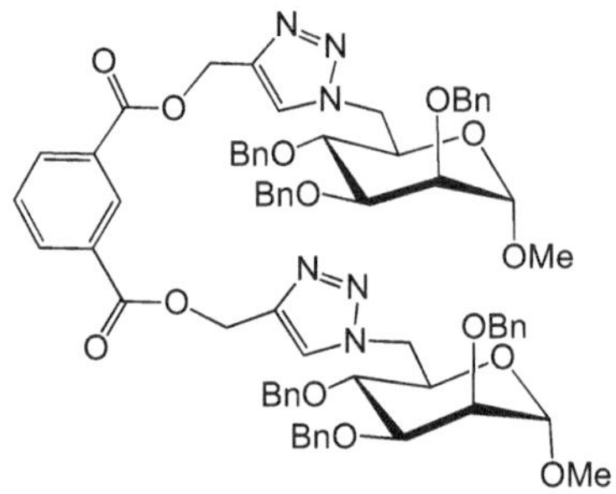

87: IC_{50} = 4.1 μM for PTP1B
= 5.2 μM for TCPTP

Figure 10.15 ***Developed triazolyl glycoconjugates as potential PTP1B inhibitor.***

Likewise, a library of monomeric and dimeric benzylated glycosyl benzenes was prepared by the click coupling of azide functionalized glucosyl, galactosyl and mannosyl sugars with mono- or di-alkyne armed ester and amide containing benzenes. Synthesized compounds were identified as PTP1B inhibitors having several fold selectivity over other homologous PTPase enzymes. The triazolyl bridged glycosyl dimers **86** and **87** exhibited inhibition activity against PTP1B (IC_{50} values of 1.5 ± 0.1 and 4.1 ± 0.03 μM, respectively) and a homologous PTP, TCPTP (T-cell protein tyrosine phosphatase) with IC_{50} values of 15.5 ± 0.1 and 5.2 ± 0.3 μM, respectively (Fig. 10.15).[114]

Owing to the structural similarity between salicylate and phosphotyrosine substrate (pTyr), Chen and co-workers envisioned development of salicylate glycosides based PTP inhibitors. Microwave assisted Cu(I)-click reaction was utilized for the synthesis of a library of salicylic acid based triazole-linked glycosides. The products showed limited inhibitory activities against PTP1B however, displayed promising selectivity over homologous PTPs, such as

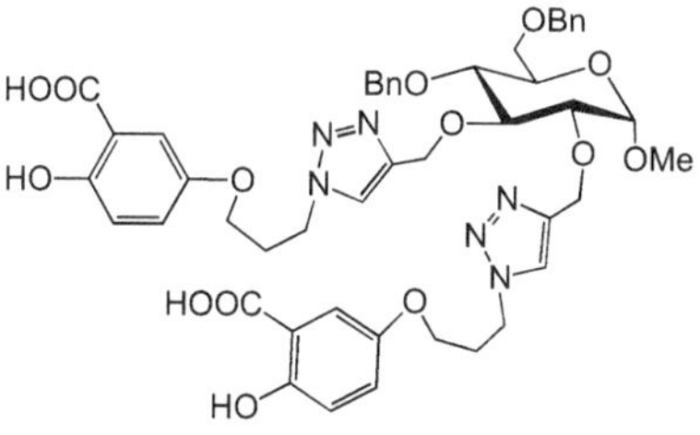

88: IC_{50} = 7.7 μM for PTP1B

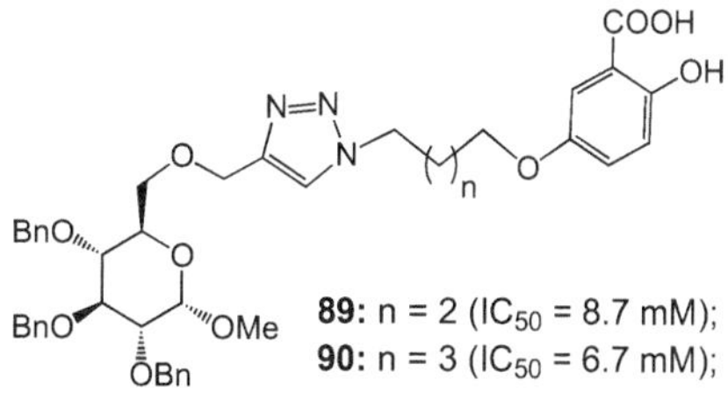

89: n = 2 (IC_{50} = 8.7 mM);
90: n = 3 (IC_{50} = 6.7 mM);

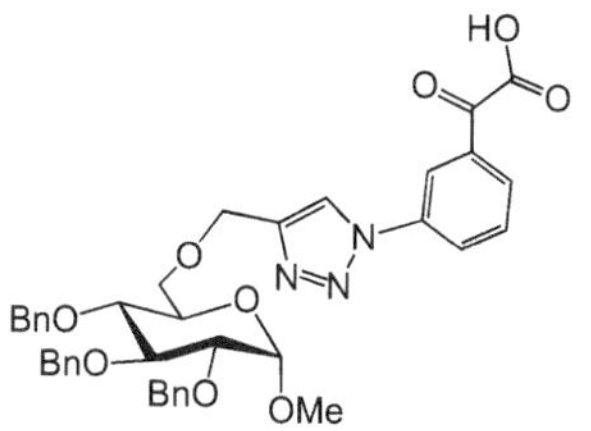

91: IC_{50} = 3.2 μM for PTP1B

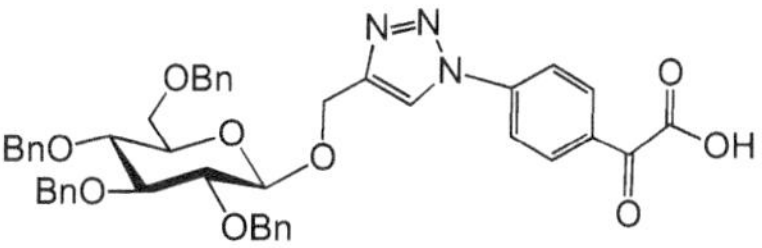

92: IC_{50} = 5.6 μM for PTP1B

Figure 10.16 ***Glycosyl-salicylic acid and a-ketocarboxylic acid linked triazoles as PTP1B inhibitor.***

TCPTP, SHP-1 (Src homology region 2 domain containing phosphatase-1), SHP-2 (Src homology region 2 containing a ubiquitously expressed tyrosine-specific protein phosphatase), LAR (leukocyte antigen-related PTP receptor).[115] Their group further extended the click methodology to access the bidentate salicylic glucosides by the coupling of azido salicylate with 2,6-, 3,4-, 4,6- and 2,3-di-*O*-propynyl-1-methoxy-*O*-glucosides. The inhibitory assay of the library established the 2,3-disubstituted salicylic 1-methoxy-*O*-glucoside **88** (IC_{50} 7.7 μM) as the structurally privileged PTP1B inhibitor among the bidentate compound series with micromole-ranged IC_{50} value and reasonable selectivity over other homologous PTPs tested (Fig. 10.16).[116]

In search of potent PTP1B inhibitors, Tang and co-workers prepared a library of benzyl 6-triazolo(hydroxy)benzoic glucosides by performing Cu(I) catalyzed 1,3 dipolar cycloaddition between azide bearing (hydroxyl)-benzoic acids with benzyl protected glucosides. The glycoconjugates bearing alkyl chain and (hydroxy)-benzoic derivatives (**89, 90**) were identified as low micro molar PTP1B inhibitors (IC_{50} = 8.7 ± 1.4 μM for **89** and IC_{50} = 6.7 ± 0.5 μM for **90**) with selectivity over T-Cell PTP (TCPTP), SH2-containing PTP-1 (SHP-1, SHP-2) and LAR (Fig. 10.16).[117]

Song et al. prepared a series of compounds by the glycoconjugation of a promising pTyr mimetic α-ketocarboxylic acids with benzylated sugars using Cu(I)-catalyzed azide–alkyne cycloaddition. Triazole-linked glycosyl α-ketocarboxylic acid derivatives **91** (IC_{50} = 3.2 μM) and **92** (IC_{50} = 5.6 μM) exhibited the maximum potency in biological assay against PTP1B with several fold selectivity over a group of homologous PTPs (Fig. 10.16).[118]

Chen and co-workers extended the microwave promoted click protocol in efficient generation of a series of triazole-linked serinyl, threoninyl, phenylalaninyl and tyrosinyl 1-*O*-gluco- or galactosides in high yields. Subsequent biological examination of pseudo-glycopeptides established them as moderate to good PTP1B and CDC25B (cell division cycle 25 homolog B) inhibitors with selectivity over TCPTP, SHP-1, SHP-2, LAR. Phenyl alanine and tyrosine based triazolyl glucosides **93** and **94** respectively, were found more potent inhibitors against PTP1B (with the IC_{50} = 5.1 μM; **93** and 8.3 μM; **94**) and CDC25B (with the IC_{50} = 9.9 μM; **93** and 3.9 μM; **94**) (Fig. 10.17).[119]

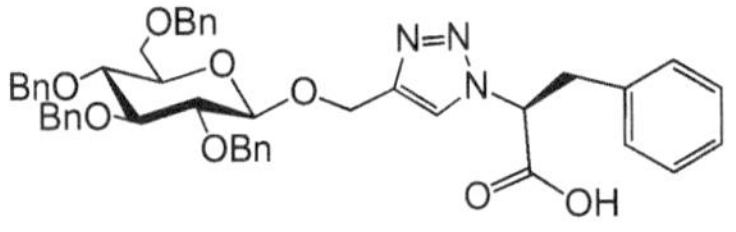

93: IC_{50} = 5.1 μM for PTP1B
= 9.9 μM for CDC25B

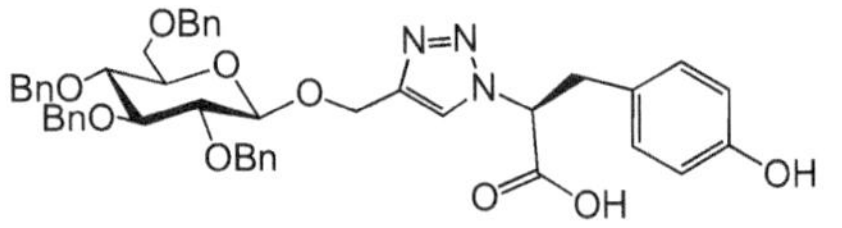

94: IC_{50} = 8.3 μM for PTP1B
= 3.9 μM for CDC25B

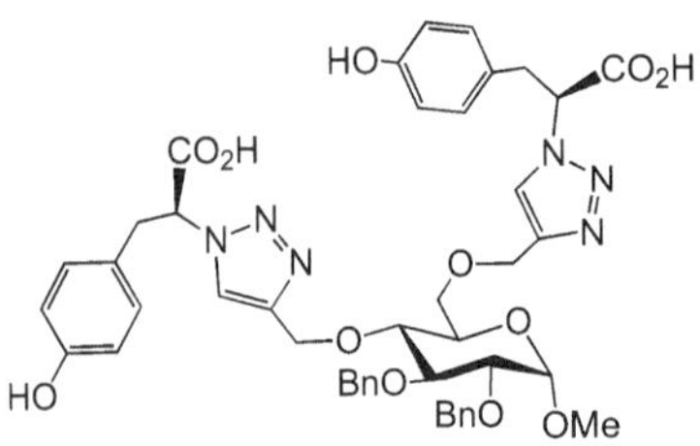

95: (IC_{50} = 10.8 μM); PTP 1B

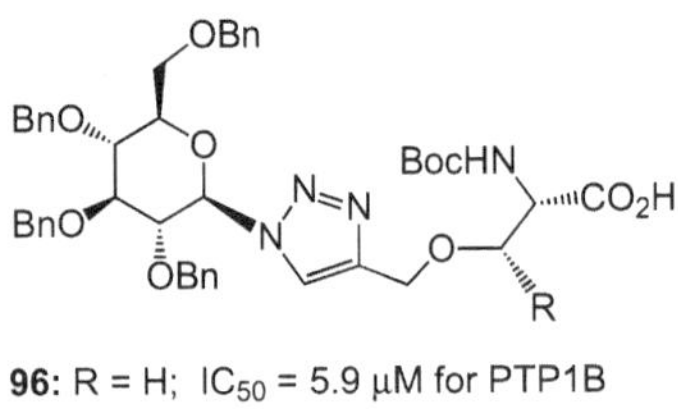

96: R = H; IC_{50} = 5.9 μM for PTP1B
97: R = Me IC_{50} = 7.1 μM for PTP1B

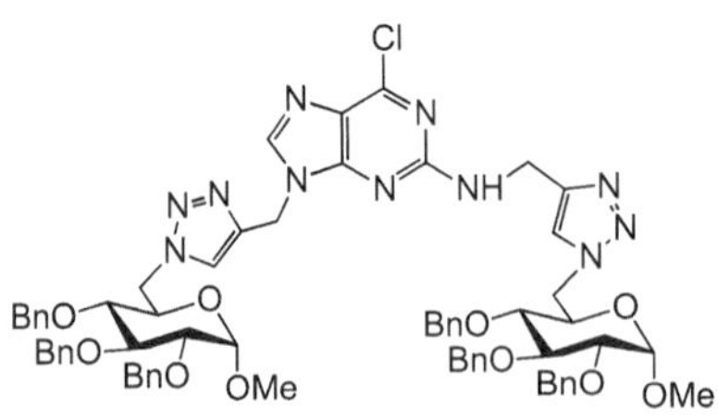

98: IC_{50} = 1.5 μM for PTP1B

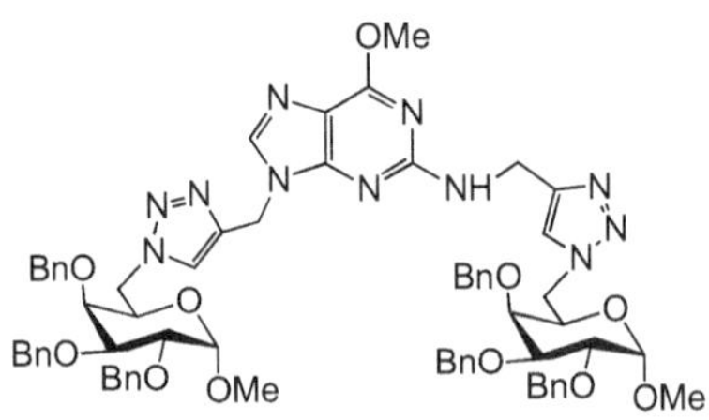

99: IC_{50} = 2.2 μM for PTP1B

Figure 10.17 *Triazole linked sugar-peptide and sugar-purine hybrids as PTP1B inhibitors.*

Their group further explored the potency of a series of mono- and *bis*-phenylalaninyl and tyrosinyl armed glucoside and galactoside derivatives towards PTP 1B inhibition. Compounds carrying one or two phenylalanine or tyrosine moieties on the 6-, 2,3-, 2,6-, 3,4-, and 4,6-positions of the glucoside template and 3,4-, position of the galactoside template were synthesized in good yields *via* MW-assisted click chemistry. In the biological study *C*3,4- and *C*2,6-disubstituted conjugates were found inactive, whereas *C*4,6-and *C*2,3- disubstituted derivatives displayed moderate inhibitory potencies (IC_{50} = 10-20 μM) on PTP1B. The *C*4,6-modified tyrosinyl glucoside **95** (IC_{50} = 10.8 μM) was found to be most potent inhibitor in the series, although was less effective than the mono-triazolyl amino acid glucosides (Fig. 10.17).[120–121]

In another effort, triazole functionalized glucosyl, galactosyl, and mannosyl based serine and threonine derivatives were efficiently synthesized by the click coupling of alkynyl amino acids onto diverse azido sugar templates. Among the developed triazolyl sugar–amino acid hybrids, **96** and **97** were identified as a potent and selective PTP1B inhibitor against a panel of homologous PTPs (IC_{50} = 5.9 ± 0.4 μM for **96** and IC_{50} = 7.1 ± 1.0 μM for **97**) (Fig. 10.17).[121]

A novel series of gluco- or galactopyranosyl-triazolyl and bis[gluco- or galactopyranosyl-triazolyl]-purine based PTP 1B inhibitors were reported by introducing gluco- or galactopyranosyl moieties to either *N*- or 9-position or both positions of the 6-*Cl* purine moiety *via* Cu(I)-catalyzed click reaction. Most of the synthesized purine derivatives showed good inhibition against PTP 1B with IC_{50} in low micromolar range (1.5-11.1 μM) and up to 5.6-fold selectivity over TC-PTP, the most homogenous enzyme of PTP1B, and at least more than tenfold selectivity over SHP1, SHP2, and LAR-PTP. Triazolyl-purine derivatives **98** and **99** exhibited the highest activity against PTP 1B with IC_{50} of 1.5 μM and 2.2 μM respectively) (Fig. 10.17).[122]

2.2 Pharmacological applications of click chemistry

2.2.1 Anti-cancer activities

Cancer is one of the major health problems and is estimated as the second leading cause of death in the world.[123] Several anticancer drugs including taxol, vinblastine, vincristine, camptothecin derivatives, opotecan and irinotecan, and etoposide developed from epipodophyllotoxin are in clinical use all over the world. Currently available medicines for cancer treatment suffer from many serious side effects for instance; nausea and vomiting, infection, hair loss, diarrhoea and in long term they may severely damage liver and kidney.[124] The development of drug candidates inspired from the natural products have been proven a useful strategy to obtain effective and better tolerated compounds. Over the past few years several groups have synthesized various derivatives of diverse natural products using click reaction.[125]

Cu(I) catalyzed click reaction was employed between C-28 acetylene functionalized triterpenes and variously substituted azides and thus obtained triazole derivatives of 3-acetylbetulin and betulone were evaluated *in vitro* for their anticancer activity against three cell lines amelanotic melanoma C-32, ductal carcinoma T47D and glioblastoma SNB-19. Betulone derivative **100** 28-[1-(1-deoxy-β-D-glucopyranosyl)-1*H*-1,2,3-triazol-4-yl]carbonylbetulone containing the 1-deoxy-β-D-glucopyranosyl substituted triazole ring exhibited anticancer activity comparable to positive control cisplatin against C-32 (IC_{50} = 0.76 ± 0.04 μM) and T47D cancer cells (IC_{50} = 32.86 ± 1.28 μM) (Fig. 10.18).[126]

Another biologically relevant triterpene, 18*β*-glycerrhetinic acid (GA) that is obtained from Licorice root, was derivatized with carbohydrate skeleton *via* click reaction. Per-*O*-acetylated glycosyl azide derivatives were clicked with propargyl ester of 18*β*-glycyrrhetinic acid (GA) to obtain a series of glycosyl triazole linked 18*β*-glycyrrhetinic acid (GA) derivatives. Evaluation of the series for their ability to block proliferation of human

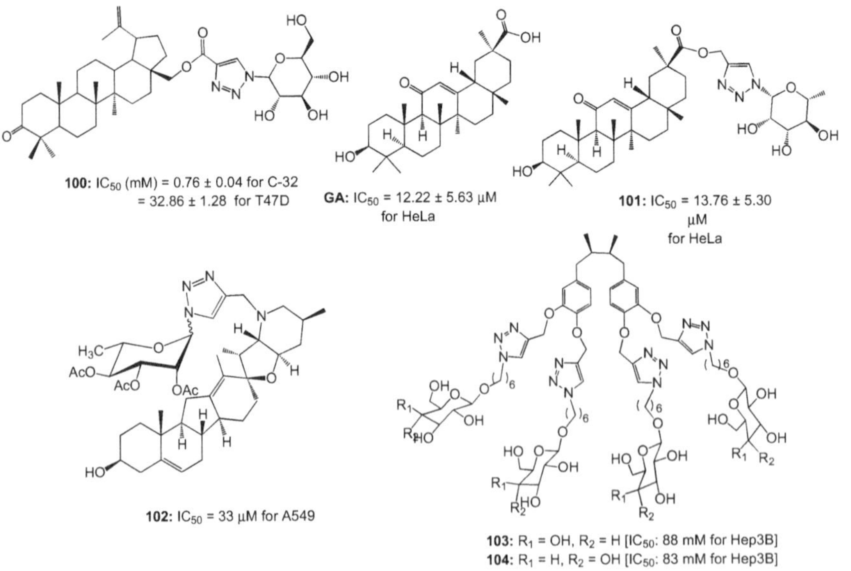

Figure 10.18 ***Triazole linked glycoconjugates as potential anti cancer agents.***

cervical cell line (HeLa) in comparison to the normal kidney epithelial cell line (NKE) disclosed GA and compound **101** (IC_{50}: 12.22 ± 5.63 μM & 13.76 ± 5.30 μM, respectively) as the most potent molecules (Fig. 10.18). Further, study of the ability to induce apoptosis of HeLa cells revealed **101** to possess higher apoptotic ability than GA and comparable to the control Resveratrol. The induction of apoptosis in cancer cells without affecting normal cells for compound **101** was found to be associated with its ability to effectively damage the mitochondria.[127]

The anticancer potential of naturally occurring cyclopamine alkaloid was explored by Chang et al. by synthesizing a library of carbohydrate-cyclopamine conjugates using click reaction. The assessment of the conjugates for their anticancer activity against lung cancer cell line (A549) identified the derivative **102** (IC_{50} = 33 μM) incorporated with L-rhamnose as more potent than the parent cyclopamine (IC_{50} = 49 μM) (Fig. 10.18).[128]

Nordihydroguaiaretic acid (NDGA), a naturally occurring lignan, well known for its potential towards antiviral and anti-tumor activity was glycosylated using Huisgen cycloaddition reaction. Thus, synthesized water soluble sugar–NDGA conjugates containing triazole-linkages possessed NDGA-(triazol-galactose) **103** (IC_{50} = 88 μM) and NDGA-(triazol-glucose) **104** (IC_{50} = 83 μM) were identified as inhibitors against human hepatocellular carcinoma Hep3B cells in culture (Fig. 10.18).[129]

Heat shock protein 90 (Hsp90) has been recognized as a promising target for the cancer therapy, as several proteins in tumor cells that are directly associated with all six hallmarks of cancer are dependent upon Hsp90 folding machinery. Novobiocin antibiotic, a modest inhibitor of Hsp90 was modified through the click reaction followed by a glycosylation reaction to synthesize novobiocin-sugar hybrids bearing a triazole ring in lieu of the amide moiety. The examination of anti-proliferative effects of the developed compounds against two breast cancer cell lines (SKBr-3 and MCF-7) manifested similar activity to that of amide counterparts. Biaryl and 3-indole substituted compounds **105** (IC_{50} = 13.16 μM, MCF-7 & IC_{50} = 21.22 μM, SKBr-3), and **106** (IC_{50} = 18.33 μM, MCF-7 & IC_{50} = 8.17 μM, SKBr-3) respectively, were determined to be the most active. In addition, western blot analyses verified that anti-proliferative action of the derivatives was resulted from Hsp90 inhibition (Fig. 10.19).[130]

Biologically relevant 8-hydroxyquinoline heterocycle was derivatized with various aromatic or protected sugars using CuAAC. The *in vitro* anti-proliferative activity of these quinoline triazolyl conjugates established protected sugar derivative **107** as the most active compound of the series

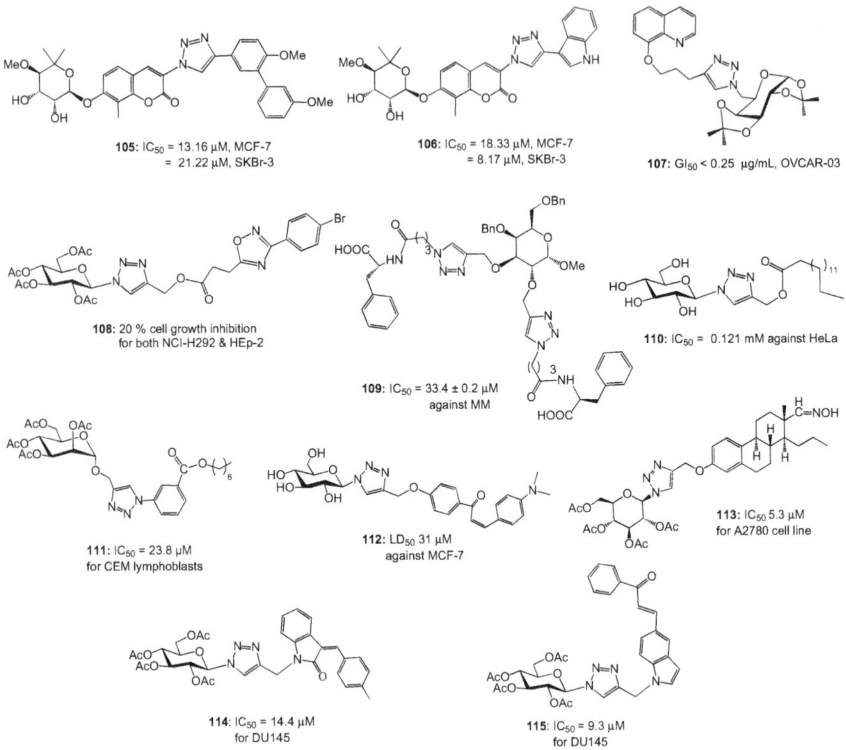

Figure 10.19 ***Potent anti-cancerous triazolyl glycohybrids.***

having high selectivity and antiproliferative activity towards ovarian cancer cells (OVCAR-03, GI_{50} < 0.25 μgmL^{-1}) better than the reference drug doxorubicin (OVCAR-03, GI_{50}: 0.43 μg mL^{-1}) (Fig. 10.19).[131]

Glycosyl-triazole linked 1,2,4-oxadiazoles synthesized by clicking 2,3,4,6-tetra-*O*-acetyl-β-D-glucopyranosyl azide with propynyl-3-[3-aryl)-1,2,4-oxadiazol-5-yl] propionates were identified as weak to moderately cytotoxic against two cell strains: NCI-H_{292} (lung carcinoma) and HEp-2 (larynx carcinoma). Most active compound **108** of the panel showed more than 20% cell growth inhibition against both cell lines (Fig. 10.19).[132]

Liu et al. explored the scope of click reaction in preparation of a series of 2,6-dichloropurine bicyclonucleoside containing a triazolyl-carbohydrates and demonstrated them as a significant inhibitors of human cancer cell lines EC109, PC-3, MGC-803, and HGC-7 through *in vitro* study.[133] A *bis*-triazolyl phenylalaninyl galactoside **109** constructed by dual-click reaction between an azido phenylalanine and a di-*O*-propynyl galactoside was identified as an anti-multiple myeloma (MM) lead compound. Interestingly,

109 exhibited selective cytotoxicity to MM cells among a panel of cancer cell lines and showed no toxicity towards the normal cells. A Western blot analysis suggested that **109** could potentiate the cleavage of poly ADP-ribose polymerase in MM cells, leading to apoptosis (Fig. 10.19).[134]

Chen and co-workers utilized a microwave induced 'click' chemistry to couple *O*-alkynyl fatty esters of varied chain length with 1-azido-tetra-*O*-benzyl-β-D-glucoside to afford a library of triazole-linked ester-type glycolipids. The cytotoxicity of the series was when tested on HeLa (cervix cancer) cell line indicated that compound **110** (IC_{50} = 0.121 mM) with an alkyl chain length of 15 carbon atoms exhibited optimum inhibition for the HeLa cells. An increase or decrease in linker length from 15 carbon atoms resulted in a gradual loss of cytotoxicity (Fig. 10.19).[135]

Polakova and their group studied the cytotoxicity of some D-mannose click conjugates with o-, m-, and *p*-substituted benzoic acid esters that were prepared by a click reaction. The *in vitro* cytotoxic study of the series activity against seven cancer cell lines including tumor and non-malignant fibroblast cells indicated the *N*-glycosyl conjugates obtained by the coupling of anomeric mannosyl azide with alkyne armed esters to be completely inactive. In contrast, *O*-glycosyl conjugates prepared by the conjugation of a propargyl mannoside with octyl esters showed increased cytotoxic activity for a wide spectrum of tumor and normal fibroblast cell lines. Also, their activity was strongly cell line dependent and was affected by the position of benzoic acid substitution as well as a length of its ester alkyl chain. The most potent compound was acetylated mannoside conjugated with octyl ester of m-substituted benzoic acid **111** (23.8 μM for CEM lymphoblasts). De-*O*-acetylation of mannose resulted in a complete loss of activity in compounds (Fig. 10.19).[136]

Recently, anti-cancer potential of novel glycosylated 1,4-substituted triazolylchalcone derivatives obtained by click coupling of D-glucosyl azides with a variety of propargylated chalcone were investigated by Misra et al. Evaluation of the series against human breast cancer cells (MDA-MB-468 and MCF-7 cells) and non-cancer cells (WI38) displayed compound **112** to be the most promising in terms of the cytotoxic activities against cancer cells and less cytotoxic against noncancer cells (WI38) (Fig. 10.19).[137] Kovács et al. synthesized fourteen new monosaccharide–D-secoestrone conjugates and studied their antiproliferative activities *in vitro* against diverse human adherent cancer cell lines (HeLa, A2780 and MCF-7) by MTT assays. The acetyl protected D-glucose-containing D-secoestrone oxime bioconjugate

113 was identified to be the most effective derivative with an IC_{50} 5.3 μM against A2780 cell line (Fig. 10.19).[138]

A novel series of triazole linked indole and oxindole glycoconjugates were evaluated for their *in vitro* anticancer activity against DU145 (prostate cancer), HeLa (cervical cancer), A549 (lung cancer) and MCF-7 (breast cancer) cell lines. In the sulforhodamine B (SRB) assay of the library, compounds **114** (indole derivative, IC_{50} = 14.4 μM) and **115** (oxindole derivative, IC_{50} = 9.3 μM) exhibited remarkably high cytotoxic activity against the DU145 cells, whereas showed no toxicity to the normal cell line NIH/3T3 (fibroblast) (Fig. 10.19). A detailed study on both the compounds further revealed that the compounds inhibit cancer cell proliferation by disrupting protein kinase B (Akt/PKB) pathway.[139]

2.2.2 Anti-leishmanial activity

Visceral Leishmaniasis (VL), also known as kala azar, threatens more than 350 million people worldwide and kills 60,000 annually. In Indian subcontinent,VL is caused by *Leishmania donovani*, a protozoan parasite transmitted by *Phlebotomus argentipes* (sand fly) in an anthroponotic cycle.[140] Despite many efforts in recent years, no effective drugs for any form of leishmaniasis has emerged and is still treated with old antimony-based drugs that are expensive and associated with significant side effects. Therefore, there is an urgent need to produce efficient well-tolerated and inexpensive drugs against this disease.[141]

With the aim to identify potent anti-leismania leads, our group glycoconjugated quercetin; a biologically potent flavanoid by employing click chemistry. *O*-benzylquercetin alkynes were prepared by the selective benzyl protection followed by propargylation. The 1,3-dipolar cycloaddition of *O*-benzylquercetin alkynes with diverse deoxy-*azido* sugars under click catalysis furnished triazole-linked *O*-benzylquercetin glycoconjugates. The *in vitro* screening of the developed glycoconjugates for anti-leishmanial activity against the promastigotes and amastigotes of *Leishmania donovani* indicated that the quercetin analogues in which triazolyl sugar entity was linked to 3,7,3́,4́-*O*-tetrabenzylquercetin skeleton at the *C*-5 position exhibited efficient activity in comparison to C-3́ linked triazolyl glycoconjugates and *bis*-triazolyl glycocojugates. Compound **116** and **117** displayed good activity against promastigotes and intramacrophase amastigote forms of *L. donovani* with the IC_{50} values of 7.76 and 6.08 μg/mL, respectively (Fig. 10.20).[142]

In continuation, our group further modified Noscapine, a naturally occurring biologically relevant opium alkaloid to design a library

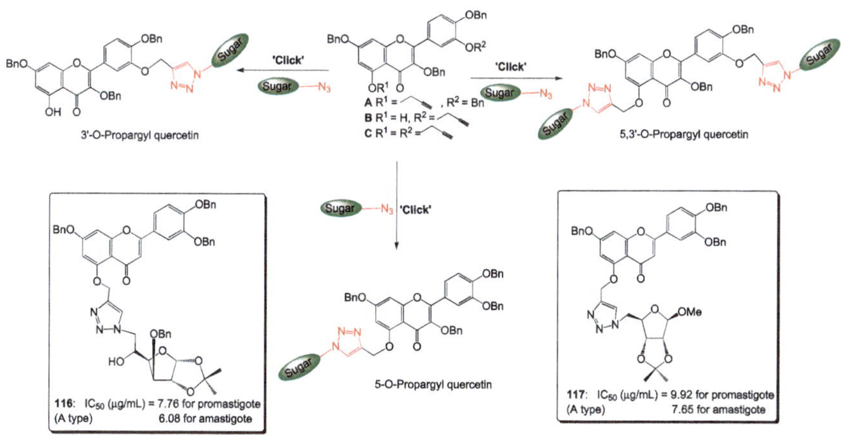

Figure 10.20 ***Synthesis of potent anti-leishmanial Triazolyl O-benzylquercetin glycoconjugates.***

of fourteen triazolyl noscapine glycoconjugates by the click coupling of 7-*O*-propargylated noscapine with various sugar azides. Four of the developed compounds of the library exhibited significant inhibition of parasite growth when screened for their anti-leishmanial properties against *L. donovani* promastigotes. None of these compounds exhibited cellular cytotoxicity against mouse macrophages. Noscapine glycoconjugates **118** and **119** were found most potent with IC_{50} value 8.36 and 15.9 μM respectively, comparable to standard drug miltefosine (IC_{50} = 10.4 μM) (Fig. 10.21).[143]

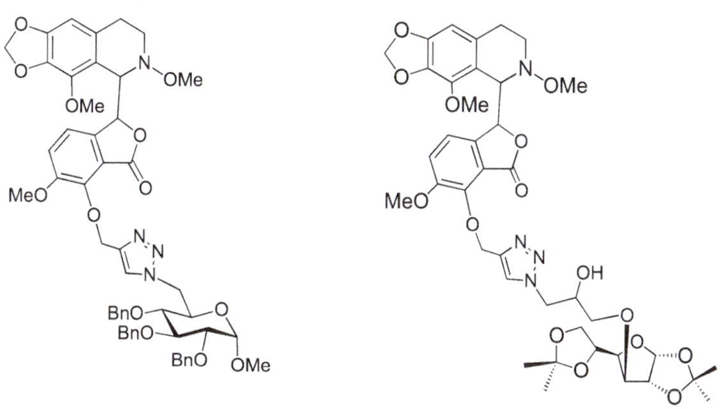

118: IC_{50} = 8.36 μM

119: IC_{50} = 15.9 μM

Figure 10.21 ***Triazole linked noscapine glycoconjugates as active anti-leishmanial agents.***

2.2.3 Anti-fungal and anti-bacterial activities

The emergence and reemergence of debilitating microbial and fungal infections and the growing resistance in microbes towards the available antibiotic treatments are the major concern that imposes the search for new antimicrobial and antifungal agents with different chemical structures and mechanism of action compared to conventional drugs.[144–145] Several biologically significant carbohydrate containing natural products and antibiotics such as Jadomycin, Neomycin-B, Kanamycin-A, Vancomycin etc. were derivatized by using facile click protocol to generate the natural product analogues with enhanced activity. Screening of the library against numerous bacterial strains showed only modest inhibition comparable to the parent skeletons in case of most of the sugar substituted derivatives; for example **120** and **121** (Fig. 10.22).[146–149]

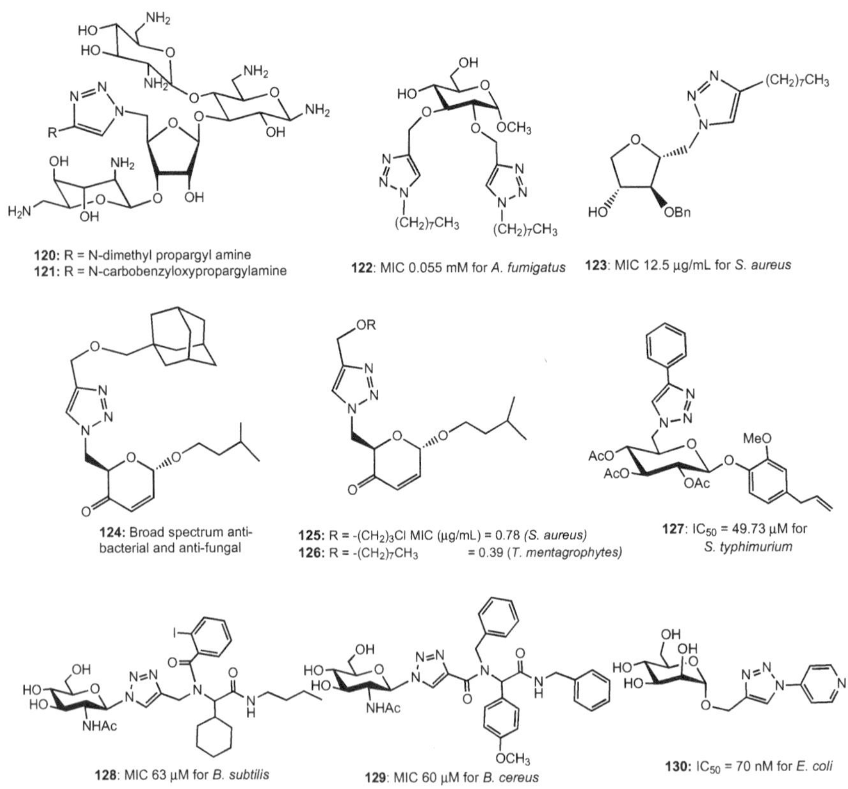

Figure 10.22 ***Click generated triazolyl glycoconjugates with potential anti-microbial and anti-fungal activity.***

A series of novel glucosyl *bis*-1,2,3-triazoles generated by the use of click reaction were studied for their *in vitro* antimicrobial and antifungal activity. Deprotected derivative **122** (0.055 mM) containing *n*-octyl chain was identified as the most potent compound of the series possessing fourteen times higher activity than the standard fluconazole against *A. fumigatus* (Fig. 10.22). Additionally, it showed good antibacterial activity against several tested bacterial strains with MIC values ranging between 0.049–0.39 mM.[150]

Click derived glycal-based novel tetrahydrofuran 1,2,3-triazoles were screened for their antimicrobial activity by Reddy et al. The most active compound **123** displayed antibacterial activity at an MIC value of 12.5 μgmL^{-1} against *S. aureus,* comparable to antibiotic gentamycin (MIC: 6.25 μgmL^{-1}) (Fig. 10.22).[151]

Shaw et al. synthesized a novel series of 2,3,6-trideoxy sugar triazole conjugates using click reaction and studied their antimicrobial and antifungal potential. The compounds showed moderate to excellent *in vitro* activities against gram positive, gram negative, resistant bacterial strains and fungi with MIC value in the range of 0.39-50 μg/mL without toxicity. Compound **124** bearing an admantyl group was found most promising with broad spectrum activity against both bacterial and fungal strains. However, compound **125** showed highest activity with MIC 0.78 μg/mL against gram positive (*S. aureus*) and gram negative (*K. pneumoniae*) bacteria and 3.12 μg/mL against methicillin- and vancomycin-resistant *S. aureus*. Compound **126** was the most potent antifungal lead with MIC 0.39 μg/mL for *Trichophyton mentagrophytes* (Fig. 10.22).[152]

The synthesis and study of 6-triazolo-6-deoxy eugenol glucosides for their antibacterial potential by de Souza et al. presented **127** as a lead compound of the series. Compound **127** bearing a phenyl group at the triazole ring exerted significant activity against the gram-negative bacteria *Salmonella typhimurium* (low IC_{50} = 49.73 μM) and low toxicity to the normal cells CC_{50} = 157.83 μM) in cytotoxicity assay (Fig. 10.22).[153]

Kuhn et al. designed triazolyl *N*-acetylglucosamine derivatives by the click coupling between 1-azido-*N*-acetylglucosamine and propionamides or *N*-propargyl diamides prepared by ugi reaction. Two compounds **128** and **129** of the series showed bacteriostatic activity against *Bacillus subtilis* with MIC values of approximately 60 μM in both cases. Their study is the first report of inhibitors of bacterial cell wall GlcNAcases that inhibited the bacterial cell growth (Fig. 10.22).[154] A library of mannosyl triazoles has been demonstrated as inhibitors of fimH, a lectin located at the bacterial

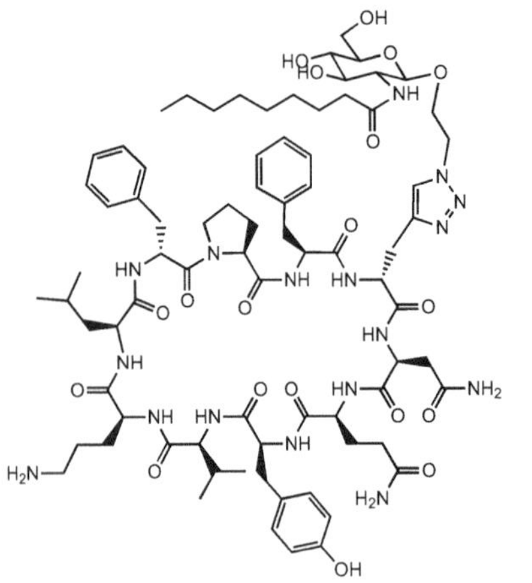

131

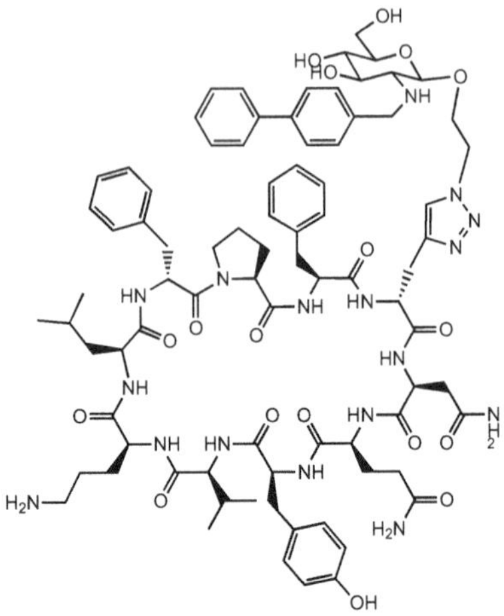

132

Figure 10.23 ***Tyrocidine based potent glycopeptide antibiotics Tyc4PG-14 and Tyc4PG-15.***

pili of the uropathogenic *E. coli.*, causative agent of urinary tract infection. The series showed nanomolar inhibition, where the most active 4-pyridiyl derivative **130** showed potency similar to the reference compound n-heptyl-α-D-mannoside.[155]

Tyrocidine (Tyc), a class of antibacterial cyclic decapeptides was modified by Walsh and Lin to afford their glycosylated analogues with the anticipation of better therapeutic index. A library of 247 carbohydrate-modified Tyc analogues was generated by CuAAC of mono-, di-, and tri-propargyl glycines with different 21 azido sugars. Screening of all the compounds for minimal inhibitory concentration (MIC) and minimal hemolytic concentration (MHC) index demonstrated the 6-times better therapeutic index of triazole-grafted glycopeptides (**131-132**) than natural tyrocidine (Tyc) (Fig. 10.23).[156,14] Novel sugar derivatives of another antibacetrial glycopeptides, ristocetin was synthesized by the click reaction and some of the derivatives exhibited dramatically improved antibacterial activity against glycopeptide-resistant enterococci. The study of antiviral activity for A/H1N1 subtype showed highest activity EC_{50} values in the range of 6–16 μM.[157]

Recently, our group have reported the click inspired synthesis of novel hexa- and octadeca-valent peripheral galactosylated dendrimers (**133**) and have studied their biological activities against multi-drug resistant bacterial isolates and human colorectal carcinoma cells (HCT116) (Fig. 10.24). The *in vitro* screening against gram positive and gram negative bacterial strains

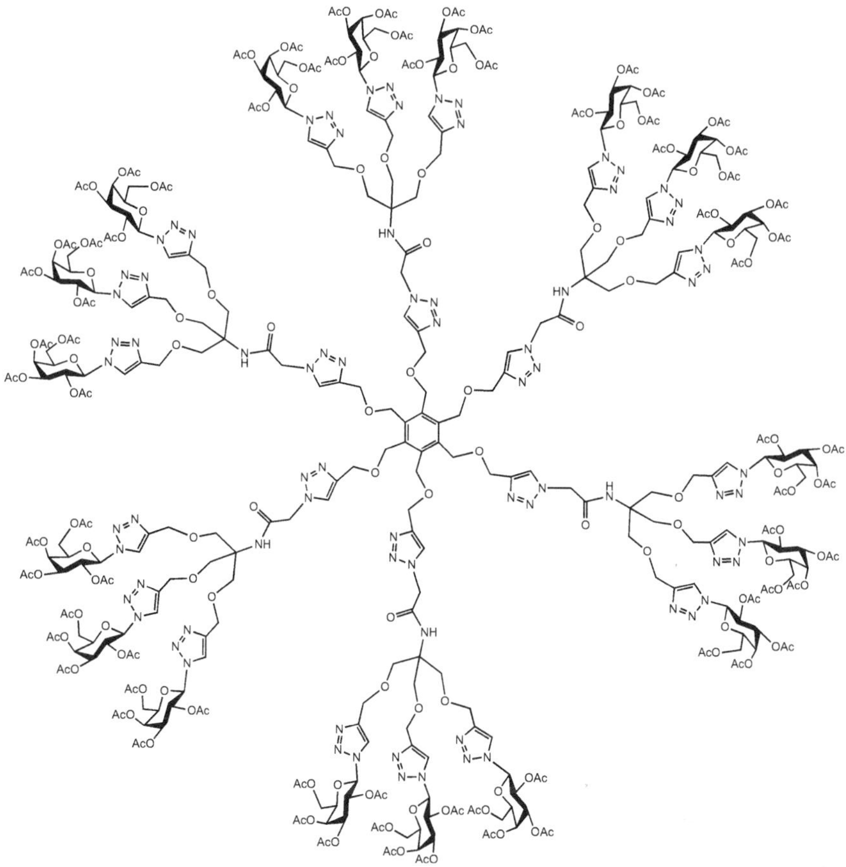

133

Figure 10.24 ***Octadecavalent glyco-dendrimers potential anti-bacterial, anti-biofilm, and anti-tumour agent.***

revealed them to be moderately active and showed a good bio film inhibition activity. Both the glycodendrimers showed significant anti proliferative activity for HCT116 cancer cell line.[158]

Recently, Benazza et al. reported the first glucocluster with efficient action against *P. aeruginosa* biofilm (a thick protective barrier in bacteria) formation. The calixarene based iron chelating glycoclusters comprised of tetrahydroxamic acid and sugar units were obtained through CuAAC. All the synthesized dendrimers (**134–136**) bearing α-L-fucose, α-D-manno- and α-D-gluco- showed strong biofilm inhibition 72%, 84% and 92% respectively, by interfering with the iron uptake process and by interacting with bacterial adhesins through glycan mimicry (Fig. 10.25).[159]

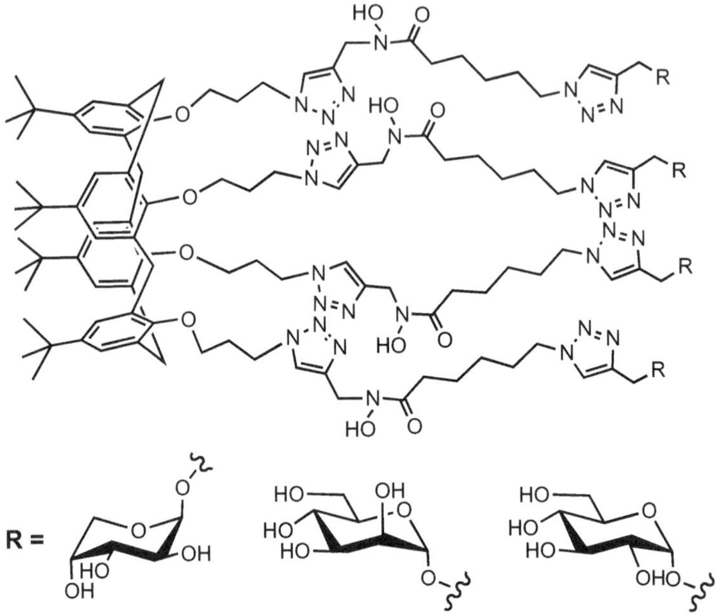

Figure 10.25 ***CuAAC derived tetrahydroxamic acid glycoclusters effective inhibitors of P. aeruginosa biofilms.***

A library of quinoline coupled 1,2,3-triazoles derived from the click precursors azidomethyl quinoline and different sugar alkynes were screened for anti-tubercular activity against *M. tuberculosis* H37Rv luciferase reporter phage (LRP) assay. Among the compounds, **137** (MIC 5 μg/mL) exhibited most promising inhibitory activity when compared to the positive control rifampicine (MIC 2 μg/mL) (Fig. 10.26).[160] Likewise, Tripathi et al. studied the antitubercular activity of 1,2,3-triazolyl glycoconjugates prepared by the coupling of 5-azido-5-deoxyxylo-, ribo-, and arabinofuranoses with different alkynes against the same strain. A mild antitubercular activity with an MIC of 12.5 μg/mL was observed for the most potent compound, *n*-pentyl substituted arbinofuranose derivative **138**; while other compounds possess MIC values >12.5 μg/mL (Fig. 10.26).[161]

In search of potential inhibitors of mycobacterial cell wall biosynthesis Fairbank's group developed a library of β-arabino glycosyl triazoles containing various hydrophobic chains as putative mimics of decaprenol phosphoarabinose (DPA). Biological assays against *M. bovis BCG* displayed weak to moderate yet selective anti-mycobacterial activity in compounds.

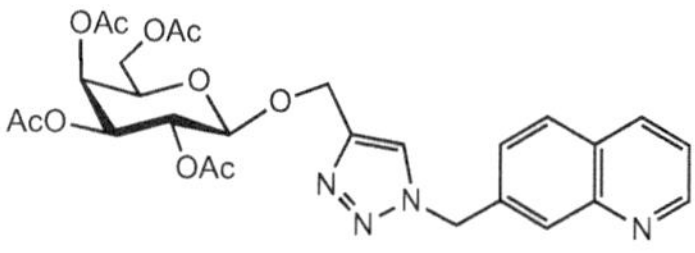

137: MIC 5 µg/mL for *M. tuberculosis H37Rv*

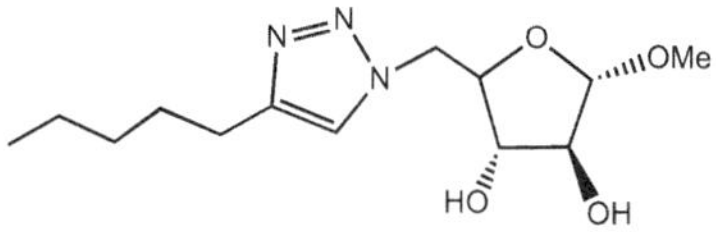

138: MIC 12.5 µg/mL for *M. tuberculosis H37Rv*

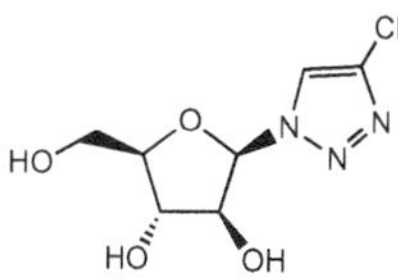

139: MIC 31 µg/mL for *M. bovis BCG*

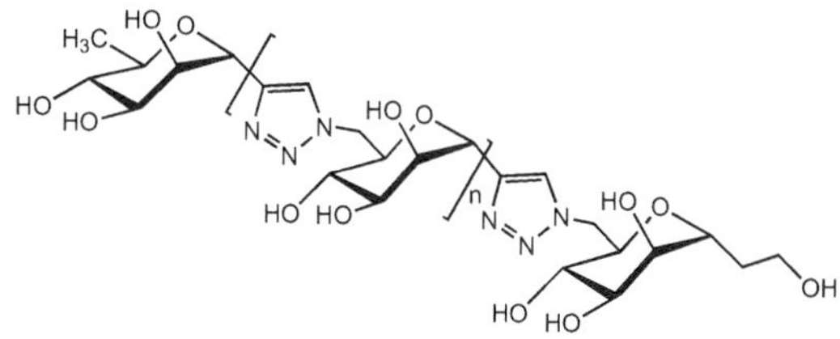

140: n = 6 IC_{50} 0.14 mM
141: n = 8 IC_{50} 0.22 mM
Mycobacterial α-(1,6)-mannosyltransferases inhibitors

Figure 10.26 ***Triazolyl glycoconjugates as anti-tubercular agent.***

Triazole glycoconjugate comprising of a tetra decyl side chain **139** was found to be the most active compound of the series (Fig. 10.26).[162] The mycobacterial α-(1,6)-mannosyltransferase enzyme that actively participates in the biosynthesis of the cell envelope of *M. tuberculosis* was targeted by Dondoni and co-workers to develop the inhibitors with the anticipation to combat tuberculosis. They utilized azide-alkyne cycloaddition to assemble triazole bridged (1,6)-oligomannosides upto hexadecamer. Suitable bioassays with oligomers (dimer, tetramer, hexamer, octamer, decamer, and hexadecamer) showed variable inhibiton against mycobacterial α-(1,6)-mannosyltransferases, out of which hexamannoside and octamannoside (**140–141**) were reported to have maximum activity (IC_{50} = 0.14-0.22 mM) (Fig. 10.26).[163]

3 Conclusion and future perspectives

In this chapter, we have explored the recent advances made in the area of carbohydrate based therapeutic development by the use of click reaction. Cu(I)-catalyzed 1,3 dipolar cycloaddition of terminal alkynes and azides (CuAAC) has been established as a versatile coupling technique that has efficiently delivered diverse glycoconjugates with enormous biological and pharmaceutical relevance. The compatibility of click reaction to carbohydrate chemistry and biological systems has speeded up the process of lead identification and drug discovery. It has allowed an easy access to the simple to complex sugar-derivatives and on the other hand facilitated their ready *in vitro* and *in vivo* evaluations.

The product of click reaction, a 1,4-disubstituted 1,2,3-triazole ring serves as a suitable linker between sugars and the functional molecule, because of its stability and potent pharmacological properties. A large number of triazolyl glycoconjugates derived from CuAAC have been identified as a selective enzyme inhibitor and also have shown wide range of bioactivities, including anticancer, anti-bacterial, antifungal, anti-viral, and anti-leishmanial, and immunostimulating effects.[15–16] In addition, many polysaccharides and multivalent glycoligands such as glycopeptides, glyco-oligonucleotides, glyco-arrays, glyco-macrocyles, glyco-clusters, glyco-dendrimers, and glycopolymers have been conveniently constructed by this reaction which finds its application in sensing materials for serum diagnosis and drug delivery etc.[13–14] The click reaction has also proven to be an ideal reaction for the bioconjugation of oligonucleotides, polypeptides, polysaccharides, viruses and bacteria to various substances.

Despite of having several virtues in Cu(I) catalyzed click reaction, this nearly perfect appearing coupling strategy faces one serious drawback. The toxicity of copper in cells restricts its direct application in biological system therefore; *in vivo* imaging studies can be done by using an alternative copper free strain promoted azide-alkyne coupling.

Acknowledgments

DK is grateful to University grants commission (UGC: SRG M-14-57), New Delhi and Banaras Hindu University for the funding and support. *VKT* thankfully acknowledge Council of Scientific & Industrial Research (CSIR), New Delhi for the funding (Grant No.: P/25/370).

References

1. Tornoe, C.W.; Christensen, C.; Meldal, M. *J. Org. Chem.* **2002**, *67*, 3057–3062.
2. Rostovtsev, V.V.; Green, L. G.; Fokin, V.V.; Sharpless, K. B. *Angew. Chem., Int. Ed.* **2002**, *41*, 2596.
3. Meldal, M.; Tornoe, C.W. *Chem. Rev.* **2008**, *108*, 2952–3015.
4. Hanni, K. D.; Leigh, D. A. *Chem. Soc. Rev* **2010**, *39*, 1240–1251.
5. Hua, Y.; Flood, A. H. *Chem. Soc. Rev* **2010**, *39*, 1262–1271.
6. Golas, P. L.; Matyjaszewski, K. *Chem. Soc. Rev* **2010**, *39*, 1338–1354.
7. Thirumurugan, P.; Matosiuk, D.; Jozwiak, K. *Chem. Rev.* **2013**, *113*, 4905–4979.
8. Wang, Q.; Chan, T. R.; Hilgraf, R.; Fokin, V.V.; Sharpless, K. B.; Finn, M. G. *J. Am. Chem. Soc.* **2003**, *125*, 3192.
9. Varki, A. *Glycobiology* **1993**, *3*, 97–130.
10. Schmidt, R. R. *Angew. Chem. Int. Ed.* **1986**, *25*, 212–235.
11. Schmidt, R. R.; Zhu, Z. *Angew. Chem. Int. Ed.* **2009**, *48*, 1900–1934.

12. Yu, H.; Cheng, J.; Ding, L.; Khedri, Z.; Chen, Y.; Chin, S.; Lau, K.; Tiwari, V. K.; Chen, X. *J. Am. Chem. Soc* **2009**, *131*, 18467–18477.
13. Tiwari, V. K.; Mishra, B. B.; Mishra, K. B.; Mishra, N.; Singh, A. S.; Chen, X. *Chem. Rev.* **2016**, *116*, 3086–3240.
14. Kushwaha, D.; Dwivedi, P.; Kuanar, S. K.; Tiwari, V. K. *Curr. Org. Synth.* **2013**, *10*, 90–135.
15. Witczak, Z. J.; Bielski, R. *Click Chemistry in Glycoscience: New Development and Strategies*. John Wiley & Sons: New York, 2013.
16. Tripathi, R. P.; Dwivedi, P.; Sharma, A.; Kushwaha, D.; Tiwari, V. K. Triazolyl glycoconjugates in medicinal chemistry. In *Click-Chemistry in glycoscience*; Witczak, Z. J., Bielski, R., Eds.; John Wiley & Sons, Inc: Hobken, New Jersy, 2013; pp. 293–324.
17. Bertozzi, C. R.; Kiessling, L. L. *Science* **2001**, *291*, 2357–2364.
18. Paulson, J. C.; Blixt, O.; Collins, B. E. *Nat Chem Biol* **2006**, *2*, 238–248.
19. Toshima, K.; Tatsuta, K. *Chem. Rev.* **1993**, *93*, 1503–1531.
20. Glycoscience, Chemistry and Chemical Biology, Eds, Fraser-Reid, Tatsuta, and Thiem, Springer, 2001.
21. Pardridge, W. M. *Adv. Drug Delivery Rev* **1995**, *15*, 5–36.
22. Wong, C. H. *Carbohydrate-based Drug Discovery*. Wiley-VCH: New York, 2003.
23. Mamidyala, S. K.; Finn, M. G. *Chem. Soc. Rev* **2010**, *39*, 1252–1261.
24. Holub, J. M.; Kirshenbaum, K. *Chem. Soc. Rev* **2010**, *39*, 1325–1337.
25. Amblard, F.; Cho, J. H.; Schinazi, R. F. *Chem. Rev.* **2009**, *109*, 4207–4220.
26. Link, A. J.; Tirrell, D. A. *J. Am. Chem. Soc.* **2003**, *125*, 11164.
27. Culp, E.; Wright, G. D. *J. Antibiotics* **2017**, *70*, 366–377.
28. Ramakrishnan, J.; Rathore, S. S.; Raman, T. *RSC Adv.* **2016**, *6*, 42387–42401.
29. Griffith, D.; Parker, J. P.; Marmion, C. J. *Anticancer Agents Med. Chem* **2010**, *10*, 354–370.
30. Wang, C. C. *J. Med. Chem.* **1984**, *27*, 1–9.
31. Supuran, C. T.; Scozzafava, A.; Casini, A. Carbonic anhydrase inibitors. *Med. Res. Rev.* **2003**, *23*, 146–189.
32. Scozzafava, A.; Owa, T.; Mastrolorenzo, A.; Supuran, C. T. *Curr. Med. Chem.* **2003**, *10*, 925–953.
33. Pastorekova, S.; Casini, A.; Scozzafava, A.; Vullo, D.; Pastorek, J.; Supuran, C. T. *Bioorg. Med. Chem. Lett.* **2004**, *14*, 869–873.
34. Wilkinson, B. L.; Bornaghi, L. F.; Houston, T. A.; Innocenti, A.; Supuran, C. T.; Poulsen, S. A. *J. Med. Chem.* **2006**, *49*, 6539–6548.
35. Wilkinson, B. L.; Bornaghi, L. F.; Houston, T. A.; Innocenti, A.; Vullo, D.; Supuran, C. T.; Poulsen, S. A. *Bioorg. Med. Chem. Lett.* **2007**, *17*, 987–992.
36. Wilkinson, B. L.; Bornaghi, L. F.; Houston, T. A.; Innocenti, A.; Supuran, C. T.; Poulsen, S. A. *J. Med. Chem.* **2006**, *49*, 6539–6548.
37. Wilkinson, B. L.; Bornaghi, L. F.; Houston, T. A.; Innocenti, A.; Vullo, D.; Supuran, C. T.; Poulsen, S. A. *J. Med. Chem.* **2007**, *50*, 1651–1657.
38. Wilkinson, B. L.; Innocenti, A.; Vullo, D.; Supuran, C. T.; Poulsen, S. A. *J. Med. Chem.* **2008**, *51*, 1945–1953.
39. Singer, M.; Lopez, M.; Bornaghi, L. F.; Innocenti, A.; Vullo, D.; Supuran, C. T.; Poulsen, S. A. *Bioorg. Med. Chem. Lett.* **2009**, *19*, 2273–2276.
40. Morris, J. C.; Chiche, J.; Grellier, C.; Lopez, M.; Bornaghi, L. F.; Maresca, A.; Supuran, C. T.; Pouysségur, J.; Poulsen, S. A. *J. Med. Chem.* **2011**, *54*, 6905–6918.
41. Mahon, B. P.; Hendon, A. M.; Driscoll, J. M.; Rankin, G. M.; Poulsen, S. A.; Supuran, C. T.; McKenna, R. *Bioorg. Med. Chem.* **2015**, *23*, 849–854.
42. Tanpure, R. P.; Ren, B.; Peat, T. S.; Bornaghi, L. F.; Vullo, D.; Supuran, C. T.; Poulsen, S. A. *J. Med. Chem.* **2015**, *58*, 1494–1501.
43. Countinho, P. M.; Deleury, E.; Davies, G. J.; Henrissat, B. *J. Mol. Biol.* **2003**, *328*, 307–317.

44. Davies, G. J.; Henrissat, B. *Structure* **1995**, *3*, 853–859.
45. Lee, L.V.; Mitchell, M. L.; Huang, S. J.; Fokin, V.V.; Sharpless, K. B.; Wong, C. H. *J. Am. Chem. Soc.* **2003**, *125*, 9588–9589.
46. Hosoguchi, K.; Maeda, T.; Furukawa, J. I.; Shinohara, Y.; Hinou, H.; Sekiguchi, M.; Togame, H.; Takemoto, H.; Kondo, H.; Nishimura, S. I. *J. Am. Chem. Soc.* **2010**, *53*, 5607–5619.
47. Kumar, R.; Nasi, R.; Bhasin, M.; Khieu, N. H.; Hsieh, M.; Gilbert, M.; Jarrell, H.; Zou, W.; Jennings, H. J. *Carb. Res.* **2013**, *378*, 45–55.
48. Li, T.; Guo, L.; Zhang, Y.; Wang, J.; Li, Z.; Lin, L.; Zhang, Z.; Li, L.; Lin, J.; Zhao, W.; Li, J. *Carb. Res.* **2011**, *346* (9), 1083–1092.
49. Giorgi, M. E.; de Lederkremer, R. M. *Carbohydr. Res.* **2011**, *346*, 1389–1393.
50. Dc-Rubin, S. S.; Schenkman, S. *Cell. Microbiol.* **2012**, *14*, 1522–1530.
51. Neres, J.; Bryce, R. A.; Douglas, K. T. *Drug. Discov. Today* **2008**, *13*, 110–117.
52. Miller, B. R.; Roitberg, A. E. *Future Med. Chem.* **2013**, *5*, 1889–1900.
53. Carvalho, I.; Andrade, P.; Campo, V. L.; Guedes, P. M.; Sesti-Costa, R.; Silva, J. S.; Schenkman, S.; Dedola, S.; Hill, L.; Rejzek, M.; Nepogodiev, S. A. *Bioorg. Med. Chem.* **2010**, *18*, 2412–2427.
54. Campo, V. L.; Sesti-Costa, R.; Carneiro, Z. A.; Silva, J. S.; Schenkman, S.; Carvalho, I. *Bioorg. Med. Chem.* **2012**, *20*, 145–156.
55. Campo, V. L.; Carvalho, I.; Da Silva, C. H.; Schenkman, S.; Hill, L.; Nepogodiev, S. A.; Field, R. A. *Chem. Sci.* **2010**, *1*, 507–514.
56. Campo, V. L.; Ivanova, I. M.; Carvalho, I.; Lopes, C. D.; Carneiro, Z. A.; Saalbach, G.; Schenkman, S.; da Silva, J. S.; Nepogodiev, S. A.; Field, R. A. *Tetrahedron* **2015**, *71*, 7344–7353.
57. Marchiori, M. F.; Riul, T. B.; Bortot, L. O.; Andrade, P.; Junqueira, G. G.; Foca, G.; Doti, N.; Ruvo, M.; Dias-Baruffi, M.; Carvalho, I.; Campo, V. L. *Bioorg. Med. Chem.* **2017**, *25*, 6049–6059.
58. Junqueira, G. G.; Carvalho, M. R.; Andrade, P. D.; Lopes, C. D.; Carneiro, Z. A.; Sesti-Costa, R.; Silva, J. S.; Carvalho, I. *J. Braz. Chem. Soc.* **2014**, *25*, 1872–1884.
59. Davies, G. J.; Gloster, T. M.; Henrissat, B. *Curr. Opin. Struct. Biol.* **2005**, *15*, 637–645.
60. Goke, B.; Herrmann-Rinke, C. *Diabetes Metab. Rev.* **1998**, *14*, S31–S38.
61. Mehta, A.; Zitzmann, N.; Rudd, P. M.; Block, T. M.; Dwek, R. A. *FEBS Lett.* **1998**, *430*, 17–22.
62. Papandreou, M. -J.; Barbouche, R.; Guieu, R.; Kieny, M. P.; Fenouillet, E. *Mol. Pharmacol.* **2002**, *61*, 186–193.
63. Sawkar, A. R.; Cheng, W. -C.; Beutler, E.; Wong, C. -H.; Balch, W. E.; Kelly, J. W. *Proc. Natl. Acad. Sci. U. S. A.* **2002**, *99*, 15428–15433.
64. Fan, J. -Q.; Ishii, S.; Asano, N.; Suzuki, Y. *Nat. Med.* **1999**, *5*, 112–115.
65. Périon, R.; Ferrières, V.; Garcia-Moreno, M. I.; Mellet, C. O.; Duval, R.; Fernández, J. M. G.; Plusquellec, D. *Tetrahedron* **2005**, *61*, 9118–9128.
66. Rossi, L. L.; Basu, A. *Bioorg. Med. Chem. Lett.* **2005**, *15*, 3596–3599.
67. Ferreira, S. B.; Sodero, A. C.; Cardoso, M. F.; Lima, E. S.; Kaiser, C. R.; Silva, F. P., Jr.; Ferreira, V. F. *J. Med. Chem.* **2010**, *53*, 2364–2375.
68. Dedola, S.; Hughes, D. L.; Nepogodiev, S. A.; Rejzek, M.; Field, R. A. *Carb. Res.* **2010**, *345* (9), 1123–1134.
69. Anand, N.; Jaiswal, N.; Pandey, S. K.; Srivastava, A. K.; Tripathi, R. P. *Carb. Res.* **2011**, *346*, 16–25.
70. Nash, R. J.; Kato, A.; Yu, C. Y.; Fleet, G. W. *Future Med Chem* **2011**, *3*, 1513–1521.
71. Chang, J.; Schul, W.; Yip, A.; Xu, X.; Guo, J. T.; Block, T. M. *Antiviral Res.* **2011**, *92*, 369.
72. Norez, C.; Noel, S.; Wilke, M.; Bijvelds, M.; Jorna, H.; Melin, P.; DeJonge, H.; Becq, F. *FEBS Lett.* **2006**, *580*, 2081.
73. Compain, P.; Martin, O. R. *Iminosugars. From Synthesis to Therapeutic Applications*. John Wiley & Sons Ltd: Chichester, 2007.

74. (a) Butters, T. D.; Dwek, R. A.; Platt, F. M. *Chem. Rev.* **2000**, *100*, 4683; (b) Howe, J. D.; Smith, N.; Lee, M. R.; Ardes-Guisot, N.; Vauzeilles, B.; Désiré, J.; Baron, A.; Blériot, Y.; Sollogoub, M.; Alonzi, D. S.; Butters, T. D. *Bioorg. Med. Chem.* **2013**, *21*, 4831-4838.
75. Diot, J.; Garcia-Moreno, M. I.; Gouin, S. G.; Mellet, C. O.; Haupt, K.; Kovensky, J. *Org. Biomol. Chem.* **2009**, 7, 357–363.
76. Zamoner, L. O. B.; Aragão-Leoneti, V.; Mantoani, S. P.; Rugen, M. D.; Nepogodiev, S. A.; Field, R. A.; Carvalho, I. *Carb. Res.* **2016**, *429*, 29–37.
77. Butters, T. D. *Curr. Opin. Chem. Biol.* **2007**, *11*, 412–418.
78. Dvir, H.; Harel, M.; McCarthy, A. A.; Toker, L.; Silman, I.; Futerman, A. H.; Sussman, J. L. *EMBO Rep.* **2003**, *4*, 704–709.
79. Barton, N. W.; Brady, R. O.; Dambrosia, J. M.; Di Bisceglie, A. M.; Doppelt, S. H.; Hill, S. C.; Mankin, H. J.; Murray, G. J.; Parker, R. I.; Argoff, C. E.; Grewal, R. P.; Yu, K. -T. *New Engl. J. Med.* **1991**, *324*, 1464–1470.
80. Cox, T.; Lachmann, R.; Hollak, C.; Aerts, J.; van Weely, S.; Hrebicek, M.; Platt, F.; Butters, T.; Dwek, R.; Moyses, C.; Gow, I.; Elstein, D.; Zimran, A. *Lancet* **2000**, *355*, 1481–1485.
81. (a) Ardes-Guisot, N.; Alonzi, D. S.; Reinkensmeier, G.; Butters, T. D.; Norez, C.; Becq, F.; Shimada, Y.; Nakagawa, S.; Kato, A.; Blériot, Y.; Sollogoub, M. *Org. Biomol. Chem.* **2011**, *9*, 5373–5388; (b) Diot, J. D.; Moreno, I. G.; Twigg, G.; Mellet, C. O.; Haupt, K.; Butters, T. D.; Kovensky, J.; Gouin, S. G. *J. Org. Chem.* **2011**, *76*, 7757-7768.
82. Díaz, L.; Bujons, J.; Casas, J.; Llebaria, A.; Delgado, A. *J. Med. Chem.* **2010**, *53*, 5248–5255.
83. Díaz, L.; Casas, J.; Bujons, J.; Llebaria, A.; Delgado, A. *J. Med. Chem.* **2011**, *54*, 2069–2079.
84. Colman, P. M. *Protein Sci* **1994**, *3*, 1687–1696.
85. Fields, B.N. in Fields Virology: Vol 1, Lippincott 1995, 1504.
86. Meindl, P.; Tuppy, H. *Monatsh. Chem* **1969**, *100*, 1295.
87. von Itzstein, M.; Wu, W. Y.; Kok, G. B.; Pegg, M. S.; Dyason, J. C.; Jin, B.; Van Phan, T.; Smythe, M. L.; White, H. F.; Oliver, S. W.; Colman, P. M.; Varghese, J. N.; Ryan, D. M.; Woods, J. M.; Bethell, R. C.; Hotham, V. J.; Cameron, J. M.; Penn, C. R. *Nature* **1993**, *363*, 418.
88. Kim, C. U.; Lew, W.; Williams, M. A.; Liu, H. T.; Zhang, L. J.; Swaminathan, S.; Bischofberger, N.; Chen, M. S.; Mendel, D. B.; Tai, C.Y.; Laver, W. G.; Stevens, R. C. *J. Am. Chem. Soc.* **1997**, *119*, 681.
89. Li, J.; Zheng, M.; Tang, W.; He, P. L.; Zhu, W.; Li, T.; Zuo, J. P.; Liu, H.; Jiang, H. *Bioorg. Med. Chem. Lett.* **2006**, *16*, 5009–5013.
90. Guillon, P.; Dirr, L.; El-Deeb, I. M.; Winger, M.; Bailly, B.; Haselhorst, T.; Dyason, J. C.; Von Itzstein, M. *Nature commun.* **2014**, *5*, 5268.
91. Fraser, B. H.; Hamilton, S.; Krause-Heuer, A. M.; Wright, P. J.; Greguric, I.; Tucker, S. P.; Draffan, A. G.; Fokin, V.V.; Sharpless, K. B. *Med. Chem. Comm.* **2013**, *4*, 383–386.
92. Zou, Y.; Albohy, A.; Sandbhor, M.; Cairo, C. W. *Bioorg. Med. Chem. Lett.* **2010**, *20*, 7529–7533.
93. Hinou, H.; Miyoshi, R.; Takasu, Y.; Kai, H.; Kurogochi, M.; Arioka, S.; Gao, X. D.; Miura, N.; Fujitani, N.; Omoto, S.; Yoshinaga, T.; Fujiwara, T.; Noshi, T.; Togame, H.; Takemoto, H.; Nishimura, S. -I. *Chem.–Asian J.* **2011**, *6*, 1048–1056.
94. Slack, T. J.; Li, W.; Shi, D.; McArthur, J. B.; Zhao, G.; Li, Y.; Xiao, A.; Khedri, Z.; Yu, H.; Liu, Y.; Chen, X. *Bioorg. Med. Chem.* **2018**, *26*, 5751–5757.
95. Shi, Y.; Si, L.; Han, X.; Fan, Z.; Wang, S.; Li, M.; Sun, J.; Zhang, Y.; Zhou, D.; Xiao, S. *Med. Chem. Commun.* **2016**, 7, 1932–1945.
96. Weïwer, M.; Chen, C. C.; Kemp, M. M.; Linhardt, R. J. *Eur. J. Org. Chem.* **2009**, 2611–2620.
97. Tollas, S.; Bereczki, I.; Borbás, A.; Batta, G.; Vanderlinden, E.; Naesens, L.; Herczegh, P. *Bioorg. Med. Chem. Lett.* **2014**, *24*, 2420–2423.

98. Kai, H.; Hinou, H.; Nishimura, S. -I. *Bioorg. Med. Chem.* **2012**, *20*, 2739–2746.
99. Mitchell, E. P.; Withers, S. G.; Ermert, P.; Vasella, A. T.; Garman, E. F.; Oikonomakos, N. G.; Johnson, L. N. *Biochemistry* **1996**, *35*, 7341–7355.
100. Baker, D. J.; Greenhaff, P. L.; Timmons, J. A. *Expert Opin. Ther. Pat* **2006**, *16*, 459–466.
101. Tracey, W. R.; Treadway, J. L.; Magee, W. P.; Sutt, J. C.; McPherson, R. K.; Levy, C. B.; Wilder, D. E.; Yu, L. J.; Chen, Y.; Shanker, R. M.; Mutchler, A. K.; Smith, A. H.; Flynn, D. M.; Knight, D. R. *Am J Physiol Heart Circ Physiol* **2004**, *286*, H1177–H1184.
102. Zois, C. E.; Harris, A. L. *J. Mol. Med* **2016**, *94*, 137–154.
103. Martin, J. L.; Johnson, L. N.; Withers, S. G. *Biochemistry* **1990**, *29*, 10745–10757.
104. Martin, J. L.; Veluraja, K.; Ross, K.; Johnson, L. N.; Fleet, G. W. J.; Ramsden, N. G.; Bruce, I.; Orchard, M. G.; Oikonomakos, N. G. *Biochemistry* **1991**, *30*, 10101–10116.
105. Chen, J.; Liu, J.; Zhang, L.; Wu, G.; Hua, W.; Wu, X.; Sun, H. *Bioorg. Med. Chem. Lett.* **2006**, *16*, 2915–2919.
106. Cheng, K.; Liu, J.; Liu, X.; Li, H.; Sun, H.; Xie, J. *Carbohyd. Res.* **2009**, *344*, 841–850.
107. Cheng, K.; Liu, J.; Sun, H.; Bokor, É.; Czifrák, K.; Kónya, B.; Tóth, M.; Docsa, T.; Gergely, P.; Somsák, L. *New J. Chem.* **2010**, *34*, 1450–1464.
108. Cheng, K.; Liu, J.; Sun, H.; Xie, J. *Synthesis* **2010**, 1046–1052.
109. Alexacou, K. M.; Hayes, J. M.; Tiraidis, Zographos, C. S. E.; Leonidas, D. D.; Chrysina, E. D.; Archontis, G.; Oikonomakos, N. G.; Paul, J. V.; Varghese, B.; Loganathan, D. *Proteins: Struct., Funct., Bioinf* **2008**, *71*, 1307–1323.
110. (a) Moller, D. E. *Nature* **2001**, *414*, 821; (b) Vang, T.; Miletic, A. V.; Arimura, Y.; Tautz, L.; Rickert, R. C.; Mustelin, T. *Annu. Rev. Immun.* **2008**, *26*, 29-55.
111. Cho, H. *Vitam. Horm* **2013**, *91*, 405–424.
112. Hendriks, W. J. A. J.; Elson, A.; Harroch, S.; Pulido, R.; Stoker, A.; Hertog, J. De. *FEBS J.* **2013**, *280*, 708–730.
113. Lin, L.; Shen, Q.; Chen, G. R.; Xie, J. *Bioorg. Med. Chem.* **2008**, *16*, 9757–9763.
114. Zhang, Y. J.; He, X. P.; Li, C.; Li, Z.; Shi, D. T.; Gao, L. X.; Qiu, B. Y.; Shi, X. X.; Tang, Y.; Li, J.; Chen, G. R. *Chem. Lett.* **2010**, *39*, 1261–1263.
115. Yang, J. W.; Li, C.; He, X. P.; Zhao, H.; Gao, L. X.; Zhang, W.; Shi, X. X.; Tang, Y.; Li, J.; Chen, G. R. *Bull. Kor. Chem. Soc.* **2010**, *31*, 3359–3365.
116. Tang, Y. H.; Hu, M.; He, X. P.; Fahnbulleh, S.; Li, C.; Gao, L. X.; Sheng, L.; Tang, Y.; Li, J.; Chen, G. R. *Bull. Kor. Chem. Soc.* **2011**, *32*, 1000–1006.
117. Li, C.; He, X. P.; Zhang, Y. J.; Li, Z.; Gao, L. X.; Shi, X. X.; Xie, J.; Li, J.; Chen, G. R.; Tang, Y. *Eur. J. Med. Chem.* **2011**, *46*, 4212–4218.
118. Song, Z.; He, X. P.; Li, C.; Gao, L. X.; Wang, Z. X.; Tang, Y.; Xie, J.; Li, J.; Chen, G. R. *Carb. Res.* **2011**, *346*, 140–145.
119. Yang, J. W.; He, X. P.; Li, C.; Gao, L. X.; Sheng, L.; Xie, J.; Shi, X. X.; Tang, Y.; Li, J.; Chen, G. R. *Bioorg. Med. Chem. Lett.* **2011**, *21*, 1092–1096.
120. He, X. P.; Li, C.; Wang, Z. Z.; Gao, L. X.; Shi, X. X.; Tang, Y.; Xie, J.; Li, J.; Chen, G. R.; Chen, K. *Glycoconj. J.* **2011**, *28*, 493.
121. (a) He, X. P.; Li, C.; Jin, X. P.; Song, Z.; Zhang, H. L.; Zhu, C. J.; Shen, Q.; Zhang, W.; Sheng, L.; Shi, X. X.; Tang, Y. *New J. Chem.* **2011**, *35*, 622–631; (b) He, X. P.; Deng, Q.; Gao, L. X.; Li, C.; Zhang, W.; Zhou, Y. B.; Tang, Y.; Shi, X. X.; Xie, J.; Li, J.; Chen, G. R. *Bioorg. Med. Chem.* **2011**, *19*, 3892-3900.
122. Luo, L.; He, X. P.; Shen, Q.; Li, J. Y.; Shi, X. X.; Xie, J.; Li, J.; Chen, G. R. *Chem. Biodiver.* **2011**, *8*, 2035–2044.
123. Bray, F.; Ferlay, J.; Soerjomataram, I.; Siegel, R. L.; Torre, L. A.; Jemal, A. *Cancer J. Clin.* **2018**, *68*, 394–424.
124. Corrie, P. G.; Pippa, G. *Medicine* **2008**, *36*, 24–28.
125. Salem, M.; Rohani, S.; Gillies, E. R. *RSC Advances* **2014**, *4*, 10815–10829.
126. Bębenek, E.; Kadela-Tomanek, M.; Chrobak, E.; Latocha, M.; Boryczka, S. *Med. Chem. Res.* **2018**, *27*, 2051–2061.

127. Parida, P. K.; Sau, A.; Ghosh, T.; Jana, K.; Biswas, K.; Raha, S.; Misra, A. K. *Bioorg. Med. Chem. Lett.* **2014**, *24*, 3865–3868.
128. Zhang, J.; Garrossian, M.; Gardner, D.; Garrossian, A.; Chang, Y. T.; Kim, Y. K.; Chang, C. W. T. *Bioorg. Med. Chem. Lett.* **2008**, *18*, 1359–1363.
129. Hwu, J. R.; Hsu, C. I.; Hsu, M. H.; Liang, Y. C.; Huang, R. C. C.; Lee, Y. C. *Bioorg. Med. Chem. Lett.* **2011**, *21*, 380–382.
130. Peterson, L. B.; Blagg, B. S. *Bioorg. Med. Chem. Lett.* **2010**, *20*, 3957–3960.
131. Luiza, B. D. O.; Borgati, T. F.; de Freitas, R. P.; Ruiz, A. L.; Marchetti, G. M.; de Carvalho, J. E.; da Cunha, E. F.; Ramalho, T. C.; Alves, R. B. *Eur. J. Med. Chem.* **2014**, *84*, 595–604.
132. dos Anjos, J. V.; Neves Filho, R. A.; do Nascimento, S. C.; Srivastava, R. M.; de Melo, S. J.; Sinou, D. *Eur. J. Med. Chem.* **2009**, *44*, 3571–3576.
133. Zhang, Q.; He, P.; Zhou, G.; Gu, Y.; Fu, T.; Xue, D.; Liu, H. M. *Carb. Res.* **2013**, *382*, 65–70.
134. Cheng, C. Y.; Feng, Y. Y.; Zang, Y.; Li, J.; He, X. P.; Chen, G. R. *Carb. Res.* **2014**, *394*, 39–42.
135. Song, S. X.; Zhang, H. L.; Kim, C. G.; Sheng, L.; He, X. P.; Long, Y. T.; Li, J.; Chen, G. R. *Tetrahedron* **2010**, *66*, 9974–9980.
136. Hradilová, L.; Poláková, M.; Dvořáková, B.; Hajdúch, M.; Petruš, L. *Carb. Res.* **2012**, *361*, 1–6.
137. Manna, T.; Pala, K.; Jana, K.; Misra, A. K. *Bioorganic & Medicinal Chemistry Letters* **2019**, *29*, 126615–126621.
138. Bodnár, B.; Mernyák, E.; Szabó, J.; Wölfling, J.; Schneider, G.; Zupkó, I.; Kupihár, Z.; Kovács, L. *Bioorg. Med. Chem. Lett.* **2017**, *27*, 1938–1942.
139. Nagarsenkar, A.; Prajapti, S. K.; Guggilapu, S. D.; Birineni, S.; Kotapalli, S. S.; Ummanni, R.; Babu, B. N. *Med Chem. Commun.* **2016**, 7, 646–653.
140. Burza, S.; Croft, S. L.; Boelaert, M. *Lancet* **2018**, *392*, 951–970.
141. Chappuis, F.; Sundar, S.; Hailu, A.; Ghalib, H.; Rijal, S.; Peeling, R. W.; Alvar, J.; Boelaert, M. *Nature Reviews Microbiology* **2007**, *5*, 873–882.
142. Dwivedi, P.; Mishra, K. B.; Mishra, B. B.; Singh, N.; Singh, R. K.; Tiwari, V. K. *Glycoconj. J.* **2015**, *32*, 127–140.
143. Mishra, K. B.; Tiwari, N.; Bose, P.; Singh, R.; Rawat, A. K.; Singh, S. K.; Mishra, R. C.; Singh, R. K.; Tiwari, V. K. *Chemistry Sel.* **2019**, *4*, 2644–2648.
144. Ventola, C. L. *Pharmacy and Therapeutics* **2015**, *40*, 277–283.
145. Walsh, C. *Nat. Rev. Microbiol.* **2003**, *1*, 65–70.
146. Zhang, J.; Chiang, F. I.; Wu, L.; Czyryca, P. G.; Li, D.; Chang, C. W. T. *J. Med. Chem.* **2008**, *51*, 7563–7573.
147. Dupuis, S. N.; Robertson, A. W.; Veinot, T.; Monro, S. M.; Douglas, S. E.; Syvitski, R. T.; Goralski, K. B.; McFarland, S. A.; Jakeman, D. L. *Chem. Sci.* **2012**, *3*, 1640–1644.
148. Bera, S.; Zhanel, G. G.; Schweizer, F. *Bioorg. Med. Chem. Lett* **2010**, *20*, 3031–3035.
149. Westermann, B.; Dörner, S.; Brauch, S.; Schaks, A.; Heinke, R.; Stark, S. L.; van Delft, F.; van Berkel, S. S. *Carb. Res.* **2013**, *371*, 61–67.
150. Wei, J. J.; Jin, L.; Wan, K.; Zhou, C. H. *Bull. Kor. Chem. Soc.* **2011**, *32*, 229–238.
151. Reddy, L. V. R.; Reddy, P. V.; Mishra, N. N.; Shukla, P. K.; Yadav, G.; Srivastava, R.; Shaw, A. K. Carb. *Res.* **2010**, *345*, 1515–1521.
152. Sharma, S.; Saquib, M.; Verma, S.; Mishra, N. N.; Shukla, P. K.; Srivastava, R.; Prabhakar, Y. S.; Shaw, A. K. *E. J. Med. Chem.* **2014**, *83*, 474–489.
153. de Souza, T. B.; Raimundo, P. O. B.; Andrade, S. F.; Hipólito, T. M. M.; Silva, N. C.; Dias, A. L. T.; Ikegaki, M.; Rocha, R. P.; Coelho, L. F. L.; Veloso, M. P.; Carvalho, D. T. *Carb. Res.* **2015**, *410*, 1–8.
154. Kuhn, H.; Gutelius, D.; Black, E.; Nadolny, C.; Basu, A.; Reid, C. *Med Chem Comm* **2014**, *5*, 1213–1217.

155. Schwardt, O.; Rabbani, S.; Hartmann, M.; Abgottspon, D.; Wittwer, M.; Kleeb, S.; Zalewski, A.; Smieško, M.; Cutting, B.; Ernst, B. *Bioorg. Med. Chem.* **2011**, *19*, 6454–6473.
156. Lin, H.; Walsh, C. T. *J. Am. Chem. Soc.* **2004**, *126*, 13998–14003.
157. Pintér, G.; Bereczki, I.; Batta, G.; Ötvös, R.; Sztaricskai, F.; Roth, E.; Ostorházi, E.; Rozgonyi, F.; Naesens, L.; Szarvas, M.; Boda, Z.; Herczegh, P. *Bioorg. Med. Chem. Lett.* **2010**, 2713–2717.
158. Agrahari, A. K.; Singh, A. S.; Singh, A. K.; Mishra, N.; Singh, M.; Prakash, P.; Tiwari, V. K. *New J. Chem.* **2019**, *43*, 12475–12482.
159. Taouai, M.; Chakroun, K.; Sommer, R.; Michaud, G.; Giacalone, D.; Maaouia, M. A. B.; Vallin-Butruille, A.; Mathiron, D.; Abidi, R.; Darbre, T.; Cragg, P. J.; Mullié, C.; Reymond, J. -L.; O'Toole, G. A.; Benazza, M. 62. *J. Med. Chem.* **2019**, *17*, 7722–7738.
160. Kumar, K. K.; Seenivasan, S. P.; Kumar, V.; Das, T. M. *Carb. Res.* **2011**, *346*, 2084–2090.
161. Singh, B. K.; Yadav, A. K.; Kumar, B.; Gaikwad, A.; Sinha, S. K.; Chaturvedi, V.; Tripathi, R. P. *Carb. Res.* **2008**, *343*, 1153–1162.
162. Wilkinson, B. L.; Long, H.; Sim, E.; Fairbanks, A. J. *Bioorg. Med. Chem. Lett.* **2008**, *18* (23), 6265–6267.
163. Lo Conte, M.; Marra, A.; Chambery, A.; Gurcha, S. S.; Besra, G. S.; Dondoni, A. *J. Org. Chem.* **2010**, *75*, 6326–6336.

CHAPTER ELEVEN

Glycohybrid molecules in medicinal chemistry: Present status and future prospective

Priti Kumari, Chintam Narayana, Ghanshyam Tiwari, Ram Sagar
Department of Chemistry, Institute of Science, Banaras Hindu University, Varanasi, India

1 Introduction

The design and synthesis of new drug molecules represents the initial and therefore probably most important step in the drug discovery process.[1] The identification of a hit and subsequently a lead structure for further development is a very challenging and expensive process.[2] There has been enormous progress made in screening technologies and assay development, and novel high-throughput screening (HTS) methods based on many biochemical and pharmacological discoveries.[3] Nevertheless, hit identification is in most cases based on random screening of compound libraries, sometimes and with increasing success rates supported by virtual screening methods.[4] Classical combinatorial chemistry approaches have not yet brought a breakthrough in the success rate of hit discovery.[5]

Natural products play an important role in the development of drugs, especially for the treatment of infections and cancer, as well as immunosuppressive agents.[6] However, the number of natural products is limited, whereas millions of hybrid molecules as combinations of parts of different natural products can be prepared.[7] This new approach seems to be very promising in the development of leads for medicinal applications, as the biological activity of several new hybrids exceeds that of the parent compounds.[8] The advantage of this concept over a combinatorial chemistry approach is the high diversity and the inherent biological activity of the hybrids.[9]

It is well recognized that numerous diseases are not caused by a defect in one specific biological target such as an enzyme or a receptor, but based on a plethora of biochemical and physiological processes that often even occur concomitantly.[10] Exogenous and/or endogenous factors can contribute to

Carbohydrates in Drug Discovery and Development. http://dx.doi.org/10.1016/B978-0-12-816675-8.00011-7

the disease state. It is obvious that even the best assay and most promising hit will only address one biological target, which is probably not sufficient to efficiently cure diseases.[11]

Hybrid molecules combine two distinct biologically active molecules that act at different targets, into one new molecule/chemical entity to combine the effects of each molecule.[12] Hybrid molecules are usually defined as a structure in which a linker, often a simple hydrocarbon chain, heterocyclic ring connects the two biologically relevant molecules that are not much altered with regard to their initial chemical structure.[13]

Carbohydrates constitute the most abundant class of biomolecules in nature. They play important roles in the natural world as diverse as energy storage, molecular recognition for intracellular trafficking or interactions between pathogenic bacteria and viruses and the surfaces of vertebrate's cells.[14] The carbohydrates present inside and at the surface of cells mediate many biological processes that are fundamentally important for both the healthy and diseased states of living organisms.[15,16]

Recent advances studies in glycochemistry and glycobiology have open new and strong interest in the study of carbohydrates for medicinal chemistry. It is evident now that oligosaccharides play fundamental roles in the development, recognition, growth, function, and survival of living cells and organisms.[17]Various natural products including antibiotics and anticancer agents also contain carbohydrates and oligosaccharides attached to it and are crucial for biological activities.

Glycoconjugates in form of carbohybrids also act as molecular probes for the interrogation of biological processes, biomolecular communication events and diseases related processes. They are rich in structural and functional diversity and involve protein–carbohydrate interactions that play important roles in mammalian biology and various disease states.[18–21] There are early events in the infectious cycles of many bacteria, viruses, mycoplasma and parasites that involve carbohydrate-mediated recognition of the host by the pathogen.[18] The ability to control or prevent these events using small carbohybrid ligands is highly desirable.

This chapter comprises the brief overview on synthesis of hybrids of carbohydrates (carbohybrids/glycohybrids) and their biological activities and their applications in medicinal chemistry.

2 Bioactive carbohybrid molecules

A brief review has been compiled here for carbohybrid molecules and categorized the findings based on their biological activities.

2.1 Anticancer activity

Ipomoeassin F, a macrolide glycoresin containing an embedded disaccharide can be considered as carbohybrid structure. It possesses potent *in vitro* antitumor activity with an unknown mechanism of function. It inhibits tumor cell growth with single-digit nanomolar IC_{50} values (2.9 nM), superior to many clinical chemotherapeutic drugs. In order to facilitate translation of Ipomoeassin bioactivity into protein function for drug development, Shi et al. reported new synthesis route for the gram-scale production of ipomoeassin F in seventeen linear steps with 3.8% overall yield starting from commercially available starting materials.[22] The conformation-controlled subtle reactivity differences of the hydroxyl groups in carbohydrates were utilized there to quickly construct the disaccharide core, which, along with judicial selection of protecting groups, made this synthesis very efficient. The same strategy they have applied for the smooth preparation of the 11R-epimer of ipomoeassin F for the first time. The cytotoxicity assays were carried out and it was noticed that, there is a crucial role of the natural 11S configuration in bioactivity. In addition, cell cycle analyses and apoptosis assays on ipomoeassin F and its 11R-epimer were also reported here. This work has laid a solid foundation for understanding the medicinal potential of the ipomoeassin family of glycolipids. The cytotoxicity screens of both compounds undoubtedly demonstrate that the 11S configuration is crucial for the cytotoxicity of ipomoeassin F. They have also confirmed for the first time that in human cells ipomoeassin F can arrest cell cycle in the G1 phase and induce remarkable apoptosis (Table 11.1).

Resende and co-workers synthesized carbohydrate-based naphthoquinones as carbohybrid molecules **2a**, **2b**, **3a** and **3b** (Fig. 11.1), derived from juglone.[23] They showed the best cytotoxicity profile against human cell lines HCT-116, A-549 and MDA-MB-435 when compared with the standard. These finding evidenced that the hydroxyl group to the pro-oxidant action of these naphthoquinone compounds is very important. Indeed, derivatives **2a** and **2b** were found more active than the clinically useful anticancer agent doxorubicin against melanoma cancer cells (MDA-MB-435 cells).

Tripathi and co-workers prepared novel class of gallic acid-based hybrid glycoconjugates **7-15** (Fig. 11.2) as potential anticancer agents.[24] Among all the compounds they have screened, compound **7a** showed potent anticancer activity against breast cancer cells. The detailed mechanistic study revealed that this compound act as inhibitor of tubulin polymerization and induced G2/M cell cycle arrest, generation of reactive oxygen species, mitochondrial depolarization and subsequent apoptosis in breast cancer cells. Further authors have shown through ultraviolet-visible spectroscopy and

Table 11.1 Anticancer activity of ipomoeassins against different cell lines.

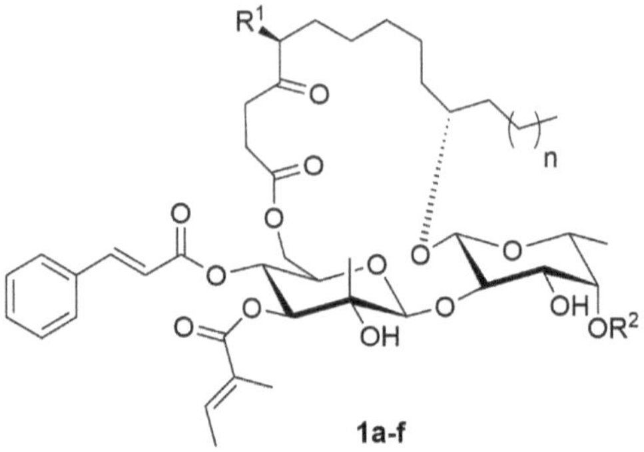

1a-f

Compounds				IC_{50}(nM)						
Ipomoeassins	R^1	R^2	n	HeLa	L-929	A2780	U937	HT-29	MDA-MB-439	H-522T1
1a	H	Ac	1	64	77.8	500	20.2	46.1	42.6	108.9
1b	H	H	1	2500	-	400	134	396	2700	1070
1c	OH	Ac	1	1500	>1000	2900	-	-	-	-
1d	OAc	Ac	1	32	135	35	7.9	11.8	19.9	23.2
1e	OAc	H	1	4300	>1000	3300	163	393	1633	967
1f	H	Ac	3	-	7.4	36	2.6	4.2	9.4	12.9

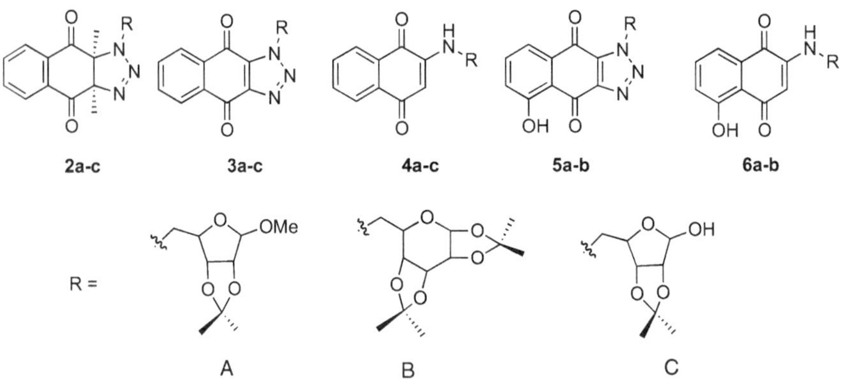

Figure 11.1 ***Anticancer naphthoquinone hybrids.***

fluorescence quenching studies of the compound with tubulin which confirmed direct interaction of compounds with tubulin. Finally, molecular modeling studies was done which revealed that it binds at the colchicine binding site in tubulin. Compound **7a** also exhibited potent *in vivo* anticancer activity in LA-7 syngeneic rat mammary tumor model.

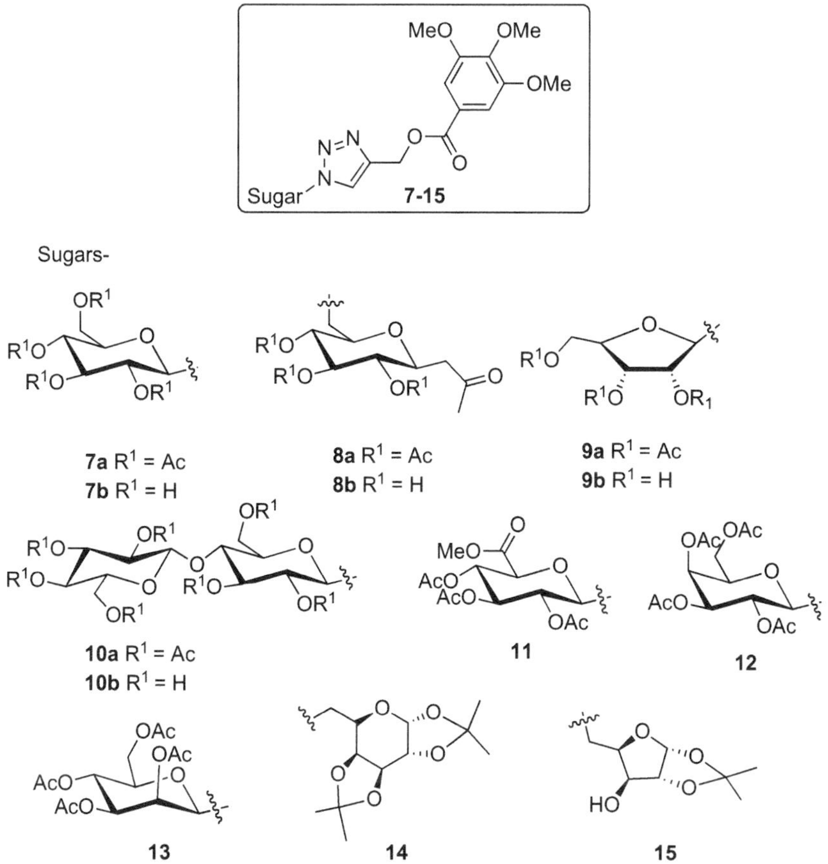

Figure 11.2 ***Anticancer carbohybrids.***

G. P. Gawolek and co-workers designed and synthesized several anticancer glycoconjugates as hybrids from small molecular quinolines and substituted gluco- and galacto- pyranosyl amines (Fig. 11.3).[25] Although the parent quinoline compounds were inactive in affordable concentrations, the glycoconjugates **16–17** that were obtained appeared to be cytotoxic against cancer cells at the micromolar concentration. When these hybrid molecules **16–17** combined with copper ions, their activity increased even further. Authors have looked for mechanism of action of these molecules and propose that this is connected to the formation of reactive oxygen species and the intercalation of DNA was responsible for anticancer activity.

G. P. Gawolek and co-workers also prepared a series of UDP-sugar analogues **18–23** (Fig. 11.4) and looked their preliminary biological activity.[26]

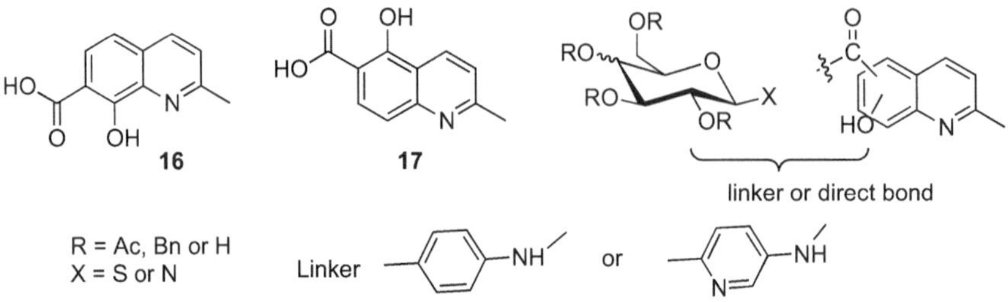

Figure 11.3 ***Quinolines hybrids as anticancer molecules.***

They have synthesized glycoconjugates of uridine **18** and **19** by condensation of uridine-5'-carboxylic acid and 1-amino sugars derivatives of *D*-glucose and *D*-galactose. The glycoconjugates **20** and **21** were synthesized by azide-alkyne 1,3-dipolar cycloaddition (CuAAC) of 1-azido sugars and propargylamide derivatives of uridine while glycoconjugates **22** and **23** were synthesized by CuAAC of propargyl β-*O*-glycosides and 5'-azido uridine (Fig. 11.4).

Evaluation of inhibitory activity of compounds **18–21** against commercially available β-1,4-galactosyltransferase I (β4GalT) show that compound **22** inhibited the enzyme in μM range. Additionally, the antitumor activity of the obtained glycoconjugates **18-23** was tested using MTT assay (Table 11.2).

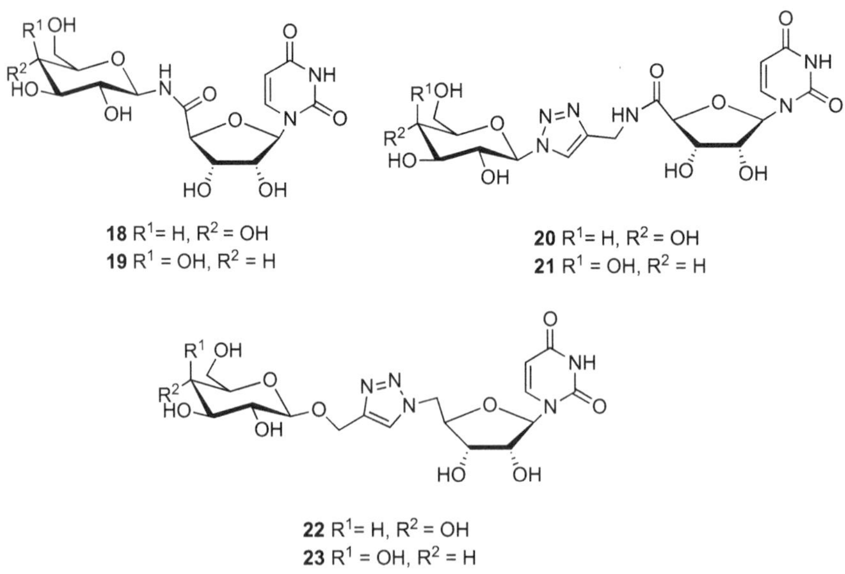

Figure 11.4 ***Antitumor glycoconjugates.***

Table 11.2 Anticancer and galactosyl transferase inhibitory activity.

Compounds	Percentage of inhibition at 0.8mM (%)	IC_{50} (mM)
18	1.3 ± 0.8	-
19	8± 0.2	-
20	1.5 ±1.1	-
21	3 ±0.5	-
22	48 ± 2.4	0.72
23	14 ± 2.6	-

2.1.1 Antiviral activity

D. J. Laever and co-workers reported synthesis of a series of bivalent 1,2,3-triazole linked galactopyranosides **24a-24b** and **25a-25d** (Fig.11.5) as potential inhibitors of cholera toxin (CT). The inhibitory activity of the bivalent series was examined (ELISA) and the series showed low inhibitory activity (millimolar IC_{50}). Conversely, the monomeric galactotriazole analogues were strong inhibitors of cholera toxin (IC_{50}= 71–75 mM).[27]

M.Yu and co-workers reported certain pentacyclic triterpenoids as glycosidic hybrids **26–28** (Fig. 11.6) displaying *in vitro* anti-influenza virus activity which was comparable and even higher in some cases than that of oseltamivir.[28]

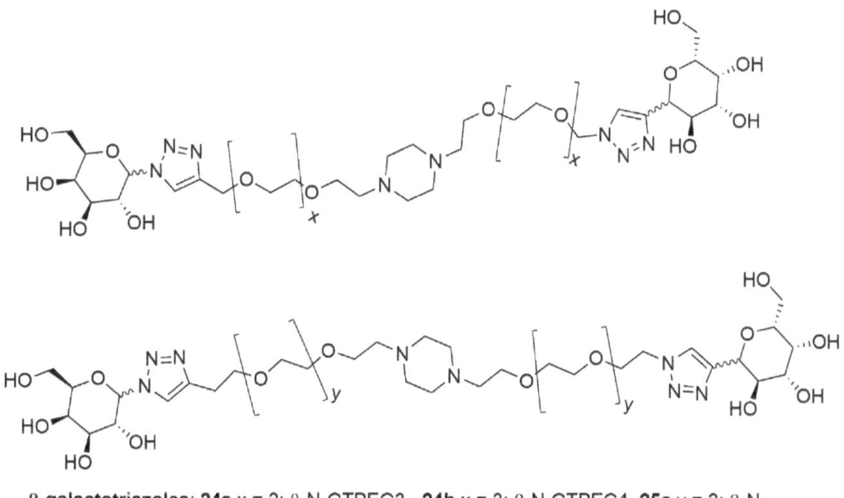

β-galactotriazoles: **24a** x = 2; β-N-GTPEG3. **24b** x = 3; β-N-GTPEG4. **25a** y = 2; β-N-GTPEG3. **25b** y = 3; β-N-GTPEG4; **α-galactotriazoles**: **25c** y = 2; α-C-GTPEG. **25d** y = 3; α-C-GTPEG.

Figure 11.5 *Potential inhibitors of cholera toxin.*

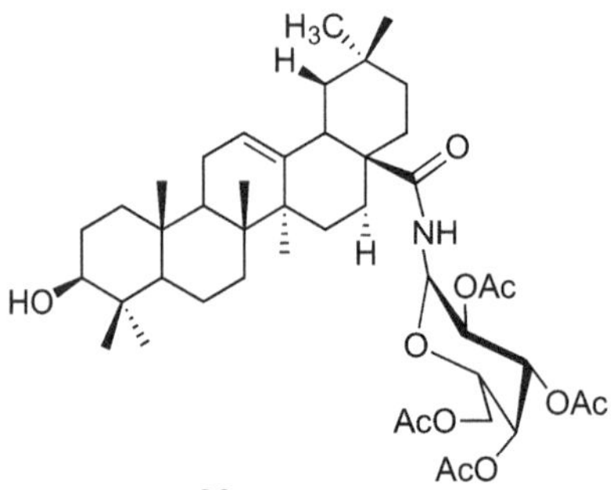

26

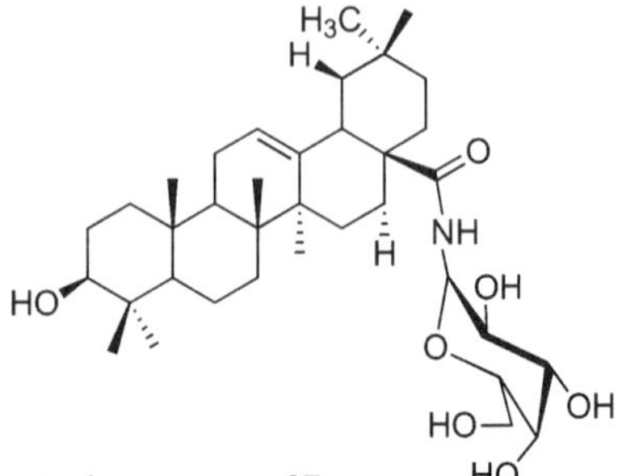

27

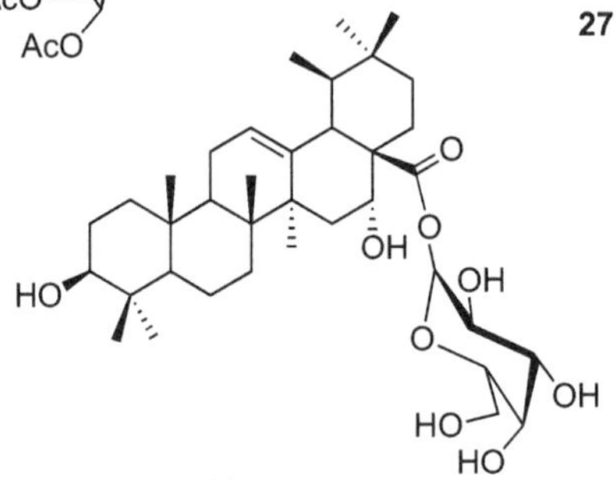

28

Figure 11.6 ***Antiviral terpenoid glycosides.***

A systemic exploration of the lead compounds clarifies the structure requirements for maintenance of high activity with oleanolic acid and acetyl galactose moieties as the optimized aglycone and glycoside. These compounds **26–28** exert broad spectrum activity even against oseltamivir- and amantadine-resistant viruses with diminished induction of viral resistance (Table 11.3). Authors have studied the molecular basis of their broad-spectrum activity and concluded that, it's due to their high affinity for HA protein, which blocks the HA – sialic acid receptor interaction and thus the attachment of influenza viruses to cells.

M. Cohen and co-workers synthesized new neuraminidase inhibitors **29–34** (Fig. 11.7).[29] Current drugs against the influenza A virus (IAV) act by inhibiting viral neuraminidase (NA) enzymes responsible for the release of budding virions from sialoglycans on infected cells. The authors approach was focused on a search for inhibitors that reinforce the protective functions of mucosal barriers that trap virus enroute to the target cells. They have generated mimetics of sialo-glycoproteins that insert into the viral envelope to provide a well-defined mucus-like environment encapsulating the virus. By introducing this barrier, which the virus must breach using its NA enzymes to infect a host cell, into a screening platform, they have been

Table 11.3 Comparisons of the anti-viral activity of lead compounds with other anti-viral drugs.

		EC_{50} (μM)				
Compds.	**CC_{50} (μM)**	**PR/8 (H1N1)**	**LN/1109 (H1N1)**	**JX/312 (H3N2)**	**HN/1222 (H3N2)**	**B/SZ/155**
26	>200	7.41 ± 0.05	6.58± 1.17	2.72 ± 0.35	3.18 ± 0.08	2.80 ± 0.74
27	>200	9.33 ± 0.60	33.25 ± 0.64	66.67 ± 0.17	26.25 ± 3.01	43.88 ± 19.38
28	>200	17.50 ± 0.43	17.8 ± 12.57	16.89 ± 5.38	25.59 ± 9.14	36.21 ± 7.21
Amantadine	>200	7.41 ± 0.21	0.44 ± 0.14	8.58 ± 1.65	>200	>83.24
OSV	>200	>200	>200	2.06 ± 0.58	8.75 ± 4.41	91.07 ± 34.51
RBV	>200	4.02 ± 1.27	5.75 ± 0.35	4.19 ± 0.35	3.56 ± 0.24	1.32 ± 0.60

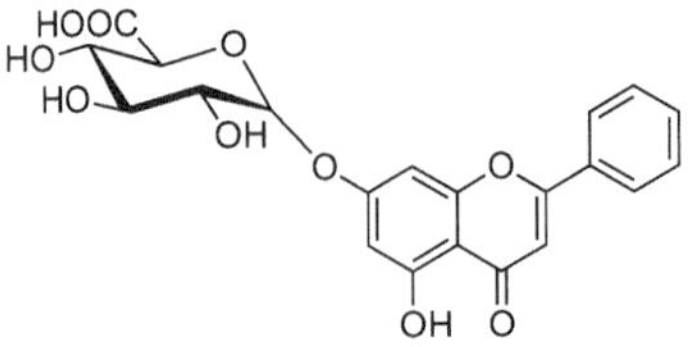

29 Baicalin IC $_{50}$ = 530 μM

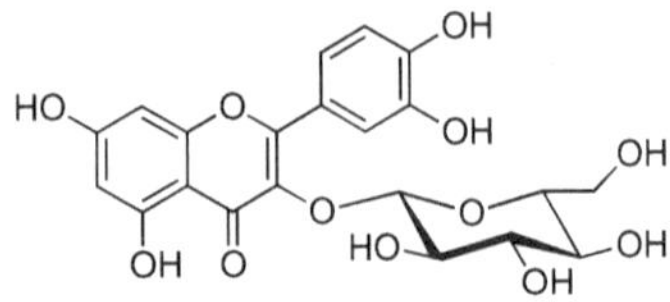

30 Hyperoside IC_{50}= 120 μM

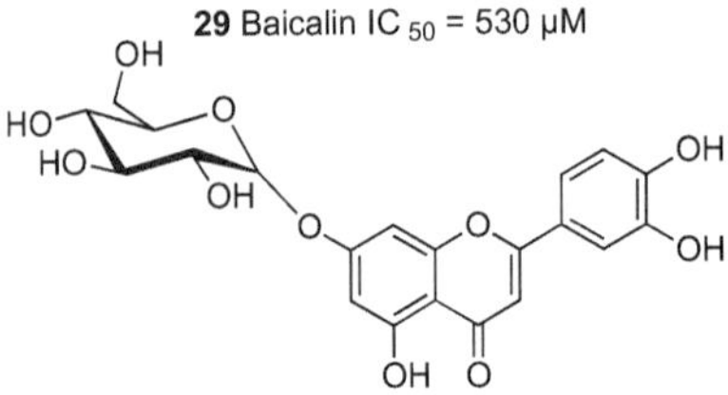

31 Luteolin-7-glucoside IC_{50}= 230 μM

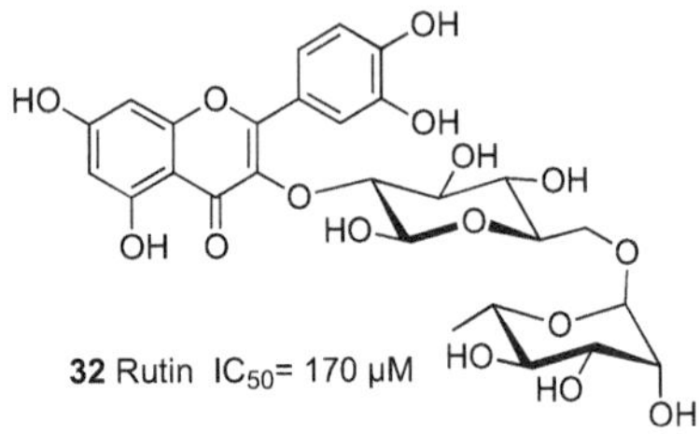

32 Rutin IC_{50}= 170 μM

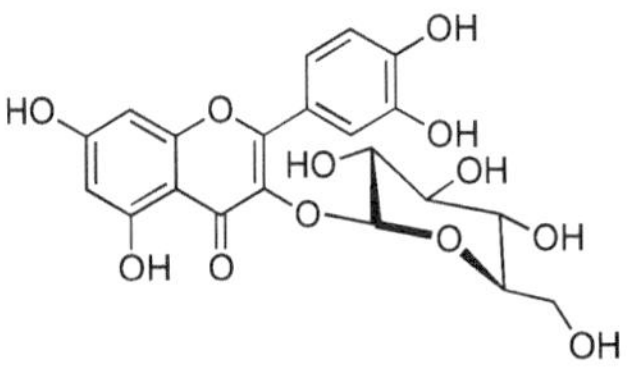

33 Isoquercitrin IC_{50} = 170 μM

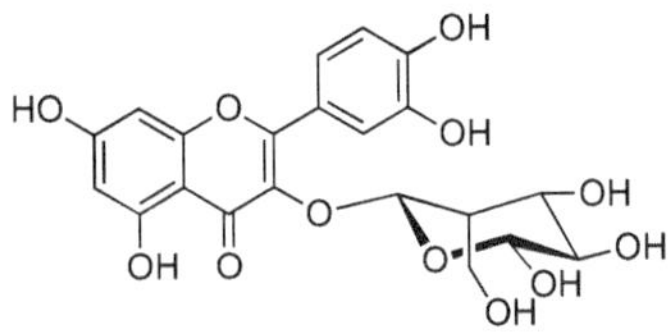

34 Quercitrin IC_{50}= 280 μM

Figure 11.7 ***Neuraminidase inhibitors.***

able to identify compounds that provide significant protection against IAV infection. Authors claimed that this approach may facilitate the discovery of potent new IAV prophylactics among compounds with NA activities too weak to emerge from traditional drug screens.

2.2 Immunomodulatory activity

Mukherjee and co-workers reported[30] therapeutic immunoadjuvants glycosides **35–44** (Fig. 11.8), which were synthesized through a straightforward glycosylation of various alcohols with unprotected and non-activated monosaccharides, under solvent free conditions using ammonium chloride as a catalyst. They have reported that, sugar acids *e.g.* glucuronic acid can be glycosylated without esterification and after acetylation novel 3,6-anhydro derivative was isolated by authors. They have subjected all the synthesized sugar glycosides **35–44** for immunomodulatory activities against the weak antigen oval albumin (OVA) to find out whether these molecules can be classified as immuno- stimulator, immuno- suppressor, or immuno- adjuvant. Most of these compounds **35–44** revealed immunosuppressive or immunostimulatory activity with reference to lymphocyte proliferation and possess potential for vaccine adjuvants (Fig. 11.8).

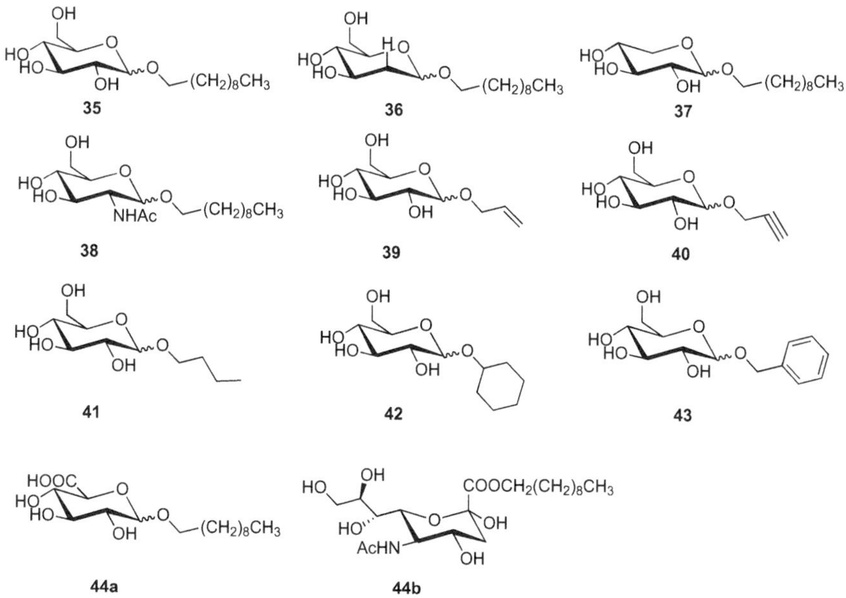

Figure 11.8 ***Immunomodulator glycosides.***

Authors have performed *in vitro* study of spleen lymphocyte transformation to Con A and LPS demonstrates that compounds **35, 36, 37, 38, 44a,** and **44b** cause a decrease in the proliferative response, while being devoid of mutagenic activity, however compound **44b** showed strong immunosuppressive activity and caused a significant decrease in T and B cell proliferation. These results indicate that varying the sugar in above hybrid molecules and keeping the alkyl chain constant led to a varying degree of immune suppression. Thus, compound **44b** has the potential for getting developed as an adjuvant for vaccines. When the T and B cell proliferations are the potency of the humoral and cell mediated immune response, compared with the control Con A and LPS; of them **39** and **40** displayed highly significant cell proliferation in studies with the specific antigen OVA, indicating their potential as adjuvants for vaccines. Thus, changing the alkyl chain length at the anomeric position of a sugar is important for the immunomodulatory activity.

2.3 PTP1B inhibitors

He and co-workers explored click chemistry for the identification of a series of mono- and *bis*-phenylalaninyl and tyrosinyl glucoside derivatives **45–46** (Fig. 11.9) as novel PTP1B inhibitors.[31] Authors have designed molecules bearing one or two phenylalanine or tyrosine derivatives on at

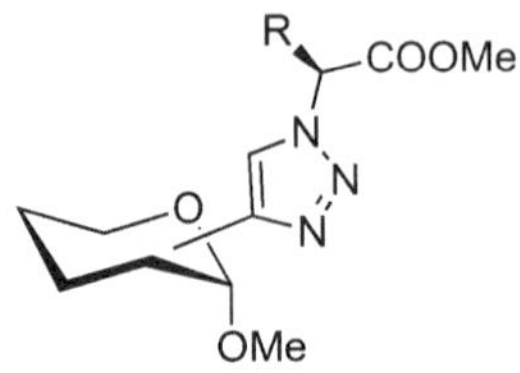

45 R = CH_2Ph
46 R = CH_2(p-OH)Ph

Figure 11.9 ***Novel PTP1B inhibitors.***

various positions of the glucosyl scaffolds and successfully constructed these via the microwave-assisted Cu(I)-catalyzed azide – alkyne cycloaddition in moderate-to-excellent yields. Subsequently they have looked for their biological activity and identified these compounds as novel PTP1B inhibitors, with the 4,6-disubstituted tyrosinyl glucoside being the most potent. They have studied kinetics and revealed that both mono- and *bis*-triazole linked glycosyl acids act as typical competitive inhibitors, whereas the *bis*-triazolyl ester that also exhibited inhibitory activity on PTP1B displayed a mixed-type inhibition pattern.

Yang and co-workers employed a microwave-accelerated Cu(I)- catalyzed azide – alkyne 1,3-dipolar cycloaddition for the preparation of a series of triazole-linked serinyl, threoninyl, phenylalaninyl, and tyrosinylon 1-*O*-glucoside **47** (Fig. 11.10) with high yields of products within 30 min only.[32] Biological assay of these molecules identified glycopeptidotriazoles as favourable PTP1B and CDC25B inhibitors with selectivity over TCPTP (T-cell protein tyrosine phosphatase). Further screening revealed compound **47** (Fig. 11.10) as a selective and potent inhibitor PTP1B over other PTPs tested with IC_{50} = 5.1 μM.

Xie and co-workers successfully synthesized dimeric acetylated and benzoylated β-*C*-D-glucosyl and β-*C*-D-galactosyl 1,4-dimethoxy benzenes or naphthalenes **48** and **50** by click chemistry.[33] These compounds were transformed into the corresponding β-*C*-D-glycosyl-1,4-quinone deriva-

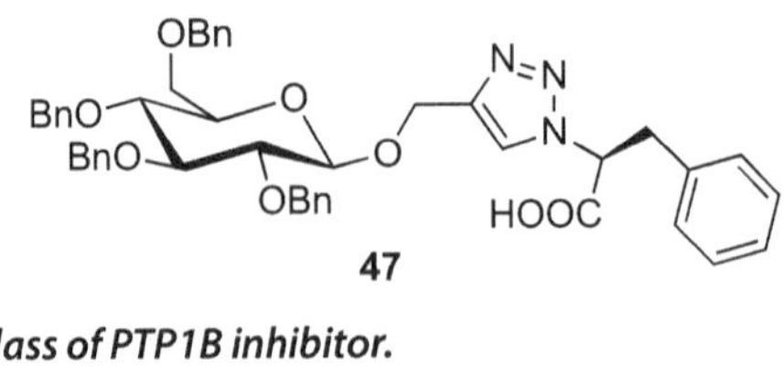

47

Figure 11.10 ***New class of PTP1B inhibitor.***

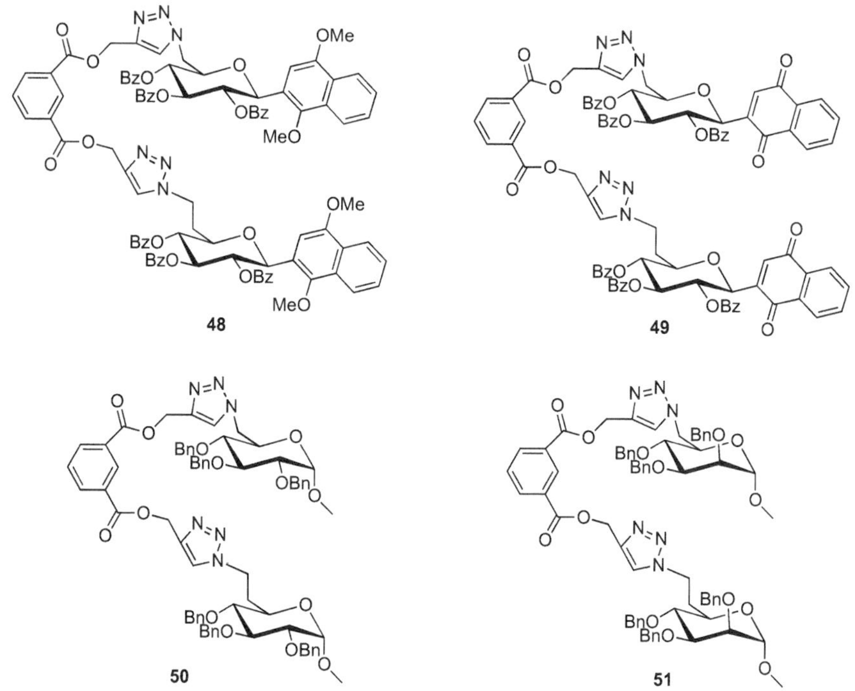

Figure 11.11 ***Structure of new PTP1B inhibitors.***

tives **49** and **51** by ceric ammonium nitrate mediated oxidation. The authors have looked for them *in vitro* inhibitory activity against PTP1B. Inhibition test showed that dimeric benzoylated β-*C*-D-glycosyl 1,4-dimethoxybenzenes (compound **48**, Fig. 11.11) or 1,4-benzoquinones (compound **49**, Fig. 11.11) were good inhibitors of PTP1B (IC_{50} = 0.62 − 0.88 μM), with no significant difference between gluco and galacto derivatives.

2.4 Carbonic anhydrase inhibitor

The zinc-containing enzymes carbonic anhydrases (CA) are very efficient catalysts for the reversible hydration of carbon dioxide to bicarbonate and hence play an important physiological role in living system. There are about sixteen carbonic anhydrases of human isoforms are considered as drug targets. The design of selective carbonic anhydrases inhibitors is a long-standing goal that has captured the attention of medicinal chemist and chemical biologist. There are carbonic anhydrases known for clinical applications against different pathologies such as glaucoma, epilepsy, and cancer. The clinical use of a highly active carbohydrate-based CA inhibitor (topiramate), constitutes an interesting demonstration of the validity of this approach. Carbohydrate-linked triazole compounds also demonstrate promising potential for the treatment of ophthalmologic diseases. Aryl and heteroaryl sulfonamides ($ArSO_2NH_2$) are therapeutically used to inhibit the catalytic activity of carbonic anhydrases. Kolb and his co-workers employed 'click chemistry' for the synthesis of sulfonamide derivatives **52–54** as carbonic anhydrase inhibitors.[34] Novel sulfonamide compounds were particularly active in inhibiting carbonic anhydrase (CA), and compound **52** (Fig. 11.12) was identified as the most potent inhibitor of hCA-II and hCA-IX (h = human) with Ki values of 0.5 and 5.0 nM, respectively. These derivatives were useful for the development of *in vivo* positron emission tomography (PET) imaging agents for the diagnosis of diseases such as cancer. Using a "click-tail" approach, a novel class of glycoconjugate benzene sulfonamides **53–54** were synthesized that contains diverse carbohydrate – triazole tails and they are found inhibitors of hCA-I, hCA-II, hCA-XII, hCA-XI, etc.

Supuran and co-workers synthesized library of compounds with sulfonamide containing neoglycoconjugates **55–78** (Fig. 11.13) and screened them for anticancer related enzymes carbonic anhydrase (CA).[35] Authors have designed CA inhibitors with a combined SAR-SPR strategy and compounds were very good CAIX inhibitors and potent CAXII inhibitors. The role of the carbohydrate fragment is of most relevance in the context of selectively targeting the extracellular active sites of CAIX and XII. The carbohydrate moiety with free hydroxyl group can takes advantage of the cell membranes lipophilic properties as a physical barrier to minimize passive membrane permeability of the polar small molecule inhibitors, this in

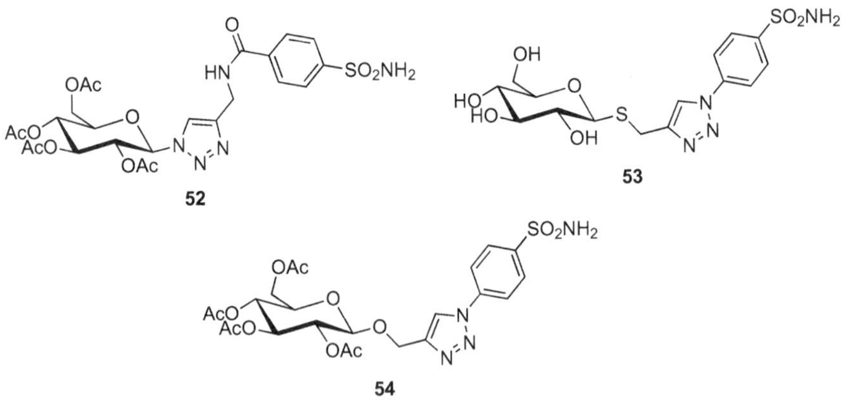

Figure 11.12 ***Humancarbonic anhydrase (hCA) inhibitors.***

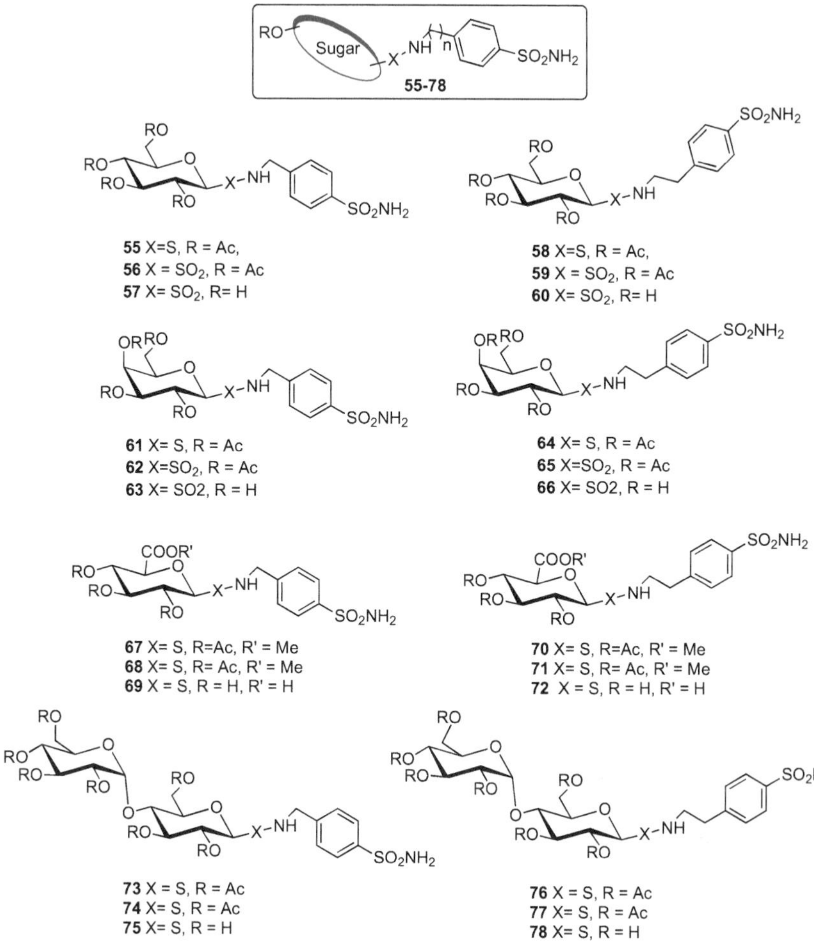

Figure 11.13 ***New humancarbonic anhydrase (hCA) inhibitors.***

turn can promote the preferential inhibition of extracellular CAs. Authors claimed that their approach has delivered neutral, water-soluble CA inhibitors that have excellent potential as isozyme selective inhibitors of cancer-associated CAs *in-vivo*. This novel finding represents an important outcome for investigating these carbohybrids for future therapeutic applications of CA inhibitors.

Fourteen of the twenty carbohybrid as glycoconjugates from this library with *O*-acetate protected carbohydrate tail moieties are low micromolar inhibitors of hCA I. The exceptional compounds are **55, 58, 61, 62, 67** and **68**, which displayed slightly stronger inhibition with *K*i's of 90–120 nM.

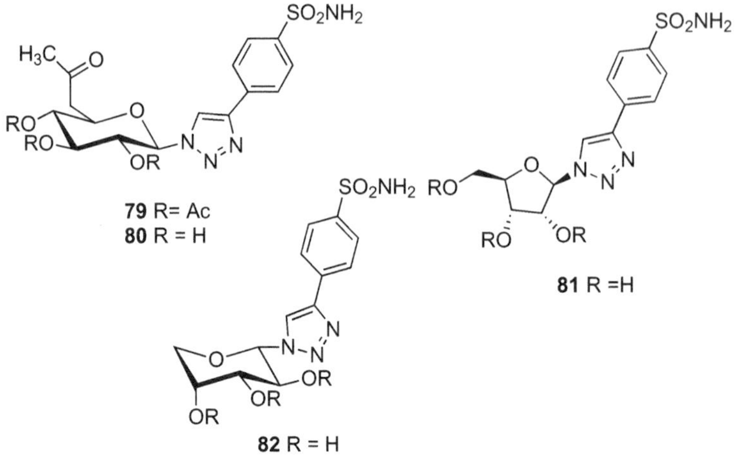

Figure 11.14 ***Selective humancarbonic anhydrase (hCA) inhibitors.***

The eight deacetylated sugar analogues (**57, 60, 63, 66, 69, 72, 75** and **78**) found hCA I inhibition with inhibition constants that ranged from 81 to 107 nM.

B. L. Wilkinson and co-workers reported library of glycoconjugate benzene sulfonamides**79-82** (Fig. 11.14) as carbohybrid and investigated their ability to inhibit the enzymatic activity of physiologically relevant human carbonic anhydrase (hCA) isozymes: hCA I, II, and tumor-associated IX.[36] In this synthetic strategy authors directly links the known CA pharmacophore ($ArSO_2NH_2$) to a sugar "tail" moiety through a rigid 1,2,3-triazole linker unit using the Cu(I)-catalyzed 1,3-dipolar cycloaddition reaction or "click chemistry". Many of the glycoconjugates were found potent CA inhibitors and exhibited some isozyme selectivity as well. In particular, the methyl-D-glucuronate triazoles **79** and **80** were potent inhibitors of hCA IX (*K*i 9.9 and 8.4 nM, respectively) with selectivity also favoring this isozyme. Other exceptional compounds included the deprotected β-D-ribofuranosyl triazole **81** and *D*-arabinosyl triazole **82**, which were inhibiting hCA II inhibitors.

Poulsen and co-workers designed and synthesized novel hybrid glycoconjugates **83–87**, primary sulfonamides (Fig. 11.15) that bind to the extracellular catalytic domain of cancer-associated human carbonic anhydrase CA IX and CA XII.[37] Authors have synthesized these compounds using variably acylated glycopyranosyl azides and either 3- or 4-ethynyl benzene sulfonamide using Cu(I)-catalyzed azide alkyne cycloaddition (CuAAC). The cancer-associated

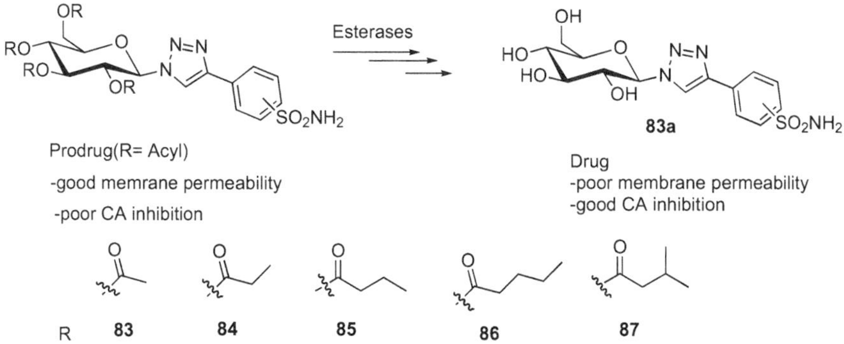

Figure 11.15 ***Sulfonamides based anticancer carbohybrids reported by Poulsen.***

human carbonic anhydrase enzyme inhibition profile for these compounds was determined, while *in vitro* metabolic stability, plasma stability, and plasma protein binding for a representative set of compounds was measured. Their findings demonstrate that the influence of the differing acyl groups on these key biopharmaceutical properties and confirmed that acyl group protected carbohydrate-based sulfonamides have potential as prodrugs for selectively targeting the extracellular cancer-associated CA enzymes.

In their research finding[37] authors have stated that the ideal profile for an oral prodrug (acyl-masked) targeting cancer-related hCA IX and XII should have good membrane permeability and poor CA inhibition. Once molecule reaches in to blood circulation it should easily get unmasked. The unmasked compound should then exhibit the opposing characteristics that of poor membrane permeability and good CA inhibition. The compounds in the glucose series of sulphonamides **83–87** displayed the needed SAR profile to fulfil these said criteria as evidenced by hCA *K*i values. The acetyl, propionyl, and butanoyl analogues **83–87** were degraded in human plasma and HLMs, consistent with esterase processing, to form the fully deacylated compound **83a** that remained stable in these environments. The propionyl analogue **84** has a cLog P value of +2.81, consistent with expected good membrane permeability, while compound **83** has a cLog P value of −1.75, consistent with poor membrane permeability. The measured membrane permeability for **83** in a Caco-2 cell model confirmed negligible membrane permeability, while compound **84** may be retained in the membrane, and future efforts needed in optimizing this compound. These findings provide a beneficial guide to the impact of these acyl groups on glycoconjugate prodrug stability and performance.

2.5 Antimalarial activity

Park and co-workers reported carbohybrid-based 2-aminopyrimidine compounds with fast-acting growth inhibitory activities against three laboratory strains of multidrug-resistant *P. falciparum*.[38] Authors studied structure-activity relationship which led to the identification of two derivatives (**88** and **89**) as the most promising anti-malarial candidates (mean EC_{50} of 0.130 and 0.096 μM against all three *P. falciparum* strains, selectivity indices >600, microsomal stabilities >80%, and mouse malaria ED_{50} values of 0.32 and 0.12 mg/kg/day, respectively), targeting all major blood stages of multidrug-resistant *P. falciparum* parasites (Table 11.4).

2.6 Antifungal activity

D. A. Ibrahim and co-workers docked thiophenyl-arabinoside conjugates against antigen Ag85C (PDB code: 1va5) using Glide.[39] Compounds **90–93** (Fig. 11.16) with good docking scores were synthesized by a Gewald synthesis followed by linking to 5-thioarabinofuranosides. The resulting thiophenyl-thioarabinofuranosides hybrids were assayed for inhibition of mycoyltransferase activity using a 4-methylumbelliferyl butyrate fluorescence assay. The conjugates showed *Ki* values ranging from 18.2 to 71.0 μM. The most potent inhibitor was soaked into crystals of *Mycobacterium tuberculosis* antigen 85C and the structure of the complex determined. The *X*-ray structure shows the compound bound within the active site of the enzyme with the thiophene moiety positioned in the putative α-chain binding site of TMM and the arabinofuranoside moiety within the known carbohydrate-binding site as exhibited for the Ag85B-trehalose crystal structure. Unexpectedly, no specific hydrogen bonding interactions were being formed between the arabinofuranoside and the carbohydrate-binding site of the active

Table 11.4 Antimalarial activity of 2-aminopyrimidine based carbohybrids.

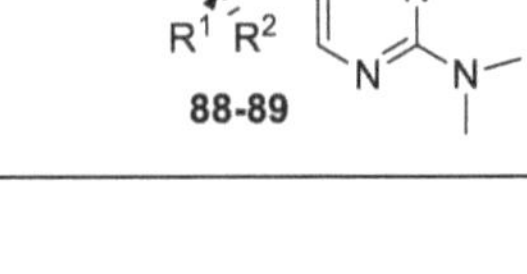

88-89

						EC_{50}(μM)
Compds.	**R¹**	**R²**	**R³**	**R⁴**	**R⁵**	**K1 Strain**
88	OPMB	H	OMs	H	Trityl	0.116
89	OPMB	H	H	N_3	Trityl	0.092

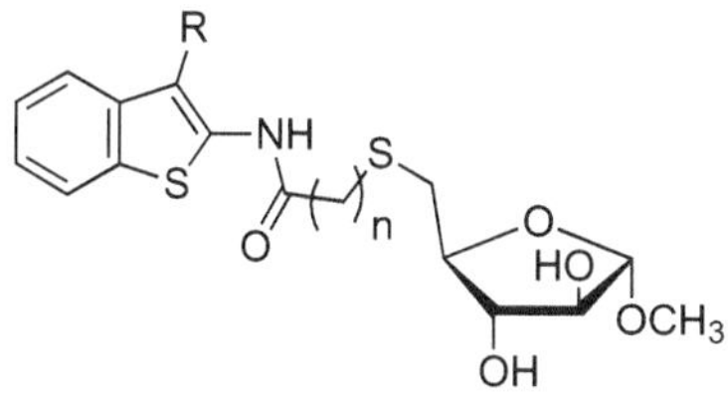

90 R = $COOC_2H_5$ n= 1
91 R = CN n = 1
92 R = $COOC_2H_5$ n =1
93 R = CN n =3

Figure 11.16 ***Mycoyltransferase inhibitors.***

site suggesting that the binding of the arabinoside within this structure is driven by shape complementarily between the arabinosyl moiety and the carbohydrate binding site.

C.S. Stauffer and co-workers investigated an as yet unknown structure-activity relationship of the Nikkomycin family of antifungal peptidyl nucleoside antibiotics. In their recent research paper the synthesis and antifungal evaluation of a carbohydrate ring-expanded pyranosyl nucleoside analogue of nikkomycin B have been reported.[40] A convergent synthetic route, independent synthesis of the *N*-terminal amino acid side chain and a stereoselective *denovo* construction of the desired pyranosyl nucleoside amino acid fragment was followed by peptidic coupling of the two components, leading to the first synthesis of a carbohydrate ring-enlarged pyranosyl nikkomycin B analogue was achieved here (**94-96**). In vitro biological evaluation of the above analogue against a variety of human pathogenic fungi demonstrated significant antifungal activity against several fungal strains of clinical significance (Table 11.5).

Antifungal activity of Nikkomycins was carried out *in vitro* to check the susceptibility testing of the pyranosyl analogue **94** (having enlarge carbohydrate ring) against clinical isolates of six different fungi, *Candida albicans, Candida glabrata, Aspergillus fumigatus, Cryptococcus neoformans, Coccidioides immitis,* and *Blastomyces dermatitidis*, were performed. They have used antifungal drug amphotericin B (Ampho-B) **95** and commercially available Nikkomycin Z (Nik-Z) **96** as reference standards in these antifungal assays. As evident from Table 11.5 that the pyranosyl Nikkomycin B analogue **94** was found to be inactive against *Candida albicans, Candida glabrata,* and *Aspergillus fumigatus.* This is not entirely unexpected, as Nikkomycin Z (the most potent among the natural Nikkomycins) itself is not a very efficient antifungal agent against most strains of *Candida* and *Aspergillus*. Gratifyingly, against

Table 11.5 Antifungal activity of peptidyl nucleoside hybrids.

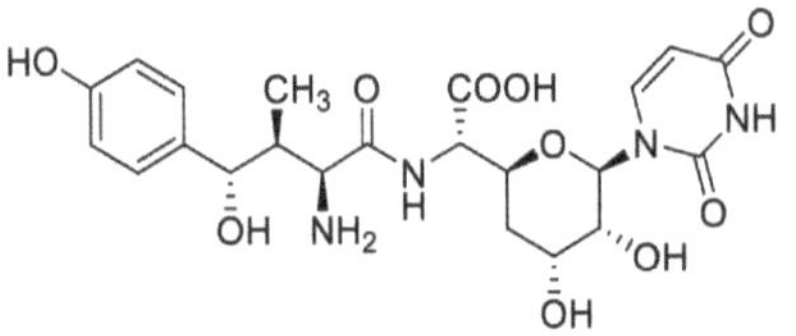

94 Pyranosyl Nikkomycin B

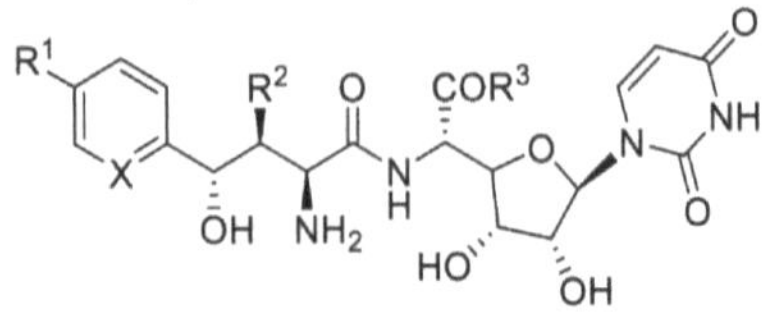

95 Nik B: X = CH;R^1= R^3 = OH, R^2 = Me
96 Nik Z X = N; R^1 = R^3 = OH; R^2 = Me

Fungal species (No. of isolates)	94Pyranosyl -Nik B MIC(μg/mL)	95AmphoB MIC (μg/mL)	96Nik-Z MIC(μg/mL)
C. albicans	>16	0.125	4,8and > 16
C. glabrata	>16	0.125 and 0.25	>16
A. fumigatus	>16	0.25-0.5	2-4
C. neoformans	≤ 0.03	0.25	≤ 0.03
C. immitis	≤ 0.03	0.25	0.25
B. dermatitidis	0.5	0.25	0.03

human pathogenic fungal strains of *Cryptococcus neoformans*and *Coccidioides immitis*, the analogue **94** exhibited strong inhibitory activity. While the antifungal activity of **94** against *Cryptococcus neoformans*was equipotent to that of Nikkomycin Z **96** (and significantly better than that of amphotericin B), the MIC of **94** against *Coccidioides immitis* was much lower than both the references amphotericin B **95** and Nikkomycin Z **96**. The pyranosyl analogue **94** also displayed some inhibitory activity against *Blastomyces dermatitidis*, although not as active as amphotericin B or Nikkomycin Z. It is worth mentioning here that, relatively less common, fungal infections caused by *C. neoformans* (cryptococcosis), *C. immitis* (coccidiodomycosis), and *B. dermatitidis* (blastomycosis) are nonetheless pathogenic systemic (lungs, brain, bone, GI tract, etc) fungal infections of increasingly serious concern. Interestingly, amphotericin B **95** was a commonly used agent for the treatment of these infections, in their antifungal assay, involving clinical isolates of the above human pathogenic strains, the newly synthesized pyranosyl Nik-B analogue **94** was found to be about ten times more active than the amphotericin B standard.

2.7 Antibacterial activity

K. Karthik Kumar and co-workers synthesized a series of quinoline coupled 1,2,3-triazoles compounds as carbohybrids **97–102** by 'click chemistry' from azidomethyl quinoline with different alkynes.[41] The efficiency and fidelity of the Cu(I)-catalyzed azide alkyne reaction are substantiated by good yields and exclusive formation of the expected 1,4-disubstituted triazole product. All the synthesized compounds were screened for anti-tubercular activity against *Mycobacterium tuberculosis* $H_{37}Rv$ by luciferase reporter phage (LRP) assay. Quinoline coupled triazole sugar hybrid, **97** is the most potent compound in the series with 76.41% and 78.37% reduction calculation based on percentage reduction in Relative Light Units at 5 and 25 lg/mL, respectively (Fig. 11.17).

L. Nagarapu and co-workers prepared novel *C*-linked imidazole derivatives **103-111** as hybrid molecules (Fig. 11.18).[42] All the newly synthesized compounds **103–107** and **108–111** were screened for antibacterial and antifungal activities. All the compounds **103–107** and **108–111** showed activity against Gram-negative and Gram-positive bacteria (Table 11.6).

Compounds **105, 106, 110,** and **111** showed moderate antibacterial activity against *Pseudomonas aeruginosa*. Compound **111** showed more activity towards Gram-positive bacteria (i.e., *Bacillus subtilis*). All the compounds were screened for antifungal activity against *Saccharomyces cerevisiae, Aspergillus niger, Rhizopus oryzae,* and *Candida albicans* by agar cup diffusion method using Amphotericin-B as standard. However, none of the compounds (Fig. 11.18) showed antifungal activity (Table 11.6).

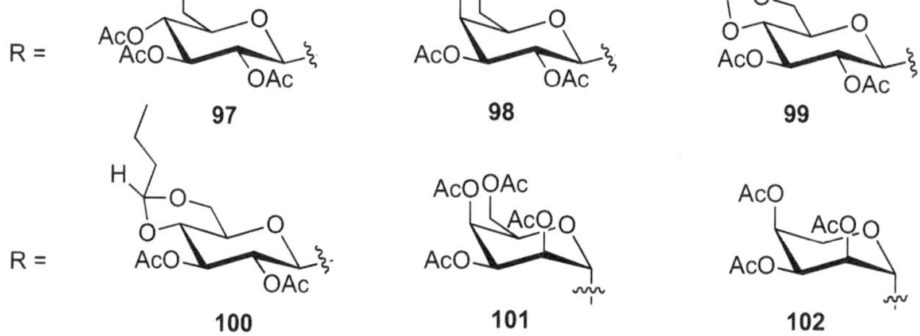

Figure 11.17 ***Anti-tubercular activity of quinoline carbohybrids.***

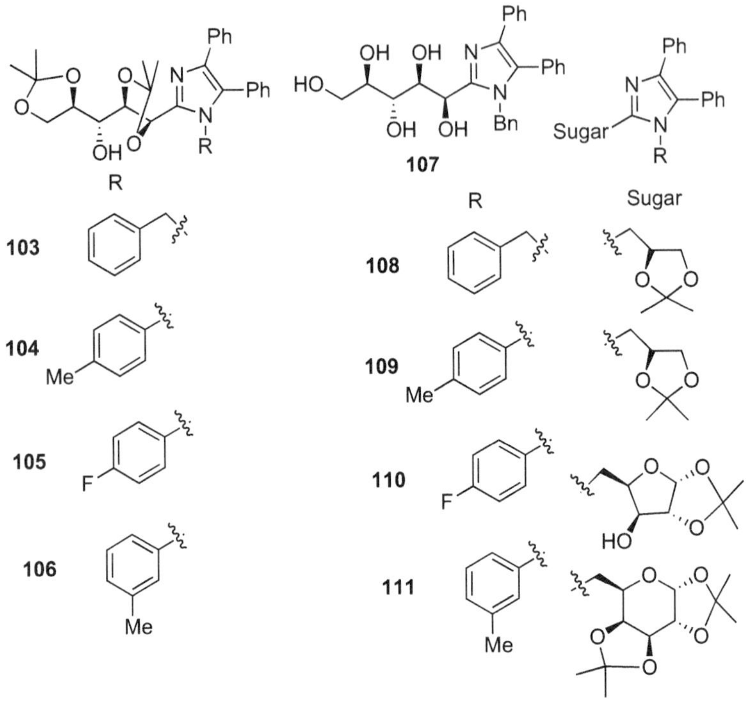

Figure 11.18 *Imidazole linked carbohybrids.*

Table 11.6 Antibacterial activity of imidazole carbohybrids.

Compounds	Gram positive bacteria			Gram negative bacteria		
	B. subtilis	*S. aureus*	*S. epidermidis*	*E. coli*	*P. aeruginosa*	*K. pneumoniae*
103	75	75	150	75	37.5	75
104	75	75	150	75	75	75
105	150	150	150	75	75	75
106	150	150	75	75	37.5	150
107	75	75	75	37.5	37.5	75
108	150	150	75	75	150	75
109	150	150	150	150	37.5	75
110	37.5	37.5	75	75	75	75
111	75	75	75	75	75	75
Streptomicin	6.25	1.562	1.562	2.35	3.125	3.125
Penicilin	1.526	6.25	3.125	7.81	12.5	6.25

2.8 Glycosidase inhibitor activity

K.K. Yeoh and co-workers prepared a series of carbohybrids as **112–115** putative mimics of UDP-sugar donors from the corresponding α-propargyl glycosides and 5-azido uridine in aqueous solution using Cu(I)-catalysed Huisgen cycloaddition.[43] Authors have observed that none of the compounds displayed significant inhibitory activity against bovine milk β-1-4-galactosyltransferase, indicating that triazole is not a good isosteric replacement here for pyrophosphate. In this respect, a *bis*-triazole moiety may be a more appropriate replacement, particularly with respect to potential ability for coordination to Mn^{2+} (Fig. 11.19).

D.R. da Rocha and co-workers have reported the synthesis often new carbohybrids as 1,2,3-triazole glycoconjugates **116–125** from *D*-glucose and evaluated *in vitro* assays for their ability to inhibit the enzyme α-glucosidase.[44] Most of the compounds had low activity or were inactive when compared with acarbose. However, the derivative 1,2-*O*-isopropylidene-3-phenyl-5-(4-phenyl-1H-1,2,3-triazole-1-yl)-α-D-ribofuranose **124** possessed activity comparable with the standard drug (Fig. 11.20).

D. E. Green and co-workers synthesized glycosylated pyridinone analogue as carbohybrids **126–133** (Fig. 11.21) and found that these compounds chelate potentially with toxic metal ions.[45]An MTT cytotoxicity assay of a selected glycosylated compound showed a relatively low toxicity of IC_{50} = 570 ± 90 μM in a human breast cancer cell line. The authors presumed that these pyridinone glycosides could be cleaved by a broad specificity β-glycosidase and tested for glycosidase inhibitory activity (β-glucosidase from *Agrobacterium sp.*). One of compound's *K*cat and *K*m

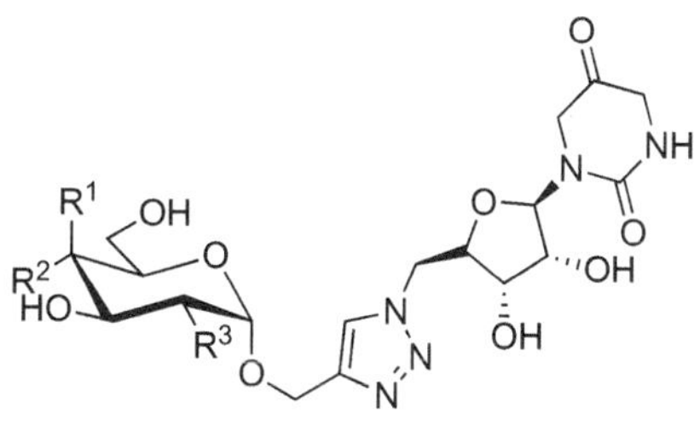

112 R¹ = OH, R² = H, R³ = OH
113 R¹ = H, R² = OH, R³ = OH
114 R¹ = H, R² = OH, R³ = NHAc
115 R¹ = OH, R² = H, R³ = NHAc

Figure 11.19 ***Compounds mimicking UDP-sugar donors.***

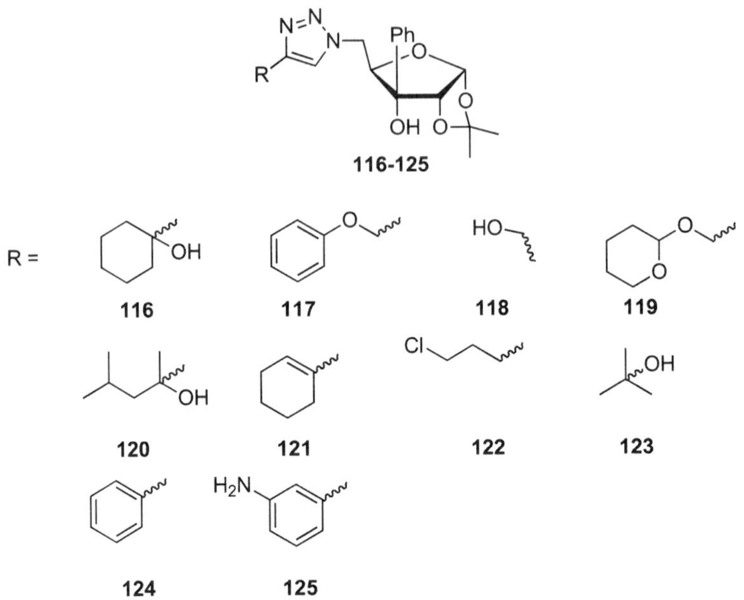

Figure 11.20 ***α-Glucosidase inhibitors.***

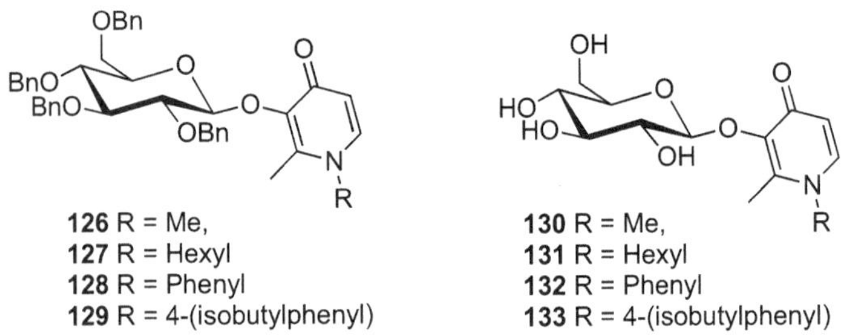

Figure 11.21 ***Glycosidase inhibitors hybrids.***

were determined to be 19.8 and 1.52 mM, respectively. Trolox Equivalent Antioxidant Capacity (TEAC) values were determined for the free pyridinones, indicating the good antioxidant properties of these compounds. Metal-Ab1-40 aggregates with zinc and copper were resolubilized by the non-glycosylated pyridinone ligands (Fig. 11.21).

2.9 Galectin-3 inhibitors

V. K. Rajput and co-workers reported the synthesis of a hybrid of doubly 3-*O*-coumaryl methyl substituted thiodigalactoside derivative **134–141**

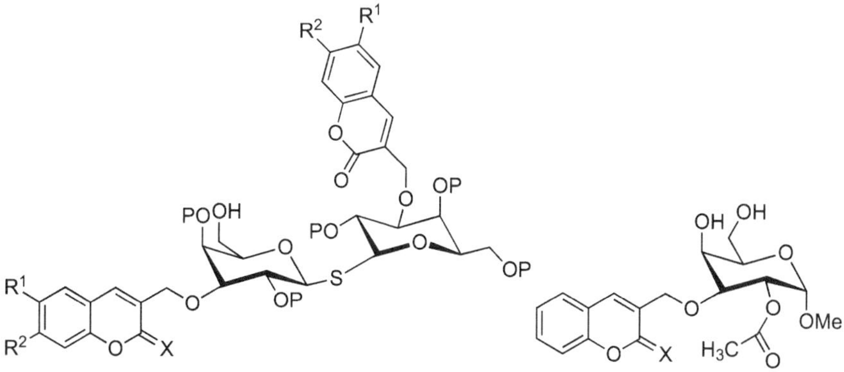

134 $R^1 = R^2$ = H,X = O, P = Ac, X = NTs
135 R^1 = H, R^2 = OMe, P = Ac, X= NTs
136 R^1 = Cl R^2 = H, P = Ac X = NTs
137 $R^1 = R^2$ = H, P = H, X = O
138 R^1 = H, R^2 = OMe, P = H, X = O
139 R^1 = Cl, R^2 = H, P = H, X = O

140 X = NTs
141 X = O

Figure 11.22 ***Galectin-3 inhibitors.***

(Fig. 11.22).[46] Compound **137** was identified as the first potent (K_d 91 nM) and selective inhibitor of galectin-3. Other galectins were also screened and found that they were inhibited with this class of molecule in the μM range. The detailed structural and mutant analysis was done by this group and it was evidenced that the affinity enhancements by the coumaryl moieties were in part due to double face-to-face stacking onto Arg144 and Arg186. They also showed water-mediated polar interaction with Lys176. Compound **137** proved efficacious in a bleomycin-induced mouse model of lung fibrosis, strongly suggesting that galectin-3-glycan interactions are limiting in progression of lung fibrosis in this model. Furthermore, search of a highly selective galectin-3 inhibitor is important not only for possible treatment of lung fibrosis with minimum side effects due to minimized cross reactivity with other galectins but possibly even more so for the discovery and development of galectin-3 inhibitors for treating fibrosis on other organs systems, and thus systemic exposure, where a different panel of galectins may be present at different levels.

2.10 Anti-inflammatory

Sah and co-workers synthesized a series of 4,6-*O*-ethylidene-β-D-glucopyranosyl amine derived glycoconjugates as hybrids **142a–142f**

Table 11.7 Anti-inflammatory glycoconjugate derivatives.

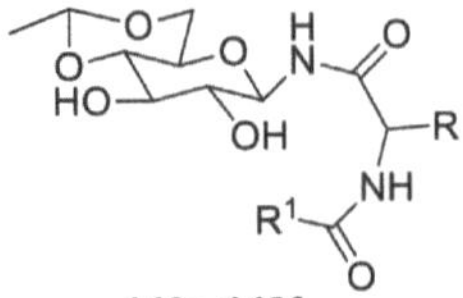

142a-142f

Compounds	R	R^1
142a	CH_3	2-hydroxybenzene
142b	CH_3	2-indole
142c	CH_3	2-hydroxy-6-methylbenzene
142d	CH_2Ph	2-hydroxybenzene
142e	CH_2Ph	2-Indole
142f	CH_2Ph	2-hydroxy-6-methylbenzene

(Table 11.7) containing amino- and aromatic acids.[47] They have screened these molecules for their anti-inflammatory and analgesic activity on Wistar rat and Swiss Albino mice respectively. The anti-inflammatory activity was explored using a carrageenan induced paw oedema model while an acetic acid induced writhing model was adapted for the analgesic studies. All of the prepared compounds **142a-142f** were found possessing anti-inflammatory and analgesic activity in the range of 63–84% and 86–94% respectively.

3 Conclusion and future prospective

From above literature reports it becomes very much clear that molecular hybridization is a powerful tool for drug discovery research. Carbohybrids is a class of hybrids which are found in lot of natural products in fused or linked form. So, carbohydrate motif with privileged scaffold, having additional advantages on Absorption Distribution Metabolism and Excretion (ADME). It's evident that, when carbohydrate molecules gets attached to privileged bioactive scaffolds it increases bioactivity of drug molecules. We have summarised here only recently reported carbohybrid molecules having bioactivity in particular anticancer, antiviral, antibacterial, antifungal, immununomodulatory, PTP1B inhibitors, carbonic anhydrase inhibitors, antimalarial, glycosidase inhibitors, galectin-3-inhibitors and anti-inflammatory activity. Out of these reported activity there are lot more biological potential exist in the carbohybrid molecules which need to be explored in near future.

In the future, we should expect the discovery of new lead bioactive molecules based on glycohybrids (carbohydrids) for neurodegenerative disorders like Alzheimer's and Parkinson's diseases along with development of some efficacious molecules from current known molecules with sub nM activities. It is believed that, increased levels of amyloid-β (Aβ) protein in brain is a critical event in triggering a wide range of molecular alterations leading to Alzheimer's diseases (AD). It is also appearing clear that, the status of the blood-brain barrier (BBB) and choroid plexus, along with hepatic functionality, is very important consideration when Aβ balance is addressed. Carbohyrate molecules have inherited advantage to cross the BBB and thus making carbohybrid molecules as an ideal scaffold to work on for AD. Authors believe that organometallic glycohybrids offer immense promise for drug design and development and this will grow into an important research field in near future. Glycohybridization with bioactive organometallic may lead some important bioactive functional molecules. This area can be pursued as new research area.

We should also expect the development of efficient and new methods to synthesize carbohybrids. Already many researchers use the multicomponent reaction and click chemistry to gain rapid access to series of carbohybrids. We have been working since a long time towards efficient synthesis of carbohybrid molecules. We have developed quite few robust route to prepare diverse classes of bio-active carbohydrids.[48]. There is lot more space to work on and develop new robust and industrially viable methods for the synthesis of such class of molecules. The fewer and most efficient chemical steps will allow for the discovery and eventual large-scale production of therapeutic hybrids in industry in shorter time in cost effective manner.

References

1. Eder, J.; Sedrani, R.; Wiesmann, C. *Nat. Rev. Drug Discov.* **2014**, *13*, 577–587.
2. Moffat, J. G.; Rudolph, J.; Bailez, D. *Nat. Rev. Drug Discov.* **2014**, *13*, 588–602.
3. Moffat, J. G.; Vincent, F.; Lee, J. A.; Eder, J.; Prunotto, M. *Nat. Rev. Drug Discov.* **2017**, *16*, 531–543.
4. Swinney, D. C.; Anthony, J. *Nat. Rev. Drug Discov.* **2011**, *10*, 507–519.
5. Macarron, R.; Banks, M. N.; Bojanic, D.; Burns, D. J.; Cirovic, D. A.; Garyantes, T.; Green, D.V.; Hertzberg, R. P.; Janzen, W. P.; Paslay, J. W.; Schopfer, U.; Sittampalam, G. S. *Nat. Rev. Drug Discov* **2011**, *10*, 188–195.
6. (a) Shen, *B. Cell,* **2015**, 163, 1297–1300; (b) Fraga, C. A. M. Expert Opin. *Drug Discov.* **2009**, 4, 605-609.
7. (a) Li, J. W.-H.; Vederas, *J. C. Science*, **2009**, 325, 161–165; (b) Tietze, L. F.; Bell, H. P.; Chandrasekhar, S. Angew. *Chem. Int. Ed.* **2003**, 42, 3996-4028.
8. (a) Li, P. H.; Zeng, P.; Chen, S. B.; Yao, P. F.; Mai, Y. W.; Tan, J. H.; Ou, T. M.; Huang, S. L.; Li, D.; Gu, L.Q.; Huang, Z. S. *J. Med. Chem.* **2016**, 59, 238–252. (b) Shah, D. K. J. Pharmacokinet. *Pharmacodyn.* **2015**, 42, 553-571.
9. Bua, S.; Mannelli, L. D. C.; Vullo, D.; Ghelardini, C.; Bartolucci, G.; Scozzafava, A.; Supuran, C. T.; Carta, F. *J. Med. Chem.* **2017**, *60*, 1159–1170.

10. Licitra, E. J.; Liu, J. O. *Proc. Natl. Acad. Sci., U.S.A.* **1996**, *93*, 12817–12821.
11. Schwinn, M. K.; Machleidt, T.; Zimmerman, K.; Eggers, C. T.; Dixon, A. S.; Hurst, R.; Hall, M. P.; Encell, L. P.; Binkowski, B. F.; Wood, K.V. *ACS. Chem. Biol.* **2018**, *13*, 467–474.
12. Lazar, C.; Kluczyk, A.; Kiyota, T.; Konishi, Y. *J. Med. Chem.* **2004**, *47*, 6973–6982.
13. V-Junior, C.; Danuello, A.; Bolzani, V. D. S.; Barreiro, E. J.; Fraga, C. A. M. *Curr. Med. Chem.* **2007**, *14*, 1829–1852.
14. Varki, A.; Cummings, R. D.; Esko, J. D.; Freeze, H. H.; Stanley, P.; Bertozzi, C. R.; Hart, G. W.; Etzler, M. E.. second edn; *Essentials of Glycobiology*,Cold Spring Harbour: New York, 2009.
15. Gabius, H. J.; Andre, S.; Jimenez-Barbero, J.; Romero, A.; Solis, D. *Trends Biochem. Sci.* **2012**, *36*, 298–313.
16. Lis, H.; Sharon, N. *Chem. Rev.* **1998**, *98*, 637–674.
17. Cipolla, L.; Araujo, A. C.; Bini, D.; Gabrielli, L.; Russo, L.; Shaikh, N. *Expert Opin. Drug Discov.* **2010**, *5*, 721–737.
18. Lundquist, J. J.; Toone, E. J. *Chem. Rev* **2002**, *102*, 555–578.
19. Galan, M. C.; Alifonso, D. B.; Watt, G. M. *Org. Biomol. Chem.* **2011**, *9*, 3598–3610.
20. Chabre, Y. M.; Roy, R. *Adv. Carbohydr. Chem. Biochem.* **2010**, *63*, 165–393.
21. Crucho, C. I.; Silva, C-D; Petrova, K. T.; Barros, M. T. *Carbohydr. Res.* **2015**, *402*, 124–132.
22. Zong, G.; Barber, E.; Aljewari, H.; Zhou, J.; Hu, Z.; Du, Y.; Shi, W. Q. *J. Org. Chem.* **2015**, *80*, 9279–9291.
23. Campos, V. R.; Cunha, A. C.; Silva, W. A.; Ferreira, V. F.; Sousa, C. S.; Fernandes, P. D.; Moreira, V. N.; Rocha, D. R.; Dias, F. R. F.; Montenegro, R. C.; Souza, M. C. B.V. D. F. D.; Boechat, C. S.; Franco, C. F. J.; Resende, J. A. L. C. *RSC Adv* **2015**, *5*, 96222–96229.
24. Upadhyaya, K.; Hamidullah, K.; Singh, A.; Arun, M.; Shukla, N.; Srivastava, R.; Ashraf, A.; Sharma, R.; Mahar, S. K.; Shukla, J.; Sarkar, R.; Ramachandran, J.; Konwar, R. L.; Tripathi, R. P. *Org. Biomol. Chem* **2016**, *14*, 1338–1358.
25. Gawołek, G. P.; Malarz, K.; Wilczkiewicz, A. M.; Musiol, M.; Serda, M.; Czaplinska, B.; Musiol, R. *Eur. J. Med. Chem.* **2016**, *112*, 130–144.
26. Gawolek, G. P.; Plesniak, M.; Komor, R.; Wyrostek, A. B.; Erfurt, K.; Szeja, W. *Bioorg. Chem.* **2017**, *72*, 80–88.
27. Leaver, D. J.; Dawson, R. M.; White, J. M.; Polyzos, A.; Hughes, A. B. *Org. Biomol. Chem.* **2011**, *9*, 8465–8474.
28. Yu, M.; Si, L.; Wang, Y.; Wu, Y.; Yu, F.; Jiao, P.; Shi, Y.; Wang, H.; Xiao, S.; Fu, G.; Tian, K.; Wang, Y.; Guo, Z.; Ye, X.; Zhang, L.; Zhou, D. *J. Med. Chem.* **2014**, *57*, 10058–10071.
29. Cohen, M.; Senaati, H. P.; Fisher, C. J.; Huang, M. L.; Gagneux, P.; Godula, K. *ACS Cent. Sci.* **2016**, *2*, 710–714.
30. Sharma, D. K.; Lambu, M. R.; Sidiq, T.; Khajuria, A.; Tripathi, A. K.; Yousufa, S. K.; Mukherjee, D. *RSC Adv.* **2013**, *3*, 11450–11455.
31. He, X. -P.; Li, C.; Jin, X. -P.; Song, Z.; Zhang, H. -L.; Zhu, C. -J.; Shen, Q.; Zhang, W.; Sheng, L.; Shi, X. –X.; Li, Y.; Chen, J.; Xie, G. -R. *New J. Chem* **2011**, *35*, 622–631.
32. Yang, J. -W.; He, X. -P.; Li, C.; Gao, L. -X.; Sheng, L.; Xie, J.; Shi, X. -X.; Tang, Y.; Li, J.; Chen, G. -R. *Bioorg. Med. Chem. Lett* **2011**, *21*, 1092–1096.
33. Lin, L.; Shen, Q.; Chen, G. -R.; Xie, J. *Bioorg. Med. Chem* **2008**, *16*, 9757–9763.
34. Kolb, H.; Joseph, J. C.; Dhanalakshmi, K.; Vani, P.; Bing, W.; Umesh, B. G.; Brian, A. D.; Kai, C.; Wei, Z.; Gang, C.; Clifton, H. P.; Farhad, K.; Peter, J. H. S.; Zhiyong, G.; Qianwa, L.; Tieming, C. T.; *Chunfang, Z. U. S. Patent US20100317842 A1*, **2010**, WO2008124703.
35. Lopez, M.; Bornaghi, L. F.; Innocenti, A.; Vullo, D.; Charman, S. A.; Supuran, C. T.; Poulsen, S. -A. *J. Med. Chem.* **2010**, *53*, 2913–2926.
36. Wilkinson, B. L.; Bornaghi, L. F.; Houston, T. A.; Innocenti, A.; Supuran, C. T.; Poulsen, S. -A. *J. Med. Chem.* **2006**, *49*, 6539–6548.

37. Carroux, C. J.; Rankin, G. M.; Moeker, J.; Bornaghi, L. F.; Katneni, K.; Morizzi, J.; Charman, S. A.; Vullo, D.; Supuran, C. T.; Poulsen, S. -A. *J. Med. Chem* **2013**, *56*, 9623–9634.
38. Lee, S.; Lim, D.; Lee, E.; Lee, N.; Lee, H.; Cechetto, J.; Liuzzi, M.; Junior, L. H. F.; Song, J. S.; Bae, M. A.; Oh, S.; Ayong, L.; Park, S. B. *J. Med. Chem* **2014**, *57*, 7425–7434.
39. Ibrahim, D. A.; Boucau, J.; Lajiness, D. H.; Veleti, S. K.; Trabbic, K. R.; Adams, S. S.; Ronning, D. R.; Sucheck, S. J. *Bioconjugate Chem.* **2012**, *23*, 2403–2416.
40. Stauffer, C. S.; Bhaket, P.; Fothergill, A. W.; Rinaldi, M. G.; Datta, A. *J. Org. Chem.* **2007**, *72*, 9991–9997.
41. Kumar, K. K.; Prabu, S.; Vasan, S.; Kumar, V.; Das, T. M. *Carbohydr. Res.* **2011**, *346*, 2084–2090.
42. Nagarapu, L.; Satyender, A.; Rajashaker, B.; Srinivas, K.; Rani, P. R.; Radhika, K.; Subhashini, G. L. *Bioorg. Med. Chem. Lett* **2008**, *18*, 1167–1171.
43. Yeoh, K. K.; Butters, T. D.; Wilkinson, L. B.; Fairbanks, A. J. *Carbohydr. Res.* **2009**, *344*, 586–591.
44. Rocha, D. R. D.; Santos, W. C.; Lima, E. S.; Ferreira, V. F. *Carbohydr. Res.* **2012**, *350*, 14–19.
45. Green, D. E.; Bowen, M. L.; Scott, L. E.; Storr, T.; Merkel, M.; Ohmerle, K. B.; Thompson, K. H.; Patrick, B. O.; Schugar, H. J.; Orvig, C. *Dalton Trans.* **2010**, *39*, 1604–1615.
46. Rajput, V. K.; MacKinnon, A.; Mandal, S.; Collins, P.; Blanchard, H.; Leffler, H.; Sethi, T.; Schambye, H.; Mukhopadhyay, B.; Nilsson, U. J. *J. Med. Chem.* **2016**, *59*, 8141–8147.
47. Soni, K.; Sah, A. K. *RSC Adv* **2014**, *4*, 6068–6073.
48. (a) Kumari, P.; Mishra, V.; Narayana, C.; Chakrabarty, A.; Sagar, R., *Sci. Rep.* **2020**, *10*, 6660; (b) Kumari, P.; Dubey, S.; Venkatachalapathy, S.; Narayana, C.; Gupta, A.; Sagar, R. *New J. Chem.* **2019**, *43,* 18590-18600; (c) Narayana, C.; Kumari, P.; Sagar, R. *Org. Lett.* **2018**, *20*, 4240–4244; (d) Kumari, P.; Gupta, S.; Narayana, C.; Ahmad, S.; Singh, S.; Sagar, R. *New J. Chem.* **2018**, *42,* 13985-13997; (e) Kumari, P.; Narayana, C.; Dubey, S.; Gupta, A.; Sagar, R. *Org. Biomol. Chem.* **2018**, *16,* 2049-2059; (f) Narayana, C.; Kumari, P.; Ide, D.; Hoshino, N.; Kato, A.; Sagar, R. *Tetrahedron* **2018**, *74,* 1957-1964; (g) Sagar, R.; Park, J.; Koh, M. and Park, S. B. *J. Org. Chem.* **2009**, *74,* 2171-2174; (h) Sagar, R.; Kim, M.-J. and Park, S. B. *Tetrahedron Lett.* **2008**, *49,* 5080-5083; (i) Sagar, R. and Park, S. B. *J. Org. Chem.* **2008**, *73*, 3270-3273.

CHAPTER TWELVE

Biologically active carbohydrate-containing macrocycles

Ashutosh K. Dash[a], Nazar Hussain[b,c], Debaraj Mukherjee[b,c]
[a]School of Pharmaceutical Sciences, Department of Pharmaceutical Chemistry, Shoolini University Solan, Himachal Pradesh, India
[b]Academy of Scientific and Innovative Research (AcSIR), CSIR-Indian Institute of Integrative Medicine, Jammu, Jammu and Kashmir, India
[c]Natural Product Chemistry Division, CSIR-Indian Institute of Integrative Medicine, Jammu, Jammu and Kashmir, India

1 Introduction

Carbohydrates are the richest class of biomolecules employed by nature for various biological tenders. Macrocycles containing carbohydrates have a great role in the areas of biologically important molecular scaffolds by exploring their structural and functional impressions. Their chemical importance is based on high density of functional groups per unit mass and the choice of stereochemical linkages at the anomeric carbon, which has always challenged the scientist toward a multitude of approaches to study this rich class of compounds. There is a huge application of this class with respect to medicinal use.[1-4] Various natural as well as non-natural macrocycles have been isolated or synthesized which have blossomed over the last decades. Carbohydrate containing macrolides are a group of molecules whose biological activity stems from the presence of the large macrocyclic ring of twelve or more atoms and belong to different classes. Such medicinally effective macrocyclic carbohydrates (MCs) which are naturally present or collected from other sources as potential products, have attracted great curiosity due to their structural convolution and the synthetic challenges associated with their total synthesis.[5-7]

Naturally occurring macrocyclic carbohydrates (MCs) are plentifully found in nature. These macrocycles have usually several monosaccharides or an oligosaccharide flanked on a hydroxylated fatty acid[8] which usually contains eighteen to forty-six- membered macrolactone rings. Some triterpene saponins named cyclic *bis*-desmosides also consists these similar types of MCs, which are composed of two oligosaccharides and a pentacyclic triterpene bridged with 3-hydroxy-3-methyl glutarate. These MCs are also

Carbohydrates in Drug Discovery and Development. http://dx.doi.org/10.1016/B978-0-12-816675-8.00012-9

involved in several biological activities *viz*, plant growth regulation, cytotoxicity against human breast cancer cell lines.[9,10]

1.1 Macrocyclic carbohydrates isolated from plants

There are numerous examples for the contribution of plants in the field of MCs. Some glycolipids which consist MCs are Calonyctin A,[11] Tricolorin A,[12] Tricolorin F,[13] Merremosides,[14] Batatoside L.[15] Calonyctin A are extracted from the leaves of *Calonyction aculeatum*, which is a reputed plant growth regulator. It is a 22-membered macrolactone composed of a 6-deoxygenated tetrasaccharide residue and 11-hydroxy fatty acid, which exists as Calonyctin A1 and A2 (Fig. 12.1).[11] Tricolorin A[13] is a MC which is isolated from *Ipomoea tricolon* (family: convolvulaceae) commonly used as a weedicide, composed of 19-membered macrolactone. Tricolorin F[13] is a subclass of Tricolorin (Fig. 12.1) consist of 21-membered trisaccharide

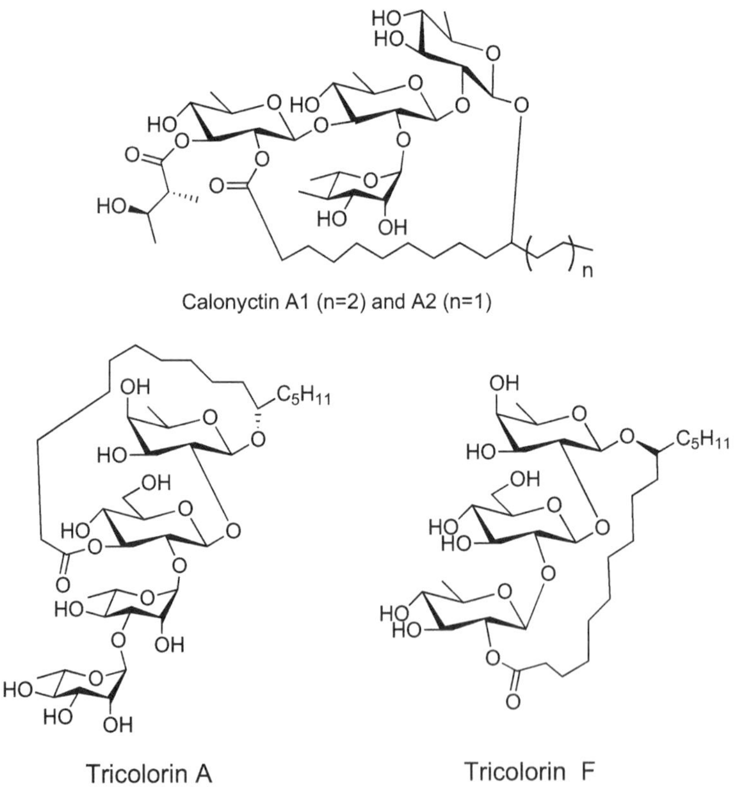

Figure 12.1 ***Structures of Calonyctin & Tricolorin glycolipids containg MCs.***

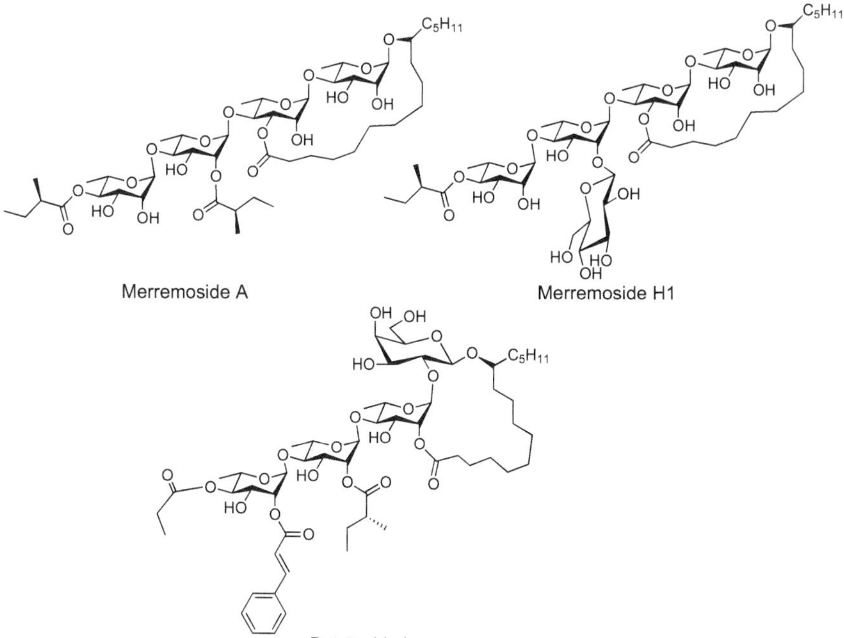

Figure 12.2 ***Structures of Meremoside & Batatoside glycolipids containg MCs.***

macrolactone. Complex MC Merremosides[14] (Fig. 12.2) were isolated from the tuber of *Merremia mammosa* (Indonesia) commonly used against various health issues such as diabetes, respiratory and throat infections. These twenty and twenty-one membered macrolactones displays ionophoretic activity which transport Na^+, K^+, and Ca^{2+} across human erythrocyte membranes.Batatoside[15] a MC is a glycoside, isolated from the tuber of *Ipomoea Batatas* (Convolvulaceae). This MC has been used as a medicinal herb for heamostasis and eliminating abnormal secretions from the ancient ages in China. Batatoside L (Fig. 12.2) has been claimed as an anticancer agent with excellent potency and significant cytotoxic activity against laryngeal carcinoma (Hep-2) cells.[15]

Some cyclic Dimers of 4-(Glycosyloxy) benzoates have been isolated from different medicinal plants[16]among them eight benzoates of MCs were reported consisting of 22-membered macrodilactones e.g., clemochinenosides A and B,[17] berchemolide, clemoarmanoside A, and clemahexapetoside A (Fig. 12.3). Clemochinenosides A and B were isolated from the roots and rhizomes of *Clematis* plant. Berchemolide was isolated from the stems of *Berchemia racemosa. Clematis* rhizomes have been used as anti-inflammatory,

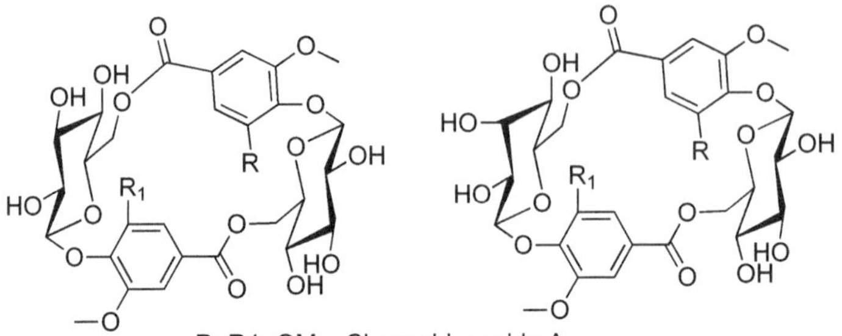

R=R1=OMe, Clemochineoside A
R=H, R1 = OMe, Clemochineoside B
R=R1=H, Clemoarmanoside / Berchemolide

Figure 12.3 ***Structure of cyclic dimers of clemochinenosides and clemoarmanoside.***

antitumor, and analgesic agents or for the treatment of headache, reported in Chinese Pharmacopoeia, and the stem of *Berchemia* was in earlier days used against gall stones and stomach-aches from ancient times in Japan.

Certain cyclic triterpene saponins, named cyclic bisdesmosides, possessing glutarate bridged macro lactones belong to the category were isolated from Chinese medicinal plants *Bolbostemma paniculatum* and *Actinostemma lobatum* (Cucurbitaceae). These saponins like biomolecules displayed antitumor activity, which was due to complex cyclic structure, Lobatoside E (Fig. 12.4).[18]

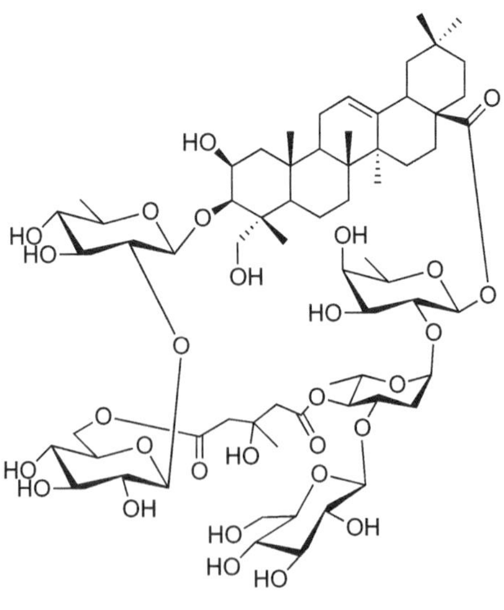

Figure 12.4 ***Structure of Lobatoside E.***

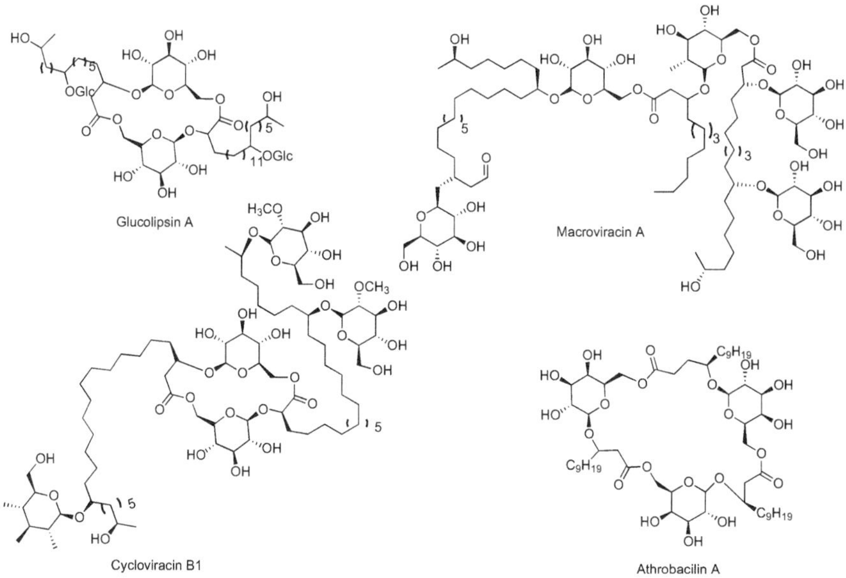

Figure 12.5 ***Some macrocyclic carbohydrates isolated from microorganisms.***

1.2 Macrocyclic carbohydrates isolated from microorganisms

Some glycolipids which were found to be MCs isolated from bacteria are Cycloviracins (Fig. 12.5). Fattiviracins, (isolated from the soil microorganisms *Kibdelosporangium albatum* so. nov. (R761-7) and *Streptomyces microflavus* no. 2445), which are 18-membered dilactone core and exhibit anti-viral activity against herpes simplex virus type 1 (HSV-1), varicellazoster virus (VZV), influenza virus A and B, and human immune deficiency virus type 1 (HIV-1).[19] Macro viracins is a forty two to forty six-membered macro dilactones isolated from the mycelium extracts of Streptomyces sp. BA-2836. These are used as antiviral agents against HSV-1 and VZV. Similarly, Arthrobacilin A, a twenty-seven-membered macrotrilactone bearing galactose as sugar unit, was isolated from the culture broth of Arthrobactorsp. NR2967 was found to be as a cell growth inhibitor.[20]

1.3 Natural carbohydrate-based macrocycles as drug candidate

From plants or microorganisms various complex carbohydrate-containing macrocycles have been isolated which usually contains several monosaccharides or an oligosaccharide having eighteen to fortysix—membered macrolactone rings. These metabolites possess unique molecular constructions which exhibits an unparalleled range of biological activities, such as plant growth regulation, cytotoxicity against human breast cancer cell lines, antiviral, as well as iontophoretic activities.[9, 10] Most of the carbohydrate-based macrocyclic drugs marketed so far are of natural origin, attained from animal extracts, bacterial products or from plant products e.g., A26771B, leucomycin A3, kitasamycin, carbomycin B, tylosin, Mithramycin, Avilamycin-A, saccharomicin, Reblastatin, Isatisine A, Thialanstatin A (Fig. 12.6). By dint of their huge size and complexity, they can fit into the targets flexibly through numerous and spatially distributed binding interactions, increasing both binding affinity and selectivity. MC is a core field from the last few years

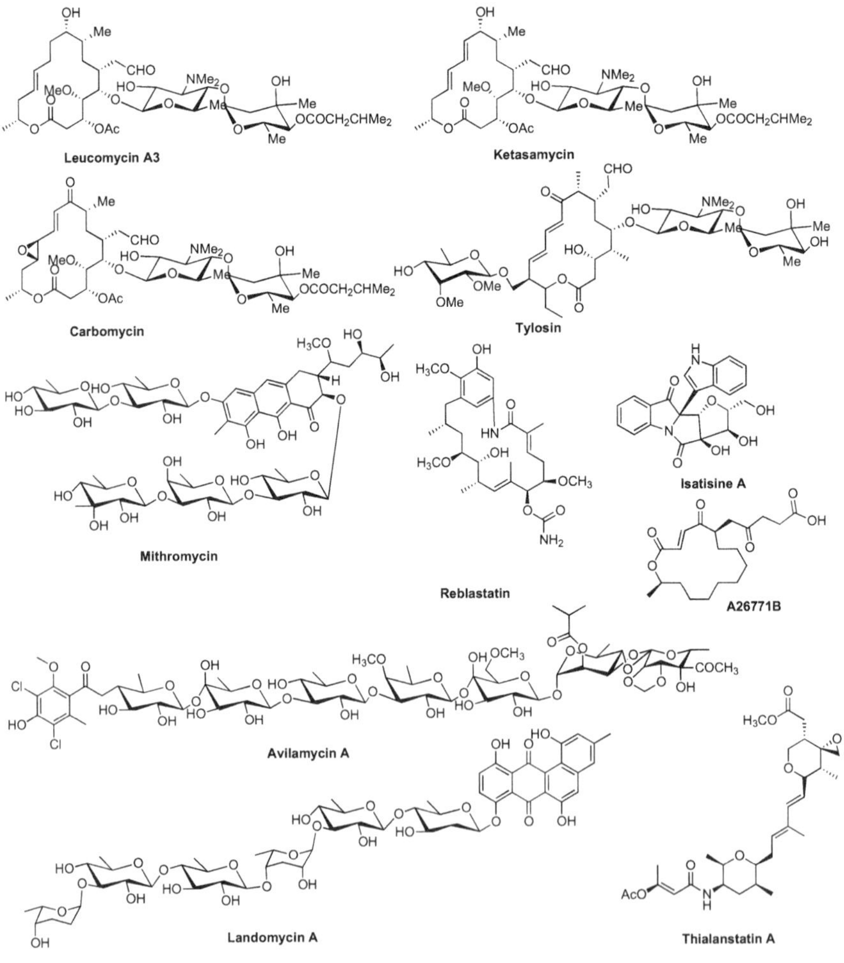

Figure 12.6 ***Macrocyclic carbohydrate as drug candidate isolated from nature.***

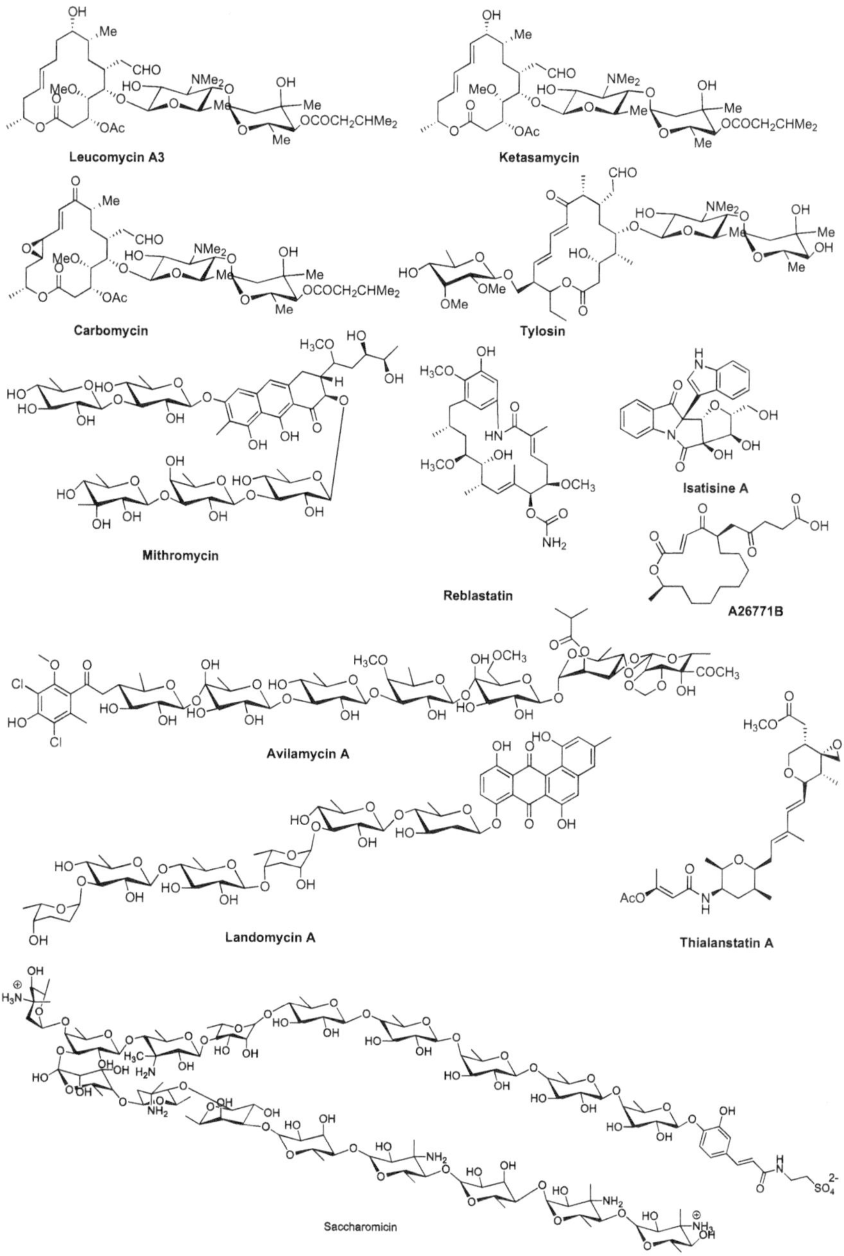

Figure 12.6 *(Cont.)*

by pharmaceutical industry and have made a continued effort aimed at the discovery of novel and clinically applicable anti-biotic, anti fungal antitumor agents. Inspecting market survey some report proclaims almost 68 MCs, are used to treat infections caused by some bacteria, fungus, or some pathogens. A very few of them are used against cancer, cardiovascular system. Some researchers say, macrocyclization has a favourable impact on pharmacokinetic properties required for drugs, such as oral bioavailability, membrane permeability, metabolic stability, etc.[21, 22]

1.4 Macrocyclic carbohydrates used for the drug delivery

In the drug delivery science cyclodextrins and dendrimers have an immense impact, where MCs, are the integral part. This topic has been reported extensively with respect to biological receptors on the corresponding organs, tissues, and cells, etc. As an example, drug delivery regarding liver has been reported where a sialoglycoprotein lectin expressed on mammalian hepatocytes lactosylated microspheres with high specificity of drugs and genes. For alveolar macrophage drug carrier dendrimers were used in the management of asthma mannose modified liposomes selectively targeting their mannose receptors. As another example LEAPT (lectin directed enzyme-activated prodrug therapy) has been developed in this context. For the regio-selective drug release by carbohydrate-lectin interaction (localization of glycosidase enzyme) these drug deliveries have been used to target different cells. In this respect cyclodextrins are amazing drug delivery tools possessing MCs which are based on the hydrophilic outside surface and the hydrophobic internal cavity, forming inclusion complexes with various hydrophobic drugs aiming to improve their water solubility. These are basically oligosaccharides produced from starch by the enzymatic conversions naturally (Fig. 12.7), which are classified by way of α-cyclodextrin (six glucose residue), β-cyclodextrin (seven glucose residue), and γ-cyclodextrin (eight glucose residue).[23] These were the first receptor molecules reported to possessed macrocyclic carbohydrates, which form complexes with many organic and inorganic compounds naturally or synthetically. However, naturally occurring such complexes have relatively low aqueous solubility and are hard to be functionalized. To resolve this issue synthetic cyclodextrins were prepared e.g., Kessler and co-workers have reported a synthetic cyclodextrin analogue, cyclohexamer based on sugar amino acid binds to *p*-nitrophenol and benzoic acid in water, but no binding constants were reported in the literature.[23a] Dendrimers are the hyperbranched complex synthetic polymers used for drug delivery purpose which possess high

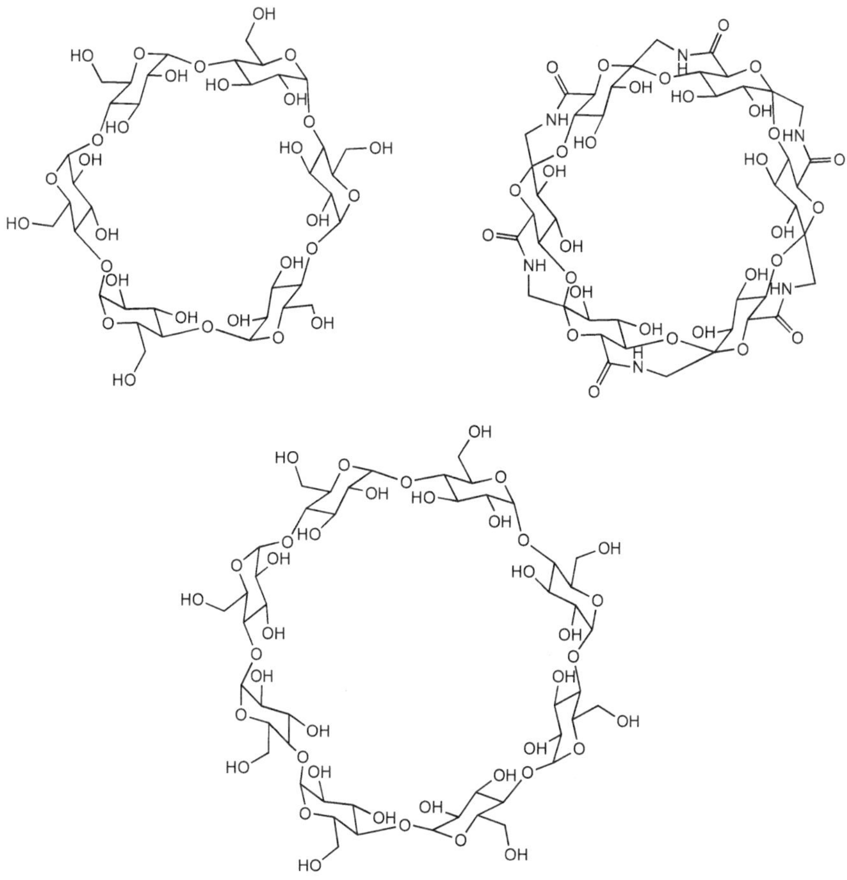

Figure 12.7 ***Structure of cyclodextrin containing macrocyclic carbohydrates.***

degree of molecular uniformity. Kushwaha and Tiwari have reported a series of porphyrin-cored glycodendrimers comprising 8, 12, 16, and 24 β-D-glucopyranose with an aid of Click chemistry.[23b] Furthermore; very recently same group reported Click inspired inspired synthesis of hexa and octadecavalent peripheral galactosylated glycodendrimer, where both glycodendrimers were evaluated for their possible therapeutic implications that resulted to promising anti-bacterial andante-biofilm activities. [23c]

1.5 Macrocyclic carbohydrates as marketed drugs

From several years of pharmaceutical research, the efficiency of carbohydrate containing macrocyclic drugs has been improved with respect to selection of biological targets for disease delivering innovative medicines. Structural information regarding MCs drugs are readily available today which provide us an ideal platform for data analysis and its commercial utility from online information e.g., Drug Bank, Thomson Reuters Integrity, GVK BIO online structure activity relationship database (GOSTAR), etc.[24-26] Some research data revealed that 50% of the MCs drugs are used against infections (both bacterial as well as fungal) and the second category on which these are used anti-cancer type. Beside these, very few proportions is being utilized against other diseases such as (cardiovascular disease, gynaecology, immunology, anaesthesiology, etc).[27] Over the last forty years, pharmaceutical industry have made a continued effort aimed at the discovery of novel and clinically utile antibiotic, antifungal, antitumor agents, many times derived from natural sources such as plant based, marine source, etc. This task has been mainly achieved through a preclinical research focused on studies with novel agents discovered to be active against specific targets, followed by clinical studies with the most promising compounds. Some of them are listed in Table 12.1.[27]

Much research has been devoted to synthesizing new derivatives or analogues with improved chemical, biological and pharmacokinetic properties. Among them if we see individual category MCs antibiotics have come forward.

1.6 Macrocyclic carbohydrates as antibiotics

Most of the MCs inhibit the peptidyl transferase, an enzyme involved in the growth of the polypeptide chain on tRNA, thus inhibiting the protein synthesis in infectious microbes. They also act by inhibiting ribosomal translocations and by causing premature dissociation of the peptidyl-tRNA from the ribosome.[39] The macrolides are broad spectrum antibiotics and possess excellent tissue infiltration and antimicrobial activity against gram-positive cocci and distinctive pathogens preventing various infections viz, respiratory tract infections, soft-tissue infections etc. Several MCs are approved by approving agencies such as US-FDA for the treatment of infectious diseases. Starting from the 1950s MCs was isolated as macrolide antibiotics, such as erythromycin (**1**), used to treat bacterial infections. Due to their safety and efficacy, these are still preferred as usual therapeutic agents in the treatment of respiratory infections. It is a 14-membered macrolide, isolated from Streptomyces cultures which were infact the first macrolide introduced into clinical practice. Erythromycin and its semi synthetic molecules like azithromycin (**2**), clarithromycin (**3**), roxithromycin (**4**), telithromycin (**5**), dirithromycin (**6**) are approved by US-FDA for the treatment of bacterial

Table 12.1 Marketed drugs containing macrocyclic alone or with carbohydrates.

Drug and its structure	Class	Disease	Biological mechanism	Administration (Major)	References
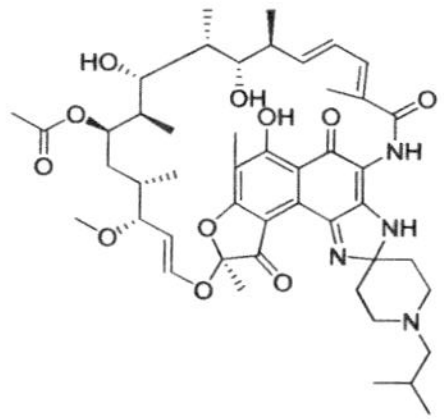 Rifabutin	Ansamycin antibiotic	Bacterial Infection (Tuberculosis)	DNA- dependent RNA polymerases	Oral	28
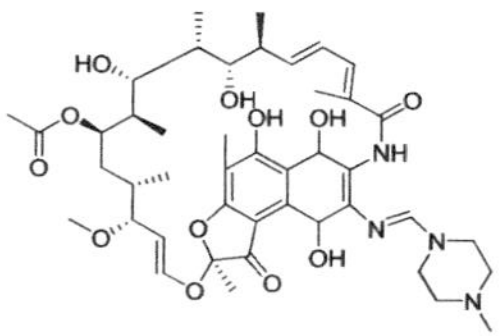 Rifampicin	Ansamycin antibiotic	Bacterial Infections e.g., Tuberculosis	DNA- dependent RNA polymerases	Oral	28
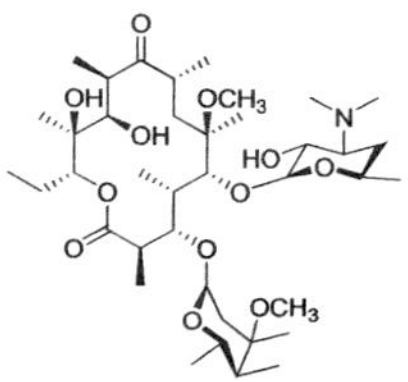 Clarithromycin	Antibiotic	Bacterial Infections, *e.g.* pneumonia	Protein synthesis inhibitor	Intravenous	29

(*Continued*)

Table 12.1 Marketed drugs containing macrocyclic alone or with carbohydrates. (*Cont.*)

Drug and its structure	Class	Disease	Biological mechanism	Administration (Major)	References
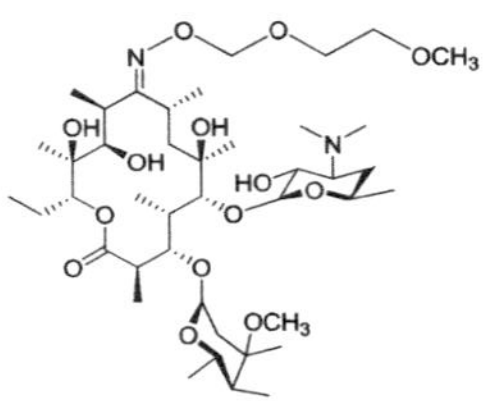 Roxithromycin	Antibiotic	Bacterial Infections, *e.g.* respiratory tract, urinary	Protein synthesis inhibitor	Oral	29
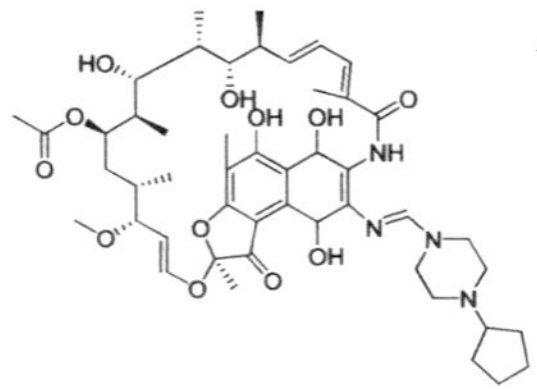 Rifapentine	Ansamycin antibiotic	Bacterial Infections e.g., latent tuberculosis	blocking DNA-dependent RNA polymerase	Oral	28
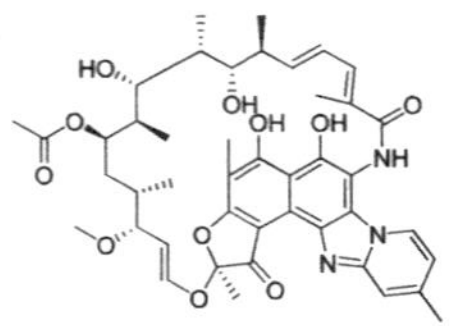 Rifaximin	Ansamycin antibiotic	Bacterial Infections e.g., irritable bowel syndrome with diarrhea.	blocking DNA-dependent RNA polymerase	Oral	28

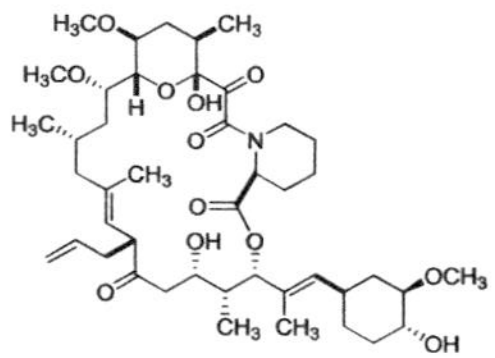 Tacrolimus	Ascomycin Macrolide Subclass	Immunosupresive	Inhibition of IL-2	Oral	30
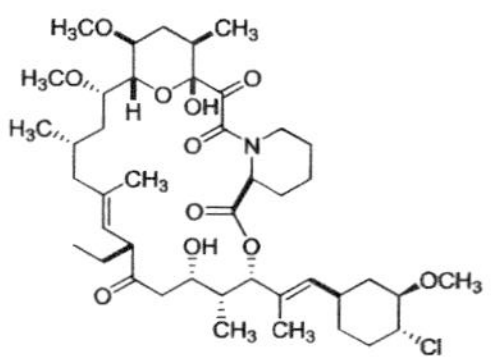 Pimecrolimus	Ascomycin Macrolide Subclass	Immunomoddulating Agent	Inhibition of Calcineurin	Topical	31
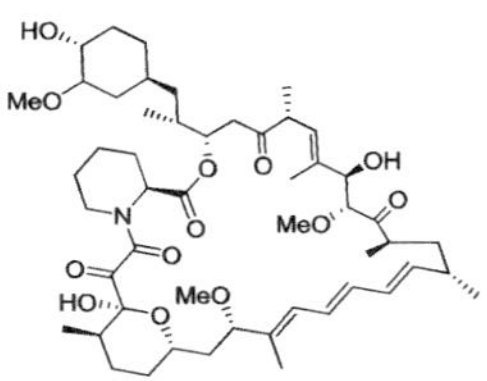 Sirolimus	Rapamycin Macrolide Subclass	Lymphangioleiomyomatosis and other diseases in preventing the rejection of kidney transplants	mTOR inhibition	Oral	32

(*Continued*)

Table 12.1 Marketed drugs containing macrocyclic alone or with carbohydrates. (*Cont.*)

Drug and its structure	Class	Disease	Biological mechanism	Administration (Major)	References
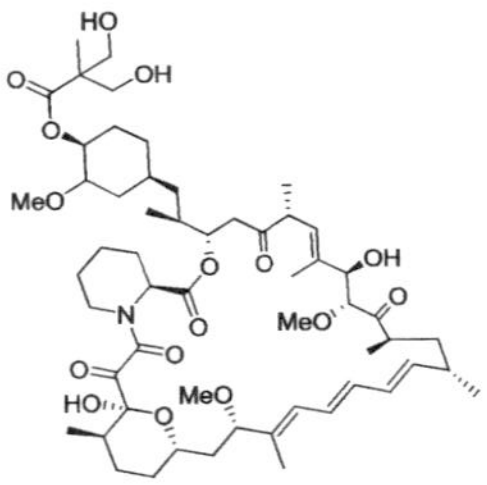 Temsirolimus	Rapamycin Macrolide Subclass	Renal cell carcinoma	mTOR inhibition	Intra venous	33
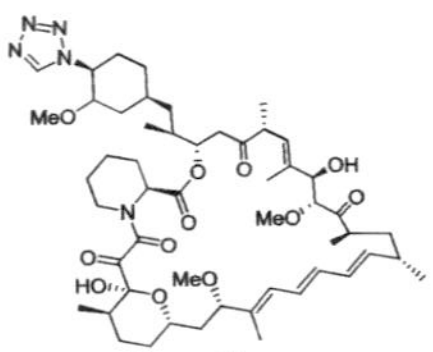 Zotarolimus	Rapamycin Macrolide Subclass	Immunosuppresant, restenosis, in stents with phosphorylcholine as a carrier	mTOR inhibition	Oral	34
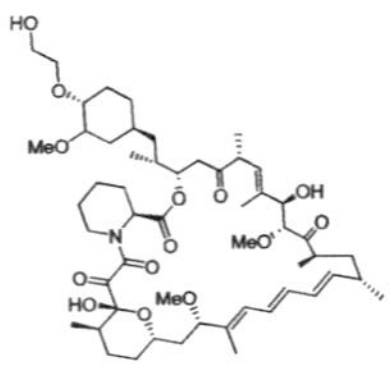 Everolimus	Rapamycin Macrolide Subclass	As immunosuppressants preventing rejection of organ transplants and against renal cell cancer and renal tumours	inhibitor of mammalian target of rapamycin (mTOR)	Oral	35

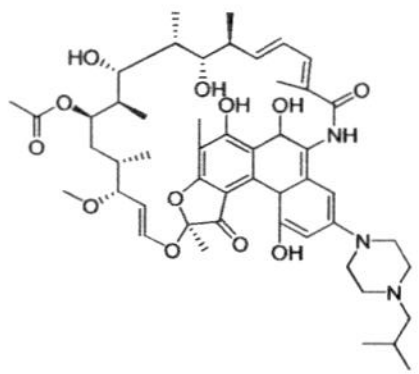 Rifalazil	Antibiotic	Chlamydia infection, Clostridium difficile associated diarrhea (CDAD), and tuberculosis (TB)	kills bacterial cells by blocking off the β-subunit in RNA polymerase	Oral	36
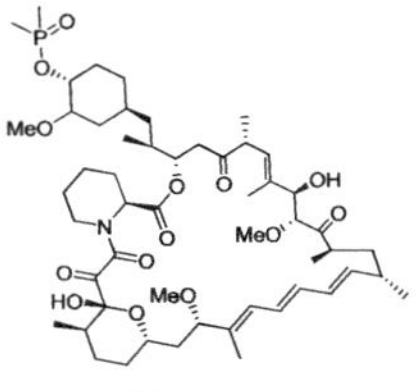 Ridaforolimus	Antibiotic	soft tissue and bone sarcoma	inhibitor of mammalian target of rapamycin (mTOR)	Oral	27
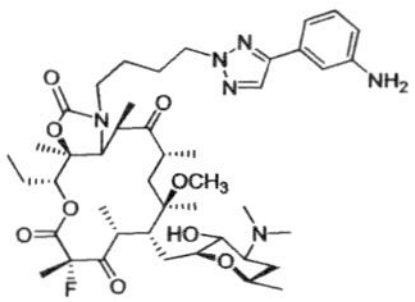 Solithromycin	ketolide antibiotic	community-acquired pneumonia and other infection	inhibits bacterial translation by binding to the 23S ribosomal RNA	Oral, IV	37
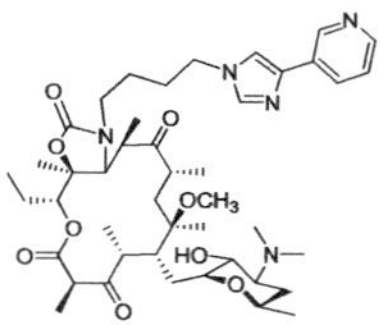 Telithromycin	ketolide antibiotic	community-acquired pneumonia and other infection	interfering with their protein synthesis	Oral, IV	38

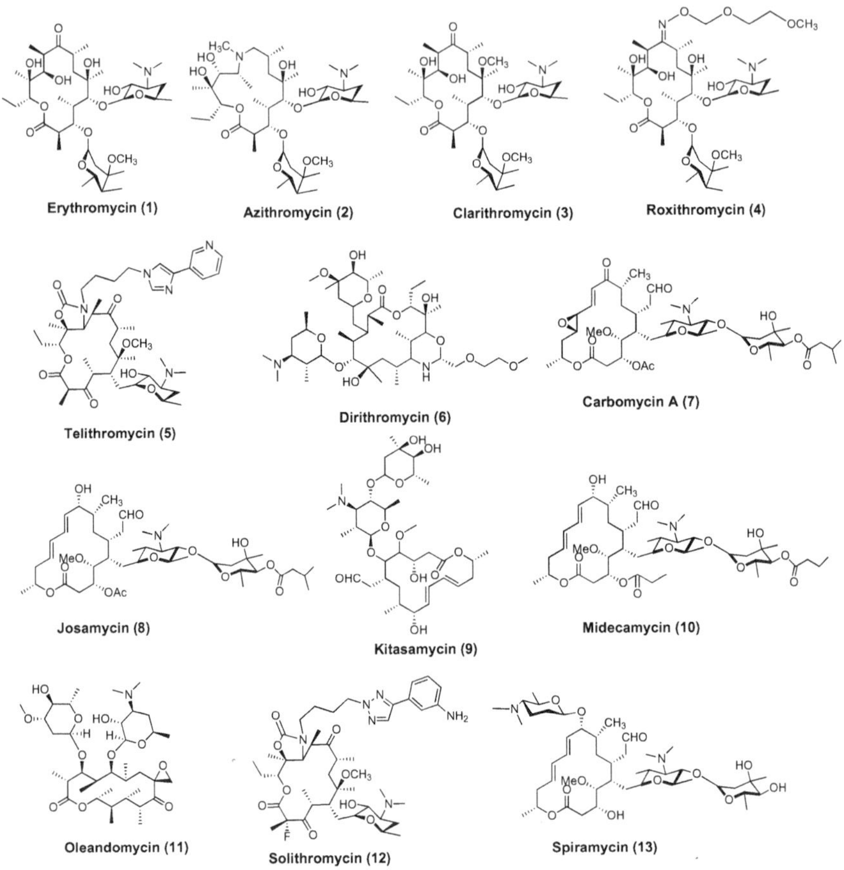

Figure 12.8 ***US-FDA-approved macrocyclic-carbohydrates as antibiotics.***

infections (**Fig. 12.10**) while carbomycin (**7**), jasomycin (**8**), kitasamycin (**9**), midecamycin (**10**), oleandomycin (**11**), solithromycin (**12**), spiramycin (**13**) are approved by other agencies (other than US-FDA) in different countries to treat various bacterial infections (Fig. 12.8). The indiscriminate use of antibiotics has developed resistance in bacteria towards many antibiotic drugs, thus, studies towards the synthesis and development of new antibiotic leads are highly demanded.

1.7 Macrocyclic carbohydrates as anti-fungal drugs

Many macrolide natural products particularly the polyene macrolides have found wide application in the treatment of fungal infections. For

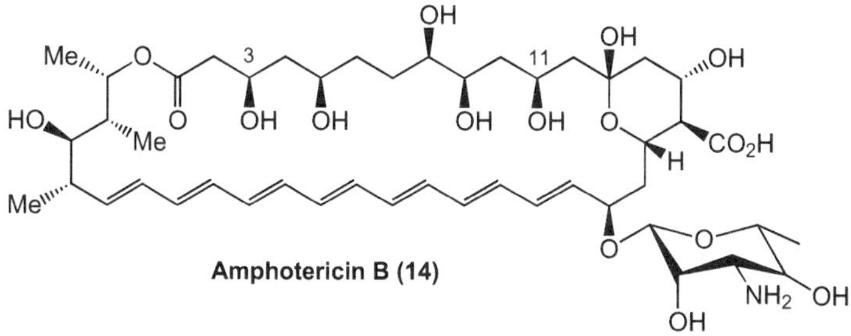

Amphotericin B (14)

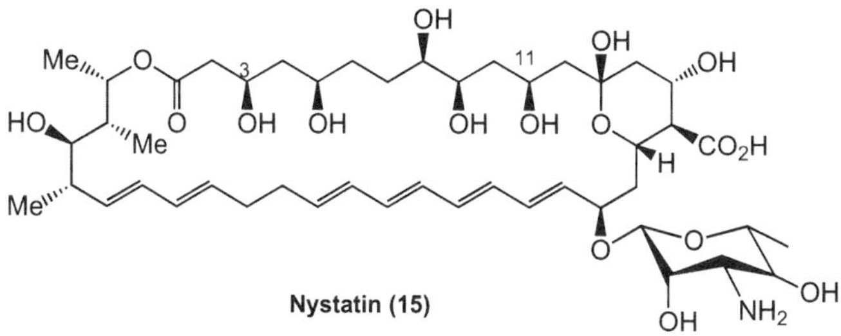

Nystatin (15)

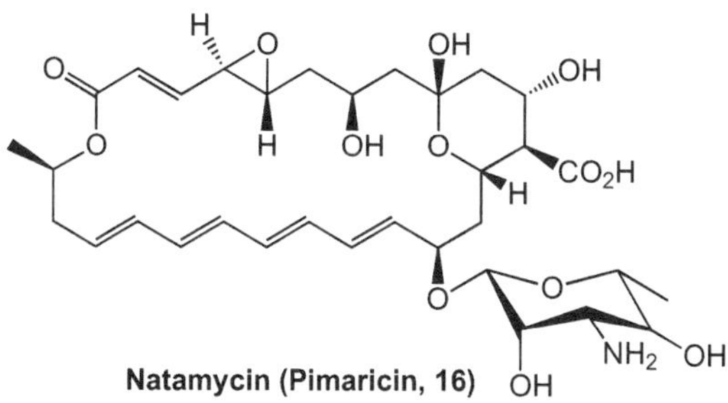

Natamycin (Pimaricin, 16)

Figure 12.9 ***Macrocyclic carbohydrates as anti-fungal drugs.***

example, the polyene macrolides like Amphotericin (**14**), nystatin (**15**), natamycin (**16**, Pimaracin) are approved drugs against fungal infections (Fig. 12.9).

1.8 Macrocyclic carbohydrates as immune suppressants or immunomodulators

In the 1960s the non-antimicrobial properties of MCs were suspected as far back,[6] but their clinical efficacy in considering diffuse panbronchiolitis (DPB) has led to their use in chronic inflammatory diseases.[40] Thus, MCs also are potent anti-inflammatory agent striating asthma, bronchiectasis,

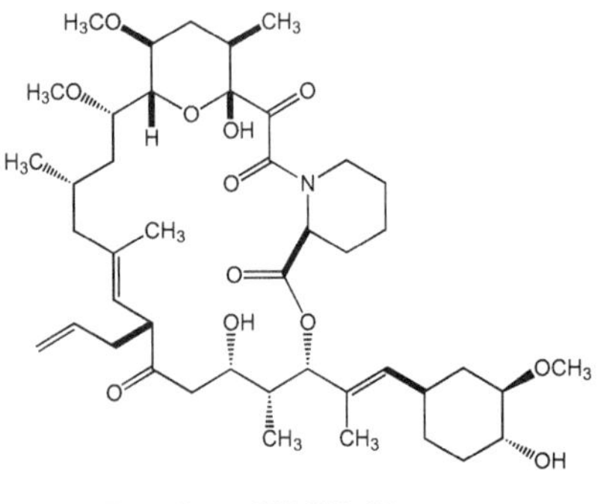

Tacrolimus (FK-506, 17)

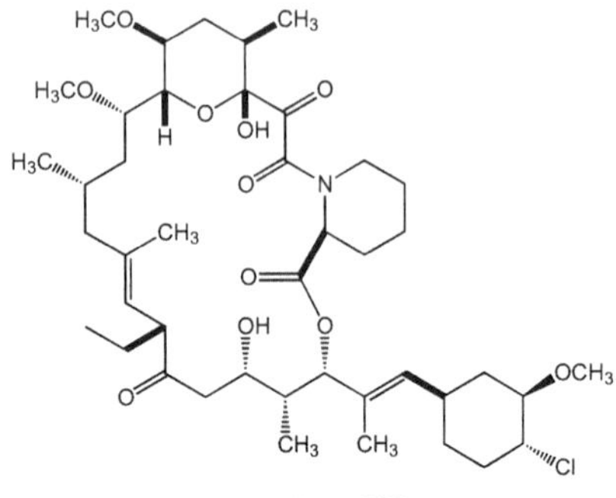

Pimecrolimus(18)

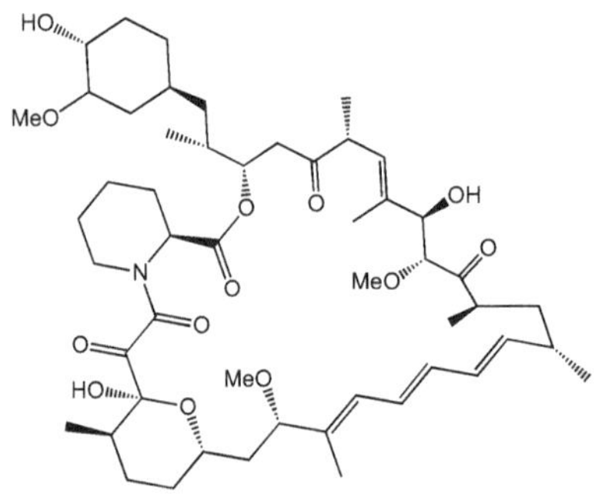

Sirolimus (Rapamycin, 19)

Figure 12.10 ***Carbohydrate macrolides as immunosuppressants or immunomodulators.***

rhinosinusitis, cystic fibrosis, etc. Recently under this category used drugs are tacrolimus (**17,** FK-506), pimecrolimus (**18**), sirolimus (**19**, rapamycin), shown in Fig. 12.10.

Many marine macrolides show promising anti-cancer activity. In this regard, bryostatin 1 (**20**, a marine macrolide natural product) has shown exceptional activity, potency, and selectivity against several cancer cell lines making it a potentially powerful agent for cancer therapy, and therefore it is currently being evaluated alone/in combination with other chemotherapeutic agents in a number of phase II/III clinical trials. MCs shown in **Fig. 12.11** under this category are laulimalide (**21**), sponistatin 1 (**22**), phorbaside A (**23**), latrunculin A (**24**), latrunculin B (**25**), muironolide A (**26**), epothilone B (**27**), swinholide (**28**), sphinholide A (**29**), E7389 (**30**) used against different human cancer lines in the micro/nano molar ranges has been shown.

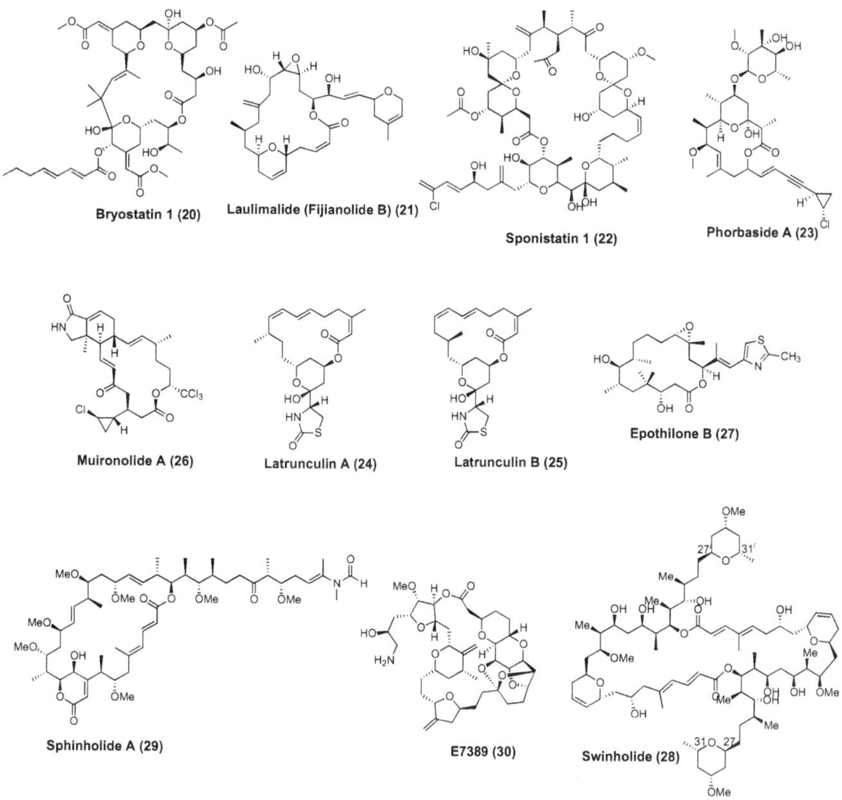

Figure 12.11 ***Carbohydrate-macrolides as promising anti-cancer drugs.***

2 Synthesis of some macrocyclic carbohydrates

Nature has created innumerable number of mysterious molecular targets having strictly defined stereochemistry and often formidable frameworks, thus providing challenges for synthetic chemists. Amongst the several chiral pool molecules prevailing in nature, carbohydrates inhabit an exceptional space, which can be endorsed to the reality that they are exuberantly available in nature, exceedingly inexpensive and above all enantiomerically pure. Another evidence for this uniqueness is the inherent exquisite arrangement of functional groups in monosaccharides and hence their utilities as chiral pool in natural product synthesis when compared to other chiral resources like amino acids, hydroxy acids and terpenes. In other words, the utility of carbohydrate as chiron is self-explanatory. This in combination with the biological importance of carbohydrates have renewed continuous interest in the synthesis and the employment of low molecular weight carbohydrates (mono- and disaccharides and their derivatives) as raw materials for providing access to rare or modified sugars in general and construction of enantiomerically pure non-carbohydrate chiral precursors. The contemporary small molecule strategy and synthesis of drugs also count on scaffolds based on MCs as these are powerful apparatuses for the study of molecular acknowledgement in structure-activity relationship studies. The mimetics are minor which maintain solubility under physiological conditions and are agreeable to detailed structural studies. The monosaccharides including hexoses, pentoses, keto sugars along with their derivatives particularly gluconolactones and unsaturated sugars called glycals have long been recognised as nonplus ultras in organic synthesis. Therefore, utility of these simple, cheap and commercially available carbohydrate derivatives in the synthesis of biologically significant macrolides will be discussed in this section. In the synthetic section of MCs, we may classify them into two parts: (i) Carbohydrate macrocycles possessing sugar as core molecule, and (ii) Glycals derived macrocycles and biologically active molecules.

2.1 Synthesis of carbohydrate macrocycles possesing sugar part as core molecule

Leucomycin A3 (**7**, Scheme 12.1) represents a clinically important 16-membered ring macrolide antibiotic. The semi-synthesis of this starts from *D*-glucose **31** by Nicolaou group in 1981, and is shown in Scheme 12.1.[41] Thus, *D*-glucose **31** was converted into the 1,2;5,6-bisacetonide, followed by C3 methylation and cleavage to form diol **32**. Formation of the ortho ester heating in the presence of benzoic acid and, finally, hydroboration of the resulted olefin with diisoamylborane supplied primary alcohol **33** in 27% overall yield. Protection with BnBr and NaH, acidic cleavage of the acetonide leads to formation of the compound **34**, olefination, acetonide formation, and addition of methyl allyl lithium furnished derivative **35**. Further elaboration of fragment **35** provided cyclic key intermediate **36** which was also derived from the natural product **37** through degradation studies. The operation was completed when compound **36** was taken to an advanced intermediate, which had previously been converted into leucomycin A3 (**37**) by Tatsuta et al.[42]

Amphotericin B (**48**, Fungilin, Fungizone, Abelcet, Ambisome, Fungisome, and Amphocil) is a polyene antifungal drug, often used intravenously against systemic fungal infections. It was originally extracted from *Streptomyces nodosus*, a filamentous bacterium.[43] Retrosynthetic analysis

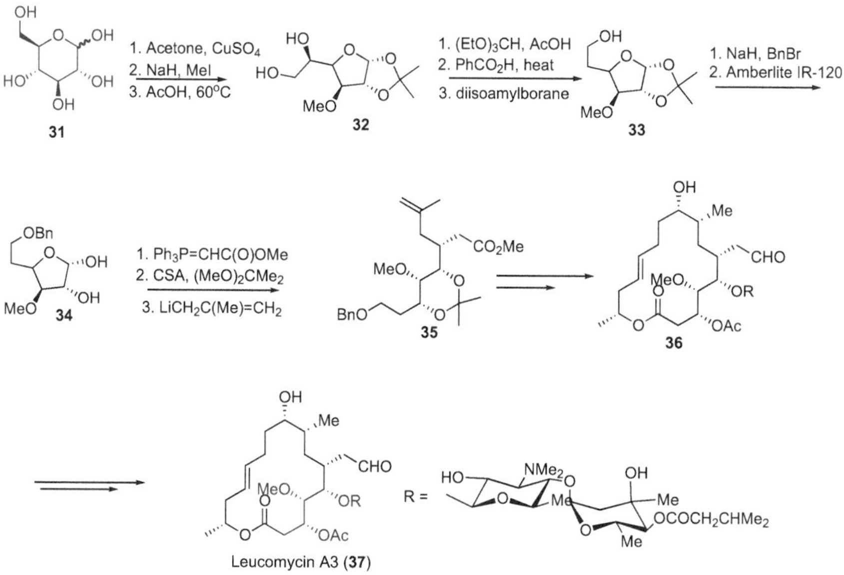

Scheme 12.1 ***Synthesis of Leucomycin A3.***

of this complex polyene macrolide suggested that certain stereochemical and symmetry elements present in amphoteronolide B could be derived from a commercially available xylose.[44] In particular, it was recognized that segments C1-C6 and C8-C13 could be obtained from the 'd' and 'l' enantiomers of xylose, respectively. Thus, *D*-xylose **38** was converted to its 1,2-monoacetonide, selectively silylated at the primary position, and deoxygenated to afford intermediate **39**. Further elaboration furnished the C1-C6 fragment of **40**, whose stereochemical elements could be readily transferred to the coupled product **44** and the final target **45**. The 'l' enantiomer of xylose (**41**) was taken through a similar sequence to afford derivative **42** and, thence, through a ten-step sequence to furnish the C8-C13 keto phosphonate **43**. Coupling of these two intermediates **40** and **43** in the presence of NaH in DME provided the larger C1-C13 fragment **44** in a highly concise and efficient manner: and the latter was additionally expanded to amphoteronolide derivative. Completion of the synthesis of the requisite derivative **45** was glycosylated with the compound **46** in the presence of PPTS to afford the compound **47**. Further manipulations, including reduction of the azide and removal of the protecting groups, proceeded smoothly to afford the target, amphotericin B (**48**) (Scheme 12.2).

Swinholide A (**53**, Scheme 12.3), a marine natural product isolated from the sponge *Theonella swinhoei*,[45] displays impressive biological properties,

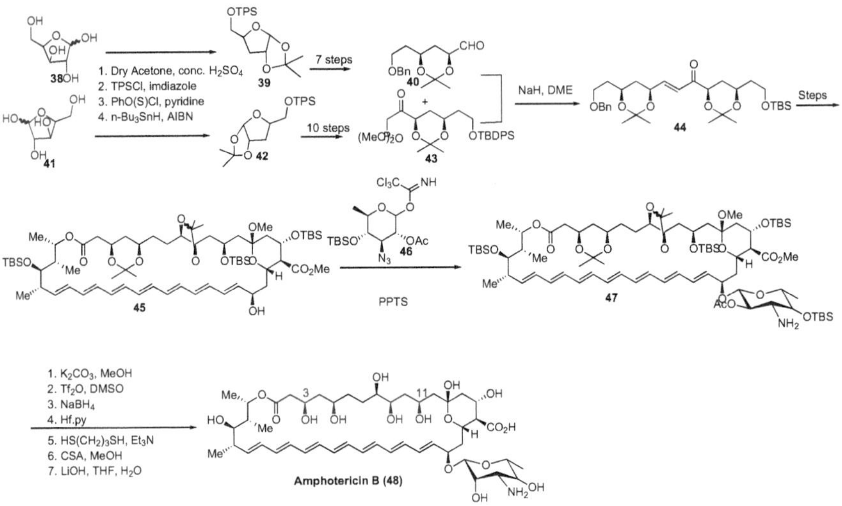

Scheme 12.2 ***Synthesis of Amphotericin B.***

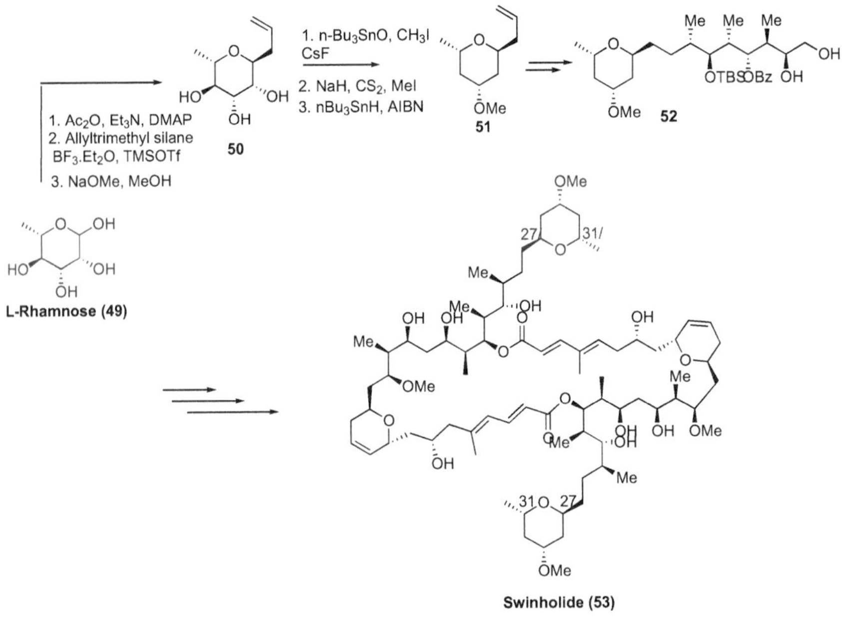

Scheme 12.3 ***Synthesis of swinholide A.***

including antifungal activity and potent cytotoxicity against several tumor cell lines. The molecular structure of **53** is distinguished by a C2-symmetric 44-membered macrolide ring, two conjugated diene systems, two trisubstituted pyran systems, two disubstituted dihydropyran systems, and a total of

30 stereocenters. Retrosynthetic analysis of this complex macrolide revealed that the C27-C32 segment could be obtained from commercially available *L*-rhamnose (**49**), which would provide the correct stereochemistry for the C27, C29, and C31 centres of this pyran moiety.[46] Thus, peracetylation of **49** followed by *C*-glycosidation with allyl trimethylsilane in the presence of TMSOTf and $BF_3.Et_2O$ and subsequent deacetylation afforded, exclusively, the desired β-glycoside **50**. Regioselective tin acetal mediated protection as a methyl ether at the C3 position was followed by Barton-McCombie deoxygenation of the remaining two alcohols to afford olefin **51**. Further elaboration of the C27-C32 segment allowed for its eventual incorporation into swinholide A (**53**).

O-Mycinosyl tylonolide (**33,** Scheme 12.4)[47] is a major degradation product of tylosin and a potential biosynthetic and synthetic precursor to this antibiotic. The synthesis of this compound also starts from *D*-glucose was converted into the 1,2;5,6-bis-acetonide, oxidation of C-3 hydroxyl group with RuO_2, reduced by $NaBH_4$ leads to formation of α − hydroxy (allose) compound, protected with benzoyl and acid catalysed 5,6-isopropylidene deprotection leads to formation of the compound **54**, this compound underwent orthoester formation and cleavage to double bond,

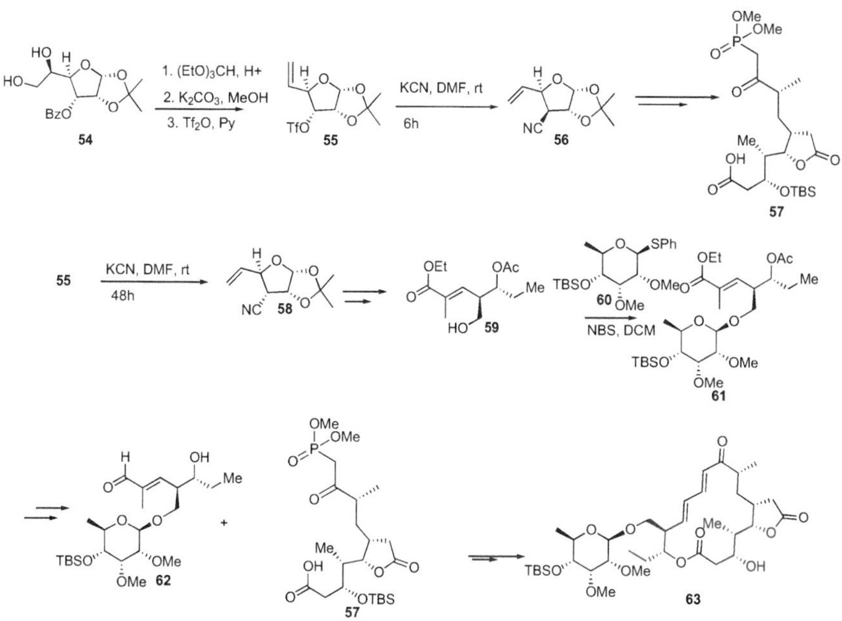

Scheme 12.4 ***Synthesis of O-Mycinosyl tylonolide.***

benzoyl deprotection and O-triflation leads to formation of the compound **55**, which was converted into kinetic (**56**) and thermodynamic (**58**) controlled products, which were subsequently converted to unsaturated ester **57** and phosphonate **59**. Thioglycoside **60** was activated with NBS in CH_2Cl_2 in the presence of **29** to afford the desired β-anomer of the expected glycoside **61**. This coupling represented the first use of NBS for activation of thiophenyl glycosides under such mild conditions. Further elaborations of these intermediates led to key building blocks **62** and **57**, which were then coupled and cyclized to afford the target, O-mycinosyl tylonolide (**63**), in high overall yield and good selectivity.

Tubelactomicin A (**74**, Scheme 12.5) was isolated from the culture broth of *Nocardia*. sp. *MK703-102F1* to show strong and specific

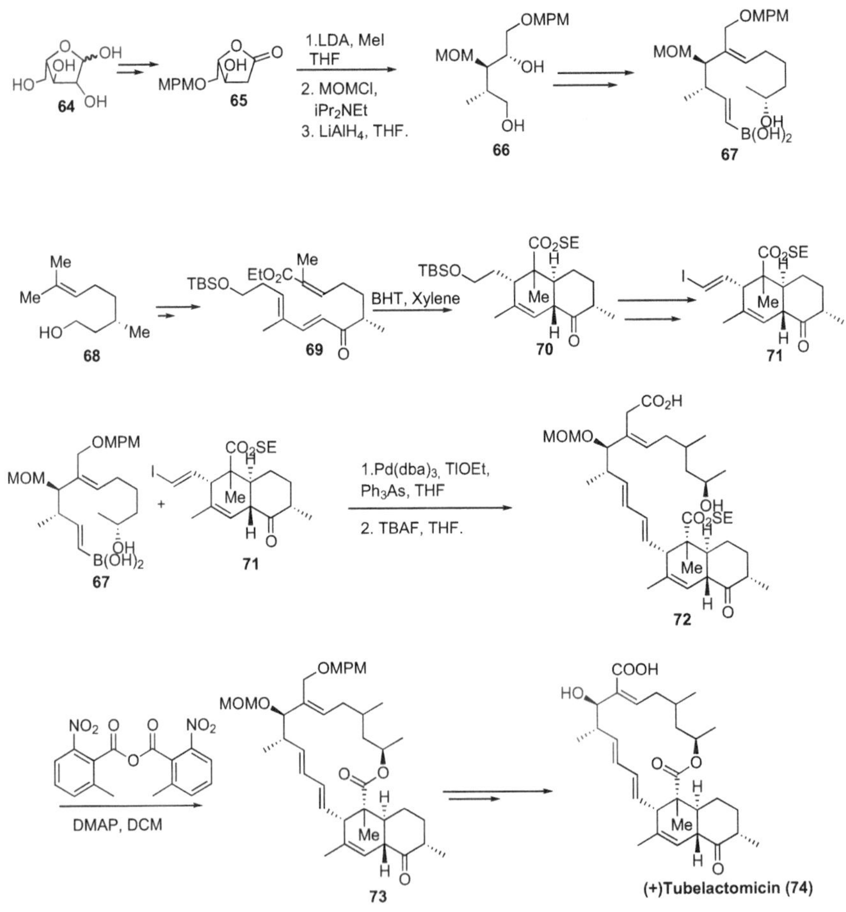

Scheme 12.5 ***Synthesis of tubelactomicin A.***

antimicrobial activities against drug resistant mycobacterium.[48] The structure was determined by *X*-ray crystallographic analysis to be the 16-membered lactone fused with a trans decalin skeleton. The stereochemical array of the northern part was derived from *L*-arabinose (**64**). Lactone **65**[49] was submitted to the stereoselective methylation and reductive ring opening to give diol **66**, possessing functionality to be the northern part **67**. The decalin moiety, the southern part of tubelactomicin A, was constructed by intramolecular Diels–Alder reaction.[50] The citronerol **68** was converted the triene **69**. The stereoselective Diels–Alder reaction to construct additional four chiral centres were realized by heating **69** in xylene, which gave **70** as a single product. The decalin **70** was converted to the alcohol **71** to couple with the northern part **67**. Treatment of the mixture of **71** and **67** under the conditions of Suzuki coupling gave the tetraene product **72**. The seco-acid derivative **72** was submitted to the macrolactonization by Shiina's method[51] to construct lactone **73**. Deprotection and selective oxidation afforded (+)-tubelactomicin A (**74**).

Stagonolides A (**69**, Scheme 12.6)[52] was 10-membered macrolide isolated in mutually liquid and solid cultures of *Stagonospora cirsii* isolated from *Cirsium arvense,* a Davis, fungal pathogen. This synthesis is anexodus from the preceding work at the readily available epoxide intermediate **66** which was synthesised from *D*-ribose. The epoxide **66** was treated with ethyl magnesium bromide in the presence of copper iodide to give alcohol, coupling of alcohol with 5-hexenoic acid under standard DCC/DMAP conditions gave ester **67** which in turn allowed to proceed further for the metathesis reaction. RCM reaction on **67** with Grubbs' first generation catalyst

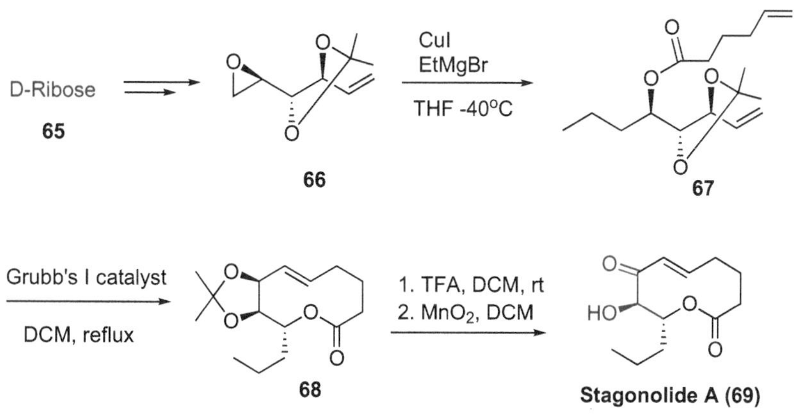

Scheme 12.6 ***Synthesis of Stagonolides A.***

working proficiently to harvest the anticipated lactone **68** as a diastereomeric mixture *E*/*Z* (8:2) which were not separated due to close R_f values. Thus, preceded further for acetonide deprotection by TFA in CH_2Cl_2 to provide the easily discrete diastereomers, further oxidation with MnO_2 yielded the stagonolide A **69** in decent yield.

Aspergillide B was isolated from marine-derived fungus *Aspergillus ostianus* strain 01F313[53] and exhibit significant cytoxic activity against mouse lymphocytic leukemica cells (L1210) with LD 50 value of 71.0 μg/mL. Jian et al.[54] reported the total synthesis of (+)-Aspergillide B from *D*-Galactose pentaacetate (**70**), Stereoselective anomeric allylation, deacetylation, selective primary silyl protection affords 2,6-*trans*-substituted pyran triol **71**. Compound **72** wasobtained through protection of *syn*-dihydroxyl with trimethyl orthoformate, protection the residual hydroxyl MOM ether, ozonolysis of olefin in methanolic NaOH, elimination of the orthoformate ester. Desilylation of **72**, Pd/C catalysed hydrogenation of olefin, IBX oxidation to give aldehyde **73**. *E*-selective Julia-Kocienski olefination of **73** with **74** to give **75**. Desilylation of **75** and base mediated deesterification to yields hydroxyl carboxylic compound **76.** Macrolactonization of **76** under *Mitsunobu* condition gives C-13 inversion of stereo center and MOM deprotection yields (+)-aspergillide B (**77**, Scheme 12.7).

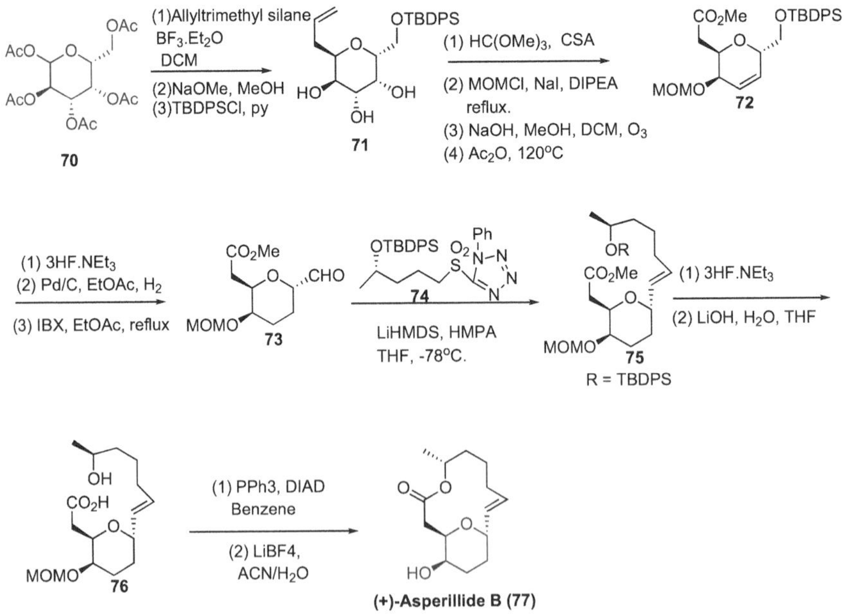

Scheme 12.7 ***Synthesis of (+)-aspergillide B.***

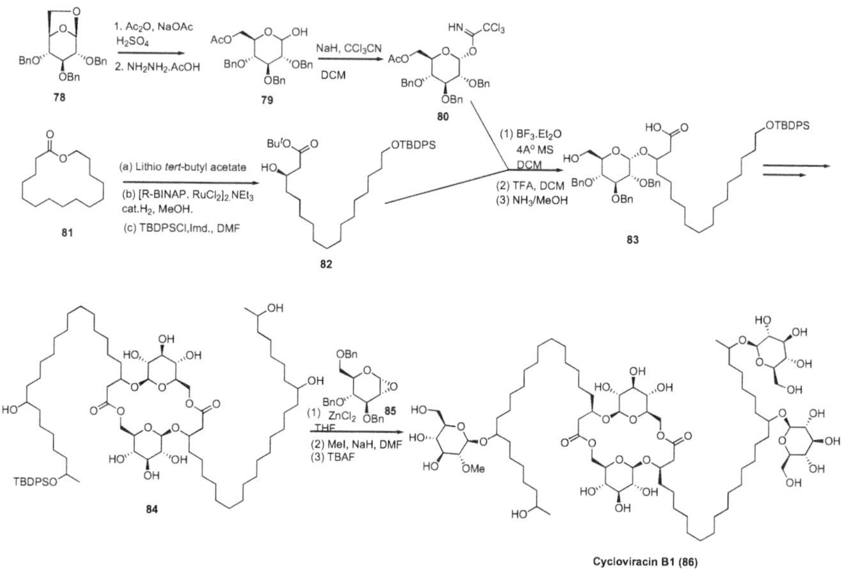

Scheme 12.8 *Synthesis of Cycloviracins B_1.*

Cycloviracins B_1 (86) was isolated from the soil microorganisms *Kibdelosporangium albatum* so. nov. (R761-7).[55] Cycloviracin B_1exhibit activity against the human pathogens herpes simplex virus type 1 (HSV-1), influenza A virus, varicella-zoster virus, and human immunodeficiency virus type 1 (HIV-1). Although less potent than acyclovir[56] (50% antivirally effective concentration of **86** against HSV-1: 5 μg/mL). Furstner et al.[57] reported the total synthesis of the antiviral glycolipid Cycloviracin B_1 by using levoglucosane **78** (Scheme 12.8).

Tri-*O*-benzyl protected levoglucosane anhydro ring opening and acetylated with Ac_2O, NaOAc in the presence of H_2SO_4, furtherselective anomeric deacetylation with hydrazine acetate to give **79.** This was protected with trichloroacetonitrile to give trichloro acetimidate donor **(80).** This was glycosylated with the acceptor **82** in the presence of $BF_3.Et_2O$ gave **83**. This undergoes certain chemical modification to give compound **84**, which was treated with Sugar epoxide (**85**) in the presence of $ZnCl_2$ to give cycloviracin B_1 **86.**

Aigialomycin D (96) was isolated in 2002 from the marine mangrove fungus *Aigialus parvus*, along with four other aigialomycins (A-C, E) and the known macrocyclic natural product hypothemycin.[58] These compounds are structurally related to other resorcylic acid macrolactones, such as radicicol, zearalenone, and the pochonins, which are mycotoxins with a diverse array of biological activities and biochemical targets.[59] Aigialomycin D displayed moderate antimalarial activity, with IC_{50} = 6.6 μg/mL for *Plasmodium falciparum*, and was cytotoxic in some human cell lines, with IC_{50} = 3.0 and 18 μg/mL in KB and BC-1 cells, respectively. More recently, studies into the molecular targets of aigialomycin D have inhibits the kinases CDK1, CDK5, and GSK3 at micromolar levels.[60] Baird et al.[61] accomplished the total synthesis of aigialomycin D by using D- ribose **65** as a starting material. One-pot methylation and isopropylidenation of *D*-ribose 86 using methanol and acetone in the presence of hydrochloric acid. The resulting primary alcohol was converted into the corresponding iodide **87** via a tosylate. Vasella reaction of compound **87**led to the ring-opened ω-enal, which underwent a Wittig reaction with the stabilized ylid methyl (triphenylphosphoranylidene) acetate to give the α,β-unsaturated ester 126 as a 4.7:1 mixture of *Z*- and *E*-isomers. Selective reduction of the α,β -unsaturated ester **88** to the saturated primary alcohol while leaving the terminal alkene intact with $NaBH_4$ in the presence of CuCl in cyclohexene, reduction of the ester to alcohol **89** was straightforward with $LiAlH_4$. Conversion of this alcohol into the coupling precursor **90** was achieved by mesylation to provide the corresponding sulfonate, followed by reaction with potassium thioacetate. Compound **90** was coupled with benzyl bromide derivative **91** in the presence of K_2CO_3 to give diphenolic compound, MOM protection of phenolic hydroxyl groups and saponification with KOH gave acid compound **92**. Esterification of the resulting carboxylic acid **130** with 4-penten-2-ol **93** using the *Mitsunobu* protocol gave **94**. Oxidation of the thioether **94**to the corresponding sulfone was performed using *m*-CPBA, macrocycle formation by ring-closing metathesis proceeded smoothly and rapidly under microwave irradiation, the obtained product **95** with only the *E*-alkene isomer. The macrocyclic sulfone **95**was subjected to a Ramberg-Backlund reaction using Meyers' conditions gave aigialomycin D (**96**, Scheme 12.9).

By utilizing the Click chemistry followed by intramolecular glycosylation, Tiwari and Schmidt recently developed a series of disaccharide-containing drug like macrocyclic carbohydrate analogs.[62] A 15 membered MCs, **97**was synthesized by Tiwari et al., from glycosyl bromide as a starting reagent resulted formation of alkyne possessing glycosylated moiety, which was furthur reacted with sugar based azide using click chemistry employing CuI/DIPEA in DCM resulted the formation of different macroyclic carbohydrates containing triazoles (Scheme 12.10).[62]

Glycals derived macrocycles and biologically active molecules: Glycals have been broadly utilized in the synthesis of structurally complex

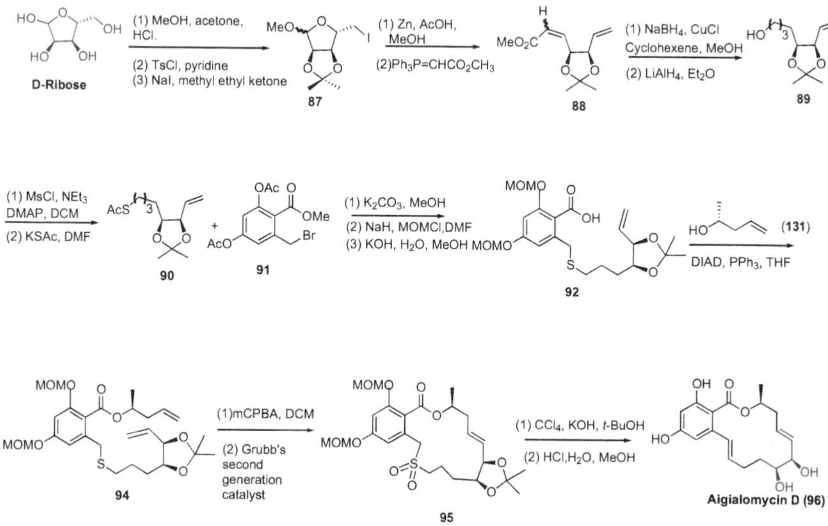

Scheme 12.9 *Synthesis of aigialomycin D.*

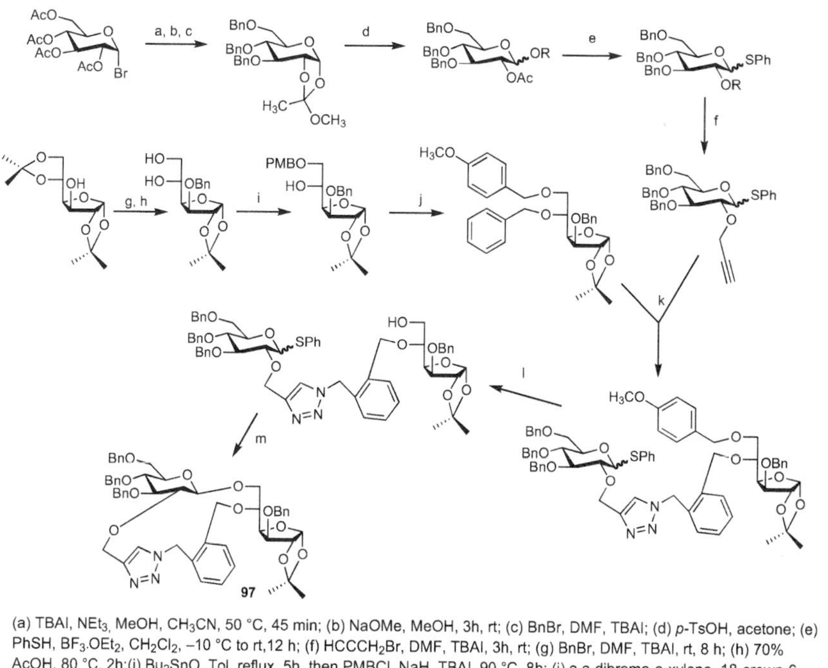

(a) TBAI, NEt_3, MeOH, CH_3CN, 50 °C, 45 min; (b) NaOMe, MeOH, 3h, rt; (c) BnBr, DMF, TBAI; (d) *p*-TsOH, acetone; (e) PhSH, $BF_3.OEt_2$, CH_2Cl_2, −10 °C to rt,12 h; (f) $HCCCH_2Br$, DMF, TBAI, 3h, rt; (g) BnBr, DMF, TBAI, rt, 8 h; (h) 70% AcOH, 80 °C, 2h;(i) Bu_2SnO, Tol, reflux, 5h, then PMBCl, NaH, TBAI, 90 °C, 8h; (j) a,a-dibromo-o-xylene, 18-crown-6, NaH, CH_2Cl_2, 0°C to rt, 14h; (k) CuI, DIPEA, CH_2Cl_2, rt, 12h; (l) DDQ, CH_2Cl_2, 0 °C, 2h; (m) NIS, TMSOTf, CH_2Cl_2, 0 °C to rt, 2h

Scheme 12.10 *Synthesis of macrocyclic carbohydrates containing triazoles.*

molecules mimicking natural products due to the readily availability with multi-functionality.[63] The inbuilt multiple chirality allows organic chemists to use carbohydrates as chiral pool in the total synthesis of natural bioactive products.[64] The wide applicability of glycals in the synthesis of various biologically important scaffolds and natural product can be attributed due the ease with which glycals can be converted into various derivatives/precursors for further derivatization as depicted in Fig. 12.12a and Fig. 12.12b.

There are numbers of biologically active known to be developed from glycols. Total synthesis of some of the representative macrocycles for eample, (+)-Aspicilin, avermectin, diplodialides, phoracantholides, Apicularen A, apicularen A, Macrolide A26771B, Aspergilide A, etc, has been established from glycols (Scheme 12.11-12.18).

The 18-membered macrocyclic lactone **(+)-aspicilin** (**103**, Scheme 12.11)[64] was first isolated in 1900 from a lichen belonging to *Lecanoraceae* family. Total synthesis starts from carbohydrate-based building block *D*-glucal. The regioselective protection of the primary alcoholic group of *D*-glucal (**64**) with TBDPS and MOM protection of secondary hydroxyl gave **98**, hydration of **98** in the company of LiBr catalyzed by amberlite IR-120 (plus) ion exchange resin to give **99**, it was Wittig olefination with methyltriphenylphosphonium bromide in the presence of *t*-BuOK, MOM protection of

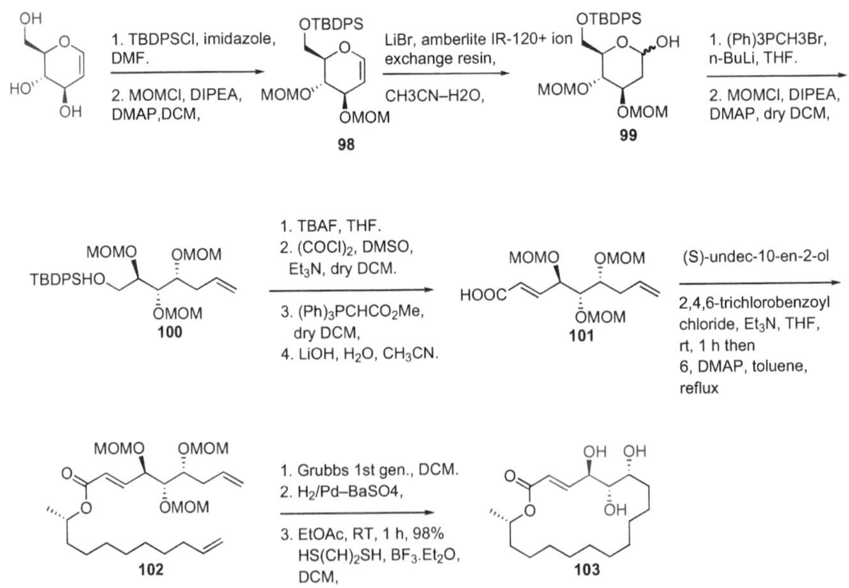

Scheme 12.11 ***Synthesis of (+)-aspicilin.***

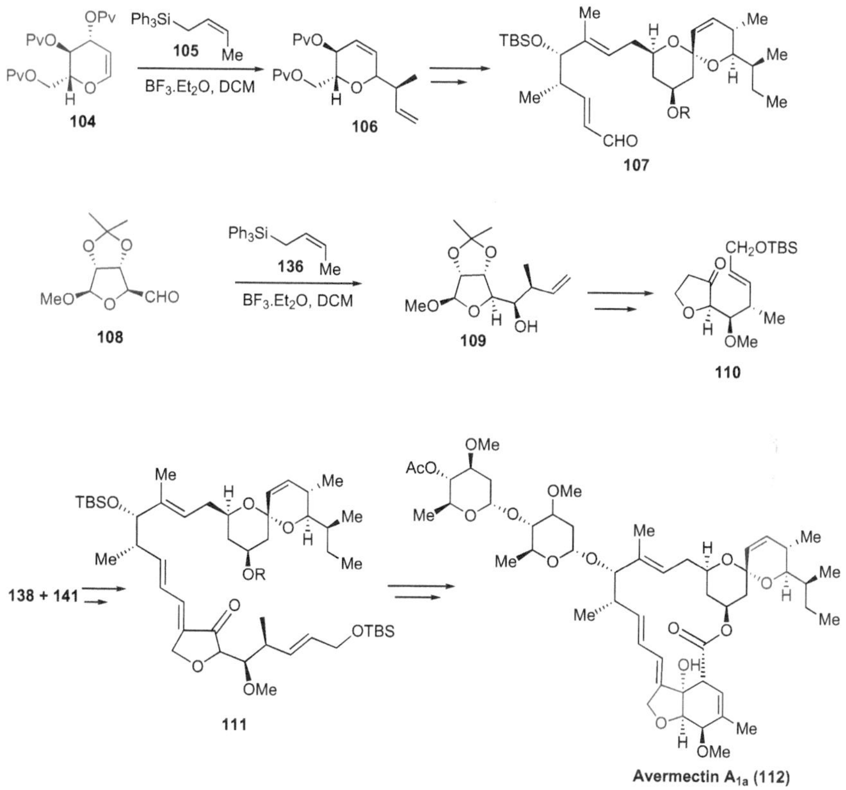

Scheme 12.12 ***Synthesis of avermectin A_{1a}.***

the alcohol gave **100**, desilylation, Swern oxidation, wittig olefination and saponification leads to formation of **101**, Yamaguchi esterification of **101** with *(S)*-undec-10-en-2-ol gave **102**, Grubb's cyclization, MOM deprotection leads to formation **103**.

The **avermectin** is a sixteen membered macrocyclic lactone derivative.It is a potent anthelmintic and insecticidal molecule[65] generated as fermentation products from *Streptomyces avermitilis* (a soil actinomycete). Danishefsky et al.[66] reported synthesis by using *D*-glucose and *D*-ribose derived compounds **104** and **108**. Synthesis of compound **107** was done from tri-O-pivaloyl-D-glucal by certain chemical transformations. Synthesis of compound **110** was done from compound **108**. Coupling of compounds **107** and **110**throughaldol condensation reaction furnished 142. Chemical transformations of compound (**111**) gave avermectin A_{1a} **112** (Scheme 12.12).

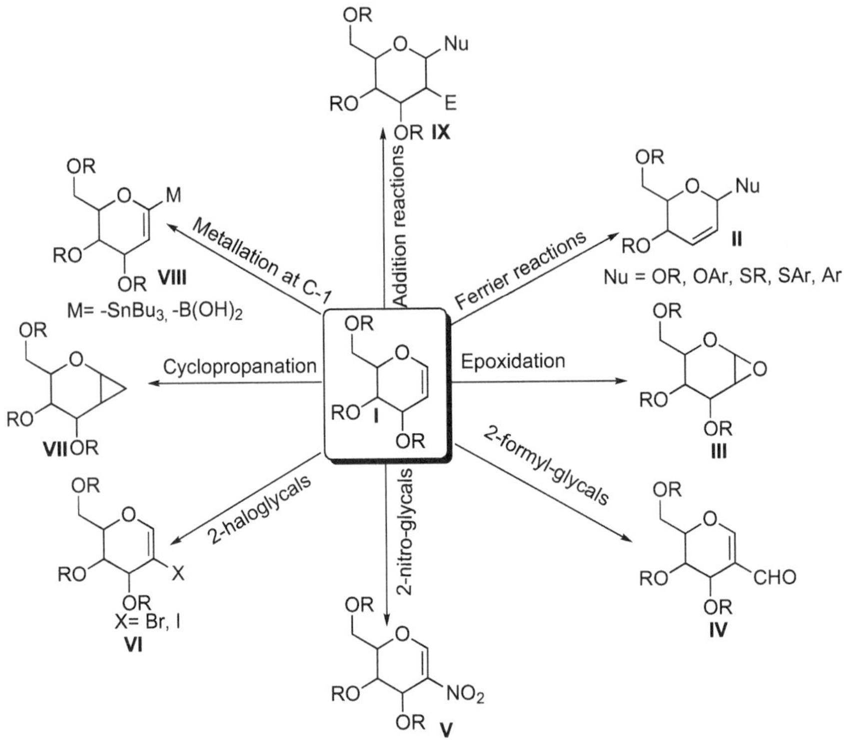

Figure 12.12a ***Transformation of glycals.***

Based on promising biological efficacy, our group was engaged for the synthesis of 13- and 14-membered macrolides related Aspergillides. We successfully developed a set of nine *trans* disubstituted dihydropyran-based medium ring macrolides using D- glucal as chiral pool (Scheme 12.13) and evaluated against a panel of three human cancer cell lines and a normal cell line.[67] Bio evaluation studies have revealed a potent cytotoxic molecule (**130**) exhibiting dose dependent growth inhibition against HL-60 cell line with IC_{50} value of 1.10± 0.75 μM which is lower than the naturally occurring molecules of this class and comparable activity to the synthetic drug fludarabine. The compound **130** inhibits PI3K/AKT signalling pathway by selectively targeting p110α subunit of PI3Kα. This leads to mitochondrial stress that causes translocation of cytochrome C from mitochondria to cytosol, which in turn activate the caspases mediated apoptotic cell death. Further *in-silico* docking simulations of four macrolides with p110α subunits have been carried out to visualize the orientation pattern. To attain trans stereochemistry at ring juncture of pyran ring we began our

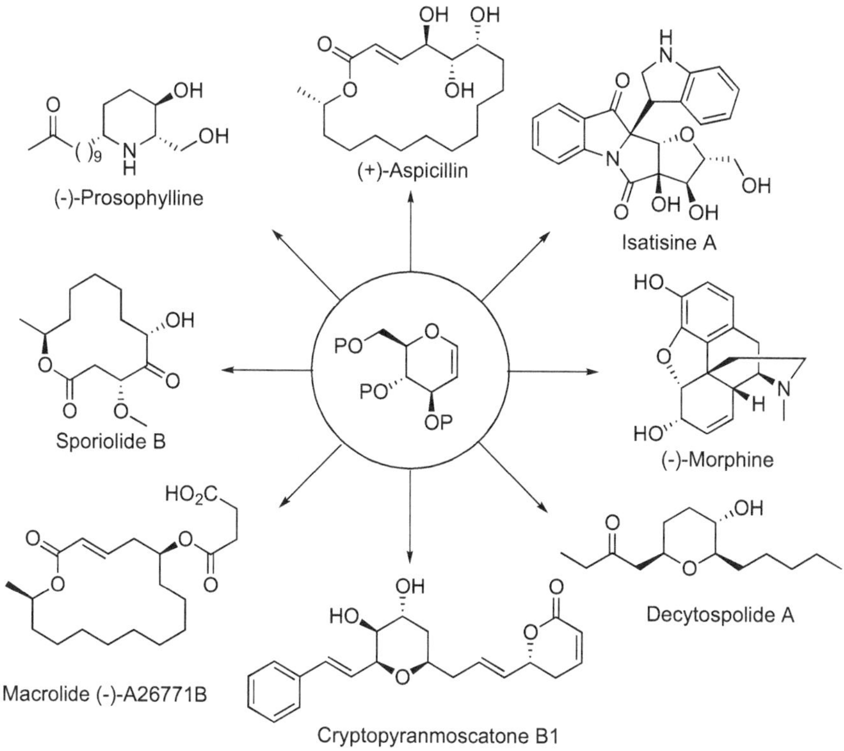

Figure 12.12b ***Glycals derived natural products.***

investigation with α-*C*-allylation of tri-*O*-acetyl-*D*-glucal, deacetylation, subsequent regioselective protection of primary hydroxyl by tosyl group, nucleophilic substitution of –OTs compound with NaCN provided cyano derivative. After protection of free hydroxyl group as ethyl ether, cyanide hydrolysis to obtain the key carboxylic acid derivatives **114–116** required for esterification. EDC-mediated esterification of acid **114** with optically pure alcohol (*S*-hex-en-2-ol) generated the di-olefinic compound **117** ready for final macrolide synthesis. Ring closing metathesis (RCM) of 20 with 5 mmol% of Gr-IIin 3 mM DCM solution gave the desired product **118**. Small coupling constant (always <12 Hz) clearly indicated the preponderance of *Z*-isomer. The ratio of E/Z isomers was deducible from the ^{1}H NMR spectrum. Encouraged by the ease in accomplishing the stereodivergent synthesis of 14-membered macrolide **118**, an aspergillide C analogue, we next directed our attention toward the synthesis of a set of medium ring macrolides for SAR. Thus, we used the same sequence of reactions but varied the alcohol partner, which was generated according to the literature

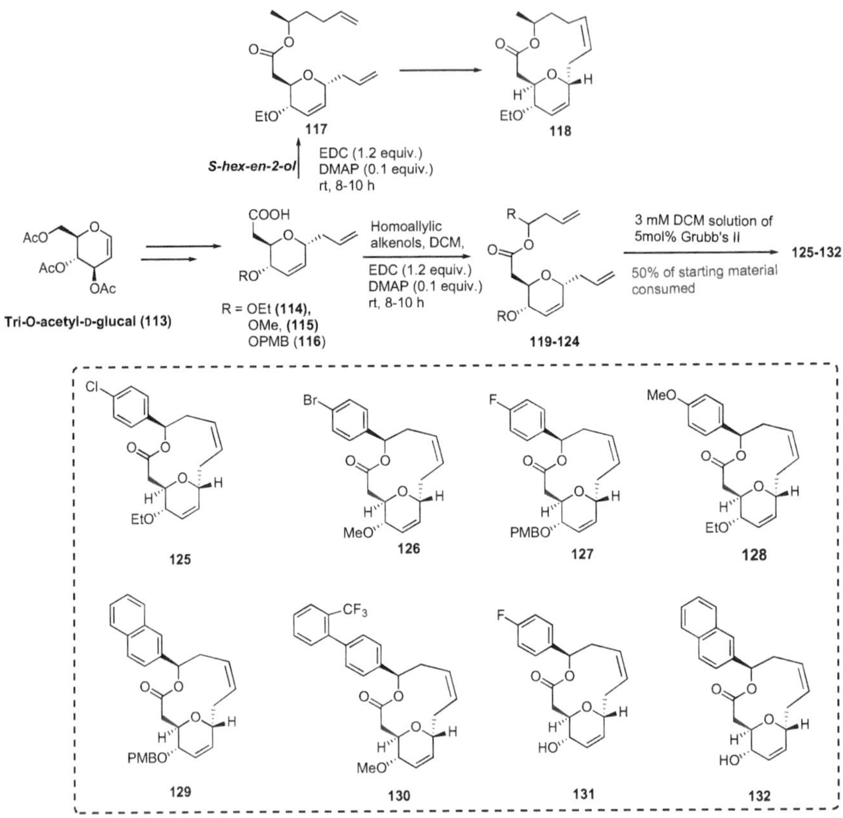

Scheme 12.13 ***Synthesis of disubstituted dihydropyran-based medium ring macrolides using D- glucal.***

procedure. It is noteworthy that changing the protection from methoxy or ethoxy to PMB leads to increased yield of acid during cyanide hydrolysis. We generated the analogues of **114** with one less carbon. Compound **125** shows a small coupling constant <12 Hz that clearly indicates the preponderance of *Z*-isomer and the stereochemistry at the C-12 position of macrolide **125** was confirmed by NOSEY correlation between H-12 and H-3. Thus, the proposed structures of the 13-membered macrolides were confirmed.

Diplodialides and **Phoracantholides** macrolides of the phoracantholide type are defensive secretion from the metasternal gland of the Eucalyptus longicorn beetle *Phoracantha synonyma*, diplodialides are metabolite of a plant pathogenic fungus *Diplodia pinea* of high hydroxylase inhibitory activity.[68] However, there invariably is a single chiral center along the carbon-chains, an (*R*)-hydroxy group tied up as a lactone, and since (5*R*)-hydroxyhexenal **136** or it is a cis - trans isomerization product **137**can readily be elaborated from tri-*O*-acetyl-D-glucal[69] in the form of the cyclic acetal **135**, sugar-based approach to natural products **138–142** of this type is not only appropriate, but efficient (Scheme 12.14).

Apicularen A (152, Scheme 12.15)[69] was formerly isolated from the myxobacterial genus *Chondromyces* by Jansen et al. in 1998 it shows remarkably high cytostatic activity against nine different human cancer cell lines. Lewis et al.[70] reported the synthesis of (+)-Apicularen A from *D*-glucal.

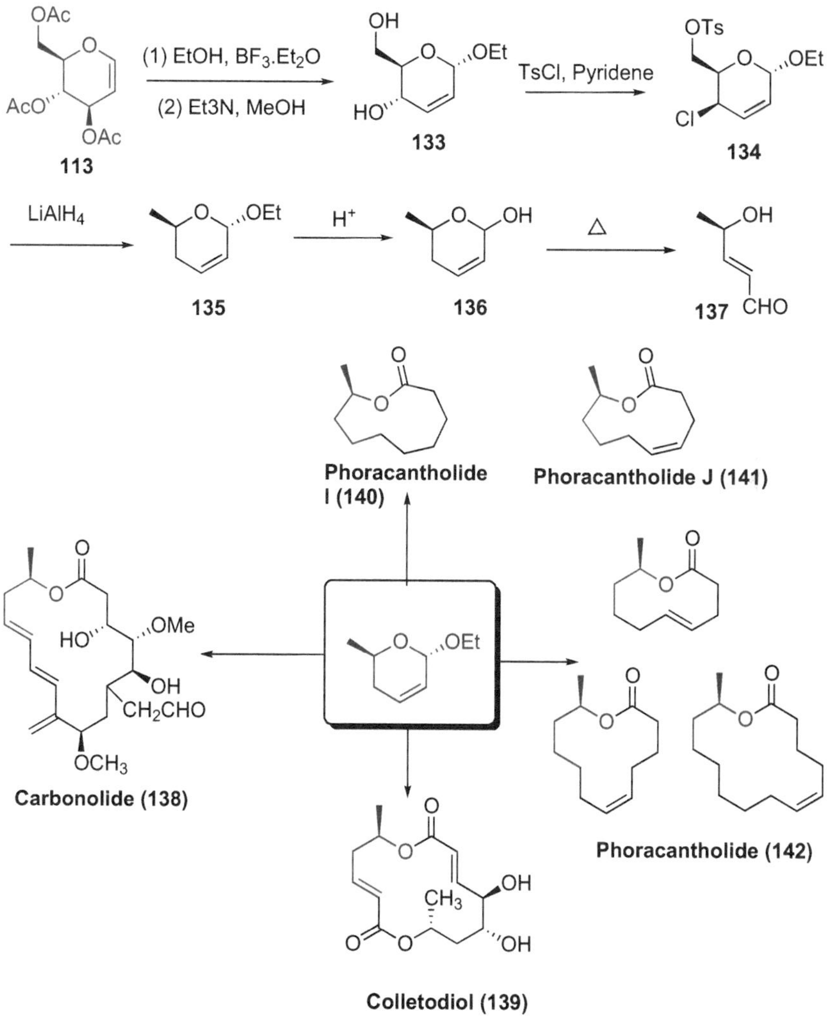

Scheme 12.14 *Synthesis of Diplodialides and Phoracantholides.*

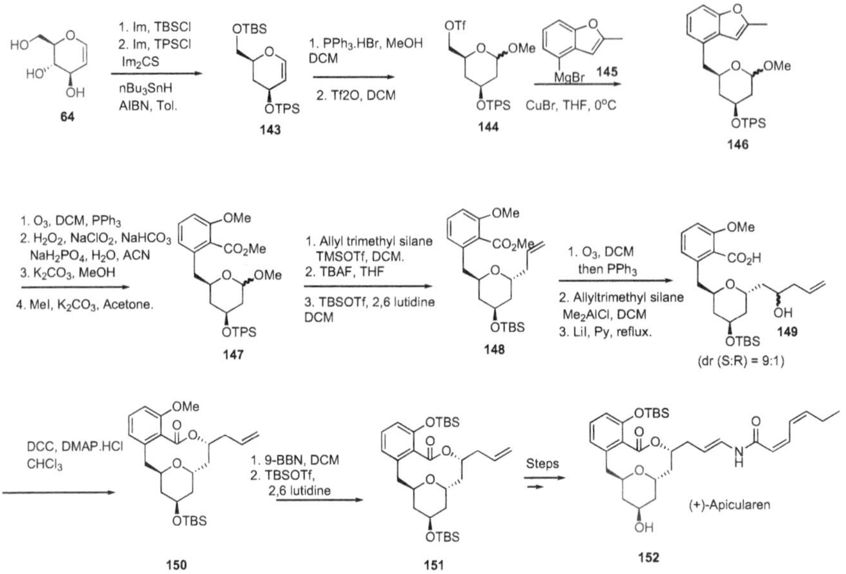

Scheme 12.15 ***Synthesis of apicularen A.***

D-Glucal (**64)** in a one-pot procedure using a three-reagent sequence (successively TBSCl, TPSCl then thiocarbonyldiimidazole) gave the protected derivative then underwent Bu_3SnH-mediated deoxygenation *via* a Barton–McCombie type process to give dihydropyran **143**. Acid-catalysed addition of methanol to the enol ether moiety, together with concomitant deprotection of the primary TBS group, was then achieved using triphenylphosphine hydrobromide in dichloromethane–methanol giving primary alcohol, which was treated with triflic anhydride in pyridine to give the compound **144.** To this Grignard reagent **145** was coupled in the presence of CuBr to afford **146,** it was ozonised, Pinnick-oxidized product was esterified with methanol, *O*-methylated to give the compound **147,** it was glycosylated with allyltrimethylsilane in the company of TMSOTf, deprotection of *O*-TPS and protected with TBS to afford **148,** it ozonised, treatment with allyltrimethylsilane, de-esertification to yield the compound **149**, DCC mediated macrolactonization furnished **150**.Further elaboration of **150** yields (+)-apicularen A (**152**).

Macrolide (−)-A26771B (**154**), isolated from the fungus Pencillium turbatum, was synthesized from acetylated glucal (**113**).[71] The synthesis started from *D*-glucal and undergoes series of steps to produce ring opened product (**153**) which was subsequently transformed into the desired compound (**154**) by using sequence of steps (Scheme 12.16).

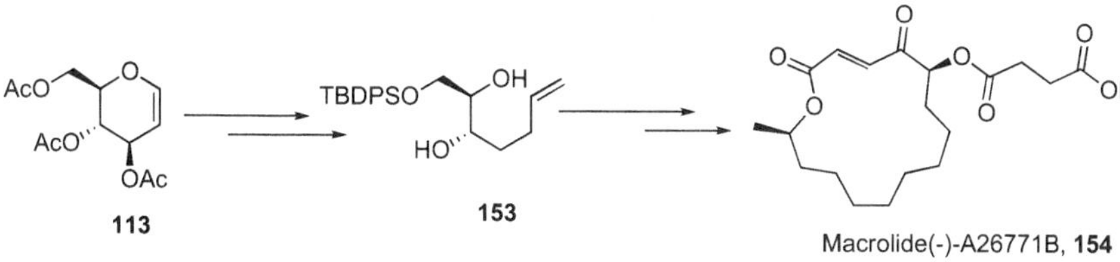

Scheme 12.16 ***Synthesis of Macrolide A26771B.***

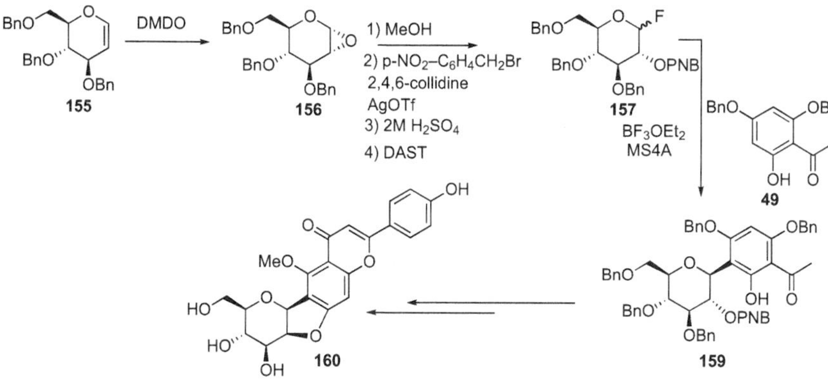

Scheme 12.17 ***Synthesis of tricyclic flavonoid.***

Tricyclic flavonoid **(160)** structurally linked to vitexin and isovitexin was isolated from oolong tea extract and displayed potent anti-inflammatory activity.[72] Nakatsuka et al. reported[73] the synthesis of tri-cyclic flavonoid using 3, 4, 6-tri-*O*-benzyl-D-glucal (**155**) when reacted with DMDO furnished the starting material 1,2 anhydro sugar by epoxidation[74] (**156**) which was latter on converted into glycosyl fluoride (**157**). *C*-glycosilation[75] of **157** with substituted acetophenone[76] (**158**) followed by various steps produced the desired tricyclic flavonoid **(160)** (Scheme 12.17).

Liu and his co-workers developed an efficient decarboxylative glycosylation method for β-glycosylation[77]using palladium catalysed reaction of glycals which was further extended in the synthesis of Aspergilide A, **169**.[78, 79, 80] Started with glucal (**161**) having an ester protecting group at C-3 position which undergo intramolecular decarboxylative coupling when treated with $Pd(OAc)_2$ to give coupling product (**162**) which underwent reduction with raney Ni to give compound (**163**).The newly formed hydroxyl group undergo elimination to produce internal alkene **(164)**. Compound (**164**) was further transformed into desired compound (**169**) using sequence of steps (Scheme 12.18).

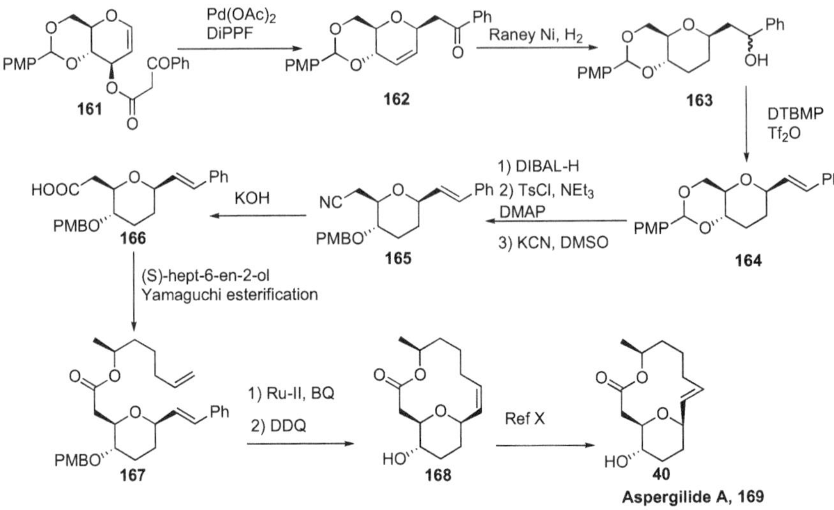

Scheme 12.18 ***Synthesis of Aspergilide A.***

3 Conclusions and future perspectives

In summary, the syntheses and pharmaceutical practicality of natural and synthetic carbohydrate-derived macrolides as drug have been appraised. Furthermore, based on the specific polysaccharides located on pathogenic microorganisms surface and tumor associated carbohydrate antigens can be specifically recognized by the host immune system and exploited as the basis for the design of carbohydrate vaccines. Employing MCs human diseases conjugated carbohydrate vaccines have been dveloped. As an example conjugation of purified capsular polysaccharide from HiB and tetanus toxoid, has been suggested for the prevention of childhood meningitis known as Haemophilus influenzaetype B (HiB) conjugate vaccine. Chemically synthesize clusters of oligosaccharide as antigen epitopes were used as MCs vaccine as promising anti-tumor and anti-HIV vaccines are intensively studied. Globo-H, a synthetic hexasaccharide from the cell surface of breast cancer, was covalently linked to KLH with QS-21 as an adjuvant and displayed favorable results in thephase I trial as MCs vaccine. Carbohydrate vaccines based on oligomannose clusters are also intensively studied for their anti-HIV activities as MCs vaccine.

Acknowledgment

Authors are thankful to Dr Ram Vishwakarma, Director IIIM Jammu and DST (GAP2155), CSIR (HCP0008) for generous funding.

References

1. Gibson, S. E.; Lecci, C. *Angew. Chem., Int. Ed.* **2006**, *45*, 1364.
2. Driggers, E. M.; Hale, S. P.; Lee, J.; Terrett, N. K. *Nat. Rev. DrugDiscovery* **2008**, 7, 608.
3. Crane, E. A.; Scheidt, K. A. *Angew. Chem., Int. Ed.* **2010**, *49*, 8316.
4. Madsen, C. M.; Clausen, M. H. *Eur. J. Org. Chem.* **2011**, *17*, 3107.
5. Furukawa, J.; Sakairi, N. *Trends Glycosci. Glycotechnol.* **2001**, *13*, 1.
6. Pereda-Miranda, R.; Rosas-Ramirez, D.; Castañeda-Gomez, J. *Fortschr. Chem. Org. Naturst* **2010**, *92*, 77.
7. Pouysegu, L.; Deffieux, D.; Malik, G.; Natangelo, A.; Quideau, S. *Nat. Prod. Rep.* **2011**, *28*, 853.
8. Xuan, X.; Nicolas, B. *Chem. Rev.* **2014**, *114*, 7678–7739.
9. Fujioka, T.; Kashiwada, Y.; Okabe, H.; Mihashi, K.; Lee, K. H. *Bioorg. Med. Chem.Lett.* **1996**, *6*, 2807–2810.
10. Furstner, A.; Albert, M.; Mlynarski, J.; Matheu, M.; DeClercq, E. *J. Am. Chem. Soc.* **2003**, *125*, 13132–13142.
11. Jiang, Z. H.; Geyer, A.; Schmidt, R. R. *Angew. Chem., Int. Ed. Engl.* **1995**, *34*, 2520–2524.
12. Larson, D. P.; Heathcock, C. H. *J. Org. Chem.* **1996**, *61*, 5208–5209.
13. Larson, D. P.; Heathcock, C. H. *J. Org. Chem.* **1997**, *62*, 8406–8418.
14. Zhu, X. M.; He, L. L.; Yang, G. L.; Lei, M.; Chen, S.-S.; Yang, J. S. *Synlett* **2006**, *20*, 3510–3513.
15. Xie, L.; Zhu, S.Y.; Shen, X. Q.; He, L. L.; Yang, J. S. *J. Org. Chem.* **2010**, *75*, 5764–5767.
16. Li, Y.; Sun, J.; Gong, Y.; Yu, B. *J. Org. Chem.* **2011**, *76*, 3654–3663.
17. Kojima, M.; Nakamura, Y.; Ito, S.; Takeuchi, S. *Tetrahedron Lett.* **2009**, *50*, 6143–6149.
18. Zhu, C.; Tang, P.; Yu, B. *J. Am. Chem. Soc.* **2008**, *130*, 5872–5873.
19. Uyeda, M.; Yokomizo, K.; Miyamoto, Y.; Habib, E. E. *J Antibiot.* **1998**, *51*, 823–828.
20. Dulce, M.; Garcia, H.; Idetoshi, Y.; Amada, S.; Usumi, H.; Atakeyama, M.; Ugio, N.; Nishizawa, M. *Tetrahedron Lett.* **1994**, *35*, 3325–3328.
21. Mc Gavin, R. S.; Gagne, R. A.; Chervenak, M. C.; Bundle, D. R. *Org Biomol. Chem.* **2005**, *3*, 2723–2732.
22. Marsault, E.; Peterson, M. L. *J. Med. Chem.* **2011**, *547*, 1961–2004.
23. (a) Rodriguez-Lucena, D.; Benito, J. M.; Álvarez, E.; Jaime, C.; Perez-Miron, J.; Ortiz Mellet, C.; Garcia Fernandez, J. M. *J. Org. Chem.* **2008**, *73*, 2967–2979; (b) Kushwaha, D.; Tiwari, V. K. *J. Org. Chem.* **2013**, *78*, 8184–8190; (c) Agrahari, A. K.; Singh, A. S.; Singh, A. K; Mishra, N.; Singh, M.; Prakash, P.; Tiwari, V. K. *New J. Chem.*, **2019**, *43*, 12475-12482.
24. Knox, C.; Law, V.; Jewison, T.; Liu, P.; Ly, S.; Frolkis, A.; Pon, A.; Banco, K.; Mak, C.; Neveu, V.; Djoumbou, Y.; Eisner, R.; Guo, A. C.; Wishart, D. S. *Nucleic Acids Res.* **2011**, *39* D1035–D1041.
25. Thomson Reuters Integrity. http://integrity.prous.com.
26. About GOSTAR. http://gostardb.com/gostar/doc/HyperlinkDownloadPDF.pdf (accessed in April, 2013).
27. Giordanetto, F.; Kihlberg, J. *J. Med. Chem.* **2014**, *57*, 278–295.
28. Zhanel, G. G.; Dueck, M.; Hoban, D. J.; Vercaigne, L. M.; Embil, J. M.; Gin, A. S.; Karlowsky, J. *Drugs* **2001**, *61*, 443–498.
29. Marchi, E.; Montecchi, L.; Venturini, A. P.; Mascellan, G.; Brufani, M.; Cellai, L. *J. Med. Chem.* **1985**, *28*, 960–963.
30. Pirsch, J. D.; Miller, J.; Deierhoi, M. H.; Vincenti, F.; Filo, R. S. *Transplantation* **1997**, *63*, 977–983.
31. Gottlieb, A. B.; Griffiths, C. E. M.; Ho, V. C.; Lahfa, M.; Mrowietz, U.; Murrell, D. F.; Ortonne, J.-P.; Todd, G.; Cherill, R.; Marks, I.; Emady-Azar, S.; Paul, C. F. *Br. J. Dermatol.* **2005**, *152*, 1219–1227.
32. Mahalati, K.; Kaha, B. D. *Clin. Pharmacokinet.* **2001**, *40*, 573–585.

33. Buckner, J. C.; Forouzesh, B.; Erlichman, C.; Hidalgo, M.; Boni, J. P.; Dukart, G.; Berkenblit, A.; Rowinsky, E. K. *Invest. New Drugs* **2010**, *28*, 334–342.
34. Burke, S. E.; Kuntz, R. E.; Schwartz, L. B. *Drug Delivery Rev.* **2006**, *58*, 437–446.
35. Kirchner, G. I.; Meier-Wiedenbach, I.; Manns, M. P. *Clin. Pharmacokinet.* **2004**, *43*, 83–95.
36. Chen, Y.-X.; Caban, B.; Robertson, L. *AAPS J.* **2004**, *6*, M1103.
37. Still, J. G.; Schranz, J.; Degenhardt, T. P.; Scott, D.; Fernandes, P.; Gutierrez, M. J.; Clark, K. *Antimicrob. Agents Chemother.* **2011**, *55*, 1997–2003.
38. Ross, D. B. *N. Engl. J. Med.* **2007**, *356*, 1601–1604.
39. (a) Sorensen, E.J. Angew. *Chem. Int. Ed.,* **2005**, *44*, 5788–5809; (b) Butler, M. S. *J. Nat. Prod.,* **2004**, *67*, 2141–2153; (c) Newman, D. J.; Cragg, G. M. *J. Nat. Prod.,* **2004**, *67*, 1216–1238; (d) Cragg, G. M.; Newman, D. J. *J. Nat. Prod.,* **2004, 67**, 232–244; (e) Mayer, A. M. S.; Gustafson, K. R. Eur. *J. Cancer,* **2004**, *40*, 2676-2689.
40. Henninger, T. C. *Expert Opin. Ther. Pat* **2003**, *13*, 787–805.
41. (a) Itkin, I.H.; M. L. Menzel. J. *Allergy,* **1970**, *157*, 146–154; (b) Kudoh, S.; Azuma, A.; Yamamoto, M.; Izumi, T.; Ando. M. Am. *J. Respir. Crit. Care Med.* **1998**, *157*,1829-1834; (c) Nagai, H.; Shishido, H.; Yoneda, R.; Yamaguchi, E.; Tamura, A.; Kurashima. *A. Respiration* **1991**, *58*, 145-156.
42. (a) Nicolaou, K. C.; Pavia, M. R.; Seitz, S. P. *Tetrahedron Lett.* **1979**, 2327–2330; (b) Nicolaou, K. C.; Seitz, S. P.; Pavia, M. R.; Petasis, N. *J. Org. Chem.* **1979**, *44*, 4011–4013; (c) Nicolaou, K. C.; Seitz, S. P.; Pavia, M. R. *J. Am. Chem. Soc.* **1981**, *103*, 1222–1224; (d) Nicolaou, K. C.; Pavia, M.R.; Seitz, S. P. *J. Am. Chem. Soc.* **1981**, *103*, 1224–1226; (e) Nicolaou, K. C.; Pavia, M. R.; Seitz, S. P. *J. Am. Chem. Soc.,* **1982**, *104*, 2027–2029.
43. (a) Tatsuta, K.; Tanaka, A.; Fujimoto, K.; Kinoshita, M.; Umezawa, S. J. Am. *Chem. Soc.* **1977**, *99*, 5826–5827; (b) Omura, S.; Katagiri, M.; Ogura, H.; Hata, T., *Chem. Pharm. Bull.* **1967**, *15*, 1529–1533; (c) Freiberg, L. A.; Egan, R. S.; Washburn, W. H., *J. Org. Chem.* **1974**, *39*, 2474-2475.
44. (a) Vandeputte, J.; Watchtel, J. L.; Stiller, E. T. *Antibiotic Annu.* **1956**, 587–597; (b) Mechinski, W.; Shaffner, C. P.; Ganis, P.; Avitabile, *G. Tetrahedron Lett.* **1970**, 3873–3876; (c) Ganis, P.; Avitabile, G.; Mechinski, W.; Shaffner, C. P. *J. Am. Chem. Sor.* **1971**, *93*, 4560-4562.
45. (a) Nicolaou, K. C.; Chakraborty, T. K.; Daines, R. A.; Simpkins, N. S. *J. Chem. Soc. Chem. Commun.* **1986**, 413–416; (b) Nicolaou, K. C.; Daines, R. A.; Uenishi, J.; Li, W. S.; Papahatjis, D. P.; Chakraborty, T. K., *J. Am. Chem. Soc.* **1987**, *109*, 2205–2208; (c) Nicolaou, K. C.; Daines, R. A.; Chakraborty, T. K., *J. Am. Chem. Soc.* **1987**, *109*, 2208–2210; (d) Nicolaou, K. C.; Daines, R. A.; Chakraborty, T. K.; Ogawa, Y. *J. Am. Chem. Soc.* **1987**, *109*, 2821–2822; (e) Nicolaou, K. C.; Chakraborty, T. K.; Daines, R. A.; Ogawa, Y. *J. Chem. Soc. Chem. Commun.,* **1987**, 686–689; (f) Nicolaou, K. C.; Chakraborty, T. K.; Ogawa, Y.; Daines, R. A.; Simpkins, N.S.; Furst, G.T., *J. Am. Chem. Soc.* **1988**, *110*, 4660–4672; (g) Nicolaou, K. C.; Daines, R. A.; Uenishi, J.; Li, W. S.; Papahatjis, D. P.; Chakraborty, T. K., *J. Am. Chem. Soc.* **1988**, *110*, 4672–4685; (*h*) Nicolaou, K. C.; Daines, R. A.; Chakraborty, T. K. *J. Am. Chem. Soc.* **1988**, *110*, 4685–4696; (i) Nicolaou, K. C.; Daines, R. A.; Ogawa, Y.; Chakraborty, T. K. *J. Am. Chem. Soc.* **1988**, *110*, 4696–4705; (j) K. C. Nicolaou, W. W. Ogilvie, Chemtracts: *Org. Chem.* **1990**, *3*, 327-349.
46. Carmely, S.; Kashman, Y. *Tetrahedron Lett* **1985**, *26*, 511–514.
47. (a) Patron, A. P.; Richter, P. K.; Tomaszewski, M. J.; Miller, R. A.; Nicolaou. K. C. *J. Chem. Soc. Chem. Commun.* **1994**, 1147–1150; (b) Richter, P. K.; Tomaszewski, M. J.; Miller, R.; Patron, A. P.; Nicolaou, K. C. *J. Chem. Soc. Chem. Commun.* **1994**, 1151-1152.
48. (a) Lewis, A.; Stefanuti, I.; Swain, S. A.; Smith, S. A.; Richard J. K. *Org. Biomol. Chem.* **2003**, *1*, 104–116; (b) Bhattacharjee, A.; Seguil O. R.; De Brabander, J. K. *Tetrahedron Lett.* **2001**, *42*, 1217-1220.

49. Igarashi, M.; Hayashi, C.; Homma, Y.; Hattori, S.; Kinoshita, N.; Hamada, M.; Takeuchi, T. *J. Antibiot* **2000**, *53*, 1096–1101.
50. Tatsuta, K.; Yasuda, S.; Araki, N.; Takahashi, M.; Kamiya, Y. *Tetrahedron Lett* **1998**, *39*, 401–402.
51. Tatsuta, K.; Itoh, M.; Hirama, R.; Araki, N.; Kitagawa, M. *Tetrahedron Lett* **1997**, *38*, 583–586.
52. Shiina, I.; Kubota, M.; Ibuka, R. *Tetrahedron Lett* **2002**, *43*, 7535–7539.
53. Saidhareddy, P.; Ajay, S.; Shaw, A. K. *RSC Adv* **2014**, *4*, 4253–4259.
54. Kito, K.; Ookura, R.; Yoshida, S.; Namikoshi, M.; Ooi, T.; Kusumi, T. *Org. Lett* **2008**, *10*, 225–228.
55. Liu, J.; Xu, K.; He, J.; Zhang, L.; Pan, X.; She, X. *J. Org. Chem* **2009**, *74*, 5063–5066.
56. Tsunakawa, M.; Komiyama, N.; Tenmyo, O.; Tomita, K.; Kawano, K.; Kotake, C.; Konishi, M.; Oki, T. *J. Antibiot* **1992**, *45*, 1467–1471.
57. Schaeffer, H. J.; Beauchamp, L.; de Miranda, P.; Elion, G. B.; Bauer, D. J.; Collins, P. *Nature* **1978**, *272*, 583–585.
58. (a) Lichtenthaler, F. W.; in New Aspects in Organic Chemistry, Yoshida Z., Shiba T., Ohshiro Y., Eds., VCH Publ.: *Weinheim/New York,* **1989**, 372, p 57; (b) Klingler, F. D. *Doctoral Dissertation, Technische Universitat Darmstadt,* **1985**.
59. Isaka, M.; Suyarnsestakorn, C.; Tanticharoen, M.; Kongsaeree, P.; Thebtaranonth, Y. *J. Org. Chem* **2002**, *67*, 1561–1566.
60. Winssinger, N.; Barluenga, S. *Chem. Commun* **2007**, *1*, 22–36.
61. Barluenga, S.; Dakas, P.Y.; Ferandin, Y.; Meijer, L.; Winssinger, N. *Angew. Chem. Int. Ed.* **2006**, *45*, 3951–3954.
62. Tiwari, V. K.; Kumar, A.; Schmidt, R. R. *Eur. J. Org. Chem* **2012**, *15*, 2945–2956.
63. (a) Xiao, M.; Wu, W.; Wei, L.; Jin, X.; Yao, X.; Xie, Z. *Tetrahedron,* **2015**, *17*, 3705–3714; (b) Saidhareddy, P.; Sama, A.; Shaw, A. K., *RSC Adv.,* **2014**, *4*, 4253–4259; (c) Chen, Q.; Du, Y. *Carbohydr. Res.* **2007**, *13*, 1405–1411; (d) Tso, H.-H.; Tsay, H. *Tetrahedron Lett.* **1999**, *40*, 6869–6870; (e) Saidhareddy, P.; Shaw, A. K. *RSC Adv.* **2015**, *5*, 29114. (f) Sabitha, G.; Siva, S.; Reddy, S.; Yadav, J. S. *Tetrahedron Lett.* **2010**, *51*, 6259–6261; (g) Tanimoto, H.; Saito, R.; Chida, N. *Tetrahedron Lett.* **2008**, *49*, 358–362; (*h*) Pazo, M.; Zuniga, A.; Perez, M.; Gomez, G.; Fall, Y. *Tetrahedron Lett.* **2015**, *56*, 3774-3776.
64. (a) Lewis, M. D.; Cha, J. K.; Kishi, Y. *J. Am. Chem. Soc.* **1982**, *104*, 4976–4978; (b) Meo, P.; Osborn, H. M. I. Best Synthetic Methods: Carbohydrates; *Elsevier Science Ltd.: New York,* **2003**, *337*; (c) Somsak, *L. Chem. Rev.* **2001**, *101*, 81–136; (d) Du, Y.; Linhardt, R. J.; Vlahov, I. *R. Tetrahedron* **1998**, *54*, 9913–9959; (e) Mc Garvey, G. J.; Schmidtmann, F. W.; Benedum, T. E.; Kizer, D. *E. Tetrahedron Lett.* **2003**, *44*, 3775–3779; (f) Harvey, J. E.; Raw, S.A.; Taylor, R. J. *K. Org. Lett.* **2004**, *6*, 2611–2614; (g) Ben, A.; Yamauchi, T.; Matsumoto, T.; Suzuki, K. Synlett. 2004, 225–230; (*h*) Gardiner, J.M.; Mills, R.; Fessard, *T. Tetrahedron Lett.* **2004**, *45*, 1215–1217; (i) Terauchi, M.; Abe, H.; Matsuda, A.; Shuto, *S. Org. Lett.* **2004**, *6*, 3751–3754; (j) Steinhuebel, D. P.; Fleming, J. J.; Du Bois, *J. Org. Lett.* **2002**, *4*, 293–295; (j) Vakalopoulos, A.; Hoffmann, H.M. R. Org. Lett. 2001, 3, 177–180.
65. Srihari, P.; Kumaraswamy, B.; Maheswara Rao, G.; Yadav, J. S. *Tetrahedron; Asymmetry* **2010**, *21*, 106–111.
66. Baird, L. J.; Timmer, M. S. M.; Teesdale-Spittle, P. H.; Harvey, J. E. *J. Org. Chem* **2009**, *74*, 2271–2277.
67. (a) Omura, S.; Shiomi, *K. Pure and Applied Chem.* **2007**, *79* (4), 581–591. (b) Pitterna, T.; Cassayre, J.; Huter, O. Bioorg. *Med. Chem.* **2009**, *17*, 4085-4095.
68. Danishefsky, S. J.; David, M. A.; Wincott, F. E.; Selnick, H. G.; Hungate, R. *J. Am. Chem. Soc* **1989**, *111*, 2967–2980.
69. Alois, F.; Albert, M.; Mlynarski, J.; Matheu, M.; DeClercq, E. *J. Am. Chem. Soc* **2003**, *125*, 13132–13142.

70. (a) Moore, B. P.; Brown, W.V. Aust. *J. Chem.* **1976**, *29*, 1365–1374; (b) Ishida, T.; Wada, K. *J. Chem. Soc., Chem. Commun.* **1975**, 205-206.
71. Nicolaou, K. C.; Seitz, S. P.; Pavia, M. R. *J. Am. Chem. Soc* **1982**, *104*, 2030–2031.
72. Saidhareddy, P.; Shaw, A. K. *RSC Adv* **2015**, *5*, 29114–29120.
73. Ishikura, Y.; Tuji, K.; Nukaya, H. *WO* **2004**, 2004–005296.
74. Nakatsuka, T.; Tomimori, Y.; Fukudaa, Y.; Nukaya, H. *Bioorg. Med. Chem. Lett* **2004**, *14*, 3201–3320.
75. Halcomb, R. L.; Danishefsky, S. J. *J. Am. Chem. Soc* **1989**, *111*, 6661–6666.
76. (a) Matsumoto, T.; Katsuki, M.; Suzuki, *K. Tetrahedron Lett.* **1988**, *29*, 6935–6938; (b) Hosoya, T.; Takashiro, E.; Matsumoto, T.; Suzuki, K., *J. Am. Chem. Soc.* **1994**, *116*, 1004–1015; (c) Mahling, J. A.; Schmidt, R. *R. Synthesis.* **1993**, 325–328, and references cited therein.
77. Kumazawa, T.; Minatogawa, T.; Matsuba, S.; Sato, S.; Onodera, J. *Carbohydr. Res* **2000**, *329*, 507.
78. Zeng, J.; Ma, J.; Xiang, S.; Cai, S.; Liu, X.-W. *Angew. Chem. Int. Ed* **2013**, *52*, 5134–5137.
79. (a) Sharma, G.V.M.; Manohar, *V. Tetrahedron Asymmetry.* **2012**, *23*, 252–263; (b) Andrea, Z.; Manuel, P.; Maria, G.; Generosa, G.; Yagamare, *F. Synthesis.* **2011**, 3301–3306; (c) Izuchi, Y.; Kanomata, N.; Koshino, H.; Hongo, Y.; Nakata, T.; Takahashi, *S. Tetrahedron Asymmetry.* **2011**, *22*, 246–251; (d) Nagasawa, T.; Nukada, T.; Kuwahara, *S. Tetrahedron.* **2011**, *67*, 2882–2888; (e) Kanematsu, M.; Yoshida, M.; Shishido, *K. Angew. Chem.* **2011**, *123*, 2666. Angew. *Chem. Int. Ed.* **2011**, *50*, 2618–2620; (f) Fuwa, H.; Yamaguchi, H.; Sasaki, *M. Tetrahedron* **2010**, *66*, 7492–7503; (g) Sabitha, G.; Reddy, D. V.; Rao, A. S.; Yadav, J. *S. Tetrahedron Lett.* **2010**, *51*, 4195–4198; (*h*) Nagasawa, T.; Kuwahara, *S. Tetrahedron Lett.* **2010**, *51*, 875–877; (i) Díaz-Oltra, S.; Angulo-Pachón, C. A.; Murga, J.; Carda, M.; Marco, J. A. *J. Org. Chem.* **2010**, *75*, 1775–1778; (j) Fuwa, H.; Yamaguchi, H.; Sasaki, *M. Org. Lett.* **2010**, *12*, 1848-1851.
80. (a) Kito, K.; Ookura, R.; Yoshida, S.; Namikoshi, M.; Ooi, T.; Kusumi, *T. Org. Lett.* **2008**, *10*, 225–228; (b) Ookura, R.; Kito, K.; Saito, Y.; Kusumi, T.; Ooi, *T. Chem. Lett.* **2009**, *38*, 384–384.

CHAPTER THIRTEEN

Carbohydrate-based antibiotics: Opportunities and challenges

Girija S. Singh
Chemistry Department, University of Botswana, Gaborone, Botswana

1 Introduction

Antibiotics are bacterial or fungal metabolites which at very low concentration inhibit the growth of bacteria, fungi, protozoa or initiate their degradation and as such are used for the treatment of infectious diseases caused by these microorganisms.[1] There are several ways of classifying antibiotics and most common classification schemes are based on their molecular structures, mode of action and spectrum of activity.[2,3] Some common classes of antibiotics based on chemical or molecular structures include β-lactams, macrolides, tetracyclines, quinolones, aminoglycosides, sulfonamides, and glycopeptides (Fig. 13.1)

Some antibiotics may kill the bacteria and some may inhibit their growth. Those that kill the bacteria are termed bactericidal while those that inhibit the bacterial growth are termed bacteriostatic. The first application of antibiotics dates back to the Chinese, 2500 years ago. Fleming's discovery of 'miracle' penicillin created a research field with immense importance in health care of mankind.[4,5] The research and development on antibiotics during the last century has been a great success story. Since the introduction of the first sulfonamides and penicillins in 1935 and 1940, respectively, the mortality rate associated with bacterial infections dropped significantly.[6] Treatment of bacterial infections got further improved by development of new generations of β-lactam antibiotics such as the cephalosporins and carbapenems or recently the macrolides/ketolides, tetracyclines, and quinolones (Fig. 13.1).[7]

The extensive overuse and/ or misuse of the antibiotics in humans, animals as well as in agriculture, however, led the bacterial species to learn about how to defend themselves against conventional antibiotics. This resulted in the development of resistant strains which entailed a worldwide appearance.[8,9] In 2013, the United States Centre for Disease Control and

Carbohydrates in Drug Discovery and Development. http://dx.doi.org/10.1016/B978-0-12-816675-8.00013-0

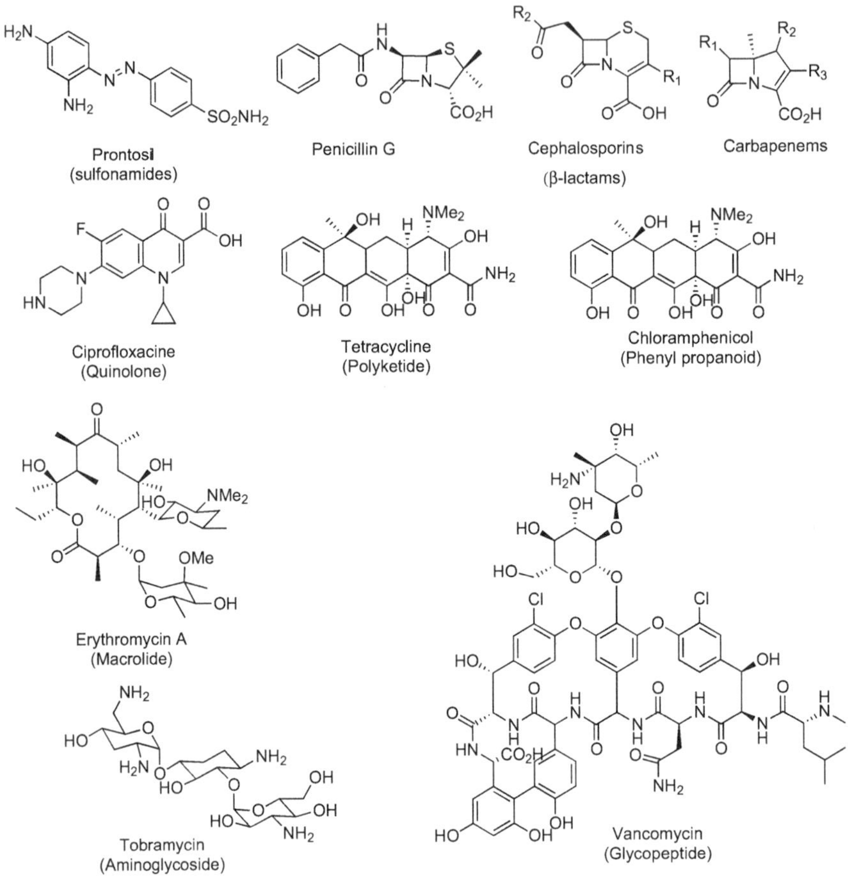

Figure 13.1 ***Some common antibiotics: structures and class.***

Prevention listed several pathogenic bacteria that are considered a serious threat.[10] The emergence and spread of multi-drug-resistant bacteria (MDR) is a global threat to public health and the economy, and urgent counter measures are required to address the problem.[11,12] The awareness about mechanism of resistance and development of knowledge about biosynthetic and metabolic pathways led to fresh initiatives in antibiotic research. The emphasis was on improving the existing drugs by chemical or enzymatic modification of the known structures to create novel semi-synthetic antibiotics against the rising danger of resistant strains and disease pattern.[13] In this context, carbohydrate-containing antibiotics were considered versatile agents to deal with this problem. However, the last decade has been challenging for the discovery of novel antibacterial drugs mostly due to technical challenges, high dose and safety requirements, regulatory difficulties and low-cost generic competition, resulting in a broken financial model. Reck, Jansen and Moser have shed light on these aspects in their recently published paper.[14]

1.1 Carbohydrates-containing antibiotics

Antibiotics belonging to different classes of organic compounds can be considered as carbohydrate antibiotics if they contain carbohydrate moiety. There are generally three types of carbohydrate-related antibiotics based on structural features. The first group consists of those in which carbohydrate moieties are part of the core structure of molecule such as aminoglycoside (Fig. 13.1) and nucleoside antibiotics. The examples include streptomycin, kanamycin, to bramycin, and amikacin, etc. The second group of antibiotics consists of those molecules in which carbohydrates may be glycosidically linked to a non-carbohydrate core structure, such as in macrolides, glycopeptides (Fig. 13.1), and amino coumarins. The common examples of these classes include erythromycin, azithromycin, vancomycin, and nystatin, etc. Additionally, there are oligosaccharide compounds with more or less complex assembling in which the individual monosaccharides are linked both glycosidically and also with one or more orthoester linkages such as orthosomycins.[15]

The target for the aminoglycosides and macrolides is protein biosynthesis while glycopeptides interfere with the cell-wall synthesis. The explanation of the role of the carbohydrate moieties in the respective antibacterial agents together with better synthetic strategies in carbohydrate chemistry as well as improvements in assay development for high throughput screening methods, carbohydrate-containing antibiotics can be used for finding the potential drug leads that can contribute to the fight against infections and diseases caused by (resistant) bacterial pathogens.

In recent years, several expert review articles have been published on development of new carbohydrate-related antibacterial agents.[15-20] This chapter encompasses around these review articles, and deals with historical development of the four main classes of carbohydrate-containing antibiotics – aminoglycosides, nucleosides, macrolides, and glycopeptides, challenges in treatment with these antibiotics and opportunities for new research. It describes the recent developments in investigations on these antibiotics.

2 Aminoglycoside antibiotics

Aminoglycoside antibiotics (Fig. 13.2) are cationic molecules six-membered aminocyclitol core structure, most commonly 2-deoxystreptamine (2-DOS) besides other central cyclitols such as streptidine, actinamine and fortimicin. The central unit is *O*-glycosylated with a varying number of different carbohydrate moieties. Based on the substitution pattern of their 2-DOS ring, aminoglycosides are classified as 4,5-disubstituted 2-DOS aminoglycosides, e.g., neamine, neomycin B, paromomycin, paromamine, and ribostamycin, 4,6-disubstituted 2-DOS aminoglycosides, e.g., arbekacin, amikacin, dibekacin, gentamicin, geneticin, isepamicin, kanamycin A and B, nebramine, netilmicin, plazomicin, formally known as ACHN-490, sisomicin, and tobramycin, and other monosubstituted 2-DOS aminoglycoside, e.g.apramycin and streptomycin. Most aminoglycosides are produced by *Streptomyces* and various *Actinomycetes* strains.

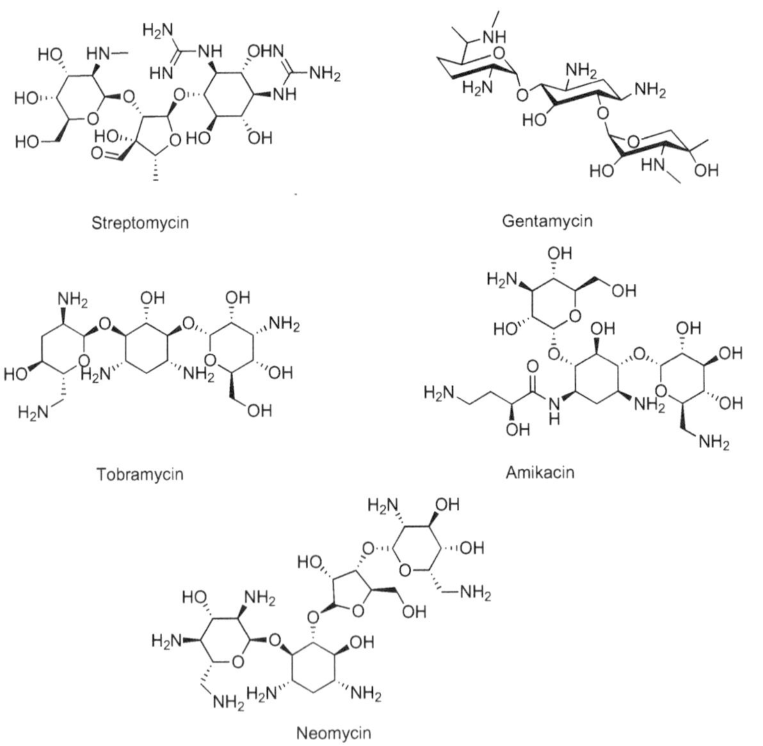

Figure 13.2 ***Some common aminoglycoside antibiotics.***

The well-known antibiotic streptomycin was isolated in 1944 from the cultures of *Streptomyces griseus* by Waksman and his group. It was the first effective drug for treatment of tuberculosis. More than two hundred types of aminoglycoside antibiotics produced by soil microorganisms (*Streptomyces, Micromonospora, Streptoalloteichus, Bacillus*, etc.) have been discovered since the discovery of streptomycin. Naturally occurring drugs in clinical use include streptomycin, neomycin, kanamycin (kanamycin A), paromomycin, spectinomycin, gentamicin, tobramycin, and ribostamycin. In the 1970s, significant progress was made in the development of effective aminoglycoside derivatives aimed at overcoming the problems of drug resistance and toxicity. As a result, semi-synthetic products such as dibekacin, amikacin, arbekacin, and isepamicin were launched onto the market. Even after seventy years of successful use of aminoglycoside antibiotics they remain powerful antimicrobial agents against Gram-negative as well as Gram-positive pathogens.

Of several modes of actions, some are protein biosynthesis inhibitors and thus compromising the structure of the bacterial cell-wall. The mechanism of resistance to aminoglycoside antibiotics has been investigated.[21] These include (1) inactivation by aminoglycoside-modifying enzymes, (2) modification of the active site by 16S rRNA methyltransferases, (3) target modification by mutation, (4) change of uptake, and (5) enhancement of efflux systems. Recent advances in structure, molecular mechanism, structure-activity relationship (SAR) of aminoglycoside mimetics have been reviewed by Silva and Carvalk.[22]

The current efforts to develop new aminoglycoside derivatives with modification and reconstruction on each sugar ring and advances in SAR have also been reviewed by Zhou and co-workers.[23] The recent emergence of infections due to Gram-(-) bacterial strains with advanced patterns of antimicrobial resistance has prompted re-evaluation of the use of aminoglycoside antibacterial agents.[24] This revived interest has brought back to light the debate on the two major issues related to these compounds, namely the spectrum of antimicrobial susceptibility and toxicity. According to recent studies, aminoglycosides retain activity against the majority of Gram-(-) clinical bacterial isolates from many parts of the world. However, the relatively frequent occurrence of nephro-toxicity and ototoxicity during aminoglycoside treatment makes physicians reluctant to use these drugs in everyday practice. Recent advances in the understanding of the effect of various dosage schedules of aminoglycosides on toxicity have provided a partial solution to this problem, although more research still needs to be done in order to overcome this problem entirely.[25]

2.1 Recent progress in design of novel aminoglycosides

The recent progress in design of aminoglycosides can be divided into two categories: a) novel drugs that improve the activity of existing aminoglycosides by providing a drug delivery system and, b) molecules with a novel structure that are effective against MDR bacteria. Examples ofthe former class include TOBI163–165, AREKACE, BAY41-6551, and ME1100. The latter class, molecules with a novel structure effective against resistant bacteria, can be exemplified by plazomicin, 2-hydroxyarbekacin, and TS3112. The chemical strategies for the generation/derivatization of novel aminoglycosides and aminoglycoside-drug conjugates include: (1) modification of aminoglycoside core, (2) preparation of alkylated and arylated aminoglycosides, (3) preparation of acylated aminoglycosides, (4) synthesis of aminoglycoside heteroconjugates, (5) synthesis of aminoglycoside homo- and heterodimers, (6) synthesis of conformationally-restricted aminoglycosides, (7) preparation of novel aminoglycosides by glycodiversification, and (8) preparation of novel aminoglycosides by chemo-enzymatic derivatization.[17] A brief description of the new aminoglycosides is given in succeeding paragraphs.

Tobramycin (TOBI) is a kanamycin-type aminoglycoside and is one component of the nebramycins produced by *Streptomyces tenebrarius.*[26] It was first isolated by Eli Lilly and company in 1967. TOBI is an inhalation solution containing tobramycin as an active ingredient, and it is inhaled into the lungs using a nebulizer or an inhaler device. TOBI significantly improves the lung function in these patients without the adverse side effects observed following parenteral aminoglycoside administration. TOBI is currently approved in more than 40 countries worldwide.

Amikacin (Fig. 13.3) is a semi-synthetic aminoglycoside produced by introducing a (*S*)-4-amino-2-hydroxybutyric acid into the amino group at the C-1 position of kanamycin.[27] Amikacin shows antibacterial activity against Gram-negative bacteria, Gram-positive bacteria (including MRSA) and acid-fast bacteria. In addition, it has the advantage of not being easily inactivated by aminoglycoside-modifying enzymes. ARIKAYCE is a form of amikacin enclosed in liposomal nano-capsules and is administered using a nebulizer system. Inhaled liposomal amikacin is expected to penetrate into bacterial biofilms and prolong the release of amikacin in the lungs while minimizing systemic exposure.[28] ARIKAYCE was approved by FDA in 2018 for the treatment of lung disease caused by *Mycobacterium avium* complex.

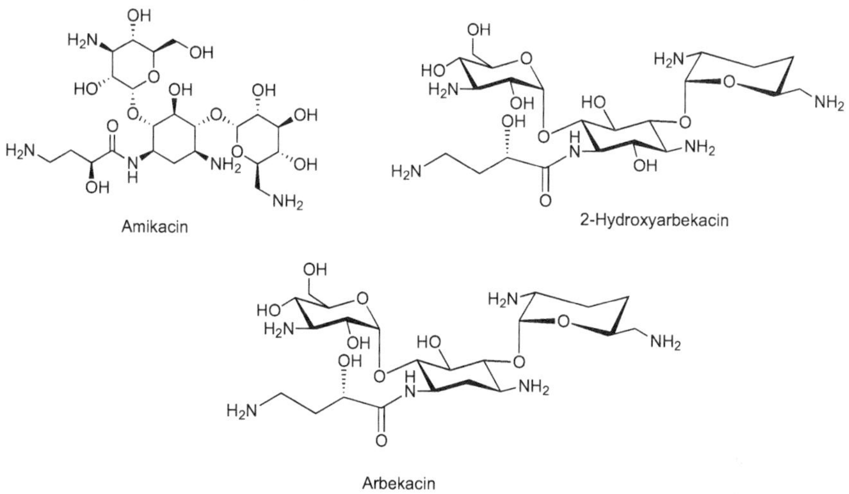

Figure 13.3 ***Structures of Amikacin, Arbekacin, and 2-hydroxyarbekacin.***

BAY41-6551 is also an inhaled amikacin solution being developed by Bayer Health Care Pharmaceuticals Inc.[29] A global phase 3 clinical trial is being conducted to evaluate the efficacy and safety of adjunctive aerosolized BAY41-6551 in the treatment of intubated and mechanically ventilated patients with Gram-negative pneumonia receiving standard-of-care intravenous antibiotics.

ME1100 is a new arbekacin formulation optimized for inhalation.[30] Arbekacin (Fig. 13.3) is a semi-synthetic antibiotic synthesized from antimicrobial drug dibekacin, derived from kanamycin B, by introducing the (*S*)-4-amino-2-hydroxybutyric acid into the amino group at the C-1 position.[31] Arbekacin is a third-generation kanamycin-type drug and is used clinically for the treatment of pneumonia and sepsis caused by MRSA. ME1100 is currently undergoing phase 1b clinical trials in the United States. It is classified as a Qualified Infectious Disease Product with Fast Track status for the adjunctive treatment of mechanically ventilated patients with bacterial pneumonia caused by MRSA and resistant bacteria such as *Klebsiella pneumoniae*, *P. aeruginosa* and *A. cinetobacter.*

2-Hydroxyarbekacin (2-OH-ABK) (Fig. 13.3) is a semisynthetic aminoglycoside specifically designed to reduce the nephrotoxicity of the anti-MRSA drug arbekacin, which is a fourth-generation kanamycin type drug.[32] As the name indicates, 2-OH-ABK is a derivative of arbekacin in which an equatorial hydroxyl group has been introduced at the C-2 position of arbekacin, resulting in a markedly lower nephrotoxicity in comparison to

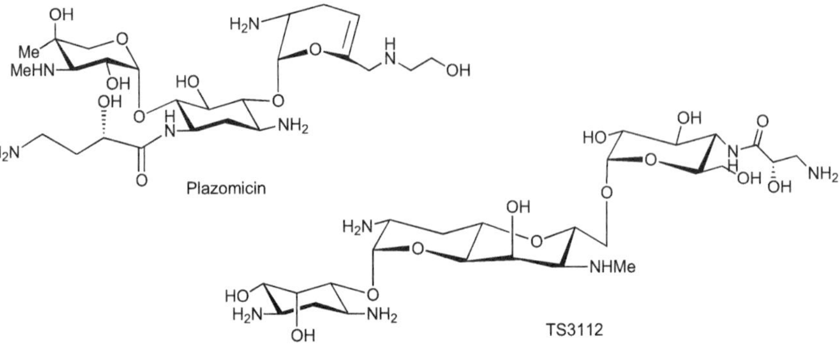

Figure 13.4 ***Structures of plazomicin and TS3112.***

arbekacin. In addition, 2-OH-ABK is effective against arbekacin-resistant bacteria and shows better antibacterial activity than arbekacin against MRSA and *P. aeruginosa*. Initially, 2-OH-ABK was synthesized through the condensation of mono-saccharide with pseudo-disaccharide but has recently been efficiently synthesized from the natural product 2-hydroxykanamycin B.[33]

Plazomicin (Fig. 13.4) is a next generation antibiotic synthetically derived from sisomicin by appending an (*S*)-4-amino-2-hydroxybutyric acid to the amino group at the C-1 position and a hydroxyethyl group to the amino group at the C-6 position.[34,35] Sisomicinis a novel unsaturated aminoglycoside antibiotic that was isolated from *Micromonosporainyoensis* in the 1970s.[18] Plazomicin exhibits excellent antibacterial activity against MRSA, *K. pneumoniae* and *Enterobacteriaceae* Also, it is stable in the presence of various clinically relevant aminoglycoside-modifying enzymes. However, similar to older parenteral AGAs, plazomicin is not effective against NDM-1-producing CRE, resistant bacteria containing a 16S rRNA methyltransferase, and MDR *P. aeruginosa*. Plazomicin was designated Fast Track by the United States Food and Drug Administration in 2012, and approved in June 2018 for treatment of adults with complicated urinary tract infections (cUTI), including pyelonephritis, caused by *Escherichia coli*, *Klebsiella pneumoniae*, *Proteus mirabilis*, or *Enterobacter cloacae*, in patients who have limited or no alternative treatment options.

TS3112 (Fig. 13.4) is an apramycin derivative developed by Takahashi and his group at the Institute of Microbial Chemistry.[36] It shows excellent activity against 16S rRNA methyltransferase-producing bacteria that are highly resistant to aminoglycosides and against MDR NDM-1-producing Enterobacteriaceae.[37] In addition, because TS3112 is effective against *P. aeruginosa* and *Acinetobacter*, including resistant strains, it is most likely to become

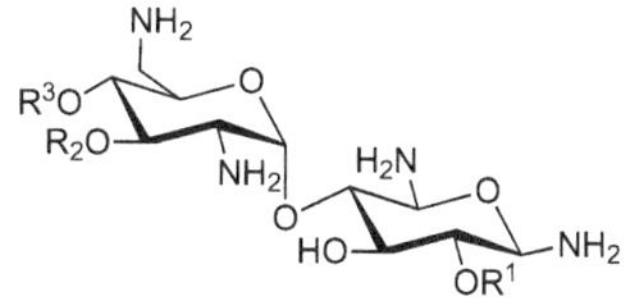

Figure 13.5 ***Structures of homodialkyl and heterodialkylneamines.***

a key drug in the treatment of infectious diseases caused by MDR Gram-negative bacteria.[38]

Recently, Zimmarmann and coworkers have discovered the 3',6-dinonyl neamine as abroad spectrum antibacterial amphiphilic aminoglycoside (Fig. 13.5).[39] They have reported the synthesis, antibacterial activity, and eukaryotic cytotoxicity. Homodialkyl and heterodialkyl neamines appeared to be more active on Gram-negative bacteria than dinaphthylalkyl neamines. However, branched dialkyl neamines or heterodialkyl derivatives were found to be more cytotoxic on mammalian cells than 3',6-dinonyl neamine. The exposure of *P. aeruginosa* over one month to half-MIC of one of the most active derivatives 3',6-dinonyl neamine demonstrated the high difficulty of resistance emergence to amphiphilic aminoglycosides.

Trehalosamine (2-amino-2-deoxy-α,α-D-trehalose) (Fig. 13.6) is an aminoglycoside with antimicrobial activity against *Mycobacterium tuberculosis*, and it is also a versatile synthetic intermediate used to access imaging

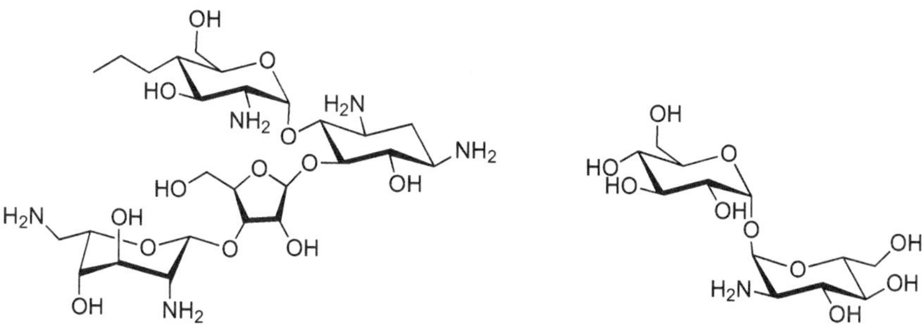

Figure 13.6 ***Structures of a 4′-C-alkyl paromycin derivative and trehalosamine.***

probes for mycobacteria. It was first isolated from *Streptomyces* in 1957.[40] Groenevelt et al. have reported a two-step chemoenzymatic synthesis of trehalosamine that features trehalose synthase (TreT)-catalyzed glycosylation as the key transformation[41].

More recently, Matsushita et al. have described the design and development of a newclass of aminoglycoside antibiotics culminating in the discovery ofpropylamycin.[42] Propylamycin is a 4'-deoxy-4'-alkyl paromomycin (Fig. 13.6) whose alkyl substituent conveys excellent activity against a broad spectrum of ESKAPE pathogens and other Gram-negative infections, including CREs, in the presence of numerous common resistance determinants, whether they are amino-glycoside modifying enzymes or rRNA methyl transferases. Importantly, propylamycin is demonstrated not to be susceptible to the action of the ArmA resistance determinant whose presence severely compromises the action of plazomicin and all other 4,6-disubstituted 2-deoxystreptamine amino-glycosides.

3 Nucleoside antibiotics

Nucleoside antibiotics are found in a large structural variety featuring nucleosides in conjugation with other moieties such as peptides, lipids, and extra carbohydrate moieties, etc.[6] The nucleoside core of these molecules is usually an aldofuranose *N*-linked to a heterocyclic base. These structural features are indicative of their potential for interacting and interfering with RNA and DNA and the related enzymes' system, which results into diverse biological activities such as antibacterial, antifungal, antitrypanosomal, antitumor, antiviral, herbicidal, insecticidal, immune stimulating as well as immunosuppressive properties.

A common strategy for the development of new compounds in this class is the design of novel sugar moieties by replacing the ring oxygen with heteroatoms such as nitrogen, sulfur, phosphorous, putting a carbocycle or changing of ribose to alternative heterocyclic systems. A complementary approach is to identify the pharmacophore of known complex nucleoside antibiotics and do further synthetic modifications on this scaffold to get new molecules with improved profiles.[43]

Many examples of nucleosides antibiotic structures show promising activities against several bacterial strains and, thus, have great potential as antibacterial agents. However, they suffer from disadvantages such as poor adsorption by oral administration, chemical liability, complexity as well as moderate cell entry ability. Furthermore, considering that the DNA as well as RNA machinery works essentially with the same molecules found in all organisms including humans, one main disadvantage of nucleoside antibiotics is the lack of selectivity and cross-activity in mammalian cells. Selected examples of nucleosides reported are described as pyrimidine analogues, fluorinated pyrimidines and thiouracils in succeeding paragraphs.

3.1 Pyrimidine analogues

Tunicamycins (Fig. 13.7), produced by several *Streptomyces* sp., are composed of an uracil moiety, a *N*-acetyl-D-glucosamine residue and an unique 11-carbon dialdose sugar called tunicamine that carries an amide-linked fatty acid residue at the amino function.[44] The compounds of this class are known to be substrate analogue inhibitors of bacterial enzyme phospho-UDP-*N*-acetylmuramoyl-pentapeptide translocase (MraY, translocase I) but they lack selectivity.[45] The MraY catalyzes the formation of lipid I by transferring the UDP-MurNAc moiety to undecaprenyl phosphate, which is a carrier molecule embedded in the cytoplasmic membrane.

Capuramycins (Fig. 13.7) are uracil nucleosides having an unsaturated uronic acid moiety connected by means of an amide linkage to a caprolactame residue. It demonstrates rapid bacterial activity against several different mycobacterial species by inhibiting translocase I.[46]

Caprazamycins and liposidomycins (Fig. 13.8) contain an uridine moiety substituted at O-5' with a 5-amino-5-deoxyribose; liposidomycins are sulfated at position O-2. Additionally, the uridine moiety carries a characteristic 1,4-diazepan-3-one heterocycle at position C-5'which is linked to lipophilic side chains.[47] Liposidomycins selectively inhibit bacterial translocase I *in vitro*. The hydrophilic sulfate group at C-2" may cause poor permeability and hence leading to less antimicrobial activity than expected. This was actually observed in studies with liposidomycin structures lacking the sulfate group at this position. Recently, it was shown that the lipophilic

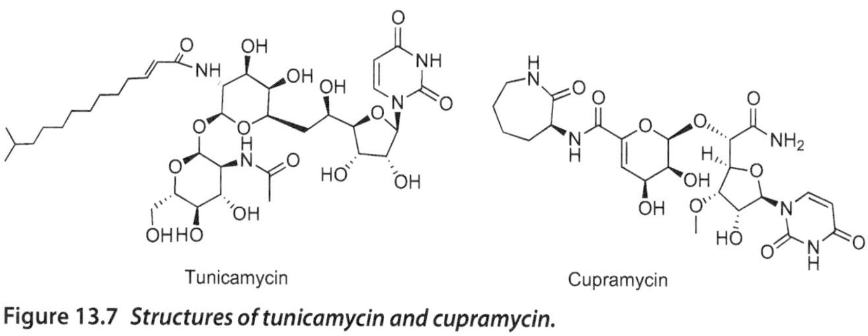

Figure 13.7 ***Structures of tunicamycin and cupramycin.***

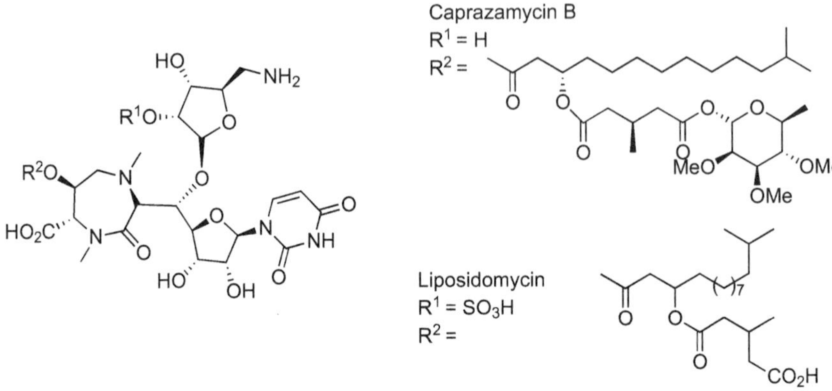

Figure 13.8 ***Structures of caprazamycins and liposidomycins.***

side chains interact with a highly hydrophobic *trans*-membrane domain of MraY.[48] Caprazamycins additionally contain a per-*O*-methylated fucosyl residue. The individual members of this class are distinguished in their lipid side chains. The compounds of this class have shown excellent *in vitro* anti-mycobacterial activity against drug-susceptible and multidrug-resistant MBT strains without significant toxicity in mice.[49] To reduce the complexity for liposidomycins and caprazamycins, acyclic analogues have been designed and synthesized.[44] These compounds are expected to be promising leads for anti-tuberculosis agents such as the semi-synthetic CPZEN-45,[50] a 4-butylaniline amide derivative of caprazamycins.

Mureidomycins are peptidyl nucleosides produced by *Streptomyces flavidovirens* and contain a 3'-deoxyuridine sugar attached by means of an enamide linkage to a peptide chain containing two *meta*-tyrosine residues, one methionine residue and an *N*-methyl-2,3-diaminobutyric acid moiety (Fig. 13.9).[51] This class also selectively and specifically inhibits bacterial

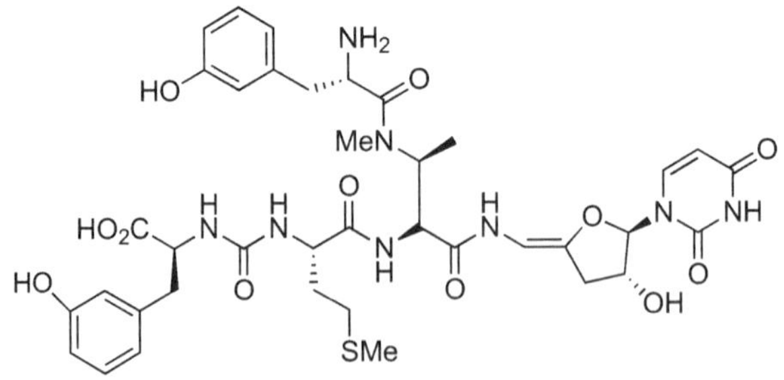

Figure 13.9 ***Structure of mureidomycin A.***

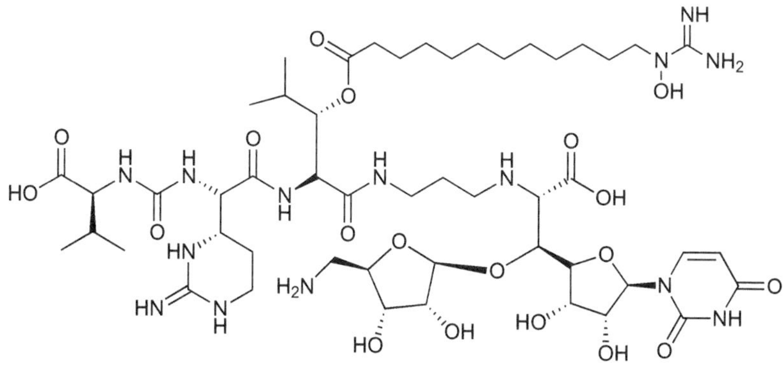

Figure 13.10 ***Structure of muraymycin A1.***

translocase I and, most importantly, does not inhibit mammalian polysaccharide or glycoprotein synthesis and, therefore, benefits from very low toxicity. With a solid-phase synthetic protocol, a simplified uridinyl branched peptide urea library with structural similarity to the mureidomycin class was synthesized, of which a high throughput screening identified several compounds with moderate anti-tuberculosis activity.[52]

Muraymycins are structurally and mechanistically related to uridyl antibiotics such as mureidomycins, and have a common peptide-appended glycosylated uronic acid core structure (Fig. 13.10).[53] They are also MraY inhibitors. Yamashita et al. have synthesized a series of muraymycin analogues and evaluated them in a SAR study.[54] Some derivatives exhibited excellent antimicrobial activities. The first total synthesis of (-)-muraymycin and its epimer was reported by Tanino and coworkers in 2010.[55] The key steps in this strategy are preparation of the urea dipeptide moiety found in the muraymycins containing an L-epi-capreomycidine via a nitrene C-H insertion of the sulfamate and the fully protected muraymycin skeleton at a late stage by an Ugi four-component reaction.

Sansanmycin (Fig. 13.11), a nucleosidyl peptide antibiotic similar to mureidomycins,isolated from an unidentified *Streptomyces* sp., has been observed to have specific inhibitory activity against *Pseudomonas aeruginosa* and MBT H37Ra as well as against multidrug-resistant strains of MBT.[56]

Gemcitabine (2′,2′-difluoro-2′-deoxycytidine) (Fig. 13.12) is an analogue of deoxycytidine bearing two fluorine atoms in place of hydrogen atoms on the 2′ carbon of the sugar component. In a study that explored the efficacy of various nucleoside analogues against clinical isolates from

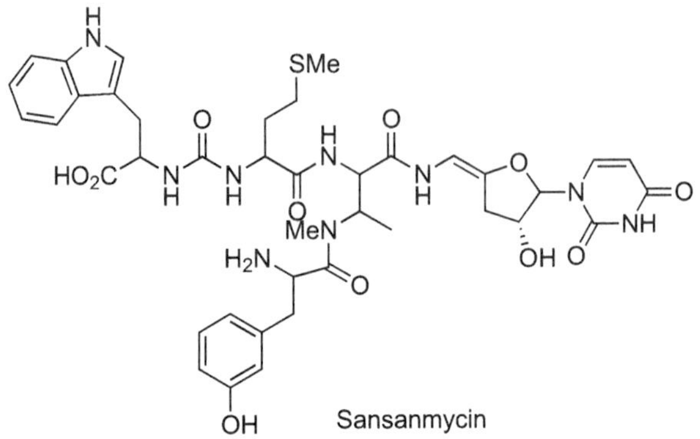

Figure 13.11 ***Structure of sansanmycin.***

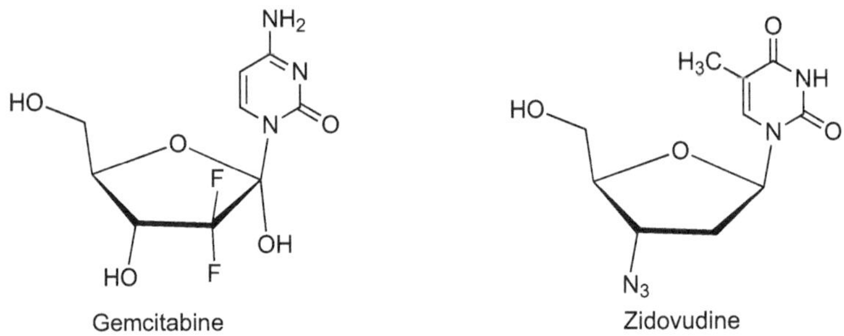

Figure 13.12 ***Structures of gemcitabine and zidovudine.***

different bacteria genera, gemcitabine was identified to have antibacterial activity.[57] Gemcitabine had a potent antibacterial activity on Gram-positive bacteria, including important pathogenic species from genera *Listeria*, *Bacillus*, *Enterococcus*, and *Staphylococcus*. It was, however, ineffective against Gram-negative bacteria. Gemcitabine has been evaluated *in vivo* on murine infection model.[58] Mice were infected with *Streptococcus pyogenes*AP1, a virulent strain responsible for causing the majority of severe *S. pyogenes* infections. Of the mice infected with a potentially fatal dose of *S. pyogenes* AP1, those treated with the control had a 100% mortality rate, whereas those treated with gemcitabine had only a 17% mortality rate. These studies, thus, revealed gemcitabine as a good candidate for repurposing as an antibiotic. Jordheim et al. have performed additional *in vitro* preclinical studies to investigate the potential of gemcitabine, especially its efficacy against clinically important multidrug-resistant strains of *Staphylococcus aureus*.[59] Gemcitabine was found to be active against 19 different strains of methicillin resistant *S. aureus*. Gemcitabine was effective against glycopeptides intermediate *S. aureus*, a strain resistant to all glycopeptides antibiotics, including vancomycin.

Zidovudine[3′azido-3′-deoxythymidine, AZT] (Fig. 13.12) is a thymidine analogue with an azido group at the C-3′ of the deoxyribose ring. Zidovudine was first discovered to have antibacterial activity in the late 1980s. It had antimicrobial activity against various Enterobacteriaceae, including *Salmonella* species.[60] Subsequently, zidovudine has been demonstrated to have potent activity against many pathogenic Gram-negative bacteria, including *Escherichia coli*, *Salmonella typhimurium*, *Klebsiella pneumoniae*, *Shigellaflexneri*, and *Haemophilus influenzae* and including isolates resistant to conventional antibiotics.[61,53,62,63] It also acts in synergy with conventional antibiotics, enhancing their effectiveness.[64-67] Zidovudine is ineffective against Gram-positive bacteria such as *Listeria* species, *Bacillus* species, Staphylococci, and *Enterococcus faecalis*as well as against *Mycobacteria* species and *Pseudomonas aeruginosa*.[53,56] It also inhibited growth of antibiotic-resistant *E. coli* and *K. pneumoniae* in a murine peritoneal infection model, acting synergistically with colistin.[62]

Plicacetin and streptcytosine A (Fig. 13.13) are disaccharide nucleosides belonging to amicetin class. Amicetin was isolated in the early 1950s as the first member of the amicetin family antibiotics which are produced in the fermentation broth of various species of Streptomyces, Arthrobacter, and Nocardia.[68] More recently, streptcytosine A(2) was isolated from the culture broth of Streptomyces sp.TPU1236A collected in Okinawa, Japan[69] and from *Streptomyces rochei* 06CM016 collected in the Mediterranean sea and named rocheicoside A.[70] These antibiotics have a common disaccharide pyrimidine nucleoside motif, namely cytosamine, in which amosamine

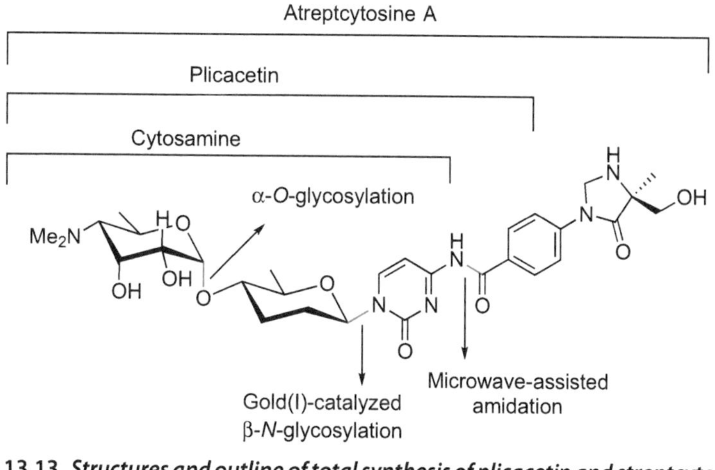

Figure 13.13 ***Structures and outline of total synthesis of plicacetin and streptcytosine. A***

is α-(1 → 4)-linked to amicetose, itself β-(1 → N1)-linked to acytosine nucleus. The principal synthetic strategies for further diversification of molecular structure are mono- or dimethylation of the amino group of amosamine, hydroxylation of C-3 of the amicetose residue, and acylation of N-7 on cytosine.The compounds of this class exhibit antibacterial and antiviral activities both *in vitro* and *in vivo*, more specifically against tuberculosis, herpes,and polio.An improved total synthesis of plicacetin and streptcytosine A has been reported recently from a common precursor cytosamine.[71] The amosamine and amicetose moieties were assembled through an α-selective *O*-glycosylation, and the cytosine nucleus was subsequently introduced through a β-selective gold (I)-catalyzed *N*-glycosylation. Finally, a microwave-assisted amidation reaction led to the formation of final products.

3.2 Fluorinated pyrimidines

Nucleosides with fluorinated pyrimidine motif exhibit antibacterial activity. Various fluorinated pyrimidines are known to inhibit the growth of the human pathogen *Ureaplasmaurealyticum*.[72] Floxuridine (Fig. 13.14) had a strong effect *in vitro*, inhibiting the growth of *U. urealyticum*for up to 72 hours. This finding was supported by a later study by Wehelie et al.[73] Subsequently, various Gram-positive and Gram-negative bacteria have been reported to be susceptible to floxuridine.[57,58,74] Gram-positive bacteria, such as *S. aureus* and *Bacillus* species, were more sensitive than the Gram-negative bacteria tested. Floxuridine has also been found to have synergic effects when used in combination with zidovudine.[64] 5-Fluorouracil (Fig. 13.14) has been reported to decrease virulence of *P. aeruginosa* by disrupting biofilm formation[75] also has activity against *Mycobacterium tuberculosis* and a number of oral microbes including the pathogen *S. pyogenes*.[76,77] Floxuridine

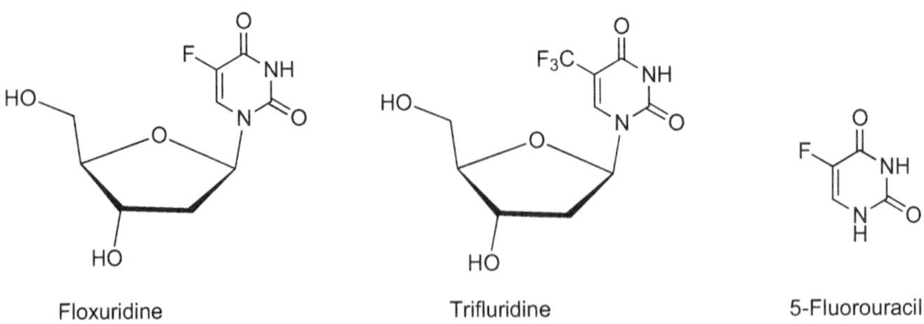

Figure 13.14 ***Structures of fluorinated pyrimidines.***

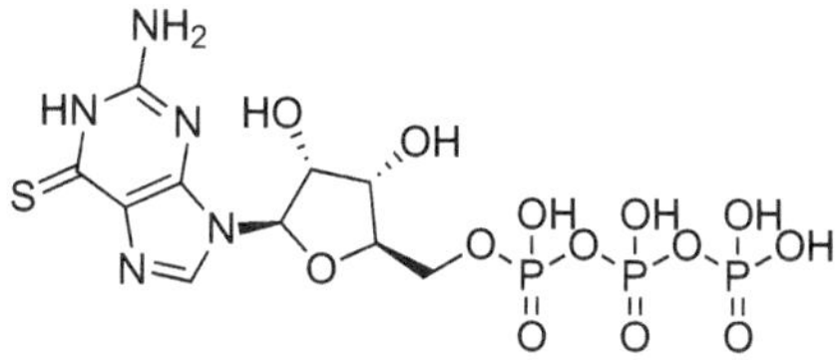

Figure 13.15 ***Structure of 6-thioguanosine 5′-triphosphate.***

and trifluridine (Fig. 13.14) both inhibited the growth of *Mycoplasma pneumoniae*, with trifluridine strongly inhibiting growth *in vitro*.[78]

3.3 Thiopurines

Currently, the thiopurines are predominantly used clinically as anticancer, immune-suppressant, and anti-inflammatory agents. In particular, they are important therapeutic agents for acute leukemias and for inflammatory bowel disease.[79,80] However, the thiopurines have also shown antibacterial activity. 6-Mercaptopurine exhibits *in vitro* activity against *Mycobacterium avium* subspecies *paratuberculosis* and *Corynebacterium* species.[81] 6-Thioguanosine 5′-triphosphate (thio-dGTP) (Fig. 13.15) inhibits spore germination of *Bacillus anthracis* in an infected macrophage cell line,[82]] and thioguanine inhibits *in vitro* growth of *Mycoplasma pneumonia*.[83]

4 Macrolide antibiotics

Macrolide antibiotics are 12-, 14-, 16- or 18-membered polyhydroxy lactones that are *O*-glycosylated at least at one position, usually with cladinose or desosamine and sometimes with other deoxy- or aminodeoxy-sugars.[84] They belong to the subgroup of polyketide natural products and constitute an important therapeutic class. Macrolide antibiotics have been used successfully to treat respiratory tract and other bacterial infections. This success is mainly due to their (i) spectrum of activity, (ii) safety, (iii) tissue and intracellular activity, and (iv) anti-inflammatory properties. These properties give macrolides an advantage over other classes of antibiotics. Their spectrum of activity in the respiratory tract is targeted to key community-acquired pathogens including *Streptococcus pneumoniae, Haemophilus influenzae, Staphylococcus aureus*, and atypical pathogens such as *Legionella pneumophila, Mycoplasma pneumoniae* and, *Chlamydophila pneumoniae*. More than 500 representatives of macrolide antibiotics are known, most of them

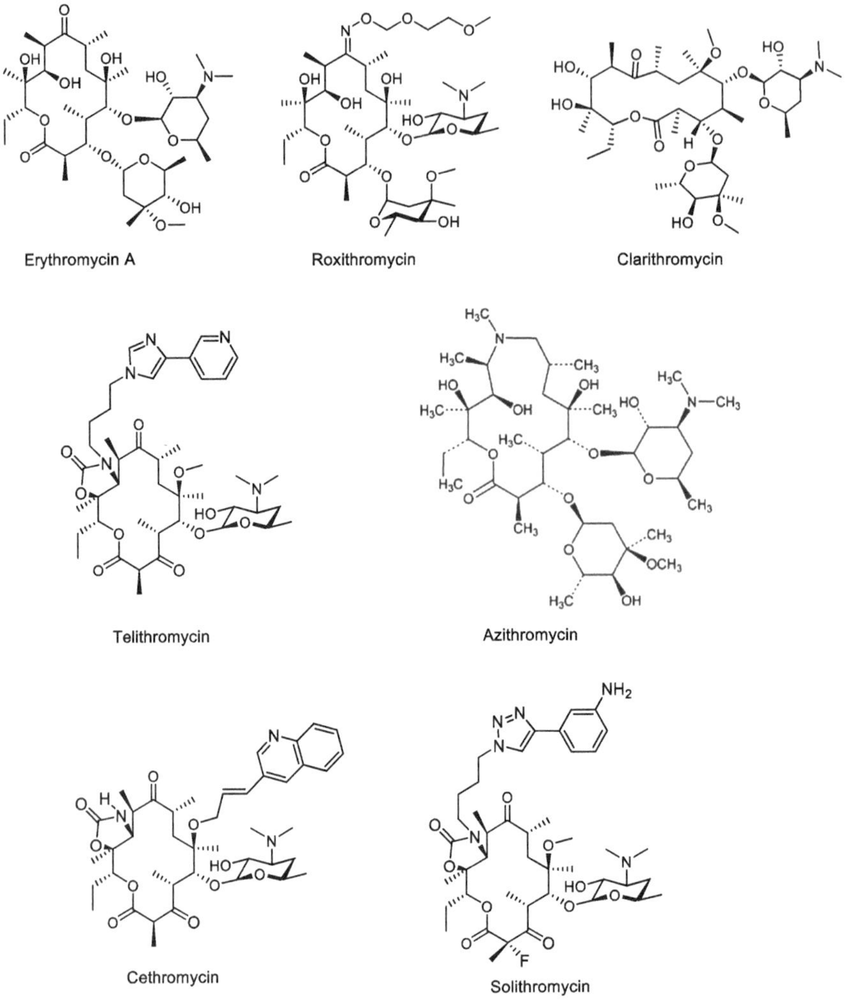

Figure 13.16 ***Structure of common macrolide antibiotics.***

showing excellent behavior against Gram-positive bacteria. Macrolides account for 20% of all the antibiotics prescribed.

The principal representative of macrolides erythromycin A (Fig. 13.16) was first isolated from *Streptomyces erythreus* at the Lilly in 1952.[85] It helps against the major respiratory pathogens and is considered safe. It is widely prescribed for treatment of children. However, it has a limited antibacterial spectrum and limited solubility in acidic medium.

The second generation macrolide antibiotics, azithromycin, clarithromycin, and roxithromycin (Fig. 13.16) have gradually replaced erythromycin

A because of their higher potency, broader spectrum of activity, improved physicochemical and pharmacokinetic profiles, and attenuated side effects.[86] However, similar to erythromycin A, the second generation variants also have poor activity against macrolide-resistant pathogens.

Ketolides are derived from 14-membered macrolides by removal of *L*-cladinone under acidic conditions and selective oxidation of the resulting 3-hydroxy group to the corresponding carbonyl group.[87,88] The first semi-synthetic ketolide RU-64004 (HMR 3004) was synthesized at Roussel Uclaf.[89] This ketolide was stable in acidic media, showed good intracellular penetration, and demonstrated potent activity against erythromycin A resistant and penicillin resistant streptococci and *H. influenzae*. Systematic SAR studies led to the discovery of several ketolide lead structures such as telithromycin, cethromycin (initially known as ABT-773) (Fig. 13.16), and EP-013420 with potent activity and improved pharmacokinetic profile. Telithromycin, was the first ketolide to be approved in Europe (2001), Japan (2003) and in the US (2004) for the once daily oral dose for treatment of respiratory tract infections. It was synthesized from clarithromycin in eight steps.[90] In fact, all the members of macrolide class approved or in trial phase for use in humans have been manufactured by chemically modifying erythromycin. Azithromycin (the first azalide antibiotic on the market) is prepared from erythromycin in four steps, telithromycin requires twelve steps, clarithromycin requires six steps, and the advanced clinical candidate solithromycin (Fig. 13.16) is currently produced from erythromycin by a 16-step linear sequence.[91-94]

Seiple et al. have presented a practical, fully synthetic route to macrolide antibiotics by the convergent assembly of simple chemical building blocks, enabling the synthesis of diverse structures not accessible by traditional semi-synthetic approaches.[16] More than 300 new macrolide antibiotic candidates, as well as the clinical candidate solithromycin, have been synthesized using these convergent approaches. Solithromycin is a new macrolide, the first fluoroketolide, which has been tested successfully in two Phase 3 trials and is undergoing regulatory review at the FDA. Solithromycin is differentiated from telithromycin chemically and biologically in that its side chain is chemically different and does not significantly block nicotinic acetylcholine receptors.[95]

EDP-420 (modithromycin) (Fig. 13.17) is a third generation analogue of erythromycin with a bridged bicyclic macrolide core structure. It has been observed to exhibit an improved pharmacological profile compared to telithromycin and cethromycin, including activity against macrolide-resistant

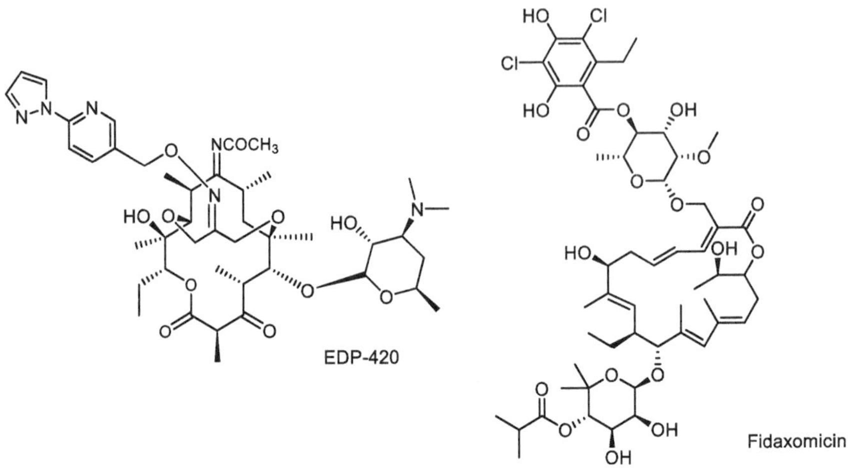

Figure 13.17 ***Structures of EDP-420 and fidaxomicin.***

Streptococci as well as MRSA and VRE.[96] Bermudez et al. have reported the activity of EDP-420 on *Micobacteriumavium*. Infection caused by *Mycobacterium avium*complex (MAC) is common in patients with immunosuppression, such as AIDS, and deficiencies of gamma interferon and interleukin-12, as well as patients with chronic lung diseases. EDP-420 showed *in vitro* inhibition at a concentration ranging from 2 to 8 μg/ml (MIC50 of 4 μg/ml and MIC90 of 8 μg/ml).[97] Xu and coworkers have reported a multikilogram-scale synthesis of EDP-420 from commercially available erythromycin A 9-oxime.[98] The synthetic procedure involves a series of transformations such as triacetylation, Pd-catalyzed *O,O*-bis-allylation (bridge formation), acid-catalyzed sugar cleavage, oxime reduction, acetylation, Os-catalyzed bridge olefin oxidative cleavage, Corey − Kim oxidation, bridge oxime formation, deprotection, and final purification.

Tiacumicins are a group of 18-membered macrolide antibiotics isolated from cultures of *Dactylosporangium aurantiacum*. Tiacumicin B (lipiarmycin A3, previously OPT-80, and PAR-101), the main component produced in this culture, contains an unsaturated macrolide core with an unusual C-5-methylated sugar at position C-11 and an aryl substituted 6-deoxy sugar moiety at C-20. This macrolide inhibits bacterial DNA-dependent RNA polymerase by blocking the formation of the first internucleotide bond. It exhibits broad-spectrum Gram-positive antibacterial activity. It is highly active against *Clostridium difficile* associated diarrhoea in low concentrations.[99,100] Tiacumicin B was approved by the FDA in 2011. It is commercially available under the generic name fidaxomicin (Fig. 13.17). Total synthesis of fidaxomicin was reported by Galus and Altmann in 2014 (published in 2015)[101] and by Carreira and Jolton in 2015.[102] Kaufmann et al. reported the first enantioselective synthesis of fidaxomicin in 2015.[103] The key features of this synthesis involve a rapid and high-yielding access to the noviose, rhamnose, and orsellinic acid precursors; the first example of a β-selective noviosylation; an effective Suzuki coupling of highly functionalized substrates; and a ring-closing metathesis reaction of a noviosylated dienoate precursor.

Bispolides were isolated from a culture of *Microbispora* sp.[104] Their core scaffold was found to be a symmetric cyclic dimeric structure of a novel unsaturated 20-membered ring macrolide that is glycosylated with 2,6-dideoxyhexopyranose and 6-deoxyhexopyranose alternatively (Fig. 13.18). These molecules exhibited good anti-MRSA activities *in vitro*.

The mode of action of macrolide antibiotics involves blocking bacterial protein biosynthesis by binding to the 23S ribosomal RNA of the 50S subunit and interfering with the elongation of nascent peptide chains during translation.[105] Located in domain V, near the peptyl transferase site, macrolide antibiotics obstruct the peptide exit tunnel without affecting peptidyl transferase activity.

Antibiotics methymycin and pikromycin, co-produced by *Streptomyces venezuelae*, represent minimalist macrolide protein synthesis inhibitors.[106] Unlike other macrolides, which carry several side chains, methymycin and

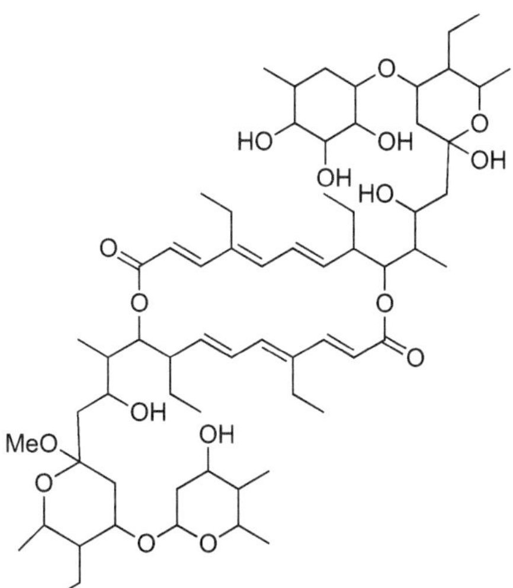

A bispolide

Figure 13.18 ***Structure of a bispolide.***

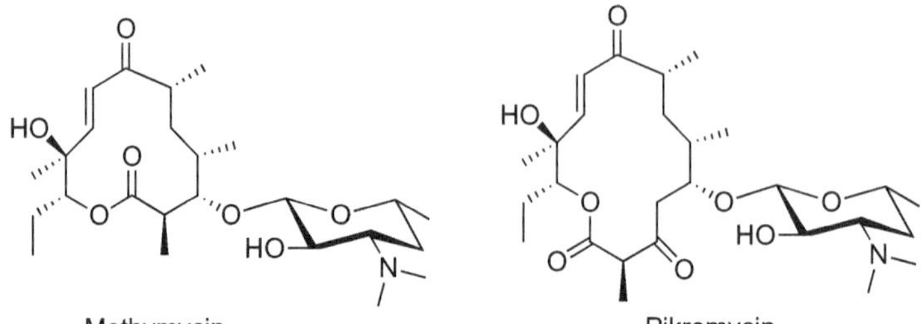

Figure 13.19 ***Structures of antibiotics methymycin and pikromycin.***

pikromycin contain a single desosamine sugar attached to their macrolactone ring. Pikromycin (Fig. 13.19), the main 14-member macrolide compound, secreted by *Streptomyces venezuelae* strain ATCC 15439, carries a C-5 desosamine and a C-3 ketone group (Fig. 13.1). Furthermore, due to an alternative translation initiation site within the polyketide synthase gene, a second, even smaller and unusual 12-membered ring ketolide, methymycin is generated via the same biosynthetic pathway. The unusual structure of these ketolides and their simultaneous secretion by *S. venezuelae* bring about the possibility that two compounds would bind to distinct ribosomal sites. However, by combining genetic, biochemical and crystallographic studies, it has been demonstrated that mythomycin and pikromycin inhibit translation by binding to overlapping sites in the ribosomal exit tunnel. Strikingly, while MTM and PKM readily arrest the growth of bacteria, ~40% of cellular proteins continue to be synthesized even at saturating concentrations of the drugs. Gel electrophoretic analysis showed that these ketolides prevented synthesis of a smaller number of cellular polypeptides compared to other ribosomal antibiotics, illustrating a unique mode of action of these molecules.

5 Glycopeptide antibiotics

Glycopeptide antibiotics[107] consist of a cyclic peptide backbone with typically seven amino acid residues, representing atropisomerism by interlinkage of amino acid side chains. These antibiotics are highly active against Gram-positive bacterial strains. The peptide backbone is *O*-glycosylated with different mono- or disaccharide moieties at up to four positions. The two most prominent and best-known representatives in this class are vancomycin and teicoplanin. Vancomycin (Fig. 13.1), the first gylcopeptide introduced into clinical practice in 1959, was isolated from *Streptomyces orientalis* from soil samples by the Lilly in 1955-56.[108] Then teicoplanin was the only additional member of this class that was available for human use. Both drugs are unaltered natural antibiotics of the large dalbaheptide group that is produced by various actinomycetes. Their common structural element is a linear heptapeptide backbone (configuration *R,R,S,R,R,S,S*) in which some aromatic amino acids residues are cross linked (biphenyl and diphenyl ether motives) leading to a rigid concave shape.

Glycopeptides inhibit bacterial cell walls biosynthesis by recognizing and strongly binding to the LLys--DAla--D-Ala termini of peptidoglycan precursor strands at the external side of the membrane. In this way, transpeptidases are prevented from executing their cross linking job.[109] Glycopeptides antibiotics are limited to treating gram-(+) infections as they cannot penetrate the outer membrane of gram-(-) bacteria. With the rise of MRSA infections in hospitals, vancomycines became the antibiotic of last resort. However, due to its frequent use, resistant gram-(+) pathogens, in particular vancomycine-resistant enterococci (VRE) emerged and worryingly spread. By 2003, more than half of the clinicalVRE isolates in the US had become resistant to glycopeptides. Consequently, research studies were conducted for the development of semisynthetic glycopeptide agents.[110-112] by *N*-alkylationor *N*-acylation of the vancosamine moiety in vancomycin and related compounds.

A new glycopeptide dalbavancin (Fig. 13.20), a semisynthetic analogue of teicoplanin that shows a different glycosylation pattern with two *O*-linked monosaccharide residues, a 2-*N*-acyl-D-glucuronic acid instead of the vancosamine and an extra *D*-mannose at residue 7, as well as modifications in the C and N terminus of the peptide part was developed and approved by FDA in 2014. Its mode of action is similar to other glycopeptides. It exerts

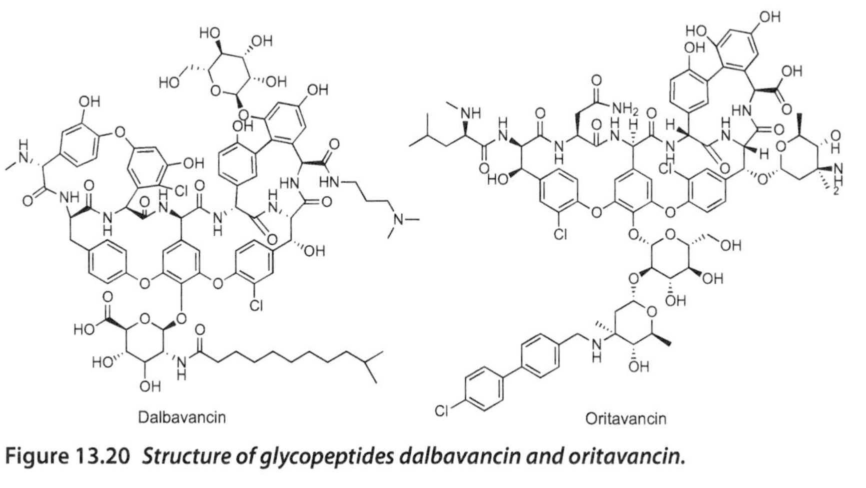

Figure 13.20 ***Structure of glycopeptides dalbavancin and oritavancin.***

its bactericidal effect by disrupting cell wall biosynthesis. It binds to the D-alanyl-D-alanyl residue on growing peptidoglycan chains and prevents transpeptidation from occurring, preventing peptidoglycan elongation and cell wall formation.

The lipoglycopeptide oritavancin (Fig. 13.20) contains a *p*-chlorophenylbenzyl substituent attached to the amino function of the disaccharide D-glucose-4-*epi*-vancosamine as the main modification, thus offering both hydrophobic as well as hydrophilic groups. It demonstrates activity similar to vancomycin but possesses improved activity against vancomycin-resistant *Staphylococcus aureus* (VRSA) *and Enterococcus* (VRE) strains. The mode of action of oritavancin involves disruption of cell membrane of Gram-positive bacteria by 4'-chlorobiphenylmethyl group. It also acts by inhibition of transglycosylation, and transpeptidation.[113] Oritavancin has been approved by the US FDA and European Medicines Agency for treatment of acute bacterial skin and skin structure infections.

Telavancin (Fig. 13.21) is a semisynthetic analogue of vancomycin, containing an *N*-decyl-aminoethylvancosamine moiety and a negatively

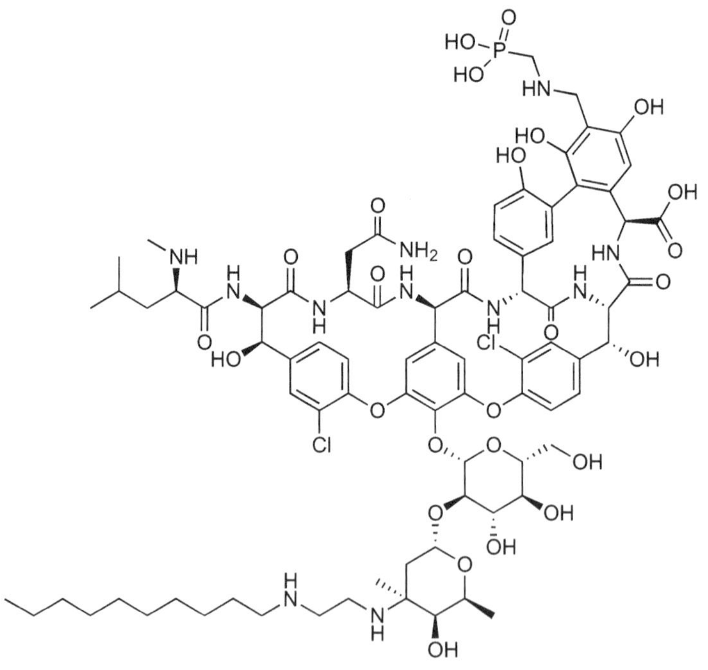

Figure 13.21 ***Structure of telavancin.***

charged *N*-(phosphonomethyl)amino-methyl group bound to the peptide backbone.[114] Telavancin shows a broad-spectrum microbial activity including VRSA as well as MRSA and is effective in the treatment of soft-tissue infections, endocarditis, meningitis and pneumonia caused by Gram-positive pathogens. Its antibacterial activity results from the common inhibition of the peptidoglycan synthesis. Additionally, it disrupts bacterial membranes by depolarization. Telavancin got approval of FDA in September 2009 for complicated skin and skin structure infections and in June 2013 for hospital-acquired and ventilator-associated bacterial pneumonia caused by *Staphylococcus aureus.*

The success of these three second generation glycopeptide antibiotics as drug candidates and their potentiated antibacterial activities in comparison to vancomycin have given impetus to further research in developing novel derivatives and studying the mechanism of action/resistance.[115,116]

Ramoplanins lipoglycopeptides, first isolated from *Actinoplanes* strain ATCC 33076, are different in structure and function compared to the vancomycin and teicoplanin families.[117] They consist of a 49-membered macrocyclic depsipeptide containing 17 L- and *D*-amino acids and are *O*-glycosylated with an α-1,2-dimannosyl moiety (Fig. 13.22). They

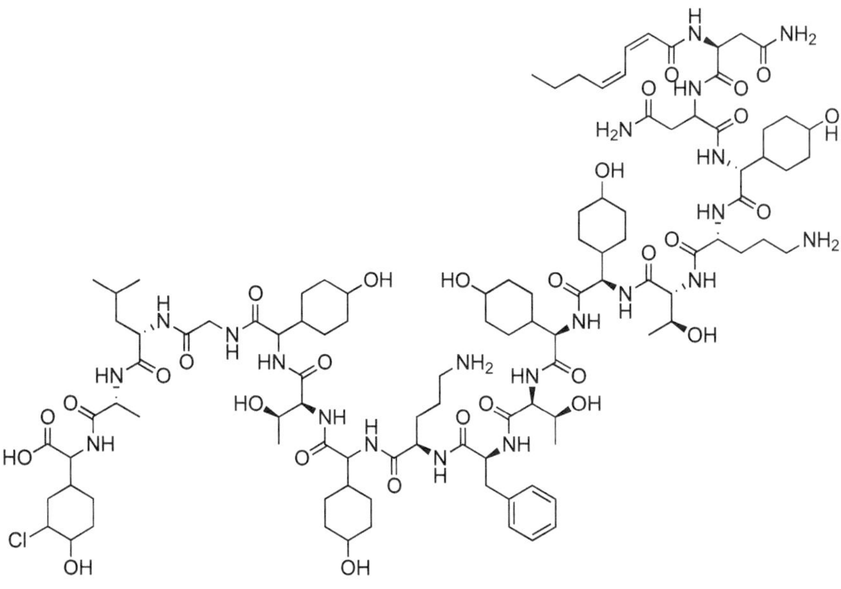

Ramoplanin

Figure 13.22 ***Structure of ramoplanin.***

inhibit two enzymes involved in the late step of peptidoglycan biosynthesis. Ramoplanin has also been observed to directly inhibit MurG, an enzyme responsible for the conversion of lipid I to lipid II, although this transformation is intracellular and would require cell membrane penetration of the agent, a pathway actually not accessible for this compound. The dimannosyl group is not required for biological activity, but clearly accounts for the noncompetitive inhibition pattern. Ramoplanin factor A2 is now undergoing Phase III clinical trials as a promising broad spectrum drug candidate against Gram-positive bacteria including VRE and MRSA and, most importantly, no resistance has been observed as yet.

Mannopeptimycins (Fig. 13.23) isolated as a mixture of compounds from the strain *Streptomyces hygroscopicus* LL-AC98,[118] is another class of cyclic hexapeptides with alternating D- and *L*-amino acids containing

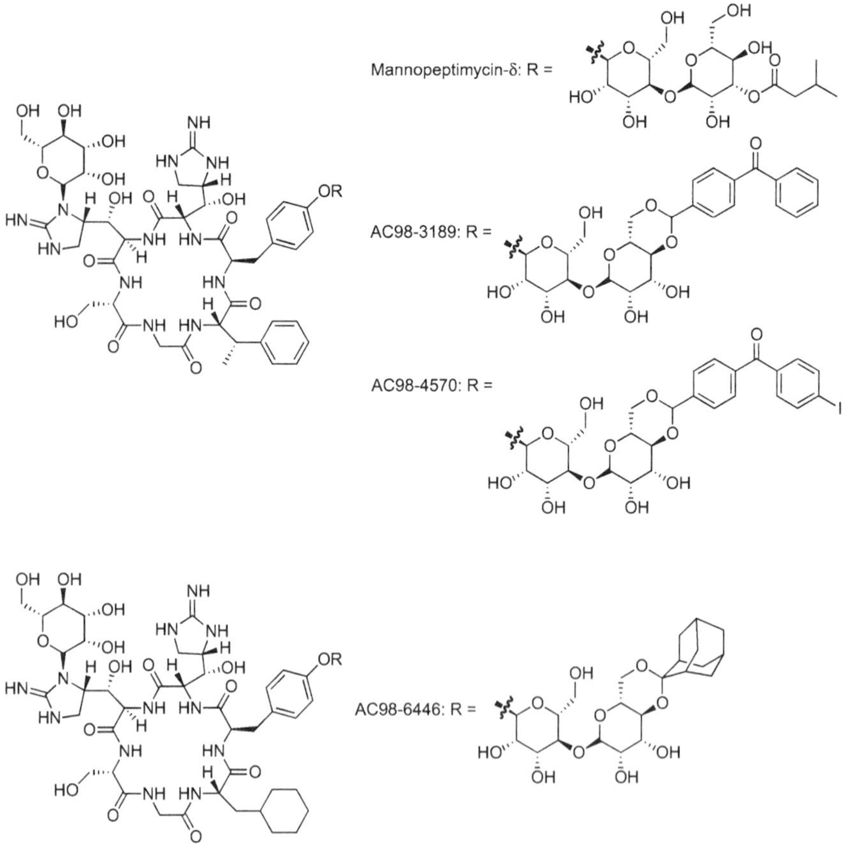

Figure 13.23 ***Structures of mannopeptimycins.***

characteristic guanidine moieties, one of them carrying an *N*-linked *D*-mannose residue. Furthermore, these compounds are *O*-glycosylated with an isovaleryl modified dimannose residue. The presence of the hydrophobic acyl substituent at the terminal mannose is relevant for the activity. These lipoglycopeptides interfere with the later stage of bacterial cell wall biosynthesis by blocking the transglycosylation reaction mediated by binding to lipid II in a different mode of action compared to vancomycin. Mannopeptimycin as well as several semisynthetic analogues obtained by chemical derivatisation show exceptional activity against multidrug-resistant pathogens such as MRSA and VRE and, therefore, clearly have potential as therapeutic candidates for clinical drugs.

A range of derivatization strategies (Fig. 13.24) have been employed in recent year to access semisynthetic glycopeptides. Okano et al. have recently reviewed the total synthesis of vancomycin-related glycopeptide antibiotics and their analogues.[119] Since glycopeptide scaffold contains many substituents, facile modifications are common. In particular, the free *C*-terminal carboxylic acid group, the vancosamine sugar primary amine (if present), and the *N*-terminal primary and secondary amines are easily transformed. Selected examples of antibacterial agent employing these modifications are described below.

There are some reports on vancomycin analogues with modification of hydroxyl groups. However, they did not show any improvement in activity.[120] A back-to-back dimerization of vancomycin and other glycopeptides

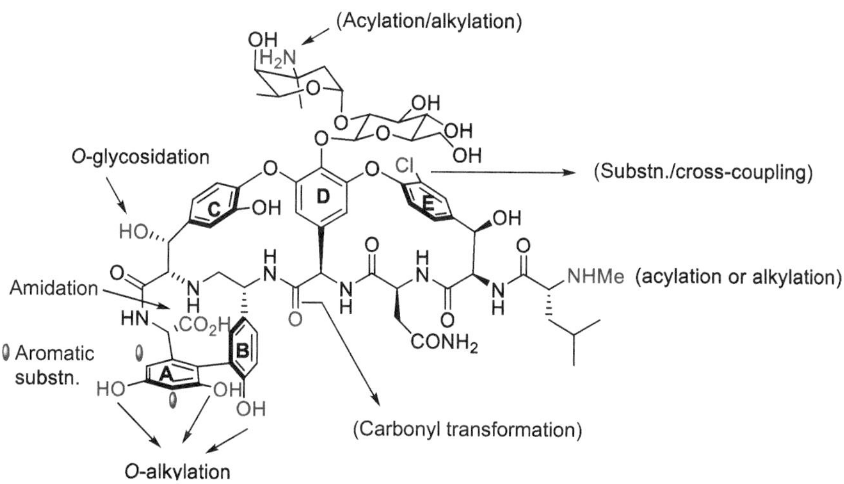

Figure 13.24 ***Derivatization strategies for new glycopeptides.***

is reported to increase their Lipid II binding affinity. Over the years, a number of groups took advantage of this characteristic by covalently linking two or more vancomycin moieties together.

Recently, the Sharpless group employed the Cu-catalyzed azide-alkyne "click" cycloaddition reaction to form a triazole bridge between two vancomycin units substituted with a variety of *C*-terminal functionalized alkynes and azides.[121] The MIC of these products against MRSA varied from 0.6 to 5.0 μg/mL for the 30 different dimers (compared to 2.5 μg/mL for vancomycin dimer) with substantial improvements in activity against VRE (0.8 – 1.6 μg/mL vs 25 μg/mL). In addition, heterodimers were prepared linking azide-functionalized *C*-terminal vancomycin derivatives with alkyne-substituted vancosamine amine vancomycin derivatives. It led to reduced activity against MRSA (5 – 30 μg/mL) but slightly improved activity against VRE (3.1 – 6.3 μg/mL).

Gu et al. synthesized new vancomycin derivatives by alkylating the amino group of vancosamine with a set of *O*-substituted glyceric acid derivatives (Fig. 13.25), in an attempt to mimic the transglycosylation inhibition properties of moenomycin and introduce lipophilic substituents while maintaining solubility. With the exception of a highly lipophilic chlorobiphenyl substituent which was 4- to >8-fold more potent than vancomycin against Newman *S. aureus* (MIC <0.125 vs 2 μg/mL) and Mu50 *S. aureus* (MIC 2 vs 8 μg/mL), the remaining analogues showed activity either similar or inferior to vancomycin.[122]

The Haldar group at the Jawaharlal Nehru Institute for advanced Scientific Research in India is actively engaged in developing new vancomycin derivatives with improved activity. Haldar and coworkers have developed another vancomycin derivative, YV54465 in which the vancosamine amine is alkylated with a decanyl group and the carboxylic acid

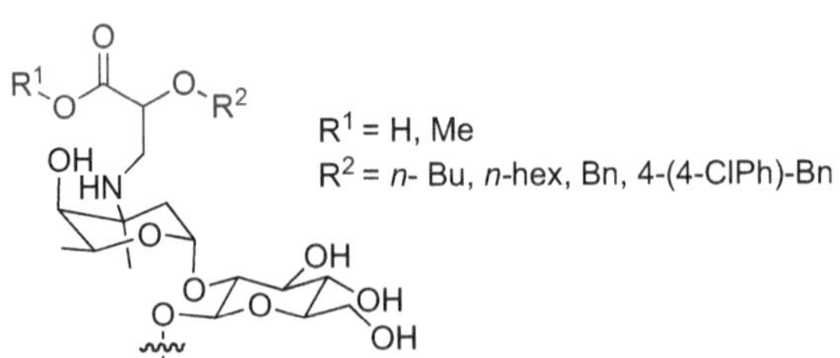

Figure 13.25 ***New vancomycin derivatives from alkylationof amino group of vancosamine with a set of O-substituted glyceric acid derivatives.***

is amidated with a cyclic-acyclic polyhydroxylated sugar moiety.[123,124] YV54465 is 2-fold more potent than vancomycin against MRSA but 35- and >100-fold more active against VISA and VRE, with an MIC of 2 μg/mL vs both VanA and VanB VRE. This group also attached a zinc-binding dipicolyl group to the C-terminus of vancomycin.[125] While activity against vancomycin-sensitiveMRSA was similar (0.5 μM), under standard testing conditions, the adduct was 10-fold more potent than vancomycin against VISA and >150-fold higher against VRE. In the presence of added Zn^{2+} ions, activity was further enhanced by a factor of 23. In further research, they attached a quaternary ammonium propylamine substituent to the *C*-terminus of vancomycin.[126]

Okano et al. synthesized the fully glycosylated version of vancomycin with the introduction of an amidinegroup in place of a carbonyl group (Fig. 13.26) along with an analogue that introduced the vancosamine-chlorobiphenylsubstituent of oritavancin.[127,128] While the compound with R = Hhad similar activity to the aglycon, the additional substituent 4-(4-chlorophenyl)benzyl improved potency by over 100-fold against VRE, with MIC of 0.005 μg/mL for VanA *E. faecalis* or *E. faecium* and 0.06 μg/mL for VanB E. faecalis. Excellent potency was also observed against MRSA (0.03 − 0.06 μg/mL for R = 4-(4-chlorophenyl)benzyl). Finally, in 2017, the carboxylic acid group was also modified by amidation with quaternary aminoalkylamine substituents somewhat similar to the dimethyl aminopropylamine group found on dalbavancin.[129] The preferred analogue with a

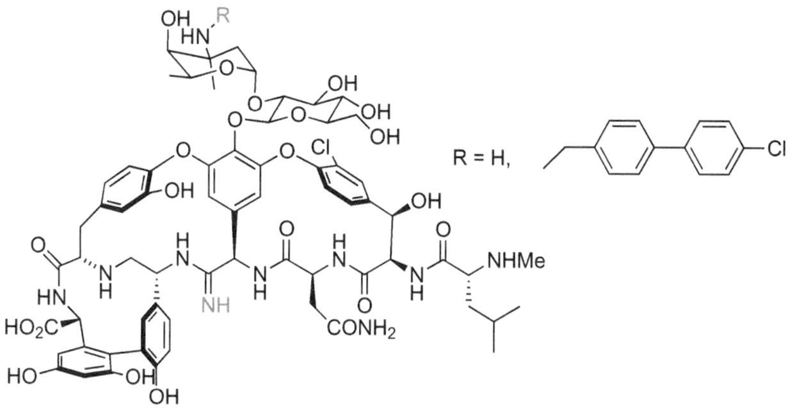

Figure 13.26 ***A fully glycosylated vancomycin with an amidine group in place of a carbonyl group.***

trimethyl aminopropylamine group showed impressive VanAVRE activity of 0.01 − 0.005 μg/mL.

Aryl ring substitution of glycopeptides is among very common approach to develop new derivatives. Several groups worldwide have reported synthesis and biological activities of new glycopeptides obtained by novel modifications of aryl groups. For example, Boger et al. haveinvestigated the influence of *E*-ring substitution on vancomycin aglycon activity, by selective functionalization of the *E*-ring aryl chloride via Pd-catalyzed conversion to a boronic acid, followed by substitution with a range of functional groups.[130] Replacement of the *E*-ring chloride with hydrogen or polar groups lead to reduced activity, while nonpolar group derivatives show activity similar to the parent chloro moiety. Permethylation of the phenolic hydroxyls and carboxylic acid of the same *E*-ring substituted aglycon series gave derivatives with up to an 8-fold improvement in activity against VanB*E. faecalis* compared to vancomycin aglycon, whereas potency was generally lost across the series (~2-fold) against vancomycin-sensitive *S. aureus* (ATCC 25923).

Guan and coworkers introduced an additional sugar residue on A-ring in combination with vancosamine alkylation (Fig. 13.27) (Scheme 13.27).[131] It resulted into enhanced activity of 128- to 1024-fold against MSSA ATCC 5904 and VISA Mu50 (MIC = 0.03 − 0.25 μg/mL), and several vanA, vanM(MIC = 1 − 8 μg/mL), and vanB (MIC = < 0.06 − 0.25 μg/mL)

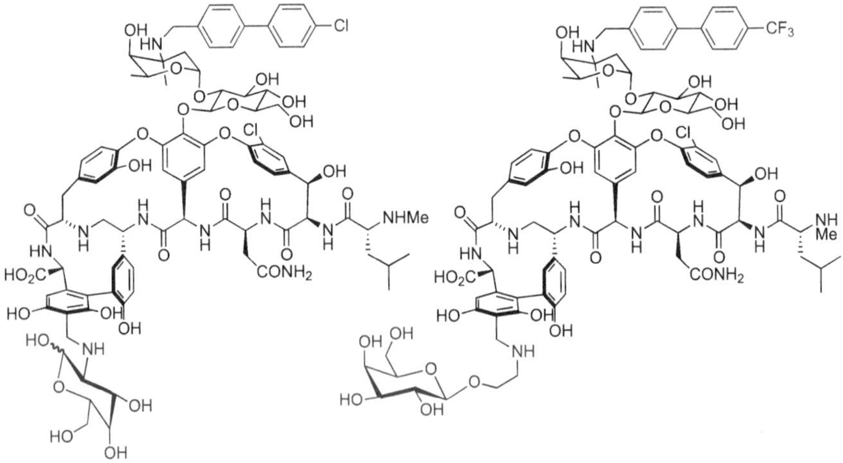

Figure 13.27 ***New glycopeptides with an additional sugar residue on A-ring in combination with vancosamine alkylation.***

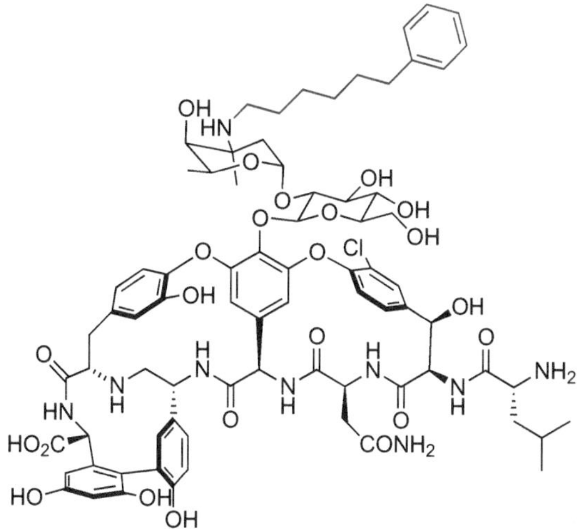

Figure 13.28 ***A vancomycin derivative with phenylhexyl group on vancosamine.***

Enterococci strains. Both analogues were equivalent to telavancin at 7 mg/ kg ip injection in a liver abscess VISA (Mu50) mouse infection mode, with approximately 2 log reduction in cfu compared to saline or vancomycin 1 at 7 mg/ kg. They also demonstrated 88 – 93% survival in a MRSA mouse lethal challenge model (again at 7 mg/kg ip), compared to 7% survival for vancomycin 1 and 93% survival for telavancin.

Zhang et al. and Chang et al. have synthesized des-methyl vancomycin analogues by selectively alkylating the vancosamine amino group with a range of aliphatic and aromatic lipophilic substituents and evaluated it for antibacterial activity.[132,133] The compounds showed a 2- to 4-fold improvement (MIC 0.78 μg/mL) for selected analogs against different strains of MRSA and *S. pneumoniae* compared to des-methyl vancomycin (MIC 1.56 – 3.13 μg/mL for both strains). A phenylhexyl derivative (Fig. 13.28) also exhibited >4-fold improvement against VRE (*E. faecalis*, MIC 12.5 μg/mL).

Vancomycin has been conjugated with a range of functional moieties, including other antibiotics, siderophores, fluorophores, and specific targeting constructs, in order to discover improved therapeutics or useful tool compounds. For example, cefilavancin (TD-1792) is a conjugate of vancomycin and a cephalosporin, leading to a dual targeting action against peptidoglycan synthesis (Fig. 13.29).[134,135]

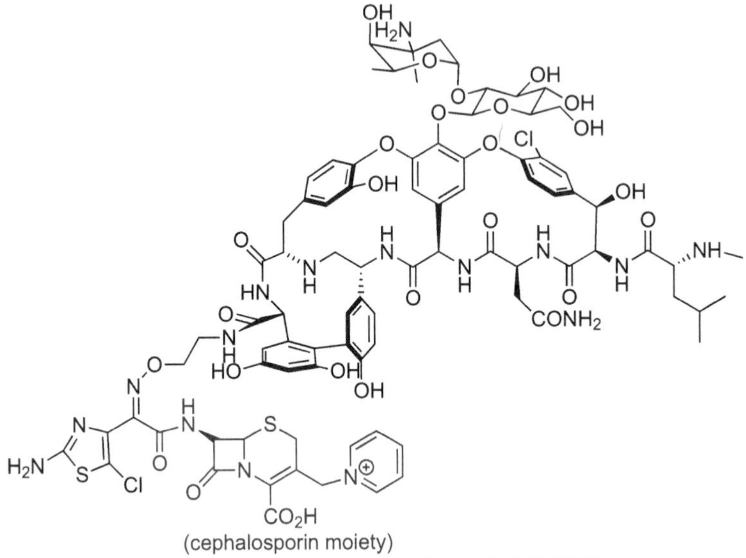

Figure 13.29 ***Cefilavancin (TD-1792)-a conjugate of vancomycin and a cephalosporin.***

6 Conclusions and future perspectives

Carbohydrate-containing antibiotics undoubtedly constitute an important class of antibiotics. Some of them contain carbohydrate scaffold in core structure while in many of them a carbohydrate moiety is attached to the core structure. Based on their molecular structure, the main carbohydrate-containing antibacterial agents can be classified as aminoglycosides, nucleosides, macrolides, and glycopeptides. The common aminoglycosides in clinical applications are streptomycin, gentamicin, kanamycin A, tobramycin, neomycin, amikacine, plazomicin. Arbekacin is a third-generation kanamycin-type drug and is used clinically for the treatment of pneumonia and sepsis caused by MRSA. Plazomicin and ARIKAYCE, an amikacin formulation have got FDA approval recently while BAY41-6551 and ME1100 are under trial. Efforts are on to improve the total synthesis of complex aminoglycosides. Nucleosides are maninly pyrimidine and fluorinated pyrimidine derivatives. Several new compounds of this class have been synthesized in recent years and evaluated for their antibacterial activity. Macrolides are among oldest known antibiotics. The second generation antibiotics like azithromycin, roxithromycin, telithromycin and clarithromycin replaced the first generation erythromycins but they are also facing challenges due to development of resistant strains. Cethromycin and selithromycin are under clinical trials. EDP-420 (modithromycin) is a third generation analogue of erythromycin. Its large scale synthesis has been reported recently from commercially available erythromycin A 9-oxime. The first enantioselective synthesis of fidaxomicin, a 18-membered macrolide antibiotic also known as tiacumicin, has been accomplished recently. Among glycopeptides, several antibiotics such as telavancin, dalbavancin, and oritavancin have been developed to tackle the vancomycin-resistant strains. Ramoplanin factor A2, a lipoglycopeptide is now undergoing Phase III clinical trials as a promising broad spectrum drug candidate against Gram-positive bacteria including VRE and MRSA and, most importantly, no resistance is reported to date. Since glycopeptide scaffold contains many such groups that can be easily transformed to other groups, facile modifications are common. In particular, the free *C*-terminal carboxylic acid group, the vancosamine sugar primary amine (if present), and the *N*-terminal primary and secondary amines are easily transformed leading to several new vancomycin derivatives of potential biological interest. A new approach in glycopeptide modification is to design conjugates with other class of antibiotics. Clearly, an extensive research is going on towards developing potential carbohydrate-containing antibiotics with broad spectrum and least toxicity. The challenges, however, will remain there because of newly developing resistant-strains.

References

1. Hodges, N. A. Antibiotics and synthetic antimicrobial agents: their properties and uses. In *Hugo and Russell's pharmaceutical microbiology,* Denyer, S. P., Hodges, N. A., Gorman, S. P., Gilmore, B. F., Eds.; 8th ed.; Wiley-Blackwell: UK, 2011; pp. 169–199.
2. Calderón C.B., Sabundayo B. P., Schwalbe R., Steele-Moore L., Goodwin A. C., (Eds.), *Antimicrobial Susceptibility Testing Protocols CRC Press (Taylor & Francis Group),* **2007**, 7–52.
3. Van Hoek, A. H.; Mevius, D.; Guerra, B.; Mullany, P.; Roberts, A. P.; Aarts, H. J. *Front. Microbiol.* **2011**, *2*, 1–27.
4. Wong, J. *Primary care update for ob/gyns.* **2003**, *10*, 124–126.
5. Bentley, R. *Persp. Biol. Med.* **2005**, *48*, 444–452.
6. Rachakonda, S.; Cartee, L. *Curr. Med. Chem.* **2004**, *11*, 775–793.
7. Nussbaum, F. V.; Brands, M.; Hinzen, B.; Weigand, S.; Häbich, D. *Angew. Chem. Int. Ed.* **2006**, *45*, 5072–5129.
8. Lambert, P. A. *Adv. Drug. Deliv. Rev.* **2005**, *57*, 1471–1485.
9. Smith A. W., Bacterial A. W., resistance to antibiotics, in, Denyer, S. P., Hodges, N. A., Gorman, S. P., Gilmore, B. F., Eds., eighth ed., *Wiley-Blackwell, UK,* **2011**, 217–229.
10. United States Centers for Disease Control and Prevention. Antibiotic resistance threats in the United States. http://www.cdc.gov/drugresistance/threat-report-2013/pdf/arthreats-2013-508.pdf (**2013**).

11. The Lancet Infectious Diseases (editorial), Time for global political action on antimicrobial resistance, *Lancet Infect. Dis.,* **2016**, *16*, 1085.
12. Barriere, S. L. *Expert Opin. Phamacother.* **2015**, *16*, 151–153.
13. Spellberg, B.; Bartlett, J. G.; Gilbert, D. N. *New Eng. J. Med.* **2013**, *368*, 299–302.
14. Reck, F.; Jansen, J. M.; Moser, H. E. *Arkivoc.* **2019**, *iv*, 227–244.
15. Schitter, G.; Wrodnigg, T. M. *Expert Opin. Drug discov.* **2009**, *4*, 315–356.
16. Seiple, I. B.; Zhang, Z.; Zacubek, P.; Langlois-Mercier, A.; Wright, P. M.; Hog, D. T.; Yabu, K.; Allu, S. R.; Fukuzaki, T.; Carlsen, P. N.; Kitamura, Y.; Zhou, X.; Condakes, M. L.; Szczypinski, F. T.; Green, W. D.; Myers, A. G. *Nature* **2016**, *533*, 338–355.
17. Chandrika, N. T.; Garneau-Tsodikova, S. *Chem. Soc. Rev.* **2018**, *47*, 1189–1249.
18. Cooper, D. J.; Jaret, R. S.; Reimann, H. *J. Chem. Soc. Chem. Commun.* **1971**, 285–286.
19. Takahashi, Y.; Igarashi, M. *J. Antibiot.* **2018**, *71*, 4–14.
20. Thomson, J. M.; Lamont, I. L. *Front. Microb.* **2019**, *10*, 1–11.
21. Santos, L. C. *Open J. Med. Microbiol.* **2012**, *2*, 24–36.
22. Silva, J. G.; Carvalho, I. *Curr. Med. Chem.* **2007**, *14*, 1101–1119.
23. Zhou, J.; Wang, G.; Zhang, L. -H.; Ye, X. S. *Med. Res. Rev.* **2007**, *27*, 279–316.
24. Falagas, M. E.; Grammatikos, A. P.; Michalopoulos, A. *Expert Rev. Anti Infect. Ther.* **2008**, *6*, 593–600.
25. Durante-Mangoni, E.; Grammatikos, A.; Utili, R.; Falagas, M. E. *Int. J. Antimicrob. Agents.* **2009**, *33*, 201–205.
26. Cheer, S. M.; Waugh, J.; Noble, S. *Drugs* **2003**, *63*, 2501–2520.
27. Olivier, K. N.; Griffith, D. E.; Eagle, G.; McGinnis second, J. P., et al. *Am. J. Respir. Crit. Care Med.* **2017**, *195*, 814–823.
28. Meers, P.; Neville, M.; Malinin, V.; Scotto, A. W., et al. *J. Antimicrob. Chemother.* **2008**, *61*, 859–868.
29. Niederman, M. S.; Chastre, J.; Corkery, K.; Fynk, J. B.; Luyt, C. E.; Garcia, M. S. *Intensive Care Med.* **2011**, *38*, 263–271.
30. Baba, N., et al. Copenhagen. Denmark. Poster P245, 2015.
31. Kondo, S.; Iinuma, K.; Yamamoto, H.; Maeda, K.; Umezawa, H. *J. Antibiot.* **1973**, *26*, 412–415.
32. Takahashi, Y.; Umemura, E.; Kobayashi, Y.; Murakami, S.; Nawa, T.; Morinaka, A.; Miyake, T.; Shibashaki, M. *J. Antibiot.* **2018**, *71*, 345–347.
33. Takahashi, Y.; Umemura, E. *WO 2017/078097* **2017**, *A1.*
34. Zhanel, G. G.; Lawson, C. D.; Zelenitsky, S.; Findlay, B., et al. *Expert Rev. Anti. Infect. Ther.* **2012**, *10*, 459–473.
35. Endimiani, A.; Hujer, K. M.; Hujer, A. M.; Armstrong, E. S.; Choudhary, Y.; Aggen, J. B.; Bonomo, R. O. *Antimicrob. Agents, Chemother.* **2009**, *53*, 4504–4507.
36. Takahashi, Y.; Umemura, E.; Ida, T.; Igarashi, M. *WO 2017/018528 A1* **2017**.
37. Sefidan, F. Y.; Mohammadzadeh-Asl, Y.; Ghotaslou, R. *Microb. Drug. Resist.* **2019**, *25*, 1261–1265.
38. Umemura, E., et al. *ASM Microbe.* Poster Sunday-335: New Orleans, LA, USA, 2017.
39. Zimmermann, L.; Kempf, J.; Briée, F.; Swain, J.; Mingeot-Leclercq, M. -P.; Décout, J. -L. *Eur. J. Med. Chem.* **2018**, *157*, 1512–1525.
40. Arcamone, F.; Bizioli, F. *Gazz. Chim. Ital.* **1957**, *87*, 896–902.
41. Groenevelt, J. M.; Meints, L. M.; Stothard, A. I.; Poston, A. W., et al. *J. Org. Chem.* **2018**, *83*, 8662–8667.
42. Matsushita, T.; Sati, G. C.; Kondasinghe, N.; Pirrone, M. G., et al. *J. Am. Chem. Soc.* **2019**, *141*, 5051–5061.
43. Clouet, A.; Gravier-Pelletier, C.; Al-Dabbag, B., et al. *Tetrahedron: Asymm.* **2008**, *19*, 397–400.
44. Price, N. P. J.; Tsvetanova, B. *J. Antibiot.* **2007**, *60*, 485–491.
45. Xu, L.; Appell, M.; Kennedy, S., et al. *Biochemistry* **2004**, *43*, 13248–13255.

46. Reddy, V. M.; Einck, L.; Nacy, C. A. *Antimicrob. Agents Chemother.* **2008**, *52*, 719–721.
47. Hirano, S.; Ichikawa, S.; Matsuda, A. *J. Org. Chem.* **2008**, *73*, 569–577.
48. Hirano, S.; Ichikawa, S.; Matsuda, A. *Bioorg. Med. Chem.* **2008**, *16*, 428–436.
49. Igarashi, M.; Takahashi, Y.; Shitara, T., et al. *J. Antibiot.* **2005**, *58*, 327–337.
50. Miyake, T.; Takahash, Y.; Igarashi, M. *New Orleans, LA, US.* **2008**.
51. Inukai, M.; Isono, F.; Takatsuki, A. *Antimicrob. Agents Chemother.* **1993**, *37*, 980–983.
52. Sun, D.; Jones, V.; Carson, E. I., et al. *Bioorg. Med. Chem. Lett.* **2007**, *17*, 6899–6904.
53. McDonald, L. A.; Barbieri, L. R.; Carter, G. T., et al. *J. Am. Chem. Soc.* **2002**, *124*, 10260–10261.
54. Yamashita, A.; Norton, E.; Petersen, P. J., et al. *Bioorg. Med. Chem. Lett.* **2003**, *13*, 3345–3350.
55. Tanino, T.; Ichikawa, S.; Shiro, M.; Matsuda, A. *J. Am. Chem. Soc.* **2010**, *75*, 1366–1377.
56. Xie, Y.; Xu, H.; Si, S., et al. *J. Antibiot.* **2008**, *61*, 237–240.
57. Sandrini, M. P. B.; Shannon, O.; Clausen, A. R.; Bjorck, L.; Piskur, J. *Antimicrob. Agents Chemother.* **2007**, *51*, 2726–2732.
58. Sandrini, M. P. B.; Clausen, A. R.; On, S. L. W.; Aarestrup, F. M.; Munch-Petersen, B.; Piskur, J. *J. Antimicrob. Chemother.* **2007**, *60*, 510–520.
59. Jordheim, L. P.; Larbi, S. B.; Fendrich, O.; Ducrot, C.; Bergeron, E.; Dumontet, C. *Int. J. Antimicrob. Agents.* **2012**, *39*, 444–447.
60. Elwell, L. P.; Ferone, R.; Freeman, G. A.; Fyfe, J. A.; Hill, J. A.; Ray, P. H., et al. *Antimicrob. Agents Chemother.* **1987**, *31*, 274–280.
61. Keith, B. R.; White, G.; Wilson, H. R. *Antimicrob. Agents Chemother.* **1989**, *33*, 479–483.
62. Doleans-Jordheim, A.; Bergeron, E.; Bereyziat, F.; Ben-Larbi, S.; Dumitrescu, O.; Mazoyer, M. A., et al. *Eur. J. Clin. Microbiol. Infect. Dis.* **2011**, *30*, 1249–1256.
63. Peyclit, L.; Baron, S. A.; Yousfi, H.; Rolain, J. M. *Int. J. Antimicrob., Agents* **2018**, *52*, 11–13.
64. Wambaugh, M. A.; Shakya, V. P. S.; Lewis, A. J.; Mulvey, M. A.; Brown, J. C. S. *PLoS Biol.* **2017**, *15*, e2001644.
65. Ng, S. M. S.; Sioson, J. S. P.; Yap, J. M.; Ng, F. M.; Ching, H. S.V.; Teo, J. W. P., et al. *Eur. J. Clin. Microbiol. Infect. Dis.* **2018**, *37*, 141–148.
66. Hu, Y.; Liu, Y.; Coates, A. *Antimicrob. Agents Chemother.* **2018**, *63*, e01630–e1718.
67. Falagas, M. E.; Voulgaris, G. L.; Tryfinopoulou, K.; Giakkoupi, P.; Kyriakidou, M.; Vatopoulos, A., et al. *Int. J. Antimicrob. Agents.* **2019**, *53*, 855–858.
68. Haskell, T. H.; Ryder, A.; Frohardt, R. P.; Fusari, S. A.; Jakubowski, Z. L.; Bartz, Q. R. *J. Am. Chem. Soc.* **1958**, *80*, 743–747.
69. Bu, Y. -Y.; Yamazaki, H.; Ukai, K.; Namikoshi, M. *Mar. Drugs* **2014**, *12*, 6102–6112.
70. Aksoy, S. C.; Uzel, A.; Bedir, E. J. *J. Antibiot.* **2016**, *69*, 51–56.
71. Fu, J.; Laval, S.; Yu, B. *J. Org. Chem.* **2018**, *83*, 7076–7084.
72. Carnrot, C.; Wehelie, R.; Eriksson, S.; Bolske, G.; Wang, L. Y. *Mol. Microbiol.* **2003**, *50*, 771–780.
73. Wehelie, R.; Eriksson, S.; Wang, L. Y. *Nucleos. Nucleot. Nucl.* **2004**, *23*, 1499–1502.
74. Zander, J.; Besier, S.; Ackermann, H.; Wichelhaus, T. A. *Antimicrob. Agents Chemother.* **2010**, *54*, 1226–1231.
75. Ueda, A.; Attila, C.; Whiteley, M.; Wood, T. K. *Microb. Biotechnol.* **2009**, *2*, 62–74.
76. Vanlancker, E.; Vanhoecke, B.; Smet, R.; Props, R.; Van De Wiele, T. *J. Med. Microbiol.* **2016**, *65*, 775–783.
77. Singh, V.; Brecik, M.; Mukherjee, R.; Evans, J. C.; Svetlikova, Z.; Blasko, J., et al. *Chem. Biol.* **2015**, *22*, 63–75.
78. Sun, R.; Wang, L. Y. *BMC Microbiol.* **2015**, *13*, 184–194.
79. Cooper, S. L.; Brown, P. A. *Clin. N. Am.* **2015**, *62*, 61–73.
80. Hanauer, S. B.; Sandborn, W. J.; Lichtenstein, G. R. *Gastroenterology* **2019**, *156*, 36–42.
81. Liu, F.; Ma, R.; Riordan, S. M.; Grimm, M. C.; Liu, L.; Wang, Y., et al. *Front. Microbiol.* **2017**, *8*, 527.

82. Alvarez, Z.; Lee, K.; Abel-Santos, E. *Antimicrob. Agents Chemother.* **2010**, *54*, 5329–5336.
83. Sun, R.; Wang, L.Y. *BMC Microbiol.* **2013**, *13*, 184.
84. Henninger, T. C. *Expert. Opin. Ther. Pat.* **2003**, *13*, 787–805.
85. McGuire, J. M.; Bunch, R. L.; Andersen, R. C.; Boaz, H. E.; Flynn, E. H.; Powell, H. M.; Smith, J.W. *Antibiot. Chemother.* **1952**, *2*, 281–283.
86. Blondeau, J. M. *Expert. Opin. Pharmacother.* **2002**, *3*, 1131–1151.
87. Zhanel, G. G.; Walters, M.; Noreddin, A.; Vercaigne, I. M., et al. *Drugs* **2002**, *62*, 1771–1804.
88. Ma, Z.; Nemoto, P.A. *Curr. Med. Chem.* **2002**, *1*, 15–34.
89. Agouridas, C.; Denis, A.; Auger, J. M.; Benedetti, Y., et al. *J. Med. Chem.* **1998**, *41*, 4080–4100.
90. Denis, A.; Agouridas, C.; Auger, J. M.; Benedetti, Y. *et. al., Bioorg. Med. Chem. Lett.* **1999**, *9*, 3075–3080.
91. Bright, G. M., et al. *J. Antibiot.* **1988**, *41*, 1029–1047.
92. Morimoto, S.; Takahashi, Y.; Watanabe, Y.; Omura, S. *J. Antibiot.* **1984**, *37*, 187–189.
93. Fernandes, P.B. WO Patent 2010/048599, 2010.
94. Putnam, S. D.; Castanheira, M.; Moet, G. J.; Farrell, D. J.; Jones, R. N. *Diagn. Microbiol. Infect. Dis.* **2010**, *66*, 393–401.
95. Fernandes, P.; Martens, E.; Bertrand, D.; Pereira, D. *Bioorg. Med. Chem.* **2016**, *24*, 6420–6428.
96. Revill, P.; Bolós, J.; Seradell, N. *Ketolide Antibio. Drugs Fut.* **2006**, *31*, 479–483.
97. Bermudez, L. E.; Motamedi, N.; Chee, C., et al. *Antimicrob. Agents Chemother.* **2007**, *51*, 1666–1670.
98. Xu, G.; Tang, D.; Gai, Y.; Wang, G.; Kim, H.; Chen, Z.; Phan, L.T.; Or, Y. S.; Wang, Z. *Org. Process. Res. Dev.* **2010**, *14*, 504–510.
99. Gerber, M.; Ackermann, G. *Expert Opin. Investig. Drugs* **2008**, *17*, 547–553.
100. Sonenshein, A. L.; Alexander, H. B. *J. Mol. Biol.* **1979**, *127*, 55–72.
101. Glaus, F.; Altmann, K. H. *Angew. Chem. Int. Ed.* **2015**, *54*, 1937–1940.
102. Carreira, E. M.; Joliton, A. *Synfacts* **2015**, *11*, 1016.
103. Kaufmann, E.; Hattori, H.; Miyatake-ondozabal, H.; Gadermann, K. *Org. Lett.* **2015**, *17*, 3514–3517.
104. Okujo, N.; Iinuma, H.; George, A., et al. *J. Antibiot.* **2007**, *60*, 216–219.
105. Hansen, L. H.; Mauvais, P.; Douthwaite, S. *Mol. Microbiol.* **1999**, *31*, 623–631.
106. Almutairi, M. M.; Svetlov, M. S.; Hansen, D. A.; Khabibullina, N. F., et al. *Nucleic Acid Res.* **2017**, *45*, 9573–9582.
107. Kahne, D.; Leimkuhler, C.; Lu, W.; Walsh, C. *Chem. Rev.* **2005**, *105*, 425–448.
108. McCormick, M. H.; McGuire, J. M.; Pittenger, G. E.; Pittenger, R. C.; Stark, W. M. *Antibiot. Ann.* **1955**, *3*, 606–611.
109. Allen, N. E.; Nicas, T. I. *Microbiol. Rev.* **2003**, *26*, 511–532.
110. Griffith, B. R.; Krepel, C.; Fu, X.; Blanchard, S.; Ahmed, A.; Edmiston, C. E.; Thorson, J. S. *J. Am. Chem. Soc.* **2007**, *129*, 8150–8155.
111. Bambeke, F.V. *Curr. Opin. Investig. Drugs* **2006**, *7*, 740–749.
112. Thayer, D. A.; Wong, C. H. *Chem. Asian J.* **2006**, *1*, 445–452.
113. Kim, S. J.; Cegelski, L.; Stueber, D.; Singh, M.; Daetrich, E.; Tanaka, K. S. E.; Parr, T. R., Jr.; Farr, A. R.; Schaefer, J. *J. Mol. Biol.* **2008**, *377*, 281–293.
114. Leadbetter, M. R.; Adams, S. M.; Bazzini, B.; Paul, R.; Fatheree, D. E.; Karr, K. M. *J. Antibiot.* **2004**, *57*, 326–336.
115. Ashford, P. A.; Bew, S. P. *Chem. Soc. Rev.* **2012**, *41*, 957–978.
116. Treviño, J.; Bayón, C.; Ardá, A.; Marinelli, F.; Gandolfi, R.; Molinari, F.; Jimenez-Barbero, J.; Hernáiz, M. J. *Chemistry* **2014**, *20*, 7363–7372.
117. Fulco, P.; Wenzel, R. P. *Expert Rev. Anti Infect. Ther.* **2006**, *4*, 939–945.
118. He, H. *Appl. Microbiol. Biotechnol.* **2005**, *67*, 444–452.

119. Okano, A.; Isley, N. A.; Boger, D. L. *Chem. Rev.* **2017**, *117*, 11952–11993.
120. Crane, C. M.; Pierce, J. G.; Leung, S. S. F.; Tirado-Rives, J.; Jorgensen, W. L.; Boger, D. L. *J. Med. Chem.* **2010**, *53*, 7229–7235.
121. Silverman, S. M.; Moses, J. E.; Sharpless, K. B. *Chem. -Eur. J.* **2017**, *23*, 79–83.
122. Gu, W.; Chen, B.; Ge, M. *Bioorg. Med. Chem. Lett.* **2014**, *24*, 2305–2308.
123. Yarlagadda, V.; Konai, M. M.; Paramanandham, K.; Nimita, V. C.; Shome, B. R.; Haldar, J. *Int. J. Antimicrob. Agents* **2015**, *46*, 446–450.
124. Yarlagadda, V.; Konai, M. M.; Manjunath, G. B.; Ghosh, C.; Haldar, J.; Tackling, J. *J. Antibiot.* **2015**, *68*, 302–312.
125. Yarlagadda, V.; Sarkar, P.; Samaddar, S.; Haldar, J. *Angew. Chem. Int. Ed.* **2016**, *55*, 7836–7840.
126. Yarlagadda, V.; Akkapeddi, P.; Manjunath, G. B.; Haldar, J. *J. Med. Chem.* **2014**, *57*, 4558–4568.
127. Okano, A.; Nakayama, A.; Schammel, A. W.; Boger, D. L. *J. Am. Chem. Soc.* **2014**, *136*, 13522–13525.
128. Okano, A.; Nakayama, A.; Wu, K. J.; Lindsey, E. A.; Schammel, A. W.; Feng, Y. Q.; Collins, K. C.; Boger, D. L. *J. Am. Chem. Soc.* **2015**, *137*, 3693–3704.
129. Okano, A.; Isley, N. A.; Boger, D. L. *Proc. Natl. Acad. Sci. U. S. A.* **2017**, *114*, E5052–E5061.
130. Pinchman, J. R.; Boger, D. L. *J. Med. Chem.* **2013**, *56*, 4116–4124.
131. Guan, D.; Chen, F.; Xiong, L.; Tang, F.; Faridoon Qiu, Y.; Zhang, N.; Gong, L.; Li, J.; Lan, L.; Huang, W. *J. Med. Chem.* **2018**, *61*, 286–304.
132. Zhang, S. J.; Yang, Q.; Xu, L.; Chang, J.; Sun, X. *Bioorg. Med. Chem. Lett.* **2012**, *22*, 4942–4945.
133. Chang, J.; Zhang, S. J.; Jiang, Y. W.; Xu, L.; Yu, J. M.; Zhou, W. J.; Sun, X. *Chem. Med. Chem.* **2013**, *8*, 976–984.
134. Tyrrell, K. L.; Citron, D. M.; Warren, Y. A.; Goldstein, E. J. *Antimicrob. Chemother.* **2012**, *56*, 2194–2197.
135. Blais, J.; Lewis, S. R.; Krause, K. M.; Benton, B. M. *Antimicrob.* **2012**, *56*, 1584–1587.

CHAPTER FOURTEEN

Carbohydrate-based anti-bacterial and anti-cancer vaccines

Rituparna Das, Balaram Mukhopadhyay
Sweet Lab, Department of Chemical Sciences, Indian Institute of Science Education and Research (IISER), Kolkata, India

1 Introduction

The introduction of vaccines in human life was a significant advancement in the history of civilisation. In the domain of medicinal and therapeutic chemistry, if anything may be considered more important than vaccines, it is only clean and hygienic water. Edward Jenner's breakthrough discovery in late 1796 about cowpox immunization[1] triggered a ripple of waves which alleviated the growing interest in vaccinology, the science behind designing vaccines to combat deadly diseases. Vaccines have since then been proved as highly efficient candidates providing protection again the foreign disease causing pathogens. The staggering vitality in the concept of vaccination can be perceived by the exponential growth in the number of vaccines approved for regular use across the globe over the years. The statement released by the American Academy of Pediatrics (AAP) stating that "most childhood vaccines are 90–99% effective in preventing diseases"[2] reveals the versatility of the concept of vaccines in the fight against diseases. According to reports by World Health Organisation (WHO), as many as 116.2 million children were vaccinated in 2017, the highest ever reported in a calendar year.[3] Fatal diseases like small pox has been completely eradicated owing to the continuous and routine administration of vaccine to the masses; the last case of small pox being reported at Somalia in Ali MaowMaal back in 1977.[4] Vaccination protocol has significantly reduced the number of polio victims which is almost on the verge of its total abolishment all over the world. Awareness drives are being conducted by WHO and UNICEF across the world to make the general people aware of the associated boons of vaccination; wherein the weekstarting from 24th April in 2019 was declared as the 'World Immunization Week'.

Vaccines, on a broader aspect have been proven as the most cost friendly and effective medium which can stimulate the formation of antibodies

Carbohydrates in Drug Discovery and Development. http://dx.doi.org/10.1016/B978-0-12-816675-8.00014-2

thereby activating the immune cells of the body to fight against deadly diseases and infections. Carbohydrate derived vaccines based on the antigenic character of the capsular bacterial cell wall polysaccharides have dominated the recent vaccine related developments. This chapter systematically illustrates the different aspects of carbohydrate vaccines implemented in the modern day pharmaceuticals and therapeutics.

2 Carbohydrates as biologically relevant molecules and their impact on carbohydrate as antigens

Carbohydrates are the most abundant bio-molecules on the earth. The exterior part of the surface membranes of all the eukaryotic living cells are decorated with extensive and varied carbohydrate moieties in a diverse and heterogeneous manner and conformity. Owing to its intrinsic location, structural complexity and plethora of characteristics associated, studies have revealed that the exterior cellular glycoconjugates are responsible for various vital biological phenomena like cell adhesion, cell-cell recognition, signal transmission, transduction etc.[5] Such glycoconjugates on the cell surface membrane acts as the actual docking site for the adjacent cells, lectins, hormones, enzymes, etc. Similarly, the progression of diseases mediated by the foreign pathogens is mainly through the unique projection of the carbohydrate epitopes on the cell wall. This phenomenon of carbohydrate recognition forms the stepping stone for the series of biological events that follow. High selectivity and affinity towards the binding motifs is the governing factor which makes the carbohydrate recognition vital for the biological activities. However, if a single carbohydrate-protein binding epitope is considered, it has been found to be extremely weak with its dissociation constant (K_d) value being mostly in the micro molar range. The seminal works by Lee[6] however explained that the living system has a complex set of functional principles wherein the disadvantage rendered due to the weak affinities of a monovalent binding motif is compensated and exemplified by a cluster of appropriately presented carbohydrate moieties binding with a large number of Carbohydrate Recognizing Domains (CRD) in the lectins. This in turn increases the avidity of the interactions and thereby being responsible for the vital biological activities of the living organisms. This phenomenon is termed as 'Glycoside Cluster Effect' of the 'Glycocalix' type cellular structures establishing 'Multivalent' interactions with the external bodies.

Bacterial polysaccharides have a thick layer of lipopolysaccharides and peptidoglycans on their capsular cell wall structures. These outer layer of polysaccharides bounded by membrane proteins and lipids are made up of repetitive sugar units in a significant amount. These exterior dense arrays of unique polysaccharide motifs completely encapsulate the organisms and malignant cells and thereby acting as the blue-print of the particular virulent species. It was back in 1923 that Heidelberg and Avery first postulated that the specific substance accounting for the specific binding of the pathogens had the intrinsic characteristics of carbohydrates.[7] Thus, the unique portray of the exposed carbohydrates enable the specific binding of the pathogens with the host cells and they pose as the most vital determinant of their virulence as well as being the probable targets for protective antibody.[8] 1930 witnessed the researchers Francis and Tillet reporting the formation of specific antibodies against capsular polysaccharides of pneumococii.[9] Previously live-attenuated or heat-killed inactive pathogens (*viz.* in the case of small pox by Edward Jenner) were introduced in the body systems into elicit antibody generation which would in turn would provide lifelong immunogenicity in the host cells. With the establishment of cellular polysaccharides being the main disease causing determinant, a number of subsequent studies revealed the phenomenon of antibody mediated immunity through vaccination of serotype specific polysaccharides.[10] The first polysaccharide vaccine composed of un-conjugated polysaccharide from pneumococii species namely Pneumo Vax was marketed in 1983 by Merck and Co.[11] However, this protocol suffered a lot of disadvantages owing to its difficulties in attenuation process and the persistent infection caused by it.

Clinical evaluation of purified polysaccharide based vaccines also proved to be inefficient as the carbohydrates intrinsically exhibit poor immunogenicity. Traditionally most carbohydrate-based antigens remain incapable of eliciting T-cell dependent immune responses[12] and thereby failing to stimulate memory responses.[13] However, they are capable of activating B-cells[14] leading to the generation of T-cell independent immunoglobulin M (IgM) antibody with low stability and affinity.[15] But in order to establish life-long antibody mediated protection, class switched antibodies need to be generated inside the body. On the other hand, peptidic antigen based $CD4^+$ epitopes are capable of generating T-cell dependent $CD4^+$T-cells which in accordance with the antigen specific B-cells go on to develop long lived class switched antibodies with high affinity.[16] Therefore, to establish effective antibody responses for the glycan layers of the pathogens, it becomes essential to introduce the participation of $CD4^+$surface protein expressed by helper T-cells (T_h).[17] It has been postulated that this can be achieved by the incorporation of external peptidic antigens or $CD4^+$ epitopes to the glycans. Avery and Goebel in 1939[18] first established the preliminary concept of glycoconjugate vaccines wherein they showed that coupling of the capsular glycans with an appropriately selected protein scaffold significantly enhancing the immunogenicity of the otherwise weak carbohydrate antigens. This initiated the path towards the development of glycoconjugate vaccines. Extensive reviews have shown the establishment and development of carbohydrate based vaccines over the years.[19]

3 Glycoconjugate vaccines

Covalent attachment of the weakly antigenic polysaccharides with suitably immunogenic carrier proteins like keyhole limpet hemocyanin (KLH),20] ovalbumin (OVA),21] bovine serum albumin (BSA)[22] or tetanus toxoid (TT)[23] renders improved antigenicity to the carbohydrate scaffolds. The $CD4^+$ epitope providing carrier proteins in association with the carbohydrate counterparts induce T-cell dependent response leading to immunoglobulin G (IgG) antibody generation.[24] The involvement of helper T-cells (T_h) invokes immunological memory and isotope switching from IgM to IgG antibody with the corresponding increase in the avidity of the generated antibody. Thus, the conjugation of peptide epitopes on polysaccharide antigens provides the efficient and lifelong protection against bacterial polysaccharides and malignant tumours in humans even in high risk age groups like in children below 2 years and also in elderly population.[25]

The steady increase in the antibiotic resistance in recent times has spurred the interest in carbohydrate vaccines even more. The corresponding studies about glycomics and carbohydrate chemistry enabled extensive research on the topic enabling the scientific community to deal with the complexities and challenges related to carbohydrate vaccines. Active research is being done by the researchers all around the globe to identify the antigenic capsular polysaccharides present in the disease causing pathogens, thereby increasing the spectrum of diseases which can be actively targeted by glycoconjugate vaccines. At present,these vaccines are claimed to provide lifelong protection and immunity towards diverse bacterial, viral and parasitic diseases as well as contribute towards solving the puzzle of cancer.

3.1 Challenges associated with carbohydrate-based vaccines

As mentioned earlier, the efficacy of carbohydrate protein binding is quite low with the dissociation values primarily in micro molar range. This poses as a serious difficulty in the design of carbohydrate based vaccines. Carbohydrates primarily lack the vital hydrophobic pockets required for efficient binding with the proteins. Moreover, there exists an unfavourable entropy penalty owing to the loss of the conformational flexibility upon complex formation. In addition to this, the solvent rearrangement also necessitates the removal of the prior H-bonding network with the bulk solvent leading to corresponding high enthalpic penalties. Thus both the conformational anomalies and de-solvation negates the favourable enthalpy conditions required for free carbohydrate-protein binding.

The introduction of multivalency to the glycoconjugates as the probable vaccine source sorted most of these challenges associated with carbohydrate based vaccines to a large extent. Multivalent interactions increased the avidity of the interactions leading to the development of more efficient glycoconjugate vaccines.

3.2 Design of glycoconjugate vaccines

It is clear that the strategy (Fig. 14.1) behind the development of glycoconjugate vaccines requires consideration of a number of factors.[26] The first

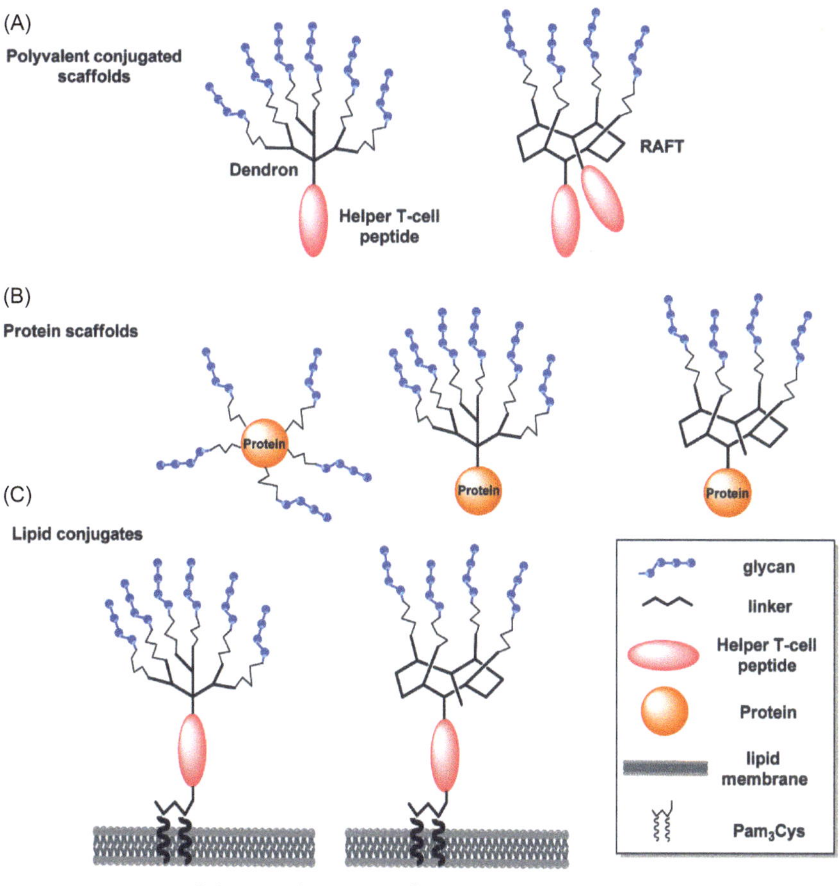

Figure 14.1 *General diagram for vaccine designing.*[28]

step is the identification of the proper antigenic carbohydrate epitope. The diversified and elaborate heterogeneous glycan arrays on the capsular cell-wall portray micro heterogeneity to the glycoproteins and glycolipids. Thus, identification and isolation of the proper antigenic source among this heterogeneous display of glycan arrays becomes the most essential step towards the design of glyconjugate vaccines. This is followed by the incorporation of a suitably immunogenic protein carrier, which is capable of providing the multivalent platform for conjugation and would in turn help in inducing strong antibody responses. It is generally done with the aid of a conjugation with a suitable immunogenic linker. Lastly, addition of an appropriate adjuvant (alum is the most preferred choice) is necessary to increase the immunogenicity of the antigenic carbohydrates.[27]

Once the appropriate antigenic carbohydrate structure is identified from the samples isolated from natural sources, chemical synthesis of those structures becomes the most prominent challenge towards developing glyconjugate vaccine candidates. Arguably the essential structures may be isolated from natural sources. However, limited natural resources and the inherent heterogeneous nature of the external glycans limit the isolation option drastically. It is extremely difficult and complicated to obtain sufficiently pure and homogenous sample of the naturally derived carbohydrate antigen which may be used as a probable vaccine candidate. Scope of impurity and batch to batch variation are also detrimental for the isolation route. Therefore, chemical synthesis of the desired oligosaccharides remains as the only option for successful glycoconjugate vaccine designing.

3.2.1 Synthesis of glycoconjugates as vaccine candidates

Although we have now learned to synthesize oligosaccharides, it should be emphasized that each oligosaccharide synthesis remains an independent problem, whose resolution requires considerable systematic research and a good deal of know-how. There are no universal reaction condition for oligosaccharide synthesis"–Hans Paulsen.[29]

Chemical synthesis of the naturally occurring carbohydrate antigens or broadly termed as 'glycomimetics' requires a high degree of synthetic exercise.[30] The past few decades have seen an explosion in the development of new protocols and methodologies to simplify the process of chemical oligosaccharide synthesis. There have been a number of reports iterating the concise synthetic protocols for the complex bacterial oligosaccharides. However, due to the diversity associated with these complex structures and requirement of appropriate stereo- and regioselectivity, chemical synthesis of a particular glycosidic bond still remains an eluding area of research. In the context, there has been an urgent need for the development of a standard and simplified means pertaining to oligosaccharide synthesis.

Reactivity based one-pot synthetic protocols[31] go a long way to implement the varying degrees of the constituent building blocks and coherently reduce the number of steps employed in the long-drawn synthesis. Solid phase protocols[32] also enable the synthetic steps to be performed without rigorous purification in each step. A computer program named Optimer[33] has also been developed which can predict the probable monosaccharide units and their corresponding sequence of coupling. However, this method witnesses high degrees of error while dealing with more than six sugar units. Recently developed automated oligosaccharide synthesis in both solution and solid phase protocols has also been generated to ease down the synthetic steps.[34] However, several technical issues *viz.* the requirement of large molar excesses of the glycoside synthons have dampened the widespread popularity. Chemo-enzymatic strategies have also been implemented to minimize the protecting group manipulations for the synthesis of oligosaccharides.[35] However, it requires the correct identification of the pathways for the natural glycan synthesis and also the systematic recombination and purification of the required enzymes. Thus, undeniably, till date, the researchers have been unable to standardize any general procedure which may be considered as the base for all oligosaccharide syntheses. Several steps are to be performed and optimized for individual glycosidic bond formation, leading to the complexity of glycomimetic synthesis. However, with the modern development towards synthetic protocols along with newer analytical techniques that enables extensive mechanistic knowhow of assembly of glycosides, it is possible to successfully replicate and modify many of the natural oligosaccharide moieties.[36] This in turn has paved the way for the generation of non-natural antigenic oligosaccharide epitopes through derivatization. These improved synthetic strategies[37] provided a large assortment of purified carbohydrates explore the domain of glycoconjugate therapeutics.

Conjugation with a suitable carrier protein *via* an intermediate linker marks the semi-synthetic approach towards vaccine designing. The carrier protein provides enhanced immunogenicity to the glycoconjugate epitope wherein it can generate high affinity antibodies. Introduction of polyvalent display in the conjugated carrier protein enhances magnitude of affinity. In addition, use of judiciously designed synthetic non-native glycomimetics with multiple loci presenting scaffolds for the generation of multivalent glycoconjugate derivatives proved to be beneficial for efficient vaccine candidates with increased affinities, enhanced bioavailability and longer life span.

In both the semi-synthetic and the fully synthetic approach, the carrier protein or the multivalent backbone of the designed scaffold remains conjugated with the antigen by means of an immunologically inactive linker.[38] The selection of the apt linker is essential enabling it to be readily accessible and providing optimum loading of the antigen with the carrier proteins. The corresponding immunogenicity and conformation of the linker should also be considered along with the knowledge of its stability relative to the original antigenic glycan epitope. Thus, the design of the entire glyconjugate vaccine candidate remains the most important step in the process of the development of carbohydrate based vaccines which would be capable of inducing the long-lived immunity.

3.3 Anti bacterial glycoconjugate vaccine

Bacterial pathogens remain completely encapsulated by a layer of dense polysaccharide moieties. The antibodies that are elicited against these disease causing pathogens recognize these exterior polysaccharide layer as the characteristic for a particular strain. The recent prevalence of antibiotic resistance has increased the spurge in the development of glycoconjugate vaccines against various bacterial strains like *Streptococcus pneumoniae*, *Haemophilus influenzae*, *Neisseria meningitidis*, *Salmonella typhi*. etc.

The first example of glycoconjugate vaccine introduced in clinical practice was against *Haemophilus influenzae*. It is a type of bacterial pathogen that has been a leading cause of meningitis in children. According to WHO reports, it has been responsible for around 386,000 deaths per year, mostly among children below 5 years of age.[39] Early 1980's saw the introduction of Hib conjugate vaccine[40] containing capsular polysaccharide (CPS) isolated from the pathogens capable of providing immunity. Although subsequent reports suggested the reduction in the occurrence of the bacterial meningitis considerably,[41] the inherent disadvantages of pathogen derived polysaccharides proved to be an obstacle towards its world-wide acceptance. Thus different types of glycoconjugates got licensure over the following years.

Success in immunisation through the purified pathogen derived CPS of Hib led to further exploration of alternate synthetic approaches. In late 1980's research was initiated in Cuba for the commercialization of the first synthetic vaccine. The initial years pertained to the standardisation of the synthetic protocols. H-phosphonate chemistry was implemented for the synthesis of the repeating unit **1**of Hib CPS. Polycondensation reaction (Fig. 14.2) further led to the production of pure size exclusive oligomers having 6–9 repeating units of **1**. This poly(ribosyl-ribitol-phosphate)

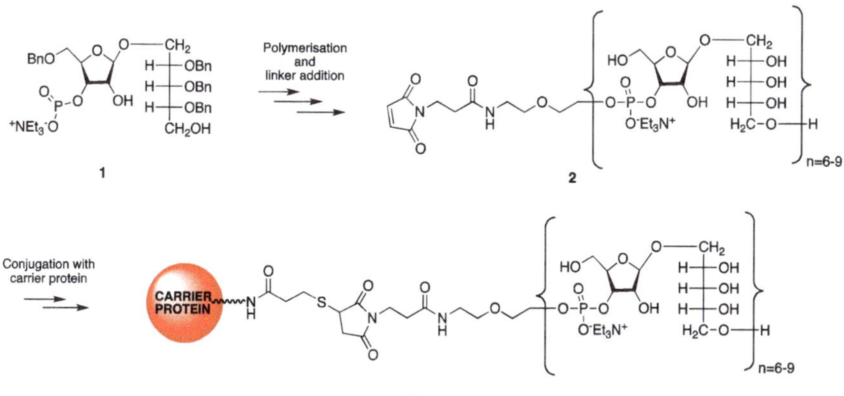

Figure 14.2 ***Synthetic protocol for the generation of the Hib capsular polysaccharide-protein conjugate for Haemophilus influenza b.***

conjugate was then activated with the help of a linker to obtain **2** and then further conjugated with a suitable carrier protein. Antigenic evaluation was done first with human serum albumin (HSA) and then with tetanus toxoid (TT). Thus, conjugation of thiolated TT protein complex with maleimide oligomer afforded the first Hib capsular polysaccharide-protein conjugate, **3**. Studies revealed that the antigenicity of this synthetic conjugate almost conformed to that of the native CPS.[42]

Subsequent clinical trials for antibody generation against this Hib conjugate was performed over mice, rats and rabbits. It was postulated through the trials that increased and more stable antibody responses were obtained after the administration of the third dose of the synthetic Hib conjugate. Human trials were then performed with people from varied age-groups, neonates, children, infants and adults[40] all of which revealed scarce adverse effects. More than a million doses were administered over almost seventeen clinical trials; of which the reports suggested 99.7% success rates, especially in children. Thus, finally in 2004, the first synthetic carbohydrate conjugate vaccine was commercialized and licensed.[43] The vaccine, Quimi Hib® (Heber Biotech) became part of Cuba's national vaccination program. Other commercialized vaccines for *Haemophilus influenzae* are Hiberix® (GlaxoSmithKline Biologicals), ActHib® (Sanofi Pasteur), HibTITER® (Pfizer).[44] Amongst the different Hib second generation vaccines, PedvaxHIB® composed of highly purified ribitolphosphate polysaccharide of Hib covalently conjugated to an outer membrane protein complex (OMPC) marketed by Merck and Co. in 1990 has proven to be the most effective (Fig. 14.3). All these vaccines in

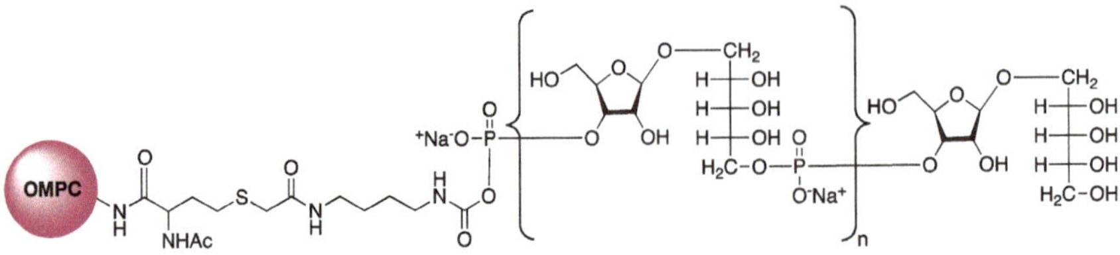

4

Figure 14.3 ***PedvaxHIB® (Merck & Co.).***

entirety have led to an enormous decrease in the occurrence of Hib disease worldwide mainly among children of age 0–4 years.[45]

Bacterial pathogens generally enjoy the advantage of considerable heterogeneity and complexity. A myriad of different capsular serotypes have been identified for the same bacterial strain in a number of instances. In such cases, 'multivalent conjugate vaccines' are deployed. These vaccines are composed of a number of clinically relevant serotypes of the antigen conjugated with an appropriate carrier protein.

Development of vaccines for *Streptococcus pneumoniae* is particularly challenging due to the presence of a large number of pathogenic serotypes. *S. pneumoniae* has been reported to cause various lower respiratory tract infections and inducive diseases like bacteraemia, meningitis, otitis media, etc.[46] Some of the relevant pathogenic serotypes of the bacteria include (Fig. 14.4):

1. Zwitterionic non-branched trisaccharide repeating unit, **4** having *N*-acetyl fucosamineand two galactouronic acid units depicting Serotype 1.
2. Serotype 2 containing hexasaccharide, **5**having three rhamnose, one glucose, one glucosamine and one glucuronic acid units.
3. Two pentasaccharide units of serotype 9A (**6**) and 9n (**7**) having difference in two constituent units wherein one galactose unit and one glucose unit of 9A are replaced by a glucose and a glucosamine unit respectively in 9n serotype.

Thus the polysaccharide vaccine commonly used to elicit antibodies against *Streptococcus pneumoniae* is PPV23 (PneumoVax 23® by Merck and Co.) accessing 23 of its different serotypes (1, 2, 3, 4, 5, 6b, 7F, 8, 9N, 9V, 10A, 11A, 12F, 14, 15B, 17F, 18C, 19F, 19A, 20, 22F, 23F and 33F). However, this vaccine could not induce high affinity antibodies in children below 2 years of age.[47] Thus for the sole purpose of improving the immunogenicity of children of this age group, a heptavalent carbohydrate conjugate vaccine was developed with seven serotype specific polysaccharide individually conjugated with CRM197, under the tradename Prevnar® (Wyeth/Pfizer Biochemicals). Other accepted vaccines include: Synflorix®, a decavalent

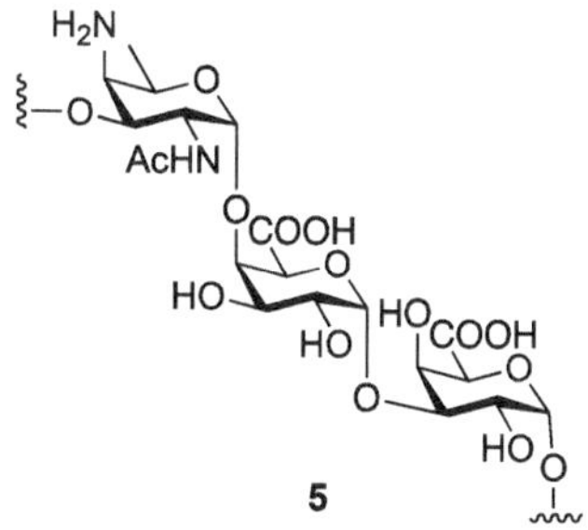

5
Serotype 1

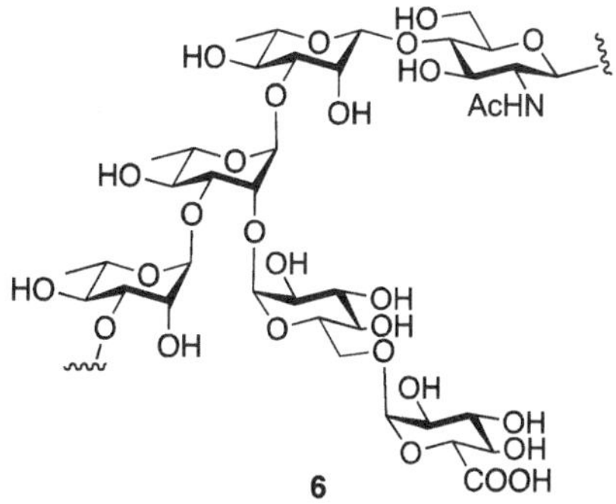

6
Serotype 2

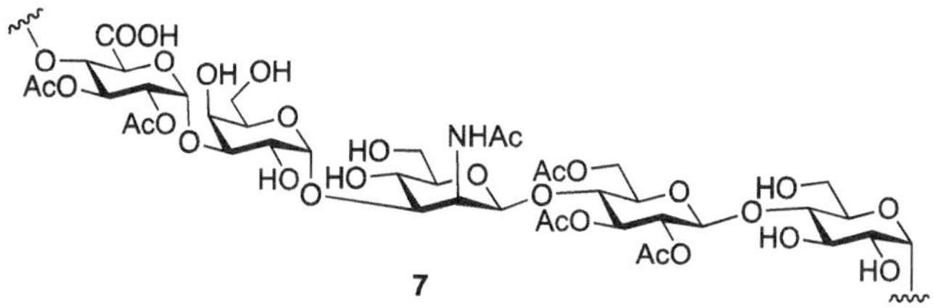

7
Serotype 9A

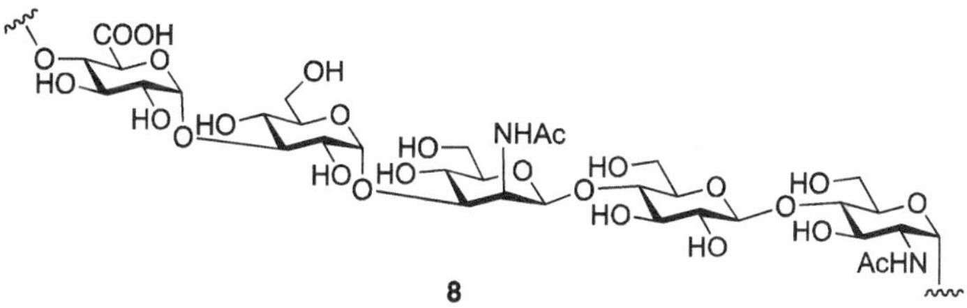

8
Serotype 9n

Figure 14.4 ***Antigenic polysaccharides of different Streptococcus pneumonia serotypes.***

conjugate (EU in 2009); Prevnar13®, a tri-decavalent vaccine, especially for children between 6 weeks and 17 years. However, research is still being carried out to develop effective conjugate vaccines for both adults as well as infants.[48]

Naturally occurring polysaccharides are usually neutral or anionic. Though they can be processedby the Antigen Presenting Cells (APC), they are incapable of binding with the Major Histo-compatibility Complex (MHC) due to the absence of inherent electrostatic interactions. Thus, T-cell independent immune responses are obtained. In this respect, serotype 1 of *S. Pneumonia* (4) presents zwitterionic polysaccharide comprising of both positive and negative charges in its trisaccharide repeating unit. This makes them sometimes capable of inducing T-cell dependent responses. Thus, the zwitterionic property of polysaccharides may sometimes replace the requirements of the carrier protein conjugation with the polysaccharides. Chemically synthesized zwitterionic polysaccharides have also been reported to induce improved immunogenicity.[49]

Another fatal disease causing pathogen that have triggered major concern is the gram negative bacteria namely *Neisseria meningitidis*. Among the different serotypes isolated for this particular pathogen, six serotypes have been identified as the main disease causing agents.[50] Of these, Group B meningocociiposes as a major health concern across the globe. However, the conjugate vaccines developed for this particular bacterial strain proved to exhibit poor immunogenicity. The polysaccharide layer encapsulating the Group B *N. meningitidis* strain comprises of an α-(2,8)-polysialic acid polymer, which has an inherent structural similarity with the human neural cell glycans found during the development stages of the central nervous system.[51] This uncharacteristic structural and antigenic similarity between the pathogenic polysaccharide and the host glycan provides immunogenic tolerance to the body; thereby leading to the vaccine formulations with poor immunogenicity.[52] Thus, this phenomenon of similarity between bacterial antigens and the host glycan leading to the generation of autoimmunity acts as one of the major concerns in vaccine development.

It may be possible to overcome this obstacle by the chemical modification of the antigenic polysaccharides and implementing them as the possible targets for administration to induce high affinity and long lived antibody response in human. For possible modifications the *N*-acetyl group of the polysialic acids in the native polysaccharide of Group B *N. meningitidis* capsule was *N*-propionylated and conjugated with TT or HAS (Fig. 14.5) leading to the formation of the synthetic glycoconjugate vaccine.[53] Further

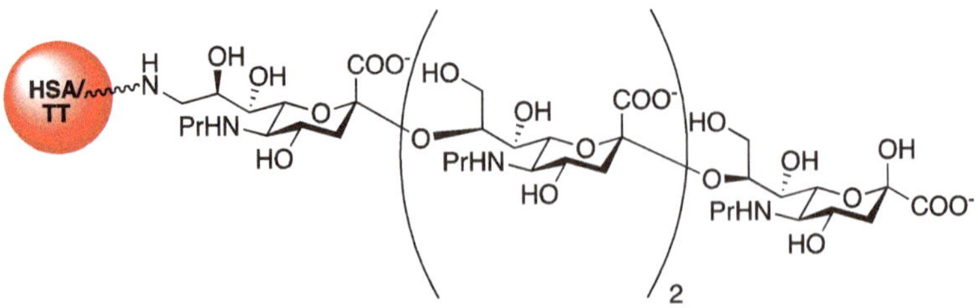

9

Figure 14.5 ***N-propionylated synthetic glyconjugate vaccine against Group B Neisseria meningitidis.***

immunological studies of the synthesized conjugate reported moderately higher antibody generation relative to the native antigen.[54]

Conjugated polysaccharide vaccines[55] for serotypes A, C, W and Y of *N. Meningitides* however have been developed and licensed as Menactra® (Sanofi Pasteur)[56] or Menveo® (Novartis Vaccines and Diagnostic).[57] Another un-conjugated polysaccharide vaccine namedMenomune® was also successful in inducing short term immunity against this virulent species.[58]

Thus, it is evident that carbohydrate conjugated with proteins pose as a promising candidate for the development of antibacterial vaccines. The development of the apt conjugate however demands the consideration of a number of important factors. The chain length of the antigenic carbohydrates is one of the most important factors in determining its immunogenicity. Reports depicting enhanced immunogenicity in rabbits and mice has been reported for di- or trisaccharide antigens of *S. pneumoniae*;[59] while in various other cases, higher lengths of oligosaccharides proved to provide better antibody eliciting property and protection in comparison to lower chain lengths of the same repeating units (in case of *S. dysenteriae*).[60] Depolymerisation of polysaccharide antigens has thus also been performed with hydrogen peroxide (*viz.* in case of *Salmonella typhi*)[61] which has also in turn facilitated efficient and successful conjugation with the carrier proteins.

Apart from the already licensed vaccines which are in practise from around the world,a great deal of work is in progress towards development of carbohydrate conjugate vaccines.[62] Maternal vaccination with CPS conjugate is recently being applied[63] to new-borns to induce immunity against Group B strain of *Streptococcus pneumoniae*. Clinical trials for multivalent vaccine conjugates with CRM are in their development phase. Trials are in the phaseII[64] for a trivalent vaccine conjugate whereas preclinical trials are underway for a pentavalent vaccine against the same strain of bacteria. A 13-valent pneumococcal vaccine is also in the pre-developmental stage for maternal immunization, being sponsored by Pfizer.[65] Recent years have also seen seminal works by Seeberger and group where they have revolutionized the synthesis and development of vaccines eliciting antibodies against various strains of S. pneumoniae. Semi-synthetic vaccine strategy with the synthesized oligosaccharide fragments of Serotype 2, 3 and 5 conjugated with CRM197[66] have been reported which have proved to induce high antibody titres in murine models.[67] Serotype 5 of *S. pneumoniae*has always proved to be a problematic component of the marketed vaccines like Prevnar13. Seeberger et.al has recently reported the synthetic procedure for the development of serotype5 strain[68] and made modifications which have thereby increased the antigen stability as well as its immunogenicity. A company named Vaxillon is presently working on the clinical trials of various such vaccine candidates. Alternative development of vaccine candidates like bioconjugate vaccines[69] are also being studied upon. Preliminary studies and Phase I clinical trials implementing the glyconjugates against *S. dysenteriae*,70] *S. flexneri*,71] *S. boydii* and *E. coli*[72] have shown promising immunogenic properties.

Vaccines against other species are also being investigated and developed during the past few years. Cryptococcal meningoencephalitis and invasive aspergillosis (IA) caused by *Cryptococcus neoformans* and *Aspergillus fumigates* respectively are the two of the major fungal infections affecting the human civilisation at present.[73] *Candida albicans* is another causative agent for widespread infections.[74] Glycoconjugate with the constituent cell-wall antigens or their synthetic repeating unit formsare being developed as a protective against these fungal strains. Glucuronoxylomnnan (GXM) of the cell wall of *Cryptococcus neoformans* are being isolated and conjugated with nontoxic mutant, CRM 197. Subsequent clinical trials are underway by various research groups.[75]

Some bacterial strains are important due to their potential use in bioterrorism weapons *viz. Burkholderia mallei, Burkholderiapseudomallei, Bacillus anthracis, Yersinia pestis* and *Francisellatularensis*.[76] *Francisellatularensis* has been the causative agent of tularemia, a category A bioterrorism agent.[77] Glycoengineered glyconjugates are being developed and implemented as vaccine candidates to test their immunological activities to combat against this deadly bacterium.[78]

There are also many research groups trying to chemically synthesize various native polysaccharide antigens[79] of the varied bacterial strains *viz. Vibrio Cholerae*,[80] *S. typhi, E. coli*,[81] *S. enterica*[82] etc. Thus, it is evident the oligosaccharide strains studied to possess the antigenic properties for the various invasive and non-invasive diseases is the primary determinant in the development of immunological vaccines. Modern methods and protocols including the synthesis of glycoarrays as well as the chemical reproduction of the rare, unnatural sugars are fast paving the way for newer and more advanced derivatives of the antigenic cell-wall repeating units. More clinical trials are being performed and pharmaceutical companies are sponsoring more of carbohydrate based vaccines owing to the exponential increase in antibody resistance in the present years. Thus, it can be safely concluded that carbohydrate conjugate based vaccines, coupled with appropriate protein carriers has the potential of solving much of the prevalent problems related to bacterial pathogenic diseases.

3.4 Anti-tumour and anti-cancer carbohydrate vaccines

The characteristic feature of cancer progression is altered glycosylation and overexpression of glycans *i.e.* tumour associated carbohydrate antigens (TACAs) generally on the surface of the tumours. Tn, sialyl Lewix X (sLe^{x}), sialyl Lewis A (sLe^{a}), Thomsen-Friedenreich (TF), sialyl 2,6-α-N-acetylgalactosamine (sTn), Globo H, GM2 etc. (Fig. 14.6) are some of the commonly found TACAs expressed on malignant cells.[83] Variation in the expression of glycosyltransferase enzyme leads to more available sites for sialic acid residues; wherein sialyl glycosyltransferases leads to increase in sialylation. Globo glycans like Globo H, Gb3, Gb5 etc are overexpressed on the surface of tumours in breast, ovaries, prostate and colon;[84] gangliosides like GM2, GM3, GD2, GD3 are overexpressed in lungs, renal and prostate cancer cells; mucin related glycans like Tn,TF etc.[82] are expressed in epithelial cancersand lastly the class of Lewis antigens are distributed in a wide range of tumours.

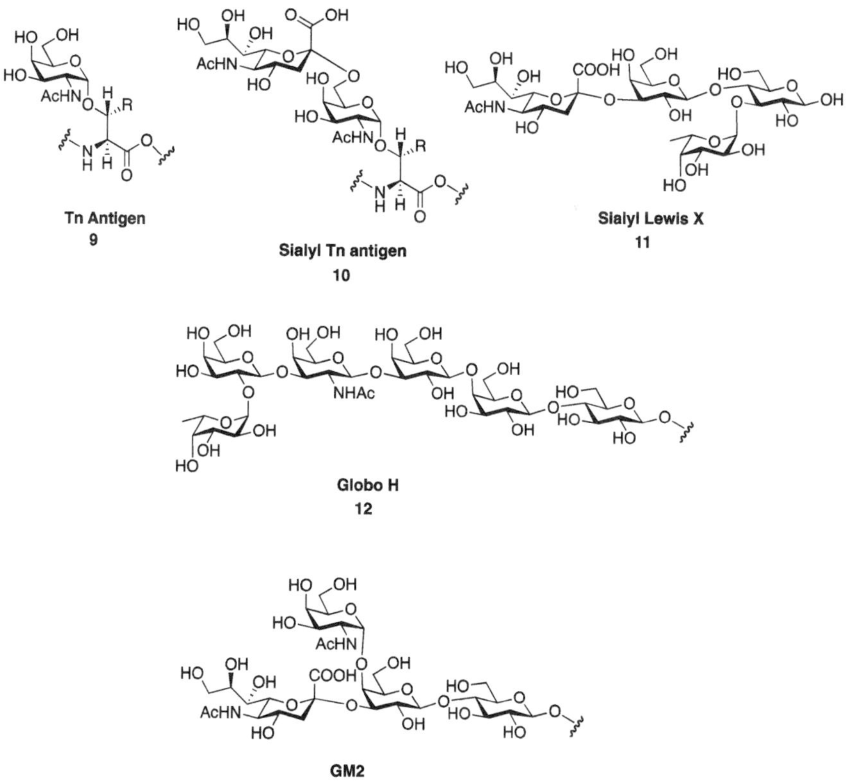

Tn Antigen
9

Sialyl Tn antigen
10

Sialyl Lewis X
11

Globo H
12

GM2
13

Figure 14.6 ***Some tumour associated carbohydrate antigens (TACAs).***

TACAs play a major role in initiating invasion and metastasis of the cells. Thus, TACAs pose as good anti-cancer epitopes with the expectation that proper presentation of these targets may induce required immune response thereby eradicating the tumour cells.[85] However, TACAs are also present on normal healthy cells which impose them the factor of being auto immune in nature.[86] But, in spite of this, TACAs may be considered as a good target for carbohydrate based vaccines as they are usually poorly expressed in normal cells.[84] However, TACAs are primarily sole B-cell activating which makes them incapable of producing long-lasting high affinity T-cell dependent antibodies. In addition to this, natural isolation of TACAs for their implementation in natural vaccines is also highly cumbersome owing to its micro heterogeneity and self-antigenic properties. Thus, conjugating synthetic TACAs with carrier proteins and their use as probable anti-cancer vaccines is being widely studied and implemented. Various research groups across the globe are working on the synthesis of these cancer antigens.

Danishefsky and co-workers have successfully synthesized many of the members of the TACAs with their subsequent conjugation with amino acid derivatives or linkers. 'Cassette hypothesis' has been extensively implemented for the synthesis of various complex oligosaccharides; the first use being with the synthesis of mucin related F1α.[87] The blood group determinant Lewis^y pentasaccharide was synthesized successfully and conjugated with KLH carrier protein[88] which was followed by highly effective preclinical immunogenic studies proving to be highly selective towards Le^y positive cells.[89] Phase I clinical trials are also underway against ovarian cancer cells. However, these TACA conjugates were primarily capable of generating IgM antibodies which belongs to the low affinity type antibody responses. Reports have also been published by both Danishefsky and Magnusson group where sLe^x and Le^x has been effectively synthesized.[90] Subsequent studies depicting their potency as anti-cancer agents are in progress. Danishefsky and co-workers have also used much of their expertise over the years for the successful convergent synthesis of Globo H.[91] They have mainly implemented glycan assembly protocols employing[3 + 3] conjugation methodology. Globo H has been long reported to have large implications in a wide range of cancers, thereby adding to its large-scale importance as a probable cancer target epitope.[92] Therefore, considering its significant contribution towards cancer therapeutics, various other groups have also reported the efficient synthetic protocols pertaining to Globo H synthesis.[93] Further conjugation of synthetic Globo H antigens with KLH or DT-CRM197 protein (Fig. 14.7) was performed and these glycoconjugates have been studied as potential

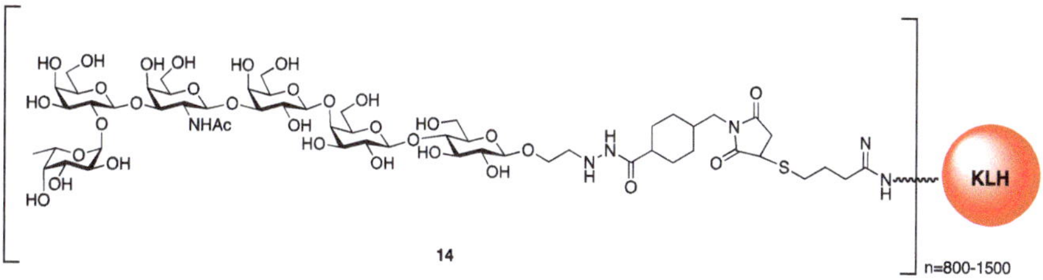

14

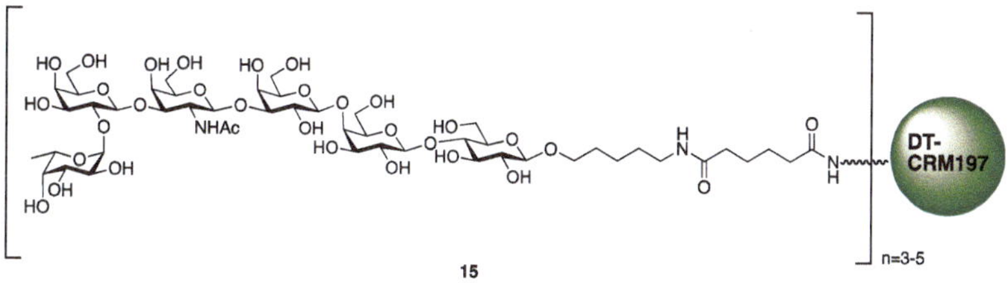

15

Figure 14.7 ***Globo-H vaccines presently in pre-clinical and clinical trials.***

vaccine candidates for both breast and prostate cancer. While breast cancer diagnosis with this conjugate has reached Phase III trials, prostate cancer studies are still in its Phase I clinical trials.

The TACAs have showed to exhibit high micro heterogeneity. Hence strategic development of multivalent vaccine candidates has also been reported. Polyvalent vaccine candidates administering a mixture of more than one monomeric antigenic epitopes have witnessed the implementation of heptavalent monomeric vaccines exhibiting high antibody titre values in the preclinical trials.[94] However, various difficulties arose along the process as the constituent mixture demanded greater degree of synthetic conjugation with each of the antigens as well as the vivid knowledge of their individual antigenic properties. Hence an alternative method for the use of multiple copies of monomeric antigens was implemented by the portrayal of multiple carbohydrate antigens attached to the same background core (Fig. 14.8). The constituent Tn, Le^x and Globo-H TACA of **16** was reported to elicit significant IgG and IgM titres when observed through enzyme-linked immunosorbent assay (ELISA).[95] A strategic pentavalent unimolecular candidate, **17** consisting of five TACAs namely Tn, STn, Le^y, TF and Globo-H conjugated with a carrier protein has also been developed and extensively studied by Danishefsky and co-workers.[96]

However, sometimes these multivalent antigenic vaccines were found to induce immune-tolerance which leads to low immunogenic response.[85] Hence antigen analogs have recently been developed and conjugated with protein carriers which may prove to be potential immunogen inducing

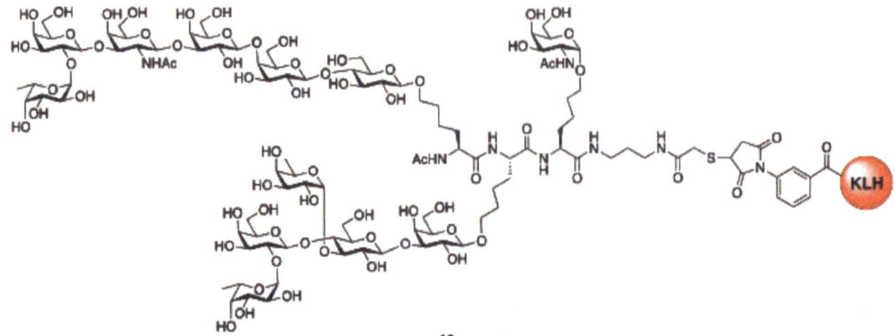

16

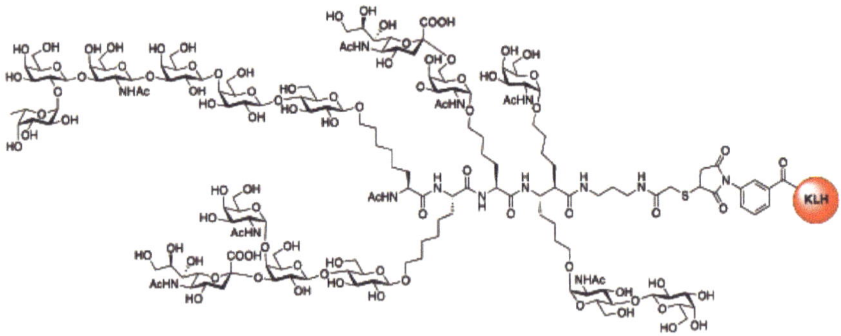

17

Figure 14.8 ***A trivalent (16) and pentavalent (17) unimolecular KLH vaccine.***

high affinity antibody. For example, GM3 analog has been developed by Guo *et. al* using glycoengineering protocol[97] for the construction of unnatural carbohydrate antigen conjugates.

In the domain of development of carbohydrate conjugate vaccines, Boons and co-workers have reported the synthesis of a three component self-adjuvating conjugate.[98] The constituent adjuvants predominantly include lipopeptide or lipo-amino acid based Toll-like receptor (TLR)2 ligands *viz*. Pam3CysSer along with a T-cell epitope.[99] Fig. 14.9 shows the developed three component vaccine candidate where glycolipopeptide was synthesized with coupling of the lipid counterpart to the *N*-terminus and the carbohydrate counterpart with the *C*-terminus of the protected peptide.

Figure 14.9 ***Three component (TLR-2 agonist, Th epitope and CPS) self adjuvating vaccine candidate.***

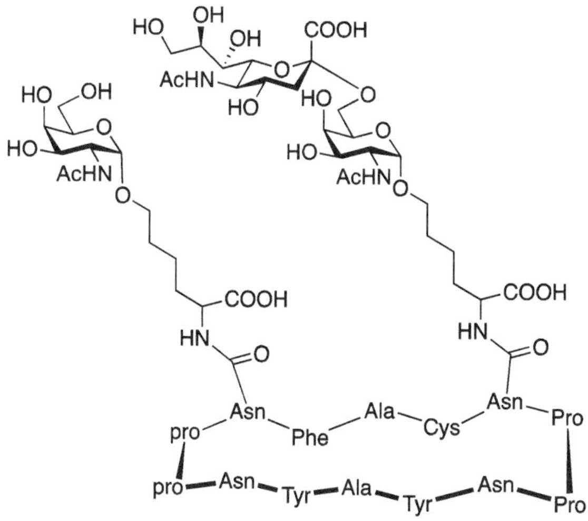

18

Figure 14.10 ***Synthetic cyclopeptide synthetic vaccine developed by Danishefsky and group.***

On the similar principle, Danishefsky has attached Pam3Cys with Le^y antigen[100] wherein multiple Le^y antigens helped in the production of stronger antibody responses. Payne *et. al* has also synthesized a multicomponent vaccine candidate conjugating a MUC1 glycopeptide with a lipopeptide adjuvant intermediated by T-helper peptide. To provide multiple copies of the antigen, anti-cancer dendrimeric vaccine candidates attached to a synthetic core in a characteristic spatial conformation have also been developed in the past years. Various TACAs like Tn, TF, STn, GM2 heterovalently conjugated with cyclopeptidic core[101] and their subsequent functionalization with T-helper cells have shown significant promise in the domain of development of anti-cancer vaccines. Cyclopeptide cores have been utilised by Danishefsky and group (Fig. 14.10) with the intention of mimicking the tumour cell glycocalyx conformation, where they have conjugated both Tn and STn antigens effectively with a strategic core.[102]

4 Conclusions and future perspectives

In conclusion, future of the carbohydrate-based vaccine development looks very promising barring the challenging synthesis of diverse class of carbohydrate antigens through chemical means. However, worldwide interest towards these molecules triggered by growing relevance considering the antibiotic resistance, breakthrough sciences is being developed. This is resulting in rapid enrichment of knowledge and development of newer techniques.

References

1. The Historical Medical Library of The College of Physicians of Philadelphia. Crookshank, E.; ed. History and Pathology of Vaccination, vol. 1. London: Lewis, H. K. 1889.
2. American Academy of Pediatrics, "Vaccine Safety: The Facts," www.aap.org, 2008.
3. (a) www.who.int/immunology WHO (2019); (ii) Vanderslott, S.; Roser, M. https://ourworldindata.org/vaccination Vaccination.
4. CDC, "Smallpox Disease Overview," www.cdc.gov, 2004.
5. (a) Varki, A. Glycobiology, 1993, 3, 97–130; (b) Dwek, R. A. **1996**, Chem. Rev, 96, 683–720; (c) Bishop, J. R.; Schukz, M.; Esko, J. D. *Nature* **2007**, 446, 1030–1037; (d) Linhardt, R. J. *J. Med. Chem.* **2003**, *46*, 2551–2564; (e) Ritchie, G. E.; Mofatt, B. E.; Sim, R. B.; Morgan, B. P.; Dwek, R. A.; Rudd, P. M. *Chem. Rev.* **2002**, *102*, 305–319 ; (f) Davis, B. G. *Chem. Rev.* **2002**, 102, 579–601; (g) Bertozzi, G. E.; Keissling, L. L. *Science* **2001**, *291*, 2357–2364; (*h*) Sears, P.; Wong, C. H. *Science* **2001**, *291*, 2344-2350.
6. (a) Chipowsky, S.; Lee, Y. C.; Roseman, S. *Proc. Natl. Acad. Sci. U. S. A.*, **1973**, *70*, 2309–2312; (b) Lee, Y. C. *Carbohydr. Res.* **1978**, *67*, 509–514; (c) Weigel, P. H.; Schmell, E.; Lee Y.C. Roseman, S.;1. *J. Biol. Chem.* **1978**, *253*, 330-333.
7. Heidelberger, M.; Avery, O.T. *J. Exp. Med.* **1923**, *38*, 73–79.
8. Daffe, M.; Etienne, G. *Tuber. Lung. Dis.* **1999**, *79*, 153–169.
9. Francis, T., Jr.; Tillet, W. S. *J. Exp. Med.* **1930**, *52*, 573–585.
10. Suna, X.; Stefanetti, G.; Berti, F.; Kaspera, D. L. *Proc. Natl. Acad. Sci.* **2019**, *116*, 193–198.
11. Merck & Co., Inc. Pneumovax23 (pneumococcal vaccine polyvalent). Merck website[online], http://www.merck.com/product/usa/pi_circulars/p/pneumovax_23/pneumovax_pi.pdf (1986).
12. (a) Mond, J. J.; Lees, A.; Snapper, C. M. *Annu. Rev. Immunol.* **1995**, *13*, 655–692; (b) Astronomo, R. D.; Burton, D. R. *Nat. Rev. Drug Discovery* **2010**, *9*, 308–324; (c) Snapper, C. M.; Mond, J. J. *J. Immunol.* **1996**, *157*, 2229–2233; (d) Payandeh, Z.; Noori, E.; Khalesi, B.; Mard-Soltani, M.; Abdolal-izadeh, J.; Khalili, S. *Biotechnol. Lett.* **2018**, *40*, 1459–1466; (e) Adams, E. W.; Ratner, D. M.; Seeberger, P. H.; Hacohen, N. *Chem.Bio.Chem.* **2008**, *9*, 294–303; (f) Duan, J.; Avci, F. Y.; Kasper, D. L. *Proc. Natl. Acad. Sci. U.S.A.* **2008**, *105*, 5183-5188.
13. (a) Vos, Q.; Lees, A.; Wu, Z. Q.; Snapper, C. M.;Mond, J. *J. Immunol. Rev.* **2000**, *176*, 154–170; (b) Stein, K.E. *J. Infect. Dis.* **1992**, *165*, Suppl. 1, 549–552; (c) Vinuesa, C. G.; MacLennan, I. C. M. In Immunobiology of Carbohydrates, Wong, S.Y. C., Arsequell, G., Eds.; Landes Bioscience/Eurekah.com: George-town, TX; *New York, NY,* **2003**, 128–147.
14. Kelly, D. F.; Pollard, A. J.; Moxon, E. R. *JAMA* **2005**, *294*, 3019–3023.
15. (a) Beuvery, E.C., van Rossum, F.; Nagel, *J. Infect. Immun.* **1982**, *37*, 15–22; (b) Barrett, D. J.; Ammann, A. J.;Stenmark, S.; Wara, D.W. *Infect. Immun.* **1980**, *27*, 411–417.
16. Carbone, F. R.; Gleeson, P. A. *Glycobiology* **1997**, 7, 725–730.
17. (a) Sun, L.; Middleton, D. R.; Wantuch, P. L.; Ozdilek, A.; Avci1, F. Y. *Glycobiology* **2016**, *26*, 1029–1040; (b) Jones, C. Anais da Academia Brasileira de Ciências **2005**, 77, 293–324.
18. Avery, O.T.; Goebel, W. F. *J. Exp. Med.* **1931**, *54*, 437–447.
19. (a) Astronoma, R. D.; Burton; D. R. *Nature* **2010**, *9*, 308–324; (b) Hever, R. *Pharmaceuticals* **2019**, *12*, 55–80; (c) Jin, K. T.; Lan, H. R.; Chen, X. Y.; Wang, S. B.; Ying, X. J.; Lin, Y.; Mou, X. Z. *Biotechnol. Lett.* **2019**, *41*, 641–650; (d) Seeberger, P. H.;

Werz, D. B. *Nature* **2007**, *446*, 1046–1051; (e) Guo, Z.; Boons, G.-J. *Carbohydrate-Based Vaccines and Immunotherapies;* Wang, B., Ed.; John Wiley Sons: Hoboken, **2009**; p. 1–416. (f) Dagan, R.; Poolman, J.; Siegrist, C. A. *Vaccine* **2010**, *28*, 5513-5523.
20. Ibrahim, N. K.; Murray, J. L. *Clin. Breast Cancer* **2003**, *3*, 139–143.
21. Gilbert, F. B.; Poutrel, B.; Sutra, L. *Vaccine* **1994**, *12*, 369–374.
22. Jacques, I.; Olivier-Bernardin, V.; Dubray, G. *Vaccine* **1991**, *9*, 896–900.
23. Finn, A. *Br. Med. Bull.* **2004**, *70*, 1–14.
24. (a) Wuorimaa, T.; Dagan, R.; Vakevainen, M.; Bailleux, F.; Haikala, R.; Yaich, M.; Eskola, J.; Kayhty, H. *J. Infect. Dis.* **2001**, *184*, 1211–1215; (b) Kasper, D. L.; Paoletti, L. C.; Wessels, M. R.;Guttormsen, H. K.; Carey, V. J.; Jennings, H. J.; Baker, C. J. *J. Clin. Invest.* **1996**, *98*, 2308–2314; (c) Westerink, M. A. J.;Giardina, P. C.; Apicella, M. A.; Kieber-Emmons, T. *Proc. Natl. Acad. Sci. U. S. A.* **1995**, *92*,4021–4025; (d) Fleuridor, R.; Lees, A.; Pirofski, L. *J. Immunol.* **2001**, *166*,1087–1096; (e) Harris, S. L.; Craig, L.; Mehroke, J. S.; Rashed, M.; Zwick, M. B.; Kenar, K.; Toone, E. J.; Greenspan, N.; Auzanneau, F. I.; Marino, A. J.; Pinto, B. M.; Scott, J. K. *Proc. Natl. Acad. Sci. U. S. A.* **1997**, *94*, 2454–2459; (f) Guttormsen, H. K.; Sharpe, A. H.; Chandraker, A. K.; Brigtsen, A. K.; Sayegh, M. H.; Kasper. D. L. *Infect. Immun.* **1999**, *67*, 6375–6384.
25. (a) Avci, F. Y.; Li, X.; Tsuji, M.; Kasper, D. L. *Nat. Med.* **2011**, *17*, 1602–1609; (b) Lai, Z.; Schreiber, J. R. *Vaccine* **2009**, *27*, 3137-3144.
26. (a) Zou, W.; Jennings, H. J. Preparation of Glycoconjugate Vaccines. In *Carbohydrate-Based Vaccines and Immunotherapies;* Guo, Z.; Boons, G.-J., Eds.; John Wiley & Sons: Hoboken, NJ, **2009**; pp. 55–88..
27. Kool, M.; Fierens, K.; Lambrecht, B. N. *J. Med. Microbiol.* **2012**, *61*, 927–934.
28. Bhatia, S.; Dimde, M.; Haag, R. *Med. Chem. Commun.* **2014**, *5*, 862–878.
29. Paulsen, H. *Angew. Chem. Int. Ed. Engl.* **1982**, *21*, 155–173.
30. Paulson, J. C.; Blixt, O.; Collins, B. E. *Nature Chem. Biol.* **2006**, *2*, 238–248.
31. Wang, Y.; Ye, X. S.; Zhang, L. H. *Org. Biomol. Chem* . **2007**, *5*, 2189–2200.
32. (a) Andrade, R.B.; Plante, O.J.; Melean, L.G.; Seeberger, P.H. *Org. Lett.* **1999**, *111*, 1811–1814; (b) Bennett, C. S. *Org. Biomol. Chem.* **2014**, *12*, 1686–1698.
33. (a) Cheng, C.-W.; Zhou, Y., Pan, W.-H.; Dey, S.; Wu, C.-Y.; Hsu, W.-L.; Wong, C.-H. *Nat. Comm.* **2018**, *9*, 5202; (b) Zhang, Z.; Ollmann, I. R.; Ye, X.-S.; Wischnat, R.; Baasov, T.; Wong, C.-H. *J. Am. Chem. Soc.* **1999**, *121*, 734–753.
34. (a) Hsu, C. H.; Hung, S. C.; Wu, C. Y.; Wong, C. H. **2011**, *50*, 11872–11923; (b) Hsu, C.-H.; Hung, S.-C.; Wu, C.-Y.; Chi-Huey Wong, C.-H. Angew. *Chem. Int. Ed.* **2011**, *123*, 12076–12129; (c) Sears, P.; Wong, C. H. *Science* **2001**, *291*, 2344–2350; (d) Seeberger, P. H. *Chem. Soc. Rev.* **2008**, *37*, 19–28; (e) Seeberger, P. H. *Chem. Commun.* **2003**, 1115–1121; (f) Seeberger, P. H.; Werz, D. B. *Nature* **2007**, *446*, 1046–1051; (g) Seeberger, P. H. *Acc. Chem. Res.* **2015**, *48*, 1450-1463.
35. McCarthy, P. C.; Saksena, R.; Peterson, D. C.; Lee, C. H.; An, Y.; Cipollo, J. F.; Vann, W. F. *Glycoconj. J.* **2013**, *30*, 857–870; (b) Fiebig, T.; Romano, M. R.; Oldrini, D.; Adamo, R.; Tontini, M.; Brogioni, B.; Santini, L.; Berger, M.; Costantino, P.; Berti, F.; Gerady-Schahn, *R. NPJ Vaccines* **2016**, 16017.
36. (a) Boh, L.; Crich, D.; Chim. C. R. C. R. *Chimie* **2011**, *14*, 3–16; (b) Zhu, X.; Schmidt, R. R. *Angew. Chem. Int. Ed.* **2009**, *48*, 1900–1934. *Angew. Chem.* **2009**, *121*, 1932–1967; (c) Das, R.; Mukhopadhyay, B. *Chem. Open* **2016**, *5*, 401–433; (d) Demchenko, A. V. *Synlett* **2003**, *9*, 1225–1240; (e) Jensen, K. J. *J. Chem. Soc. Perkin Trans. 1*, **2002**, 2219–2233; (f) Mulani, S. K.; Hung, W. C.; Ingle, A. B.; Shiau, K. S.; Tony Mong, K.-K. *Org. Biomol. Chem.* **2014**, *12*, 1184–1197; (g) Davis, B. G. *J. Chem. Soc. Perkin Trans. 1* **2000**, 2137–2160.
37. Guerrini, M.; Raman, R.; Venkataraman, G.; Torri, G.; Sasisekharan, R.; Casu, B. *Glycobiology* **2002**, *12*, 713–719.
38. Buskas, T.; Li, Y.; Boons, G. J. *Chemistry* **2004**, *10*, 3517–3524.

39. World Health Organization. Haemophilus influenzae type B (HiB). World Health Organization website[online], http://www.who.int/mediacentre/factsheets/ fs294/en/ (2005).
40. https://www.cdc.gov/vaccines/pubs/pinkbook/hibHTML(online) Epidemiology and Prevention ofVaccine-Preventable Diseases Chapter 8: Haemophilus influenzae type b.
41. Peltola, H. *Clin. Microbiol. Rev.* **2000**, *13*, 302–317.
42. Toraño, G.; Toledo, M. E.; Baly, A.; Fernandez-Santana, V.; Rodriguez, F.; Alvarez, Y.; Serrano, T.; Musachio, A.; Hernandez, I.; Hardy, E.; Rodríguez, A.; Hernandez, H.; Aguilar, A.; Sánchez, R.; Diaz, M.; Muzio, V.; Dfana, J.; Rodríguez, M. C.; Heynngnezz, L.; Verez-Bencomo, V. *Clin. Vaccine Immunol.* **2006**, *13*, 1052–1056.
43. Verez-Bencomo, V.; Fernandez-Santana, V.; Hardy, E.; Toledo, M. E.; Rodriguez, M. C.; Heynngezz, L.; Rodriguez, A.; Baly, A.; Herrera, L.; Izquierdo, M.; Villar, A.; Valdes, Y.; Cosme, K.; Deler, M. L.; Montane, M.; Garcia, E.; Ramos, A.; Aguinar, A.; Medina, E.; Torano, G.; Sosa, I.; Hernandez, I.; Martinez, R.; Muzachio, A.; Carmenates, A.; Costo, L.; Cardoso, F.; Campa, C.; Diaz, M.; Rene, R. *Science* **2004**, *305*, 522–525.
44. (a) Schneerson, R.; Barrera, O.; Sutton, A.; Robbins, J, B. *J. Exp. Med.* **1980**, *152*, 361–372; (b) Anderson, P. W.; Pichichero, M. E.; Insel, R. A.; Betts, R.; Eby, R.; Smith, D. H. *J. Immunol.* **1986**, *137*, 1181–1186.
45. Peltola, H. *Clin. Microbiol. Rev.* **2000**, *13*, 202–207.
46. (a) Mesnzner, Z. *Orv. Hetil.* **2014**, *155*, 243–247; (b) Degan, R. *Vaccine* **2010**, *19*, S9-S16; (b) Dowell, S. F.; Butler, J. C.; Giebink, G. S.; Jacobs, M. R.; Jernigan, D.; Musher, D. M.; Rakowsky, A.; Schwartz, B. *Pediatr. Infect. Dis.* **1999**, *18*, 1–9; (c) Whitney, C. G.; Moore, M.; Pilishvili, T. MMWR. *Morbidity and Mortality Weekly Report* **2012**, *61*, 816–819.
47. (a) Melegaro, A.; Edmunds, W.J. *Eur. J. Epidemiol.* **2004**, *19*, 353–363; (b) Bricks, L.F. *J. Pediatr.* **1994**, *70*, 75–81.
48. Burns, T.; Abadi, M.; Pirofski, L. A. *Infect. Immun.* **2005**, *73*, 4530–4538.
49. (a) Gallorini, S.; Berti, F.; Parente, P.; Baronio, R.; Aprea, S.; D'Oro, U.; Pizza, M.; Telford, J. L.; Wack, A. *J. Immunol.* **2007**, *179*, 8208–8215.
50. (a) Stoof, S. P.; Rodenburg, G. D.; Knol, M. J.; Rumke, L. W.; Bovenkerk, S.; Berbers, G. A. Spanjaard, L.; van der Ende, A.; Sanders, E. A. *Clin Infect Dis.* **2015**, *61*, 1281–1292; (b) Tirani, M.; Meregaglia, M.; Melegaro, A. *PLoS ONE* **2015**, *10*, 0123383; (c) Goldschneider, I.; Gotschlich, E.C.; Artenstein, M.S. *J Exp Med.* **1969**, *129*, 1307–1326; (d) Stephens, D. S.; Greenwood, B.; Brandtzaeg, P. *Lancet* **2007**, *369*, 2196–2210.
51. Finne, J.; Leinonen, M.; Makela, P. H. *Lancet* **1983**, *2*, 355–357.
52. (a) Wyle, F. A. et al. *J. Infect. Dis.* **1972**, *126*, 514–521; (b) Finne, J.; Leinonen, M.; Makela, P. H. *Lancet* **1983**, *2*, 355-357.
53. (a) Jennings, H. J. *Contrib. Microbiol. Immunol.* **1989**, *10*, 151–165; (b) Jennings, H. *J. Int. J. Infect. Dis.* **1997**, *1*, 159–164; (b) Jennings, H. J.; Lugowski, C.; Ashton, F. E. *Infect. Immun.* **1984**, *43*, 407–412; (c) Jennings, H. J.; Lugowski, C. J. *Immunol* **1981**, *127*, 1011–1018; (d) Beuvery, E. C.; Roy, R.; Kanhai, V.; Jennings, *J. Dev. Biol.* **1986**, *65*, 197–204.
54. (a) Bruge, J.; Bouveret-Le Cam, N.; Danve, B.; Rougon, G.; Schulz, D. *Vaccine* **2004**, *22*, 1087–1096; (b) Krug, L. M.; Ragupathi, G.; Hood, C.; George, C.; Hong, F.; Shen, R.; Abrey, L.; Jennings, H. J.; Kris, M. G.; Livingston, P. O. *Cancer Immunol. Immunother.* **2012**, *61*, 9–18.
55. Trotter, C. L.; Andrews, N. J.; Kaczmarski, E. B.; Miller, E.; Ramsay, M. E. *Lancet* **2004**, *364*, 365–367.
56. Noya, F.; McCormack, D.; Reynolds, D. L.; Neame, D.; Oster, P. *Can. J. Infect. Dis. Med. Microbiol.* **2014**, *25*, 211–216.
57. Pina, L. M.; Bassily, E.; Machmer, A.; Hou, V.; Reinhardt, A. *Pediatr. Infect. Dis. J.* **2012**, *31*, 1173–1183.
58. (a) King, W. J.; MacDonald, N. E.; Wells, G.; Huang, J.; Allen, U.; Chan, F.; Ferris, W.; Diaz-Mitoma, F.; Ashton, F. *J. Pediatr.* **1996**, *128*, 196–202; (b) Snape, M. D.; Pollard, A. J. *The Lancet Infectious Diseases,* **2005**, *5*, 21-30.

59. (a) Jansen, W. T. M.; Hogenboom, S.; Thijssen, M. J. L.; Kamerling, J. P.; Vliegenthart, J. F. G.; Verhoef, J.; Snippe, H.; Verheul, A. F. M. A. *Infect. Immun.* **2001**, *69*, 787–793; (b) Cavallari, M.; Stallforth, P.; Kalinichenko, A.; Rathwell, D. C.; Gronewold, T. M.; Adibekian, A.; Mori, L.; Landmann, R.; Seeberger, P. H.; De Libero, G. *Nat. Chem. Biol.* **2014**, *10*, 950–956.
60. Pozsgay, V.; Chu, C.; Pannell, L.; Wolfe, J.; Robbins, J. B.; Schneerson, R. *Proc. Natl Acad. Sci. USA* **1999**, *96*, 5194–5197.
61. Arcuri, M.; Benedetto, R. D.; Cunningham, A. F.; Saul, A.; MacLennan, C. A. F. *PLoS ONE* **2017**, *12*, 0189100.
62. Kay, E. J.; Yates, L. E.; Terra, V. S.; Cuccui, J.; Wren, B. W. *Open Biol.* **2016**, *6*, 150243; (b) Micoli, F.; Rondini, S.; Alfini, R.; Lanzilao, L.; Necchi, F.; Negrea, A.; Rossi, O.; Brandt, C.; Clare, S.; Mastroeni, P.; Rappuoli, R.; Saul, A.; MacLennan, C. A. *Proc. Natl Acad. Sci.* **2018**, 115, 10428 10433; (c) Sun, P.; Pan, C.; Zeng, M.; Liu, B.; Liang, H.; Wang, D.; Liu, X.; Wang, B.; Lyu, Y.; Wu, J.; Zhu, L.; Wang, H. NPJ *Vaccine* **2018**, *3*, 4; (d) Szijártó, V.; Guachalla, L. M.; Hartl, K.; Varga, C.; Banerjee, P.; Stojkovic, K.; Kaszowska, M.; Nagy, E,; Lukasiewicz, J.; Nagy, G. Int. *J. Med. Microbiol.* **2016**, *306*, 89–98; (e) Theilacker, C.; Kaczynski, Z.; Kropec, A.; Fabretti, F.; Sange, T.; Holst, O.; Huebner, *J. Infect Immun.* **2006**, *74*, 5703–5712.
63. Heath, P. T. *Vaccine* **2016**, *34*, 2876–2879.
64. (a) Madhi, S. A.; Cutland, C. L.; Jose, L.; Koen, A.; Govender, N.; Wittke, F.; Olugbosi, M.; Meulen, A. S.; Baker, S.; Dull, P. M.; Narasimhan, V.; Slobod, K. *Lancet Infect. Dis.* **2016**, *16*, 923–934; (b) Leroux-Roels, G.; Maes, C.; Willekens, J.; De Boever, F.; de Rooij, R.; Martell, L.; Bedell, L.; Wittke F.; Slobod K.; Dull P. *Vaccine* **2016**, *34*, 1786–1791; (c) Heyderman, R. S.; Madhi, S. A.; French, N.; Cutland, C.; Ngwira, B.; Kayambo, D.; Mboizi, R.; Koen, A.; Jose, L.; Olugbosi, M.; Wittke, F.; Slobod, K.; Dull, P. M. *Lancet Infect. Dis.* **2016**, *16*, 546–555.
65. Donders, G. G.; Halperin, S. A.; Devlieger, R.; Baker, S.; Forte, P.; Wittke, F.; Slobod, K. S.; Dull, P. M. *Obstet. Gynecol.* **2016**, *127*, 213–221.
66. (a) Emmadi, M.; Khan, N.; Lykke, L.; Reppe, K.; Parameswarappa, S. G.; Lisboa, M. P.; Wienhold, S.-M.; Martin Witzenrath, Pereira, C. L.; Seeberger, P. H. *J. Am. Chem. Soc.* **2017**, *139*, 14783–14791; (b) Parameswarappa, S. G.; Reppe, K.; Geissner, A.; Nová, P.-M.; Govindan, S.; Calow, A. D. J.; Wahlbrink, A.; Weishaupt, M.W.; Monnanda, B. P.; Bell, R. L.; Pirofski, L.-A.; Suttorp, N.; Sander, L.-E.; Witzenrath, M.; Pereira, C. L.; Anish, C.; Seeberger, P. H. *Cell Chem. Biol.* **2016**, *23*, 1407–1416.
67. Cavallari, M.; Stallforth, P.; Kalinichenko, A.; Rathwell, D. C. K.; Gronewold, T. M. A.; Alexander Adibekian Mori, L.; Landmann, R.; Seeberger, P. H.; Libero, G. D. *Nat. Chem. Biol.* **2014**, *10*, 950–956.
68. Lisboa, M. P.; Khan, N.; Martina, C.; Xua, F. -F.; Reppe, K.; Geissnera, A.; Govindana, S.; Witzenrath, M.; Pereira, C. L.; Seeberger, P. H. *Proc. Natl. Acad. Sci.* **2017**, *114*, 11063–11068.
69. Ollis, A. A.; Zhang, S.; Fisher, A. C.; DeLisa, M. P. *Nat. Chem. Biol.* **2014**, *10*, 816–822.
70. Hatz, C. F. R.; Ballya, B.; Rohrer, S.; Steffena, R.; Kramme, S.; Siegrist, C.-A.; Wacker, M.; Alaimoc, C.; Fonck, V. G. *Vaccine* **2015**, *33*, 4594–4601.
71. (a) Riddle, M. S.; Kaminski, R. W.; Paolo, C. D.; Porter, C. K.; Gutierrez, R. L.; Clarkson, K. A.; Weerts, H. E.; Duplessis, C.; Castellano, A.; Alaimo, C.; Paolino, K.; Gormley, R.; Fonck, V. G. *Clin. Vaccine Immunol.* **2016**, *23*, 908–917; (b) Mitra, A.; Mukhopadhyay, B. *Carbohydr. Res.* **2016**, *426*, 1-8.
72. Huttner, A.; Hatz, C.; van den Dobbelsteen, G.; Abbanat, D.; Hornacek, A.; Frölich, R.; Dreyer, A. M.; Martin, P.; Davies, T.; Fae, K.; van den Nieuwenhof, I.; Thoelen, S.; de Valliére, S.; Kuhn, A.; Bernasconi, E.; Viereck, V.; Kavvadias, T.; Kling, K.; Ryu, G.; Hülder, T.; Gröger, S.; Scheiner, D.; Alaimo, C.; Harbarth, S.; Poolman, J.; Fonck, V. G. *Lancet Infect. Dis.* **2017**, *17*, 528–537.

73. Ellerbroek, P. M.; Walenkamp, A. M.; Hoepelman, A. I.; Coenjaerts, F. E. *Curr. Med. Chem.* **2004**, *11*, 253–266.
74. Han, Y.; Ulrich, M. A.; Cutler, J. E. *J. Infect. Dis.* **1999**, *179*, 1477–1484.
75. (a) Oscarson, S., Alpe, M., Svahnberg, P., Nakouzi, A.; Casadevall, A. *Vaccine* **2005**, *23*, 3961–3972; (b) Guazzelli, L.; Ulc, R.; Oscarson, S. *ChemistryOpen* **2015**, *4*, 729–739; (c) Guazzelli, L.; Ulc, R.; Rydner, L. et al. *Org. Biomol. Chem.* **2015**, *13*, 6598–6610; (d) Guazzelli, L.; McCabe, O.; Oscarson, S. *Carbohydr. Res.* **2016**, *433*, 5–13; (e) Torosantucci, A. et al. *J. Exp. Med.* **2005**, *202*, 597–606.
76. (a) Gilad, J.; Harary, I.; Dushnitsky, T.; Schwartz, D.; *Amsalem, Y. IMAJ* **2007**, *9*, 499–503; (b) Dingle, T. C.; Butler-Wu, S. M.; Abbott, A. N. *J. Clin. Microbiol.* 2014, 52, 3490–3491.
77. Sebastian, S., et al. *Infect. Immun.* **2007**, *75*, 2591–2602.
78. (a) Stefanetti, G.; Okana, N.; Finkd, A.;Gardnera, E.; Kaspera, D. L. *Proc. Natl. Acad. Sci. USA* **2019**, *116*, 7062–7070; (b) Conlan, J. W.; Shen, H.; Webb, A.; Perry, M. *B. Vaccine* **2002**, *20*, 3465–3471.
79. (a) Pifferi, C.; Bertheta, N.; Renaudet, O. *Biomater. Sci.* **2017**, *5*, 953–965; (b) Pistorio, S. G.; Geringer, S. A.; Stine, K. J.; Demchenko, A. V. *J. Org. Chem.* **2019**, *84*, 6576–6588; (c) Nagasaki, M.; Manabe, Y.; Minamoto, N.; Tanaka, K.; Silipo, A.; Molinaro, A.; Fukase, K. *J. Org. Chem.* **2016**, *81*, 10600–10616; (d) Groneck, L.; Schrama, D.; Fabri, M.; Stephen, T. L.; Harms, F.; Meemboor, S.; Hafke, H.; Bessler, M.; Becker, J. C.; Kalka-Moll, W. M. *Infect. Immun.* **2009**, 77, 3705–3712; (e) NíCheallaigh, A.; Oscarson, S. Canad. *J. Chem.* **2016**, *94*, 940–960; (f) Sundgren, A.; Lahmann, M.; Oscarson, S. Beilstein *J. Org. Chem.* **2010**, *6*, 704–708; (g) Zhang, H.; Shao, L.; Wang, X.; Zhang, Y.; Guo, Z.; Gao, *J. Org. Lett.* **2019**, *21*, 2374–2377.
80. Soliman, S. E.; Kovác, P. *Angew. Chem. Int. Ed.* **2016**, *55*, 12850–12853.
81. (a) Mitra, A.; Mukhopadhyay, B. Eur. *J. Org. Chem.* 2019, *30*, 4869–4878; (b) Pal, K. B.; Mukhopadhyay, B. *ChemistrySelect* 2017, *2*, 7378–7381; (c) Budhadev, D.; Mukhopadhyay, B. *RSC Advances* **2015**, *5*, 98033–98040; (d) Bhaumik, I.; Misra, A. K. *Glycoconj. J.* **2016**, *33*, 887–896; (e) Si, A.; Misra, A. K. *Tetrahedron* 2016, *72*, 4435–4444; (f) Si, A.; Misra, A. K. *Chemistry Open* **2015**, *5*, 47–50.
82. Das, R.; Mukhopadhyay, B. *J. Carbohydr. Chem.* **2015**, *34*, 247–262.
83. (a) Zhang, S. et al. *Int. J. Cancer* **1997**, *73*, 42–49; (b) Zhang, S. et al. *Int. J. Cancer* **1997**, *73*, 50–56; (c) Kannagi, R.; Levery, S.B.; Ishijamik, F.; Hakomori, S.; Schevinsky, L. H.; Knowles, B. B.; Solter, D. *J. Biol. Chem.* **1983**, *258*, 8934–8942.
84. (a) Chang, W-W.; Lee, C. H.; Lee, P.; Lin, J.; Hsu, C-W.; Hung, J-T.; Lin, J-J.; Yu, J-C.; Shao, L-e.; Yu, J.; Wong, C-H.; Yu, A. L. *Proc. Natl. Acad. Sci. U.S.A.* **2008**, *105*, 11667; (b) Balic, M.; Lin, H.; Young, L.; Hawes, D.; Giuliano, A.; McNamara, G.; Datar, R. H.; Cote, R. *J. Clin. Cancer Res.* **2006**, *12*, 5615.
85. (a) Livingston, P. O.; Wong, G.Y. C.; Adluri, S.; Tao, Y.; Padavan, M.; Parente, R.; Hanlon, C.; Calves, M. J.; Helling, F.; Ritter, G.; Oettgen, H. F.; Old, L. J. *J. Clin. Oncol.* **1994**, *12*, 1036; (b) Feng, F.; Shaikh, A. S.; Wang, F. ACS *Chem. Biol.* **2016**, *11*, 850–863.
86. Monzavi-Karbassi, B.; Pashov, A.; Kieber-Emmons, T. *Vaccines (Basel)* **2013**, *1*, 174–203.
87. Chen, X. -T.; Sames, D.; Danishefsky, S. J. *J. Am. Chem. Soc.* **1998**, *120*, 7760–7769.
88. Behar, V.; Danishefsky, S. J. *Angew. Chem., Int. Ed. Engl.* **1994**, *33*, 1468.
89. Kudryashov, Y.; Kim, H. M.; Ragupathi, G.; Danishefsky, S. J.; Livingston, P. O.; Lloyd, K. O. *Cancer Immunol. Immunother.* **1998**, *45*, 281.
90. (a) Danishefsky, S. J.; Koseki, K.; Griffith, D. A.; Gervay, J.; Peterson, J. M.; McDonald, F. E.; Oriyamalf, T. *J. Am. Chem. Soc.* **1992**, *114*, 8331–8333; (b) Ellervik, U.; Magnusson, G. *J. Org. Chem.* **1998**, *63*, 9314–9322; (c) Ellervik, U.; Grundberg, H.; Magnusson, G. *J. Org. Chem.* **1998**, *63*, 9323–9338; (d) Tsukida, T.; Yoshida, M.; Kurokawa, K.; Nakai, Y.; Achiha, T.; Kiyoi, T.; Kondo, H. *J. Org. Chem.* **1997**, *62*, 6876–6881.

91. (a) Danishefsky, S. J.; Shue, Y-K.; Chang, M. N.; Wong, C-H. *Acc. Chem. Res.* **2015**, *48*, 643–652; (b) Park, T. K.; Kim, I. J.; Hu, S.; Bilodeau, M. T.; Randolph, J. T.; Kwon, O.; Danishefsky, S. J. *J. Am. Chem. Soc.* **1996**, *118*, 11488–11500; (c) Jeon, I.; Iyer, K.; Danishefsky, S. J. *J. Org. Chem.* **2009**, *74*, 8452–8455.
92. Kannagi, R.; Levery, S. B.; Ishijamik, F.; Hakomori, S.; Schevinsky, L. H.; Knowles, B. B.; Solter, D. *J. Biol. Chem.* **1983**, *258*, 8934.
93. (a) Zhu, T.; Boons, G-J. *Angew. Chem. Int. Ed.* **1999**, *38*, 3495–3497; (b) Tsai, T-I.; Lee, H-Y.; Chang, S-H.; Wang, C-H.; Tu, Y-C.; Lin, Y-C.; Hwang, D-R.; Wu, C-Y.; Wong, C-H. *J. Am. Chem. Soc.* **2013**, *135*, 14831–14839.
94. (a) Kaiser, A.; Gaidzik, N.; Westerlind, U.; Kowalczyk, D.; Hobel, A.; Schmitt, E.; Kunz, H. *Angew. Chem. Int. Ed. Engl.* **2009**, *48*,7551–7555; (b) Ragupathi, G.; Cappello, S.; Yi, S. S.; Canter, D.; Spassova, M.; Bornmann, W. G.; Danishefsky, S. J.; Livingston, P. O. *Vaccine* **2002**, *20*, 1030–1038.
95. Ragupathi, G.; Coltart, D. M.; Williams, L. J.; Koide, F.; Kagan, E.; Allen, J.; Harris, C.; Glunz, P. W.; Livingston, P. O.; Danishefsky, S. J. *Proc. Natl. Acad. Sci. USA* **2002**, *99*, 13699–13704.
96. (a) Keding, S.J.; Danishefsky, S. *J. Proc. Natl. Acad. Sci. USA* **2004**, *101*, 11937–11942; (b) Schwarz, J. B.; Kuduk, S. D.; Chen, X. T.; Sames, D.; Glunz, P. W.; Danishefsky, S. J. *J. Am. Chem. Soc.* **1999**, *121*, 2662–2673.
97. (a) Wang, Q., Zhang, J.; Guo, Z. *Bioorg. Med. Chem.* **2007**, *15*, 7561–7567; (b) Arcangeli, A.; Toma, L.; Contiero, L.; Crociani, O.; Legnani, L.; Lunghi, C.; Nesti, E.; Moneti, G.; Richichi, B.; Nativi, C. *Bioconjate Chem.*, **2010**, *21*, 1432–1438.
98. (a) Yin, X-G.; Chen, X-Z.; Sun, W-M.; Geng, X-S.; Zhang, X-K.; Wang, J.; Ji, P-P.; Zhou, Z-Y.; Baek, D. J.; Yang, G-F.; Liu, Z.; Guo, *J. Org. Lett.* **2017**, *19*, 456–459; (b) Buskas, T.; Ingale, S.; Boons, G. *J. Angew. Chem. Int. Ed.* **2005**, *44*, 5985–5988; (c) McDonald, D. M.; Byrne, S. N.; Payne, R. *J. Front. Chem.* **2015**, *3*, 60.
99. Li, Q.; Guo, Z. *Molecules* **2018**, *23*, 1583.
100. (a) Glunz, P. W; Hintermann, S.; Williams, L. J.; Schwarz, J. B.; Kuduk, S. D.; Kudryasov, V.; Lyoyd, K. O.; Danishefsky, S. J. *J. Am. Chem. Soc.* **2000**, *122*, 7273–7279; (b) Kudryasov, V.; Glunz, P. W.; Williams, L. J.; Hintermann, S.; Danishefsky, S. J.; lyoyd, K. O. *Proc. Natl. Acad. Sci. USA*, **2001**, *98*, 3264–3269.
101. (a) Pifferi, C.; Thomas, B.; Goyard, D.; Berthet, N.; Renaudet, O. *Chem. Eur. J.* **2017**, *23*, 16283–16296; (b) Renaudet, O.; Dumy P. *Bioconjugate Chem.* **2005**, 16, 1149-1159.
102. Jeon, I.; Lee, D.; Krauss, I. J.; Danishefsky, S. J. *J. Am. Chem. Soc.* **2009**, *131*, 14337–14344.

CHAPTER FIFTEEN

Opportunity of plant oligosaccharides and polysaccharides in drug development

Vineet Kumar[a], **Shipra Nagar**[a,b], **Pradeep Sharma**[a]
[a]Chemistry and Bioprospecting Division, Forest Research Institute, Dehradun, India
[b]National Centre for Polar and Ocean Research, Ministry of Earth Sciences, Goa, India

1 Introduction

Although carbohydrate-based drugs have been developed during the last few decades, a meticulous study of their structure based activity is yet to be thoroughly explored. Irrespective of size and chain length, their ability to undergo physico-chemical transformations as a backbone or attached to aglycan moiety makes them the most versatile candidates amongst all biomacromolecules and natural products.[1] They exhibit assorted diversity at the levels of conformation, rheology, reducing and non-reducing nature, branching, hydrodynamic volume, molecular weight (MW), etc. The vast array of carbohydrate polymers may vary from a simple molecule as 7-membered cyclodextrin (MW 1.1 kDa) to huge xanthan gums (MW range 2000 kDa), encompassing linear (e.g. pullulans, inulin) or branched (e.g. arabinogalactans, glycogen) or globular (e.g. cyclodextrins) structures, decorated with numerous functional moieties like uronic acids, sulfates, pyruvates, methyls, acetyls, 6-deoxy sugars, 3,6-deoxy sugars, amines, acetylated amines, etc. thereby, giving the polymer hydrophobic and/or hydrophilic attributes.

The use of medicinal plants since 4000 B.C., as a main source of bioactive polysaccharides and oligosaccharides in Ayurvedic formulations called as 'Rasayanas' to cure dreaded diseases, has been documented precisely in Ancient Indian Medicinal Literature including *Sushruta Samhita*,[2] *Charak Samhita*, *Bhav Prakash*, *Nighntu* etc. The Rasayanas containing polysaccharides have been known for various ethnotherapeutic effects *viz.* gastro-protective, lipid lowering, antioxidant, tumor preventive and cytotoxic,

Carbohydrates in Drug Discovery and Development. http://dx.doi.org/10.1016/B978-0-12-816675-8.00015-4

macrophage activation, immunomodulating, free radical scavenging, anti-diabetic, etc.[3]

In recent years, significant phytochemical and pharmacological interventions have revealed a wide range of carbohydrates including acetylated glucomannans,[4] fructans,[5] glucans,[6] arabinogalactans,[7] xylans,[8] and numerous other glycoconjugate based drugs for treating influenza, cancer, inflammation, blood coagulation, tumor, bacterial and viral infections, colitis, human immunodeficiency virus (HIV), Alzheimer, etc.[9,10] The chapter presents a comprehensive account of the status of oligosaccharides and polysaccharides based drug development and underlines the challenges and future perspectives.

2 Biological and pharmacological significance

Carbohydrates are one of the most ubiquitous and principal life supporting molecules. Glucose, the primary source of energy in all living beings is the most abundant monosaccharide synthesized by plants and algae. The monosaccharide occurs naturally in fruits and other parts of plant in free state as D-glucose. The other enantiomer L-glucose does not occur naturally and can only be obtained by synthetic routes. Interestingly, it is not metabolized in the living organisms and therefore less significant. *D*-Glucose, being the smallest energy molecule in biotic world constitutes the major biomass in the form of distinctive biopolymers which possess immense biological and industrial importance *viz.* starch, cellulose, hemicellulose, glycogen etc. It is stored as amylose and amylopectin in plants and as glycogen in animals. The biopolymers are formed through biosynthetic routes involving series of steps including polymerization. Plants synthesize cellulose and starch as the major components in the nature's combi lab for their own structural and storage necessities, respectively. The physico-chemical properties and stability of these biopolymers depend upon the glycosidic linkage among glucose units (1,4-α or 1,4-β linkage). Intriguingly, these biopolymers have diverse industrial applications and brief details of their usage are as follows:

2.1 Starch and its products

Starch is synthesized by majority of green plants and is normally used as staple food throughout the world. The biopolymer is a white, tasteless and odourless powder encompassing variable proportion of amylose (20 to 25%) and amylopectin (75 to 80%) depending upon the species from which it is extracted (Fig. 15. 1). It is largely insoluble in cold water in raw state. Starch is hydrolysed to oligosaccharides and monosaccharides by malting

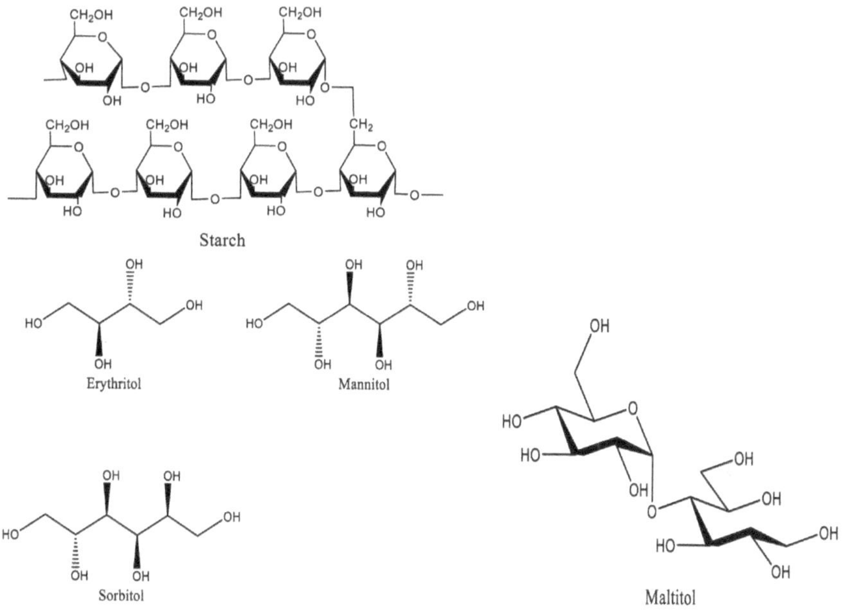

Figure 15.1 ***Chemical structure of starch and alditols.***

and fermented to produce ethanol as raw material for usage in alcoholic beverages (beer, whisky, rum) and as bio-fuel. It is a widely demanded polysaccharide in food, feed, pharmaceutical, beverage, etc. Maltodextrin, one of the partially hydrolysed products of starch, is an easily digestible, moderately sweet or almost flavourless oligosaccharide and polysaccharide having DP 2 to 20. It is used as filler and thickener in food and pharmaceuticals due to its absorption and digestibility equivalent to glucose. Further, the completely hydrolysed starch commonly called dextrose, is used in baking products as a sweetener, and is commonly used in processed foods and corn syrup. Additionally, dextrose has significant pharmaceutical applications. The sterile, non-pyrogenic solution of dextrose (5% and 10%) is given in a single dose intravenously for fluid replenishment and calorie supply. The dextrose solution can be combined with other drugs for intravenous administration or used to increase blood sugar in humans. Another fully hydrolysed form of starch is glucose syrup which is used as sweetener and thickener in food industry.

Resistant starch, another form of starch which evades digestion due to its resistance to hydrolysis by α-amylase in the small intestine of healthy persons, holds various benefits as a dietary supplement for metabolic health. It can improve insulin sensitivity and is very effective in lowering blood sugar level, improve satiety and colon functions. The resistant starch also contributes immensely towards health benefits of intact whole grains.

2.2 Monosaccharide based natural polyols

Natural polyols are another class of carbohydrates that are gaining the status of artificial sweeteners. They are reduced form of monosaccharides, also called sugar alcohols or alditols *viz.* mannitol, maltitol, erythritol, xylitol, sorbitol, etc. The most common sugar alcohols are sorbitol and mannitol, and their availability in some plant parts and exudates may be very high. Like D-Glucose, these sugar alcohols also occur as *D*-enantiomers. *D*-Mannitol is used as a sweetener in diabetic food due to its lesser absorption ability in the biological system. In pharmaceuticals, mannitol is commonly employed to decrease pressure in the eyes in case of glaucoma, and to reduce the intra-cranial pressure. Further, it is also used in certain cases of low urine output as cholecysto kinetic agent to prevent kidney failure. Xylitol, another sugar substitute, though not a common household sweetener, is widely used in pharmaceuticals, dietary supplements, toothpaste, confectionery, chewing gum, etc. Interestingly, xylitol is metabolized independent of insulin and hence has insignificant impact on blood sugar. It provides 40% lesser calories than sucrose due to its slower metabolism in biological system than glucose and is an approved food additive in the United States. On the other hand, erythritol possesses 60–70% sweetness as compared to sucrose and is not metabolized in biological system. Consequently, it does not affect blood sugar or tooth decay. In nature, erythritol occurs in fruit and fermented foods, but industrially it is produced by enzymatic degradation of starch with yeast (*Moniliella* sp.) followed by fermentation of glucose with bacteria (*Yarrowia lipolytica*). It is stable over a broader pH range (2 to 10) and temperature (up to 160 °C) and employed in freshly made artificially sweetened beverages for improving taste. It is valuable for functional beverages as it acts as an antioxidant along with non-caloric and non-glycemic additive.[11–13]

D-Sorbitol or D-Glucitol is obtained by reduction or hydrogenation of glucose. It is known for slow absorption in humans and employed as sweetener in food industry. It is largely obtained from potato starch but also occurs in fruits such as apples, pears and prunes. It is used in preparation of soft gel capsules, mouthwash, toothpaste, transparent gels owing to its laxative and thickening ability and high refractive index. Being a humectant, it is widely used to retain the moisture content of skin powders and is also demanded in bakery products, peanut butter and fruit preservatives. During baking, it act as a plasticizer and reduces salting process.[14–16] Maltitol is obtained by hydrogenation of maltose. It is mixed with other sweeteners in sugar-free hard candies, chocolates, confectioneries, chewing gums and ice creams. It finds application in pharmaceutical industry as an excipient as well as plasticizer in gelatin capsules, emollient and humectant.[17]

2.3 Cellulose and its derivatives

Cellulose, one of the most abundant naturally occurring polysaccharide on earth, is the major structural polymer in the cell walls of higher plants, green algae, fungi, grasses etc. Being a high molecular weight polymer of (1→4)-linked β-D-glucopyranosyl residues, it has an extended ribbon-like conformation which are stacked parallel to each other in Cellulose I and anti-parallel in Cellulose II. Due to strong hydrogen bonding between different chains (chain stacking, Fig. 15.2A), cellulose neither melts nor dissolves readily in common solvents and hence of lesser pharmaceutical importance. However, cellulose can be functionalized by introducing desired functional groups, leading to steric hindrance between the chains and resulting increase in hydrodynamic volume (Fig. 15.2B). Consequently, different derivatives of cellulose are soluble in aqueous and organic solvents and are currently in use for diverse applications including assorted pharmaceutical products as given in Table 15.1[18–20]

2.4 Exudate gums

Exudate gums are one of the oldest industrial polysaccharides, usually exuded from trees, grown in arid regions that exhibit high water retention capacity. The industrially important exudate gums *viz.* gum arabic, gum ghatti, gum karaya, gum tragacanth have wide applications in food and feed, pharmaceuticals, etc. (Table 15.2). They are characterized by relatively high amount of uronic acids which are mainly present as calcium and magnesium salts. Gum arabic and gum ghatti are water soluble gums and demanded in applications where low viscosity solutions are required. However, gum karaya and gum tragacanth result in higher viscosity and low solubility in aqueous solutions. These gums absorb large volume of water and form gels. Plant exudates have been known for their remarkable pharmacological activities *viz.* anthelmintic, antimicrobial, anti-inflammatory, antinociceptive, anti-ulcerogenic, antioxidant, wound healing, etc.[21]

2.5 Glycoconjugates

Many bioactive oligosaccharides and polysaccharides are produced as exopolymers by several fungi and algal species in response to defence mechanism

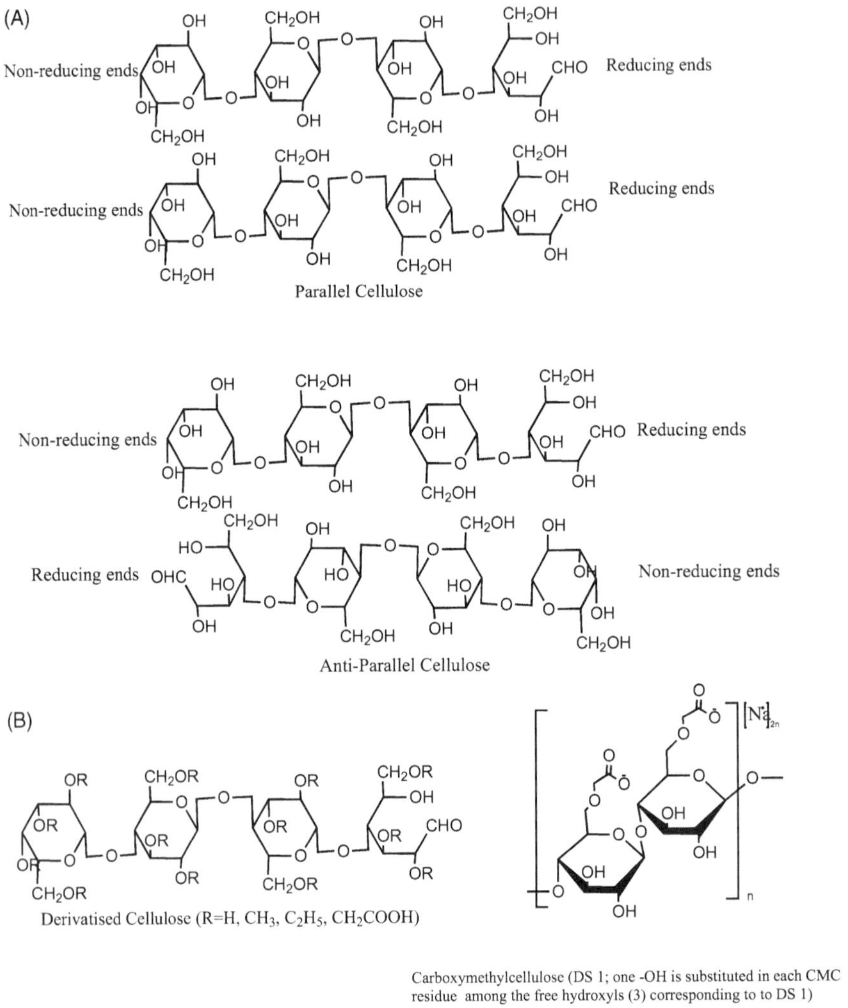

Figure 15.2 (A) Cellulose chains (symbolic representation). (B) Derivatized cellulose (symbolic representation).

against the surroundings, thereby, exhibiting anticancer, anti-ulcer, anti-thrombotic, cardioprotective, anti-dengue, hepatoprotective properties, etc. These bioactive carbohydrates demonstrate different mechanisms of action, which include either exhibition of biological activity by the biopolymer itself or its involvement in complex reactions as a precursor, leading to formation of a product illustrating the particular activity.[22] The present market is flooded with glycoconjugates as antibacterial, anticancer, antimicrobial,

Table 15.1 Diverse applications of cellulose derivatives.

Product	Functional Group	DS (Degree of Substitution)	Solubility	Application
Carboxymethyl cellulose	$-CH_2COONa$	0.5–2.9	Water	Pharmaceuticals, Coatings, paints, adhesives
Ethyl cellulose	CH_2CH_3-	0.5–0.7	4% aq. NaOH	Pharmaceutical industry
		0.8–1.7	Cold water	
		2.3–2.6	Organic solvents	
Methyl cellulose	CH_3-	0.4–0.6	4% aq. NaOH,	Bioadhesives, Films, food and tobacco industry
		1.3–2.6	Cold water	
		2.5–3.0	Organic solvents	
Hydroxyethyl cellulose	CH_2CH_2OH-	0.1–0.5	4% aq. NaOH	Bioadhesives, films, coatings, and cosmetics
		0.6–1.5	Water	
Cellulose acetate	–OAc	0.6–0.9	Water,	pH sensitive membranes and controlled release coating of pharmaceuticals
		1.2–1.8	2-methoxy ethanol,	
		2.2–2.7	Acetone,	
		2.8–3.0	Chloroform	
Cellulose nitrate	NO_2-	1.8–2.0	Ethanol,	topical anti-wart solution and micro-porous membrane filters in pharmaceutical industries
		2.0–2.3	Methanol, acetone	
		2.2–2.8	Acetone	

Table 15.2 Pharmaceutical applications of exudate gums.

Application	Gum arabic	Gum ghatti	Gum karaya	Gum tragacanth
Drug tablets	Encapsulation agent	Encapsulation agent	Encapsulation agent	-
Pharmaceutical emulsion	Emulsifier	-	Emulsifier	Emulsifier
Denture adhesive	-	-	Denture adhesive	Denture adhesive
Colostomy seal	-	-	Colostomy seal	-

antiviral and antiparasitic drugs (Table 15.3) and it has been established that the efficacy of the drug is due to presence of glycan part.[9,23]

3 Plant oligosaccharide based molecules

Since past few decades, oligosaccharides have gained the status of biologically active molecules and extended their application as prebiotics, drug carriers, plant elicitors, immunomodulators, antioxidants, antiangiogenic, antithrombotic, anti-infectives, blood sugar level regulators in diabetes etc.[24,25]. They are isolated from plants, fungi and algae as well as synthesized either by degradation of higher polysaccharides or by polymerization of required monosaccharide moieties. The oligosaccharides utilized as prebiotics are non-digestible oligosaccharides (NOD) [e.g. galacto-oligosaccharides (GOS, Stachyose and Raffinose; Fig. 15.3A and B), gluco-oligosaccharides, fructo-oligosaccharides (FOS), xylo-oligosaccharides (XOS), β-glucan oligosaccharides, gentio-oligosaccharides (GeOS), pectin derived acidic oligosaccharides (PaOS), arabino-oligosaccharides (AOS)] obtained from soyabean, oatmeal starch, sugar beet arabinan, gram husk, barley hulls, wheat bran, almond shells, straw, aspen wood, brewery spent grains, bamboo, corn cob, etc.[26] Prebiotics cover a class of dietary fibres that resists hydrolysis by human alimentary enzymes but are fermented by colonic microflora. The main physiological effects of dietary fibre are: (1) primarily on gastric emptying and small intestinal transit time, resulting in an improved glucose tolerance and a decreased digestion of starch, (2) on colonic transit time and large bowel functions due to fermentation by ceco-colonic microbial flora or bulking action. The soluble dietary fibres are fermented to a large extent by a wide variety of anaerobic bacteria that result in an increase in bacterial biomass, an increase in fecal mass, a change in intracolonic pH, and production of short chain fatty acids and various gases as metabolic end products. The insoluble fibres are only marginally fermented; they serve almost exclusively as bulking agents that result in shorter transit time and increased fecal mass. The short chain fatty acids resulting from the colonic fermentation of

Table 15.3 Mode of actions of bioactive glycoconjugates.

Glycoconjugates	Mode of action	Activity	Source	Drugs
Amicetins	Interact with RNA	Universal antibiotics (against herpes virus and poliovirus)	*Streptomyces fasciculatus*	Amicetin
Aminoglycosides	Interact with RNA	Anti-infectives	*Streptomyces and Streptoverticillis strains*	Streptomycin, Gentamicin, Kanamycin
Angucyclines	Interact with Proteins	Antibiotics	*Streptomyces sp.*	Urdamycin A, Landomycin A
Anthracyclines	Interact with DNA	Antitumor antibiotics	*Streptomyces sp., Micromonospora sp., Actinomadura sp., Nomadura sp. Actinosporangium sp., Chaetomium sp., Actinoplanes sp., Ampullariella sp., Nocardia sp.*	Daunomycin, Doxorubicin, Aclacinomycin A, Isoquinocycline A Kosinostatin, Viriplanins A & D
Aureolic Acids	Interact with DNA	Antimicrobials (against Gram-positive and mycobacteria)	*Streptomyces* sp., *Actinoplanes sp.*	Chromomycin A3, Mithramycin, Chromomycins, Olivomycins, Durhamycins, Chromocyclomycin, Variamycins
Avermectins	Interact with Proteins	Against endo- and ectoparasites	*Streptomyces avermitilis, Streptomyces hygroscopicus, Amycolatopsis* sp., *Nocardia sp.*	Avermectins, Milbemycins
Benzoisochromanequinones	Interact with Proteins	Antibacterial	*Streptomyces olivaceus*	Granaticin

(*Continued*)

Table 15.3 Mode of actions of bioactive glycoconjugates. (*Cont.*)

Glycoconjugates	Mode of action	Activity	Source	Drugs
Bleomycins	Interact with DNA	Against malignant lymphomas, pulmonary fibrosis	*Streptomyces verticillus*	Phleomycins, Tallysomycins, Zorbamycin, Cleomycins
Cardiac Glycosides	Interact with Proteins	Treatment of congestive heart failure	*Digitalis sp.*	Cardenolides, Digitoxin, Digoxin,
Coumarins	Interact with Proteins	Antibiotics	*Streptomyces spheroides*	Novobiocin, Isonovobiocin, Clorobiocin, Vanillobiocin, Novenamine, Isovanillobiocin, Declovanillobiocin
Diazobenzofluorenes	Interact with DNA	Anticancer	*Micromonospora lomaivitiensis*	Lomaiviticins
Enediynes	Interact with DNA	Antitumor antibiotics	*Micromonospora echinospora* *Streptomyces sp.*, *Actinomadura* sp., *Streptoalloteichus* sp., *Salinospora* sp., *Nocardiopsissp.*	Calicheamicin γ_1^{I}, Neocarzinostatin, Maduropeptin, Kedarcidin
Ginsenosides	Interact with Proteins	Anticancer	*Panax ginseng*	Ginsenosides
Indolocarbazoles	Interact with Proteins	Antitumor, antiviral	*Streptomyces* sp., *Staurosporeus* sp., *Saccharothrix aerocolonigenes*	Staurosporine, Rebeccamycin, Tjipanazoles, Akashins

Macrolides	Interact with RNA	Antibacterial (against Gram-positive bacteria)	*Streptomyces* sp., *Nocardia* sp., *Saccharopolyspora* sp., *Amycolatopsis sp.*	Methymycin, Pseudoerythromycin, Erythromycin A1, Oleandromycin, Spiramycin, Josamycin, Midecamycin
Non-Ribosomal Peptides	Interact with Cell Walls and Cell Membranes	Glycopeptide antibiotics	*Amycolatopsis orientalis*	Vancomycin
Orthosomycins	Interact with RNA	Antibacterial (against Gram-positive and some Gram-negative bacteria)	*Streptomyces* sp., *viridochromogenes Tu57 sp.*	Avilamycin A
Pluramycin	Interact with DNA	Antitumor antibiotics	*Streptomyces* sp., *Actinomyces* sp., *Actinomadura* sp., *Saccharotrix sp.*	Altromycin B, Hedamycin
Polyenes	Interact with Cell Walls and Cell Membranes	Antifungal	*Streptomyces nodosus*	Candicidin, Nystatin A1, Fattiviracins, Moenomycins, Arthrobacillins
Saccharomicins	Interact with Cell Walls and Cell Membranes	Bacteriocidal antibiotics	*Saccharothrix espanaensis*	Saccharomicins A and B

dietary fibre are largely absorbed via the portal blood and reach both the liver and the peripheral tissues. They induce changes in glucose and fat metabolism leading to post-prandial hypoglycemia and long-term hypolipidemia.

Most oligosaccharides are non-toxic and can be utilized in food, feed and pharmaceuticals like proteins and polysaccharides. However, there are some facts that make oligosaccharides exceptionally different from proteins and polysaccharides. Firstly, they are of smaller size with enough hydroxyl groups making it highly polar. Secondly, they possess high thermal stability and remain unaltered in presence of alcohol unlike proteins and polysaccharides, which implies that they can withstand higher temperatures and can be easily sterilized without any loss of bioactivity.[25] These properties of oligosaccharides have been widely utilized as excipients and in drug delivery systems due to their higher aqueous solubility over drugs. Cyclodextrins (Fig. 15.3C)

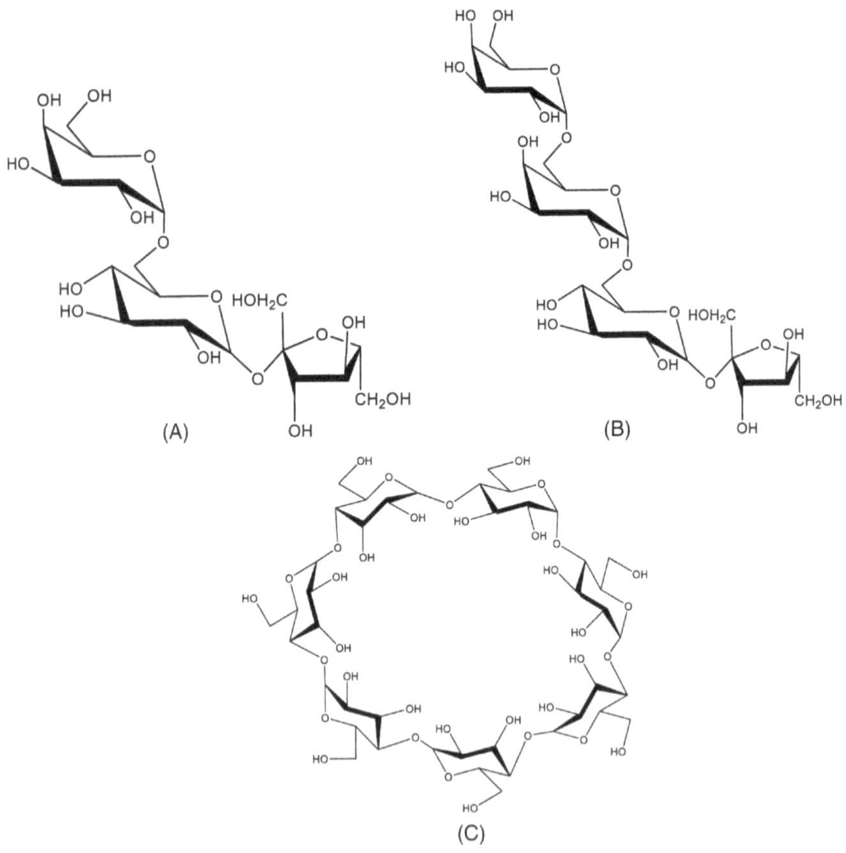

Figure 15.3 (A) Raffinose (B) Stachyose (C) Cyclodextrans

and their derivatives exhibit cyclic cage like structures that aid in stabilization and solubilization of active molecule and allow the drug penetration through polar cellular membranes and enhance drug efficacy.[27] Another remarkable property of cyclodextrins is their ability to form self-aggregations and non-inclusion complexes that allow encapsulation of even bigger drug molecules as well as vitamins and prevent their interaction with surroundings, thereby, preventing their deterioration and enhancing their shelf life.

4 Plant polysaccharide based drugs

Plant polysaccharides are renewable, biodegradable primary metabolites, which are utilized enormously in food and pharmaceuticals. Though there are several other bioactive secondary metabolites that possess equivalent bioactivity as polysaccharides, however, what make polysaccharides distinct from them is their attributes of hydrophilicity, biocompatibility and presence of free hydroxyl groups that enable them to bind with receptors more efficiently. Numerous species have been screened and investigated for bioactive polysaccharides *in vitro* and *in vivo* and some of the polysaccharides have been established for potent anti-cancer, hepatoprotective, antithrombotic, antidiabetic, immunomodulatory, anti-dengue drugs such as β-glucans, λ-carrageenan, fucoidans, galactomannans (Fig. 15.4) etc. to name a few as illustrated in Tables 15.4 and 15.5.[28–34] Further, numerous polysaccharides extracted from plants, used in folk and traditional systems of medicine, have been investigated for complement modulating properties. Intriguingly, pectins, type II arabinogalactans, arabinans and other heteroglycans like glucuronoarabinoxylan were found to stimulate the complement system.[35] The complement system is a significant part of the immune defence against infections, and proteolytic cleavage of the complement components lead to generation of biologically active complement activation products which may increase local vascular permeability, attract leucocytes, mediate immune adherence and modulate antibody production.[36]

4.1 Antidiabetic activity

Plant polysaccharides have been reported to exhibit antidiabetic activity. The study carried out in our laboratory using polysaccharide isolated from *Acacia tortilis* gum exudates revealed α-glucosidase inhibitory activity. During investigation, α-glucosidases from *Saccharomyces cerevisiae* and rat small intestine were used as *in vitro* model to evaluate α-glucosidase inhibitory activity of *A. tortilis* polysaccharide against yeast as well as mammalian

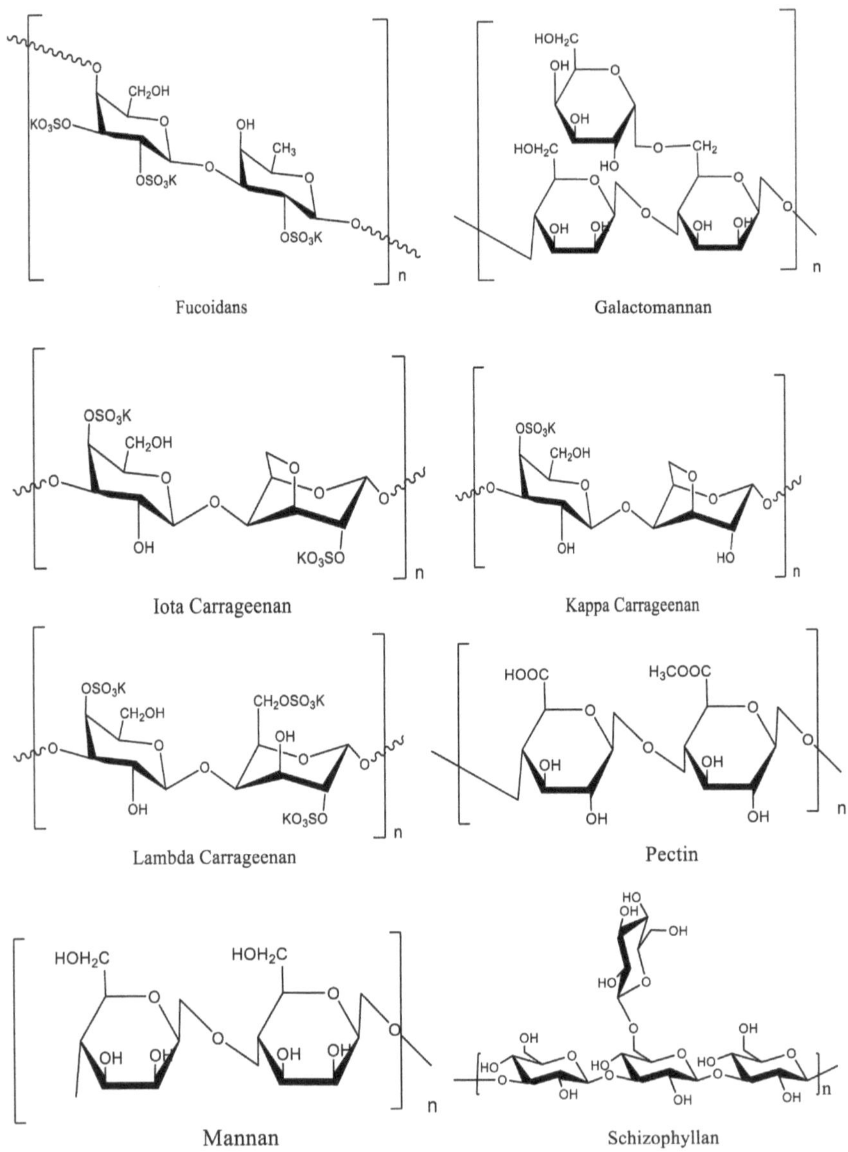

Figure 15.4 ***Bioactive polysaccharides.***

enzyme. The reduction in postprandial blood glucose level after carbohydrate rich diet fed to albino wistar rats was employed as *in vivo* model of α-glucosidase inhibition. The findings demonstrated the α-glucosidase inhibitory activity of isolated polysaccharide in *in vitro* both as well *in vivo* as models. Interestingly, unlike acarbose as standard drugs, the *Acacia tortilis*

Table 15.4 *In-vitro* activity of polysaccharides from different plant species.

S.No.	Species	Type	Polysaccharides	Bioactivity	References
1.	*Ascophyllum nodosum*	Anticancer	Fucose containing sulfated polysaccharide [Fuc:Xyl:Glc:Man:Gal as 1:0.95:0.02:0.2:0.04; 21.4% GlcA; 9.6% sulfate content]	Inhibition of U937 cells proliferation; Stimulated DNA-fragmentation and apoptotic nuclear morphological changes	40
2.	*Ascophyllum nodosum*	Anticoagulant/ anti-thrombotic	Fucoidan	Prolongation of activated partial thromboplastin time	41
3.	*Asparagus officinalis*	antioxidant and anti-tumor activities	Heteropolysaccharide Mw: 6.18×10^4 Da. glucose, fucose, arabinose, galactose and rhamnose in a ratio of 2.18:1.86:1.50:0.98:1.53.	Significant hydroxyl radical scavenging activity; Inhibition of HeLa and BEL-7404 cells growth	42
4.	*Buckwheat*	Anticancer	Not mentioned	THP-1 cell differentiation and maturity; Exhibited phagocytic activity and superoxide anion production	43
5.	*Cladosiphon okamuranus*	Anti-dengue	Fucoidan	Inhibition of DENV-2 multiplication in BHK-21 cells	44
6.	*Cladosiphon okamuranus tokida*	Anti-ulcer	Fucoidan	Possess anti-peptic activity without inflammation thereby causing gastric mucosal protection	45
7.	*Cordyceps gunnii*	Anticancer	Heteroglucan Mw:3.72×10^6 Da Rha:Ara:Xyl:Man:Glu:Gal = 3.0:2.6:1.0:1.3:106.0:2.8	Inhibition of tumor growth in K562 cells	46

(*Continued*)

Table 15.4 *In-vitro* activity of polysaccharides from different plant species. (*Cont.*)

S.No.	Species	Type	Polysaccharides	Bioactivity	References
8.	*Cordyceps sphecocephala*	Anticancer	polysaccharide-peptide complexes 88.3% Protein content; Mw: 1.38 kDa; Man (56.1%), Gal (37.5%), Glc (5.4%).	Apoptosis of Human Hepatocarcinoma (HepG2) and Neuroblastoma (SKN- SH) Cells via DNA fragmentation, caspase-3 activation, Bcl-2 and Bax modulation	47
9.	*Costaria costata*	Anticancer	Sulfated galactofucan [18.9% sulfate content; Fuc:Gal:Man:Rha:Xyl::1:0.83:0.01:0.05:0.06]	Inhibition of colony formation of DLD-1 cells as well as SK-MEL-28 SK-MEL-28 cells, DLD-1 cells inhibited cell colony formation	48
10.	*Cryptonemia crenulata*	Anti-dengue	Sulfated Galactan	Inhibition of DENV-2 multiplication in Vero cells	49
11.	*Curcuma kwangsiensis*	Anticancer	Heteropolysaccharide with fructose, xylose, mannose, glucose and galacose in mole % of 25.0, 25.0, 10.0, 12.5 and 12.5, respectively	Inhibition of CNE-2 cells proliferation via apoptosis mediated by Bcl-2 expression attenuation and p53 expression promotion	50
12.	*Ecklonia cava*	Anti-inflammation	Fucoidan	Suppression of prostaglandin-E2 (PGE2)and NO production; Inhibition of nitric oxide synthase and cyclooxygenase-2 expression in LPS-stimulated RAW 264.7 cells	51
13.	*Ecklonia cava.*	Anti-proliferation/ anticancer	Fucoidan (sulfated polysaccharide)	Selective inhibition of tumor growth in HL-60 and U-937cells	52

14.	*Ecklonia cava.*	Anticoagulant	Sulfated polysaccharides [Fuc (82.1%), GlcN (0.52%), Gal (16.7%), Glc (0.52%), Man (0.07); Degree of sulfation (0.92)]	Preservation of activated partial thromboplastin time, thrombin time and prothrombin time	53
15.	*Eclonia cava*	Anticancer	Sulfated rhamnogalacto-fucan [19.1% sulfate content; Fuc:Gal:Man:Rha::1:0.21:0.05:0.16]	Inhibition of colony formation of DLD-1 cells as well as SK-MEL-28 SK-MEL-28 cells, DLD-1 cells inhibited cell colony formation	48
16.	*Flammulina velutipes*	Anticancer	Heteroglucans- FVP-1 [Mw: 28 kDa, 1.56% UA, 0.09% sulfate, Glc (81.3%), Fuc (3.0%), Man (3.6%), Gal (12.1%)] FVP-2 [Mw: 268 kDa, 3.42% UA, 0.14% sulfate, Glc (57.9%), Fuc (5.5%), Xyl (9.5%), Man (15.1%), Gal (12%)]	Inhibition of growth of BGC-823 cells proliferation	54
17.	*Fomes fomentarius*	Anticancer	Not mentioned	Anti-proliferation of human gastric cancer cell lines SGC-7901 and MKN-45	55
18.	*Fucus evanescens*	Anticancer	Fucoidan	Apoptosis enhancement of MT-4 cells by fucoidans	56
19.	*Fucus vesiculosus*	Anticancer	Fucoidan	Enhancement of binding ratio of NY-ESO-1 (expressing human cancer cells) to human dendritic cells; Stimulation of CD8+ T cells by fucoidan treated dendritic cells	57

(*Continued*)

Table 15.4 *In-vitro* activity of polysaccharides from different plant species. (*Cont.*)

S.No.	Species	Type	Polysaccharides	Bioactivity	References
20.	*Fucus vesiculosus*	Antitumor	Fucoidan	Induced apoptosis of Human leukemic cells *i.e.* HL-60, NB4, and THP-1 cells via activation of MEKK1, MEK1, ERK1/2, and JNK and production of NO	58
21.	*Fucus vesiculosus*	Anticancer	Fucoidan	Inhibition of growth of AGS cells via autophagy and apoptosis through downregulation of antiapoptotic Bcl-2 and Bcl-xL expression	59
22.	*Ganoderma lucidum*	Antitumor	GP-1 [Mw: 1.926 kDa; Man(3.10%), Rha(0.53%), Glc (60.11%) Gal (30.58%), Fuc (5.67%)] GP-2 [Mw: 1086 kDa; Man(9.89%), Rha(0.35%), GlcA(1.46%), Glc (68.04%), Gal (15.81%), Fuc (4.45%)]	Inhibition of cell proliferation in human breast cancer cell (MDA-MB-231) and activation of macrophage cell (RAW 264.7)	60
23.	*Ganoderma lucidum*	Anti- angiogenesis	Not mentioned	Anti-angiogenesis through the inhibition of secretion of VEGF and TGF-β1 from prostate cancer cells	61
24.	*Gastrodia elata Bl.*	Anticancer	Glucan [Mw: 7.0×10^5 Da; α-(1 → 4)-glc*p* & α-(1 → 4,6)-glc*p*]	Anti-proliferation of PANC-1 pancreatic cancer cells	62

25.	*Ginkgo biloba sarcotesta*	Antitumor	4 Heteropolysaccharide fractions GBEPP11 [Mw: 3.4×10^3 Da; Rha:Glc:: 1.9:1], GBEPP22 [Mw: 4.8×10^4 Da; Rha:Glc:Gal:: 37.01:1:8.46], GBEPP33 [Mw: 3.1×10^5 Da , Rha:Glc:Gal:: 2.64:1:1.43], GBEPP44 [Mw: 9.5×10^5 Da; Rha:Glc:Gal:: 16:10:0.96]	Inhibition of growth of U937 cells Inhibition rate increased with decreasing MW Acetylated GBEPP11 exhibited higher inhibition rate than that of GBEPP11	43
26.	*Grifola frondosa*	Antitumor	β- glucan	Inhibition of growth of prostatic cancer cells via oxidative stress and induced apoptosis	63
27.	*Gymnogongrus griffithsiae*	Anti-dengue	Kappa carrageenan	Inhibition of DENV-2 multiplication in Vero cells	49
28.	*Gymnogongrus torulosus*	Anti-dengue	Galactan	Inhibition of DENV-2 multiplication in Vero cells	64
29.	*Gynostemma pentaphyllum Makino*	Antitumor	Rhamnoxylan artificially Sulfated [Mw: 8.96 kDa; Rha:Xyl::1:12.25; C(17.21%), H (3.37%), S(13.49%)]	Inhibition of growth of HepG2 cells and Hela cells and mediating cell cycle arrest	65
30.	*Hedysarum polybotrys Hand.-Mazz*	Antitumor	Glucan [Mw: 9.4×10^4 Da; α-(1 → 4)-glc*p* & α-(1 → 4,6)-glc*p*]	Inhibition of human hepatocellular carcinoma HEP-G2 cells and human gastric cancer MGC-803 cells proliferation	66

(*Continued*)

Table 15.4 *In-vitro* activity of polysaccharides from different plant species. (*Cont.*)

S.No.	Species	Type	Polysaccharides	Bioactivity	References
31.	*Hizikia fusiforme*	Anticoagulant/ anti-thrombotic	Two Fucose containing sulfated polysaccharide **pD-I** [Mw: 42 kDa; 2.3% GlcA; 26.2% Sulfate content; Fuc (67%), Man (2%), Gal (31%)] **pD-II** [Mw: 95 kDa; 1.0% GlcA; 31.6% Sulfate content; Fuc (77%), Xyl (2%), Man (11%), Gal (11%)]	Prolongation of activated partial thromboplastin time	67
32.	*Hypsizigus marmoreus*	Anticancer and immunomodulatory	Heteroglucan artificially Sulfated [Mw: 7×10^6 Da Glc (88.4%), Gal (11.6%)]	Inhibition of growth of AGS cells and increase in nitric oxide (NO) and cytokine (IL-1β and TNF-α) production	68
33.	*Laminaria cichorioides*	Anticoagulant/ anti-thrombotic	Heterogeneous sulfated fucan [2,3-disulfated, 4-linked α-fucose unit]	Inhibition of thrombin by heparin cofactor II	69
34.	*Laminaria guryanovae*	Anticancer	Fucoidan	Inhibition of phosphorylation of epidermal growth factor receptor (EGFR) and EGF-induced the *c-fos* and *c-jun* transcriptional activities in JB6Cl41 cells	70
35.	*Laminaria japonica*	Anticancer	WPS-2-1[Mw: 80 kDa; Man:Rha:Fuc:: 1.0:2.3:1.2]	Higher antitumor activities against A375 and BGC823 cells and low cytotoxicity	71

36.	*Lentinula edodes*	Anticancer and immunomodulatory	Neutral Polysaccharide [Rib (2.3%), Ara (1.5%), Xyl (1.5%), Man (28.6%), Glc (55.9%), Gal (10.1)]	Suppression of proliferation of human breast cancer cells; Exhibit mitogenic and co-mitogenic activity	72
37.	*Lepista inversa*	Antitumor and antioxidant	Not mentioned	Inhibition of cell proliferation of Lung cancer cell line; Inhibition of lipid peroxidation along with radical-scavenging capacity and reducing power	73
38.	*Lycium barbarum*	hypoglycemic	Acidic polysaccharide [Mw: 2.25×10^6 Da; Rha:Ara:Xyl:Gal: Man:GalA:: 1.00:7.85:0.37:0.65:3.01:8.16]	Inhibition of oxidative damage in pancreatic islets cells Prolongation of cell survival ratio	74
39.	*Melia toosendan Sieb. Et Zucc*	Anticancer & antioxidant	pMTPS-3 [Mw:26100 Da; Ara:Glu:Man:Gal::17.3:28.3:41.6:12.6]	Inhibition of growth of human gastric cancer BGC-823 cells; Exhibit Superoxide, DPPH, hydroxyl & free radicals scavenging activities	75
40.	*Meristiella gelidium*	Anti-dengue	Kappa carrageenan Mw:425.6–956.7 kDa.	Inhibition of DENV-2 multiplication in Vero cells	76
41.	*Monostroma nitidum*	Anticancer & immunomodulatory	Sulfated polysaccharide [Mw: 1103 kDa; 13.5% Sulfate content; 16.8% Uronic acid; Rha (51%), Glc (48.2%), Xyl (0.8%)]	Exhibit direct cytotoxic effects on human cancer (AGS) cells; Macrophages & Raw 264.7 cells stimulation; Considerable NO and PGE_2 production	77
42.	*Orostachys japonicus*	Anticancer	Polysaccharides (Mw: 30-50 kDa)	Inhibition of proliferation and growth of HT29 human colon cancer cells along with apoptosis stimulation	78

(*Continued*)

Table 15.4 *In-vitro* activity of polysaccharides from different plant species. (*Cont.*)

S.No.	Species	Type	Polysaccharides	Bioactivity	References
43.	*Padina gymnospora*	Anticoagulant	Heterofucan [Mw: 18 kDa; Fuc:Xyl:UA:Gal:Sulfate::1:0.3:1.3:0.2:0.4]	Exhibit anticoagulant activity due to the presence of 3-*O*-sulfation at C-3 of fucose units	79
44.	*Phellinus ribis*	Anticancer	Polysaccharide artificially sulfated PRP-SIII [Mw: 21.42 kDa; C(16.29%), H (2.48%), S (13.05%)]	Inhibition of growth of HepG2 cells; Blocking of new angiogenic vessel formation in zebrafish assay	80
45.	*Pholiota dinghuensis Bi*	Antitumor	Heteroglucan [2.31% UA; 3.81% sulfate content; Rha (1.75%), Fuc (4.35%), Xyl (7.67%), Man (11.11%), Glc (67.83%), Gal (7.30%)]	Inhibition of human gastric cancer BGC-823 cells proliferation	81
46.	*Pleurotus geesteranus*	Anticancer	Glucan [Mw: 22.3 kDa]	Inhibition of growth of Human breast cancer cell line MCF-7	82
47.	*Pleurotus tuber-regium*	Antitumor	polysaccharide-protein complex further sulfated [55.1% carbohydrate content, 32.3% protein content, 2.3% sulfate content 10.7 x 10^4 Da]	Inhibition of tumor growth in S180 bearing mice	83
48.	*Pleurotus tuber-regium*	Antitumor	β-glucan carboxymethylated	Inhibition of tumor growth in human breast carcinoma MCF-7 cells along with Cell-cycle arrest and apoptosis	84

49.	*Punica granatum*	Anticancer & antioxidant	Galactomannan [Mw:110 kDa]	Inhibition of growth of MCF-7 (breast cancer) and K562 (leukemia) cells; Exhibit Superoxide, DPPH, hydroxyl, nitric oxide & free radicals scavenging activities	85
50.	*Saccharina japonica*	Anticancer	Acetylated sulfated galactofucan composed of (1→3)-α-l-fucose	Inhibition of Tumor growth in human breast cancer T-47D and melanoma SK-MEL-28 cell lines	86
51.	*Sargassum fulvellum*	Anticoagulant	Sulfated polysaccharide [8 kDa > Mw > 20 kDa; 29.70 μg mL^{-1} sulfate content]	Inhibition of both intrinsic and extrinsic blood coagulation pathways	87
52.	*Sargassum horneri*	Anticancer	Homofucan [fuc:Rha::1:0.11]; Sulfated rhamnofucan [16.9% sulfate content; Fuc:Rha:: 1:0.44]	Inhibition of colony formation of DLD-1 cells as well as SK-MEL-28 SK-MEL-28 cells, DLD-1 cells inhibited cell colony formation	48
53.	*Sargassum pallidum*	Anticancer & antioxidant	Polysaccharide (Mw < 50 kDa; 22.6% sulfate content)	Inhibition of Tumor growth in HepG2 cells, A549 cells, and MGC-803 cells; Exhibit DPPH, free radicals scavenging activity	88
54.	*Taxus yunnanensis*	Antitumor	Heteropolysaccharide [Mw: 17.37 kDa; Rha:Gal:Xyl::2.01:6.74:1]	Inhibition of growth of Hela and HT1080 cells	89

(*Continued*)

Table 15.4 *In-vitro* activity of polysaccharides from different plant species. (*Cont.*)

S.No.	Species	Type	Polysaccharides	Bioactivity	References
55.	*Undaria pinnatifida*	Antitumor/anti-proliferation/anticancer	*O*-acetylated sulfated galactofucan [Mw: 1246 kDa; α-fucose (50.9 mol%) and β-galactose (44.6 mol%); Sulfate (0.97 mol mol − 1); acetate (0.24 mol mol − 1) esters]	Exhibited antitumor activity against HepG2 (hepatocellular carcinoma), HeLa (cervical cancer), A549 (alveolar carcinoma) and PC-3 (prostate cancer), cells.	90
56.	*Undaria pinnatifida*	Immunomodulation	Fucoidan	Suppression in proliferation of murine macrophage RAW 264.7 cells	91
57.	*Undaria pinnatifida*	Anticancer	Acetylated sulfated galactofucan composed of (1→3)- or (1→3);(1→4)-α-l-fucose	Inhibition of Tumor growth in human breast cancer T-47D and melanoma SK-MEL-28 cell lines	86
58.	*Undaria pinnatifida*	Anticancer	Fucoidan	Inhibition of cell proliferation in A549Hu549Hu49Human Lung Carcinoma Cells along with induced apoptosis	51
59.	*Undaria pinnatifida*	Antivirus	Galactofucan Sulfates [Mw:150 kDa; Gal:Fuc::1:1; 10.1% UA; Degree of sulfation:0.41]	Inhibition of cell proliferation in herpes viruses HSV-1, HSV-2 and HCMV cells	92
60.	*Zizyphus jujuba*	Anticancer	Deproteinized polysaccharide -2 fractions A (Mw: 1,43,108 Da) and B (Mw: 67,633 Da)	Inhibition of cell proliferation of melanoma cells; Production of apoptotic bodies and increase in caspase-3 and caspase-9 activity	93

Table 15.5 *in-vivo* activity of polysaccharides from different plant species.

S.No.	Species	Type	Polysaccharides	Bioactivity	References
1.	*Camellia sinensis (green tea)*	antidiabetic	Two Heteropolysaccharides **TPC-W** [Rha:Fuc:Ara:Xyl: Man:Glc:Gal:: 8.74:4.69:29.04:0.42: 7.11:14.10:35.89] **TPC-A** [Rha:Fuc:Ara:Xyl: Man:Glc:Gal:: 13.81:1.43:36.07:5.2 4:4.89:6.28:32.27]	Decrease in blood glucose and anti-glutamic acid decarboxylase antibody level in non-obese diabetic mice	107
2.	*Cladosiphon okamuranus*	Cardioprotection	Fucoidan	Reduction in myocardial damage in isoproterenol induced myocardial infarction in rats; Enhancement in creatinine phosphokinase, lactate dehydrogenase, alanine transaminase and aspartate transaminase levels	108
3.	*Isaria farinose B05*	Antitumor	WIPS1 [Mw:42 kDa; Man:Gal:Glc:: 8.0:4.8:1.0; 8% uronic acid], WEPS1 [Mw: 208 kDa; Man:Gal:Glc:: 21.6:4.7:1.0; 8.06% uronic acid]	Inhibition of tumor growth in S180 bearing mice	109
4.	*Lactarius deliciosus Gray*	Anticancer	Heteropolysaccharide [Mw: 1.1×10^4 Da; Man:Xyl::3:1] Composed of →6)-α-L-Man*p*-(1→, →2,6)-α-L-Man*p*-(1→, α -D-Xyl*p*-(3→	Inhibition of tumor growth in S180 bearing mice	110

(Continued)

Table 15.5 *in-vivo* activity of polysaccharides from different plant species. (*Cont.*)

S.No.	Species	Type	Polysaccharides	Bioactivity	References
5.	*Liriope spicata var. prolifera*	antidiabetic	Fructans [Mw: 4.29 kDa; β-(1→2)-fruc, β-(1→2,6)-fruc]	Reduction in decrease of the fasting blood glucose (FBG); Glucose tolerance enhancement in type-2 diabetic mice	111
6.	*Ophiopogon japonicus*	antidiabetic	β-D-fructan [Fru*f* (2→1), Fru*f* (2→6), Fru*f* (2→)]	Reduction of hyperglycemia, hyperinsulinemia and hyperlipidemia in the KKAy mice via InsR/IRS-1/PI3K/Akt/GSK-3/Glut-4 signaling pathway	112
7.	*Padina gymnospora*	Cardioprotection	Fucoidan	Cardioprotection in myocardial ischemia–reperfusion (I/R) bearing mice via regulation of the inflammation response through high-mobility group box 1 and NF-κB inactivation.	113
8.	*Sparassis crispa*	Antitumor	β-1,3-D-Glucan	Inhibition of growth of metastatic lung tumor in C57BL/6J mice; Inhibition of B16-F10 cell-induced angiogenesis in ICR mice	114
9.	*Trametes versicolor*	Antitumor	Not mentioned	Inhibition of interferon gamma expression in orthotopic CRPC PC3-LN4 tumors bearing C57BL/6 mice; Upregulation of T-regulatory cell; Enhancement of NK cell activity and CD4 count	115

polysaccharide maintained lipid profile, liver enzymes and was devoid of gastrointestinal side effects.[33] The case study also infers that basically, the polysaccharides in food are degraded by salivary and pancreatic amylases into oligosaccharides which are hydrolyzed by α-glucosidase in the brush borders of the small intestine to form glucose. The resulting postprandial hyperglycemia could induce the non-enzymatic glycosylation of various proteins leading to development of chronic complication. Therefore, control of postprandial plasma glucose levels is essentially required at early stages of Diabetes mellitus to reduce chronic vascular complications. Inhibition of enzymes involved in metabolism of carbohydrates including α-glucosidase is one of the therapeutic approaches for reducing postprandial hyperglycemia.[33] Based on such observations, a series of studies have been carried out to develop effective therapeutics for the control of Diabetes mellitus by isolation of polysaccharides from different plant species.[32] In further quest to understand the therapeutic potential of polysaccharides, the antidiabetic mechanism of dietary polysaccharides based on their gastrointestinal functions was proposed.[37] It was observed that the dietary polysaccharides isolated from natural sources, namely plants, fungi, algae, etc. were resistant to human digestion and absorption alongwith complete or partial fermentation in the large bowel, thereby, demonstrating antidiabetic ability. Further, dietary polysaccharides could relieve diabetes due to modification of gastrointestinal viscosity, satiety, anti-inflammation effect and large bowel fermentation.

4.2 Immunomodulatory activity

Polysaccharides have been reported to possess strong immunomodulatory activity. The major role of plant polysaccharides is to augment and/or activate immune responses of macrophage leading to immunomodulation and related therapeutic effects. Actually, human system possess innate (non-specific) and acquired (specific) immunity. Natural killer (NK) cells, complement system, macrophages, antigen presenting cells (APCs) and neutrophils make up the innate immune system, and mount an immediate non-specific response to foreign microbial agents.[38] The therapeutic effect of polysaccharides may be due to direct or indirect interactions with immune system components, activating assorted cellular and molecular events and is dependent upon a number of structural features like chemical structure, molecular weight, conformation, functional groups (acetyl and sulfate), and branching. In addition, the activity is also dependent upon the triple-helix conformation of the polysaccharide backbone.[34] Despite numerous polysaccharides reported to exhibit very good immunomodulatory activity as reported in Tables 15.4, 15.5 and 15.6, further studies of well-characterized polysaccharides are required to comprehend the immunomodulatory effects and develop precise immunomodulators for the benefit of mankind from medicinally important plants and fungi.[39]

Polysaccharides isolated from mushrooms exhibited potent antitumor and immunomodulating properties.[94–102] Decoctions and essences obtained from the medicinal mushrooms are mostly hot water extracts and polysaccharide rich fractions, which had been historically used as medicine in the Far East.[103] The very first scientific studies on antitumor activities of decoction and essences obtained from fruiting bodies of mushrooms belonging to the family *Polyporaceae* (*Aphyllophoromycetideae*) and a few other families, demonstrated host-mediated activity against grafted cancer like Sarcoma 180 in animals.[104] Consequently, the first three polysaccharide based major drugs developed from medicinal mushrooms consisted of β-glucans, *i.e.* lentinan from fruiting bodies of *Lentinus edode,* krestin from cultured mycelial biomass of *Trametes versicolor* (Turkwey Tail), and schizophyllan from the liquid cultured broth product of *Schizophyllum commune* (Split Gill) (Table 15.5). In continuum, mushrooms containing anticancer β-glucans were thoroughly investigated for *in vitro* and *in vivo* model systems in drug discovery[94–96,100,104] and polysaccharides possessing antitumor and immunomodulatory properties had been identified and developed as anticancer agents.

Biologically active polysaccharides isolated from higher Basidiomycetes mushrooms possess unique structural diversity in different species. The different strains of same Basidiomycetes species may have polysaccharides with diverse properties due to the presence of distinct protein molecules bound to the polysaccharide.[105] The proteoglucan krestin from the strain *Trametes (Coriolus) versicolor* CM-101 (Japan) has different structure and properties with that of polysaccharide-peptide (PSP) (China) isolated from submerged culture of the strain Cov-1 of the same species due to the dissimilar protein residues linked to same polysaccharide unit.[105] Further, the bio-active polysaccharides are present mostly as glucans having different types of glycosidic linkages, such as (1→3), (1→6)-β-glucans and (1→3)-α-glucans, but some are true heteroglycans. Many antitumor polysaccharides contain either only (1→3)-β-glucans or (1→3)-β-glucans in combination with (1→6)-β-glucans branching. The others mostly bind to protein residues as PSPcomplexes (PSPC).[106] The main source of antitumor polysaccharides appears to be fungal cell wall polysaccharides. A triple-helical tertiary

Table 15.6 *In vitro and in-vivo* activity of polysaccharides from different species.

S.No.	Species	Type	Polysaccharides	Bioactivity	References
1.	*Aconitum coreanum*	Anti-hepatocarcinoma	Crude polysaccharides	Inhibition of H22 tumor cells growth *in vitro* and *in vivo*; Significant lower cytotoxicity; H22 cells were arrested in G0/G1 phage	138
2.	*Actinidia eriantha*	Antitumor and immunomodulatory activity	Neutral Heteropolysaccharide with galactose, arabinose and fucose as major monosaccharides Mw: 1.13 x10^6 Da	Inhibition of tumor growth; Splenocytes proliferation; Incease in NK cell and CTL activity, IL-2 and IFN-γ production from splenocytes	139, 140
3.	*Agaricus blazei*	Antitumor	(1→3)-β-D-glucan	Inhibition of tumor growth in prostate cancer cells and in immunodeficient mice with PC3 tumor xenograft via an apoptotic pathway; Suppression of prostate tumor growth via antiproliferative and antiangiogenic mechanisms	117
4.	*Agaricus brasiliensis*	Anticancer	β-glucan	Inhibition of tumor growth in Ehrlich tumor-bearing mice by reducing tumor-induced IL-10 production and changing microenvironment of subcutaneous Ehrlich adenocarcinoma	116
5.	*Anemone raddeana*	Anti-hepatoma activity	Homogenous polysaccharide Mw: 83 kDa; 8.2% uronic acid	Inhibition of H22 tumor cells growth; Reversal of 5-FU induced hematological parameters to near normal level	141

(*Continued*)

Table 15.6 *In vitro and in-vivo* activity of polysaccharides from different species. (*Cont.*)

S.No.	Species	Type	Polysaccharides	Bioactivity	References
6.	*Angelica sinensis*	Antitumor	APS-1d [Mw: 5.1 kDa; Glc:Ara:: 13.8:1]	Inhibition of Hela cells proliferation and tumor growth in athymic nude mice; Exhibit apoptosis via Bcl-2 family protein expression regulation, mitochondrial membrane potential decline and cytosolic cytochrome c levels enhancement	142
7.	*Astraeus hygrometricus*	Antitumor	Heteroglucan with 96% glc	Inhibition of tumor growth in Daltons lymphoma-bearing mice via immunoactivation and cell apoptosis	143, 144
8.	*Astragalus membranaceus (Radix Astragali)*	Antidiabetic	Two heteroglucans APS-I [Mw:1,699,100 Da; Ara:Glc:: 1:3.45] APS-II [Mw: 1,197,600 Da; Rha:Ara:Glc:: 1:6.25:17.86]	Upregulation of cholesterol-7-α-hydroxylase and LDL-receptor gene expressions	145
9.	*Auricularia auricular-judae*	Antitumor	β-D-glucan; Mw: 2.88×10^5 Da (1 → 4)-β-D-glcp & (1 → 4,6)-β-D-glcp	Exhibited inhibition against Acinar cell carcinoma (ACC) proliferation; Inhibition of tumor growth in S180-bearing mice via cell apoptosis and up-regulation of Bax and down-regulation of Bcl-2 expression	118

10.	*Auricularia polytricha*	Anticancer	β-D-glucan Mw: 259kDa (1 → 3)- β-D-glcp & (1→ 3, 6)-β-D- glcp	Inhibition of tumor growth in S180-bearing mice	119
11.	*Boschniakia rossica*	Antitumor	Homogeneous polysaccharide Mw:2.2 × 10^4 Da	Inhibition of tumor growth in Sarcoma180 bearing mice; Exhibited synergistic effects with 5-Fluorouracil	146
12.	*Brassica napus L.*	Antitu mor	Heteropolysaccharide 2.5 × 10^4 Da glucose, galactose, mannose, arabinose and xylose, whose molar ratio was 1.0:1.45:0.44:3.22:0.03 uronic acid (8.09%)	Inhibition of tumor growth in Sarcoma180 and B16 melanoma-bearing mice. Activity mediated by immunomodulation, leukogenic and antianemic actions.	147
13.	*Camellia oleifera Abel.*	Antitumor and antioxidant activities	Heteropolysaccharide [w: 362 kDa; Rha:Fuc:Ara:Man:Gal:Glc as 4.05:11.62: 1.78:3.91:8.76:27.06]	Inhibition of tumor growth in Sarcoma180 bearing mice; Exhibited hydroxyl radical and superoxide radical anion scavenging effect	148
14.	*Chondrus ocellatus*	Anticancer	λ-Carrageenan 27.8% sulfate content, Mw: 15 kDa	Synergistic inhibition activity of 5-Fu on tumor growth in H22 bearing mice by improving the immunocompetence damaged by 5-Fu	127
15.	*Cladosiphon okamuranus Tokida*	Anticancer	Fucoidan	Inhibition of tumor growth in Sarcoma180 bearing mice; Stimulation of Murine RAW264.7 macrophages	124

(*Continued*)

Table 15.6 *In vitro and in-vivo* activity of polysaccharides from different species. (*Cont.*)

S.No.	Species	Type	Polysaccharides	Bioactivity	References
16.	*Codonopsis pilosula*	Antitumor and immunomodulatory	Not mentioned	Inhibition of tumor growth in H22 bearing mice; Stimulation of lymphocyte proliferation and activation of immune system	149
17.	*Cordyceps militaris*	Anticancer	selenium-protein polysaccharide	Inhibition of tumor growth in S180-bearing mice; Reduction in toxic effect of chemotherapy drug	150
18.	*Cordyceps militaris*	Immunostimulating	Polysaccharide [Mw: 127 kDa]	Enhancement of ROS, TNF-α, NO and phagocytic uptake in RAW264.7 and mouse peritoneal macrophages; Suppression of tumor growth in Melanoma bearing mice along with Macrophages activation	151
19.	*Cordyceps militaris*	Anticancer	Not mentioned	Inhibition of cell proliferation in A549Hu549Hu49Human Lung Carcinoma Cells along with induced apoptosis	152
20.	*Dahlia tubers _-inulin*	Anticancer	γ-Inulin Mw: 16kDa , β-D-(2→1) polyfructofuranosyl α-D-glucose	γ-Inulin established as photodynamic therapy adjuvant; Inhibition of tumor growth in B16BL6, MCA205 and FsaR-bearing mice	132

21.	*Dendrobium nobile Lindl*	Anticancer	Neutral polysaccharide with Mw: 92kDa Man, Glc,Gal,Ara in mol % 64.2 28.1 6.6 & 1.1% respectively.	Inhibition of tumor growth in HL-60 cells, S180 bearing mice	22
22.	*Dimocarpus longan Lour.*	antioxidant, immunity-modulatory and antitumor activities	Not mentioned	Hydroxyl radicals and a,a-diphenyl-1-picrylhydrazyl (DPPH) radicals scavenging activity; Inhibition of tumor growth in S180-bearing mice activated delayed-type hypersensitivity (DTH) response, macrophage phagocytosis and ConA-stimulated splenocyte proliferation	153
23.	*Fucus evanescens*	Antitumor and antimetastatic activities	Fucoidan, Mw: 20–40 kDa	Potentiation of antimetastatic activity in Lewis lung carcinoma bearing C57Bl/6 mice at dose 10 mg/kg; Potentiation of the toxic effect of Cyclophosphamide at dose 25 mg/kg	125
24.	*Fucus vesiculosus*	Anticancer	Fucoidan	Inhibition of growth of Lewis lung carcinoma cells (LCC) and melanoma B16 cells; Enhanced cytolytic activity of natural killer (NK) cells in C57BL/6JJCL mice	126

(*Continued*)

Table 15.6 *In vitro and in-vivo* activity of polysaccharides from different species. (*Cont.*)

S.No.	Species	Type	Polysaccharides	Bioactivity	References
25.	*Ganoderma lucidum*	Antitumor	Polysaccharide peptide [Mw: 512,500 Da; Rha:Xyl:Fruc:Gal:Glc:: 0.549: 3.614: 3.167: 0.556: 6.89]	Anti-proliferation of Human umbilical vein endothelial cell along with induced Vascular endothelial growth factor (VEGF) in human lung cancer cell	154
26.	*Ganoderma lucidum*	Antioxidant	Mannans	Elevation of antioxidant enzyme activities (SOD, CAT and GPx), Decline of IL-1β, IL-6 and TNF-α in cervical cancer bearing rats	155
27.	*Gracilaria lemaneiformis*	Antitumor and immunomodulatory	Sulfated galactan [Mw: 1.37 × 106 Da; 6.13% sulfate content; 4.5% UA]	Inhibition of growth of tumor cells in H22 hepatoma bearing ICR mice via splenocytes proliferation, macrophage phagocytosis and increased the level of IL-2 and CD8 + T cells	156
28.	*Grateloupia longifolia*	Antiangiogenic and antitumor	Galactans [Mw: 1000 ±100 kDa; 3,6-AG:6-methyl-Gal:2-methyl-Gal:Gal:: 0.96:0.35:0.05:7.48]	Inhibition of growth of human microvascular endothelial cells (HMEC1) and human umbilical vein endothelial cells (HUVEC); Inhibition of tumor growth in S180 bearing mice by reducing vascular density of tumor mass; Anti-angiogenesis induced by down-regulation of tissue factor.	131

29.	*Hizikia fusiforme*	Neuroprotection	Fucoidan	Inhibition of MPTP (1-methyl-4-phenyl-1, 2, 3, 6-tetrahydropyridine) -induced lipid peroxidation and decrease in antioxidant enzyme activity	157
30.	*Inonotus obliquus*	Anticancer and immuno-stimulating	α- fucoglucomannan [Mw: 1000 kDa]	Inhibition of tumor growth in B16F10-bearing mice Prolonged the survival rate	158
31.	*Leucaena leucocephala*	Anti-dengue	Galactomanan	In vitro Inhibition of DENV-1 in C6/36 cell culture assays; In vivo activity tested against yellow fever virus YFV-infected mice	159
32.	*Lycium barbarum*	Anticancer & anti-oxidant	LBPF5 [Mw: 5.3 × 10^4 Da; ara, man, xyl, glc and rha]	Inhibition of growth of human bladder carcinoma cell line BIU87 and BIU87 apoptosis stimulation; Exhibit Superoxide and DPPH radical scavenging activities	135
33.	*Mimosa scabrella*	Anti-dengue	Galactomanan	In vitro Inhibition of DENV-1 in C6/36 cell culture assays; In vivo activity tested against yellow fever virus YFV-infected mice	159

(Continued)

Table 15.6 *In vitro and in-vivo* activity of polysaccharides from different species. (*Cont.*)

S.No.	Species	Type	Polysaccharides	Bioactivity	References
34.	*Ornithogalum caudatum Ait*	Anticancer	4 heteropolysaccharides (OCAP-2-1 [Mw:102.1kDa], OCAP-2-2 [Mw:62.3kDa], OCAP-3-1 [Mw:46.4 kDa], OCAP-3-3 [Mw: 22.8kDa]) with monosaccharides Glc, Gal, Ara, Man, Xyl, GlcA & GalA.	Inhibition of tumor growth in S180 bearing mice and K562 cell line	136
35.	*Passiflora edulis*	Anticancer	Acidic Polysaccharide [Mw: 6.0 × 104 Da; GalA (44.2%), Ara (11.8%), Rha (10.6%), Glc (11.8%), Man (9.0%), Gal (6.1%), Xyl (3.6%), Rib (1.3%), Fuc (1.6%)]	Inhibition of tumor growth in S180 bearing mice	137
36.	*Patrinia scabra Bunge*	Antitumor	Not mentioned	Inhibition of tumor growth in cervical cancer cell U14 -bearing mice with reduced serum lactate dehydrogenase (LDH) activity	160

37.	*Phellinus gilvus*	Anticancer	Not mentioned	Inhibition of cell proliferation along with enhanced cell apoptosis; Inhibition of tumor growth and metastasis in TMK-1-bearing mice	161
38.	*Phellinus gilvus*	Antitumor	Not mentioned	Inhibition of cell proliferation in B16F10 melanoma cell line and melanoma bearing mice along with enhanced cell apoptosis	162
39.	*Phellinus igniarius*	Anticancer	Heteropolysaccharide [Xyl:Man:Fuc: Glc:Gal:: 2.3: 1: 6.4: 22.1: 19.83]	Inhibition of tumor growth in S180 and Hepatoma 22 bearing mice along with increased life span	133
40.	*Phellinus linteus*	Antitumor	Acidic polysaccharide	Inhibition of cell metastasis and cancer cell adhesion in melanoma bearing mice; Elevated macrophage NO production	163
41.	*Phellinus rimosus (Berk.)*	Radioprotective effect	Polysaccharide Protein Complex	Increase in leukocyte count, bone marrow cellularity, glutathione content in γ Radiation exposed Swiss albino mice	164
42.	*Pleurotus ostreatus*	Immunomodulating and antitumor	β- Glucans [→6)-β-D-Glcp-(1→, →3)-β-D-Glcp-(1→]	Suppression of tumor growth in Melanoma B16 bearing mice; Prolongation of life span	165

(*Continued*)

Table 15.6 *In vitro and in-vivo* activity of polysaccharides from different species. (*Cont.*)

S.No.	Species	Type	Polysaccharides	Bioactivity	References
43.	*Porphyridium cruentum*	Immunomodulation and antitumor	Sulfated polysacharide [Mw: 6.53kDa; 14.37% sulfate content]	in S180 bearing mice Activation of peritoneal macrophage along with increased lymphocyte proliferation	166
44.	*Prunella vulgaris L.*	Anticancer	Heteropoylsaccharide [Mw: 58kDa; Rha:Ara:Xyl:Man :Glc:Gal:: 3.46:49.32: 58.91:0.43:2.64: 3.11]	Inhibition of tumor growth in adenocarcinoma tumor-bearing C57BL/6 mice; Enhanced thymus and spleen indices	134
45.	*Pulsatilla chinensis*	Anticancer	Not mentioned	Inhibition of tumor growth in C6- glioma bearing mice as well as cells Prolongation of life span along with enhanced thymus and spleen indices	167
46.	*Rice Bran Fermented with Lentinus edodes*	Anticancer	Heteropolysaccharide [Ara:Gal:Glc:Man:Xyl:: 0.95:0.81:0.96: 1.00: 0.39; 32.2% UA]	Activation of natural killer (NK) cells and inhibition of tumor growth in S180 and B16/Bl6 melanoma bearing C57/Bl6 mice Prolongation of life span	168

47.	*Sargassum sp.*	Anticancer	Fucoidan	Inhibition of growth of Lewis lung carcinoma cells (LCC) and melanoma B16 cells; Enhanced cytolytic activity of natural killer (NK) cells in C57BL/6JJCL mice	126
48.	*Schisandra chinensis(Turcz.) Baill*	antitumor and immunomodulating	SCPP11 [Mw: 3.4 × 103 Da]	Inhibition of growth of Heps cells bearing mice; Enhanced immune system by elevating phagocytic activity and production of NO of RAW264.7	169
49.	*Thuja occidentalis*	Anticancer	Not mentioned	Elevation of enhanced the NK cell activity, antibody-dependent cell-mediated cytotoxicity and antibody-dependent complement-mediated cytotoxicity in B16F-10 melanoma bearing mice	170

conformation of medicinal mushroom β-(1→3)-glucans is known to be important for their immune-stimulating activity.

5 Structure-activity relationship

Recent polysaccharide investigations have contributed significantly in characterization of various antibiotic and antitumor agents. Bioassay guided isolation of active compounds showed that in many cases, polysaccharides are responsible for biological activity. It has been further observed that a wide range of polysaccharides with completely distinct structural features have been reported to possess equivalent activity and this fact unfolds diverse aspects of structure-activity relationship of polysaccharides. For instance, polysaccharides with antitumor action differ greatly in their chemical composition and configuration, as well as their physical properties. Antitumor activity is exhibited by a wide range of glycans extending from homo-polymers to highly complex heteropolymers (Tables 15.4–15.6).[99] Although it is difficult to correlate the structure and antitumor activity of complex polysaccharides, some relationships can be inferred. For instance, β-glucans[116–119] have been established as anticancer biopolymers since decades and lentinan is one such anticancer drug with backbone of (1→3)-linked β-D-glc*p* branching points at *O*-6 of (1→6)-linked β-D-glc*p* (Fig. 15.5).[120]

Beside the well-known antitumor β-(1→3)-glucans, a wide range of biologically active glucans with diverse structures have been described. The list includes laminarins and other β-glucans with similar linkage for the activity.[117,119,121] It is obvious that structural features such as β-(1→3) linkages in the main chain of the glucan and additional β-(1→6) branch points are needed for antitumor action. β-Glucans containing mainly (1→6) linkages have less activity. Earlier it was believed that the linkage and branching points might have been responsible for the activity, which was later on found contradictory when (1→4)-linked β-glucans with (1→6)-linked β-D-glc*p* branching were reported to possess potent anti-cancer activity.[118] If the linkage had been the reason for bioactivity, then cellulose and barley glucans with similar linkage should also have exhibited anticancer efficacy. Nevertheless, bioactivity was found to be a function of helical conformations of β-glucans and hydrophilic hydroxyl groups located exterior to the helix.[121] While studying the biological activity, it was observed that in the structure of schizophyllan tertiary conformation, active β-D-glucan was found to have a triple-strand right-winding structure.[122] Acidic

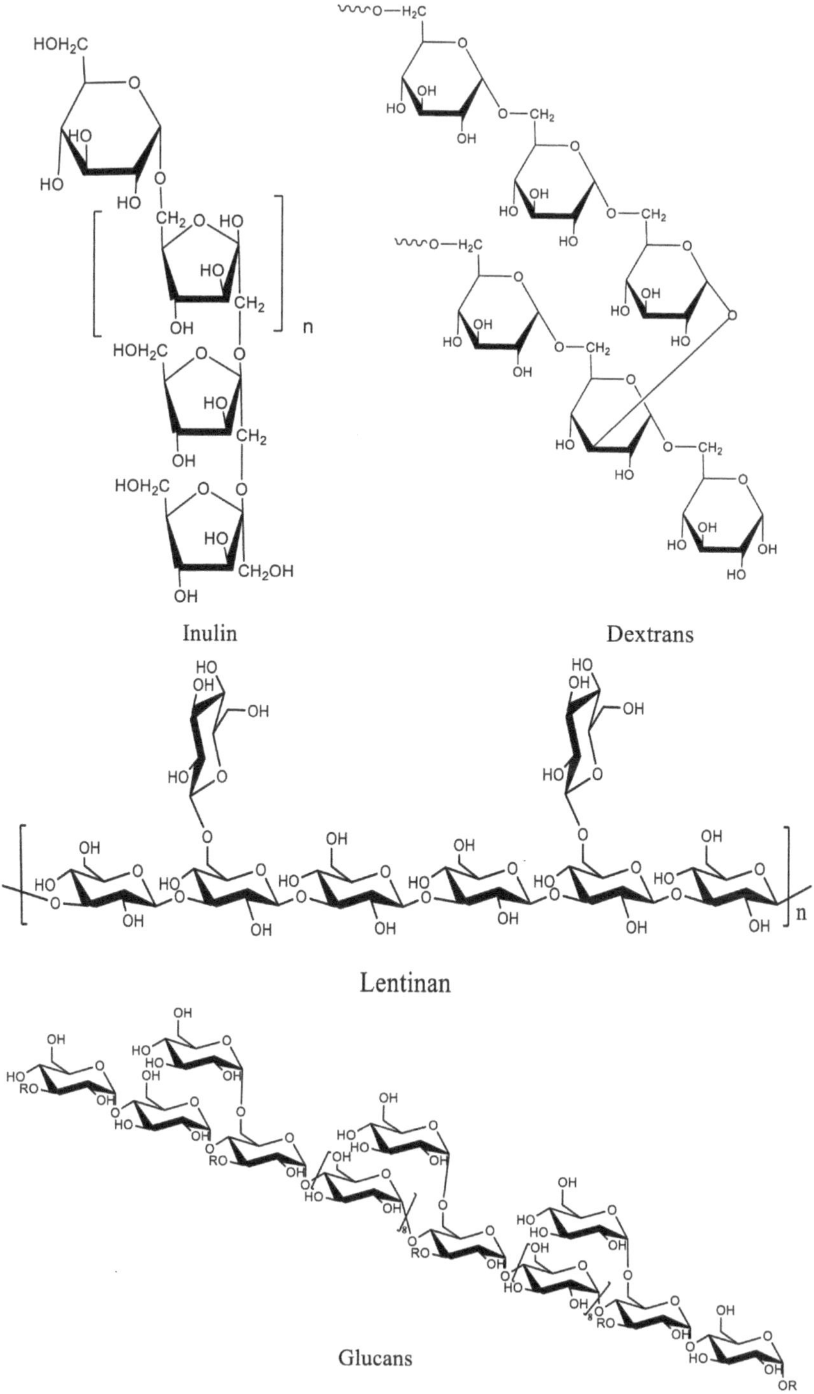

Figure 15.5 ***Structure of some neutral bioactive polysaccharides.***

glucuronoxylomannan isolated from the fruit body of *Tremella fuciformis* also demonstrated a left-handed, threefold helical backbone conformation.[123] However, obvious variation in antitumor polysaccharides have also been noted. In such cases, differences in activity can be correlated with solubility of polymer in water, size of the molecules, branching rate and form. High molecular weight glucans appear to be more effective than those of low molecular weight.[94–96]

In case of sulfated carbohydrate polymers *viz.* fucoidan,[40,59,124–126] λ-carrageenan,[127] sulfated rhamnofucan, sulfated galactofucan, sulfated rhamnogalactofucan[48], [86] and other sulfated heteropolysaccharides,[53,68,77,87] it was considered that the bioactivity is due to poly-anionic structure owing to the presence of sulfate content and glucuronyl branching. However, on one hand where sulfate content enhanced the bioactivity, uronic acid residues correlated inversely with activity.[128] On the contrary, several other polysaccharides show no direct correlation between sulfate content and anticancer activity and it has been found that fucoidan fractions devoid of sulfate groups have been reported to be equally potent as their sulfated counterparts.[48] Further, other neutral polysaccharides such as hetero-β-glucans,[46,54,97] heteroglycan,[129] β-glucan-protein,[130] α-manno-β-glucan, α-glucan-protein,[97] and heteroglycanprotein complexes,[94] galactans,[131] galactomannans,[85] γ-Inulin,[132] and heteropolysaccharides[72,75,133–135] have been found as peer contenders for anticancer activities. Some of these bioactive polysaccharides have been given in Fig. 15.5. These polysaccharides have linear or branched molecules in a backbone composed of α- or β-linked glucose units, and contain side chains that are attached to varied degree of branching. Heteroglucan side chains may contain glucuronic acid, xylose, galactose, mannose, arabinose or ribose as a main component or in different combinations. Further, contrary to the studies performed on fucoidans, it has been revealed that acidic polysaccharides containing only uronic acid illustrated inhibition of tumor cell growth and biopolymer conjugates with higher uronic acid content demonstrated higher anticancer activity.[136, 137] Hence, unlike proteins and secondary metabolites, the structural attributes of polysaccharides are so complex that it's extremely difficult to correlate that specific feature, which may justify a particular structure activity relationship. Though, development of *in silico* studies have aided in making the task simpler, nevertheless properties such as high molecular weight, complex geometry of the three dimensional structure and its interaction within the system (between constituent moieties) and with surroundings in terms of electrostatic attraction/repulsion, dispersion forces, hydrogen bonding, coiling and uncoiling due to hydrophobic and hydrophilic interactions etc. still need to be meticulously investigated to observe their effect on bioactivity of a carbohydrate polymer.

6 Polysaccharides in drug delivery

Though several potent drugs have been formulated against numerous diseases, they can play their role effectively only if they reach the target intact and conserved prior to their action. This awareness has shifted the focus from drugs to drug delivery systems and established their significance in making the drug more efficient. Carbohydrates are perfect models for controlled drug delivery systems owing to their biodegradability, biocompatibility, safety, low cost, low toxicity, high stability, amphiphillic nature and ability to undergo conformational alterations. Various drug delivery systems such as photosensitive, thermo-sensitive, timed release, pH-sensitive, microbially triggered, pressure regulated and target specific drug systems have been developed.[171] The target drug is attached to excipient/carrier *via* cross linking and then carried to the specific site in the body. Natural polysaccharides are cross linked to synthetic biodegradable polymers such as acrylic acid, polyethylene glycol, polyvinyl alcohol etc. to make systems like hydrogels, microcapsules, nanoparticles, microgels, nanogels that can encapsulate the required drug and then release it at the target in a controlled manner.[172] The cross-linking can be performed by a chemical reaction, irradiation, ionization of polymer or through ionic moieties. Some of the established drug delivery systems are modified hemicellulose–*N*-isopropylacrylamide, psyllium–acrylamide and 2-acrylamido-2-methylpropanesulfonicacid,[173] psyllium–acrylic acid,[174] xanthan–poly(acrylic acid)–poly(vinyl alcohol),[175] tara gum–acrylic acid,[176] etc.

7 Bioactive polysaccharides: Structural aspects

Bioassay guided isolation and fine structural characterization of active polysaccharides pave the way to structure activity relationship. Optimizing functions and understanding the biological roles of polysaccharides are directly dependent upon knowledge of their composition, structure, molecular weight distribution and the extent to which these properties affect polymer behaviour. However, the structure elucidation of polysaccharides involves series of steps *viz.* isolation, purification, fractionation and chemical characterization including degradation and derivatization. It is worth mentioning that same standard protocol cannot be applied to all polysaccharides

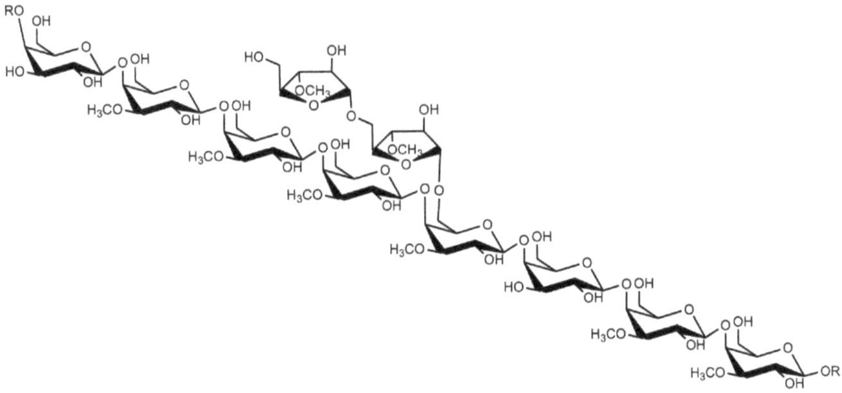

Figure 15.6 ***Chemical structure of methylated arabinogalactans.***

due to huge structural variability and therefore, one has to optimize each and every protocol depending upon the diverse structural requirements of sample. Isolation, purification and fractionation of polysaccharides are very complex and many times these objectives work contrary to each other. After isolation, as the sample is subjected to further stages of purification and fractionation, the yield decreases. Hence, one has to maintain a balance between yield, purity and homogeneity of sample.[177, 178] Intriguingly, these steps are critical and these parameters are essential attributes, leading to meticulous structural characterization. Chemical investigation is a next big step that includes determination of monosaccharide composition (through acidic and enzymatic degradation),[179,180] uronic acid content (by colorimetry and reduction)[181–183] linkage analyses (through methylation, hydrolysis and further derivatization),[184] conformation determination and linkage (by 1D and 2D homo and heteronuclear NMR measurements), sequencing and repeating unit of constituent monosaccharides in biopolymer (by partial hydrolysis and ESI-MS^n analyses).[185] Some of the complex structures of polysaccharides isolated and characterized in our laboratory are given in Fig. 15.6 and Fig. 15.7.

8 Conclusions

The transformation of a plant polysaccharide to a drug requires interdisciplinary amalgamation of polysaccharide chemistry for isolation, purification and structural characterization of polysaccharide; pharmacology to test the bioactivity *in vitro* and *in vivo* and define the dosage; biochemistry for structure activity relationship; toxicology to check whether the chemical moiety has toxicity; clinical trials to scrutinize the efficacy of drug

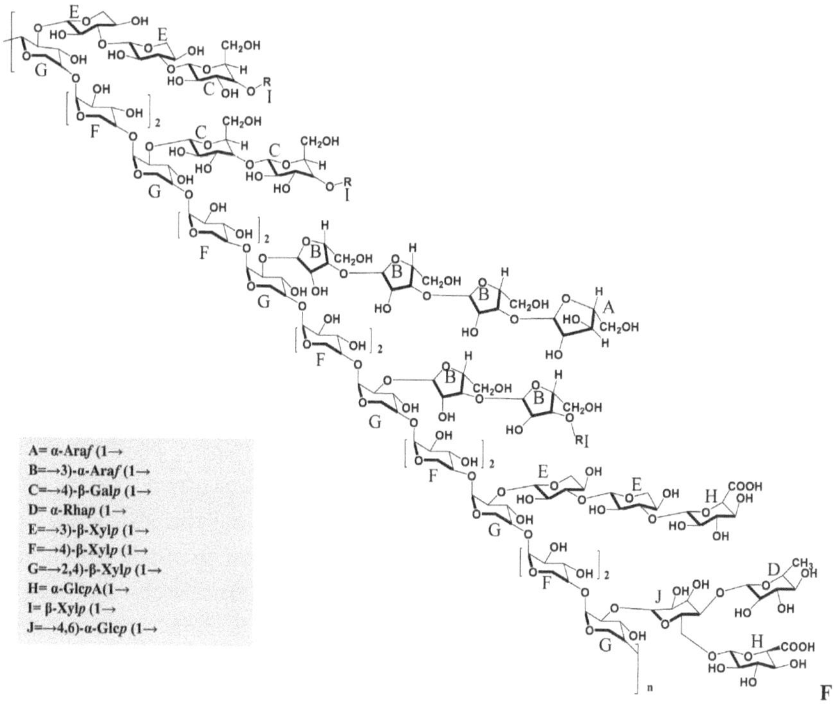

Figure 15.7 ***Chemical structure of xylan.***

and finally pharmaceuticals to scale up the process and shape the polysaccharide residue to tablet/capsule with suitable additive. Once the polysaccharide/oligosaccharide qualifies all these stages, the possibility of drug manufacturing and release depends upon the cost of investment required to scale up the isolation, purification and packaging processes. With passage of time, a decline in interest of screening and scrutiny of plants for drugs has occurred due to several reasons. Firstly, plant containing polysaccharides with established bioactivities (e.g. *Tinospora cordifolia, Aloe vera, Panax ginseng, Ganoderma lucidum, Acacia senegal),* are directly harvested from wild, which may lead to decline in their population as well as accessibility. Secondly, the yield of bioactive residues is usually low compared to dried mass of starting material and hence much large quantities of plant materials may be needed. Thirdly, as the bioactivity of any polysaccharide depends upon its functionalized structure and biosynthesis in plants at certain age, maturity and abiotic conditions including soil, weather, altitude etc., obtaining polysaccharides of exactly identical structure of its bioactive analogue may not be always possible unless the growth conditions of plants have been meticulously monitored and precisely optimized. It further requires a systematic study of variation of bioactivity and structure of polysaccharide with plant age and other changing parameters. Due to stated reasons, research efforts have been swung in the direction of synthesis of drugs. However, polysaccharide or even oligosaccharide structures are far tricky and complicated, first to understand and then to synthesize owing to number of stereogenic centres, conformational flexibility/rigidity, stereospecific linkages, different chain length, molecular weights with varying polydispersity indices and their ability to undergo chemical and biotechnological transformations etc. For instance pectins are bioactive acidic hetero polysaccharides and occur naturally with a wide array of varying galacturonic acid contents, neutral sugars (xylose, arabinose, rhamnose) and degree of *O*-methylation and *O*-acetylation.[10] Further, unlike natural products, synthesized drugs may be non-biodegradable with high/unknown toxicity levels and may lack diversity in their structures and hence the efficacies. On the contrary, there is a huge spectrum of plant polysaccharides which are bioactive in nature and yet unexplored. Therefore, there is still a vast scope of investigating flora and discovering new polysaccharides with unique structures and distinctive properties.

9 Future perspectives

In view of the above discussion, the following activities may be pragmatic in drug development from plant resources:

1. *Taxus, Undaria, Saccharina, Orostachys, Monostroma, Hedysarum, Gynostemma,* etc. are some of the plant genus, known for their significant anticancer and immunomodulatory bioactive polysaccharides like α-glucans, fucoidans, *O*-acetylated sulfated galactofucan, rhamnoxylan, sulfated heteropolysaccharides, etc. There is a scope to validate the structures and set some functional parameters (e.g., uronic acid or sulfate content, molecular weight range, polydispersity index, viscosity, degree of substitution) that could be experimented to categorize the polysaccharides as bioactive. Once such parameters are set, they could be modified to enhance their efficacy.
2. Medicinal mushrooms from different species of *Agaricus, Trametes, Ganoderma, Lentinus, Pleurotus, Schizophyllum* etc. have significant bioactive polysaccharides like lentinan, schizophyllan, krestin, etc. with well-defined

structures and potent biological activities leading to their usage in food and pharmaceuticals. Such leads may be extended to other unexplored mushroom species.

3. The structure-activity relationship of polysaccharides from higher plant species (gymnosperms and angiosperms) did not yield commercial drugs despite their vital biological activities. The impediment is primarily due to variability in structure with age, season, soil type, moisture and many other geographical parameters. Hence, there is a strong need to study variability in polysaccharide structure from same species and the effect of such variability on biological activity.
4. The species having potential polysaccharides demand their availability on sustainable basis. It can be realized by domestication of the important species. The diverse applications of seed polysaccharides, namely galactomannans from guar and *Cassia tora* gum in food and feed, pharmaceuticals, etc. has been possible, due to fulfilling the demand by farming. The approach needs to be extended to other potential species for their wider utilization.
5. Meticulous strategies may be developed toward conversion of abundant and relatively low-cost polysaccharides (namely, cellulose, starch, galactomannans, xyloglucans, etc.) to high valued carbohydrate polymers by chemical modification (carboxymethyl, hydroxypropyl, methyl, ethyl derivatives, etc.) or blending with other homogeneous and compatible polysaccharides for specific end use applications in pharmaceuticals.
6. Marine Microbial polysaccharides from extreme environments, namely, pelagic domains, polar regions hydrothermal vents, etc. owing to their unique properties may be explored for biomedical and pharmaceutical applications.
7. There is scope to develop facile strategies for culturing biologically active and/or chemically novel polysaccharides in the laboratory and scale-up facile productions for sustainable utilization as pharmaceutical drugs and other cosmeceutical products.
8. Structural analysis of interesting and novel polysaccharides from medicinal plant species has led to better understanding of the biological activity. However, in-depth polysaccharide structure analysis is tricky, lengthy and still ambiguous due to enormous complexity in structure. There is an exigency to develop facile approaches that are fast, reliable, and unambiguous along with development of novel, consistent, and fast biological activity protocols.

References

1. Stick, R.V.; Williams, S. Carbohydrates: the essential molecules of life. *Elsevier*: 2010.
2. Ali, M. *Bull. Ind. Inst. History of Medicine* **1998**, *28* 95-110.3.
3. Thakur, M.; Weng, A.; Fuchs, H.; Sharma, V.; Bhargava, C. S.; Chauhan, N. S.; Dixit, V. K.; Bhargava, S. *Carbohydr. Polym.* **2012**, *87*, 3–15.
4. Paulsen, B. S. *Phytochem. Rev.* **2002**, *1*, 379–387.
5. Thakur, M.; Dixit, V. K. *Ind. Pharmacist* **2005**, *4*, 7–12.
6. Nair, P. R.; Rodriguez, S.; Ramachandran, R.; Alamo, A.; Melnick, S. J.; Escalon, E.; Garcia, P. I., Jr.; Wnuk, S. F.; Ramachandran, C. *Int. Immunopharmaco.* **2004**, *4*, 1645–1659.
7. Deters, A. M.; Lengsfeld, C.; Hensel, A. *J. Ethnopharmacol.* **2005**, *102*, 391–399.
8. Akpinar, O.; Erdogan, K.; Bostanci, S. *Carbohydr. Res.* **2009**, *344*, 660–666.
9. Thorson, J. S.; Vogt, T.; Wong, C. *Carbohydrate-based drug discovery*. Wiley Online Library, 2003.
10. Liu, J.; Willför, S.; Xu, C. *Bioact. Carbohydr. Diet. Fibre* **2015**, *5*, 31–61.
11. De Cock, P.; Bechert, C. L. *Pure Appl. Chem.* **2002**, *74*, 1281–1289.
12. Moon, H. J.; Jeya, M.; Kim, I. W.; Lee, J. K. *Appl. Microbiol. Biotechnol.* **2010**, *86*, 1017–1025.
13. Kawanabe, J.; Hirasawa, M.; Takeuchi, T.; Oda, T.; Ikeda, T. *Caries Res.* **1992**, *26*, 358–362.
14. Kearsley, M. W.; Deis, R. C. *In Sweeteners and Sugar Alternatives in Food Technology; Ames: Oxford*. **2008**, 249–261.
15. Campbell, J.; Goheen, S.; Donald, P. In Recent trends for enhancing the diversity and quality of soybean products, *IntechOpen*: **2011**.
16. Hayes, C. *J. Dent. Educ.* **2001**, *65*, 1106–1109.
17. Rodríguez Trabado, A.; Cámara Hijón, C.; García-Trujillo, J. A.; Magriz Trascón, I.; Fernández Pereira, L. M. *Case Reports in medicine* **2017**, *2017*.
18. Kumar, ENVIS Forestry Bulletin 2010, pp 76-83.
19. Swatloski, R. P.; Spear, S. K.; Holbrey, J. D.; Rogers, R. D. *J. Am. Chem. Soc.* **2002**, *124*, 4974–4975.
20. Klemm, D.; Heublein, B.; Fink, H. P.; Bohn, A. *Angew. Chem. Int.* **2005**, *44*, 3358–3393.
21. Licá, I. C. L.; dos Santos Soares, A. M.; de Mesquita, L. S. S.; Malik, S. *Food Res. Int.* **2018**, *105*, 1039–1053.
22. Wang, X.; Quinn, P. J. *Progress in Lipid Res.* **2010**, *49*, 97–107.
23. Elshahawi, S. I.; Shaaban, K. A.; Kharel, M. K.; Thorson, J. S. *Chem. Soc. Rev.* **2015**, *44*, 7591–7697.
24. Courtois, J. *Curr Opin Microbiol.* **2009**, *12*, 261–273.
25. Zopf, D.; Roth, S. *The Lancet* **1996**, *347*, 1017–1021.
26. Patel, S.; Goyal, A. *World J. Microbiol. Biotechnol.* **2011**, *27*, 1119–1128.
27. Jansook, P.; Ogawa, N.; Loftsson, T. *Int. J. Pharm.* **2018**, *535*, 272–284.
28. Zong, A.; Cao, H.; Wang, F. *Carbohydr. Polym.* **2012**, *90*, 1395–1410.
29. Kadir, S. L. A.; Yaakob, H.; Zulkifli, R. M. *J. Nat. Med.* **2013**, *67*, 677–689.
30. Wijesinghe, W.; Jeon, Y. J. *Carbohydr. Polym.* **2012**, *88*, 13–20.
31. Ren, L.; Perera, C.; Hemar, Y. *Food Funct.* **2012**, *3*, 1118–1130.
32. Wang, P. C.; Zhao, S.; Yang, B. Y.; Wang, Q. H.; Kuang, H. X. *Carbohydr. Polym.* **2016**, *148*, 86–97.
33. Bisht, S.; Kant, R.; Kumar, V. *Int. J. Biol. Macromol.* **2013**, *59*, 214–220.
34. Ferreira, S. S.; Passos, C. P.; Madureira, P.; Vilanova, M.; Coimbra, M. A. *Carbohydr. Polym.* **2015**, *132*, 378–396.
35. Yamada, H.; Kiyohara, H., Complement-activating polysaccharides from medicinal herbs. In *Immunomodulatory agents from plants, Springer:* **1999**; pp 161–202.

36. Wagner, H.; Jurcic, K. *Methods Plant Biochem.* **1991**, *6*, 195–217.
37. Hu, J. L.; Nie, S. P.; Xie, M.Y. *J. Agric. Food Chem.* **2018**, *66*, 4781–4786.
38. Tan, B. K.; Vanitha, J. *Curr. Medicinal Chem.* **2004**, *11*, 1423–1430.
39. Ramberg, J. E.; Nelson, E. D.; Sinnott, R. A. *Nutr. J.* **2010**, *9*, 54.
40. Nakayasu, S.; Soegima, R.; Yamaguchi, K.; Oda, T. *Biosci., Biotechnol., Biochem.* **2009**, *73*, 961–964.
41. Chevolot, L.; Foucault, A.; Chaubet, F.; Kervarec, N.; Sinquin, C.; Fisher, A. M.; Boisson-Vidal, C. *Carbohydr. Res.* **1999**, *319*, 154–165.
42. Zhao, Q.; Xie, B.; Yan, J.; Zhao, F.; Xiao, J.; Yao, L.; Zhao, B.; Huang, Y. J. *Carbohydr. Polym.* **2012**, *87*, 392–396.
43. Wu, X.; Mao, G.; Zhao, T.; Zhao, J.; Li, F.; Liang, L.; Yang, L. *Carbohydr. Polym.* **2011**, *86*, 1073–1076.
44. Hidari, K. I.; Takahashi, N.; Arihara, M.; Nagaoka, M.; Morita, K.; Suzuki, T. *Biochem. Biophys. Res. Commun.* **2008**, *376*, 91–95.
45. Shibata, H.; Kimura-Takagi, I.; Nagaoka, M.; Hashimoto, S.; Aiyama, R.; Iha, M.; Ueyama, S.; Yokokura, T. *Biofactors* **2000**, *11*, 235–245.
46. Zhu, Z.Y.; Liu, N.; Si, C. I.; Liu, Y.; Ding, L. N.; Jing, C.; Liu, A. J.; Zhang, Y. M. *Carbohydr. Polym.* **2012**, *88*, 1072–1076.
47. Oh, J. -Y.; Baek, Y. M.; Kim, S. W.; Hwang, H.-J.; Hwang, H. S.; Lee, S. H.; Yun, J. W. J. *J. Microbiol. Biotechnol.* **2008**, *18*, 512–519.
48. Ermakova, S.; Sokolova, R.; Kim, S. -M.; Um, B.-H.; Isakov, V.; Zvyagintseva, T. *Appl. Biochem. Biotechnol.* **2011**, *164*, 841–850.
49. Talarico, L.; Pujol, C.; Zibetti, R.; Faria, P.; Noseda, M.; Duarte, M.; Damonte, E. *Antivir. Res.* **2005**, *66*, 103–110.
50. Zeng, J.; Dai, P.; Ren, L.; Song, B.; Chen, X.; Wang, X.; Wang, J.; Zhang, T.; Zhu, W. *Carbohydr. Polym.* **2012**, *89*, 1067–1072.
51. Boo, H. J.; Hyun, J. H.; Kim, S. C.; Kang, J. I.; Kim, M. K.; Kim, S.Y.; Cho, H.; Yoo, E. S.; Kang, H. K. *Phytother. Res.* **2011**, *25*, 1082–1086.
52. Athukorala, Y.; Ahn, G. N.; Jee, Y. -H.; Kim, G.Y.; Kim, S.; Ha, J. H.; Kang, J. S.; Lee, K.W.; Jeon, Y. -J. J. *Appl. Phycol.* **2009**, *21*, 307–314.
53. Athukorala, Y.; Jung, W. -K.; Vasanthan, T.; Jeon, Y. -J. *Carbohydr. Polym.* **2006**, *66*, 184–191.
54. Yang, W.; Pei, F.; Shi, Y.; Zhao, L.; Fang, Y.; Hu, Q. *Carbohydr. Polym.* **2012**, *88*, 474–480.
55. Chen, W.; Zhao, Z.; Li, Y. J. *Carbohydr. Polym.* **2011**, *85*, 369–375.
56. Philchenkov, A.; Zavelevich, M.; Imbs, T.; Zvyagintseva, T.; Zaporozhets, T. *Exp. Oncol.* **2007**, *29*, 181–185.
57. Hu, Y.; Cheng, S. C. S.; Chan, K. T.; Ke, Y.; Xue, B.; Sin, F. W.Y.; Zeng, C.; Xie, Y. *Biochem. Biophys. Res. Commun.* **2010**, *392*, 329–334.
58. Jin, J. O.; Song, M. G.; Kim, Y. N.; Park, J. I.; Kwak, J. Y. *Mol. Carcinogenesis* **2010**, *49*, 771–782.
59. Park, H. S.; Kim, G.Y.; Nam, T. J.; Deuk Kim, N.; Hyun Choi, Y. *J. Food Sci.* **2011**, *76*, T77–T83.
60. Zhao, L.; Dong, Y.; Chen, G.; Hu, Q. *Carbohydr. Polym.* **2010**, *80*, 783–789.
61. Stanley, G.; Harvey, K.; Slivova, V.; Jiang, J.; Sliva, D. *Biochem. Biophys. Res. Commun.* **2005**, *330*, 46–52.
62. Chen, X.; Cao, D.; Zhou, L.; Jin, H.; Dong, Q.; Yao, J.; Ding, K. *Carbohydr. Polym.* **2011**, *86*, 1300–1305.
63. Fullerton, S.; Samadi, A.; Tortorelis, D.; Choudhury, M.; Mallouh, C.; Tazaki, H.; Konno, S. *Mol Urology* **2000**, *4*, 7–13.
64. Pujol, C.; Estevez, J.; Carlucci, M.; Ciancia, M.; Cerezo, A.; Damonte, E. *Antiviral Chem. Chemother.* **2002**, *13*, 83–89.
65. Chen, T.; Li, B.; Li, Y.; Zhao, C.; Shen, J.; Zhang, H. *Carbohydr. Polym.* **2011**, *83*, 554–560.

66. Li, S. G.; Wang, D. G.; Tian, W.; Wang, X. X.; Zhao, J. -x.; Liu, Z.; Chen, R. *Carbohydr. Polym.* **2008**, *73*, 344–350.
67. Dobashi, K.; Nishino, T.; Fujihara, M.; Nagumo, T. *Carbohydr. Res.* **1989**, *194*, 315–320.
68. Bao, H.; Choi, W. -S.; You, S. *Biosci., Biotechnol., Biochem.* **2010**, *74*, 1408–1414.
69. Yoon, S. J.; Pyun, Y. R.; Hwang, J. K.; Mourão, P. A. *Carbohydr. Res.* **2007**, *342*, 2326–2330.
70. Lee, N. Y.; Ermakova, S. P.; Zvyagintseva, T. N.; Kang, K. W.; Dong, Z.; Choi, H. S. *Food Chem. Toxicol.* **2008**, *46*, 1793–1800.
71. Peng, Z.; Liu, M.; Fang, Z.; Wu, J.; Zhang, Q. *Carbohydr. Polym.* **2012**, *89*, 1022–1026.
72. Israilides, C.; Kletsas, D.; Arapoglou, D.; Philippoussis, A.; Pratsinis, H.; Ebringerova, A.; Hříbalová, V.; Harding, S. *Phytomed.* **2008**, *15*, 512–519.
73. Vaz, J. A.; Heleno, S. A.; Martins, A.; Almeida, G. M.; Vasconcelos, M. H.; Ferreira, I. C. *Food Chem. Toxicol.* **2010**, *48*, 2881–2884.
74. Zou, S.; Zhang, X.; Yao, W.; Niu, Y.; Gao, X. *Carbohydr. Polym.* **2010**, *80*, 1161–1167.
75. He, L.; Ji, P.; Gong, X.; Li, W.; Cheng, J.; Qian, H.; Song, X. *Int. J. Biol. Macromol.* **2011**, *49*, 422–427.
76. De Sf-Tischer, P. C.; Talarico, L. B.; Noseda, M. D.; Guimarães, S. M. P. B.; Damonte, E. B.; Duarte, M. E. R. *Carbohydr. Polym.* **2006**, *63*, 459–465.
77. Karnjanapratum, S.; You, S. *Int. J. Biol. Macromol.* **2011**, *48*, 311–318.
78. Ryu, D. S.; Baek, G. O.; Kim, E. Y.; Kim, K. H.; Lee, D. S. *BMB reports* **2010**, *43*, 750–755.
79. Silva, T.; Alves, L.; De Queiroz, K.; Santos, M.; Marques, C.; Chavante, S. F.; Rocha, H. A. O.; Leite, E. J. *Braz. J. Med. Biol. Res.* **2005**, *38*, 523–533.
80. Liu, Y.; Liu, C.; Tan, H.; Zhao, T.; Cao, J.; Wang, F. *Carbohydr. Polym.* **2009**, *77*, 370–375.
81. Gan, D.; Ma, L.; Jiang, C.; Xu, R.; Zeng, X. *Carbohydr. Polym.* **2011**, *84*, 997–1003.
82. Zhang, M.; Zhu, L.; Cui, S. W.; Wang, Q.; Zhou, T.; Shen, H. *Int. J. Biol. Macromol.* **2011**, *48*, 5–12.
83. Tao, Y.; Zhang, Y.; Zhang, L. *Int. J. Biol. Macromol.* **2009**, *45*, 109–115.
84. Zhang, M.; Cheung, P. C. K.; Chiu, L. C. -M.; Wong, E. Y. L.; Ooi, V. E. C. *Carbohydr. Polym.* **2006**, *66*, 455–462.
85. Joseph, M. M.; Aravind, S.; Varghese, S.; Mini, S.; Sreelekha, T. *Mol. Med. Rep.* **2012**, *5*, 489–496.
86. Vishchuk, O. S.; Ermakova, S. P.; Zvyagintseva, T. N. *Carbohydr. Res.* **2011**, *346*, 2769–2776.
87. De Zoysa, M.; Nikapitiya, C.; Jeon, Y. J.; Jee, Y.; Lee, J. *J. Appl. Phycol.* **2008**, *20*, 67–74.
88. Ye, H.; Wang, K.; Zhou, C.; Liu, J.; Zeng, X. *Food Chem.* **2008**, *111*, 428–432.
89. Yin, Y.; Yu, R.; Yang, W.; Yuan, F.; Yan, C.; Song, L. *Carbohydr. Polym.* **2010**, *82*, 543–548.
90. Synytsya, A.; Kim, W. J.; Kim, S. M.; Pohl, R.; Synytsya, A.; Kvasnička, F.; Čopíková, J.; Park, Y. I. *Carbohydr. Polym.* **2010**, *81*, 41–48.
91. Yoo, Y. C.; Kim, W. J.; Kim, S. Y.; Kim, S. M.; Chung, M. K.; Park, J. W.; Suh, H. H.; Lee, K. B.; Park, Y. I. *Algae* **2007**, *22*, 333–338.
92. Hemmingson, J. A.; Falshaw, R.; Furneaux, R.; Thompson, K. *J. Appl. Phycol.* **2006**, *18*, 185.
93. Hung, C. F.; Hsu, B. Y.; Chang, S. C.; Chen, B. H. *Nutrition* **2012**, *28* (1), 98–105.
94. Mizuno, T.; Ykohlui, P.; Kinoshita, T.; Zhuang, C.; Ito, H.; Mayuzumi, Y. *J. Biosci. Biotechnol. Biochem.* **1996**, *60*, 30–33.
95. Mizuno, T.; Zhuang, C.; Abe, K.; Okamoto, H.; Kiho, T.; Ukai, S.; Leclerc, S.; Meijer, L. *Int. J. Med. Mushrooms* **1999**, *1*, 301–316.
96. Mizuno, M.; Minato, K. i.; Ito, H.; Kawade, M.; Terai, H.; Tsuchida, H. *Biochem. Mol. Biol. Int.* **1999**, *47*, 707–714.
97. Mizuno, T.; Saito, H.; Nishitoba, T.; KaWagishi, H. *Food Rev. Int.* **1995**, *11*, 23–61.
98. Borchers, A. T.; Stern, J. S.; Hackman, R. M.; Keen, C. L.; Gershwin, M. E. *Soc. Exp. Biol. Med.* **1999**, *22*, 281–293.
99. Ooi, V. E. C.; Liu, F. *Int. J. Med. Mushrooms* **1999**, *1*, 195–206.

100. Wasser, S. P.; Weis, A. L. *Int. J. Med. Mushrooms* **1999**, *1*, 31–62.
101. Tzianabos, A. O. *Clin. Microbiol. Rev.* **2000**, *13*, 523–533.
102. Reshetnikov, S.V.; Tan, K. K. *Int. J. Med. Mushrooms* **2001**, *3*, 361–394.
103. Ikekawa, T.; Uehara, N.; Maeda, Y.; Nakanishi, M.; Fukuoka, F. *Cancer Res.* **1969**, *29*, 734–735.
104. Ikekawa, T. *Int. J. Med. Mushrooms* **2001**, *3*, 291–298.
105. Sakagami, H.; Takeda, M. In Diverse biological activity of PSK (Krestin), a protein-bound polysaccharide from Coriolus versicolor (Fr.) Quél, Proceedings of the First International Conference on Mushroom Biology and Mushroom Products, The Chinese University Press, *Hong Kong*, **1993**; pp 23–26.
106. Gorin, P.; Barreto-Bergter, E., The chemistry of polysaccharides of fungi and lichens. *In The polysaccharides, Elsevier:* **1983**; pp 365–409.
107. Chen, X.; Lin, Z.; Ye, Y.; Zhang, R.; Yin, J.; Jiang, Y.; Wan, H. *Carbohydr. Polym.* **2010**, *82*, 28–33.
108. Thomes, P.; Rajendran, M.; Pasanban, B.; Rengasamy, R. *Phytomedicine* **2010**, *18*, 52–57.
109. Jiang, Y. H.; Jiang, X. L.; Wang, P.; Mou, H. J.; Hu, X. K.; Liu, S. Q. *Microbiol. Res.* **2008**, *163*, 424–430.
110. Ding, X.; Hou, Y.; Hou, W. J. *Carbohydr. Polym.* **2012**, *89*, 397–402.
111. Chen, X.; Bai, X.; Liu, Y.; Tian, L.; Zhou, J.; Zhou, Q.; Fang, J.; Chen, J. *J. Ethnopharmacol.* **2009**, *122*, 205–209.
112. Wang, L. Y.; Wang, Y.; Xu, D. S.; Ruan, K. F.; Feng, Y.; Wang, S. *J. Ethnopharmacol.* **2012**, *143*, 347–354.
113. Li, C.; Gao, Y.; Xing, Y.; Zhu, H.; Shen, J.; Tian, J. *Food Chem. Toxicol.* **2011**, *49*, 2090–2095.
114. Yamamoto, K.; Kimura, T.; Sugitachi, A.; Matsuura, N. *Biol. Pharm. Bull.* **2009**, *32*, 259–263.
115. Slaton, J.; Verneris, M.; Lu, H.; Wenner, C. *J. Urol.* **2011**, *185*, e295.
116. da Silva Pinto, A.V. F.; Martins, P. R.; Romagnoli, G. G.; Campanelli, A. P.; Terezan, A. P.; Rodrigues Filho, E.; da Eira, A. F.; Kaneno, R. *Cell. Immunol.* **2009**, *256*, 27–38.
117. Yu, C. H.; Kan, S. F.; Shu, C. H.; Lu, T. J.; Sun-Hwang, L.; Wang, P. S. *J. Nutr. Biochem.* **2009**, *20*, 753–764.
118. Ma, Z.; Wang, J.; Zhang, L. *Biopolymers: Original Research on Biomolecules* **2008**, *89*, 614–622.
119. Song, G.; Du, Q. *J. Chromatogr. A* **2010**, *1217*, 5930–5934.
120. Sasaki, T.; Takasaka, N. *Carbohydr. Res.* **1976**, *47*, 99–104.
121. Bohn, J. A.; BeMiller, J. N. *Carbohydr. polym.* **1995**, *28*, 3–14.
122. Marchessault, R.; Deslandes, Y.; Ogawa, K.; Sundararajan, P. *Can. J. Chem.* **1977**, *55*, 300–303.
123. Yui, T.; Ogawa, K.; Kakuta, M.; Misaki, A. *J. Carbohydr. Chem.* **1995**, *14*, 255–263.
124. Takeda, K.; Tomimori, K.; Kimura, R.; Ishikawa, C.; Nowling, T. K.; Mori, N. *Int. J. Oncol.* **2012**, *40*, 251–260.
125. Alekseyenko, T.; Zhanayeva, S. Y.; Venediktova, A.; Zvyagintseva, T.; Kuznetsova, T.; Besednova, N.; Korolenko, T. *B. Exp. Biol. Med.* **2007**, *143*, 730–732.
126. Ale, M. T.; Maruyama, H.; Tamauchi, H.; Mikkelsen, J. D.; Meyer, A. S. *Int. J. Biol. Macromol.* **2011**, *49*, 331–336.
127. Zhou, G.; Sheng, W.; Yao, W.; Wang, C. *Pharmacol. Res.* **2006**, *53*, 129–134.
128. Cumashi, A. *Glycobiology* **2007**, *17*, 541–552.
129. Gao, Q.; Seljelid, R.; Chen, H.; Jiang, R. *Carbohydr. Res.* **1996**, *288*, 135–142.
130. Kawagishi, H.; Kanao, T.; Inagaki, R.; Mizuno, T.; Shimura, K.; Ito, H.; Hagiwara, T.; Nakamura, T. *Carbohydr Polym.* **1990**, *12*, 393–403.
131. Zhang, C.; Yang, F.; Zhang, X. W.; Wang, S. C.; Li, M. H.; Lin, L. P.; Ding, J. *Br. J. Pharmacol.* **2006**, *148*, 741–751.

132. Korbelik, M.; Cooper, P. *Br. J. Cancer* **2007**, *96*, 67–72.
133. Chen, L.; Pan, J.; Li, X.; Zhou, Y.; Meng, Q.; Wang, Q. *Int. Immunopharmacol.* **2011**, *11*, 255–259.
134. Feng, L.; Jia, X. B.; Shi, F.; Chen, Y. J. *Molecules* **2010**, *15*, 5093–5103.
135. Ke, M.; Zhang, X. J.; Han, Z. H.; Yu, H. Y.; Lin, Y.; Zhang, W. G.; Sun, F. H.; Wang, T. J. *Carbohydr. Polym.* **2011**, *86*, 136–141.
136. Chen, R.; Meng, F.; Liu, Z.; Chen, R.; Zhang, M. *Carbohydr. Polym.* **2010**, *80*, 845–851.
137. Silva, D. C.; Freitas, A. L. P.; Barros, F. C. N.; Lins, K. O.; Alves, A. P. N.; Alencar, N. M.; de Figueiredo, I. S.; Pessoa, C.; de Moraes, M. O.; Costa-Lotufo, L.V.; Feitosa, J. P. *Carbohydr. Polym.* **2012**, *87*, 139–145.
138. Liang, M.; Li, S.; Shen, B.; Cai, J.; Li, C.; Wang, Z.; Li, X.; Gao, J.; Huang, H.; Zhang, X. *Carbohydr. Polym.* **2012**, *88*, 973–976.
139. Xu, H.; Yao, L.; Sun, H.; Wu, Y. *Carbohydr. Polym.* **2009**, *78*, 316–322.
140. Xu, H. S.; Wu, Y. W.; Xu, S. F.; Sun, H. X.; Chen, F.Y.; Yao, L. *J. Ethnopharmacol.* **2009**, *125*, 310–317.
141. Liu, Y.; Li, Y.; Yang, W.; Zhang, L.; Cao, G. *Int. J. Biol. Macromol.* **2012**, *50*, 632–636.
142. Cao, W.; Li, X. Q.; Wang, X.; Fan, H. T.; Zhang, X. N.; Hou, Y.; Liu, S. B.; Mei, Q. B. *Phytomedicine* **2010**, *17*, 598–605.
143. Mallick, S.; Maiti, S.; Bhutia, S.; Maiti, T. *Food Chem. Toxicol.* **2010**, *48*, 2115–2121.
144. Mallick, S. K.; Bhutia, S. K.; Maiti, T. K. *Int. J. Med. Mushrooms* **2009**, *11*, 237–248.
145. Cheng, Y.; Tang, K.; Wu, S.; Liu, L.; Qiang, C.; Lin, X.; Liu, B. *PloS one* **2011**, *6*, e27437.
146. Wang, Z.; Wu, B.; Zhang, X.; Xu, M.; Chang, H.; Lu, X.; Ren, X. *Carbohydr. Polym.* **2012**, *89*, 31–35.
147. Yang, X.; Guo, D.; Zhang, J.; Wu, M. *Int. Immunopharmacol.* **2007**, 7, 401–408.
148. Jin, X. *Carbohydr. Polym.* **2012**, *87*, 2198–2201.
149. Xu, C.; Liu, Y.; Yuan, G.; Guan, M. *Int. J. Biol. Macromol.* **2012**, *50*, 891–894.
150. Zhong, M.; Wang, L.; Ma, H.; Guo, Z.; Wang, C.; Liu, S. *Zhongguo Zhong yao za zhi= Zhongguo zhongyao zazhi= China journal of Chinese materia medica* **2008**, *33*, 2120–2123.
151. Lee, J. S.; Hong, E. K. *Int. Immunopharmacol.* **2011**, *11*, 1226–1233.
152. Park, S. E.; Yoo, H. S.; Jin, C. -Y.; Hong, S. H.; Lee, Y. -W.; Kim, B. W.; Lee, S. H.; Kim, W.-J.; Cho, C. K.; Choi, Y. H. *Food Chem. Toxicol.* **2009**, *47*, 1667–1675.
153. Zhong, K.; Wang, Q.; He, Y.; He, X. *Int. J. Biol. Macromol.* **2010**, *47*, 356–360.
154. Cao, Q. Z.; Lin, Z. B. *Life Sci.* **2006**, *78*, 1457–1463.
155. XiaoPing, C.; Yan, C.; ShuiBing, L.; YouGuo, C.; JianYun, L.; LanPing, L. *Carbohydr. Polym.* **2009**, *77*, 389–393.
156. Fan, Y.; Wang, W.; Song, W.; Chen, H.; Teng, A.; Liu, A. *Carbohydr. Polym.* **2012**, *88*, 1313–1318.
157. Luo, D.; Zhang, Q.; Wang, H.; Cui, Y.; Sun, Z.; Yang, J.; Zheng, Y.; Jia, J.; Yu, F.; Wang, X. *Eur. J. Pharmacol.* **2009**, *617*, 33–40.
158. Kim, Y. O.; Park, H. W.; Kim, J. H.; Lee, J. Y.; Moon, S. H.; Shin, C. S. *Life Sci.* **2006**, *79*, 72–80.
159. Ono, L.; Wollinger, W.; Rocco, I. M.; Coimbra, T. L.; Gorin, P. A.; Sierakowski, M. -R. *Antivir. Res.* **2003**, *60*, 201–208.
160. Lu, W. Z.; Geng, G. X.; Li, Q. W.; Li, J.; Liu, F. Z.; Han, Z. S.; Gao, D. W.; Yan, X.; Yang, X. L. *Am. J. Chin. Med.* **2009**, *37*, 933–944.
161. Bae, J. S.; Jang, K. H.; Jin, H. K. J. *Cancer Lett.* **2006**, *235*, 60–68.
162. Bae, J. S.; Jang, K. H.; Yim, H.; Jin, H. K. *Cancer Lett.* **2005**, *218*, 43–52.
163. Han, S. B.; Lee, C. W.; Kang, J. S.; Yoon, Y. D.; Lee, K. H.; Lee, K.; Park, S. K.; Kim, H. M. *Int. Immunopharmacol.* **2006**, *6*, 697–702.
164. Joseph, J.; Smina, T. P. P.; Janardhanan, K. K. *Cancer Biother. Radiopharm.* **2011**, *26*, 299–308.

165. Kokina, M.; Frioui, M.; Shamtsyan, M.; Sibirtsev, V.; Krasnikova, L.; Konusova, V.; Simbirtsev, A. *Chemistry & Chemical Engineering, Biotechnology, Food Industry* **2018**, *19*, 465.
166. Sun, L.; Wang, L.; Zhou, Y. *Carbohydr. Polym.* **2012**, *87*, 1206–1210.
167. Zhou, F.; Lv, O.; Zheng, Y.; Wang, J.; Hu, P.; Wang, Z.; Yang, L. *Int. J. Biol. Macromol.* **2012**, *50*, 1322–1326.
168. Kim, H. Y.; Kim, J. H.; Yang, S. B.; Hong, S. G.; Lee, S. A.; Hwang, S. J.; Shin, K. S.; Suh, H. J.; Park, M. H. *J. Med. Food* **2007**, *10*, 25–31.
169. Zhao, T.; Mao, G.; Mao, R.; Zou, Y.; Zheng, D.; Feng, W.; Ren, Y.; Wang, W.; Zheng, W.; Song, J.; Chen, Y. *Food Chem. Toxicol.* **2013**, *55*, 609–616.
170. Sunila, E.; Hamsa, T.; Kuttan, G. *Pharm. Biol.* **2011**, *49*, 1065–1073.
171. Philip, A. K.; Philip, B. *Oman Med. J.* **2010**, *25*, 79.
172. Pushpamalar, J.; Veeramachineni, A. K.; Owh, C.; Loh, X. *J. Chem. Plus Chem.* **2016**, *81*, 504–514.
173. Singh, B.; Bala, R. *Radiat. Phys. Chem.* **2014**, *103*, 178–187.
174. Kumar, K.; Kaith, B.; Mittal, H. *J. Appl. Polym.* **2012**, *123*, 1874–1883.
175. Bhattacharya, S. S.; Mishra, A.; Pal, D.; Ghosh, A. K.; Ghosh, A.; Banerjee, S.; Sen, K. *Polym. -Plast. Technol.* **2012**, *51*, 878–884.
176. Alla, S. G. A.; Sen, M.; El-Naggar, A. W. M. *Carbohydr. Polym.* **2012**, *89*, 478–485.
177. Nagar, S. Isolation and Structural investigation of polysaccharide from *Tinospora sinensis*. PhD thesis, Forest Research Institute (Deemed University), **2017**.
178. Rana, V.; Kumar, V.; Soni, P. L. *Carbohydr. Polym.* **2012**, *90*, 243–250.
179. Kumar, V.; Nagar, S. *Carbohydr. Polym.* **2014**, *99*, 291–296.
180. Kumar, V.; Rana, V.; Soni, P. L. *Phytochem Analysis* **2013**, *24*, 75–80.
181. Kumar, V.; Nagar, S.; Tripathi, Y. C. *Int. J. Biol. Macromol.* **2014**, *70*, 360–363.
182. Lakhera, A. K.; Kumar, V. *Int. J. Biol. Macromol.* **2017**, *94*, 45–50.
183. Kumar, P.; Kumar, V. *Carbohydr. Polym.* **2017**, *165*, 205–212.
184. Nagar, S.; Lakhera, A. K.; Kumar, V. *J. Biol Active Prod. from Nat.* **2020**, *10*, 70–85.
185. Nagar, S.; Hensel, A.; Mischnick, P.; Kumar, V. *Carbohydr. Polym.* **2018**, *193*, 326–335.

CHAPTER SIXTEEN

N-acetylgalactosamine (GalNAc)-conjugates: Delivering oligonucleotide drugs to the liver

Vivek K. Sharma[a], Balaji Olety[b], Ashok K. Prasad[c]
[a]RNA Therapeutics Institute, University of Massachusetts Medical School, Worcester, MA, United States
[b]Department of Molecular Cell and Cancer Biology, University of Massachusetts Medical School, Worcester, MA, United States
[c]Bioorganic Laboratory, Department of Chemistry, University of Delhi, Delhi, India

1 Introduction

Oligonucleotide therapeutics is an attractive drug development platform since in principle it can potentially modulate the expression of any gene or gene product.[1,2] The drug design process is relatively straightforward –the nucleotide sequence of the oligonucleotide drug is complementary to the target gene or gene product. Owing to this fact, oligonucleotide-based therapeutics has a great advantage over other drug development platforms such as drugs based on small molecules or biologics. Despite its ease of design, however, conventional oligonucleotides cannot be readily used as drugs because of several limitations such as stability, specific delivery and its pharmacokinetic properties. For example, conventional oligonucleotides are prone to be degraded by nucleases affecting its stability. Additionally, because target sequences are sometimes redundant across genes or gene families, strong and stable interactions with specific targets are therefore necessary to minimize "off-target" effects. Since gene or gene products are expressed in multiple tissues and cell types, specific delivery of oligonucleotides to the site of interest poses another challenge. In addition, the size/length of the oligonucleotides are also important determinants for their uses as drugs. Therefore, to overcome these limitations, chemical modifications are often necessary to improve their drug-like properties.[2–6] One such modification that has been successfully used is the conjugation of *N*-acetylgalactosamine (GalNAc).[7–8] The discovery that *N*-acetylgalactosamine (GalNAc) conjugation which can specifically deliver the oligonucleotide-GalNAc conjugate to the liver has addressed an important limitation of specific delivery and

Carbohydrates in Drug Discovery and Development. http://dx.doi.org/10.1016/B978-0-12-816675-8.00016-6

has revolutionized the field.[9–12] Within five years of this discovery, more than twenty-five GalNAc based oligonucleotide conjugates have entered into advanced clinical trials to treat liver diseases suggesting the immense promise this holds.[10,13] In November 2019, the first GalNAc-conjugated siRNA drug, Givosiran (Givlaari) was approved for the treatment of acute hepatic porphyria (AHP). This chapter aims to describe recent developments in oligonucleotide therapeutics with an emphasis on GalNAc-conjugation strategy to target a variety of liver ailments.

1.1 Oligonucleotide therapeutics

The concept of oligonucleotide therapeutics was first demonstrated against Rous sarcoma virus (RSV) in 1978. In that pioneering work, Zamecnik and Stephenson found that a synthetic 13-mer oligonucleotide complementary to viral RNA could inhibit its translation as well as viral replication.[14–15] Soon it was realized that only chemically modification of oligonucleotides can exert the therapeutic impact *in vivo*.[16] As a result, Fomivirsen, the first phosphate backbone modified antisense drug was approved by the FDA in 1998 for the treatment of cytomegalovirus (CMV) retinitis.[17]

Since then oligonucleotide therapeutics has made tremendous progress in the last four decades and is no longer limited to antisense or sugar-phosphate backbone modifications. Oligonucleotide therapeutics now includes diverse class of drugs, *e.g.* antisense oligonucleotides (ASOs), small interfering RNAs (siRNAs), microRNAs (miRNAs) and aptamers (Table 16.1).[4,18,19] This has been made possible due to advances in the medicinal chemistry of synthetic oligonucleotides.[2,20]

1.2 Oligonucleotide therapeutics: Mode of action

There are two prevalent mechanisms of therapeutic action of oligonucleotides, antisense oligonucleotides (ASOs) and small interfering RNA (siRNA).[21] The antisense oligonucleotides (ASOs) are ~ 12–22 nucleotide long oligonucleotides that are fully complementary to the target mRNA (Fig. 16.1). Depending on the chemistry of ASOs, the fate of target mRNA can be predicted. The ASO:mRNA heteroduplex can recruit enzymes such as RNase H which can degrade the target mRNA. Alternatively, the ASO can bind and modulate the splicing of mRNA or simply act as a steric blocker for ribosomes during protein synthesis.[18,19,22]

The siRNAs, which are delivered as double-stranded RNAs, need enzyme AGO2 for therapeutic action (Fig. 16.1).[23] The AGO2 can selectively recruit the antisense siRNA strand to form the RNA induced

Table 16.1 Approved oligonucleotide drugs.[1,3]

Oligonucleotide drug	Indication	Chemical modification	Mode of action	Year of approval
Fomivirsen (Vitravene)	Cytomegalovirus (CMV) retinitis	Fully PS	antisense	1998
Pegaptanib (Macugen)	Neovascular age-relatedmacular degeneration(AMD)	2'-OMe purine and 2'-F pyrimidine nucleotides, 5'-end is covalently linked to two 20KDa monomethoxy polyethylene glycol (PEG)	aptamer	2004
Mipomersen (Kynamro)	Homozygous familialhypercholesterolemia (HoFH)	Fully PS, 5-10-5 2'-OMOE gapmer, all pyrimidines are 5-methylated	Antisense	2013
Nusinersen (Spinraza)	spinal muscular atrophy (SMA)	Fully PS, fully 2'-OMOE, all pyrimidines are 5-methylated	antisense	2016
Eteplirsen[26] (Exondys 51)	Duchenne muscular dystrophy (DMD)	Fully phosphorodiamidate morpholino (PMO)	antisense	2016
Inotersen (Tegsedi)	Hereditary transthyretin (hATTR) amyloidosis	Full PS, 5-11-4OMOE gapmer, all pyrimidines are 5-methylated	antisense	2018
Patisiran (Onpattro)	hATTR amyloidosis	dT at 3'-end, some pyrimidine nucleotides are 2'-OMe	siRNA	2018
Volanesorsen (Waylivra)	Familial Chylomicronemia Syndrome (FCS)	Fully PS, 5-10-5 2'-OMOE gapmer, all cytosine are 5-methylated	antisense	2019
Golodirsen (Vyondys53)	Duchenne muscular dystrophy (DMD)	PMO	Antisense	2019
Givosiran (Givlaari)	Acute Hepatic porphyrias	2'-F, 2'-OMe and PS modifications, triantennary GalNAc (*N*-acetylgalactosamine) conjugate	siRNA	2019

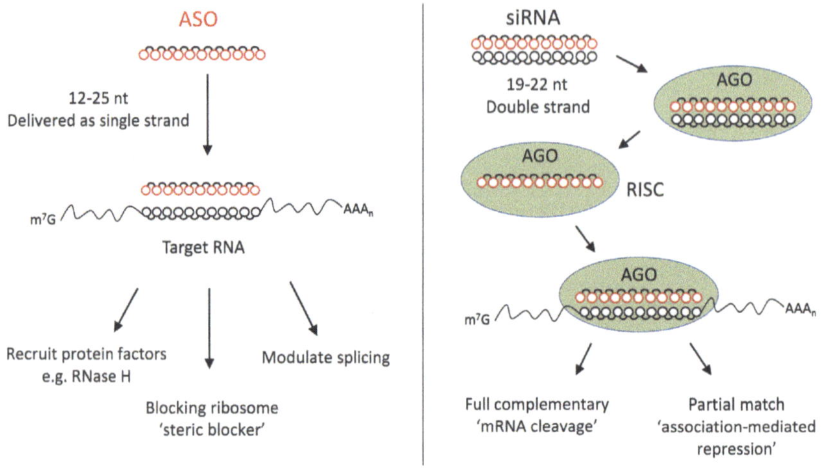

Figure 16.1 ***Schematic presentation of therapeutic action of antisense oligonucleotides (ASOs) and small interfering RNAs (siRNAs).***[21]

silencing complex (RISC). Depending on the extent of complementarities between siRNA and target mRNA, RISC can lead to repression or mRNA cleavage.[18–19,22]

The GalNAc conjugation has been employed in both ASO and siRNA drug candidates as evidenced in the clinical pipelines of Ionis Pharmaceuticals and Alnylam Pharmaceuticals, respectively.[13] The advantages of GalNAc conjugation can be better understood by comparing two antisense drugs, Miravirsen and RG-101, both targeting miR-122, a liver specific micro RNA that has been shown to be essential for Hepatitis C replication.[24] While Miravirsen is an ASO that is unconjugated, RG-101 is covalently conjugated with triantennary N-acetylgalactosamine[13] (GalNAc). While both similarly target miR122, RG-101 is significantly more potent where a single subcutaneous dose of just 2 mg/Kg has resulted in liver-specific uptake followed by anti-HCV activities possibly due to the conjugation.[25] The reason for this higher potency of RG-101 is due to its covalent tethering with triantennary *N*-acetylgalactosamine ligand.[13] The mechanism by which GalNAc leads to efficient liver-specific uptake of oligonucleotides will be discussed in the following sections in detail.

1.3 Oligonucleotide therapeutics: Challenges

As eluded to earlier, the mechanism of action of oligonucleotides is a straightforward process. They bind to their target complementary to their

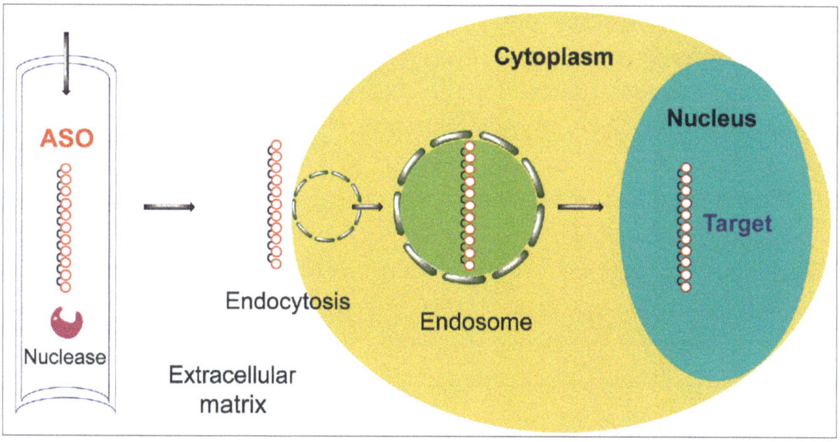

Figure 16.2 ***Schematic representation of inter- and intra-cellular trafficking of therapeutic oligonucleotides (siRNAs are delivered mainly as double-stranded oligonucleotides).***

sequence. The stability of this binding determines its potency to recruit cellular factors that eventually promote target degradation (Fig. 16.1).[19] However, unlike their mode of action, their utility as a therapeutic agent is not straightforward and depends also on a number of factors. To access their target site, oligonucleotides must undergo complex inter and intra-cellular trafficking pathways (Fig. 16.2).[4] For example, oligonucleotides must be stable to withstand degradation by various nucleases both prior to or after they enter cells. Furthermore, their entry into the cells that are mainly mediated through endocytosis through the endosomes which are eventually degraded by its fusion with the lysosome and/or recycled back outside the cells. Only a small fraction of oligonucleotides (< 1%) can escape from the endosomes. Depending on the localization of the target, oligonucleotides may further need to access particular sub-cellular compartment and/or specific cellular factors to exert its function (Fig. 16.1).

In summary, unmodified oligonucleotides lack the "drug-like" properties. These oligonucleotides are unable to enter cells, have poor serum stability and low binding affinities. Therefore, this necessitates chemical modifications in oligonucleotides to improve their "drug-like" properties.

1.4 Medicinal chemistry of therapeutic oligonucleotides

Therapeutic oligonucleotide candidates are almost always chemically modified to improve drug-like properties.[2,20] When compared to other classes of drugs, oligonucleotide-based drugs have shown considerable promise.

Research in medicinal chemistry of therapeutic oligonucleotides has made tremendous progress in the past three decades. As a result, seven oligonucleotide drugs have been approved in the last four years (Table 16.1).[1] This pace of drug approval is likely to only further increase as the available chemical toolbox is continuously growing in its size and sophistication.[20]

Oligonucleotides are polynucleotide chains having a specific sequence of nucleotide residues. The sequences specificity comes from the nucleobases (either a purine or a pyrimidine)) which are held by the pentose sugar (either a ribose or a deoxyribose) linked through phosphate backbone (Fig. 16.3). Nearly every structural component of the oligonucleotide can be modified to improve the therapeutic utility of oligonucleotides such as chemical modifications to the sugar, backbone and/or the nucleobase

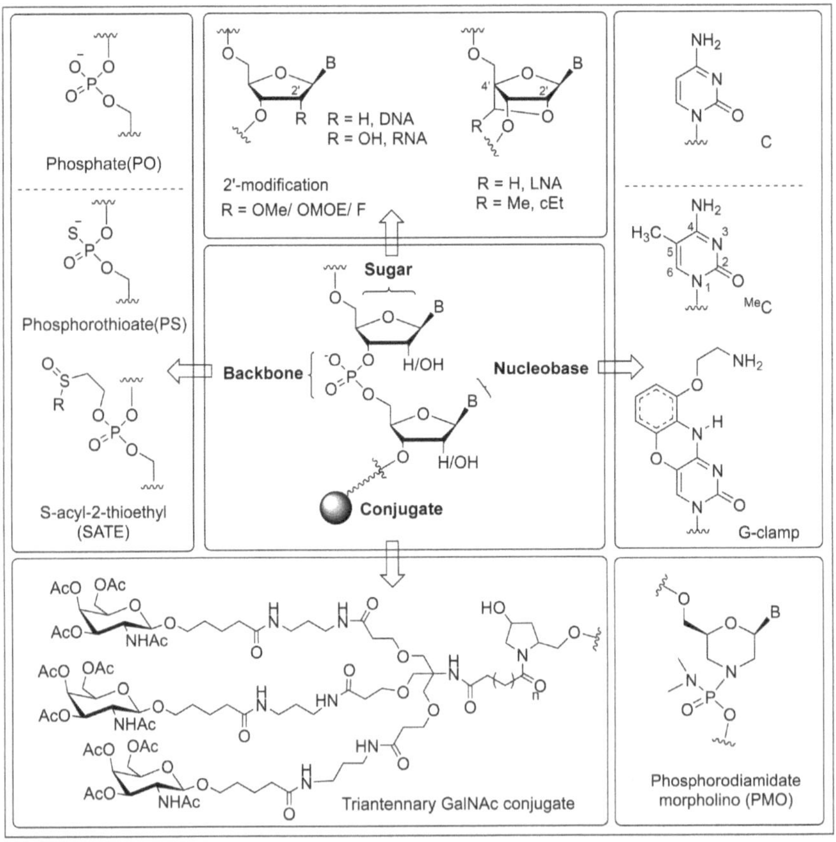

Figure 16.3 ***Potential modification sites and some common chemical modifications in therapeutic oligonucleotides.***[3]

moiety of the oligonucleotide (Fig. 16.3, Table 16.1).[3–6,10,13,27] Recently, it has been shown that covalent tethering of small molecules such as GalNAc can enhance potency as well as specificity in delivery of the oligonucleotide drug (Fig. 16.3).[8,28,29] Chemical modifications are also needed to improve cellular uptake, serum stability, binding affinity, potency and specificity of the oligonucleotide drugs.[20,30] Since targets differ in their properties, it is important therefore that the best available tools are picked. In the following section, we aim to discuss the latest developments to the oligonucleotide chemical toolbox.

1.5 Phosphate-backbone modifications

The natural phosphodiester linkage is highly sensitive to cleavage from ubiquitous nucleases. Therefore, the phosphate backbone is one of the primary sites of modification in therapeutic oligonucleotides.[27] Phosphorothioate (PS) backbone was one of the earliest and by far the most successful modification to date.[31] Replacement of one of the non-bridging oxygen with sulphur in PS backbone (Fig. 16.3) provides greater stability against both endo- and exo- nucleases.[31–33] Furthermore, PS backbone has a higher affinity to plasma proteins and cell surface receptors which facilitates cellular uptake.[34–35] The PS-ASOs can recruit RNase H and direct the cleavage of the target mRNA.[36] Due to these advantages, PS modification have been utilized successfully. Indeed, all the approved antisense drugs so far Formivirsen, Mipomersen, Nusinersen, Inotersen, and Volanesorsen are fully phosphorothioated (Table 16.1). The GalNAc-conjugated siRNA drug, Givosiran is partially modified with PS backbone (Table 16.1).

The removal of negative charge from the phosphate backbone has been shown to improve cellular uptake and binding affinity of the therapeutic oligonucleotides. Highly robust RNAi responses were demonstrated by short interfering ribonucleic neutrals (siRNNs) which contains a number of neutral phosphotriester groups, *S*-acyl-2-thioethyl (SATE) (Fig. 16.3).[37] Here, the negative charges are neutralized by masking some of the phosphates with SATE-phosphate modifications. This potentially increases the siRNNs affinity to serum albumins and thereby has better pharmacokinetic properties over natural siRNAs which quickly gets cleared by kidney (half-life < 5 minutes). The neutral SATE-phosphate modification acts as prodrug, enables cellular entry and also releasesfully charged siRNAs upon the action of cytoplasmic thioesterases (Fig. 16.4A), and induces robust RNAi responses.[37] For hepatocyte-specific delivery, triantennary

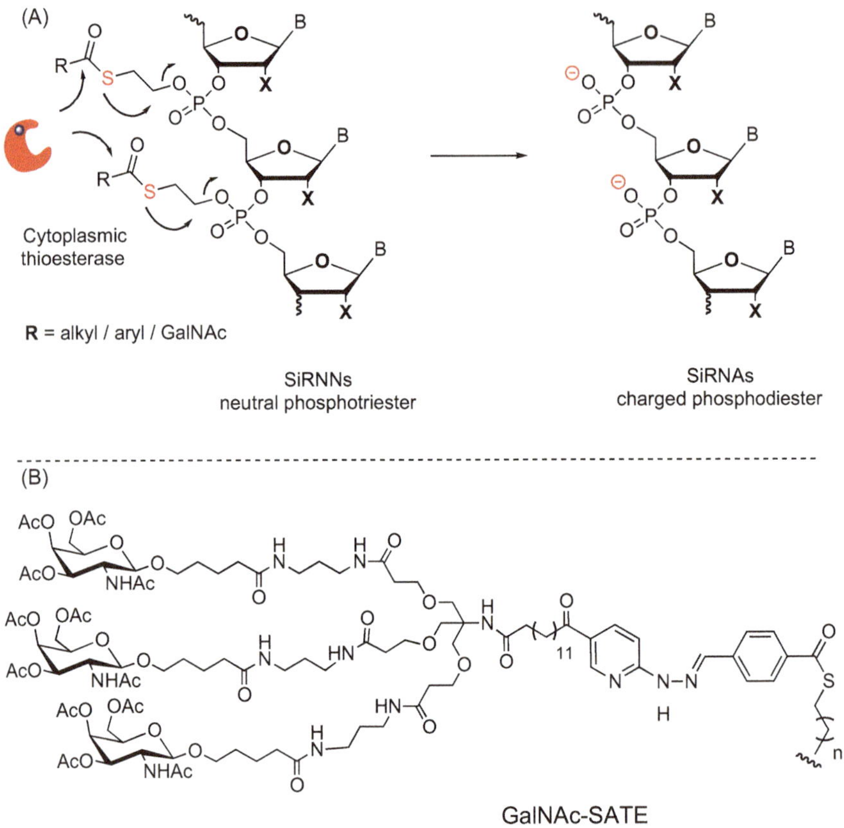

Figure 16.4 (A) Short interfering ribonucleic neutrals (siRNNs), and mechanism of release of charged siRNAs by cytoplasmic thioesterase.[4,37] (B) GalNAc-SATE modification in siRNNs. X = OH or 2′- modification (F/OMe).

Gal*N*Ac conjugates were introduced with SATE backbone modification (Fig. 16.4B). The systemic delivery of GalNAc-SATE siRNNs provide sextended dose-dependent robust RNAi responses *in vivo* (> 12 days).[37]

1.6 Sugar modifications

In addition to the phosphate backbone, the 2′-position of nucleotides is also an attractive site for modifications in therapeutic oligonucleotides.[4,38] 2′-*O*-alkyl or -alkoxy modifications, 2′-fluorinated analogues and constrained analogues where the 2′-substituent makes an extra bridge to the sugar moiety (Fig. 16.3) have all been successfully performed.[38–39] The common sugar modifications in therapeutic oligonucleotides are 2′-*O*-methyl (OMe), 2′-*O*-methoxyethyl (OMOE), 2′-fluoro (F) and locked nucleic acid (LNA) (Fig. 16.3).[38,40]

Substitution of an electronegative atom/group at the 2′-position pre-organizes the sugar moiety in a way that favors A-form (RNA-like) conformation in oligonucleotides.[38] This leads to high binding-affinity for the complementary mRNA targets. The 2′-OMe, 2′-F, and 2′-OMOE modifications have been used in the approved oligonucleotide drugs, Pegaptanib, Mipomersen, Nusinersen, Inotersen,Volanesorsen and Patisiran (Table 16.1).[1,3] The Locked nucleic acid (LNA) or Bridged nucleic acid (BNA) that is obtained after connecting 2′-O and 4′-C of ribose sugar has also been found to have very high binding affinities (ΔT_m ~ 4–8°C per modification) towards complementary RNA (Fig. 16.3).[40–44]

1.7 Combining the sugar and phosphate-backbone modifications

Although these sugar-modified oligonucleotides have demonstrated improved pharmacokinetics and toxicological profiles, these modifications are not capable of mediating target degradation through RNase H. This is because of its RNA-like conformation as RNase H can only cleave a DNA:RNA heteroduplexes.[45,38] To overcome this, antisense oligonucleotides are therefore designed as 'Gapmers' which have a fully PS backbone with 2'-modification only at the terminal wings.[46,47] Such gapmer 'wings' provide high binding affinity and the stretch of PS DNA in the middle ensures RNase H activity. Such strategy has been used in the approved drug Mipomersen, a 5-10-5 'gapmer' antisense oligonucleotide drug which has five 2′-*O*-methoxyethyl (2′-OMOE) nucleotides wing at each end (Fig. 16.5).

Gapmers are designed when RNase H activity is required to degrade the target mRNA. But antisense oligonucleotides have also other modes of therapeutic actions (Fig. 16.1).[19,48] To enable other modes of action, 'chimera' design where multiple modifications are used in a single oligonucleotide have been tested.[49,50] For example, the phosphorothioated anti-HCV oligonucleotide drug RG-101 contains 2′-OMOE and cEt (constrained ethyl) sugar modifications, as well as triantennary GalNAc ligand linked through a bio-cleavable PO linker (Fig. 16.5).[25] The BNA sugar modification in RG-101 is constrained ethyl (cEt) nucleotide that has a methyl substituent on the methylene bridge of parent LNA (Fig. 16.3).[51–52] Such chimeras are more potent than normal antisense oligonucleotides simply because different chemical modifications augmentmultiple drug-like attributes. For example, in case of RG-101, the PS backbone provides stability against nuclease, the cEt BNA ensures high binding affinity for miR-122

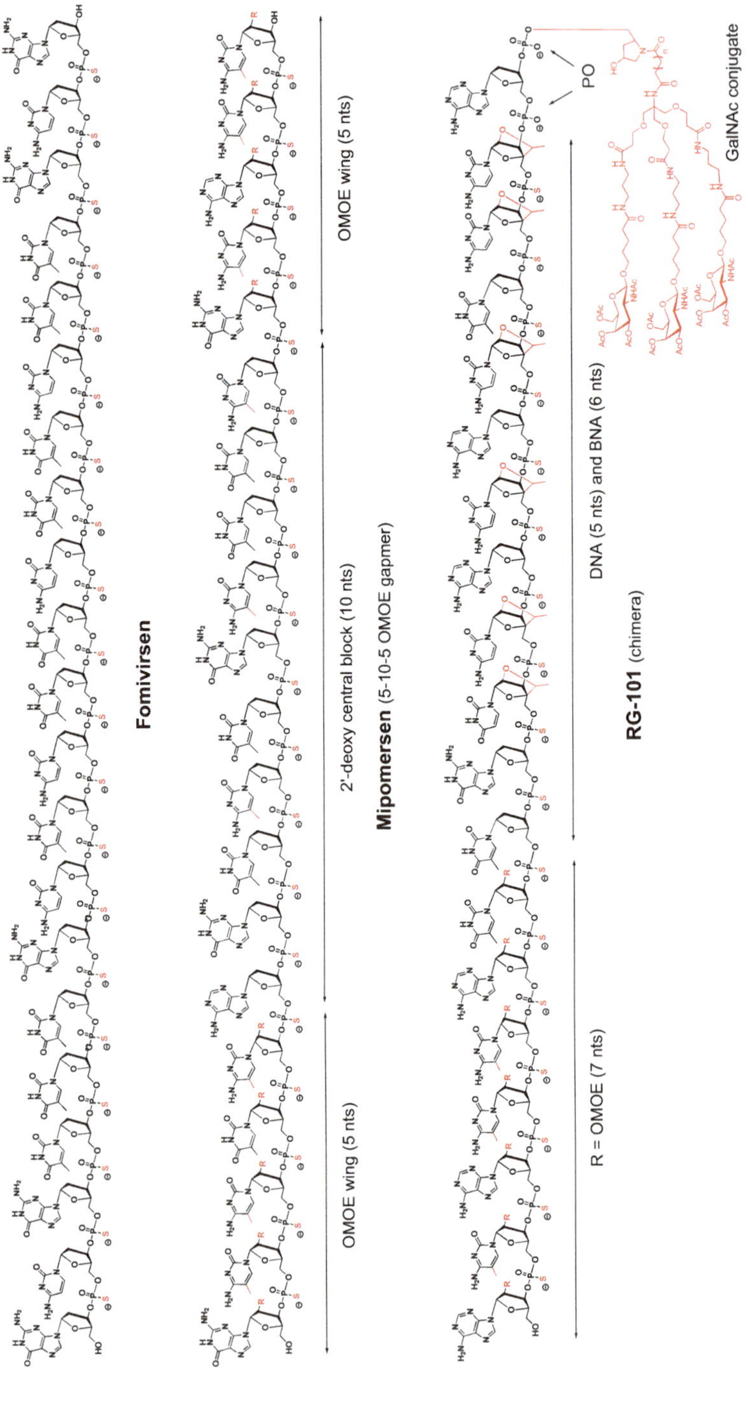

Figure 16.5 ***Structures of Fomivirsen, Mipomersen and RG-101. Chemical modifications are colored red.***

where as GalNAc lug the whole cargo specifically to the hepatocytes inside the liver.[25]

1.8 Complete replacement of the sugar-phosphate backbone

In addition to the sugar alone or the sugar-phosphate backbone modifications, the entire sugar-phosphate backbone can be replaced with alternatives such as in phosphorodiamidate morpholinos (PMO, Fig. 16.3).[53] In PMO's, the five-membered ribose sugar is replaced with six-membered morpholine. The morpholine rings bearing the nucleo bases are connected through uncharged phosphorodiamidate linkages and therefore, are highly stable to nucleases.[53–55] While PMO's can't recruit RNase H, these have found applications as steric blockers or splicing modulators (Fig. 16.1). The fully PMO modified oligonucleotide drug, Eteplirsen, has been recently approved for the treatment of spinal muscular atrophy (SMA) (Table 16.1).[26,56]

1.9 Nucleobase modifications

The Watson-Crick nucleobase paring is the mode in which a therapeutic oligonucleotide binds its specific target. Therefore, there is limited scope for nucleobase modification in oligonucleotides.[57] The most tolerated sites in nucleoside bases are those that are exposed to the solution in the major groove, *i.e.* 5-positions of pyrimidines (Fig. 16.3).[57–59] The methyl substituent at 5-position of cytosine or uracil nucleobase do not interfere with base pairing.[57] The binding affinity of ^{Me}C is higher than the unmethylated nucleobase ($\Delta T_m \sim +0.5°C$ per modification).[60] Another advantage of ^{Me}C modification is that it mitigates innate immune response triggered by TLR9 recognition of unmethylated CpG motifs (5′-CG-3′ dinucleotides).[61–63] All pyrimidine bases are methylated at 5-position in the recently approved antisense drugs, Mipomersen, Nusinersen and Inotersen (Table 16.1).

Another cytosine derivative, the tricyclic G-clamp (Fig. 16.3), pairs specifically with guanosine with very high affinity (ΔT_m up to +18°C).[64] G-clamp modified oligonucleotides are able to recruit RNase H and have demonstrated ~ 25-fold increase in potency *in vitro*.[65] However, poor uptake of G-clamp limits it's *in vivo* use.[46]

1.10 Covalent-conjugate approach: GalNAc-conjugates

As briefly mentioned earlier, the delivery of oligonucleotides to specific target sites is a big hurdle limiting its therapeutic utility.[7] Therefore, to address this limitation, a targeting moiety that improves targeting and/or uptake is usually covalently conjugated.[12,66] GalNAc-conjugation strategies has emerged as a powerful platform to deliver oligonucleotide drugs to the liver.[10,13] It has also been shown that GalNAc-conjugated oligonucleotides are more efficient in endosomal escape in a yet unknown mechanism.[10] We have earlier discussed two examples, GalNAc-SATE siRNNs (Fig. 16.4) and RG-101 (Fig. 16.5), where GalNAc ligands have a beneficial role in improving the specificity and potency of the oligonucleotide drugs. We will now discuss the mechanism of specific uptake of GalNAc-conjugated oligonucleotides.

1.11 GalNAc-conjugates: mechanism of liver-specific delivery

As previously discussed, covalent conjugation of GalNAc can specifically deliver oligonucleotide drugs into the hepatocytes inside the liver.[10,13] This hepatocyte-specific targeting is mediated by a sialoglycoprotein receptor (ASGPR), which is highly expressed ($\sim 5 \times 10^5$ per cell vs insulin receptor, $1x10^5$/cell) on the hepatocytes (Fig. 16. 6).[67–68] ASGPR's are involved in clearing serum glycoproteins containing terminal galactose (Gal) or *N*-acetyl galactosamine (GalNAc) moieties *via* clathrin-mediated endocytosis (affinity for GalNAc > Gal).[69]

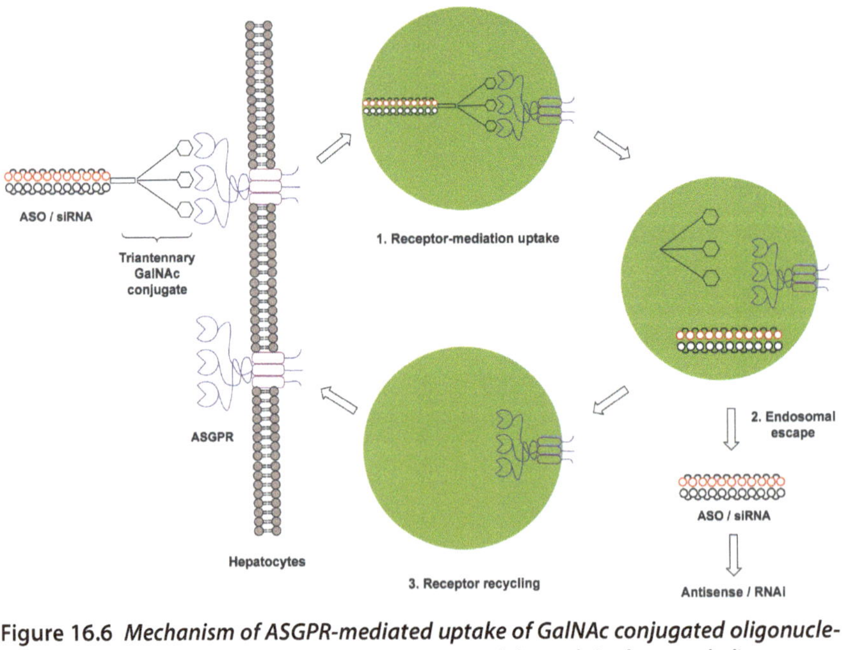

Figure 16.6 ***Mechanism of ASGPR-mediated uptake of GalNAc conjugated oligonucleotides to the hepatocytes inside the liver (ASOs are delivered single-stranded).***

In addition to glycoproteins, ASGPR also has high avidity for trivalent GalNAc-bound oligonucleotides.[11,70–71] The ASGPR-mediated endocytosis results into engulfment of trivalent GalNAc-oligonucleotides into hepatocytes. Inside the endosome, the cargo is released from the ASGPR due to pH drop, and the receptor is recycled to the hepatocyte surface. It has been shown that such a transport mechanism is a highly efficient process wherein a single ASGPR can be recycled about 200 times with a turnover time of ~15 minutes.[67,68,72,73] However, the mechanism by which a small fraction of the oligonucleotide escapes the endosomes to reach its target has still remained unknown.[10]

1.12 GalNAc-conjugates: design

The landmark papers describing GalNAc-mediated liver-targeted delivery of ASOs[9] and siRNAs[70] confirms that the triantennary GalNAc design (Fig. 16.3) is superior to other designs in terms of efficacy and potency. However, GalNAc designs other than the triantennary design have also been explored.[71,74-77]

There are mainly two synthetic strategies to achieve multivalent GalNAc conjugation. The first strategy is to pre-assemble a triantennary GalNAc ligand on the solid-support (Fig. 16.7),[9,70] whereas the other strategy is based on multiple incorporations of a monomeric GalNAc phosphoramidite during solid-phase oligonucleotide synthesis (Fig. 16.8).[75,78] The later strategy based on monomeric GalNAc phosphoramidite allows more flexibility in terms of design, valency and synthesis of GalNAc ligands for oligonucleotides.[75,77,78] For example, linear multivalent GalNAc conjugates (mono, di, tri, tetra and penta-valent) can be readily assembled from monomeric

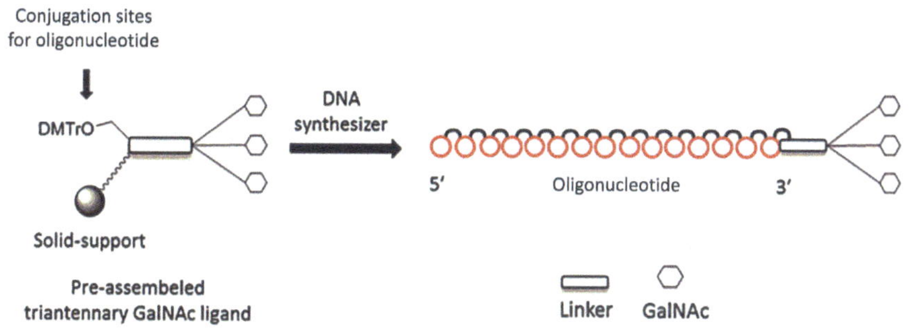

Figure 16.7 ***Synthetic strategy for GalNAc-oligonucleotide conjugates starting from a pre-assembled triantennary GalNAc ligand on solid-support.***

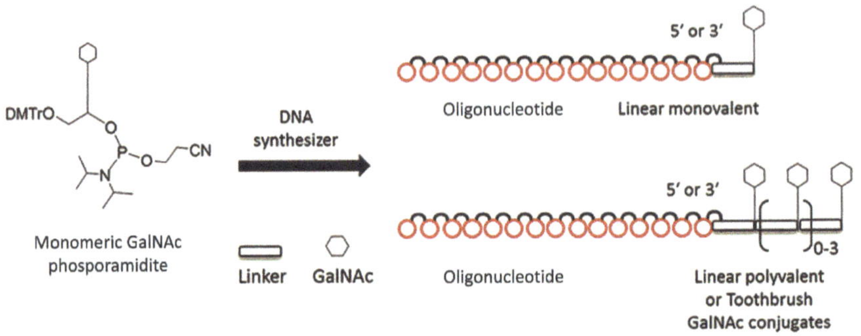

Figure 16.8 ***Synthetic strategy for linear polyvalent (toothbrush) GalNAc-oligonucleotide conjugates starting from a monomeric GalNAcphosphoramidite.***

GalNAc phosphoramidite (Fig. 16.8).[75,78] For simplicity we will refer linear polyvalent design as 'toothbrush'.

1.12.1 Triantennary GalNAc

Nair et al. found that conjugation of GalNAc ligand results in robust gene silencing in the liver without the need for any delivering agents.[70] These GalNAc-siRNA conjugates were synthesized using solid-phase oligonucleotide method. The GalNAc ligands were pre-assembled on a solid-support which was used to synthesize siRNA oligonucleotides. In that landmark work, the trivalent GalNAc arranged in triantennary fashion (**7**$_{(m=9)}$, Scheme 16.1) was found to be the optimal design for ASGPR-mediated siRNA uptake. Further, silencing of *TTR* gene in liver was achieved with a single-dose median effective dose (ED_{50}) of ~1 mg/kg when these conjugates were delivered subcutaneously in mice.[70]

Prakash et al. found that the ASOs can also be covalently conjugated with triantennary GalNAc (**7**$_{(m = 7)}$, Scheme 16.1) to provide robust gene silencing.[9] The ED_{50} of GalNAc conjugated 5-10-5 OMOE gapmer targeting SRB1 mRNA in mouse liver was 3.3 mg/Kg, much higher compared to the unconjugated gapmer ASO (ED_{50}= 18.3 mg/Kg). When the OMOE nucleotide sugar wings in gapmer ASO were replaced by constrained ethyl (cEt)-bridged ribonucleotides[52], ~60 fold enhancement in potency was observed over parent GalNAc-ASO.[9] The low charge and low molecular weight of shorter 2-10-2 cEt LNA gapmeris believed to help in escaping from endosomes.

The above gene silencing studies using GalNAcligands were conjugated to the 3'-end of the oligonucleotides (siRNA/ASOs). However, 5'-GalNAc conjugated ASO's was shown to be more effective (ED_{50}= 2.1 mg/Kg) than 3'-end (ED_{50}= 3.4 mg/Kg), both *in vitro* and *in vivo* for targeting SRB1.[79]

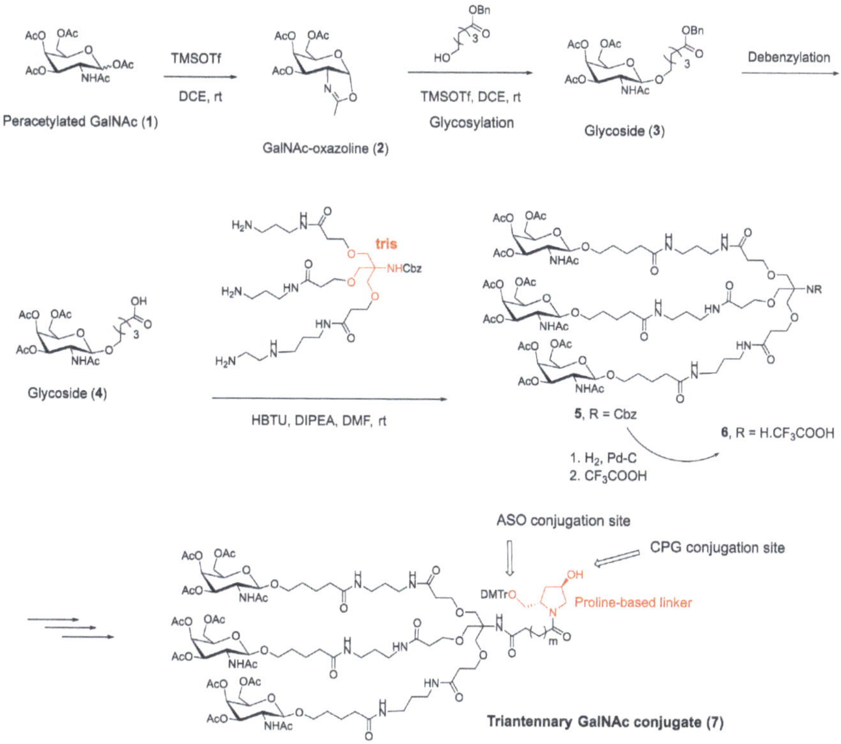

Scheme 16.1 General strategy for the synthesis of triantennary GalNAc conjugate (**7**) for liver-specific delivery of oligonucleotide drugs.

Such 5'-GalNAc-ASO conjugates were synthesized using a solution-phase approach.[79] To overcome this limitation of solution-phase synthesis, Prakash et al. developed a GalNAc phosphoramidite (**10**) which was coupled to 5'-trebler branched ASO during solid-phase oligonucleotide synthesis (Scheme 16.2).[80] This monomer-based triantennary assembly at the 5'-end of ASO (**11**) results in 50-fold improvement in potency in hepatocytes compared to 3'conjugated ASO.

Prakash et al. did a comprehensive structure-activity relationship study of triantennary GalNAc for the hepatocyte-specific delivery of ASOs.[11] Several GalNAc clusters were synthesized based on six different scaffolds and conjugated to ASOs targeting various targets such as SRB-1, A1AT, FXI, TTR, and ApoC III. This comprehensive study resulted in the identification of a simplified *tris*-based GalNAc cluster (n = 6, Fig. 16.9), which can be easily assembled from readily available materials using solution-phase conjugation methods.[11]

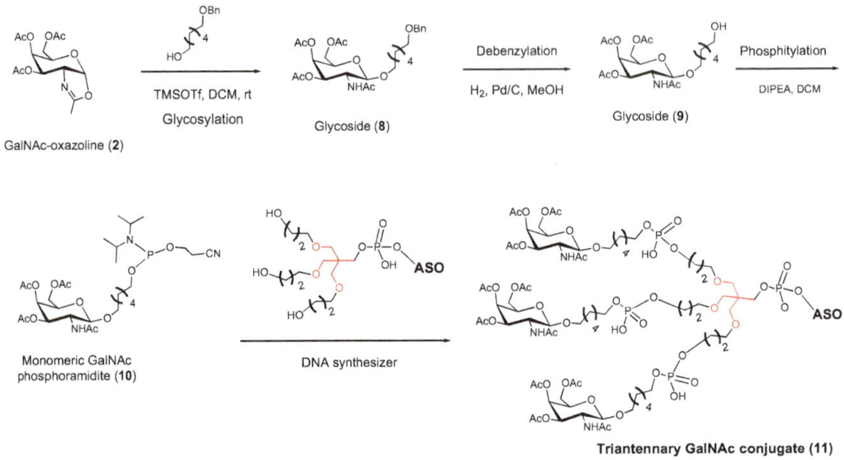

Scheme 16.2 ***Synthesis of monomeric GalNAc phosphoramidite and solid-phase coupling with 5′-trebler-functionalized ASO.***[80]

1.12.2 Monomeric linear GalNAc

The monomeric approach involves the synthesis of monovalent GalNAc building block and its incorporation during solid-phase oligonucleotide synthesis (Scheme 16.3). This approach is straightforward and requires fewer synthetic steps than pre-assembly of triantennary GalNAc clusters on solid support. Further, the GalNAc phosphoramidite monomers allow the rapid screening of different designs due to flexibility in the conjugation site and linkers connecting GalNAc ligand to oligonucleotides.[81]

Recently, in addition to triantennary GalNAc design, it has been shown that three GalNAc moieties in a linear sequential manner at the 3'-end of siRNA are also effective in binding to ASGPR.[81] The hydroxymethylpyrrolidine-based monomeric GalNAc phosphoramidite (**12**)[78,81] was synthesized (Scheme 16.4) and incorporated during solid-phase oligonucleotide synthesis (Scheme 16.3). It is important that the *in vitro* and *in vivo* potency of these linear trivalent GalNAc conjugates (**13**) are similar to triantennary GalNAc conjugates (**7**).

Yamamoto et al., coupled the hydroxymethylpyrrolidine based monomeric GalNAc phosphoramidite to ASOs during the solid-phase oligonucleotide synthesis (Scheme 16.4).[78] The serial incorporation up to five GalNAc, either at 3'- or 5'-terminus of ASOs was achieved in linear fashion (toothbrush design, Fig. 16.8). It is interesting that the gene-silencing activity of ASOs was found to increase with the increase in the number of GalNAc ligands.

Sharma et al., developed a threoninol-based monomeric GalNAc ligand which was synthesized from monomeric GalNAc phosphoramidite (**16**) (Scheme 16.5).[82]The sequential linear incorporation of GalNAc ligands

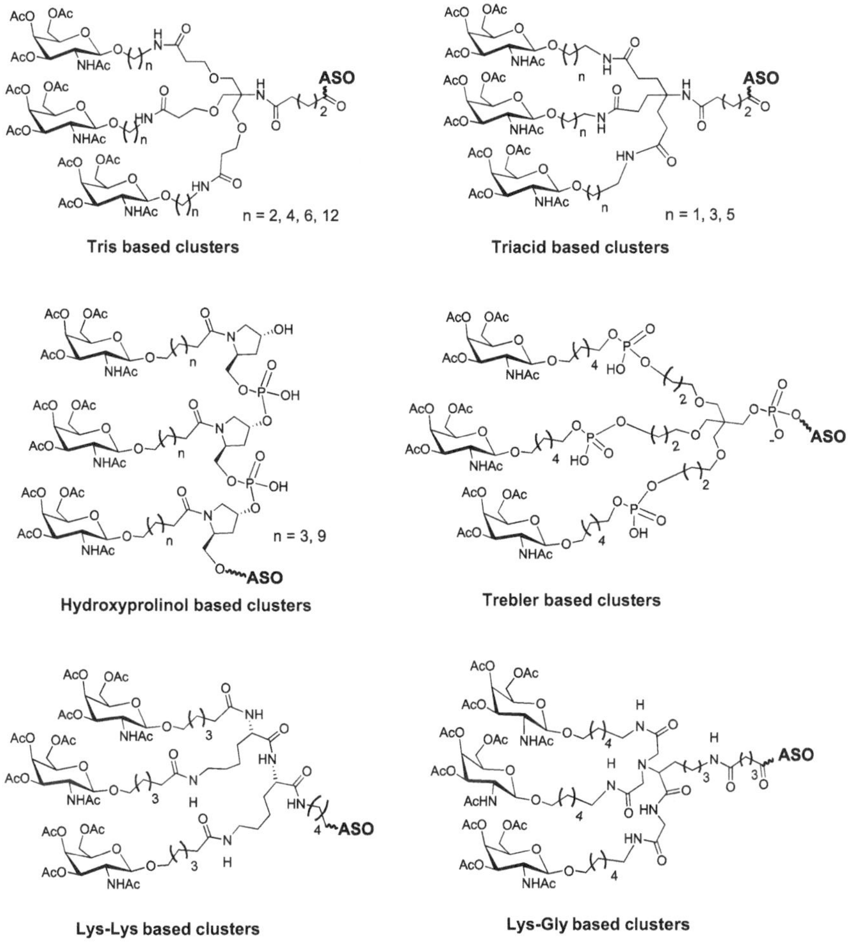

Figure 16.9 ***Structures of the different scaffolds used for structure-activity relationship of triantennary GalNAc clusters conjugated to ASOs.***[11]

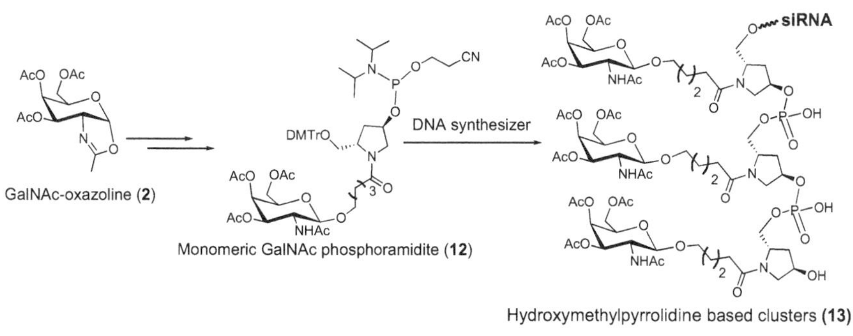

Scheme 16.3 ***General strategy for the synthesis of linear monomeric GalNAc clusters (13).***

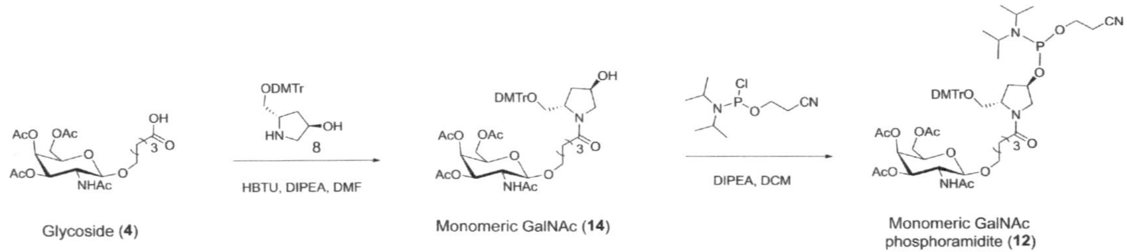

Scheme 16.4 ***Synthesis of monomeric GalNAc for polyvalent GalNAc conjugate approach for liver-specific delivery of oligonucleotide drugs.***[78]

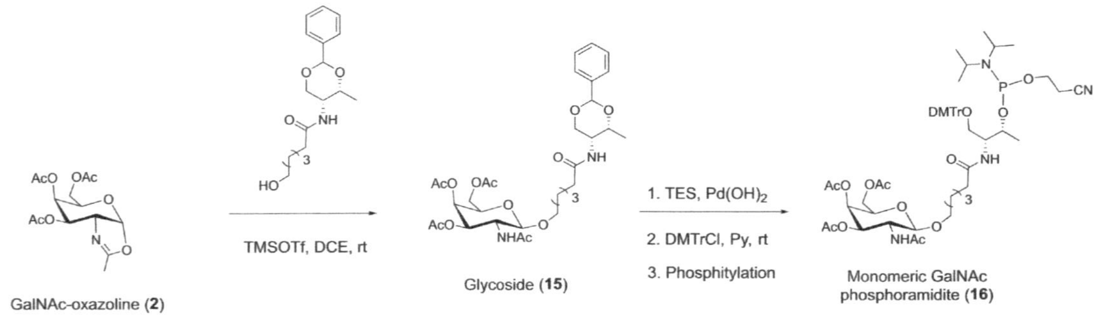

Scheme 16.5 ***Synthesis of monomeric GalNAc for polyvalent GalNAc conjugate approach for liver-specific delivery of oligonucleotide drugs.***

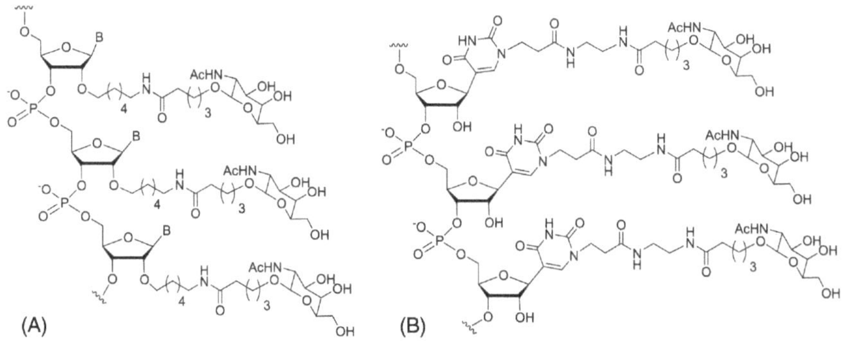

Figure 16.10 ***Nucleosidic GalNAc linear monomer approach.*** [74]

(Fig. 16.8) lead to increased potency in mice. In this study, the potency of four linear GalNAc ligands (tetra-valent) was found to be more than three or two GalNAc ligands. However, surprisingly the tetra-valent threoninol-based GalNAc conjugate demonstrated low potency compared to the tri-antennary GalNAc conjugate (**7**).

Matsuda et al., synthesized the nucleoside based monomeric GalNAc building blocks. The GalNAc moiety was linked either through 2' or 3' oxygen atom of nucleosides (Fig. 16.10A) or N1 of pseudouridine (Fig. 16.10B).[74] The linear incorporation of three nucleosidic GalNAc moieties at or near 3'-end of the siRNA retained *in vitro* and *in vivo* silencing activity.

Following the linear polyvalent approach, Dicerna Pharmaceuticals is developing siRNA drugs based on the GalXC platform.[83] The GalXC siRNA drugs can be delivered subcutaneously to hepatocytes in the liver. In this unique platform, folded tetra loop structures are used to connect two strands of siRNA (Fig. 16.11). Now four GalNAc ligands can be attached, one GalNAc per nucleotide in the tetraloop (Fig. 16.11). The GalNAc moieties are introduced at the 2' position of tetra loop nucleotide ssimilar to Fig. 16.10A. Due to asymmetry, two strands in the GalXC molecule can be discriminated by Dicer enzyme, resulting in loading of antisense siRNA strand to RNAi machinery (Fig. 16.11).

1.13 Further improvements in GalNAc-oligonucleotide conjugates

One of the biggest challenges for the therapeutic GalNAc-oligonucleotide conjugates is hepatotoxicity possibly due to intracellular accumulation of

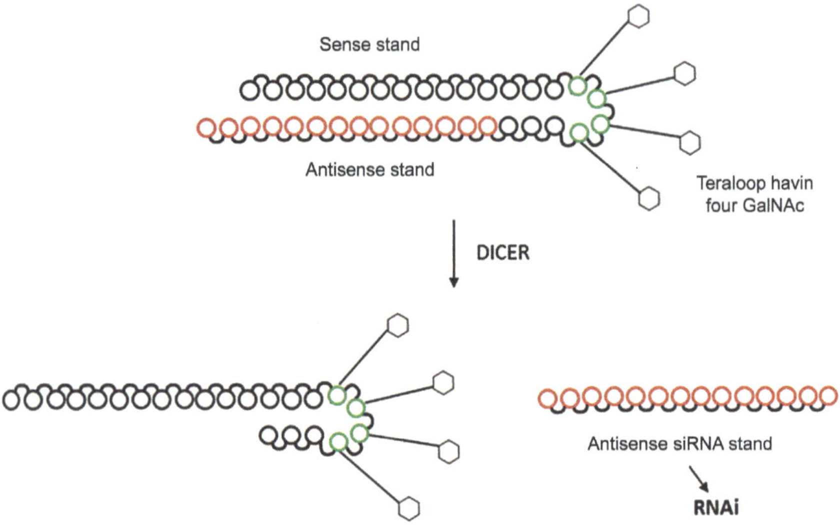

Figure 16.11 ***The GalXC siRNA platform.***[83]

the oligonucleotides or their metabolites or due to other off-target effects of these oligonucleotides.[84–85] Therefore it is necessary to overcome this challenge to make safer and efficient GalNAc-oligonucleotide conjugates. Zlatev et al., recently reported a novel approach termed as REVERSIR, to reduce the hepatotoxity observed with GalNAc-conjugated siRNAs.[85] The REVERSIR is a short LNA-modified oligonucleotide (9-mer) complementary to the GalNAc-siRNA. The high affinity REVERSIR template in a single subcutaneous administration (3mg/kg)was capable of reversing the TTR-siRNA response in mice. This approach will provide fine control over the duration of action of GalNAc-siRNAs.

The GalNAc conjugation strategy is continuously evolving with the next generation nucleic acid modifications called 'enhance stabilization chemistry' (ESC).[86] The ESC-GalNAc platform further improves the efficacy and potency of oligonucleotide drugs.[13,86] The ESC platform is about using chemical modifications such as PS backbone, 2'-OMe or 2'-F modifications, 5'-vinylphosphonate (VP), 5'-morpholino (Mo), inverted abasic (iB) and Glycol nucleic acid (GNA) at specific positions.[84] Today, the most common siRNA design is the combination of 2'-OMe and 2'-F modifications in siRNA with terminal phosphorothioate (PS) linkages (Fig. 16. 12).[70] The field has progressed from ESC to ESC+ platform by reducing the 2'-F content with 2'-OMe nucleotides which significantly enhance the metabolic stability of siRNA drugs which leads to increase

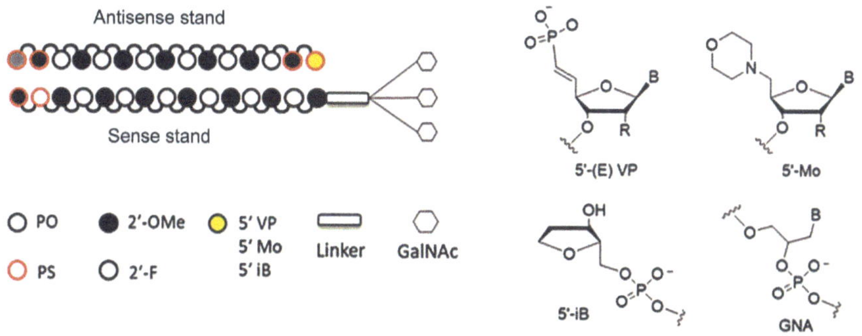

Figure 16.12 ***Examples of the Enhance Stabilization Chemistry (ESC) platform for siRNA drugs.***

in the potency and duration of RNAi.[87] The metabolic stability of siRNA drugs is further enhanced with 5'-end modifications such as 5'-vinylphosphonate (VP), morpholino (Mo), inverted a basic (iB) (Fig. 16.12) which provide protection from 5'-exonucleases.[84] 5'-Vinylphosphonate (VP) modification alone has been found to improve the potency of siRNA drugs upto 10 folds.[88–90]

The Glycol nucleic acid (GNA) modification in the antisense siRNA strand at positions 7 and 8 (seed region) reduces the off-target effects without compromising the on-target activity.[84] The acyclic GNA modification has much lower binding-affinity than the natural nucleic acids.[91–92] These results prove that the hybridization-based off-target effects regulated by the seed region of the antisense strand are the major cause for the observed hepatotoxicity with siRNAs which provide protection against 3' and 5'-exonucleases.

2 Clinical status of GalNAc-conjugated oligonucleotide drugs

The GalNAc-conjugation has emerged as a powerful technology to deliver oligonucleotide drugs into the liver. As discussed in previous sections, the medicinal chemistry efforts have made GalNAc technology even more powerful by enhancing the potency and specificity and reducing toxicity profiles of the GalNAc-oligonucleotide drugs. All these advances have pushed a significant number of GalNAc-conjugated oligonucleotides in the advanced clinical trials (Table 16. 2 and Table 16.3).

Table 16.2 The clinical status of GalNAc-ASO drugs being developed by Ionis Pharmaceuticals.[93]

Drug	Indication(s)	Target	Clinical Phase
AKCEA-APO(a)-L_{Rx}[94–95]	High lipoprotein(a)	Apo(a)	II
AKCEA-ANGPTL3-L_{RX}[96]	Rare Hyperlipidemias	ANGPTL3	II
IONIS-FB-L_{Rx}[97]	Complement-Mediated Diseases	Complement Factor B	II
IONIS-HBV-L_{Rx}[98]	Hepatitis B virus infection	HBV genome	II
AKCEA-APOCIII-L_{Rx}[99]	Severe high triglycerides / High triglycerides with type 2 diabetes	APOCIII	II
AKCEA-ANGPTL3-L_{Rx}	NAFLD/Metabolic Complications	ANGPTL3	II
AKCEA-TTR-L_{Rx}	TTR Amyloidosis (ATTR)	Transthyretin (TTR)	I
IONIS-TMPRSS6-L_{Rx}[100]	Beta-thalassemia and Iron Toxicity	β-Thalassemia	I
IONIS-AGT-L_{Rx}[101]	Treatment-Resistant Hypertension	AGT	I
IONIS-GHR-L_{Rx}[102]	Acromegaly	GHr	II
IONIS-GSK4-L_{Rx}	Ocular disease	Undisclosed targets	I
IONIS-FXI-L_{RX}	Clotting disorder	Factor XI	I

Table 16.3 The clinical status of GalNAc-siRNA drugs being developed by Alnylam Pharmaceuticals[103] and Arrowhead Pharmaceuticals[104], and micro-RNA targeting GalNAc-anti-miR developed by Regulus Therapeutics[104] and Dicerna Pharmaceuticals.[104]

Drug	Indication(s)	Target	Clinical Phase
ALN-PCSsc/ Inclisiran[105]	Hypercholesterolemia	PCSK9	III
ALN-AT3/ Fitusiran[106]	Hemophilia and rare bleeding disorders	Antithrombin (AT)	III
ALN-AS1/ Givosiran[107]	Acute Hepatic porphyrias	ALAS1	Approved
ALN-CC5/ Cemdisiran[108]	Paroxysmal nocturnal hemoglobinuria	C5	II
ALN-GO1/ Lumasiran[109]	Primary hyperoxaluria type 1	GO (HAO1)	II
ALN-HBV02[110]	Hepatitis B virus infection	conserved region of HBV	II

(Continued)

Table 16.3 The clinical status of GalNAc-siRNA drugs being developed by Alnylam Pharmaceuticals[103] and Arrowhead Pharmaceuticals[104], and micro-RNA targeting GalNAc-anti-miR developed by Regulus Therapeutics[104] and Dicerna Pharmaceuticals.[104] (*Cont.*)

Drug	Indication(s)	Target	Clinical Phase
ALN-TTRsc02	Hereditary ATTR amyloidosos	TTR	II
DCR-PHXC	Primary Hyperoxaluria (PH)	LDHA	II
ARO-HBV	Hepatitis B virus infection (HBV)	integrated DNA of HBV genome	II
ARO-AAT	Alpha-1 antitrypsin deficiency	AATD	II
RG-101[25]	Hepatitis C virus infection	microRNA-122	I
RG-125	NASH/type 2 diabetes/ pre-diabetes	microRNA-103/107	I
ALN-AGT	Hypertension	angiotensinogen (AGT)	I

3 Conclusions and future perspectives

Seven drug approvals in the last four years show that the field of oligonucleotide therapeutics has made great strides. This is mainly due to advances in the medicinal chemistry of synthetic oligonucleotides which began from phosphorothiaote (PS) backbone modification and has expanded to include a wide range of tools having different degrees of specificity and sophistication. Current efforts are aimed at solving the long-standing delivery problems of oligonucleotides.

One successful approach is the *N*-acetylgalactosamine (GalNAc)-based therapeutic oligonucleotides which can be specifically delivered to hepatocytes. In addition to specific delivery, GalNAc-oligonucleotide conjugates are also found to be up to 100 folds more potent than its unconjugated counterpart. Further developments in the medicinal chemistry have led to Enhance Stabilization Chemistry (ESC) and ESC+ platforms for GalNAc-conjugated oligonucleotides. These advances have resulted in more than 25 oligonucleotide drugs into advanced clinical trials for various liver diseases and have sparked the quest for conjugates targeting other cell and tissue types.

References

1. Stein, C. A.; Castanotto, D. *Mol. Ther.* **2017**, *25*, 1069–1075.
2. Khvorova, A.; Watts, J. K. *Nat. Biotechnol.* **2017**, *35*, 238–248.
3. Sharma, V. K.; Sharma, R. K.; Singh, S. K. *Med. Chem. Comm.* **2014**, *5*, 1454–1471.
4. Sharma, V. K.; Watts, J. K. *Future Med. Chem.* **2015**, 7, 2221–2242.
5. Deleavey, Glen F.; Damha, M.J. Chem. Biol., 2012, 19, 937-954.
6. Wan, W. B.; Seth, P. P. *J. Med. Chem.* **2016**, *59*, 9645–9667.
7. Juliano, R. L. *Nucleic Acids Res.* **2016**, *44*, 6518–6548.
8. Kim, Y.; Jo, M.; Schmidt, J.; Luo, X.; Prakash, T. P.; Zhou, T.; Klein, S.; Xiao, X.; Post, N.; Yin, Z.; MacLeod, A. R. *Mol. Ther.* **2019**, *27*, 1547–1557.
9. Prakash, T. P.; Graham, M. J.; Yu, J.; Carty, R.; Low, A.; Chappell, A.; Schmidt, K.; Zhao, C.; Aghajan, M.; Murray, H. F.; Riney, S.; Booten, S. L.; Murray, S. F.; Gaus, H.; Crosby, J.; Lima, W. F.; Guo, S.; Monia, B. P.; Swayze, E. E.; Seth, P. P. *Nucleic Acids Res.* **2014**, *42*, 8796–8807.
10. Springer, A. D.; Dowdy, S. F. *Nucleic acid Ther.* **2018**.
11. Prakash, T. P.; Yu, J.; Migawa, M. T.; Kinberger, G. A.; Wan, W. B.; Ostergaard, M. E.; Carty, R. L.; Vasquez, G.; Low, A.; Chappell, A.; Schmidt, K.; Aghajan, M.; Crosby, J.; Murray, H. M.; Booten, S. L.; Hsiao, J.; Soriano, A.; Machemer, T.; Cauntay, P.; Burel, S. A.; Murray, S. F.; Gaus, H.; Graham, M. J.; Swayze, E. E.; Seth, P. P. *J. Med. Chem.* **2016**, *59*, 2718–2733.
12. Rajeev, K.G.; Manoharan, M. Liver-targeted RNAi Therapeutics: Principles and Applications. In Advances in Nucleic Acid Therapeutics, *The Royal Society of Chemistry:* **2019**; pp 233-265.
13. Huang, Y. *Nucleic Acids* **2017**, *6*, 116–132.
14. Zamecnik, P. C.; Stephenson, M. L. *Proc. Nat. Acad.* Sc. *U.S.A.* **1978**, *75*, 280–284.
15. Stephenson, M. L.; Zamecnik, P. C. *Proc. Nat. Acad.* Sc. *U.S.A.* **1978**, *75*, 285–288.
16. Zamecnik, P. C.; Goodchild, J.; Taguchi, Y.; Sarin, P. S. *Proc. Nat. Acad.* Sc. *U.S.A.* **1986**, *83*, 4143–4146.
17. Roehr, B. *J. Int. Assoc. Physicians in AIDS Care* **1998**, *4*, 14–16.
18. Sharma, V. K.; Rungta, P.; Prasad, A. K. *RSC Adv.* **2014**, *4*, 16618–16631.
19. Bennett, C. F.; Swayze, E. E. *Ann. Rev. Pharmac. Toxicol.* **2010**, *50*, 259–293.
20. Watts, J.K. The Medicinal Chemistry of Antisense Oligonucleotides. In Oligonucleotide-Based Drugs Ther., 2018; pp 39-89.
21. Watts, J. K.; Corey, D. R. *J. Pathology* **2012**, *226*, 365–379.
22. Shen, X.; Corey, D. R. *Nucl. Acids Res.* **2018**, *46*, 1584–1600.
23. Lima, W. F.; Prakash, T. P.; Murray, H. M.; Kinberger, G. A.; Li, W.; Chappell, A. E.; Li, C. S.; Murray, S. F.; Gaus, H.; Seth, P. P.; Swayze, E. E.; Crooke, S. T. *Cell* **2012**, *150*, 883–894.
24. Gebert, L. F. R.; Rebhan, M. A. E.; Crivelli, S. E. M.; Denzler, R.; Stoffel, M.; Hall, J. *Nucleic Acids Res.* **2014**, *42*, 609–621.
25. van der Ree, M. H.; de Vree, J. M.; Stelma, F.; Willemse, S.; van der Valk, M.; Rietdijk, S.; Molenkamp, R.; Schinkel, J.; van Nuenen, A. C.; Beuers, U.; Hadi, S.; Harbers, M.; van der Veer, E.; Liu, K.; Grundy, J.; Patick, A. K.; Pavlicek, A.; Blem, J.; Huang, M.; Grint, P.; Neben, S.; Gibson, N. W.; Kootstra, N. A.; Reesink, H. W. *Lancet* **2017**, *389*, 709–717.
26. Charleston, J. S.; Schnell, F. J.; Dworzak, J.; Donoghue, C.; Lewis, S.; Chen, L.; Young, G. D.; Milici, A. J.; Voss, J.; DeAlwis, U.; Wentworth, B.; Rodino-Klapac, L. R.; Sahenk, Z.; Frank, D.; Mendell, J. R. *Neurology* **2018**, *90*, e2146–e2154.
27. Jason, M. *Cur. Med. Chem.* **2001**, *8*, 1157–1179.
28. Wang, Y.; Yu, R. Z.; Henry, S.; Geary, R. S. *Exp. Opin. Drug Metab. Toxicol.* **2019**, *15*, 475–485.
29. Crooke, S. T.; Baker, B. F.; Xia, S.; Yu, R. Z.; Viney, N. J.; Wang, Y.; Tsimikas, S.; Geary, R. S. *Nucleic Acid Therap.* **2018**, *29*, 16–32.
30. Sharma, V. K.; Watts, J. K. *Fut. Med. Chem.* **2015**, 7, 2221–2242.

31. Fritz, E. *Nucleic Acid Therap.* **2014**, *24*, 374–387.
32. Eckstein, F. *Antisense Nucl. Acid Drug Dev.* **2000**, *10*, 117–121.
33. Stein, C. A.; Subasinghe, C.; Shinozuka, K.; Cohen, J. S. *Nucleic Acids Res.* **1988**, *16*, 3209–3221.
34. Geary, R. S. *Exp. Opin. Drug Metab. Toxicol.* **2009**, *5*, 381–391.
35. Pirollo, K. F.; Rait, A.; Sleer, L. S.; Chang, E. H. *Pharmacol. Therap.* **2003**, *99*, 55–77.
36. Furdon, P. J.; Dominski, Z.; Kole, R. *Nucleic Acids Res.* **1989**, *17*, 9193–9204.
37. Meade, B. R.; Gogoi, K.; Hamil, A. S.; Palm-Apergi, C.; Berg, A. v. d.; Hagopian, J. C.; Springer, A. D.; Eguchi, A.; Kacsinta, A. D.; Dowdy, C. F.; Presente, A.; Lönn, P.; Kaulich, M.; Yoshioka, N.; Gros, E.; Cui, X. -S.; Dowdy, S. F. *Nature Biotechnol.* **2014**, *32*, 1256.
38. Prakash, T. P. *Chem. Biodiv.* **2011**, *8*, 1616–1641.
39. Wengel, J. *Acc. Chem. Res.* **1999**, *32*, 301–310.
40. Ito, K.R.; Obika, S. Recent Advances in Medicinal Chemistry of Antisense Oligonucleotides. *In Reference Module in Chemistry, Molecular Sciences and Chemical Engineering, Elsevier:* **2016**.
41. Lundin, K. E.; Hojland, T.; Hansen, B. R.; Persson, R.; Bramsen, J. B.; Kjems, J.; Koch, T.; Wengel, J.; Smith, C. I. *Adv. Genetics* **2013**, *82*, 47–107.
42. Watts, J. K. *Chem. Commun.* **2013**, *49*, 5618–5620.
43. Hagedorn, P. H.; Persson, R.; Funder, E. D.; Albaek, N.; Diemer, S. L.; Hansen, D. J.; Moller, M. R.; Papargyri, N.; Christiansen, H.; Hansen, B. R.; Hansen, H. F.; Jensen, M. A.; Koch, T. *Drug Disc. Today* **2018**, *23*, 101–114.
44. Sharma, V. K.; Kumar, M.; Olsen, C. E.; Prasad, A. K. *J. Org. Chem.* **2014**, *79*, 6336–6341.
45. Yu, R. Z.; Grundy, J. S.; Geary, R. S. *Exp. Opin. Drug Metab. Toxicol.* **2013**, *9*, 169–182.
46. Stanton, R.; Sciabola, S.; Salatto, C.; Weng, Y.; Moshinsky, D.; Little, J.; Walters, E.; Kreeger, J.; DiMattia, D.; Chen, T.; Clark, T.; Liu, M.; Qian, J.; Roy, M.; Dullea, R. *Nucleic Acid Therap.* **2012**, *22*, 344–359.
47. Monia, B. P.; Lesnik, E. A.; Gonzalez, C.; Lima, W. F.; McGee, D.; Guinosso, C. J.; Kawasaki, A. M.; Cook, P. D.; Freier, S. M. *J. Biol. Chem.* **1993**, *268*, 14514–14522.
48. Crooke, S. T.; Witztum, J. L.; Bennett, C. F.; Baker, B. F. *CellMetab.* **2018**, *27*, 714–739.
49. Nishina, T.; Numata, J.; Nishina, K.; Yoshida-Tanaka, K.; Nitta, K.; Piao, W.; Iwata, R.; Ito, S.; Kuwahara, H.; Wada, T.; Mizusawa, H.; Yokota, T. *Mol. Therapy. Nucleic Acids* **2015**, *4*, e220.
50. Fujino, T.; Suzuki, T.; Okada, K.; Kogashi, K.; Yasumoto, K. -i.; Sogawa, K.; Isobe, H. *J. Org. Chem.* **2016**, *81*, 8967–8976.
51. Pallan, P. S.; Allerson, C. R.; Berdeja, A.; Seth, P. P.; Swayze, E. E.; Prakash, T. P.; Egli, M. *Chem. Commun.* **2012**, *48*, 8195–8197.
52. Seth, P. P.; Siwkowski, A.; Allerson, C. R.; Vasquez, G.; Lee, S.; Prakash, T. P.; Wancewicz, E.V.; Witchell, D.; Swayze, E. E. *J. Med. Chem.* **2009**, *52*, 10–13.
53. Iversen, P. L. *Curr. Opin. Mol. Therap.* **2001**, *3*, 235–238.
54. Arora, V.; Devi, G. R.; Iversen, P. L. *Current Pharm. Biotech.* **2004**, *5*, 431–439.
55. Amantana, A.; Iversen, P. L. *Curr. Opin. Pharmac.* **2005**, *5*, 550–555.
56. Lim, K. R. Q.; Maruyama, R.; Yokota, T. *Drug Design, Devel. Therapy* **2017**, *11*, 533–545.
57. Herdewijn, P. *Antisense Nucleic Acid Drug Dev.* **2000**, *10*, 297–310.
58. Verma, S.; Eckstein, F. *Ann. Rev. Biochem.* **1998**, *67*, 99–134.
59. Peacock, H.; Kannan, A.; Beal, P. A.; Burrows, C. J. *J. Org. Chem.* **2011**, *76*, 7295–7300.
60. Freier, S. M.; Altmann, K. -H. *Nucleic Acids Res.* **1997**, *25*, 4429–4443.
61. Takeda, K.; Akira, S. *Int. Immun.* **2005**, *17*, 1–14.
62. Dalpke, A.; Frank, J.; Peter, M.; Heeg, K. *Infection Immunity* **2006**, *74*, 940–946.
63. Henry, S.; Stecker, K.; Brooks, D.; Monteith, D.; Conklin, B.; Bennett, C. F. *J. Pharm. Exp. Ther.* **2000**, *292*, 468–479.
64. Lin, K. -Y.; Matteucci, M. D. *J. Am. Chem. Soc.* **1998**, *120*, 8531–8532.
65. Flanagan, W. M.; Wagner, R. W.; Grant, D.; Lin, K. -Y.; Matteucci, M. D. *Nature Biotech.* **1999**, *17*, 48.

66. Benizri, S.; Gissot, A.; Martin, A.; Vialet, B.; Grinstaff, M. W.; Barthélémy, P. *Bioconjugate Chem.* **2019**, *30*, 366–383.
67. Schwartz, A. L.; Fridovich, S. E.; Lodish, H. F. *J. Biol. Chem.* **1982**, *257*, 4230–4237.
68. Spiess, M. *Biochemistry* **1990**, *29*, 10009–10018.
69. Ashwell, G.; Harford, J. *Ann. Rev. Biochem.* **1982**, *51*, 531–554.
70. Nair, J. K.; Willoughby, J. L.; Chan, A.; Charisse, K.; Alam, M. R.; Wang, Q.; Hoekstra, M.; Kandasamy, P.; Kel'in, A.V.; Milstein, S.; Taneja, N.; O'Shea, J.; Shaikh, S.; Zhang, L.; van der Sluis, R. J.; Jung, M. E.; Akinc, A.; Hutabarat, R.; Kuchimanchi, S.; Fitzgerald, K.; Zimmermann, T.; van Berkel, T. J.; Maier, M. A.; Rajeev, K. G.; Manoharan, M. *J. Am. Chem. Soc.* **2014**, *136*, 16958–16961.
71. Schmidt, K.; Prakash, T. P.; Donner, A. J.; Kinberger, G. A.; Gaus, H. J.; Low, A.; Ostergaard, M. E.; Bell, M.; Swayze, E. E.; Seth, P. P. *Nucleic Acids Res.* **2017**, *45*, 2294–2306.
72. Bider, M. D.; Spiess, M. *FEBS Lett.* **1998**, *434*, 37–41.
73. Bridges, K.; Harford, J.; Ashwell, G.; Klausner, R. D. *Proc. Natl. Acad. Sci. U.S.A.* **1982**, *79*, 350–354.
74. Matsuda, S.; Keiser, K.; Nair, J. K.; Charisse, K.; Manoharan, R. M.; Kretschmer, P.; Peng, C. G.; A, V. K. i.; Kandasamy, P.; Willoughby, J. L.; Liebow, A.; Querbes, W.; Yucius, K.; Nguyen, T.; Milstein, S.; Maier, M. A.; Rajeev, K. G.; Manoharan, M. ACS *Chem. Biol.* **2015**, *10*, 1181–1187.
75. Sharma, V. K.; Osborn, M. F.; Hassler, M.; Echeverria, D.; Ly, S.; Ulashchik, E.; Martynenko-Makaev, Y.; Shmanai, V.V.; Zatsepin, T. S.; Khvorova, A.; Watts, J. *Bioconjugate Chem.* **2018**, *29*, 2478–2488.
76. Kinberger, G.A.; Prakash, T. P.; Yu, J.; Vasquez, G.; Low, A.; Chappell, A.; Schmidt, K.; Murray, H. M.; Gaus, H.; Swayze, E. E.; Seth, P. P. *Bioorg. Med. Chem. Lett.* **2016**, *26*, 3690–3693.
77. Wada, F.; Yamamoto, T.; Ueda, T.; Sawamura, M.; Wada, S.; Harada-Shiba, M.; Obika, S. *Nucleic Acid Ther.* **2018**, *28*, 50–57.
78. Yamamoto, T.; Sawamura, M.; Wada, F.; Harada-Shiba, M.; Obika, S. *Bioorg. Med. Chem.* **2016**, *24*, 26–32.
79. Ostergaard, M. E.; Yu, J.; Kinberger, G. A.; Wan, W. B.; Migawa, M. T.; Vasquez, G.; Schmidt, K.; Gaus, H. J.; Murray, H. M.; Low, A.; Swayze, E. E.; Prakash, T. P.; Seth, P. P. *Bioconjug. Chem.* **2015**, *26*, 1451–1455.
80. Prakash, T. P.; Brad Wan, W.; Low, A.; Yu, J.; Chappell, A. E.; Gaus, H.; Kinberger, G. A.; Ostergaard, M. E.; Migawa, M. T.; Swayze, E. E.; Seth, P. P. *Bioorg. Med. Chem. Lett.* **2015**, *25*, 4127–4130.
81. Rajeev, K. G.; Nair, J. K.; Jayaraman, M.; Charisse, K.; Taneja, N.; O'Shea, J.; Willoughby, J. L.; Yucius, K.; Nguyen, T.; Shulga-Morskaya, S.; Milstein, S.; Liebow, A.; Querbes, W.; Borodovsky, A.; Fitzgerald, K.; Maier, M. A.; Manoharan, M. *ChemBioChem.* **2015**, *16*, 903–908.
82. Sharma, V. K.; Osborn, M. F.; Hassler, M. R.; Echeverria, D.; Ly, S.; Ulashchik, E. A.; Martynenko-Makaev, Y. V.; Shmanai, V. V.; Zatsepin, T. S.; Khvorova, A.; Watts, J. K. *Bioconjugate Chem.* **2018**, *29*, 2478–2488.
83. Marc Abrams, G. C. Jihye Park Compositionsand Methods for Inhibiting HMGB1 Expession, US 2019/0000870 A1.
84. Janas, M. M.; Schlegel, M. K.; Harbison, C. E.; Yilmaz, V. O.; Jiang, Y.; Parmar, R.; Zlatev, I.; Castoreno, A.; Xu, H.; Shulga-Morskaya, S.; Rajeev, K. G.; Manoharan, M.; Keirstead, N. D.; Maier, M. A.; Jadhav, V. *Nature Commun.* **2018**, *9*, 723.
85. Zlatev, I.; Castoreno, A.; Brown, C. R.; Qin, J.; Waldron, S.; Schlegel, M. K.; Degaonkar, R.; Shulga-Morskaya, S.; Xu, H.; Gupta, S.; Matsuda, S.; Akinc, A.; Rajeev, K. G.; Manoharan, M.; Maier, M. A.; Jadhav, V. *Nature Biotech.* **2018**, *36*, 509.
86. Nair, J. K.; Attarwala, H.; Sehgal, A.; Wang, Q.; Aluri, K.; Zhang, X.; Gao, M.; Liu, J.; Indrakanti, R.; Schofield, S.; Kretschmer, P.; Brown, C. R.; Gupta, S.; Willoughby, J. L. S.; Boshar, J. A.; Jadhav, V.; Charisse, K.; Zimmermann, T.; Fitzgerald, K.; Manoharan, M.; Rajeev, K. G.; Akinc, A.; Hutabarat, R.; Maier, M. A. *Nucleic Acids Res.* **2017**, *45*, 10969–10977.

87. Foster, D. J.; Brown, C. R.; Shaikh, S.; Trapp, C.; Schlegel, M. K.; Qian, K.; Sehgal, A.; Rajeev, K. G.; Jadhav, V.; Manoharan, M.; Kuchimanchi, S.; Maier, M. A.; Milstein, S. *Mol. Ther.* **2018**, *26*, 708–717.
88. Parmar, R.; Willoughby, J. L.; Liu, J.; Foster, D. J.; Brigham, B.; Theile, C. S.; Charisse, K.; Akinc, A.; Guidry, E.; Pei, Y.; Strapps, W.; Cancilla, M.; Stanton, M. G.; Rajeev, K. G.; Sepp-Lorenzino, L.; Manoharan, M.; Meyers, R.; Maier, M. A.; Jadhav, V. *ChemBioChem.* **2016**, *17*, 985–989.
89. Parmar, R. G.; Brown, C. R.; Matsuda, S.; Willoughby, J. L. S.; Theile, C. S.; Charisse, K.; Foster, D. J.; Zlatev, I.; Jadhav, V.; Maier, M. A.; Egli, M.; Manoharan, M.; Rajeev, K. G. *J. Med. Chem.* **2018**, *6*, 734–744.
90. Elkayam, E.; Parmar, R.; Brown, C. R.; Willoughby, J. L.; Theile, C. S.; Manoharan, M.; Joshua-Tor, L. *Nucleic Acids Res.* **2017**, *45*, 3528–3536.
91. Schlegel, M. K.; Foster, D. J.; Kel'in, A. V.; Zlatev, I.; Bisbe, A.; Jayaraman, M.; Lackey, J. G.; Rajeev, K. G.; Charissé, K.; Harp, J.; Pallan, P. S.; Maier, M. A.; Egli, M.; Manoharan, M. *J. Am. Chem. Soc.* **2017**, *139*, 8537–8546.
92. Zhang, L.; Peritz, A.; Meggers, E. *J. Am. Chem. Soc.* **2005**, *127*, 4174–4175.
93. http://www.ionispharma.com/pipeline/. (accessed 2019)
94. Yu, R. Z.; Gunawan, R.; Post, N.; Zanardi, T.; Hall, S.; Burkey, J.; Kim, T. W.; Graham, M. J.; Prakash, T. P.; Seth, P. P.; Swayze, E. E.; Geary, R. S.; Henry, S. P.; Wang, Y. *Nucleic Acid Therap.* **2016**, *26*, 372–380.
95. Viney, N.; Marcovina, S.; Xia, S.; Geary, R.; Witztum, J.; Tsimikas, S. *J. Am. College of Cardiology* **2016**, *67*, 2081.
96. Yamamoto, T.; Wada, F.; Harada-Shiba, M. *J. Atheroscler. Thromb.* **2016**, *23*, 1011–1025.
97. Grossman, T. R.; Johnson, R. B.; Shen, L.; Hettrick, L. A.; Henry, S. P.; Monia, B. P.; Mc-Caleb, M. *Investigative Ophthal. Visual* Sc. **2016**, *57*, 5000.
98. Boucle, S.; Bassit, L.; Ehteshami, M.; Schinazi, R. F. *Clin. Liver Dis.* **2016**, *20*, 737–749.
99. Jorgensen, A. B.; Frikke-Schmidt, R.; Nordestgaard, B. G.; Tybjaerg-Hansen, A. *New Eng. J. Med.* **2014**, *371*, 32–41.
100. Makis, A.; Hatzimichael, E.; Papassotiriou, I.; Voskaridou, E. *Am. J. Hematology* **2016**, *91*, 1135–1145.
101. Williams, B. *Therap. Adv. Cardiovascular Dis.* **2016**, *10*, 118–125.
102. Maffezzoni, F.; Formenti, A. M.; Mazziotti, G.; Frara, S.; Giustina, A. *Exp. Opin. Pharmac.* **2016**, *17*, 1631–1642.
103. http://www.alnylam.com/alnylam-rnai-pipeline/ (accessed 2019).
104. http://arrowheadpharma.com/pipeline/. (accessed 2019)
105. Nishikido, T.; Ray, K. K. *Exp. Opin. Investigat. Drugs* **2018**, *27*, 287–294.
106. Sehgal, A.; Barros, S.; Ivanciu, L.; Cooley, B.; Qin, J.; Racie, T.; Hettinger, J.; Carioto, M.; Jiang, Y.; Brodsky, J.; Prabhala, H.; Zhang, X.; Attarwala, H.; Hutabarat, R.; Foster, D.; Milstein, S.; Charisse, K.; Kuchimanchi, S.; Maier, M. A.; Nechev, L.; Kandasamy, P.; Kel'in, A. V.; Nair, J. K.; Rajeev, K. G.; Manoharan, M.; Meyers, R.; Sorensen, B.; Simon, A. R.; Dargaud, Y.; Negrier, C.; Camire, R. M.; Akinc, A. *Nature Med.* **2015**, *21*, 492–497.
107. Chan, A.; Liebow, A.; Yasuda, M.; Gan, L.; Racie, T.; Maier, M.; Kuchimanchi, S.; Foster, D.; Milstein, S.; Charisse, K.; Sehgal, A.; Manoharan, M.; Meyers, R.; Fitzgerald, K.; Simon, A.; Desnick, R. J.; Querbes, W. *Mol. Ther. Nucl. Acids* **2015**, *4*, e263.
108. Borodovsky, A.; Yucius, K.; Sprague, A.; Banda, N. K.; Holers, V. M.; Vaishnaw, A.; Maier, M.; Kallanthottathil, R.; Charisse, K.; Kuchimanchi, S.; Manoharan, M.; Salant, D. J.; Fitzgerald, K.; Meyers, R.; Sorensen, B. *Blood* **2014**, *124*, 1606.
109. Liebow, A.; Li, X.; Racie, T.; Hettinger, J.; Bettencourt, B. R.; Najafian, N.; Haslett, P.; Fitzgerald, K.; Holmes, R. P.; Erbe, D.; Querbes, W.; Knight, J. *J. Am. Soc. Nephrology* **2017**, *28*, 494–503.
110. Petersen, J.; Thompson, A. J.; Levrero, M. *J. Hepatol.* **2016**, *65*, 835–848.

Index

Note: Page numbers followed by "f" indicate figures, "t" indicate tables.

H

I

J

K

N

T

FSC
www.fsc.org
MIX
Papier | Fördert
gute Waldnutzung
FSC® C083411

Kolibri 360 GmbH
Ferdinand-Jühlke-Straße 7
99095 Erfurt, Deutschland
produktsicherheit@kolibri360.de